# Advances in Intelligent Systems and Computing

Volume 542

**Series editor**

Janusz Kacprzyk, Polish Academy of Sciences, Warsaw, Poland
e-mail: kacprzyk@ibspan.waw.pl

*About this Series*

The series "Advances in Intelligent Systems and Computing" contains publications on theory, applications, and design methods of Intelligent Systems and Intelligent Computing. Virtually all disciplines such as engineering, natural sciences, computer and information science, ICT, economics, business, e-commerce, environment, healthcare, life science are covered. The list of topics spans all the areas of modern intelligent systems and computing.

The publications within "Advances in Intelligent Systems and Computing" are primarily textbooks and proceedings of important conferences, symposia and congresses. They cover significant recent developments in the field, both of a foundational and applicable character. An important characteristic feature of the series is the short publication time and world-wide distribution. This permits a rapid and broad dissemination of research results.



More information about this series at http://www.springer.com/series/11156

Suresh Chandra Satapathy
Vikrant Bhateja · K. Srujan Raju
B. Janakiramaiah
Editors

# Data Engineering and Intelligent Computing

## Proceedings of IC3T 2016

*Editors*
Suresh Chandra Satapathy
Department of Computer Science
and Engineering
Anil Neerukonda Institute of Technology
and Sciences
Visakhapatnam, Andhra Pradesh
India

Vikrant Bhateja
Shri Ramswaroop Memorial Group
of Professional Colleges (SRMGPC)
Lucknow, Uttar Pradesh
India

K. Srujan Raju
Department of Computer Science
and Engineering
CMR Technical Campus
Hyderabad
India

B. Janakiramaiah
DVR and Dr. HS MIC College
of Technology
Kanchikacherla, Andhra Pradesh
India

ISSN 2194-5357 ISSN 2194-5365 (electronic)
Advances in Intelligent Systems and Computing
ISBN 978-981-10-3222-6 ISBN 978-981-10-3223-3 (eBook)
DOI 10.1007/978-981-10-3223-3

Library of Congress Control Number: 2016963322



Printed on acid-free paper

This Springer imprint is published by Springer Nature
The registered company is Springer Nature Singapore Pte Ltd.
The registered company address is: 152 Beach Road, #21-01/04 Gateway East, Singapore 189721, Singapore

# Preface

This volume contains the papers that were presented at the 3rd International Conference on Computer and Communication Technologies (IC3T 2016) held in Department of Computer Science and Engineering, Devineni Venkata Ramana & Dr. Hima Sekhar MIC College of Technology, Vijayawada, during November 5–6, 2016. IC3T 2016 aims to provide a forum for exchanges of research results and ideas, and experience of applications among researchers and practitioners involved with all aspects of computing and communication technologies. Previous editions of the conference were held in CMR Group of Colleges, Hyderabad, under the guidance of their management and Dr. Srujan Raju. The IC3T 2016 received total 398 submissions. Each of which was peer reviewed by at least two members of the Program Committee. Finally, a total of 63 papers were accepted for publication in this proceeding. The IC3T 2016 was technically supported by Div-V of Computer Society of India. Several Special sessions were offered by eminent professors in many cutting edge technologies. A preconference workshop on multi-disciplinary research challenges and trends in computer science was the highlight of this meet. Topics on sensor networks, wireless communications, big data, and swarm optimization were discussed at length for the benefits of the participants, and guidance to research direction was provided. An excellent author workshop on "How to write for and get published in Scientific Journal" was conducted by Mr. Aninda Bose, Senior Publishing Editor, Springer India Pvt. Ltd, Delhi, India.

We would like to express our appreciation to the members of the Program Committee for their support and cooperation in this publication. We are also thankful to team from Springer for providing a meticulous service for the timely production of this volume. Our heartfelt thanks to secretary and other management members of Devineni Venkata Ramana & Dr. Hima Sekhar MIC College of Technology for extending wholehearted support to host this in their campus. Special thanks to all guests who have honored us in their presence in the inaugural day of the conference. Our thanks are due to all special session chairs, track managers, and reviewers for their excellent support. Last, but certainly not the least, our special thanks go to all the authors who submitted papers and all the attendees for their contributions and fruitful discussions that made this conference a great success.

| | |
|---|---|
| Visakhapatnam, India | Suresh Chandra Satapathy |
| Lucknow, India | Vikrant Bhateja |
| Hyderabad, India | K. Srujan Raju |
| Kanchikacherla, India | B. Janakiramaiah |
| November 2016 | |

# Organising Committee

## Patrons

**Chief Patron**
Dr. M.V. Ramana Rao M.E., Ph.D., *Chairman, MIC College of Technology, Vijayawada, India.*

**Patrons**
Sri N. Srinivasa Rao, *Vice Chairman, MIC College of Technology, Vijayawada, India.*
Sri D. Panduranga Rao, *CEO, MIC College of Technology, Vijayawada, India.*
Sri M. Srinivasa Rao, *Director (P&D), MIC College of Technology, Vijayawada, India.*
Sri K. Janardhan, *Director, MIC College of Technology, Vijayawada, India.*
Prof. N. Krishna, *Director (Academics), MIC College of Technology, Vijayawada, India.*
Dr. K.B.K. Rao, *MIC College of Technology, Vijayawada, India.*
Dr. Y. Sudheer Babu, *Principal, MIC College of Technology, Vijayawada, India.*

**Honorary Chairs**
Dr. Swagatam Das, *Indian Statistical Institute Kolkata, India.*
Dr. B.K. Panigrahi, *Indian Institute of Technology, Delhi, India.*

**Advisory Committee**
Sri C. Gopal Reddy, *Chairman, CMRTC, & Secretary CMR Group Hyderabad, India.*
Smt C. Vasanth Latha, *Secretary, CMR Technical Campus, Hyderabad, India.*
Dr. A. Raji Reddy, *Director, CMR Technical Campus, Hyderabad, India.*

## Organising Committee

**Organizing Chair**
Dr. B. Janakiramaiah, *Professor of CSE, MIC College of Technology, Vijayawada, India.*

**Program Chairs**
Prof. Vikrant Bhateja, *SRMGPC, Lucknow, India.*
Dr. Nilanjan Dey, *TICT, Kolkata, India.*

**Special Session Chairs**
Dr. R.T. Goswami, *BITS Mesra, Kolkata campus, India.*
Dr. Manas Sanayal, *University of Kalyani, West Bengal, India.*

**Track Chairs**
Dr. Seerisha Rodda, *GITAM, Visakahapatnam, India.*
Prof. Pritee Parwekar, *ANITS, Visakahapatnam, India.*
Dr. M. Ramakrishna Murty, *Vignan Institute of Information Tech, Visakahapatnam.*

**Technical Chairs**
Dr. A. Jayalakshmi, *Professor & HOD of CSE, MIC College of Technology, Vijayawada, India.*
Dr. P. Srinivasulu, *Professor of CSE, MIC College of Technology, Vijayawada, India.*

**Technical Co-chairs**
Mr. A. Rama Satish, *Assoc. Professor of CSE, MIC College of Technology, Vijayawada, India.*
Mr. D. Varun Prasad, *Assoc. Professor of CSE, MIC College of Technology, Vijayawada, India.*
Ms. G. Kalyani, *Assoc. Professor of CSE, MIC College of Technology, Vijayawada, India.*
Mr. *D. Prasad, Assoc. Professor of CSE, MIC College of Technology, Vijayawada, India.*

**Steering Committee**
Dr. Suresh Chandra Satapathy, *Professor of CSE, ANITS, Visakhapatnam, India.*
Dr. K. Srujan Raju, *Professor of CSE, CMR Technical Campus, Hyderabad, India.*

**Publicity Chairs**
Dr. J.K. Mandal, *Professor of CSE, University of Kalyani, Kolkata, India.*
Mr. D. Prasad, *Assoc. Professor of CSE, MIC College of Technology, Vijayawada, India.*
Mr. C.S. Pavan Kumar, *Asst. Professor of CSE, MIC College of Technology, Vijayawada, India.*

**Web Committee**
Mr. K. Mahanthi, *Asst. Professor of CSE, MIC College of Technology, Vijayawada, India.*

**Advisory Committee**
Dr. T.S. Nageswara Rao, *Professor & HOD of CIVIL, MIC College of Technology, Vijayawada, India.*
Dr. A. Guruva Reddy, *Professor & HOD of ECE, MIC College of Technology, Vijayawada, India.*
Dr. T. Vamsee Kiran, *Professor & HOD of EEE, MIC College of Technology, Vijayawada, India.*
Dr. M. Srinivasa Rao, *Professor & HOD of Mechanical, MIC College of Technology, Vijayawada, India.*
Mr. C.V.V.D. Srikanth, *Assoc. Professor & HOD of MBA, MIC College of Technology, Vijayawada, India.*
Dr. A.V. Naresh Babu, *Professor of EEE, MIC College of Technology, Vijayawada, India.*
Dr. Sarath Babu, *Professor of Chemistry, MIC College of Technology, Vijayawada, India.*
Dr. B. Seshu, *Professor of Physics, MIC College of Technology, Vijayawada, India.*
Dr. K.V. Rao, *Professor of MBA, MIC College of Technology, Vijayawada, India.*
Dr. M. Kaladhar, *Professor of Mechanical, MIC College of Technology, Vijayawada, India.*
Dr. T. Sunil Kumar, *Professor of Mechanical, MIC College of Technology, Vijayawada, India.*
Dr. G. Chenchamma, *Professor of ECE, MIC College of Technology, Vijayawada, India.*
Dr. P.V. Srinivasa Rao, *Assoc. Professor of Mathematics, MIC College of Technology, Vijayawada, India.*
Dr. K. Trimula Prasad, *Assoc. Professor of Chemistry, MIC College of Technology, Vijayawada, India.*
Dr. R. Durga Prasad, *Assoc. Professor of Mathematics, MIC College of Technology, Vijayawada, India.*
Dr. K. Praveen, *Assoc. Professor of Chemistry, MIC College of Technology, Vijayawada, India.*
Dr. Prasanna Kumar, *Assoc. Professor of MBA, MIC College of Technology, Vijayawada, India.*
Mr. Ch. Vijaya Kumar, *Assoc. Professor & HOD of BED, MIC College of Technology, Vijayawada, India.*
Mr. R.J. Lakshmi Narayana, *Chief Librarian, MIC College of Technology, Vijayawada, India.*

**International Advisory Committee**
Maurice Clerc, France.
Roderich Gross, England.

L. Perkin, USA.
Sumanth Yenduri, USA.
Carlos A. Coello Coello, Mexico.
Dipankar Dasgupta, USA.
Peng Shi, UK.
Saman Halgamuge, Australia.
Jeng-Shyang Pan, Taiwan.
X.Z. Gao, Finland.
Dr. Kun-lin Hsieh, NTU, Taiwan.
Dr. Ahamad J. Rusumdar, KIT, Germany.
Dr. V.R. Chirumamilla, EUT, the Netherlands.
Dr. Halis Altun, MU, Turkey.
Juan Luis Fernández Martínez, California.
Oscar Castillo, Mexico.
Leandro Dos Santos Coelho, Brazil.
Heitor Silvério Lopes, Brazil.
Rafael Stubs Parpinelli, Brazil.
Prof. A. Govardhan, SIT, JNTUH, India.
Prof. A. Rama Mohan Reddy, SVU, India.
Prof. A. Ananda Rao, JNTUA, India.
Prof. B. Eswara Reddy, JNTUA, India.
Prof. P.V.G.D. Prasad Reddy, AU, India.
Prof. J.V.R. Murthy, JNTUK, India.
Prof. K. Usha Rani, SPMVV, Tirupati, India.
Prof. S. Satyanarayana, JNTUK, India.
Prof. S. ViswanadhaRaju, JNTUHCEJ, India.
Prof. E. Sreenivasa Reddy, ANU, India.
Prof. P. Bala Krishna Prasad, India.
Dr. Ch. Satyananda Reddy, AU, India.
Gerardo Beni, USA.
Namrata Khemka, USA.
G.K. Venayagamoorthy, USA.
K. Parsopoulos, Greece.
Zong Woo Geem, USA.
Lingfeng Wang, China.
Athanasios V. Vasilakos, Athens.
S.G. Ponnambalam, Malaysia.
Pei-Chann Chang, Taiwan.
Ying Tan, China.
Chilukuri K. Mohan, USA.
M.A. Abido, Saudi Arabia.
Saeid Nahavandi, Australia.
Almoataz Youssef Abdelaziz, Egypt.
Hai Bin Duan, China.
Delin Luo, China.

Oscar Castillo, Mexico.
John MacIntyre, England.
Frank Neumann.
Rafael Stubs Parpinelli, Brazil.
Jeng-Shyang Pan, Taiwan.
P.K. Singh, India.
M.K. Tiwari, India.
Sangram Samal, India.
K.K. Mohapatra, India.
Sachidananda Dehuri, India.
P.S. Avadhani, India.
G. Pradhan, India.
Anupam Shukla, India.
Dilip Pratihari, India.
P.K. Patra, India.
T.R. Dash, Kambodia.
Kesab Chandra Satapathy, India.
Amit Kumar, India.
Srinivas Sethi, India.
Lalitha Bhaskari, India.
V. Suma, India and many more.

**National Advisory Committee**
Dr. G.V.S.N.R.V. Prasad, *Professor of CSE, GEC, Gudlavalleru.*
Dr. M.V.P. Chandra Sekhara Rao, *Professor of CSE, RVR&JC, Guntur.*
Dr. K. Subba Ramaiah, *Professor, YITS, Tirupati.*
Dr. M. Babu Rao, *Professor & HOD of CSE, GEC, Gudlavalleru.*
Dr. M. Suneetha, *Professor & HOD of IT, VRSEC, Vijayawada.*
Dr. B. Narendra, *Professor & HOD of CSSE, Sree Vidyanikethan, Tirupati.*
Dr. M. Sunil Kumar, *Professor of CSE, Sree Vidyanikethan, Tirupati.*
Dr. B. Thirumala Rao, *Professor of CSE, KLU, Vijayawada.*
Dr. N. Ravi Shankar, *Professor & HOD of CSE, LBRC, Mylavaram.*
Dr. D. Naga Raju, *Professor & HOD of IT, LBRC, Mylavaram.*
Dr. K. Hima Bindu, *Professor of CSE, Vishnu, Bhimavaram.*
Dr. V.V.R. Maheswara Rao, *Professor of CSE, SVECW, Bhimavaram.*
Dr. B. Srinivasa Rao, *Professor & HOD of CSE, Dhanekula, Vijayawada.*
Dr. P. Harini, *Professor & HOD of CSE, SACET, Chirala.*
Dr. A. Yesu Babu, *Professor & HOD of CSE, CRR, Eluru.*
Dr. S. Krishna Rao, *Professor & HOD of IT, CRR, Eluru.*
Dr. D. Haritha, *Professor & HOD of CSE, SRK, Vijayawada.*
Dr. S.N. Tirumala Rao, *Professor & HOD of CSE, NEC, Narasaraopeta.*
Dr. G.V. Padma Raju, *Professor & HOD of CSE, SRKR, Bhimavaram.*
Dr. G. Satyanarayana Murty, *Professor & HOD of CSE, AITAM, Tekkali.*

Dr. G. Vijay Kumar, *Professor of CSE, VKR,VNB & AGK, Gudivada.*
Dr. P. Kireen Sree, *Professor of CSE, SVECW, Bhimavaram.*
Dr. C. Nagaraju, *Professor of CSE, Y.V.U, Proddatur.*

## Local Organizing Committee

**Chair**: Dr. A. Jayalakshmi, *Professor & HOD of CSE, MIC College of Technology, Vijayawada, India.*

**Registration Chair**: Ms. L. Kanya Kumari, *Assoc. Professor of CSE, MIC College of Technology, Vijayawada, India.*

**Reception, Registration Committee**
Ms. A. Anuradha, *Asst. Professor of CSE, MIC College of Technology, Vijayawada, India.*
Ms. K. Vijaya Sri, *Asst. Professor of CSE, MIC College of Technology, Vijayawada, India.*
Ms. G. Rama Devi, *Asst. Professor of CSE, MIC College of Technology, Vijayawada, India.*
Ms. B. Lalitha Rajeswari, *Asst. Professor of CSE, MIC College of Technology, Vijayawada, India.*
Ms. R. Srilakshmi, *Asst. Professor of CSE, MIC College of Technology, Vijayawada, India.*
Ms. P. Reshma, *Asst. Professor of CSE, MIC College of Technology, Vijayawada, India.*

**Stage Management Chair**: Dr. B. Seshu, *Professor of Physics, MIC College of Technology, Vijayawada, India.*

**Decoration, Stage Management Committee**
Ms. Sujata Agarwal, *Asst. Professor of English, MIC College of Technology, Vijayawada, India.*
Ms. T.P. Ann Thabitha, *Asst. Professor of CSE, MIC College of Technology, Vijayawada, India.*
Ms. K. Vinaya Sree, *Asst. Professor of CSE, MIC College of Technology, Vijayawada, India.*

**IT Services Chair**: D. Varun Prasad, *Assoc. Professor of CSE, MIC College of Technology, Vijayawada, India.*
Mr. J.V. Srinivas, *Asst. Professor of CSE, MIC College of Technology, Vijayawada, India.*
Mr. P. Srikanth, *Asst. Professor of CSE, MIC College of Technology, Vijayawada, India.*

Mr. Y. Narayana, *Asst. Professor of CSE, MIC College of Technology, Vijayawada, India.*

**Chair**: Dr. P. Srinivasulu, *Professor of CSE, MIC College of Technology, Vijayawada, India.*

**Technical Sessions Chair**: Ms. G. Kalyani, *Assoc. Professor of CSE, MIC College of Technology, Vijayawada, India.*

**Technical Sessions Committee**
Mr. J. Ranga Rajesh, *Asst. Professor of CSE, MIC College of Technology, Vijayawada, India.*
Ms. V. Sri Lakshmi, *Asst. Professor of CSE, MIC College of Technology, Vijayawada, India.*
Ms. N.V. Maha Lakshmi, *Asst. Professor of CSE, MIC College of Technology, Vijayawada, India.*
Mr. R. Venkat, *Asst. Professor of CSE, MIC College of Technology, Vijayawada, India.*
Mr. A. Prashant, *Asst. Professor of CSE, MIC College of Technology, Vijayawada, India.*
Ms. M. Madhavi, *Asst. Professor of CSE, MIC College of Technology, Vijayawada, India.*
Mr. G.D.K. Kishore, *Asst. Professor of CSE, MIC College of Technology, Vijayawada, India.*
Ms. Lakshmi Chetana, *Asst. Professor of CSE, MIC College of Technology, Vijayawada, India.*
Mr. N. Srihari, *Asst. Professor of CSE, MIC College of Technology, Vijayawada, India.*
Ms. G. Prathyusha, *Asst. Professor of CSE, MIC College of Technology, Vijayawada, India.*

**Refreshment Chair**: Mr. D. Varun Prasad, *Assoc. Professor of CSE, MIC College of Technology, Vijayawada, India.*

**Refreshment Committee**
Mr. T. Krishnamachari, *Asst. Professor of CSE, MIC College of Technology, Vijayawada, India.*
Mr. B. Murali Krishna, *Asst. Professor of CSE, MIC College of Technology, Vijayawada, India.*
Mr. Kamal Rajesh, *Asst. Professor of CSE, MIC College of Technology, Vijayawada, India.*

**Transport Chair**: D. Prasad, *Assoc. Professor of CSE, MIC College of Technology, Vijayawada, India.*

**Transport and Accommodation Committee**
Mr. D. Durga Prasad, *Assoc. Professor of CSE, MIC College of Technology, Vijayawada, India.*
Mr. C.S. Pavan Kumar, *Asst. Professor of CSE, MIC College of Technology, Vijayawada, India.*

**Finance Chair**: A. Rama Satish, *Assoc. Professor of CSE, MIC College of Technology, Vijayawada, India.*

**Sponsoring Chair**: D. Prasad, *Assoc. Professor of CSE, MIC College of Technology, Vijayawada, India.*

**Press and Media Chair**: G. Rajesh, *Assoc. Professor of ME, MIC College of Technology, Vijayawada, India.*

# Contents

# About the Editors

**Dr. Suresh Chandra Satapathy** is currently working as a Professor and Head of the Department of Computer Science and Engineering, Anil Neerukonda Institute of Technology and Sciences (ANITS), Andhra Pradesh, India. He received his Ph.D. in Computer Science and Engineering from Jawaharlal Nehru Technological University (JNTU), Hyderabad, and his Master's degree in Computer Science and Engineering from the National Institute of Technology (NIT), Rourkela, Odisha. He has more than 27 years of teaching and research experience. His research interests include machine learning, data mining, swarm intelligence studies, and their applications to engineering. He has more than 98 publications to his credit in various reputed international journals and conference proceedings. He has edited many volumes from Springer AISC and LNCS in the past. In addition to serving on the editorial board of several international journals, he is a senior member of the IEEE and a life member of the Computer Society of India, where he is the National Chairman of Division-V (Education and Research).

**Prof. Vikrant Bhateja** is an associate professor at the Department of Electronics and Communication Engineering, Shri Ramswaroop Memorial Group of Professional Colleges (SRMGPC), Lucknow, and also the Head of Academics & Quality Control at the same college. His research interests include digital image and video processing, computer vision, medical imaging, machine learning, pattern analysis and recognition, neural networks, soft computing, and bio-inspired computing techniques. He has more than 90 quality publications in various international journals and conference proceedings to his credit. Professor Vikrant has been on TPC and chaired various sessions from the above domain in international conferences of IEEE and Springer. He has been the track chair and served on the core-technical/editorial teams for numerous international conferences: FICTA 2014, CSI 2014, and INDIA 2015 under the Springer-ASIC Series, and INDIACom-2015 and ICACCI-2015 under the IEEE. He is an associate editor for the International Journal of Convergence Computing (IJConvC) and also serves on the editorial board of the International Journal of Image Mining (IJIM) under Inderscience Publishers. At present, he is the guest editor for two special issues of the International Journal of Rough Sets and Data Analysis (IJRSDA) and the International Journal of System Dynamics Applications (IJSDA) under IGI Global publications.

**Dr. K. Srujan Raju** is a Professor and Head of the Department of Computer Science and Engineering (CSE), CMR Technical Campus. Professor Srujan earned his Ph.D. in the field of network security, and his current research interests include computer networks, information security, data mining, image processing, intrusion detection, and cognitive radio networks. He has published several papers in referred international conferences and peer-reviewed journals. He was also on the editorial board of CSI 2014 Springer AISC series 337 and 338 volumes. In addition, he served as a reviewer for many indexed journals. Professor Raju has been honored with the Significant Contributor and Active Member Awards by the Computer Society of India (CSI) and is currently the Hon. Secretary of the CSI's Hyderabad Chapter.

**Dr. B. Janakiramaiah** is currently working as a professor in the Department of Computer Science and Engineering, Devineni Venkata Ramana and Dr. Hima Sekhar MIC College of Technology (DVR & Dr. HS MIC College of Technology), Kanchikacherla, Vijayawada, Andhra Pradesh, India. He received his Ph.D. in Computer Science and Engineering from the Jawaharlal Nehru Technological University (JNTU), Hyderabad, and Master's degree in Computer Science and Engineering from the Jawaharlal Nehru Technological University, (JNTU), Kakinada. He has more than 15 years of teaching experience. His research interests include data mining, machine learning, studies, and their applications to engineering. He has more than 25 publications to his credit in various reputed international journals and conference proceedings. He is a life member of the Computer society of India.

# Analysis of Genetic Algorithm for Effective Power Delivery and with Best Upsurge

**Azeem Mohammed Abdul, Srikanth Cherukuvada, Annaram Soujanya, R. Srikanth and Syed Umar**

**Abstract** Wireless network is ready for hundreds or thousands of nodes, where each node is connected to one or sometimes more sensors. WSN sensor integrated circuits, embedded systems, networks, modems, wireless communication and dissemination of information. The sensor may be an obligation to technology and science. Recent developments underway to miniaturization and low power consumption. They act as a gateway, and prospective clients, I usually have the data on the server WSN. Other components separate routing network routers, called calculating and distributing routing tables. Discussed the routing of wireless energy balance. Optimization solutions, we have created a genetic algorithm. Before selecting an algorithm proposed for the construction of the center console. In this study, the algorithms proposed model simulated results based on parameters depending dead nodes, the number of bits transmitted to a base station, where the number of units sent to the heads of fuel consumption compared to replay and show that the proposed algorithm has a network of a relative.

**Keywords** LEACH-GSA · Genetic algorithm · Dead nodes · Upsurge

## 1 Introduction

Distributed sensor system can certainly follow the area. In this system, unlike traditional wired system, on the other hand, reduce the installation, configuration and network configuration instead of thousands of meters of cable, on the other

A.M. Abdul (✉)
Department of ECE, K L University, Vaddeswaram, India
e-mail: mohammedazeem123@gmail.com

S. Cherukuvada · A. Soujanya · R. Srikanth
Department of CSE, Institute of Aeronautical Engineering, Hyderabad, India

S. Umar
Department of CSE, Gandhiji Institute of Science and Technology, Jaggaihpet, India
e-mail: umar332@gmail.com



S.C. Satapathy et al. (eds.), *Data Engineering and Intelligent Computing*,
Advances in Intelligent Systems and Computing 542,
DOI 10.1007/978-981-10-3223-3_1

hand have only a small device, the size of the same coin. Wireless sensor networks are wireless sensors are spread out, and measuring by us to say that a group of a number of physical quantities or to the environment, such as temperature, sound, vibration, pressure, motion or polluting substances in different places [1]. Sensor Networks motivate used in military applications such as battlefield surveillance, to be developed. But now, using wireless sensor networks in the industry and many non-military purposes, such as monitoring and control of industrial processes, health monitoring, environmental monitoring, and applications for home care, smart home and traffic [2].

Except addition, one or more sensors, such as each network node is usually equipped with a radio transmitter (or other wireless communication devices), small and micro-power source (usually a battery) [3, 4]. Depending on the size of the entire sensor node packaging is a little sand is the microscopic parts of the nest is still under development. The network sensor access networks usually down (ad hoc network) is present, which means that each node is a multi-hop algorithm. (Many of the nodes and the central station is a package). Currently, wireless sensor networks, computing and communications are held numerous seminars and conferences active research each year in this regard. In this article we will have a considerable impact on reducing energy consumption algorithm communication will focus on the network. According to our research, direct transfer methods, and is not optimized for use in sensor networks [5, 6]. We leach clustering protocol based on the proposal that we do not turn the head of the local pastries; Power is shared by all nodes. Scalability and reliability algorithm that provides local coordination and data transfer, reduce uses [7]. This algorithm reduces the energy consumption by a factor of [8]. Breakdown of life. Direct access to each node to send data directly to the transmitter. If the plant is small nodules, it takes a lot of energy network nodes and reduces life expectancy significantly. However, when the cells close to each other, this method is acceptable, and probably better. Least energy transfer nodes between nodes cells. Another way to make. The nodes in the group, which at each node is connected to the local station and the local station data to the public stations ultimately reach the hands of the users. When the low-energy cluster [9, 10].

## 2 Literature Survey

In the Paper [8] routing multi path effect and streams studied treatment was wireless networks. Three criteria are used to measure the effectiveness of the network is the network bandwidth minimum requirements for the popular fairs and acceptable QoS bandwidth. Each of these criteria "show in any case in many directions and ongoing clinical trials without" multi-beam calculated route. The calculation of these measures, but if the level of protection for solving optimization model of these standards. In both cases, the comparison of these measures, it has been found that even "with multi-path routing to the fixed network can, wireless, or is not climbing twice." Moreover, showed in comparison to the two-state solution of the model with ensures the help of multi-beam routing optimization model for a much simpler solution. The confrontation between the two forms of the above results were obtained.

(1) Multi-way routing and distribution. You can use it to network efficiency, wireless and the use of complex algorithms and protocols plants prospects increase.
(2) A multi-beam routing and power allocation optimization model reduces the complexity of flora.

Optimal size designation, a group shredding algorithms are tools online use, such as cultural model of cultural values of the results differ. Article [9], because the limited flexibility of wireless communication channels are divided into large number of sensor nodes is a serious problem that collisions, packet collisions factors that increase latency end-end network throughout the plant. Collision nodes try to write an increasing number of pseudo miss the beginning of the end of the period. In this article, we describe a method which graphics, network, increases network capacity XTC algorithm built, and the modified distance vector algorithm that makes up the crane, so that the delay is eliminated by the supply end area. XTC algorithm is one of the most realistic and practical control algorithm limits of the sensor. Algorithms and many other algorithms have network nodes, the exact cause of its geographical neighbor's day. This algorithm in difficult circumstances, called wireless applications know it in this environment. In general, this algorithm gives three steps building all nodes in the network, the network, and these three steps the development of others is a numbered list of the roadmap to a network diagram each growing network.

Many search queries obligations tools to give advice. Working with minimal cost routing using some genetic PMX crossings to drive each population. Evaluation of the total population in the pretreatment liquid to obtain only results "has always been a part of the population." comprehensive research analysis to find the optimal solution for better targeting problem respectively 15. The work I often temperature logarithmic function, leading to the best solution to this problem [10]. PMX operator routing problem of the junction caused by repeated time and characteristic of her equipment grows gradually logarithmic rate convergence algorithm of the search algorithm. Paper [11] a wealth of wireless network routing optimization and compression for maximum durability. They solve the problem, it was the evaluation function, which is a necessary condition and optimal algorithm proposed slope compensation [12]. Routing algorithm on the basis of the geometry selected from acyclic graph consists quote for instructions on the basis of the routing algorithm is one of the modern culture can send traffic to neighboring nodes adjacent text in the distance of the text culture. Using data compression each culture, there are two features about culture or obtain. Since the original data are grown near the cultural and local knowledge is encoded transmission nodes nearest neighbor nodes for information directly screwed.

Wood was purpose brought to the nodes for a low price in [13], respectively. The complexity of routing tables using sending results of the search engine optimization of information on genetic algorithms. Environmental problems latex nodes connected to a plurality of the large losses in the target nodes. Latex once factors, such as the link between the nodes surrounding text nodes to a collection of objects to break, the addition of target nodes in the group [14].

## 3 Various Methodologies and Flow of Execution of Proposed Algorithm

In this study is to optimize the routing of the balance between the wireless energy was examined genetic algorithm. Here, examples of the resulting applications. Given the complexity of this solution in order polynomial. So, this model solve with great optimization.

### *3.1 Algorithm Proposed for*

- The first sensors are randomly distributed in space.
- Classification of sensors or groups so that each group has a leader.
- The number of clusters selected in accordance with the temperature distribution of all the sensors and to save energy.
- If you are part of the cultural calendar on the map, at the end of the cluster.
- Select the part closest to the data center of the use of a genetic algorithm:
- The creation of the indigenous population.

  1. Crossover online
  2. Mutations Internet
  3. Controversial Internet

- Go to step 5, all nodes in the cluster to their optimum.

### *3.2 The First Group of Algorithms*

The values below are based on the best escape. We run the simulation several times until the parameter in order to adapt the algorithm to achieve the best conditions. The values of these parameters: number of nodes in the network 50, the number of repetitions, 3000, Pico network nodes [0100] packages sent to 6400 bits, the probability of selection of the head of Culture 00:05 team.

Key factors and Formulation used as below

$$\text{Distance} = \sqrt{(((s(i)^*xd - S(n+1)^*xd)^\wedge 2) + ((S(i).yd - S(n+1)^*yd)^\wedge 2))} \quad (1)$$

In the above Eq. (1) S stands for as follows

S(C (min_dis_cluster).id) = {S(C (min_dis_cluster).id). E (ETX*(ctrPacketLength) + Efs*ctrPacketLength*

$$((\text{min\_dis})^\wedge 2)) \quad (2)$$

# 4 Experimental Analysis

In order to evaluate the algorithm, I introduced four of the key associated with a number of rides, let me. The first argument to the number of nodes in the network is dead. The second parameter is the number of bits in the down. The third parameter is the number of devices in the hands of the head and the fourth parameter, power consumption is played online.

Figure 1 shows the number of repetitions of the dead volume of 1100 is repeated up to 50 minimum. The scientific results or proposal is better than the standard method.

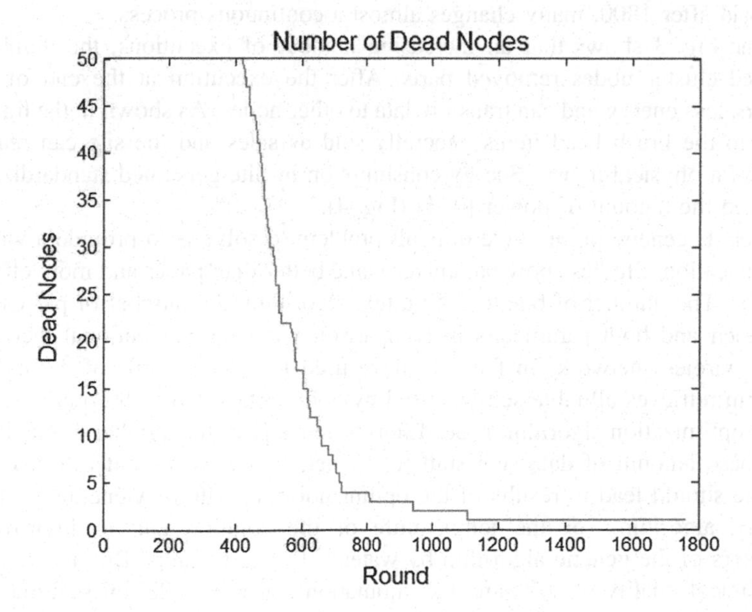


**Fig. 1** The number of dead nodes on which the algorithm proposed

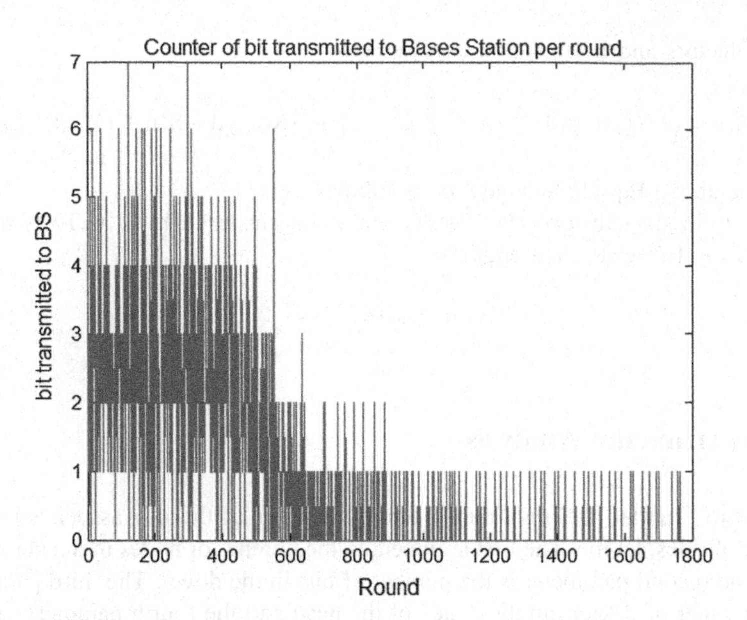


**Fig. 2** The number of bits transmitted from the base station, the algorithm

In the Fig. 2 many other embodiments, the reduced number of bits in the base station. The number of bits transmitted in a given time, the network is a good argument for quality. It is known for finishing Run, energy and other affected nodes can send data. The simulation was able to move running in the morning, but the taxes paid after 1800, many changes almost a continuous process.

In the Fig. 3 shows that an increasing number of executions, the number of delivered cluster nodes removed parts. After the execution at the end of 1100 numbers, less energy and can transmit data to other nodes. As shown in the figure, it is sent to the brush head items generally sold as sales and the site can read the results of a physical injury. Energy consumption in Site generated standardization; calculated the amount of power [0, 1] (Fig. 4).

Article L genetic progressive difficult problem to solve is to provide a wireless communication. Studies show parameter value better than paper and more effective solutions. The number of bits to send a text algorithm, the number of pieces GSA sent Leach and both parameters in comparison with the conventional method of routing wireless network. In fact, we have tried to answer some of the methods algorithm retrieves all table settings used by both parties. How all criteria network routing optimization algorithm is dead sensor nodes [15], the number of bits for use in the base amount of data sent staff to the net. Below is presented a chart, and therefore should lead to results of the optimization algorithm? Generation of evolutionary algorithms or the temperature of the solution and to improve the parameters of the genetic algorithm by water with good results. But above all test the technical quality of software and simulation run in parallel on multiple computers [16].

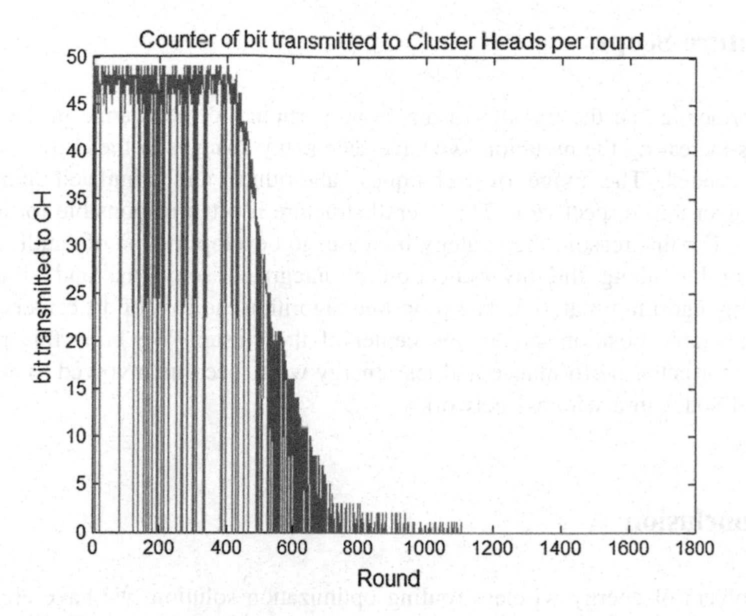


**Fig. 3** The number of transmitted bits of the proposed algorithm

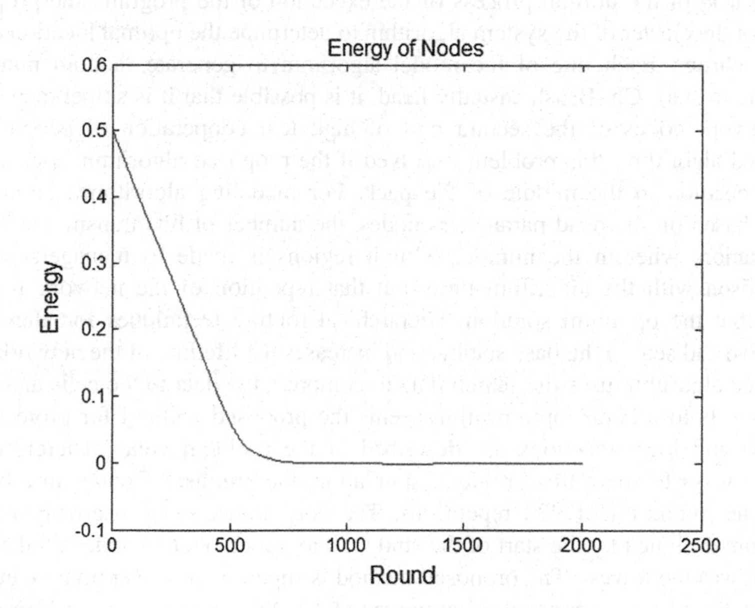


**Fig. 4** Total energy fuel by increasing the frequency|theater

## 5 Future Scope

Is the principle that the transfer means is proportional to the change in the population is increasing the mutation, we have seen many changes in the chromosomes. Recommended. The range of techniques, algorithms and simulated annealing search algorithm respectively. The overall structure is often responsible for a large research. For this reason, the strategy increases to develop the list of candidates in the algorithm along the diversification of integrated solutions, and simulated annealing algorithm material. The proposed algorithm the role of art centers and a more favorable position can to the center of the cluster. We hope this project solution for better performance and less energy waste become involved to develop our total nodes in a wireless network.

## 6 Conclusion

The problem of energy wireless routing optimization solution, we have created a genetic algorithm. Before selecting a proposal for a cluster design algorithm. In fact, instead of the normal process of the execution of the program, the proposed center of the cluster of the system algorithm to determine the optimal location of the central cluster. Each line of the model algorithm to generate random numbers, deciding to start, Ch. Brush casually head, it is possible that it is sometimes not to the network nodes of the second part of high text cooperation of people. The proposed algorithm, this problem is solved if the proposed algorithm in charge to find a position in the middle of the pack. For modeling algorithms, simulation results based on the dead parameters nodes, the number of bits transmitted by the base station, wherein the number of unit regions is made to managers fuel in comparison with the algorithm turns out that repetition of the network is for a report, but the optimum solution. Hierarchical routing techniques for alleviating obsessive and sent to the base station, and increases the lifetime of the network. The proposed algorithm uses the search data to compress the data to the cells and send, which leads to a better optimization seems the proposed method for protecting a scalable and local reactions are described in the problem areas. Therefore, we propose ways to solve this problem, simulation, the number of nodes in a bin 50 when the minimum of 990 repetitions. The cost model sensor network routing algorithm designed by the start of the study, 50 to the number of nodes on the label in 1100 was the lowest. The proposed method is superior to conventional methods. The number of nodes parameters compared to 1050 is a result of natural, resulting in better performance. The assessment of energy use repetition of standard algorithms, and sets the parameter proposed is a good performance, and the lowest 550 MPs. As the proposed algorithm is not much changed in comparison with conventional algorithms shows one possible solution to change the proposed algorithm to change the parameters of the genetic algorithm for the simulation.

## References


1. Yang, H., Ye, F., Sikdar, B: A written warning is based on swarm intelligence telephone hub. Faculty of Electrical Engineering, Computer and Systems Engineering, Rensselaer Polytechnic Institute, Troy, NY 12180 (2004)
2. Toosizadeh, D.S.: Multi-Niche impact on the stable position multicast dynamic routing EAS. International Res. Organ. Sci. (IROCS-IJCEE) **1**(1) (2013) (IROCS published online magazine)
3. Vijayalakshmi, K., Radhakrishnan, S.: Dynamic routing to different locations over an IP network using a hybrid genetic algorithm (DRHGA). Int. J. Inf. Technol. **4** (2007)
4. Umar, S.: A review of limitations and challenges in wireless sensor networks for environmental research. Int. J. Appl. Eng. Res. **9**(16), 3309–3318 (2014). ISSN 0973-4562
5. Zahmatkesh, A., Yaghmaei, H.: Genetic algorithm approach to how groups of energy efficiency in wireless sensor networks. In: ICNCC Conference Network and Sport Communication (2011)
6. Umar, S.: A review of limitations and challenges in wireless sensor networks for environmental research. Int. J. Appl. Eng. Res. **9**(6). 3309–3318 (2014). ISSN 0973-4562
7. Zahhad Abo, M., Ahmed Triumph No, S., Sasaki, S.: The new energy-efficient protocol for adaptive genetic algorithm to the collection month, and improve wireless sensor networks. Int. J. Energy Inf. Commun. **5**(3), 47–72 (2014)
8. Treatment of S.K., Abdalla, T.Y.: High-efficiency routing protocol for wireless sensor networks in order to optimize the core by means of fuzzy logic. Int. J. Comput. Appl. (0975-8887) **122**(1) (2015)
9. Bakhshi, B., Khorsand, S.: Optimal routing and multicast network woven net effect on the performance of the network. CSICC, Paper_96 (2009)
10. Zahrani, M.S., Loomes, M.J., Malcolm, J.A., Ullah, A.D., Steinhöfel, K., Albrecht, A.A.: Genetic study multicast simulated annealing pretreated logarithmic routing. Comput. Oper. Res. **35**, 2049–2070 (2008)
11. Ding, W.: S. S. R. portfolio has Rummler energy watt unparalleled routing wire. Microprocess. Ciulli 467–475 (2004)
12. Rostam network of 11 Dr. H. Mottaret number of wireless sensors, particle swarm optimization to reduce the excess energy. Int. J. Inf. Technol. Manag. (IJMIT) **6**(4) (2014)
13. Kumar, K., Bhavani, S.: Italian recording routing optimized energy for wireless sensor networks. Middle East J. Sci. Res. **23**(5):915–923 (2015). ISSN 1990-9233 © 2015 idose Publications
14. Lady, D.: Evolutionary game theory QoS routing in hybrid wireless networks. Int. J. Sports Sci. Eng. **4**(9) (2013)
15. Sharma, N.R.: 15 fm, WSN: basic optimization maximize rumor routing protocol and evaluation of nervous system. Article 266–279 no.BJMCS.122 2015. Br. J. Math. Comput. Sci. **7**(4) (2015)
16. Iqbal, M., Naeem Anpalagan, A., Ahmed Azam, A.: Optimization of wireless sensor networks: multi-function paradigm. Sensors 15:17572–17620 (2015). doi:10.3390/s150717572

## References


1. [illegible], Ye [illegible], Sit [illegible] [illegible] routing based on swarm intelligence [illegible] Faculty of Electrical Engineering, Computer and Systems Engineering, Rensselaer Polytechnic Institute, Troy [illegible] (2003)
2. [illegible] D.S., [illegible] impact on [illegible] routing [illegible] using EMS. International [illegible] (IJCCS [illegible]) (2013) [illegible] published online
3. [illegible] K., Padeski [illegible] Dynamic routing [illegible] locations over an IP network using a hybrid genetic algorithm (DRHGA). Int. J. [illegible] (2007)
4. [illegible], [illegible] wireless sensor network [illegible]. Int. J. Appl. Eng. Res. 9(16), 3309–3316 (2014). ISSN 0973-4562
5. Zaheeruddin, [illegible] approach to form groups of [illegible] in wireless sensor networks. In: [illegible] Conference Networks and [illegible] Communication (2013)
6. [illegible] A. [illegible] limitations and challenges of wireless sensor networks over [illegible]. Int. J. Appl. Eng. Res. 9(20), [illegible] (2014). ISSN 0973-4562
7. Zaheeruddin, M., Ahmed, [illegible] The new energy efficient protocol for adaptive genetic algorithm for optimum routing and improve wireless sensor network. [illegible] Energy Int. Comput. Stat. 4(3), [illegible] (2014)
8. [illegible] S.S., [illegible] S.Y. [illegible] routing protocol for wireless sensor network in order to optimize lifetime by means of [illegible]. Int. J. Comput. Appl. 0975-8887, 122(1) (2015)
9. Baldini, [illegible] Optimal routing and [illegible] network [illegible] effect on the performance of the network. [illegible] (2005)
10. Zahari, M.S., [illegible], M., [illegible] Genetic and [illegible] simulated annealing [illegible] routing. Comput. Oper. Res. 35, [illegible] (2008)
11. [illegible] W. [illegible] S. [illegible] (2001)
12. [illegible] network [illegible] H. [illegible] wireless sensor [illegible] particle swarm optimization [illegible]. Int. J. [illegible] (2013)
13. [illegible], K., Bhavsi, S. [illegible] energy for wireless sensor networks. Middle East J. Sci. Res. [illegible] (2015). ISSN 1990-9233 © 2015 IDOSI Publications
14. [illegible] D. [illegible] routing [illegible] network [illegible] (2000)
15. Shamsul [illegible] optimization [illegible] routing protocol and [illegible] systems. Article [illegible] J. Mob. Commun. [illegible] (2015)
16. [illegible] Rajakumar, A., Ahmed [illegible] Optimization [illegible] of [illegible] Sensors [illegible] (2015)

# Edge Detection of Degraded Stone Inscription Kannada Characters Using Fuzzy Logic Algorithm

**B.K. RajithKumar, H.S. Mohana, J. Uday and B.V. Uma**

**Abstract** Digital India is an initiative by the Government of India. This initiative encourages digitization and analysis in all walks of life. Digitization will preserve any historical document and that information can access by any individuals by his finger tip from any place. Stone inscriptions are one of the key historical evidences of literature and culture of that region in the passage of time. Recognition and analysis of stone inscriptions play a pivotal role in deciding the era/age it belongs ad to understand the content. A proper digitization and recognition technique is pre-requisite and desired. Here, in this work digitization of characters has been done by using ordinary digital camera. Further, the captured images are pre-processed in order to extract features. In this proposed algorithm, gradient analysis is carried out at every pixel in the x and y directions, based on the result it defines an edge using Fuzzy Inference System. The experiment was conducted on twenty set of analogous degraded stone inscriptions Kannada characters and result obtained was magnificent with better time efficiency compared to prior methods.

**Keywords** Digitalization · Edge detection · Fuzzy inference system

B.K. RajithKumar (✉) · B.V. Uma
Department of Electronics and Communication Engineering, RV College of Engineering, Bengaluru, India
e-mail: rajith.bkr@rvce.edu.in

B.V. Uma
e-mail: umabv@rvce.edu.in

H.S. Mohana
Department of Electronics and Instrumentation Engineering, Malnad College of Engineering, Hassan, India
e-mail: hsm@mce.ac.in

J. Uday
Department of Electronics and Communication Engineering,
Srinivas Institute of Technology, Mangaluru, India
e-mail: uday.j@gmail.com



S.C. Satapathy et al. (eds.), *Data Engineering and Intelligent Computing*,
Advances in Intelligent Systems and Computing 542,
DOI 10.1007/978-981-10-3223-3_2

## 1 Introduction

Feature extraction plays a vital role in stone inscription character recognition, for better feature extraction an enhanced pre-processing is required. Comprehensible feature extraction for any Kannada stone inscriptions is challengeable, especially in the stone inscriptions each character are not uniformly inscripted and some characters are degraded due to some natural calamities. Edge detection is imperative step in preprocessing; the prior methods like Sobel, Canny are deficient to find the edges of old stone inscription characters so new-fangled edge detection technique is required for better feature extraction. In this present work, the captured image is converted into gray scale and Fuzzy algorithm works only on double precision data so the gray scale image was again converted into double-precision. To determine image gradient in $x$-axis and $y$-axis, we define Fuzzy Inference System (FIS) using triangular membership function and defined rule such that, any pixel belongs to a uniform region is make it as white else make same pixel as black. Finally we compared the recognition results of proposed method with Sobel edge detection method.

## 2 Related Works

The review of the literature pertaining to the present topic is presented to the readers. In [1] authors concentrate on recognition of old Hoysala, Ganga characters, but in that work they achieved only 90% recognition rate for few set of Kannada stone inscriptions special characters and here that work was extended and that recognition rate was increased using advanced edge detection method. In [2] author concentrate on Fuzzy interface rule, here that work is extended and defined Fuzzy "logical or" for detect stone inscriptions image edges. In [3] author concentrate on Design of Fuzzy interface system and that work was tailored to construct a fuzzy interface system for extraction of old stone inscriptions Kannada characters. In [4] author concentrate on extraction of old stone inscriptions Kannada characters using Gaussian filter and some morphological operation but this approach doesn't extract exact features of degraded Special stone inscriptions Kannada characters and pre-processing algorithm used was simple and that can't be extended to large dataset.

## 3 Algorithm

**Step 1:** The Kannada Stone inscriptions characters are capture by using ordinary Camera of 16 Mega pixel Resolution.

**Step 2:** The Captured images are 3-Dimensional so in this step that images are converted into gray scale image (2-D array).
**Step 3:** Salt and pepper noise of captured image is remove by using Mean shift filter.
**Step 4:** Fuzzy logic tool box operates on double-precision data so the grayscale unit 8 array is converting into double array.
**Step 5:** Calculation of image gradient in X and Y direction and define Specify input and output to the Fuzzy interface system using Membership function. Specify Fuzzy Interface rule using Fuzzy Interface system and display of Edge detection image.
**Step 6:** The edge detected image features were extracted and Feed into Advance Recognition Algorithm for recognition.

# 4 Methodology and Implementation

The Block diagram of proposed work is shown in Fig. 1.

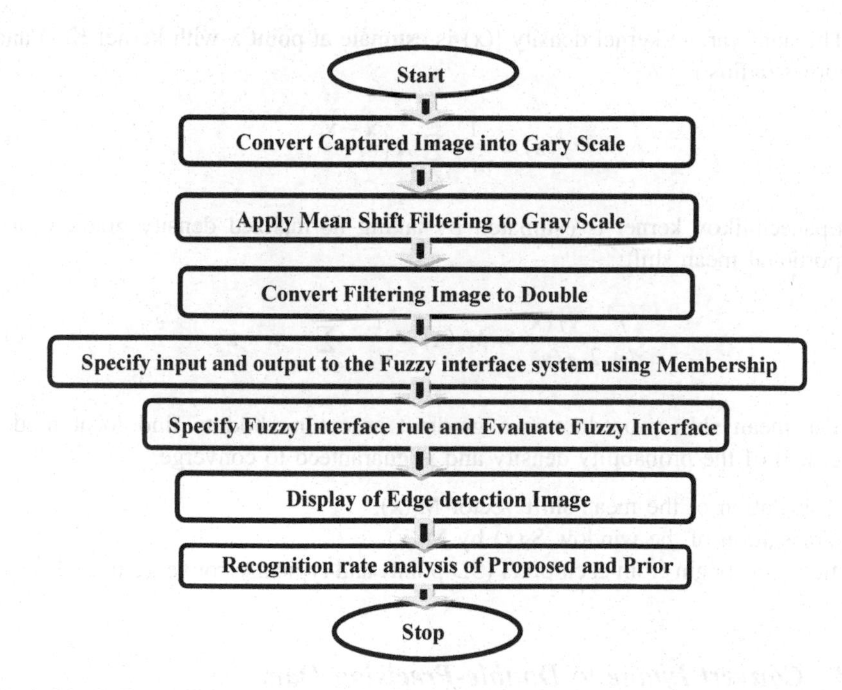


**Fig. 1** Block diagram of proposed work

## 4.1 Image Conversion

Computation with 2-D array is simple than computation with 3-D array, the captured image contains red, green, blue intensities so standard NTSC conversion formula used to calculate the effective luminance of each pixel. Instead rgb2gray function can also be used.

$$\text{Igray} = 0.2989{*}\text{Irgb}(,1) + 0.5870{*}\text{Irgb}(,,2) + 0.1140{*}\text{Irgb}(,,3) \tag{1}$$

## 4.2 Mean Shift Filtering

Apply mean shift filtering for to remove salt and pepper noise and for image smoothing mean shift filtering used. Mean shift filtering basically works on non-parametric probability density estimation method in which each pixel in an image is replaces by probable local v

$$\{\mathrm{x}\}_{\mathrm{i=1..n}} = \mathrm{x_i} € \mathrm{R^d} \tag{2}$$

The multivariate kernel density f(x) is estimate at point x with kernel K(x) and windows radius r

$$\hat{f}(\mathbf{x}) = \frac{1}{nr^d} \sum_{i=1}^{n} K\left(\frac{\mathbf{x} - \mathbf{x_i}}{r}\right) \tag{3}$$

Epanechnikov kernel is estimated by taking normalized density gradient and proportional mean shift:

$$\frac{r^2}{d+2} \frac{\nabla f(\mathbf{x})}{\hat{f}(\mathbf{x})} = M_r(\mathbf{x}) = \frac{1}{n_{\mathbf{x}}} \sum_{\mathbf{x}_i \in S_r(\mathbf{x})} \mathbf{x_i} - \mathbf{x} \tag{4}$$

The mean shift procedure is a gradient ascent method to find local modes (maxima) of the probability density and is guaranteed to converge.

1. Calculation of the mean shift vector $M_r(\mathbf{x})$.
2. Translation of the window $S_r(\mathbf{x})$ by $M_r(\mathbf{x})$.
3. Iterations begin from each pixel (5D point) and typically converge in 2–3 steps.

## 4.3 Convert Image to Double-Precision Data

Fuzzy logic tool box operates on double-precision data so the grayscale unit 8 array is converting into double array.

## *4.4 Calculate Image Gradient*

The image gradient measures the varying information of magnitude and direction along X and Y axis in an image. The image gradient vector is obtain by combining derivates of X and Y direction as shown in Eq. 5

$$\Delta I = \left( \frac{\partial I}{\partial X}, \frac{\partial I}{\partial Y} \right) \quad (5)$$

For a continuous function, the I(x,y) can calculate by taking the partial derivative of I with respect to X and determining how rapidly the image intensity changes as X changes, by using Eq. 6

$$\frac{\partial I(X,Y)}{\partial X} = \frac{\lim}{\Delta X \to 0} \frac{I(X+\Delta X,Y) - I(X,Y)}{\Delta X} \quad (6)$$

For discrete case, a differences between I(x,y) and the pixel before or after it could be taken shown in Eq. 7.

$$\frac{\partial I(X,Y)}{\partial X} = \frac{I(X+1,Y) - I(X-1,Y)}{2} \quad (7)$$

## *4.5 Define Fuzzy Inference System (FIS)*

Fuzzy Inference System is a process of obtain an output value from an input on basis of fuzzy inference rules. Fuzzy Inference operation involves FIS Editor, membership functions, Fuzzy interface rule, Evaluate FIS.

In this work, FIS is created by specifying X, Y image gradient and zero-mean Gaussian membership function for each input. If the gradient value for a pixel is 0, then it belongs to the zero membership function with a degree of 1.

The standard deviation Sx and Sy are the zero membership function for the Ix and Iy inputs. These values can change; increasing the values of Sx, Sy makes the algorithm less sensitive to the edges in the image and decreases the intensity of the detected edges. Specify triangular membership function for black, white pixel in Iout.

### 4.5.1 Specify FIS Rules

The rule editor is used for editing list of rules that defines the behavior of the system. In proposed work "logical or" rules is used as shown in Table 1. $I_X$, $I_Y$ are

**Table 1** Fuzzy inference system rules

| $I_X$ | $I_Y$ | $I_{OUT}$ ($I_X$ or $I_Y$) |
|---|---|---|
| 0 | 0 | 0 |
| 0 | 1 | 1 |
| 1 | 0 | 1 |
| 1 | 1 | 1 |

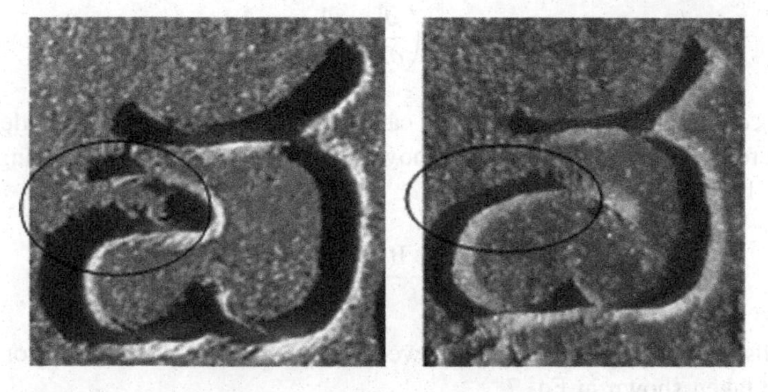

**Fig. 2** Captured Kannada stone inscription character 'SHA' and 'VA'

image gradients in X and Y direction. $I_{out}$ is the output of the system. Logical one and zero is represent white Pixel and black pixel respectively.

### 4.5.2 Evaluate FIS

Evaluate the output of the edge detector for each row of pixels in $I_{out}$ using corresponding rows of $I_x$ and $I_y$ as inputs.

## *4.6 Advance Recognition Algorithm (ARA)*

The pre-processed characters were passed into Advance Recognition algorithm (ARA) [2]. The ARA algorithm recognizes each Kannada character in an image by two steps, first it calculates Mean and Sum of absolute difference value of featured extraction character and database characters. Based on its shape and size it recognizes stone inscriptions Kannada characters in a test image [6].

The Mean and Sum of Absolute difference value of an image is calculate by using Eqs. 8, 9 and 10.

$$x = f^{-1}\frac{1}{n}\sum_{i=1}^{n} f(xi) \tag{8}$$

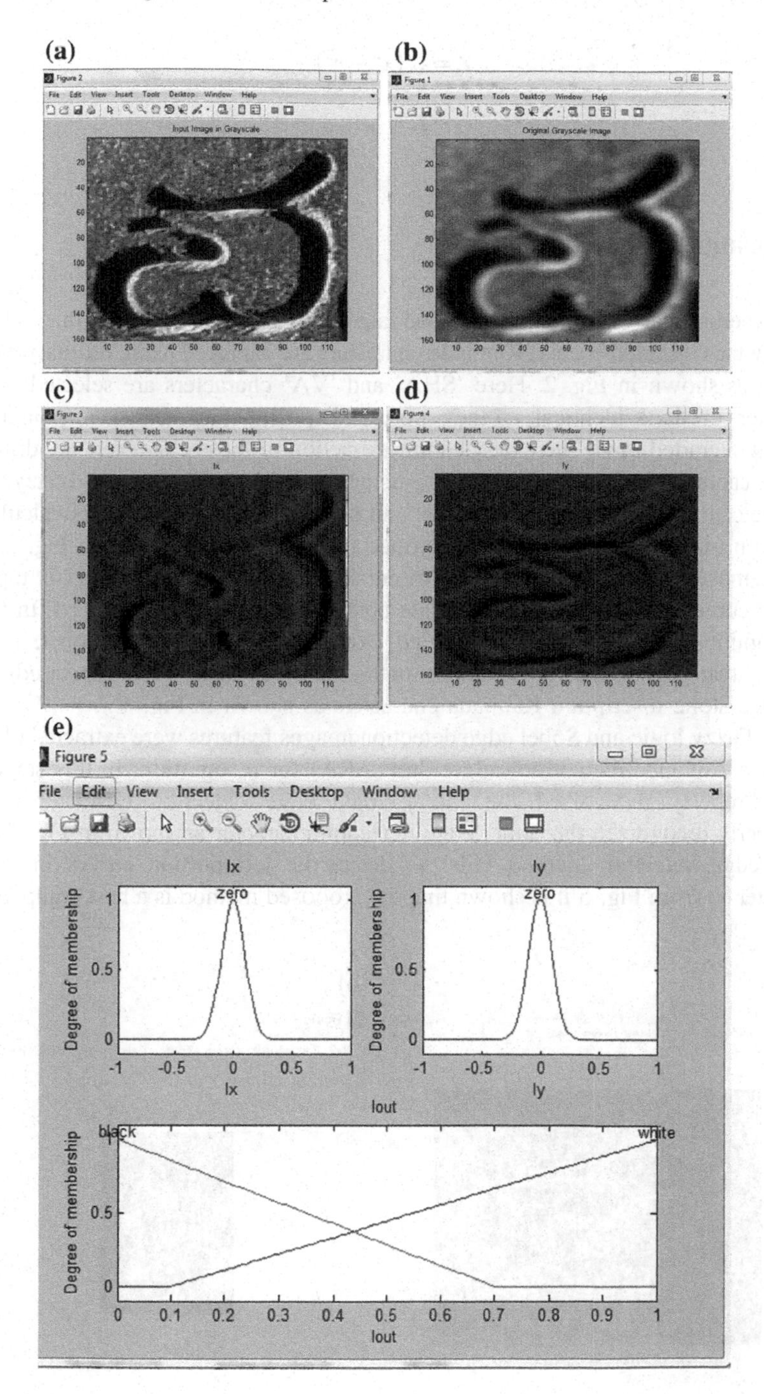


**Fig. 3** **a** Input image, **b** gray scaled image, **c** gradient Ix, **d** gradient IY, **e** degree membership $I_x$, $I_y$

$$Z = \sqrt{(x^2 - y^2)} \tag{9}$$

$$A = |Z| \tag{10}$$

## 5 Results and Discussion

The experimental results of developed algorithm are discussed in this section. Initially the degraded special Kannada characters are captured using ordinary digital camera as shown in Fig. 2. Here 'SHA' and 'VA' characters are selected. These characters are look identical, so they are called special characters. Due to identical look, in degraded mode the exact feature extraction of these characters are difficult.

The captured image was filtered by using mean shift filter. Using Fuzzy logic edge detection, the $I_X$, $I_Y$, coordinators and degree of Membership was calculated. Based on fuzzy rule each edge is identified (white or black) shown in Fig. 3.

The most important deference between 'SHA' and 'VA' characters is top left portion curves, these curves differentiate both characters shown in Fig. 4. In fuzzy this significant portion was indentified exactly when compared Sobel, it was affirmed that proposed edge detection method is preeminent for to detect edges of degraded stone inscription Kannada characters as shown in Fig. 4.

The Fuzzy logic and Sobel edge detection images features were extracted [4] and feed into Advance recognition algorithm (ARA) for recognition. In this stage the ARA Properly recognized the Fuzzy logic edge detection character and it improperly recognized the Sobel edge detection character as shown in Fig. 5. The Sobel edge detection method fails' to detect the left portion curves in 'SHA' character so from Fig. 5 it is shown that the proposed method is a best compared to

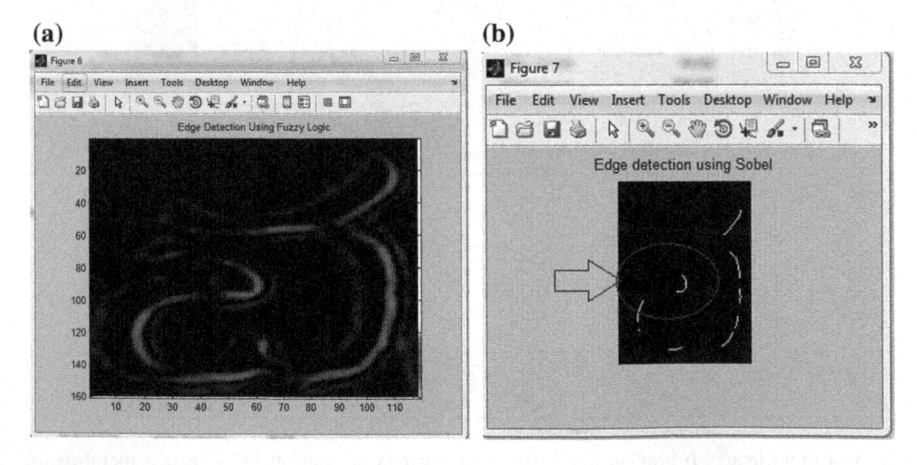


**Fig. 4** Comparison of fuzzy logic edge detection image (*left*) and Sobel edge detection (*right*) image

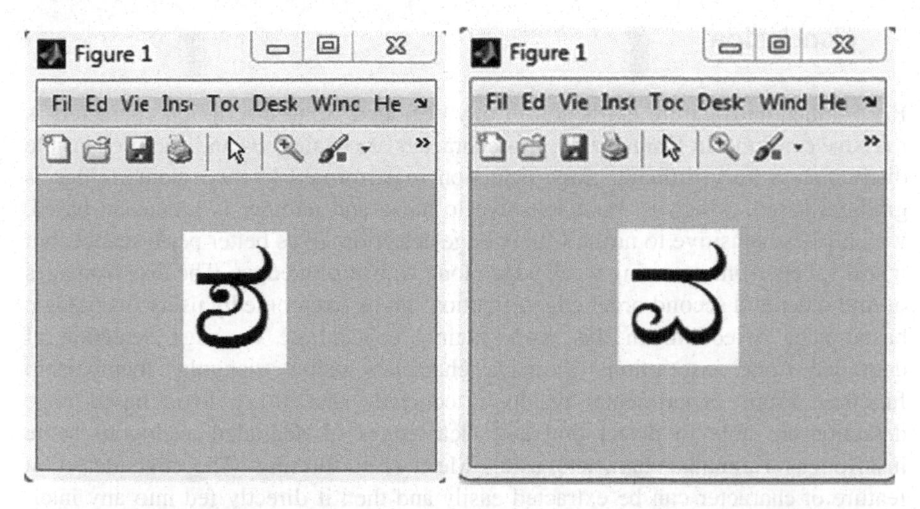


**Fig. 5** Accurate recognition of 'SHA' Character (*left*) and inaccurate recognition of 'SHA' Character (*right*)

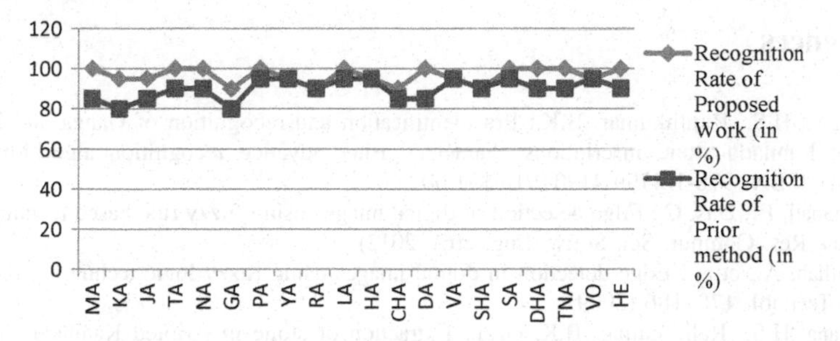


**Fig. 6** Results comparison

Sobel detection algorithm in recognition of old degraded Kannada stone Inscriptions characters.

## 6 Results Validation

The extracted feature from twenty degraded stone inscription characters are feed into Advance Recognition algorithm (ARA) [1] and Characters recognition rate is compared with Sobel edge detection method [5]. Figure 6 it is shown that the proposed work have 99.8% accuracy when compare to prior method (Sobel).

## 7 Conclusion

Recognition and feature extraction of any degraded stone inscription characters is massive confront. In Kannada 20 plus characters are analogous and comprehensible distinction is very difficult. Edge detection is performed by two methods that is gradient based, which is more sensitive to noise and another is Laplacian based, which is less sensitive to noise. Canny edge detection gives better performance, but it still suffers from detecting weak edge along with strong edges. The disadvantages of first order and second order edge detection can be overcome by using fuzzy logic based edge detection. In this work mainly concentrate on edge detection of degraded stone inscriptions Kannada characters using triangular membership function. From experimental results it conclude that fuzzy logic based edge detection are able to detect thin and clear edges of degraded analogous stone inscriptions Kannada characters using Mean shift filtering. The edge detection feature of character can be extracted easily and then it directly fed into any intelligence network for recognition.

## References

1. Mohana, H.S., Rajithkumar, B.K.: Era identification and recognition of Ganga and Hoysala phase Kannada stone inscriptions characters using advance recognition algorithm. IEEE (2014). ISBN: 978-1-4799-4190-2/14/$31.00
2. Bhagabati, B., Das, C.: Edge detection of digital images using fuzzy rule based technique. Int. J. Adv. Res. Comput. Sci. Softw. Eng. **2**(6) (2012)
3. Abdullah, A., et al.: Edge detection in digital images using fuzzy logic technique. IEEE Sci. Eng. Technol. 178–186 (2009)
4. Mohana, H.S., Rajithkumar, B.K. et al.: Extraction of stone in-scripted Kannada characters using sift algorithm based image mosaic. Int. J. Electr. Commun. Technol. (IJECT) **5**(2) (2014). ISSN: 2230-7109
5. Rajithkumar, B.K. et al.: Read and recognition of old Kannada stone inscriptions characters using novel algorithm. IEEE 308–312 (2015). ISBN: 978-1-4673-9825-1/15/$31.00
6. Rajithkumar, B.K. et al.: Template matching method for recognition of stone inscripted Kannada characters of different time frames based on correlation analysis. Int. J. Comput. Electr. Res. (IJCER) **3**(3) (2014). ISSN: 2278-5795

# Design of Alpha/Numeric Microstrip Patch Antenna for Wi-Fi Applications

**R. Thandaiah Prabu, R. Ranjeetha, V. Thulasi Bai and T. Jayanandan**

**Abstract** The Microstrip Patch Antenna's (MPA's) find usability in several day to day applications. They hold several advantages like low-profile structure, low fabrication cost, and they support both circular as well as linear polarizations, etc. This paper proposes MPA for Wi-Fi applications where it is used within the 2.4 GHz range of frequencies (IEEE 802.11). Although various antenna designs are currently prevalent to support the existing systems, in-order to overcome the bandwidth limitations, Alpha-Numeric (Alphabets and Numbers) Microstrip Patch Antennas are proposed in this paper. The MPA's are designed with 2.4 GHz as their resonant frequency. The simulation results based on the essential antenna performance analysis parameters like Return Loss, Gain, Radiated Power and Directivity are discussed. The antenna's are designed with a thickness of 1.5 mm, height of 70 mm and width of 60 mm, the substrate material of the antenna's is Flame Retardant 4 (FR4) and its relative permittivity is 4.4. The designs for all the Microstrip Patch Antennas are simulated using Advanced Design System 2009 (ADS) software.

**Keywords** Wi-Fi · ADS · FR4 · Microstrip patch antenna (MPA) · Radiation loss · Directivity

---

R. Thandaiah Prabu (✉)
Anna University, Chennai, India
e-mail: thandaiah@gmail.com

R. Thandaiah Prabu · R. Ranjeetha
Prathyusha Engineering College, Thiruvallur, India
e-mail: ranji.ae001@gmail.com

V. Thulasi Bai
KCG College of Technology, Chennai, India
e-mail: thulasi_bai@yahoo.com

T. Jayanandan
Waveguide Technologies, Chennai, India
e-mail: jayanandan@waveguidetech.com



S.C. Satapathy et al. (eds.), *Data Engineering and Intelligent Computing*,
Advances in Intelligent Systems and Computing 542,
DOI 10.1007/978-981-10-3223-3_3

## 1 Introduction

Wireless Fidelity technology is an essentiality in LAN topologies. Wi-Fi mainly uses 2.4 GHz frequency range with IEEE 802.11 standard. Wi-Fi is one of the rapidly growing wireless communication technologies, whose need is drawn based on the requirement to provide QoS guarantee for high traffic end user connectivity. Proper antenna design is the fundamental necessity to ensure high device performance. Antenna's designed with high radiation; low return loss and high efficiency

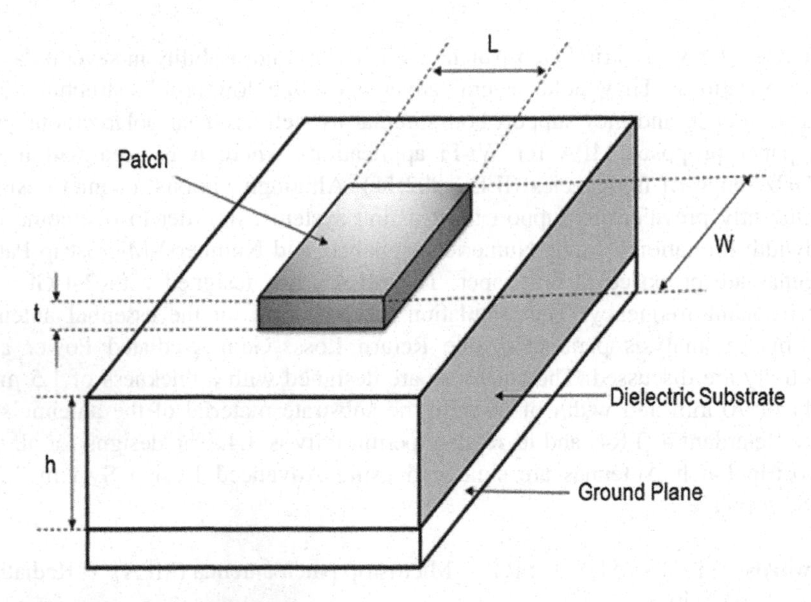


**Fig. 1** Structure of microstrip patch antenna

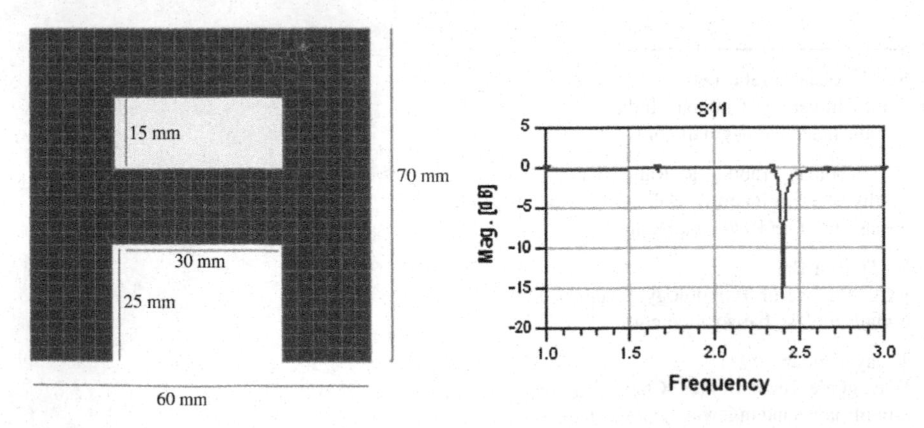


**Fig. 2** A shape antenna and return loss of antenna

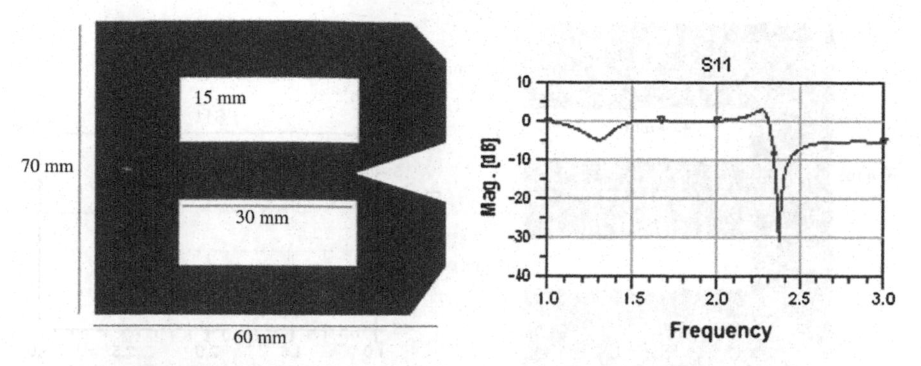


**Fig. 3** B shape antenna and return loss of antenna

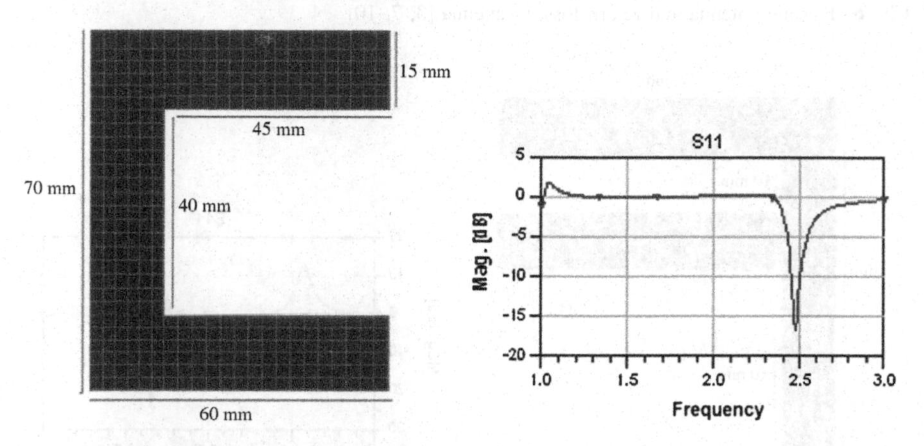


**Fig. 4** C shape antenna and return loss of antenna [2]

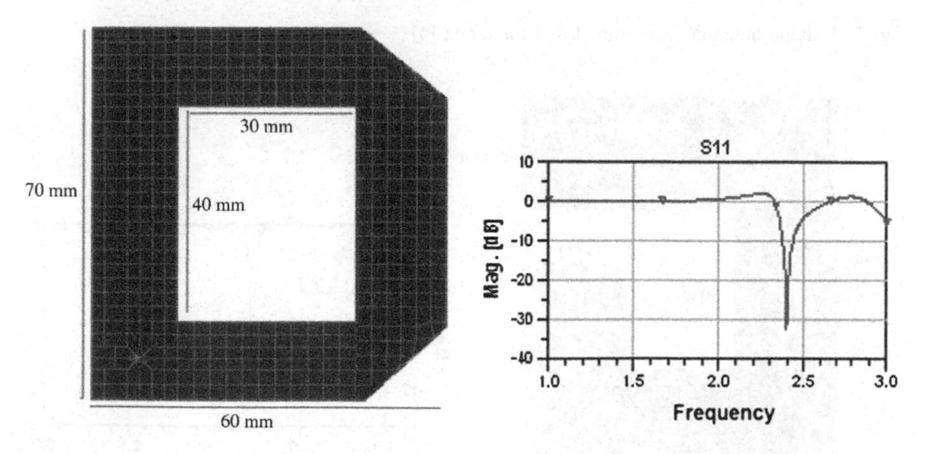


**Fig. 5** D shape antenna and return loss of antenna

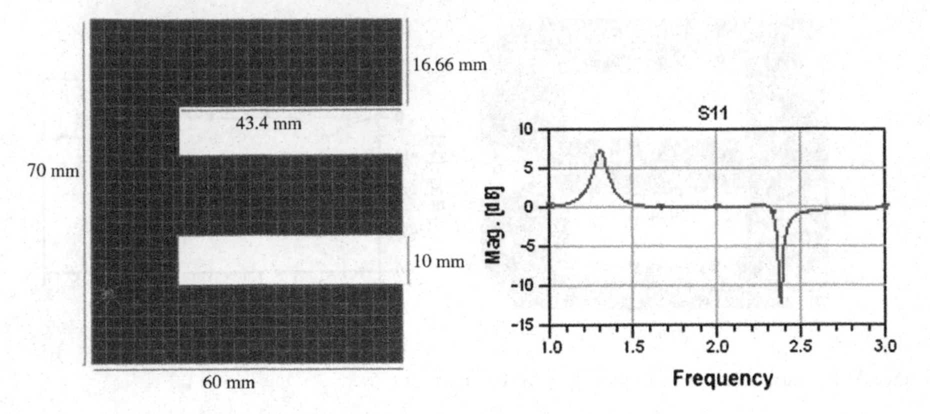


**Fig. 6** E shape antenna and return loss of antenna [3, 7, 10]

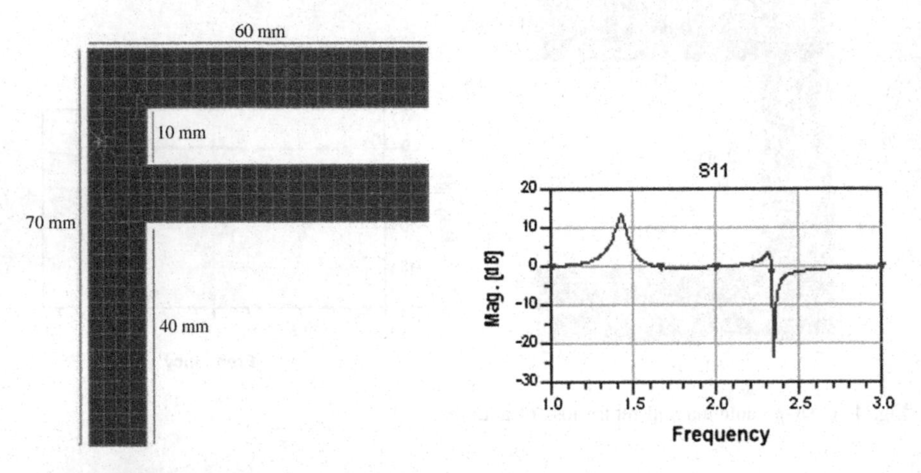


**Fig. 7** F shape antenna and return loss of antenna [4]

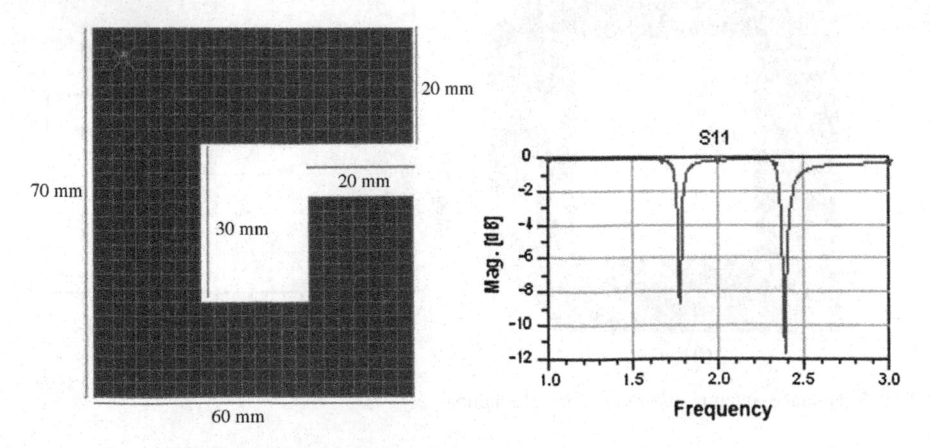


**Fig. 8** G shape antenna and return loss of antenna

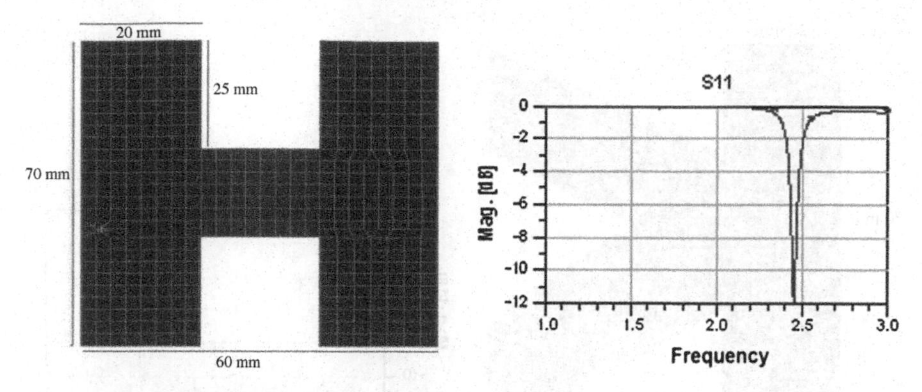


**Fig. 9** H shape antenna and return loss of antenna [5, 10]

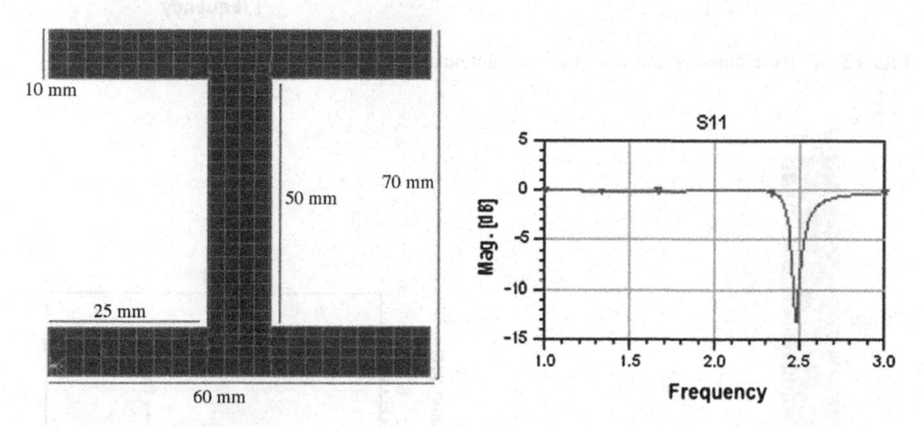


**Fig. 10** I shape antenna and return loss of antenna [6]

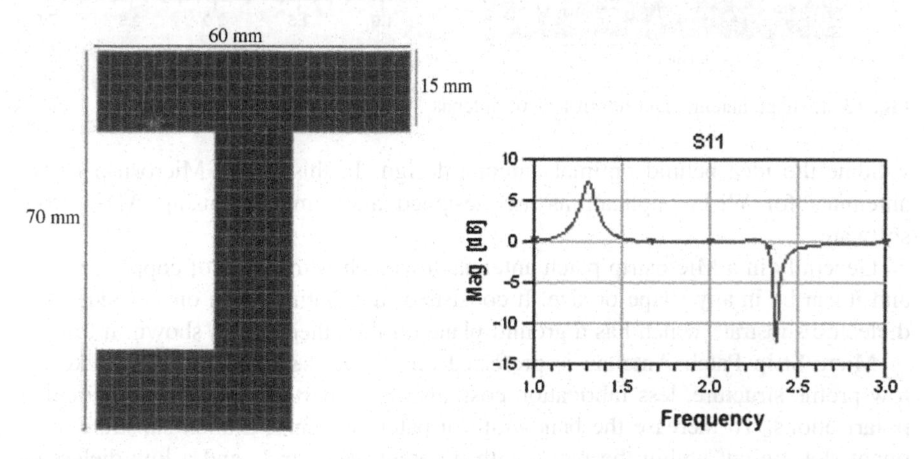


**Fig. 11** J shape antenna and return loss of antenna

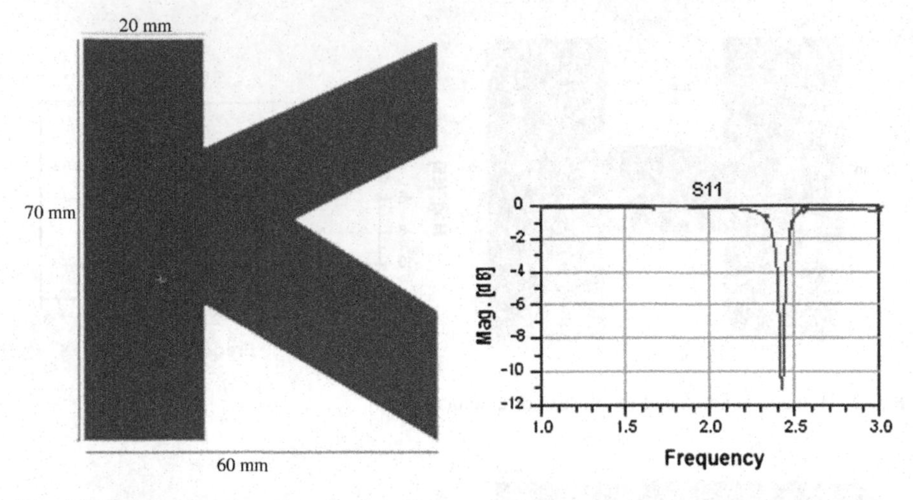


**Fig. 12** K shape antenna and return loss of antenna

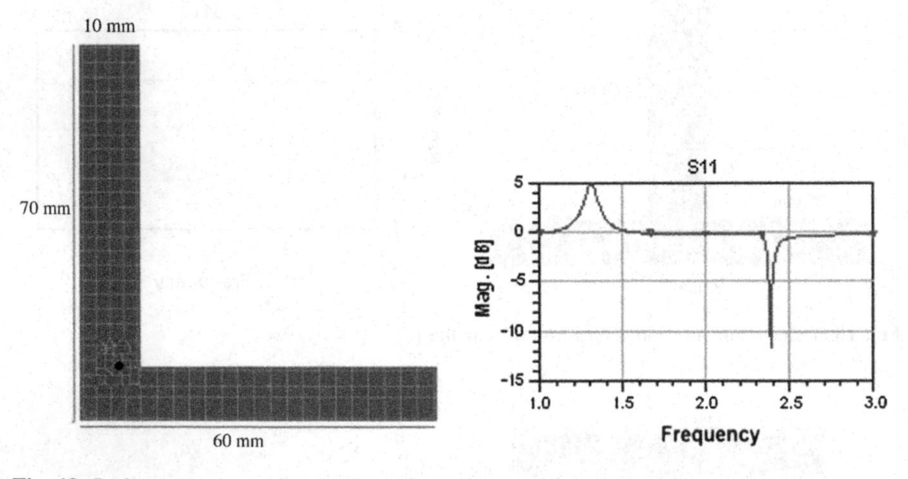


**Fig. 13** L shape antenna and return loss of antenna

validate the idea behind optimal antenna design. In this paper, Microstrip patch antenna's for Wi-Fi applications are designed and simulated using ADS 2009 software.

Generally in a Microstrip patch antenna, the patch is made up of copper or gold and it can be in any shape or size. It consists of a radiating patch on one side of a dielectric substrate which has a ground plane on the other side as shown in Fig. 1.

Micro-Strip Patch Antenna is preferred due to the several advantages like its low-profile structure, less fabrication cost; also it supports both linear and circular polarizations. To increase the bandwidth of patch antennas, a thick substrate, resonant slot are cut within the patch with different geometry's and a low dielectric substrate have to be used [1].

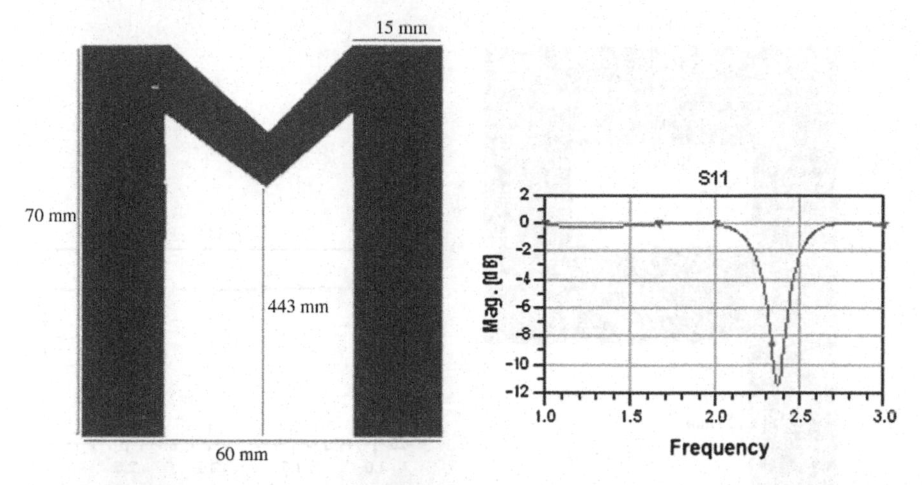


**Fig. 14** M shape antenna and return loss of antenna

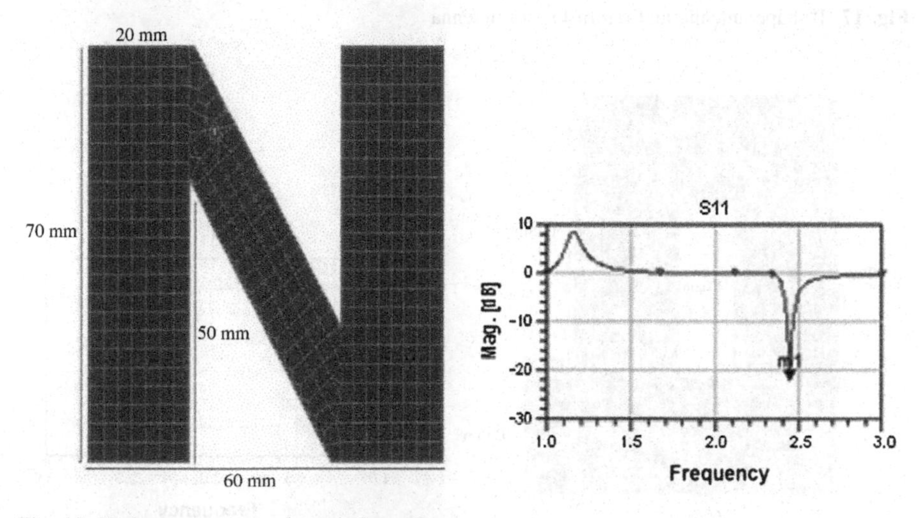


**Fig. 15** N shape antenna and return loss of antenna

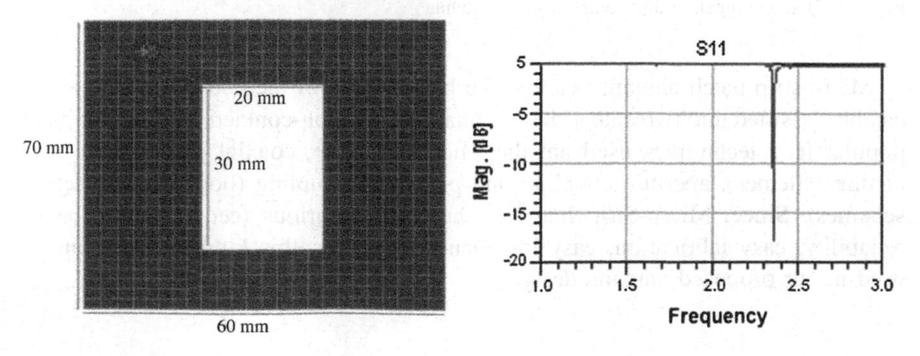


**Fig. 16** O shape antenna and return loss of antenna

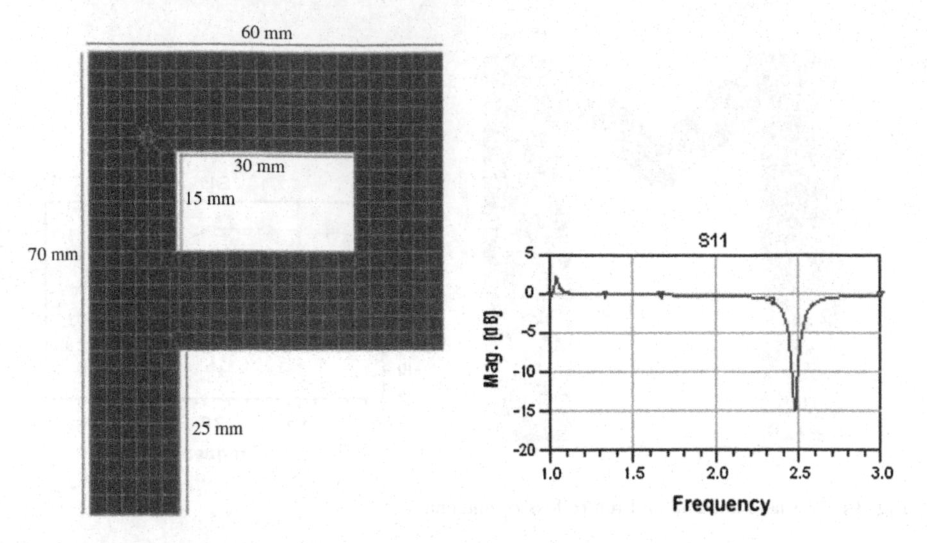


**Fig. 17** P shape antenna and return loss of antenna

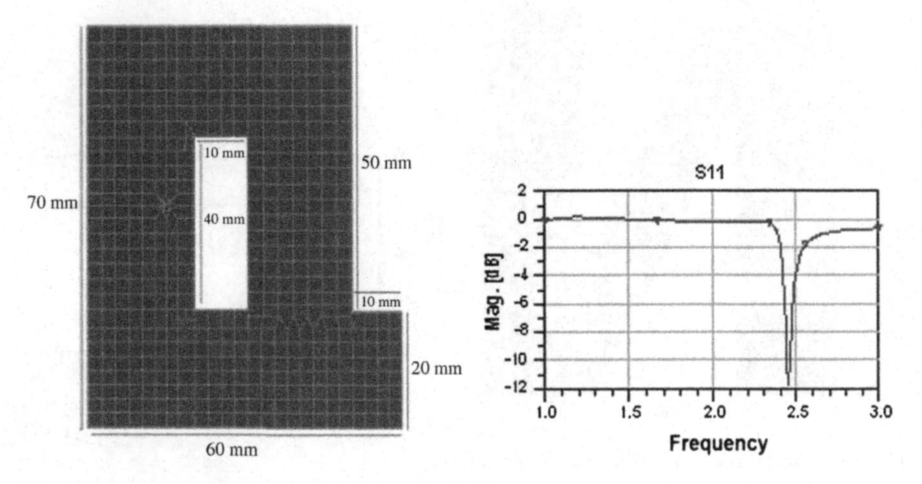


**Fig. 18** Q shape antenna and return loss of antenna

Micro-strip patch antennas can be fed by a variety of methods. These methods can be classified into two categories—contacting and non-contacting. The four most popular feed techniques used are the Micro-strip line, coaxial probe (both contacting schemes), aperture coupling and proximity coupling (both non-contacting schemes). Since, Micro-Strip line feed has more spurious feed radiation, better reliability, easy fabrication, easy impedance matching, this kind of feed point is used in our proposed antenna designs.

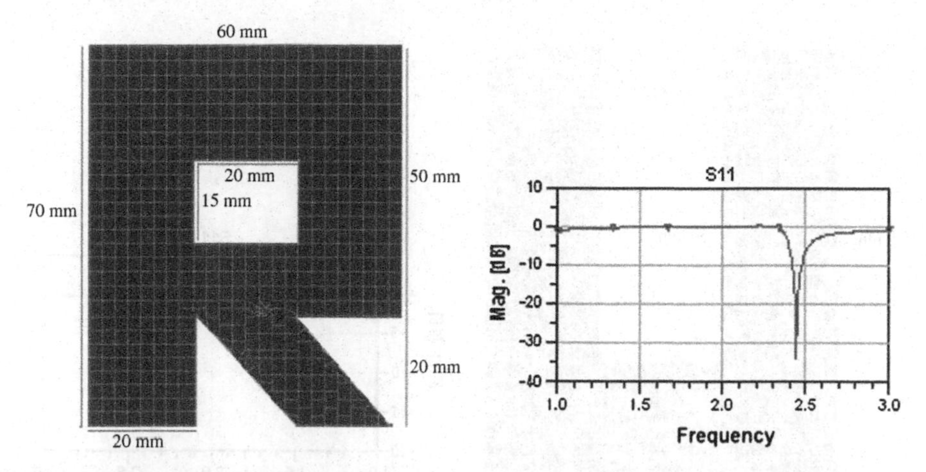


**Fig. 19** R shape antenna and return loss of antenna

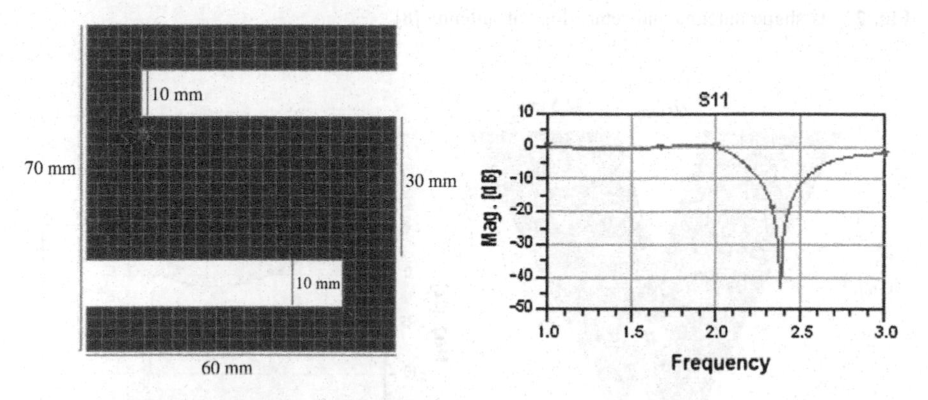


**Fig. 20** S shape antenna and return loss of antenna [7]

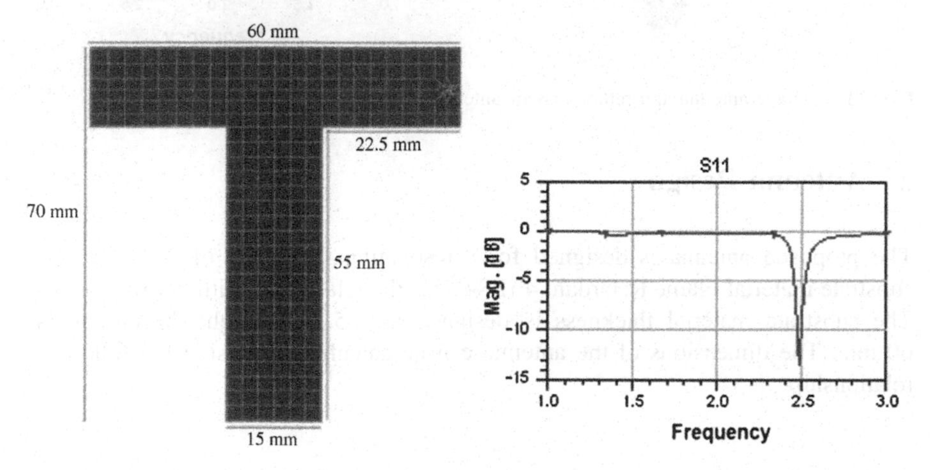


**Fig. 21** S shape antenna and return loss of antenna

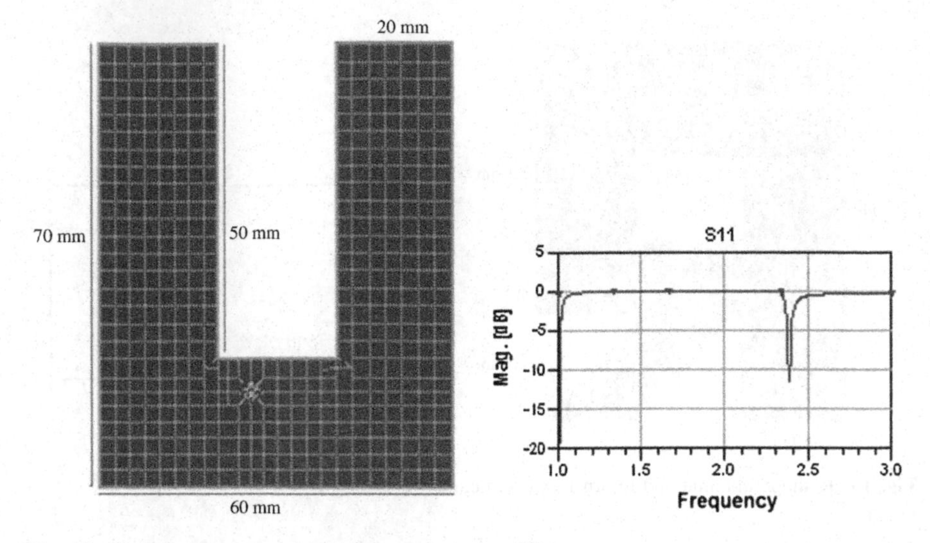


**Fig. 22** U shape antenna and return loss of antenna [8]

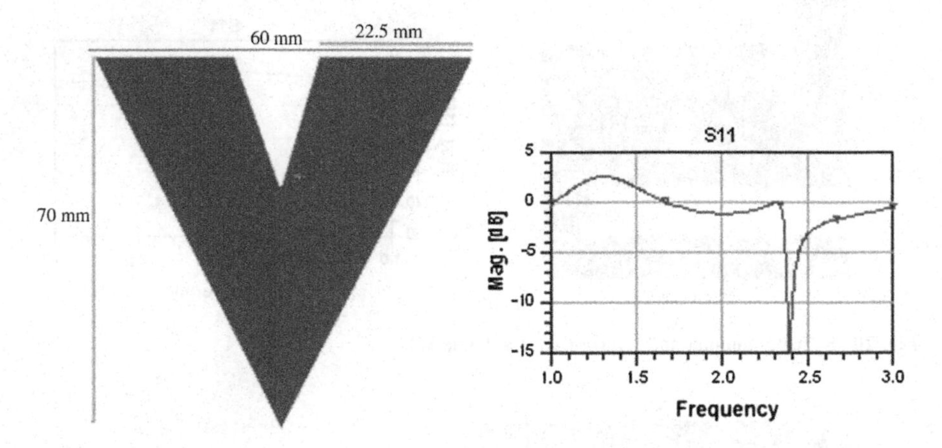


**Fig. 23** V shape antenna and return loss of antenna

## 2 Antenna Design

The proposed antenna is designed for a resonating frequency of 2.4 GHz. The substrate material Flame Retardant 4 (FR4) has the relative permittivity of $\varepsilon_r = 4.4$. The substrate material thickness is designed as 1.5 mm, height 70 mm, width 60 mm. The dimensions of the antenna can be calculated by using the following relationship.

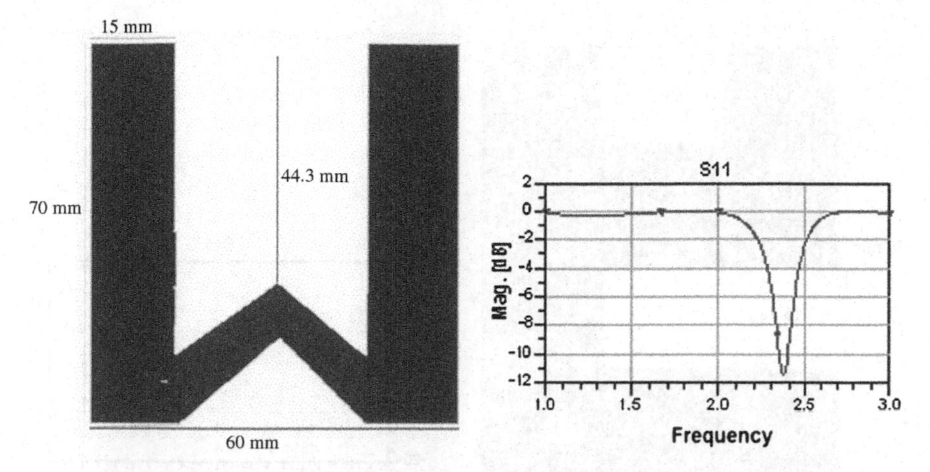


**Fig. 24** W shape antenna and return loss of antenna

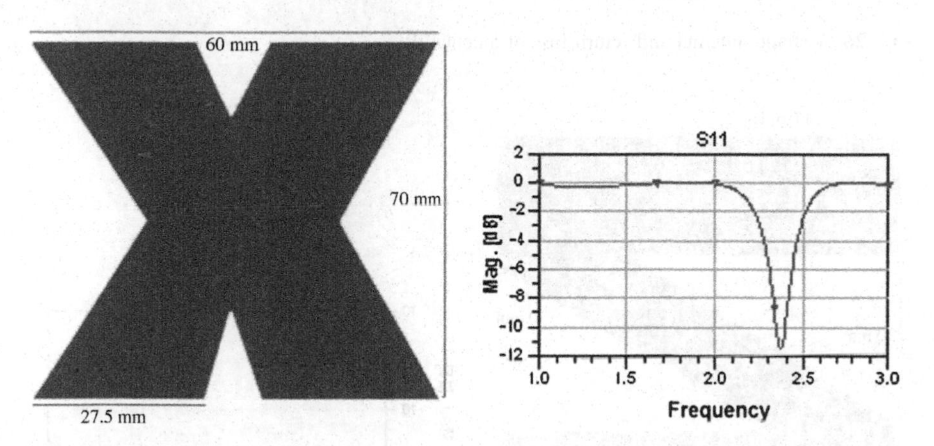


**Fig. 25** X shape antenna and return loss of antenna

The width of the patch is given by

$$W = \frac{c}{2f_o}\sqrt{\frac{2}{\epsilon_r + 1}} \tag{1}$$

The effective dielectric constant $\varepsilon_{\text{reff}}$ is given as

$$\varepsilon_{reff} = = \frac{\varepsilon_r + 1}{2} + \frac{\varepsilon_r - 1}{2}\left[1 + 12\frac{h}{w}\right]^{\frac{-1}{2}} \tag{2}$$

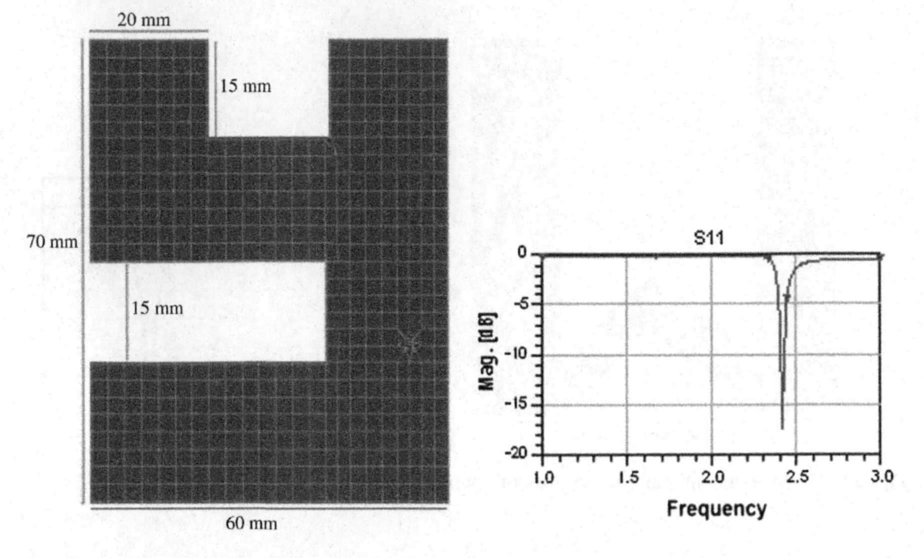


**Fig. 26** Y shape antenna and return loss of antenna [9]

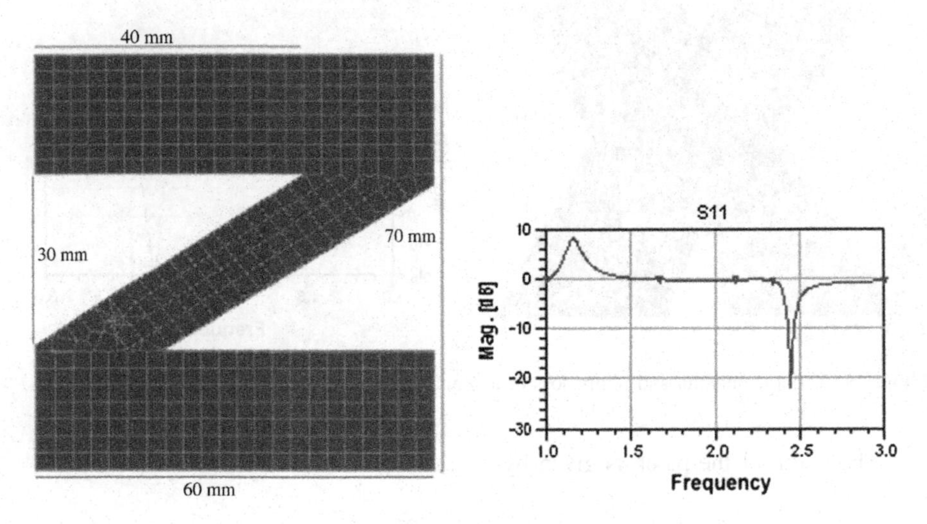


**Fig. 27** Z shape antenna and return loss of antenna [10]

The length extension of the patch Δl is given as

$$\Delta L = 0.412\text{h}\left[\frac{(\varepsilon_{reff} + 0.3)\left(\frac{w}{h} + 0.264\right)}{(\varepsilon_{reff} - 0.258)\left(\frac{w}{h} + 0.8\right)}\right] \tag{3}$$

**Table 1** Comparison of alphabet shape antennas

| Shape | Frequency | Return loss (dB) | Gain (dB) |
|---|---|---|---|
| A | 2.4 | −17 | 5.13 |
| B | 2.39 | −31 | 6.11 |
| C | 2.48 | −17 | 6.03 |
| D | 2.39 | −34 | 3.02 |
| E | 2.39 | −13 | 5.57 |
| F | 2.43 | −20 | 5.84 |
| G | 2.39 | −12 | 5.30 |
| H | 2.45 | −12 | 4.13 |
| I | 2.48 | −13 | 4.28 |
| J | 2.39 | −13 | 5.57 |
| K | 2.4 | −11.5 | 4.23 |
| L | 2.4 | −12 | 3.08 |
| M | 2.38 | −11 | 4.93 |
| N | 2.42 | −22 | 3.23 |
| O | 2.38 | −18 | 6.15 |
| P | 2.46 | −15 | 5.77 |
| Q | 2.45 | −12 | 5.55 |
| R | 2.41 | −35 | 5.96 |
| S | 2.39 | −44 | 4.95 |
| T | 2.47 | −13 | 4.56 |
| U | 2.39 | −12 | 5.71 |
| V | 2.39 | −15 | 5.59 |
| W | 2.37 | −11.21 | 4.96 |
| X | 2.38 | −11 | 4.93 |
| Y | 2.4 | −18 | 5.26 |
| Z | 2.4 | −22 | 3.24 |

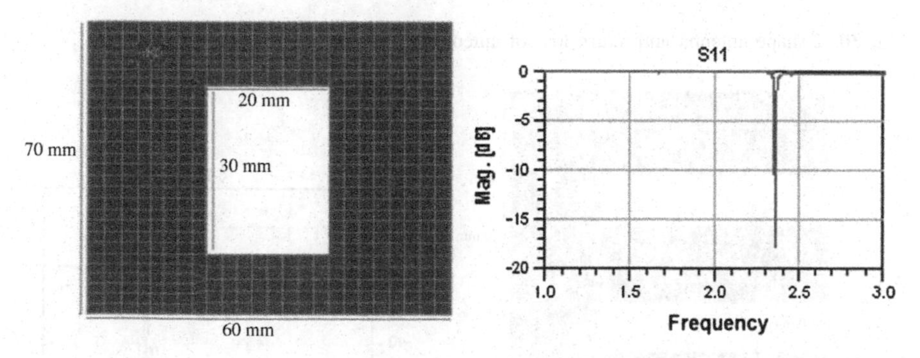


**Fig. 28** 0 shape antenna and return loss of antenna

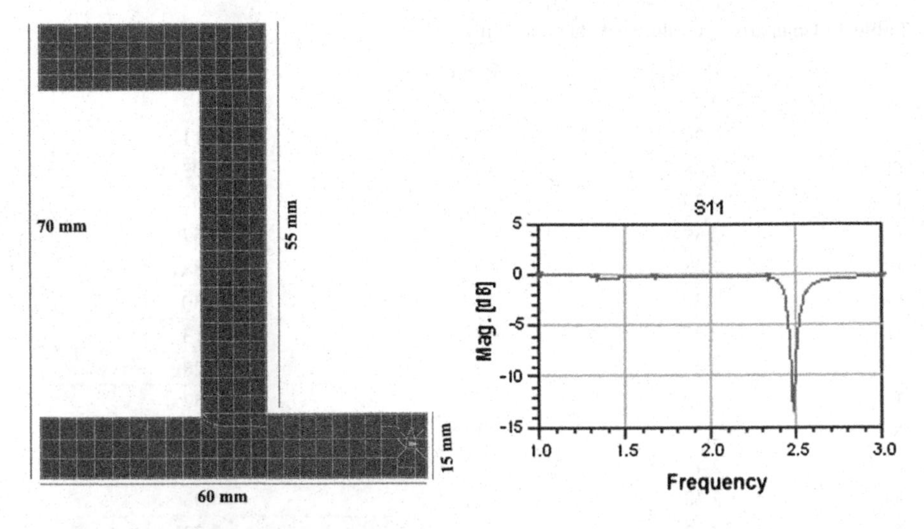


**Fig. 29** 1 shape antenna and return loss of antenna

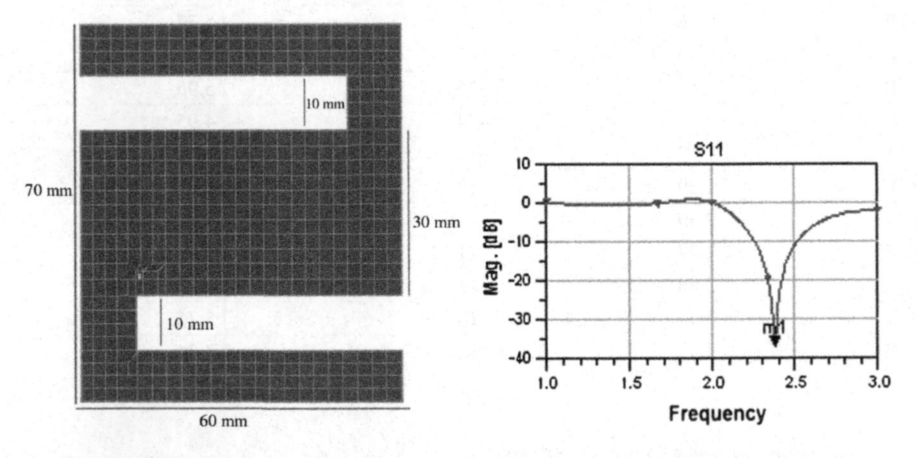


**Fig. 30** 2 shape antenna and return loss of antenna

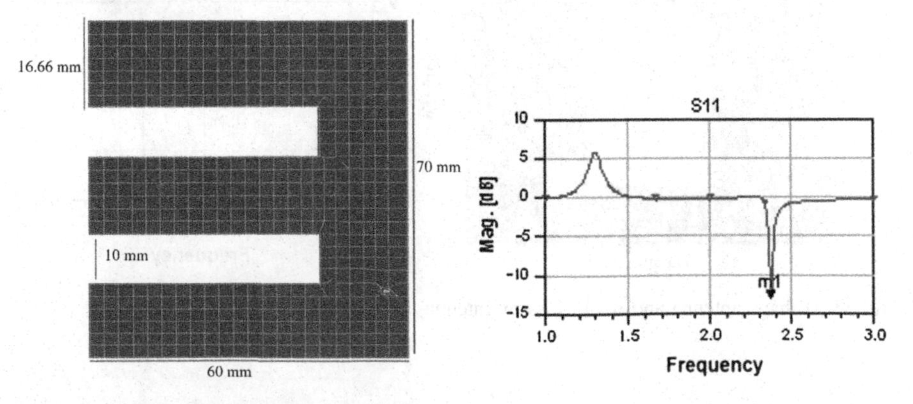


**Fig. 31** 3 shape antenna and return loss of antenna

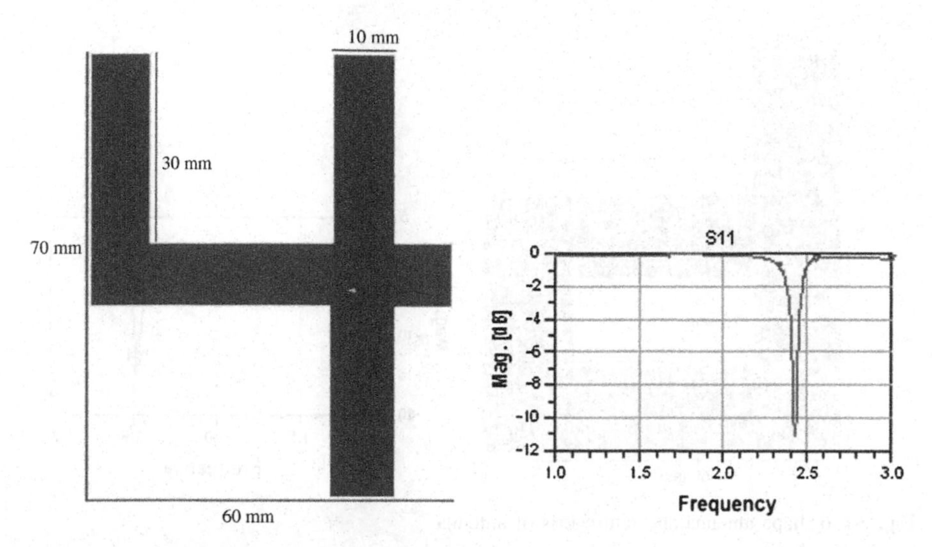


**Fig. 32** 4 shape antenna and return loss of antenna

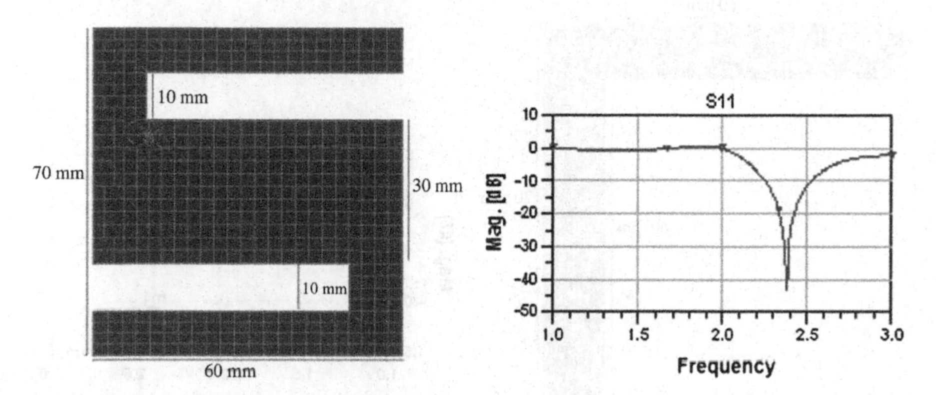


**Fig. 33** 5 shape antenna and return loss of antenna

The effective Length of patch $L_{eff}$ is given by

$$L_{reff} = \frac{C}{2f_o\sqrt{\varepsilon_{reff}}} \tag{4}$$

The final patch length L is given by

$$L = L_{eff} - 2\Delta L \tag{5}$$

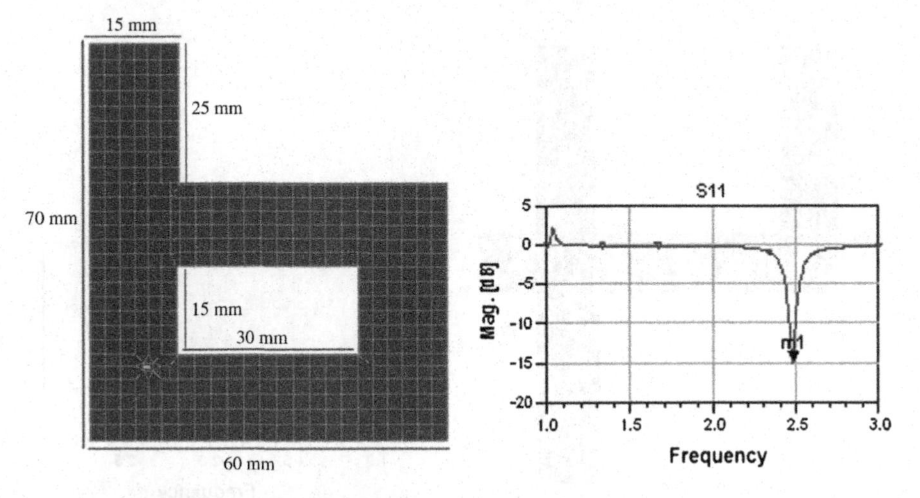


**Fig. 34** 6 shape antenna and return loss of antenna

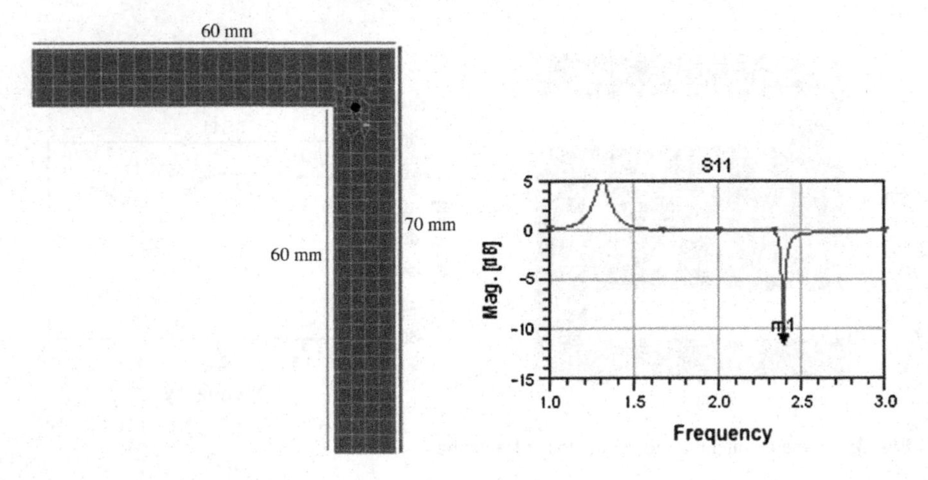


**Fig. 35** 7 shape antenna and return loss of antenna

The ground plane dimension is given by

$$W_g = 6h + W \tag{6}$$

$$L_g = 6h + W \tag{7}$$

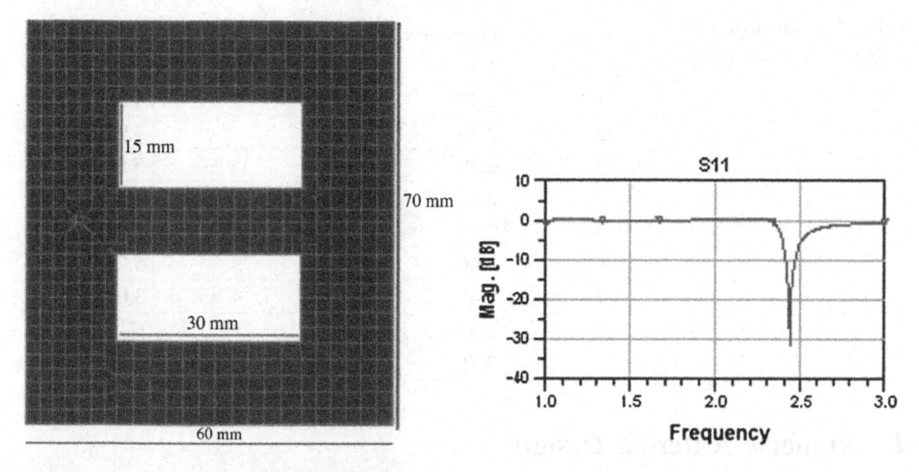


**Fig. 36** 8 shape antenna and return loss of antenna

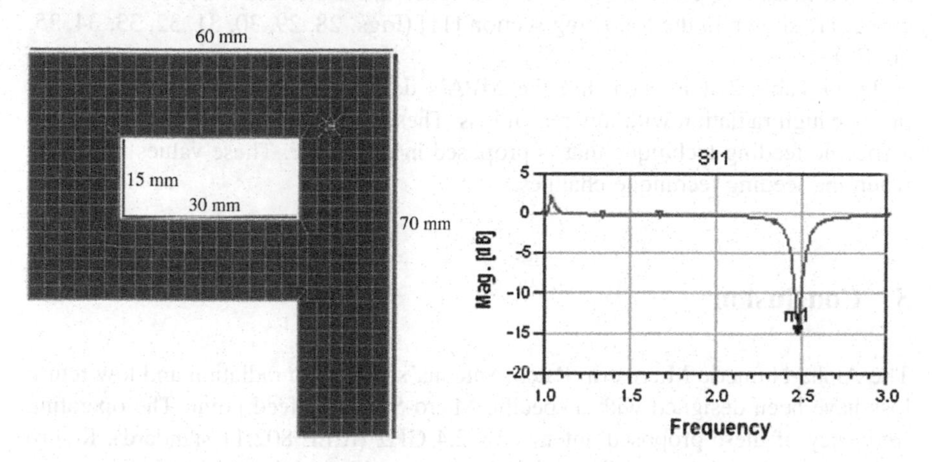


**Fig. 37** 9 shape antenna and return loss of antenna

# 3 Alphabet Antenna Design

The following section holds the pictorial representation of the various MPA's designed based on alphabetical shapes. Each alphabet shaped MPA's are substantiated using their return loss curves (Figs. 2, 3, 4, 5, 6, 7, 8, 9, 10, 11, 12, 13, 14, 15, 16, 17, 18, 19, 20, 21, 22, 23, 24, 25, 26, 27).

The Table 1 relates the antenna frequency, return loss and gain values of the corresponding MPA's. Amongst the 23 antennas considered, the MPA's designed based on alphabet shapes B, R and S give high radiation with less return loss.

**Table 2** Comparison of numeric shape antennas

| Shape | Frequency | Return loss (dB) | Gain (dB) |
|---|---|---|---|
| 0 | 2.38 | −18 | 6.15 |
| 1 | 2.47 | −13 | 4.56 |
| 2 | 2.39 | −44 | 4.95 |
| 3 | 2.39 | −13 | 5.57 |
| 4 | 2.4 | −11.5 | 4.23 |
| 5 | 2.39 | −44 | 4.95 |
| 6 | 2.46 | −15 | 5.77 |
| 7 | 2.4 | −12 | 3.08 |
| 8 | 2.39 | −31 | 6.11 |
| 9 | 2.46 | −15 | 5.77 |

## 4 Numeric Antenna Design

The numerical shape based MPA's are designed and simulated and their return loss curves are shown in the following section [11] (Figs. 28, 29, 30, 31, 32, 33, 34, 35, 36, 37).

From Table 2 it is seen that the MPA's designed based on shapes 2 and 5 provide high radiation with low return loss. These characteristics are obtained using a specific feeding technique that is proposed in this paper. These values may vary when the feeding technique changes.

## 5 Conclusion

The Alpha-Numeric Microstrip Patch Antenna's with high radiation and low return loss have been designed with a specific Micro-Strip line feed point. The operating frequency of these proposed antennas is 2.4 GHz (IEEE 802.11 standard). Return loss, gain, radiated power, directivity are discussed based on the simulation results obtained using ADS 2009 software. This Alpha-Numeric antenna's can be fabricated to analyse the overall system performance by studying the antenna system characteristics when used in tandem with an appropriately designed filter and amplifier.

## References

1. Prabu, R.T., Reddy, G., Reddy, K.: Design and implementation of E-shaped antenna for GSM/3G applications. IJISET **2**(4) (2015)
2. Bagal, K.M.: C-shaped microstrip patch for dual band on different shape ground plane. IJCA **121**(16) (2015)

3. Ramesh, B., Lakshmi, V.R., Raju, G.S.N.: Design of E-shaped triple band microstrip patch antenna. IJERA **3**(4) (2013)
4. Gupta, S., Arora, R.: A novel miniaturized, multiband F shape fractal antenna. IJARCSSE **5**(8) (2015)
5. Majumdar, A.: Design of an H shaped microstrip patch antenna for bluetooth applications. IJIAS **3**(4), 987–994 (2013)
6. Trivedi, N., Gurjar, S., Singh, A.K.: Design and simulation of novel I shape fractal antenna. IJEST **4**(11) (2012)
7. Arora, H., Jain, K., Rastogi, S.: Review of performance of different shapes (E, S, U) in micro-strip patch antenna. IJSRET **4**(3), 178–181 (2015)
8. Kamble, R., Yadav, P., Trikolikar, V.U.: Design of U-shape microstrip patch antenna at 2.4 GHz **4**(3) (2015)
9. Singh, S., Rani, S.: Design and development of Y shape power divider using improved DGS. IJESTTCS **2**(6) (2013)
10. Venkatrao, K., Reddy, M.: Design and performance analysis of microstrip patch antenna using different dielectric substrates. IIJEC **2**(10) (2014)
11. Kriti, V., Abishek, A., Ray, T.B.: Development and design of compact antenna on seven segment pattern. IJERGS **3**(3) (2015)

8. [illegible] (2015)

# Performance Evaluation of Leaf Disease Measure

**D. Haritha**

**Abstract** Leaf disease is a state where I find the abnormality observation in the growth of the plant. Most of the diseases I can easily find out by observing the conditions on leaf at regular intervals of time. If I found the diseases in early stage, then, I can save the plant without further growth of the disease by taking necessary actions. So, here made an attempt to find the leaf disease measure. Local Binary pattern, median filter, Morphological operations and edge detection are used for analyzing the disease in leaf images. For comparing the disease level, Rank table is considered. Finally, calculates the execution time for measuring the leaf disease in a leaf by using various edge detection techniques. This is further extended by using other techniques like hierarchical clustering.

**Keywords** Edge detection techniques · Morphological operations · Local binary pattern

## 1 Introduction

Agriculture has become an important area for growing populations. Plants are playing an important role for solving many problems like global warming etc. There is an emergency need of saving the plants by not effecting with different diseases. Plant disease is a state where I find the abnormality observation in the growth of the plant. This plant disease is a visible reaction. With this, there is an abnormal coloration or interferes in the development or growth of the plant. Plant diseases are divided into two groups by considering their causes. They are Biotic and Abiotic. Biotic disease is caused due to bacteria, viruses, fungi …etc. Abiotic disease is caused due to the variation in temperature, Moisture, nutrition …etc. or I found the detectable changes in leafs like color, shape …etc. The Abiotic diseases

D. Haritha (✉)
Jawaharlal Nehru Technological University Kakinada,
Kakinada 533003, Andhra Pradesh, India
e-mail: harithadasari9@yahoo.com



S.C. Satapathy et al. (eds.), *Data Engineering and Intelligent Computing*,
Advances in Intelligent Systems and Computing 542,
DOI 10.1007/978-981-10-3223-3_4

are also called as symptoms. The Biotic diseases are also called as signs. The Diseases won't occur instantly and it is a time taking process [1, 2].

Most of the diseases I can easily find out by observing the conditions on leaf at regular intervals of time. If I found the diseases in early stage, then, I can save the plant without further growth of the disease by taking necessary actions.

Many researchers are developed several techniques and methods for identifying the leaf diseases. Shen Weizheng et al. developed a new method based on computer image processing. They considered the Otsu method for image segmentation for leaf regions. For disease spot edges, they used Sobel operator. They calculated the plant disease by calculating the quotient of disease spot and leaf areas [3]. Revathi and Hemalatha, they expressed a new technological strategies for cotton leaf spot images. They found the disease level by using the HPCCDD algorithm. They used the RGB feature ranging techniques for identifying the disease. Color segmentation techniques used for getting the target region and Homogenize techniques used for identifying the edges. Finally Extracted edges are used for classification of the disease spot [4].

Dheeb Al Bashish et al. they proposed the fast, cheap and accurate image processing based solution for leaf diseases. They segmented the image using the K-means algorithm. They used neural networks for Training the images. They proved that the leaf detection system gives the accurate results. Using this Neural network classifier based on statistical classification could detect successfully. And the detection rate is 93% [5]. In Yan-cheng-Zhang et al., a fuzzy feature selection approach, fuzzy curves and surfaces are proposed for selecting features of cotton disease leaves image. This method reduces the dimensionality of the feature space. The features of the leaf image selected by the FC and FS method is giving better performance than the features selected by human randomly [6].

Abdullah et al. studied leaf diseases in rubber trees for digital RGB color extraction. They identified regions of interest of the diseases then they further processed it to quantify the normalized data. They used artificial neural networks for classification. They found that using PCA model was giving loir network size [7]. In Sankaran et al., presented a review for recognizing the basic need for development in the agriculture. They compared the benefits and limitation of the methods [8]. In Ying et al., discussed and compared two filters that are simple filter and median filter. They found that using median filter, noise is removed effectively. They used two methods like edge detection and snake model for separating disease image from background. By using snake model, they found the desired results [9].

In Kurniawati et al., concentrated on extracting paddy features through off-line image. After image acquisition, they converted the RGB image into binary image. For the binary image, they applied the Otsu method and local entropy threshold. Morphological algorithm is used for removing noise and region filling. By using production rule technique, the paddy diseases are identified [10]. By using Support Vector Machine, in Jian and Wei, they recognized the leaf disease. They used the Radial Basis Function, polynomial and Sigmoid kernel function [11]. By using the BP neural network classifier in Liu and Zhou, they separated the healthy and diseased parts in the leaf images [12].

In Husin et al., they discussed early detection of the diseases in chili leafs. Leaf images are acquired and further processed for the knowing the health status. The image processing techniques are used for detecting the chili diseases. This system was inexpensive and very useful for chili diseases [13]. In Hashim et al., presented the classification methods for the five types of rubber leaf diseases. They used the spectrometer and SPSS. By using spectrometer, they clearly discriminating the diseases of five types in the rubber leaf [14].

Several researchers are worked for the identification of the leaf diseases. But, there is still a gap in the preprocessing of the leaf images. Here, in this paper, applied median filter for removing of the noise. The preprocessing technique local binary pattern (LBP) was used for elimination of illumination effects and to identify the minor changes among the pixels. These minor changes have significant effects on the leaf images for the identification of the leaf diseases. With this, the disease spots are easily identified in the leaf images. Different edge detection techniques by using the Sobel operator and Prewitt ...etc. are used for comparing the disease levels in the leaf images. Hence, in this paper, an attempt is made to find the leaf diseases and percentage of leaf that was affected due to disease.

## 2 Modeling

The overview of the leaf disease measure system is shown in the Fig. 1.

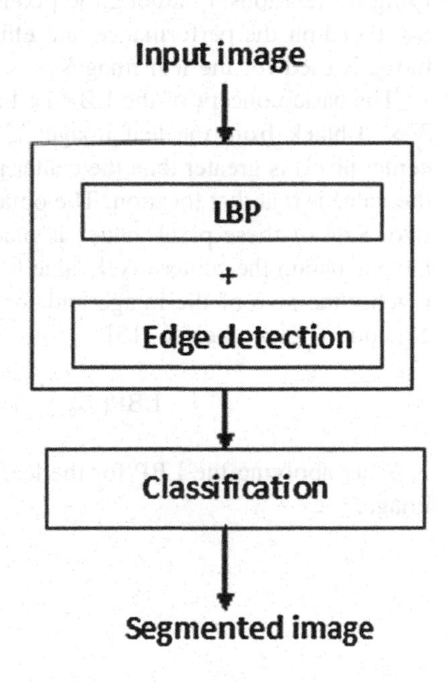


**Fig. 1** System overview

Here, the disease level is identified by two methods. In the first case, LBP was applied for the leaf image and then the edge detection techniques are applied for the LBP image which is obtained by applying LBP. In second case, edge detection techniques are applied first then for the resultant image is further applied with the technique of LBP. For the obtained resultant images, in two cases, are separately segmented using morphological operations. Finally, the results are compared with the rank table for identifying the disease measure.

# 3 Preprocessing

Preprocessing is done by two ways. In the first step, image acquisition was done. In the next step, illuminations in the leaf image are eliminated. For this, Local Binary Pattern method applied. This method eliminates the illumination effects and identifies the smaller changes in the pixels region, which is very useful for the disease identification.

## *3.1 LBP Definition*

LBP is a mainly used for the texture analysis and for the classification of the images. In face recognition system, the LBP classifier has given accurate and efficient results. For identifying the minor changes in leaf images and for identifying the relationship among the pixels LBP is one of the best method for images. So, based on the performance and efficiency of the LBP classifier, this LBP technique is used for the leaf images.

The basic concept of the LBP is: First I have to apply the LBP operator for the $3 \times 3$ block from the leaf image. If the pixel value at any position (except the center pixel) is greater than the center pixel then give the value as 1, otherwise give the value is 0 at that location. The obtained Pixel value is multiplied by the poirs of two. Sum of these pixel values is placed at the center pixel position. Like this, I have to obtain the center pixel value for every $3 \times 3$ block in the leaf image until it reaches the size of the image and for calculating the center pixel LBP value, the equation is given below [15].

$$LBP(x_{ce}, y_{ce}) = \sum_{q=0}^{q-1} S(i_q - i_{ce}) 2^q \quad (1)$$

After applying the LBP for the leaf image, the resultant image is called as LBP image.

## 4 Edge Detection Techniques

Edge Detection is for identifying the discontinuities in images. I have used the Sobel, Prewitt and Robert operators for detecting edges in the images.

By taking various approximations of the 2D-gradient, I obtain the first order derivatives of the digital image. The magnitude of the 2D gradient is calculated by using the following Eq. [16]:

$$f = \left[G_x^2 + G_y^2\right]^{1/2} \tag{2}$$

For a sample of the image, the gradient is obtained by using the equations of Robert, Sobel, Prewitt and Canny operators. Compared to Sobel, Prewitt masks are simple to implement. But, Sobel masks are superior in noise suppression [16]. The obtained image is further processed by applying the median filter for sharpening the edges in the leaf image and for noise reduction.

## 5 Segmentation Using Morphological Operations

Here, mainly used two morphological operations for segmentation. They are dilation and erosion. The main reason of using dilation is for bridging gaps. The Broken segments are easily joined by using the dilation. Compared to the low pass filtering, the dilation is giving better results for the binary images. The erosion is mainly used for eliminating the irrelevant content from the given binary image. By taking the structuring element of the required size, I can simply eliminate the unwanted portions of the image [16].

By using these dilation and erosion, I can able to segment the required disease part of the leaf image which was shown in the implementation part. By applying these threshold, dilation and erosion, I obtain the segmented image.

## 6 Implementation

Experimentation is carried using MATLAB. Sample of leaf images are shown in Fig. 2. The leaf images approximate size is 3500 × 2500 pixels. The leaf disease measure is calculated by considering the total area of the leaf, segmented diseased area and the rank table [3].

Here, the experimentation was done by two ways. One is, by applying LBP first then the edge detection. In second case, edge detection techniques are applied first then the LBP. By considering the procedures discussed in Sects. 3 and 4, experimentation is carried. For the obtained resultant image using two cases are further

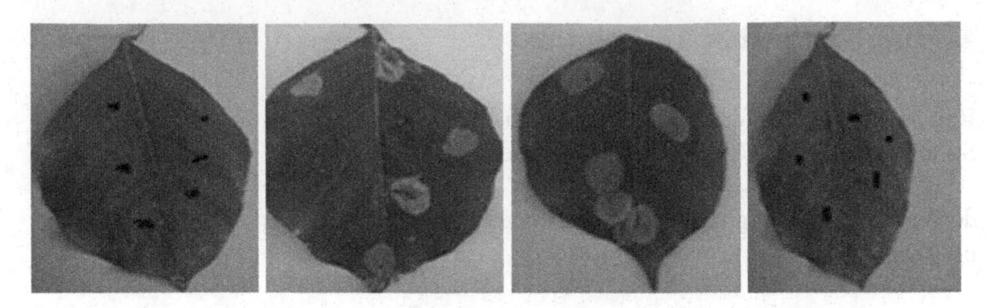

**Fig. 2** Sample of leaf images

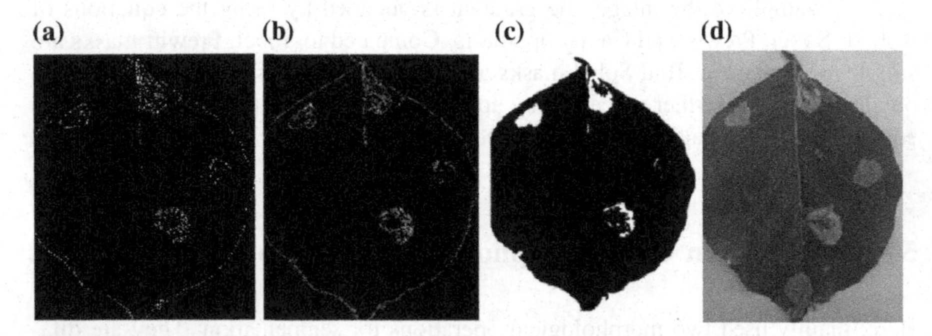


**Fig. 3** **a** Leaf image after applying LBP, **b** leaf image after applying edge detection using Sobel, **c** leaf image after segmentation, **d** final outlined leaf image

processed by applying the morphological operations for finding the segmented image, which I have discussed in Sect. 5. Finally, results are compared for identifying the disease measure by using the rank table which was discussed in Shen et al. [3].

Based on the procedures discussed in Sects. 3 and 4, LBP is applied for the leaf images and then edge detection. After this method, the obtained image is called the gradient image, which is shown in the Fig. 3b. Figure 3a shows the Leaf image after applying the LBP. Figure 3b shows Leaf image after edge detection technique using Sobel operator. The obtained image is having more clarity for identifying the diseases in the leaf images. For the resultant image, the dilation and erosion are applied for segmenting the image. The resultant segmented image is shown in Fig. 3c. Finally in Fig. 3d, the outline of the disease is shown in red lines on the original leaf image.

In the second case, edge detection is applied first, then the LBP. Figure 4a shows the leaf image after edge detection using Sobel operator. Figure 4b shows the resultant image after edge detection using Sobel operator and LBP. The segmented image and final outlined image is shown in Fig. 4c, d. Figure 5a, b shows the resultant segmented image and outlined images using Robert and Prewitt operators. By observing these 4 and 5 figures, found that the edge detection technique using

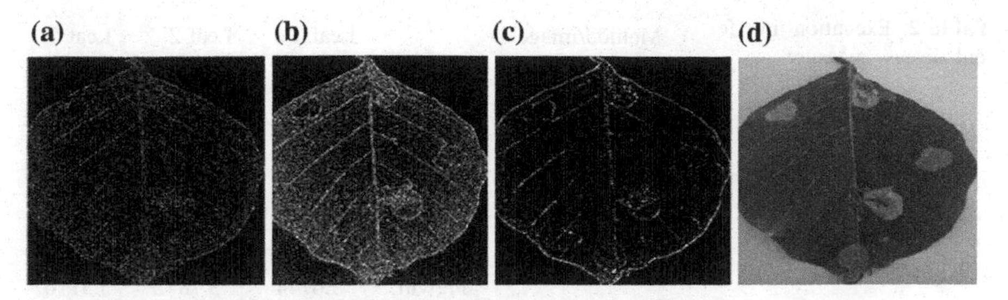

**Fig. 4** **a** After applying edge detection using Sobel, **b** after applying edge detection using Sobel and LBP, **c** after applying segmentation, **d** final outlined image

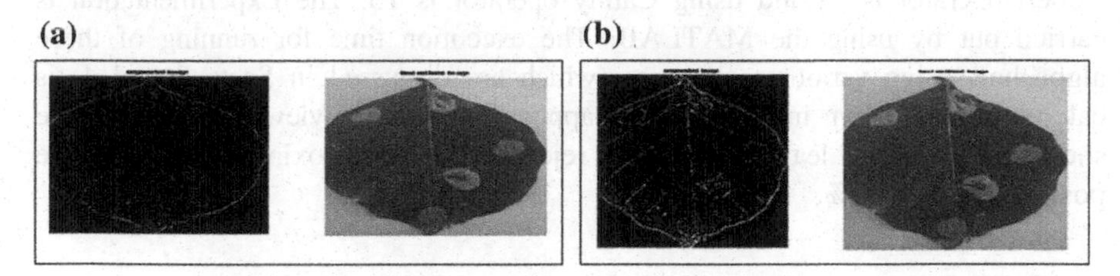

**Fig. 5** **a** Segmented and final outlined image after edge detection using Robert and LBP. **b** Segmented and final outlined image after edge detection using Prewitt and LBP

**Table 1** Leaf disease measure table

| Method/images | | Leaf 1 | Leaf 2 | Leaf 3 |
|---|---|---|---|---|
| Edge detection then LBP | Sobel | 0.4825 | 0.3217 | 0.4357 |
| | Robert | 0.6502 | 0.4330 | 0.4410 |
| | Prewitt | 0.4796 | 0.3149 | 0.4352 |
| | Canny | 0.3073 | 0.1863 | 02053 |
| LBP then edge detection | Sobel | 0.0052 | 0.0015 | 0.0111 |
| | Robert | 0.0054 | 0.0016 | 0.0113 |
| | Prewitt | 0.0051 | 0.0014 | 0.0111 |
| | Canny | 0.0237 | 0.0031 | 0.0312 |

Sobel was giving the better clarity in identifying the diseases in leaf image. Next using Prewitt operator is giving better clarity compared to the Robert operator in edge detection.

The disease measure value using various techniques was shown in Table 1. By observing the Table 1 and based on the rank table, found that using Sobel operator for measuring the leaf disease is giving the better performance compared to the remaining techniques. Here, further found that by applying edge detection first and then the LBP obtained the accurate disease measures in the leaf images. Whereas, in case of using the LBP first and then the edge detection, the disease spots are clearly identified in the Leaf images compared to the first method. By observing the

**Table 2** Execution time for leaf disease measure

| Method/images | | Leaf 1 | Leaf 2 | Leaf 3 |
|---|---|---|---|---|
| Edge detection then LBP | Sobel | 4.2883 | 4.2944 | 3.5607 |
| | Robert | 4.1654 | 4.2562 | 3.4972 |
| | Prewitt | 4.3922 | 5.0069 | 3.5809 |
| | Canny | 10.437 | 11.022 | 8.6802 |
| LBP then edge detection | Sobel | 3.7066 | 3.8139 | 3.2293 |
| | Robert | 3.7409 | 3.8457 | 3.2442 |
| | Prewitt | 3.6744 | 3.7872 | 3.3116 |
| | Canny | 5.2031 | 4.8720 | 5.3183 |

Table 1, the obtained disease level using Sobel and Prewitt operator are $T_4$, using Robert operator is $T_5$ and using Canny operator is T3. The Experimentation is carried out by using the MATLAB. The execution time for running of these algorithms using various techniques (which are discussed in Sects. 3 and 4) is calculated and shown in Table 2. For application point of view, here, results are shown only for three leaf images. False rejection rate is approximately 2% and true positive rate is 97.7%.

## 7 Conclusion

In this paper, proposed a method for leaf disease measure using two ways. One is by applying the Local Binary Pattern to the leaf image and for the obtained image applying the edge detection techniques like Sobel, Prewitt, Canny and Robert operators. In the second case, edge detection techniques are applied first then the LBP. For the resultant images, by using the two methods are further processed using morphological operators, as a result, obtained the segmented image. Finally, the results are compared and found that the leaf disease measure was accurately found by the edge detection technique using Sobel than the other edge detection techniques like Prewitt and Robert. Further, this method using the edge detection then the LBP was giving the better performance for disease measure in leaf images compared to the method LBP then edge detection. By using the LBP and next edge detection, only the disease spots are clearly seen on the leaf images, which are shown by red lines on the original leaf image. The time complexity of these algorithms calculated and studied.

## References

1. Clement, D.L.: An introduction to plant diseases. http://www.gardening.cornell.edu/education/mgprogram/mgmanual/04diseases.pdf
2. Introduction to plant pathology. http://www.ipm.iastate.edu/files/curriculum/05-introduction-to-Plant-Pathology_0.pdf

3. Weizheng, S., Yachun, W., Zhanliang, C., Hongda, W.: Grading method of leaf spot disease based on image processing. In: International Conference on Computer Science and Software Engineering, vol. 6, pp. 491–494. IEEE Computer Society, Washington (2008)
4. Revathi, P., Hemalatha, M.: Classification of cotton lead spot diseases using image processing edge detection techniques. In: International Conference on Emerging Trends in Science, Engineering and Technology, pp. 169–173. IEEE, Tiruchirappalli (2012)
5. Al Bashish, D., Braik, M., Bani-Ahmad, S.: A framework for detection and classification of plant leaf and stem diseases. In: International Conference on Signal and Image Processing, pp. 113–118. IEEE, Chennai (2010)
6. Zhang, Y.C., Mao, H.-P., Hu, B., Li, M.-X.: Features selection of cotton disease leaves image based on fuzzy feature selection techniques. In: International Conference on Wavelet Analysis and Pattern Recognition, vol. 1, pp. 124–129. IEEE, Beijing (2007)
7. Abdullah, N.E., Alam, S., Rahim, A.A., Hashim, H., Kamal, M.M.: Classification of rubber tree leaf diseases using multilayer perceptron neural network. In: 5th Student Conference on Research and Development, pp. 1–6. IEEE, Selangor (2007)
8. Sankaran, S., Mishra, A., Ehsani, R., Davis, C.: A review of advanced techniques for detecting plant diseases. Int. J. Comput. Electr. Agric. **72**(1), 1–13 (2010). (Elsevier, Great Britain)
9. Ying, G., Miao, L., Yuan, Y., Zelin, H.: A study on the method of image pre-processing for recognition of crop diseases. In: International Conference on Advanced Computer Control, pp. 202–206. IEEE, Singapore (2009)
10. Kurniawati, N.N., Abdullah, S.N.H.S., Abdullah, S.: Investigation on image processing techniques for diagnosing paddy diseases. In: International Conference of Soft Computing and Pattern Recognition, pp. 272–277. IEEE, Malacca (2009)
11. Jian, Z., Wei, Z.: Support vector machine for recognition of cucumber leaf diseases. In: 2nd International Conference on Advanced Computer Control, vol. 5, pp. 264–266. IEEE, Shenyang (2010)
12. Liu, L., Zhou, G.: Extraction of the rice leaf disease image based on BP neural network. In: International Conference on Computational Intelligence and Software Engineering, pp. 1–3. IEEE, Wuhan (2009)
13. Husin, Z.B., Shakaff, A.Y.B.M., Aziz, A.H.B.A., Farook, R.B.S.M.: Feasibility study on plant chili disease detection using image processing techniques. In: 3rd International Conference on Intelligent Systems, Modelling and Simulation, pp. 291–296. IEEE, Kota Kinabalu (2012)
14. Hashim, H., Haron, M.A., Osman, F.N., Al Junid, S.A.M.: Classification of rubber tree leaf disease using spectrometer. In: 4th Asia International Conference on Mathematical/Analytical Modelling and Computer Simulation, pp. 302–306. IEEE, Kota Kinabalu (2010)
15. Ojala, T., Pietikäinen, M., Harwood, D.: A comparative study of texture measures with classification based on feature distributions. Pattern Recoginit. **29**(1), 51–59 (1996). (Elsevier, Great Britain)
16. Gonzales, R.C., Woods, R.E.: Digital Image Processing, 2nd edn. Addison-Wesley (1992)

# Ear Recognition Using Self-adaptive Wavelet with Neural Network Classifier

**Jyoti Bhardwaj and Rohit Sharma**

**Abstract** We present a novel approach to ear recognition that utilizes ring-projection method for reducing the dimensions of two-dimensional image into one-dimensional information, that consists of the summation of all pixels that lie on the boundary of a circle with radius ‘r’ and center at the centroid of the image. As a 2-D image is transformed into a 1-D signal, so less memory is required and it is faster than existing 2-D descriptors in the recognition process. Also, ring-projection technique is rotation-invariant. The 1-D information, obtained in the ring-projection method, is normalized so as to make a new wavelet which is named as self-adaptive wavelet. Features are extracted using this wavelet by the process of decomposition. Neural Network based classifiers such as Back Propagation Neural Network (BPNN) and Probabilistic Neural Network (PNN) are used to obtain the recognition rate. A survey of various other techniques has also been discussed in this paper.

**Keywords** Wavelet · Ring-Projection · Neural network · BPNN · PNN

## 1 Introduction

This Ear recognition plays a very important role in the field of human authentication [1]. The geometry and shape of an ear have been found to be unique for every individual. The applications of the ear biometric include identification in an unconstrained environment, i.e., application in smart surveillance, border control system and identification of people (perpetrators, thieves, visitors, etc.) on CCTV images, etc. Ear biometrics individually or along with the face, i.e., multi-modal

---

J. Bhardwaj (✉)
DCRUST, Murthal, Sonepat, India
e-mail: bhardwajjyoti1010@gmail.com

R. Sharma
AMITY University, Noida, Uttar Pradesh, India
e-mail: r25sharma@gmail.com

 
S.C. Satapathy et al. (eds.), *Data Engineering and Intelligent Computing*,
Advances in Intelligent Systems and Computing 542,
DOI 10.1007/978-981-10-3223-3_5

biometrics, provides valuable information for identification of individuals even in the presence of occlusion, unconstrained environment and pose variations. Cameras are located overhead at public places, so as to capture ear images of a large number of people. Therefore, it becomes quite difficult to capture the face image and recognize the person. The ear biometrics plays a significant role in this situation.

In recent years, pertinent research activities have increased significantly and much progress has been made in this direction.

However, existing systems perform well only under constrained environments, requiring subjects to be highly cooperative. Furthermore, it is also observed that the variations between the images of the same ear due to the rotation and viewing direction are often larger than those caused by changes in ear identity. Also, 2-D image recognition needs more memory for storage of information. So, there is a requirement of such technique that is rotation-invariant and requires less memory for storage of information.

## 2 Previous Work

To recognize people by their ear has attained significant attention in the literature and a number of relative techniques have been published. An overview of various latest techniques employing two-dimensional (2D) ear recognition and ear detection is discussed below.

### *2.1 Methods for Ear Detection*

Generally, in surveillance cameras or in CCTV images, the human body is captured in multiple poses and with lots of occlusions. That is why it is necessary to detect the ear from human face and from the complex background too. Segmentation is done to locate ear from the face image. Segmentation method also helps to get rid of the unwanted objects like eyeglass frame, hat or hair, etc.

**Subregion based techniques.**
Marsico et al. [2] considered the problem arising due to pose, illumination, and partial occlusion while matching two images to identify a person in a database. For this, they proposed a fractal based technique, Human Ear Recognition against Occlusions (HERO), to classify human ear.

**Template based techniques.**
Ansari and Gupta [3] fetched the features from ear image using their outer helix curve. Canny edge operator was used to detecting the ear from an image. These edges may be of two forms either concave or convex. From these shapes, outer helix edges were decided.

Pflug et al. [4] introduced a new ear detection approach for 3D profile images based on surface curvature and semantic analysis of edge patterns.

## 2.2 *Methods for Ear Recognition*

In this section, various ear recognition methods are summarized and categorized based on classification and feature extraction. Recognition can further be of two types. One is recognition, which uses the ear as a single unit. This type of method is very much feasible when ear image is free of occlusion. Other is, working on the ear in separate regions. In the case of multiple poses and occlusion problem, separate regions method works well. Different methods of ear recognition are as follows:

**Neural network based techniques.**
Alaraj et al. [5] used the neural network as a classifier based on two criteria, i.e., recognition accuracy and computation time. They demonstrated the relation between accuracy and computation time which depends on the no. of eigenvectors obtained from feature extraction using PCA (Principal Component Analysis) approach.

Wang et al. [6] used high order moment invariants having the feature of translation, rotation, and scale change for ear identification along with Back Propagation (BP) artificial neural network with an accuracy of 91.8%.

**Principal component analysis based techniques.**
This technique is quite easy to use and implement, provided that the images are registered accurately and extra information has not indulged with the important one (but having poor invariance). That's why this process depends entirely on the preprocessing stage and alignment of data.

Kus et al. [7] presented a fully embedded ear recognition system using PCA based on an ARM microcontroller which can be programmed with Micro-C language.

Xie and Mu [8] used LLE (locally linear embedding) which deals with non-linear dimensionality reduction but having less labeled information in the dataset. LLE observed the neighboring relationship using the structure of data.

Xie and Mu [9] proposed two techniques for multiple poses ear recognition namely LLE (locally linear embedding) and IDLLE (improved locally linear embedding algorithm) and compare these techniques with PCA and KPCA to prove their betterment.

**Wavelet-based techniques.**
Chen and Xie [10] proposed a new descriptor based techniques for invariant pattern recognition by the use of ring-projection and dual-tree complex wavelets. The ring-projection sums up all the pixels that lie on the circle having radius ‘r’ and center at the centroid of the pattern. This converts the pattern from a 2-D image to a 1-D image signal that requires less memory than 2-D image. Also, it is faster than

current 2-D descriptors in the process of recognition. Due to approximate shift invariant property, Dual-tree complex wavelet has been applied to the ring projection technique that is very important in pattern recognition.

Tang et al. [11] proposed a novel approach for recognizing optical characters by the use of ring-projection-wavelet-fractal signatures (RPWFS). This proposed method is effective in reducing the dimensionality of a 2-D pattern with the help of ring-projection method.

In image processing, wavelets play an important role as these are used to derive necessary information from an image. Earlier, wavelets were used to extract features from the image/data by decomposition and then classifiers are used to classify those features. Till now, this is the strategy, one used for recognition of a person with the use of wavelets. But, in this research work, a new wavelet is designed through a ring-projection technique. This wavelet is designed according to the ear image and extracts similar information from that ear image. The main motive behind designing a new wavelet is to approximate the given information by using least square optimization under constraints that leads to an admissible wavelet. This admissible wavelet uses continuous wavelet transform for the detection of information. And in this way, we can extract the desired information from each ear image and thus the task of recognizing a person becomes comparatively easy and efficient. The proposed methodology is shown in Fig. 1.

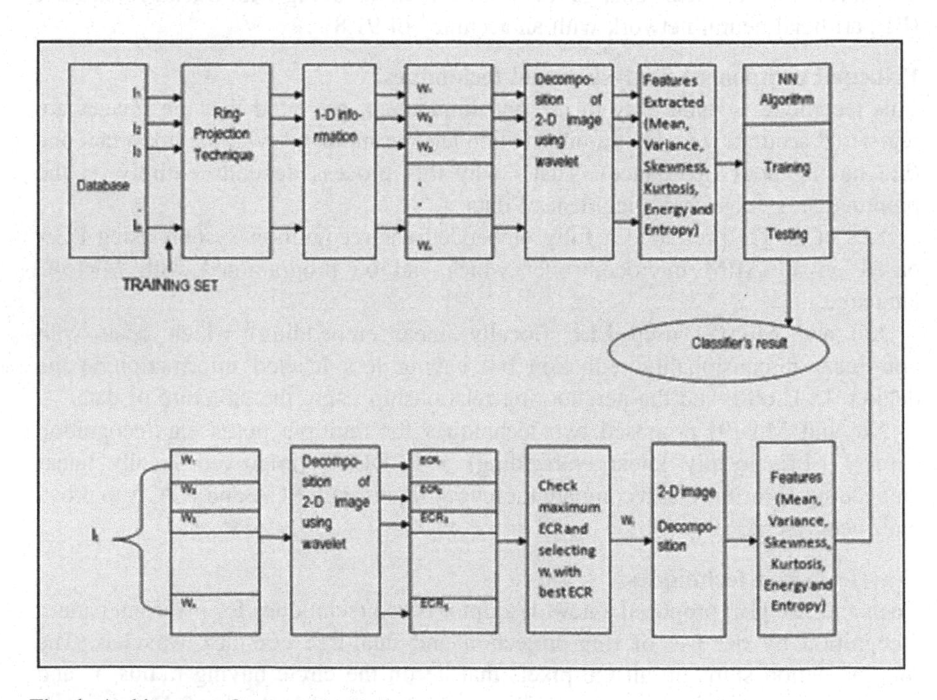


**Fig. 1** Architecture of proposed technique

# 3 Proposed Methodology

## 3.1 Training

For training the data, ear image is used as input to the classifier. In our work, we used ear images from two databases, one of them is IIT Delhi Database and other is a West Pommeranian University of Technology Database (WPUT-DB) database.

**IIT Delhi Database.**
In IIT Delhi database, more than three images per subject are captured with the help of simple imaging center from a distance in the indoor environment. It consists of total 493 grayscale images of the ear. All the data is provided online free of cost [12].

**WPUT-DB Database.**
The WPUT-DB database consists of 501 subjects of all ages and 2071 images in total. There are 4 to 8 images for each subject which are taken under different lightning conditions. The database also contains images that are occluded by headdresses, earrings, hearing aids, hair, etc. This database is also available online free of cost [13].

## 3.2 Pre-processing

There are two ways to acquire ear image for further stages. Either enter the ear image directly or crop the ear image from the face image. In our work, we manually cropped the ear part from the side face image and the portions of the image which do not constitute the ear are removed so as to remove noise from the image. The colored cropped ear image is converted into gray scale image.

## 3.3 Ring-Projection

Ring-Projection technique is used for reducing the dimensions of two-dimensional image into one-dimensional signal [10, 11]. The ring-projection takes the summation of all pixels that lie on the circle with radius r and center at the centroid of the image. As the 2-D image is transformed into a 1-D information, so less memory is needed and it is faster than existing 2-D descriptors in the recognition process. The various steps involved in this technique are as follows:

STEP 1: Initially, the RGB image is converted into gray scale p(x, y) and if the image is already gray scaled then no change is required. We have taken both, i.e., colored and gray scaled images in our databases.

STEP 2: Now, the boundary points of the considered image in Cartesian plane are defined, which are taken as ($x_1$, $y_1$), ($x_2$, $y_2$), ($x_3$, $y_3$) and ($x_4$, $y_4$) in our work.

STEP 3: Now based on the above boundary points, the centroid ($x_c$, $y_c$) and radius (r) of the image are calculated by using the following formulae:

$$xc = \left(\frac{\mathrm{xmax} + \mathrm{xmin}}{2}\right) \quad (1)$$

$$yc = \left(\frac{\mathrm{ymax} + \mathrm{ymin}}{2}\right) \quad (2)$$

$$r = \sqrt{(xc - xi)^2 + (yc - yi)^2} \quad (3)$$

Where, $i = 1, 2, 3, 4$

$$r = (max(r)) \quad (4)$$

STEP 4: Further, we transform the original reference Cartesian frame into a polar frame based on the following relations:

$$x = \gamma cos\theta \quad (5)$$

$$y = \gamma sin\theta \quad (6)$$

where, $\gamma \epsilon [0, \infty)$; $\theta \epsilon (0, 2\Pi]$

For any fixed $\gamma \, \epsilon [0, r]$, we then compute the following integral:

$$f(\gamma) = \int_0^\theta p(\gamma cos\theta, \gamma sin\theta) d\theta \quad (7)$$

The resulting $f(\gamma)$ is in fact equal to the total mass as distributed along circular rings.

The boundary points on the defined range of circle are given by:

$$x = xc + r * cos(\theta * pi/180) \quad (8)$$

$$y = yc + r * sin(\theta * pi/180) \quad (9)$$

The derivation of f(γ) is termed as Ring-Projection as it returns a circle containing a summation of all the points lying on it.

STEP 5: The radius decides how many circles are obtained, and each obtained circle consists of the summation of all points lying on it. The number of points on the circle depends on the range of θ, e.g., if θ = [0:10:360], then there are total 36 points lying on the circle. The total number of circles combine together to return 1-D information. The single-variate function f(γ); γ = [0, r], sometimes also denoted as f(x); x = [0, r], can be viewed as a 1-D information that is directly transformed from the original 2-D pattern through a ring-projection. Owing to the fact that the centroid of the mass distribution is invariant to rotation and that the projection is done along circular rings, the derived 1-D information will be invariant to the rotations of its original 2-D image. In other words, the ring-projection is rotation-invariant.

## 3.4 Information Normalization

The 1-D information we get in above stage needs to be normalized so as to confine it to a smaller range. Normalization of the signals is done by using the following formula:

$$pNi = (pi - p)/p^* \quad (10)$$

Where pi is the ith element of the 1-D information, P, before normalization, p and p* are the mean and standard deviation of the vector P, respectively. pNi is the ith element of the normalized 1-D information obtained. This new 1-D information obtained after normalization is then used to design a new wavelet that is used for further feature extraction from the 2-D image.

## 3.5 Wavelet Design

After normalization of 1-D information, a new wavelet is then designed for each ear image using wavelet toolbox. The algorithm used for the design of adaptive wavelet is described as follows:

1. Go to wavelet toolbox or wave menu in Matlab.
2. Select 'new wavelet for CWT'.
3. Load 1-D information (obtained after normalization) from the workspace in the command window.
4. The 1-D information is in the range [0, 1], so it is not a wavelet but it is a good candidate since it oscillates like a wavelet.
5. To synthesize a new wavelet adapted to the given 1-D information, use the least squares polynomial approximation of degree 6 with constraints of continuity at the beginning and the end of 1-D information.

6. Now approximate the 1-D information signal obtained.
7. Adaptive wavelet is obtained which can be compared with other existing wavelets in the wavelet toolbox.
8. This wavelet is applied to the 2-D image for decomposition.

### 3.6 Decomposition

The adaptive wavelet, we get in above stage, is used for the decomposition of ear image into four sub-bands: LL_LL, LH_LL, HL_LL, HH_LL.

### 3.7 Features Extraction

To feed the classifier information to use for classification, we need to extract mathematical measurements (features) from that object. There are two categories of features: shape (features of the shape's geometry captured by both the boundary and the interior region) and texture (features of the grayscale values of the interior). Discriminative features are extracted separately from image blocks of the LL_LL, LH_LL, HL_LL, HH_LL sub-bands. In our work, we compute six different types of features (Energy, Entropy, Mean, Variance, Skewness, Kurtosis) that are total 24 in number, four from each filter.

### 3.8 Classification

Classifiers are designed for classification task of the vectors obtained after extracting the features. Classifiers play an important role in finding the accuracy of the method designed. For classification, we use two classifiers that are Neural Network (NN) classifier and Probabilistic Neural Network (PNN) Classifier. The features obtained are stored in a matrix that is given as input to these classifiers.

## 4 Results

### 4.1 Results of Ring Projection Technique

Results for both the databases are shown separately. A number of circles obtained from ring-projection technique are 181, i.e., 'r' is equal to 181 which are

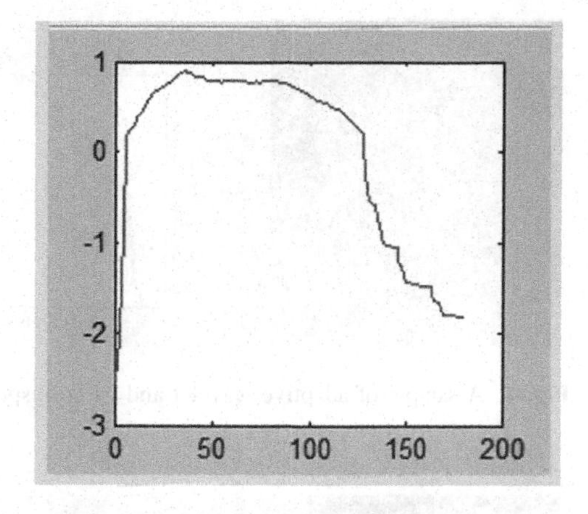


**Fig. 2** 1-D information obtained after normalization from ring-projection technique for a sample of ear image from IIT Delhi database

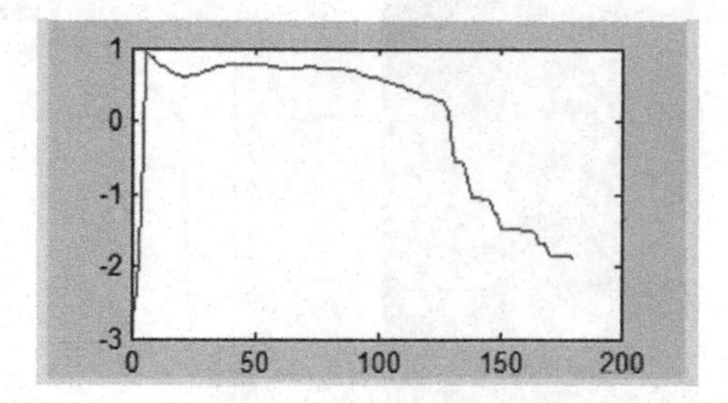


**Fig. 3** 1-D information obtained after normalization from ring-projection technique for a sample of ear image from WPUT-DB database

represented by x-axis of the figures below. And summation of pixel points lie on the boundary of each circle lies is represented on the y-axis of the 1-D signal. Only a sample of results obtained from both the databases is represented in the results.

Figures 2 and 3 shows 1-D information obtained after normalization from Ring-Projection technique for a sample of ear image from IIT Delhi database and WPUT-DB database. Here, x-axis (ranges from 0–200) denotes the number of circles and y-axis (ranges from −3–1) denotes the summation of pixels that lie on the boundary of circles.

## *4.2 Adaptive Wavelet*

This wavelet is obtained by inserting the coefficients obtained after normalization into wave menu (wavelet toolbox in Matlab) for constructing a new wavelet, i.e., adaptive wavelet. Figures 4 and 5 shows a sample of results for adaptive wavelets and their corresponding ears for IIT Delhi Database and WPUT-DB Database.

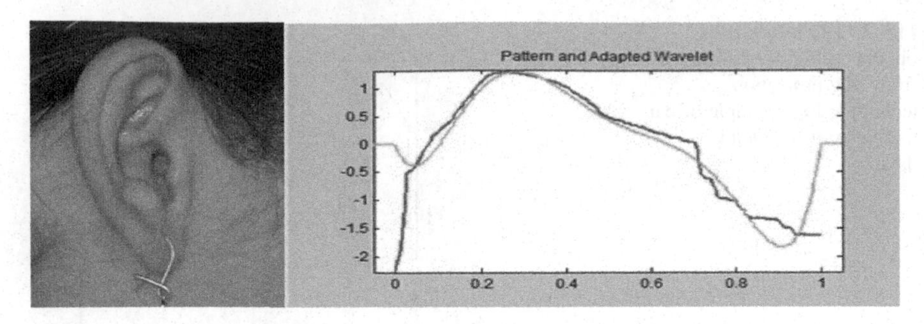


**Fig. 4** A sample of adaptive wavelet and its corresponding ear for IIT Delhi database

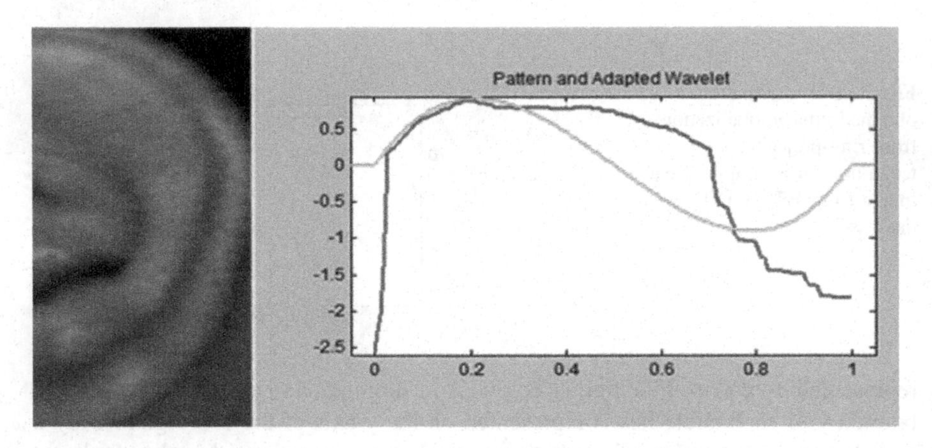


**Fig. 5** A sample of adaptive wavelet and its corresponding ear for WPUT-DB database

## 4.3 Feature Extraction

With this adaptive wavelet, obtained above, the 2-D ear image is decomposed to two levels for extracting features from the image. The various features extracted from both the databases are shown in Table 1 (mentioned after references).

## 4.4 Neural Network Classification

Neural networks are much in demand in the pattern recognition problems. Neurons, elements of the neural network, can classify data with good accuracy and are best suited for decision boundary problems with multiple variables. Therefore, neural networks are considered as a good candidate to solve the ear classification

**Table 1** Extracted features for a sample of IIT Delhi database and WPUT-DB database

| Features | Extracted features for a sample of IIT Delhi database | Extracted features for a sample of WPUT-DB database |
|---|---|---|
| | | |
| Energy | 1.45489811217982e + 17 | 1.30096462879746e + 17 |
| | 9.18973043603821e + 16 | 1.06171406206065e + 17 |
| | 1.75447098482279e + 17 | 2.86663000891790e + 16 |
| | 3.87271816403285e + 17 | 1.73717509453974e + 17 |
| Entropy | −1.59992555841891e + 17 | −2.59077793134260e + 17 |
| | −4.41467483149918e + 16 | −4.20587684185622e + 17 |
| | −7.07389209520546e + 16 | −5.30172787338413e + 16 |
| | −8.21143320664034e + 17 | −6.89714603026766e + 17 |
| Skewness | 0.377861771272962 | 0.653634898959975 |
| | −0.379168279747595 | −1.14585883758366 |
| | 0.0495658918386443 | 0.341071557114864 |
| | −0.0429563209130982 | −0.592553934658552 |
| Mean | 29061225.8489115 | 21266597.6231361 |
| | −32331135.0106542 | −12828627.4424261 |
| | −32168191.7659871 | −27005213.2642575 |
| | 36001946.3885917 | 21554730.4888384 |
| Variance | 1.37449259873164e + 15 | 1.20924613453403e + 15 |
| | 1.70326376219486e + 15 | 587693409589778 |
| | 4.40677484719567e + 15 | 1.83478758751571e + 15 |
| | 5.45973234510193e + 15 | 881558676993522 |
| Kurtosis | 1.35454784195563 | 1.75612383677717 |
| | 1.35422431014704 | 2.75672324177030 |
| | 1.63567470973403 | 1.95551179446882 |
| | 1.63043847600484 | 2.23927883519164 |

problems. The following results illustrate how neural network can identify a human from a database with the help of ear biometrics using Ring-Projection technique.

Confusion plot matrix (Fig. 6) is a tool to study the performance of the neural network. The confusion matrix is plotted across all samples and shows the correct and incorrect classifications percentage in the green squares and red squares

**Confusion Matrix**

| Output Class | 1 | 2 | 3 | 4 | 5 | 6 | 7 | 8 | 9 | 10 | |
|---|---|---|---|---|---|---|---|---|---|---|---|
| 1 | 0<br>0.0% | 0<br>0.0% | 0<br>0.0% | 0<br>0.0% | 0<br>0.0% | 0<br>0.0% | 0<br>0.0% | 0<br>0.0% | 0<br>0.0% | 0<br>0.0% | NaN%<br>NaN% |
| 2 | 0<br>0.0% | 0<br>0.0% | 0<br>0.0% | 0<br>0.0% | 0<br>0.0% | 0<br>0.0% | 0<br>0.0% | 0<br>0.0% | 0<br>0.0% | 0<br>0.0% | NaN%<br>NaN% |
| 3 | 0<br>0.0% | 0<br>0.0% | 0<br>0.0% | 0<br>0.0% | 0<br>0.0% | 0<br>0.0% | 0<br>0.0% | 0<br>0.0% | 0<br>0.0% | 0<br>0.0% | NaN%<br>NaN% |
| 4 | 0<br>0.0% | 0<br>0.0% | 0<br>0.0% | 1<br>100% | 0<br>0.0% | 0<br>0.0% | 0<br>0.0% | 0<br>0.0% | 0<br>0.0% | 0<br>0.0% | 100%<br>0.0% |
| 5 | 0<br>0.0% | 0<br>0.0% | 0<br>0.0% | 0<br>0.0% | 0<br>0.0% | 0<br>0.0% | 0<br>0.0% | 0<br>0.0% | 0<br>0.0% | 0<br>0.0% | NaN%<br>NaN% |
| 6 | 0<br>0.0% | 0<br>0.0% | 0<br>0.0% | 0<br>0.0% | 0<br>0.0% | 0<br>0.0% | 0<br>0.0% | 0<br>0.0% | 0<br>0.0% | 0<br>0.0% | NaN%<br>NaN% |
| 7 | 0<br>0.0% | 0<br>0.0% | 0<br>0.0% | 0<br>0.0% | 0<br>0.0% | 0<br>0.0% | 0<br>0.0% | 0<br>0.0% | 0<br>0.0% | 0<br>0.0% | NaN%<br>NaN% |
| 8 | 0<br>0.0% | 0<br>0.0% | 0<br>0.0% | 0<br>0.0% | 0<br>0.0% | 0<br>0.0% | 0<br>0.0% | 0<br>0.0% | 0<br>0.0% | 0<br>0.0% | NaN%<br>NaN% |
| 9 | 0<br>0.0% | 0<br>0.0% | 0<br>0.0% | 0<br>0.0% | 0<br>0.0% | 0<br>0.0% | 0<br>0.0% | 0<br>0.0% | 0<br>0.0% | 0<br>0.0% | NaN%<br>NaN% |
| 10 | 0<br>0.0% | 0<br>0.0% | 0<br>0.0% | 0<br>0.0% | 0<br>0.0% | 0<br>0.0% | 0<br>0.0% | 0<br>0.0% | 0<br>0.0% | 0<br>0.0% | NaN%<br>NaN% |
| | NaN%<br>NaN% | NaN%<br>NaN% | NaN%<br>NaN% | 100%<br>0.0% | NaN%<br>NaN% | NaN%<br>NaN% | NaN%<br>NaN% | NaN%<br>NaN% | NaN%<br>NaN% | NaN%<br>NaN% | 100%<br>0.0% |

Target Class

Fig. 6 Confusion matrix

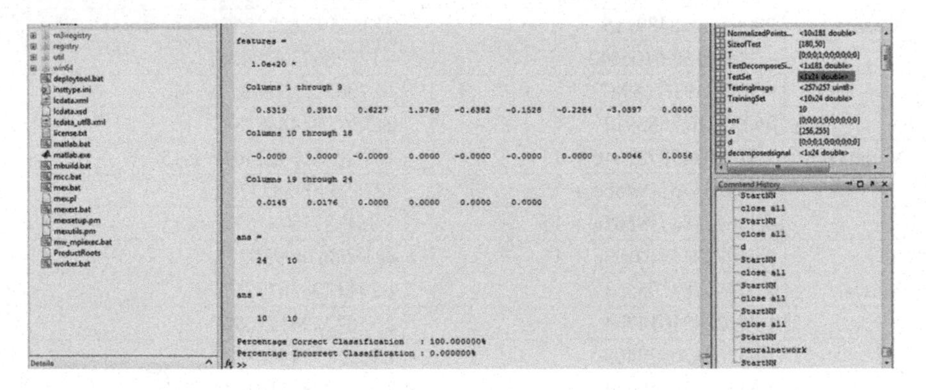


**Fig. 7** MATLAB results for BPNN

respectively. Value '1' against 4th output class shows that test image '4' is classified accurately. Matlab results for BPNN are shown in Fig. 7.

MATLAB results as shown in Fig. 7 indicate:

1. Features extracted from test image
2. Correct classification rate and incorrect classification rate by NN (100%, 0.0% resp.).

### 4.5 PNN Classifier

PNN, used for classification, is a type of radial basis network. In Matlab, 'newpnn' manages classification task using two-layer network. 10 images are used from the database to classify the required test image. These images are used for training with 10 classes as shown in Fig. 7. After classification, the result is 'c1', i.e., the test set belongs to class 1 as shown in red color in Fig. 8. MATLAB results as shown in Fig. 9 indicate the class of the matching image within the database and the time taken by the PNN to classify the test image.

The X-axis, i.e., P(1,:) in Fig. 8 indicates 1st element of the training vector from each class and Y-axis, i.e., P(2,:) indicates 2nd element of the training vector from each class. In Figs. 8 and 9 respectively, P(1,:) indicates 1st element of the training vector and P(1) indicates 1st element of the test vector from each class on X-axis and P(2,:) indicates 2nd element of the training vector and P(2) indicates 2nd element of the test vector from each class on Y-axis.

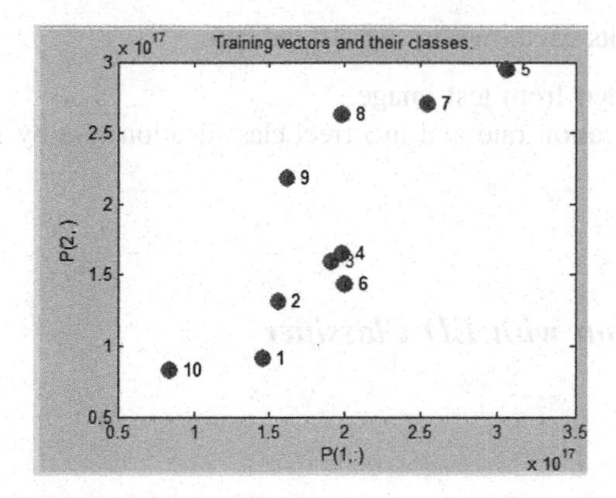


**Fig. 8** Trained PNN network

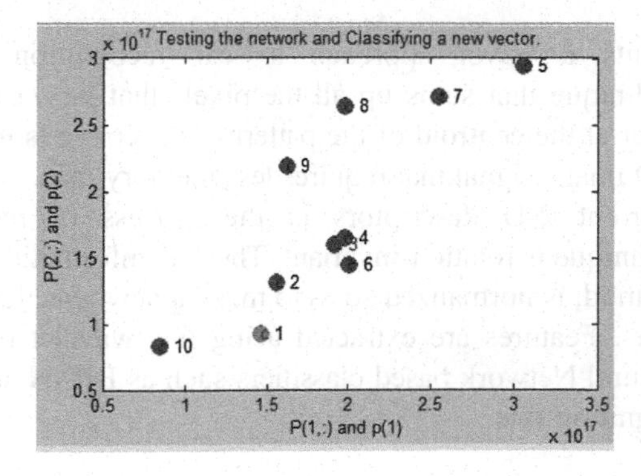


**Fig. 9** Classified image

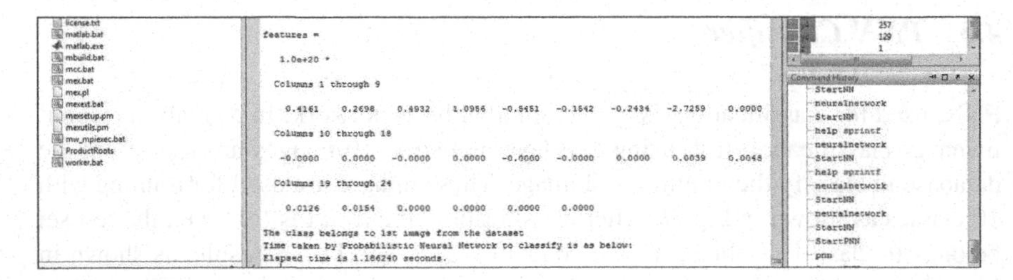


**Fig. 10** MATLAB results for PNN

**Table 2** Comparison between ED and BPNN classifier

| Classifier | Number of subjects | Recognition rate (%) |
|---|---|---|
| Euclidean distance [14, 15] | 150 | 86.67 |
| Current technique (BPNN, PNN) | 150 | 100.00 |

MATLAB results as shown in Fig. 10 indicate:

1. Features extracted from test image.
2. Correct classification rate and incorrect classification rate by NN (100%, 0.0% resp.).

### *4.6 Comparison with ED Classifier*

Table 2.

## 5 Conclusion

This work presents a novel approach to ear recognition that utilizes a ring-projection technique that sums up all the pixels that lie on the circle having radius ‘r’ and center at the centroid of the pattern. This converts the pattern from a 2-D image to a 1-D image signal that requires less memory than 2-D image. Also, it is faster than current 2-D descriptors in the process of recognition. Also, ring-projection technique is rotation-invariant. The 1-D information, obtained in the ring-projection method, is normalized so as to make a new wavelet which is named as adaptive wavelet. Features are extracted using this wavelet by the process of decomposition. Neural Network based classifiers such as BPNN and PNN are used to obtain the recognition rate.

## References


1. David, W.K.: Court of appeals of Washington, kryminalistyka.wpia.uw.edu.pl/_les/2012/10/oto2.pdf (1999)
2. Marsico, M.D., Michele, N., Riccio, D.: HERO: human ear recognition against occlusions. In: IEEE Computer Society Conference on Computer Vision and Pattern Recognition Workshops (CVPRW), pp. 178–183 (2010)
3. Ansari, S., Gupta, P.: Localization of ear using outer helix curve of the ear. In: Proceedings of the International Conference on Computing: Theory and Applications, pp. 688–692. Barcelona, Spain (2007)
4. Pflug, A., Winterstein, A., Busch, C.: Ear detection in 3D profile images based on surface curvature. In: 8th International Conference on Intelligent Information Hiding and Multimedia Signal Processing (IIH-MSP), pp. 1–6 (2012)
5. Alaraj, M., Hou, J., Fukami, T.: A neural network based human identification framework using ear images. In: IEEE Region 10th Conference, Fukuoka, Japan, pp. 1595–1600. TENCON (2010)
6. Wang, X.Q., Xia, H.Y., Wang, Z.L.: The research of ear identification based on improved algorithm of moment invariant. In: 3rd IEEE International Conference on Information and Computing, China, pp. 58–60 (2010)
7. Kus, M., Kacar, U., Kirci, M., Gunes, E.O.: ARM Based Ear Recognition Embedded System, pp. 2021–2028. IEEE, EUROCON, Zagreb, Croatia (2013)
8. Xie, Z.X., Mu, Z.C.: Improved locally linear embedding and its application on multi-pose ear recognition. In: Proceedings of the International Conference on Wavelet Analysis and Pattern Recognition, pp. 1367–1371. Beijing, China (2007)
9. Xie, Z., Mu, Z.: Ear recognition using LLE and IDLLE algorithm. In: Proceedings of the 19th International Conference on Pattern Recognition ICPR, pp. 1–4. Tampa-FL, USA (2008)
10. Chen, G.Y., Xie, W.F.: Invariant pattern recognition using ring-projection and dual-tree complex wavelets. In: Proceedings of the International Conference on Wavelet Analysis and Pattern Recognition, Guilin, 10–13 July (2011)
11. Yuan, Y.T., Bing, F.L., Hong, M., Jiming, L.: Ring-projection-wavelet-fractal signatures: a novel approach to feature extraction. In: IEEE Transactions on Circuits and Systems—Ii: Analog and Digital Signal Processing, vol. 45(8) (1998)
12. http://www4.comp.polyu.edu.hk/~csajaykr/myhome/database_request/ear/
13. http://ksm.wi.zut.edu.pl/wputedb
14. Specht, D.F., Romsdahl, H.D.: Experience with adaptive probabilistic neural network and adaptive general regression neural network. In: IEEE International Conference on Neural Networks, Vol. 2, pp. 1203–1208. Orlando, FL (1994)
15. Wang, Y., Mu, Z., Zeng, H.: Block-based and multi-resolution methods for ear recognition using wavelet transform and uniform local binary patterns. In: Proceedings of the 19th International Conference on Pattern recognition. Tampa, FL (2008)

# An Efficient Algorithm for Mining Sequential Patterns

**Gurram Sunitha, M. Sunil Kumar, B. Sreedhar and K. Jeevan Pradeep**

**Abstract** The temporal component of the spatio-temporal databases is the key factor that leads to large accumulation of data. It can be said that continuous collection of spatial data, leads to spatio-temporal databases. A event type sequence is called as an sequential pattern and extracting spatio-temporal sequential patterns from spatio-temporal event data sets paves way to define causal relationships between event types. In this paper, a data structure has been proposed to support efficient mining of sequential patterns.

**Keywords** Data mining · Sequential patterns · Spatio-Temporal databases · Voronoi diagram · Sequential graph

## 1 Introduction

An event in the real-world object which inherently has spatial and temporal extent. Along with the object, if its location and time of occurrence are also recorded, the database is called as a spatio-temporal database. Hence, a spatio-temporal database contains both spatio-temporal as well as non-spatio-temporal attributes. Spatio-temporal extent of the data provides dynamic view of the underlying phenomenon. Algorithms handling extraction of sequential patterns from

---

G. Sunitha (✉) · M.S. Kumar · B. Sreedhar · K. Jeevan Pradeep
Department of Computer Science and Engineering, Sree Vidyanikethan Engineering College, A. Rangampet, Chittoor, Andhra Pradesh, India
e-mail: gurramsunitha@gmail.com

M.S. Kumar
e-mail: sunilmalchi1@gmail.com

B. Sreedhar
e-mail: sreedharburada@gmail.com

K. Jeevan Pradeep
e-mail: jeevanpradeep33@gmail.com



S.C. Satapathy et al. (eds.), *Data Engineering and Intelligent Computing*, Advances in Intelligent Systems and Computing 542,
DOI 10.1007/978-981-10-3223-3_6

spatio-temporal databases has to mainly concentrate on: (1) scalability of algorithms as spatio-temporal databases are inherently large. (2) design of efficient data structures to store and retrieve data quickly. (3) design of good significance measure to prove the significance of sequential patterns.

The work done in this paper concentrates on construction of sequential graph to reduce the size of the given spatio-temporal database. Voronoi diagram is used for the purpose.

## 2 Problem Statement

Consider spatio-temporal database D, which consists of a set of events of various event types. Each event defines a real-world event that has occurred at a definite space and time. Each event in the database is uniquely identified by its event ID and is described by its event type, space and time of event occurrence, and other non-spatio-temporal attributes. Let all the event types in the given spatio-temporal database be $E = \{E_1, E_2, \ldots, E_n\}$ where n specifies the number of event types. A spatio-temporal sequential pattern is an event type sequence of form $E_i \rightarrow E_j \rightarrow \ldots \rightarrow E_k$ where $E_i, E_j, \ldots E_k \in E$.

The main problem approached in this paper is to extract all significant sequential patterns from the given spatio-temporal database.

## 3 Related Work

Since the idea of sequential pattern mining is proposed in [1], there has been colossal research done in this area. Substantial number of algorithms has been proposed to work with a variety of spatio-temporal databases such as trajectory databases, geographical databases, scientific databases, web traffic databases etc.

Work in [2–4] has concentrated on mining frequent trajectory patterns by extending sequential pattern mining paradigm. [5, 6] have proposed their theories on finding evolution patterns in satellite image time series. [7] have proposed significant work on extracting significant sequential patterns from spatio-temporal event datasets.

## 4 Mining Spatio-Temporal Sequential Patterns

Extracting significant sequential patterns from the spatio-temporal event data set goes through 2 stages: constructing sequential graph and mining the sequential patterns.

## 4.1 Overview of Voronoi Diagram

Voronoi Diagram takes its origin from computational geometry. It has its applications in various areas of computer science which implicitly have problems related to geometry such as computer graphics, computer-aided design, robotics, pattern recognition etc.

A Voronoi diagram is a way of dividing space into a number of regions. Given a set of input data points, the voronoi diagram is formed by dividing space into discrete regions, where each region is called as voronoi region or voronoi cell. Each voronoi region is bounded by the voronoi edges, also called as segments. Voronoi diagram also consists of voronoi vertices, where a voronoi vertex is the point where three or more voronoi edges meet. i.e., a voronoi vertex is surrounded by three or more voronoi regions.

To construct voronoi diagram, a set of points called as site points or seed points are to be selected from the input data set. This can be done randomly or by a using particular method. The number of site points decides the number of voronoi regions to be constructed in the voronoi diagram. Each input data point is added to a voronoi region if it is closer to the site point of that region than any other site points i.e., nearest neighbor rule are used to construct voronoi diagram. One input data point is associated with only one voronoi region.

Consider a input data point q, a set of voronoi regions P and Euclidean distance ED. Point $q$ lies in the region corresponding to a seed $p_i \in P$ *iff* $\mathrm{ED}(q, p_{i)} < \mathrm{ED}(q, p_{j)}$, for each $p_i \in P, j \neq i$.

## 4.2 Space-Time

For the given spatio-temporal database D, for constructing the sequential graph as well as for extracting sequential patterns only space and time attributes are considered. Events in the database D are considered as points in space-time which are specified by their location and time of occurrence. The given database D is represented visually using space-time diagram. A space-time diagram is normally 4-dimensional, since space has 3-dimensions and time has 1-dimension. Here, we are considering space and time both as 1-dimensional. Hence, the space-time diagram here is 2-dimensional. Considering space as x-axis and time as y-axis on the co-ordinate grid, each point in the diagram specifies an occurrence of the event.

Once, the space-time diagram is formed, the events are grouped using nearest-neighbor rule. This is done by dividing space-time plane into cells or regions. Events belonging to one region form one group. Space-time plane can be divided into regular equal-sized rectangular cells by applying rectangular grid onto the plane. The plane can also be divided into irregular cells by constructing voronoi diagram on the plane.

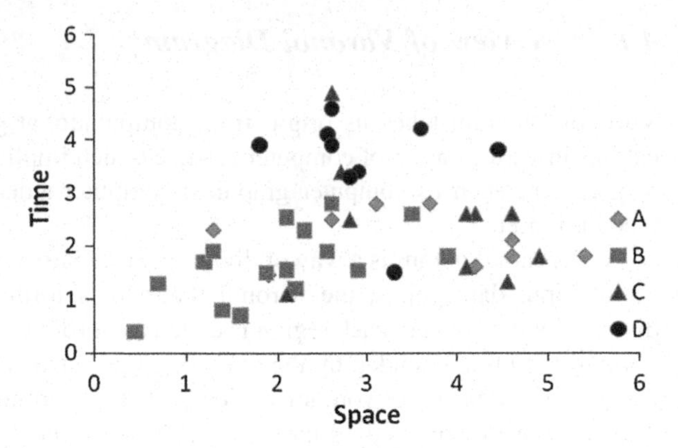


**Fig. 1** Space-time diagram of the spatio-temporal database D

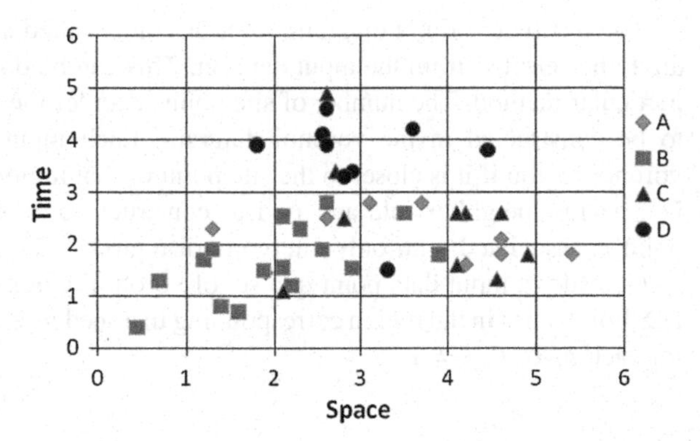


**Fig. 2** Space-time diagram divided into equal-sized grid cells

Consider a database with 4 event types A, B, C, D. Figure 1 shows the space-time diagram of the given spatio-temporal database D. Figure 2 shows the space-time diagram divided into equal-sized cells or regions by applying rectangular grid onto the plane. Figure 3 shows the space-time diagram divided into irregular sized and shaped regions by constructing voronoi diagram. Each cell may contain events of more than one event type. The dots shown in each region are the site points.

### *4.3 Construction of Sequential Graph with Grid-Based Approach*

Once the events in the database are divided into grid cells, these cells are used to construct the sequential graph. Before constructing the graph effort is made to reduce the size of the graph by adding only significant cells to the graph. Density is calculated for each cell and the cells with density greater than the minimum cell density threshold is said to be significant.

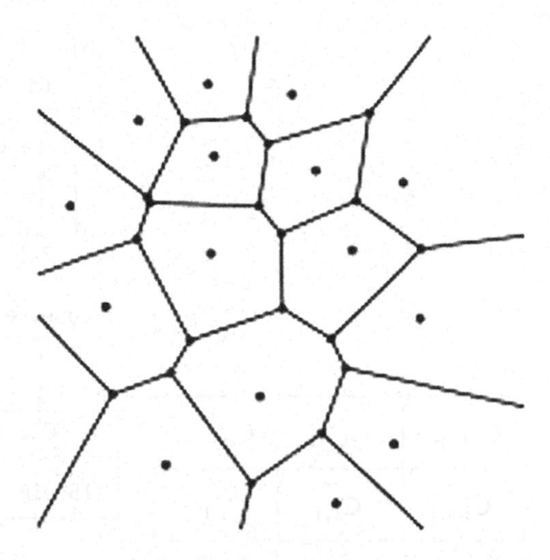

**Fig. 3** Space-time diagram divided into voronoi regions

Cell's density can be calculated by using any applicable density function. A simple density function can be defined as the ratio of number of events in the cell to the total number of events in the database. Once the significant cells are identified, sequential graph can be constructed. Each significant cell forms a node in the graph. Each node stores: Cell ID and information of all events belonging to that cell.

For each significant cell $C_{ij}$, the neighborhood cells are determined. The left, right-side cells and upper-side cells of $C_{ij}$ form the neighborhood cells of $C_{ij}$. The bottom-side cells are not considered as neighborhood because of the unidirectional property of the sequential patterns. i.e., a sequential pattern A → B says that A is followed by B not vice-a-versa. Figure 4a shows the neighborhood of a grid cell $C_{ij}$. $C_{i-1,\, j}$ is the left side cell, $C_{i+1,\, j}$ is the right side cell and $C_{i-1,\, j+1}$, $C_{i,\, j+1}$, $C_{i+1,\, j+1}$ are the upper side cells. So, for every cell there can be a maximum of five neighborhood cells. But, among these five cells, only those that are significant are only considered as the neighborhood of the $C_{ij}$.

Once neighborhood cells of significant cells are determined, each cell connects itself to its neighborhood cell by directed edges. i.e., there is an edge from the significant cell to each of its significant neighborhood cell. Figure 4b shows an example sequential graph where all cells is significant cells.

## 4.4 *Mining Sequential Patterns*

Once the sequential graph, sequential patterns can be mined using the information available in the graph. Hence, the original database can be discarded from the memory. As the sequential graph is smaller in size when compared with the original database, this approach is efficient in terms of memory usage.

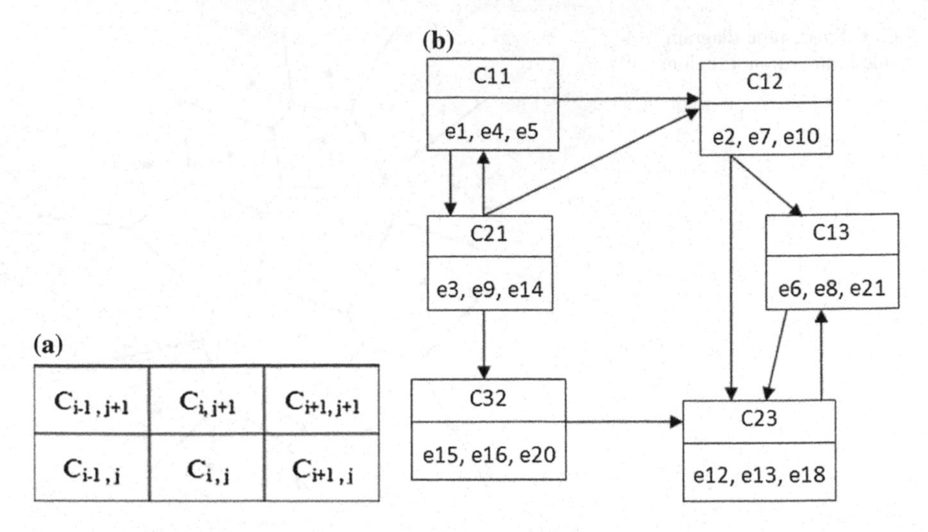


**Fig. 4** **a** Neighborhood cells of cell $C_{i,\,j}$. **b** Example sequential graph

The mining procedure is as follows:

Step (A)

Let the sequential graph be $S_G$. Consider all the event types available in $S_G$. Let the set of event types be $E_1$. Follow step (B) to generate all sequential patterns of length 2 i.e., 2-sequential patterns.

Step (B)—Generating 2-Sequential Patterns

(1) Let the set of 2-Sequential Patterns be $SP_2$. Initially $SP_2 = \emptyset$.
(2) For each event type $E_{1i}$ in $E_1$ do
    (a) Find all nodes that contain one or more events of event type $E_{1i}$ from $S_G$. Let such nodes be $N_{1i}$.
    (b) Find all neighborhood nodes of nodes in $N_{1i}$ from $S_G$. Let such nodes be $NN_{2i}$.
    (c) List out all event types in nodes $NN_{2i}$. Let such event types be $E_2$.
    (d) For each event type $E_{2i}$ in $E_2$ do
        (i) Form a new 2-Sequential Pattern as $E_{1i} \rightarrow E_{2i}$ and add it to $SP_2$.

$$SP_2 = SP_2 \cup \{E_{1i} \rightarrow E_{2i}\}$$

Step (C)—Generating (n + 1)-Sequential Patterns from n-Sequential Patterns

(1) Let the set of n-Sequential Patterns be $SP_n$. Let (n + 1)-Sequential Patterns be $SP_{n+1}$. Initially $SP_{n+1} = \emptyset$.
(2) For each sequential pattern $SP_{ni}$ in $SP_n$ *do*

(a) Consider $NN_{ni}$ generated while forming sequential pattern $SP_{ni}$. Let $N_{ni} = NN_{ni}$.
(b) Find all neighborhood nodes of nodes in $N_{ni}$ from $S_G$. Let these neighborhood nodes be $NN_{(n+1)i}$.
(c) List out all event types in nodes $NN_{(n+1)i}$. Let such event types be $E_{n+1}$.
(d) For each event type $E_{(n+1)i}$ in $E_{n+1}$ do

(i) Form a new (n + 1)-Sequential Pattern as $SP_{ni} \rightarrow E_{(n+1)i}$ and add it to $SP_{n+1}$.

$$SP_{n+1} = SP_{n+1} \cup \{SP_{ni} \rightarrow E_{(n+1)i}\}$$

At each step, once all sequential patterns of length $n$ are generated, pruning of patterns can be done to identify significant patterns and only forward them to next step to generate $(n + 1)$-sequential patterns. This will avoid unnecessary processing of insignificant patterns. Number of significance measures has been proposed in the literature. Any measure appropriate to the situation and application can be applied to identify significant patterns. [7] have proposed a significance measure called as *sequence index* that can be used for the purpose.

## 5 Conclusion and Future Work

An efficient algorithm has been proposed to extract sequential patterns from spatio-temporal event datasets using grid-based approach. A new concept called as sequential graph has been proposed to efficiently store event dataset along with all the information needed to generate sequential patterns. As sequential graph occupies less storage when compared to the original database, it is efficient. The proposed algorithm has yet to be compared with the approach proposed in [7] to prove the efficiency. Defining neighborhood of regions when space-time diagram is divided using voronoi diagram is a complex task and needs to be worked upon in future.

## References

1. Agrawal, R., Srikant, R.: Mining sequential patterns. In: Proceedings of 1995 International Conference on Data Engineering, Mar (1995)
2. Cao, H., Mamoulis, N., Cheung, D.W.: Mining frequent spatio-temporal sequential patterns. In: Proceedings of Fifth IEEE International Conference on Data Mining (2005)
3. Giannotti, F., Nanni, M., Pinelli, F., Pedreschi, D.: Trajectory pattern mining. In: KDD' 07, pp 330–339. ACM NY, USA
4. Lee, A.J.T., Chen, Y.-A., Ip, W.-C.: Mining frequent trajectory patterns in spatial–temporal databases. Int. J. Inf. Sci. **179**(13) (2009)

5. Julea, A., et al.: Unsupervised spatiotemporal mining of satellite image time series using grouped frequent sequential patterns. IEEE Trans. Geosci. Remote Sens. **49**(4) (2011)
6. Petitjean, F., Masseglia, F., Gançarski, P., Forestier, G.: Discovering significant evolution patterns from satellite image series. Int. J. Neur. Syst. **21**(475) (2011)
7. Huang, Y., Zhang, L., Zhang, P.: A framework for mining sequential patterns from spatio-temporal event data sets. IEEE Trans. Knowl. Data Eng. **20**(4), 433–448 (2008)
8. Sunitha, G., Rama Mohan Reddy, A.: A grid-based algorithm for mining spatio-temporal sequential patterns. Int. Rev. Comput. Soft. **9**(4), 659–666 (2014)
9. Sunitha, G., Rama Mohan Reddy, A.: WRSP-miner algorithm for mining weighted sequential patterns from spatio-temporal databases. In: Proceedings of the Second International Conference on Computer and Communication Technologies, pp. 309–317. Springer, India (2016)

# Comparison of SVM Kernel Effect on Online Handwriting Recognition: A Case Study with Kannada Script

**S. Ramya and Kumara Shama**

**Abstract** Proposed research work is aimed at investigating the issues specific to online Kannada handwriting recognition and design an efficient writer independent Online Handwriting recognizer. The proposed system accepts continuous Kannada online handwriting from pen tablet and produces recognized Kannada text as the system output. System comprises of pre-processing, segmentation, feature extraction and character recognition units. SVM classifier is implemented to test its efficiency with the Kannada handwritten characters. The recognition rates are analyzed for different SVM kernels.

**Keywords** Online handwriting recognition · Support vector machine · Pre-processing · Pattern recognition

## 1 Introduction

Now a days, human and machine interaction attained more importance. Commonly used interaction media are the keyboard or pointing devices. Keyboard is very inconvenient when the machine size is comparable with the human palm. Thus pen-based interfaces gained the interest and playing easy and efficient medium of human-computer interaction. The challenge with pen enabled PC is to recognize user handwriting. Handwriting Recognition (HR) is a field of Pattern recognition which aims to assign objects into a set of categories [1]. HR is classified into offline handwriting recognition and Online Handwriting Recognition (OHR). In offline handwriting systems, written information is optically scanned and converted into images once the

S. Ramya (✉) · K. Shama
Department of Electronics and Communication Engineering,
Manipal Institute of Technology,
Manipal, Karnataka, India
e-mail: ramya.lokesh@manipal.edu

K. Shama
e-mail: shama.kumar@manipal.edu



S.C. Satapathy et al. (eds.), *Data Engineering and Intelligent Computing*,
Advances in Intelligent Systems and Computing 542,
DOI 10.1007/978-981-10-3223-3_7

writing process is over. In OHR, handwritten data are acquired during the process of writing, which is represented by a sequence of coordinate points, known as digital ink. Hence digital ink represents the dynamic information preserving the order of the strokes. This representation can be converted into character codes. The applications of the OHR are e-mail, e-chat, online form filling in hospital and colleges etc.

A survey on OHR [2] reports, work carried out in English, German, Chinese and Japanese language by using feature extraction, time zone direction and stroke code based. But, OHR in Indian languages in general is underway. OHR for Indian languages is more challenging due to the large number of consonants, vowels and conjuncts. Efforts are made for Indian languages like Devanagari, Tamil and Telugu etc. But not many efforts are reported for the Kannada language which is the state language of Karnataka.

### *1.1 Kannada Language*

Kannada has history of more than 2000 years. Kannada language is inherited from the Dravidian languages and is one among the twenty-two other official languages recognized by the constitution of India [3]. The script of Kannada language is evolved from Kadamba script which is syllabic in nature. The language has sixteen vowels known as Swaragalu, thirty four consonants known as Vyanjanagalu as shown in Fig. 1(a) and (d) respectively.

Consonant when combines with vowel will modify the base consonant giving rise to distinct symbols called Consonant-Vowel (CV) modifiers. Figure 1b shows symbols for respective vowel and when it is applied to the consonant, it modifies the base consonant as shown in Fig. 1c for consonant *ka*. The number of these modifiers symbol is equal to that of the consonant. Therefore, the number of possible combinations of Kannada characters is $34 \times 16$ consonant-vowel combinations. When a consonant combines with another consonant-vowel combination, it will lead to $34 \times 34 \times 16$ Consonant Consonant-Vowel (CCV) combinations. Figure 1e shows some samples of consonant-consonant-vowel modifiers.

## 2 Literature Review

Work on Kannada OHR has been few. In 2000, the Kannada OHR was reported by Kunte [4] where neural network classifier is used with the wavelet features. Divide and Conquer Technique [5] is implemented which provides 81% accuracy. Statistical Dynamic Space Warping classifier was applied on Online Kannada word level recognition [6] by considering first derivatives of x and y co-ordinates and normalized x and y co-ordinates as features. 88% classification accuracies obtained for the character level and 80% at the word level. Rakesh [7] recently reports fusion of offline and online character recognition for better recognition efficiency of 89.7%. Dexterous technique for Kannada akshara was implemented [8] which improves accuracy

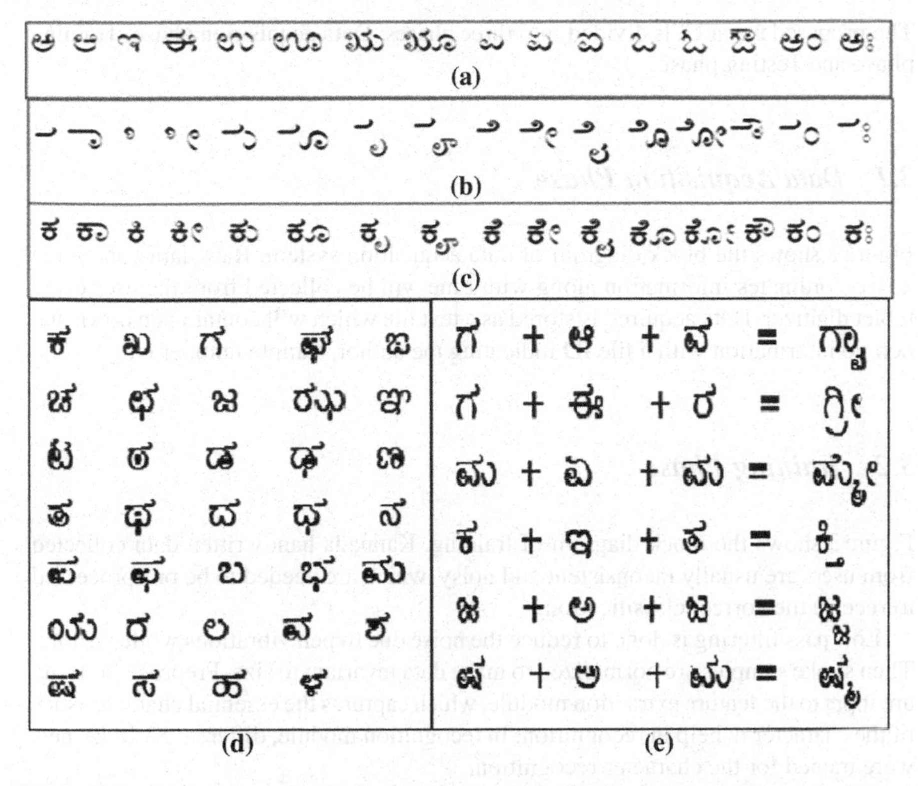

**Fig. 1** Kannada characters **a** vowels **b** respective vowel modifiers **c** vowel modifiers as applied to a consonant '*Ka*' **d** consonants **e** examples of consonant consonant-vowel combinations

from 77% to 92%. In summary there is a scope to develop efficient, robust Kannada OHR system which provides easy human computer interactions.

# 3 Methodologies

Support Vector Machines (SVM) are widely used for pattern recognition field. SVM uses the different kernel function to classify. To compare the capability of different kernel, it is necessary to have online Kannada character database. Since there is no publicly available Kannada character database, we have built a database. Since Kannada characters are very vast, work is tested for vowels and consonants and it can be extended for other combinations of vowels and consonants. Following are our objectives

- To build online Kannada character database.
- To analyze dominant features in the stroke.
- To identify online strokes using SVM based technique.
- To compare the efficiency of different kernel functions for Kannada Characters.

The proposed research is divided into three phases: Data acquisition phase, Training phase and Testing phase.

### 3.1 Data Acquisition Phase

Figure 2 shows the block diagram of data acquisition system. Raw data containing x, y co-ordinates information along with time will be collected from the user using tablet digitizer. Data acquired is stored as a text file which will contain pen down and pen up information with a file ID indicating the author, sample number etc.

### 3.2 Training Phase

Figure 3 shows the block diagram of training. Kannada handwritten data collected from users are usually inconsistent and noisy, which are needed to be pre-processed to receive the correct classification.

Low pass filtering is done to reduce the noise due to pen vibrations while writing. Then stroke samples are normalized to make data invariant to size. Preprocessed data are input to the feature extraction module, which captures the essential characteristics of the character to help in recognition. In recognition module, different SVM kernels were trained for the character recognition.

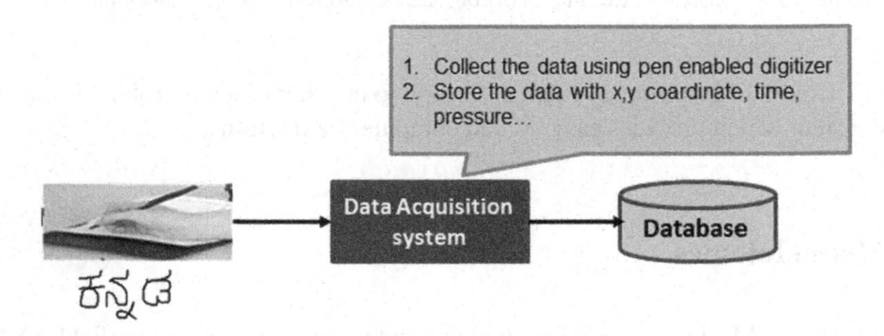


**Fig. 2** Block diagram of data acquisition system

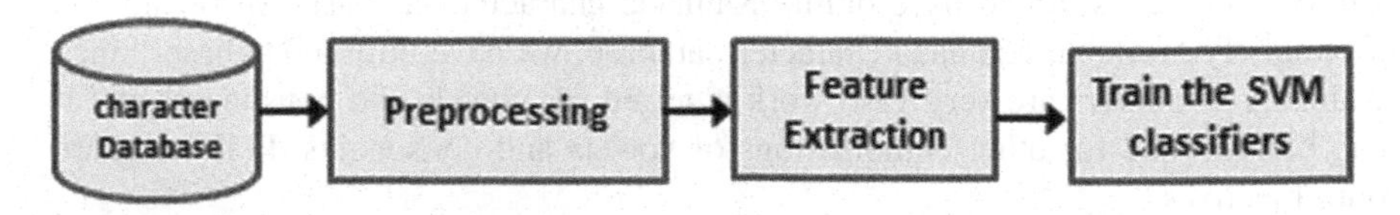


**Fig. 3** Block diagram of training procedure

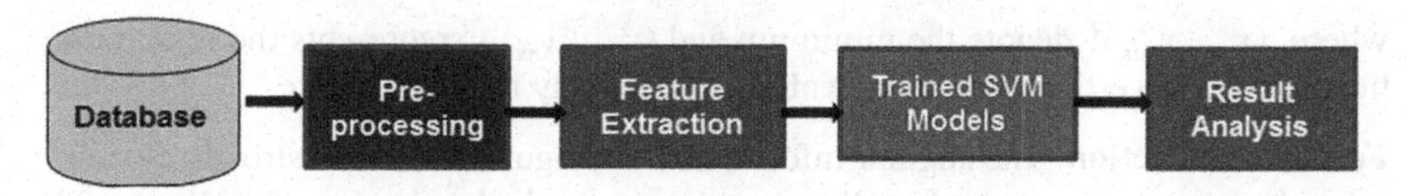


**Fig. 4** Block diagram of testing procedure

## *3.3 Testing Phase*

The testing procedure is as shown in Fig. 4. Raw data collected from data acquisition system is preprocessed, feature extracted and recognized by different SVM models and recognition rates are evaluated for the same.

## *3.4 Implementation*

A graphical user interface is been developed to collect data using Genius Mouse Pen i608X. Native users of Kannada were selected between the age group 12–55 years by visiting the nearby school and colleges. The x and y coordinates of characters written on the tablet PC along with the user information is saved into a text file. The database includes varying size and shapes with different speed of writing. Validity of the characters is tested in semiautomatic way to ensure data integrity.

**Pre-processing**: The online handwritten data consists of sequence *x* and *y* sequence representing horizontal and vertical coordinate points of the pen movements. Due to the writing, data may contain jitter which will be eliminated by subjecting to noise removal [8] techniques. Moving average filter of window span of three was used to remove the noise. The written data has varying in length of x, y sequence. Hence the data is resampled [9] using linear interpolation method. By empirical testing, resample size was set to 40 points.

Let D represents the resampled data sequence:

$$D = [d_1, d_2, \ldots, d_{40}] \quad (1)$$

where the vector $di = (xi, yi)^T$. The resampled data are normalized to get a new sequence:

$$N = [n_1, n_2, \ldots, n_{40}] \quad (2)$$

where the vector $n_i = (p_i, q_i)^T$ is given by,

$$p_i = (x_i - x_{min})/(x_{max} - x_{min}) \quad (3)$$

$$q_i = (y_i - y_{min})/(y_{max} - y_{min}) \quad (4)$$

where, $(x_{min}, y_{min})^T$ denote the minimum and $(x_{max}, y_{max})^T$ represents the maximum horizontal and vertical coordinate values respectively for a pen stroke.

**Feature Extraction**: The angular information distinguishes the two strokes. So tangent slope angle at point i for all $x$, $y$ points are calculated using (5). We get 39 angular data from 40 resampled points.

$$\theta_i = \tan^{-1}(y_{i+1} - y_i)/(x_{i+1} - x_i) \tag{5}$$

We considered writing direction as another feature. The local writing direction at a point (x(i), y(i)) is represented by cosine and sine [9] angles as given by Eq. 6 and Eq. 7 respectively.

$$\cos\alpha(i) = \delta x(i)/\delta s(i) \tag{6}$$

$$\sin\alpha(i) = \delta y(i)/\delta s(i) \tag{7}$$

where $\delta x(i)$, $\delta y(i)$ and $\delta s(i)$ are defined by Eq. 8, Eq. 9 and Eq. 10 respectively.

$$\delta x(i) = (x_{(i-1)} - x_{(i+1)}) \tag{8}$$

$$\delta y(i) = (y_{(i-1)} - y_{(i+1)}) \tag{9}$$

$$\delta s(i) = (\delta x^2(i) + \delta y^2(i))^{1/2} \tag{10}$$

**Character Recognition**: Character recognition is a sub class of pattern recognition which aims to classify data based on a prior knowledge. We implemented SVM classifier which is faster and efficient for the character recognition. We tested with different SVM kernel functions namely Linear, Polynomial and Radial Bias Function (RBF). The basic SVM takes a set of input data and predicts maximum marginal Hyper Plane (MMH) for the separation of two classes leading to the binary classification. In non-linear SVM, classification is performed by kernel function mapping the data into a higher dimension whereas in RBF classifier, margin is nonlinear and Gaussian in nature.

### *3.5 Result Analysis*

Handwritten data before normalization and after normalization are as seen in Figs. 5 and 6 respectively. We can observe, normalized data is invariant to size and displacement.

Output of each pre-processing methods namely, smoothing, normalization and resampling methods are depicted in Fig. 7.

SVM classifier with kernel functions namely, Linear, Polynomial and RBF are trained with 70% of the character samples from the database. 30% of the remaining character data are used for testing the efficiency of the classifier. Recognition rates are as compared in Table 1.

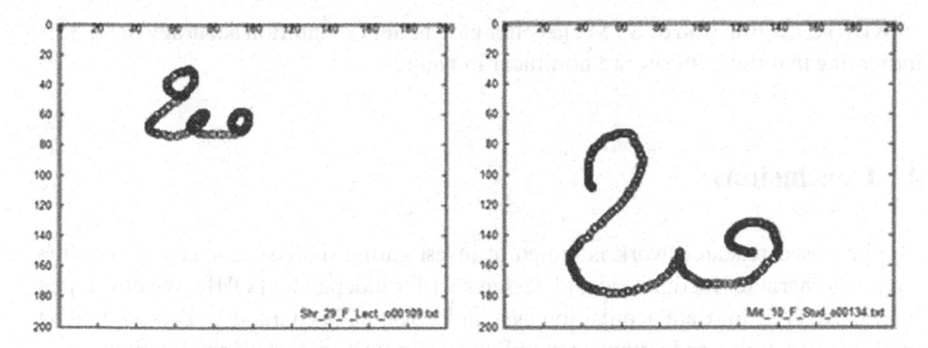

Fig. 5 Raw data before normalization

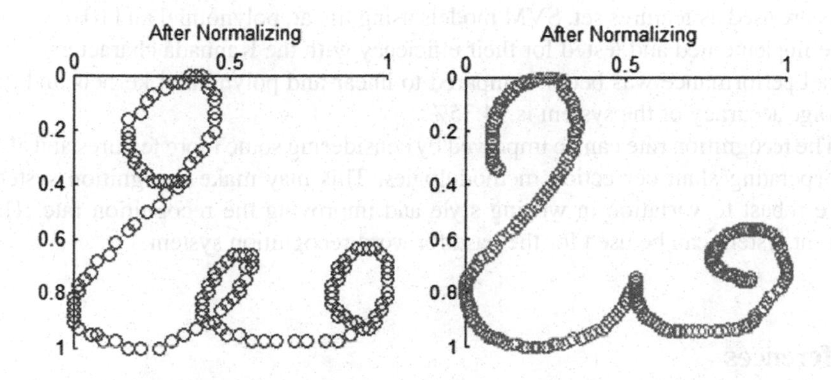


Fig. 6 Output of normalization module

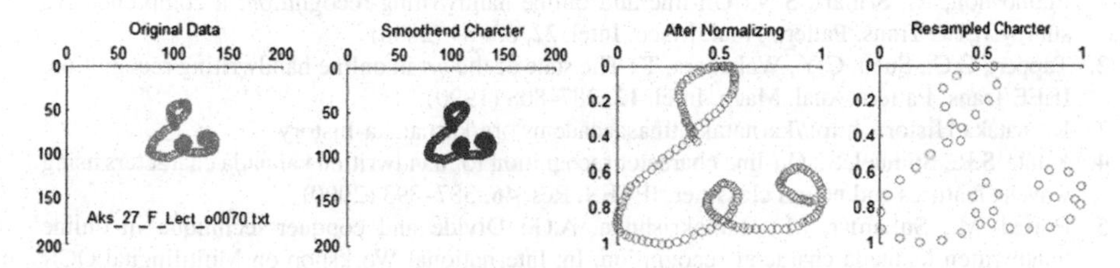


Fig. 7 Output of various pre-processing module

Table 1 Recognition result with SVM classifier

| Method | Classifier | Recognition rate in % |
|---|---|---|
| Prasad et al. [5] | PCA with K-NN | 81 |
| Kunwar et al. [6] | Statistical DSW | 87.9 |
| Proposed method | Linear kernel | 86.07 |
| | Polynomial kernel | 93.44 |
| | RBF kernel | 94.35 |

RBF kernel function of SVM classifier gave better recognition accuracy of 94.35% indicating that the patterns are nonlinear in nature.

## 4 Conclusions

The proposed research work is aimed at investigating the issues specific to online Kannada character recognition and design a writer independent OHR. We developed interactive GUI for data acquisition system for Kannada characters. Pre-processing methods like noise reduction, resampling of points and size normalizations were used. Writing direction, angular difference and normalized horizontal vertical positions are used as features set. SVM models using linear, polynomial and RBF kernel were implemented and tested for their efficiency with the Kannada characters. RBF kernel performance was better compared to linear and polynomial kernels and the average accuracy of the system is 94.35%.

The recognition rate can be improved by considering some more features and also incorporating slant correction methodologies. This may make recognition system more robust to variation in writing style and improving the recognition rate. The current system can be used for the base for word recognition system.

## References

1. Plamondon, R., Srihari, S.N.: On-line and offline handwriting recognition: a comprehensive survey. IEEE Trans. Pattern Anal. Mach. Intel. **22**, 63–84 (2000)
2. Tappert, C.C., Suen, C.Y., Wakahara, T.: The state of the art in online handwriting recognition. IEEE Trans. Pattern Anal. Mach. Intel. **12**, 787–808 (1990)
3. Karnataka History. http://karnatakaitihasaacademy.org/karnataka-history
4. Kunte, S.R., Samuel, S.: On-line character recognition for handwritten kannada characters using wavelet features and neural classifier. IETE J. Res. **46**, 387–393 (2000)
5. Prasad, M., Sukumar, M., Ramakrishnan, A.G.: Divide and conquer technique in online handwritten kannada character recognition. In: International Workshop on Multilingual OCR (MOCR '09), pp. 11:1–11:7. ACM, New York (2009)
6. Kunwar, R., Shashikiran, K., Ramakrishnan, A.G.: Online Handwritten Kannada Word Recognizer with Unrestricted Vocabulary. In: International Conference Frontiers in Handwriting Recognition (ICFHR), pp. 611–616, IEEE Computer Society, Washington, DC, USA (2010)
7. Rakesh, R., Ramakrishnan, A.G.: Fusion of complementary online and offline strategies for recognition of handwritten kannada characters. J. Univ. Compu. Sci. **17**, 81–93 (2011)
8. Venkatesh, N.M., Ramakrishnan, A.G.: Choice of classifiers in hierarchical recognition of online handwritten kannada and tamil aksharas. J. Univ. Comput. Sci. **17**, 94–106 (2011)
9. Deepu, V., Madhvanath, S., Ramakrishnan, A.G.: Principal component analysis for online handwritten character recognition. In: 17th IEEE International Conference on Pattern Recognition, pp. 327–330. IEEE Computer Society, Washington, DC, USA (2004)

# Feature Extraction of Cervical Pap Smear Images Using Fuzzy Edge Detection Method

**K. Hemalatha and K. Usha Rani**

**Abstract** In Medical field Segmentation of Medical Images is significant for disease diagnose. Image Segmentation divide an image into regions precisely which helps to identify the abnormalities in the Cancer cells for accurate diagnosis. Edge detection is the basic tool for Image Segmentation. Edge detection identifies the discontinuities in an image and locates the image intensity changes. In this paper, an improved Edge detection method with the Fuzzy approach is proposed to segment Cervical Pap Smear Images into Nucleus and Cytoplasm. Four important features of Cervical Pap Smear Images are extracted using proposed Edge detection method. The accuracy of extracted features using proposed method is analyzed and compared with other popular Image Segmentation techniques.

**Keywords** Image segmentation · Edge detection · Fuzzy logic · Feature extraction · Cervical cancer

## 1 Introduction

Cervical Cancer is one of the most frequent cancers that occur in women. Unlike remaining cancers it does not cause pain and show symptoms. Cervical Cancer usually takes a long period of 10–15 years to become dangerous [1]. Early diagnosis will reduce the risk. Pap smear test introduced by Papanicolaou in 1940 is the most common screening test to detect precancerous and cancerous cervix cells using stained Pap smear slides. Cytotechnicians detect the abnormal changes in the cervix cells in laboratory using Pap smear test. In Pap smear test trained technicians

K. Hemalatha (✉) · K. Usha Rani
Department of Computer Science, Sri Padmavati Mahila Visvavidyalayam, Tirupati, India
e-mail: hemalathakulala@gmail.com

K. Usha Rani
e-mail: usharanikuruba@yahoo.co.in



S.C. Satapathy et al. (eds.), *Data Engineering and Intelligent Computing*,
Advances in Intelligent Systems and Computing 542,
DOI 10.1007/978-981-10-3223-3_8

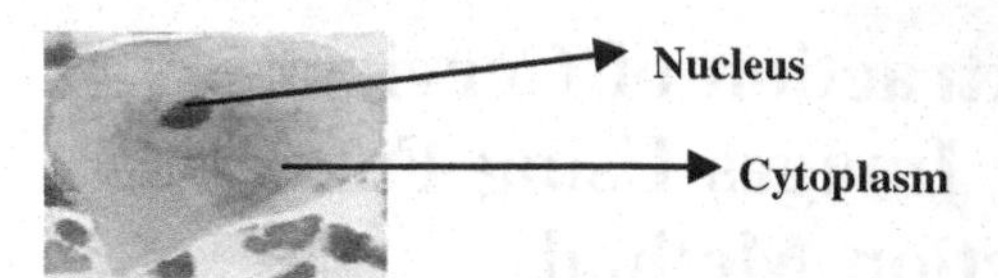


**Fig. 1** Pap smear image

collect cervix cells using a speculum and then the cell samples are prepared by examining under a microscope [2] (Fig. 1).

The morphology and abnormalities of the cells are examined by the cytologists. In this process time and errors are critical [3]. Human error may occur in the Pap smear test which leads to wrong assessment of the cervical cells. Hence to reduce the errors and for quick diagnosis computer aided automated screening system is essential. Image Processing is vital in automated Pap smear slide screening.

Image Segmentation, a critical step in Image Processing divides an image into regions based on the interest. Selection of Image Segmentation technique is based on the problem considered [4]. Edge detection is a commonly used Image Segmentation technique. Based on the grey tones variation in the image edges are formed in Edge detection [5]. In Edge Detection uniform regions are not defined crisply so that the small intensity difference between pixels did not represented by the edge always. That may represent a shading effect to the edge.

The Fuzzy Logic approach for Edge Detection is necessary to define the degree of pixel relevance to a uniform region. Membership functions [6] consider the values in the interval [0, 1]. The important features can be extracted from the resulted edges. In this paper four commonly used features [7] such as Size of Nucleus, Size of Cytoplasm, Grey Level of Nucleus and Grey Level of Cytoplasm are extracted using Fuzzy Logic based Edge Detection which are important to classify cervical cells into normal and abnormal.

## 2 Review of Literature

Mat-Isa et al. [8] segmented Pap Smear Cytology images and isolated nucleus and cytoplasm from the background using segmentation process using Non-adaptive k-means and fuzzy c-means clustering algorithms.

N. Senthil kumaran et al. [9] extracted the features: Nucleus size, Cytoplasm size, Nucleus grey level, Cytoplasm grey level from Pap Smear Images using Seeded Region Growing Features Extraction Algorithm.

Khang Siang Tan et al. [10] analyzed the efficiency of Fuzzy based, Genetic Algorithm based and Neural Network based approaches for Image Segmentation.

J. Mehena et al. [11] proposed an approach based on Histogram Thresholding technique to acquire all identical regions in color image. After that Fuzzy C-Means algorithm is used to improve the compactness of the clusters formed.

Koushik Mondal et al. [12] presented Neuro-Fuzzy approach to detect more fine edges of noisy Medical Images.

Mahesh Prasanna et al. [13] proposed a feature based fuzzy rule guided novel technique and the performance was analyzed using Mean Squared Error (MSE), Mean Absolute Error (MAE), Peak Signal to Noise Ratio (PSNR) measures.

Abhishek Raj et al. [14] enhanced cerebral MRI features using enhancement approaches related to frequency and spatial domain.

Anushk Gupta et al. [15] proposed an edge detector for color images with Gaussian and Speckle noises.

## 3 Proposed Methodology

The theoretical structure of the proposed Fuzzy Edge Detection Method (FEDM) for feature extraction of Cervical Pap Smear Images based on Fuzzy Logic approach is presented in the Fig. 2.

Cervical Pap Smear Images sometimes contains menstrual discharge, vaginal discharge, air artifacts, etc., which may lead to wrong classification from normal to abnormal cells. Hence, Image Pre Processing is necessary to obtain better results. Four important features of Cervical Pap Smear images are automatically extracted using

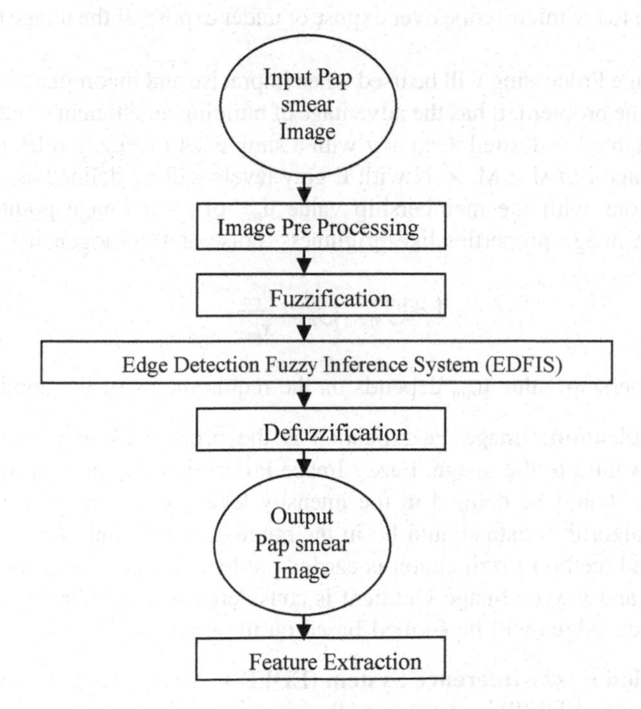


**Fig. 2** Proposed fuzzy edge detection method (FEDM) structure

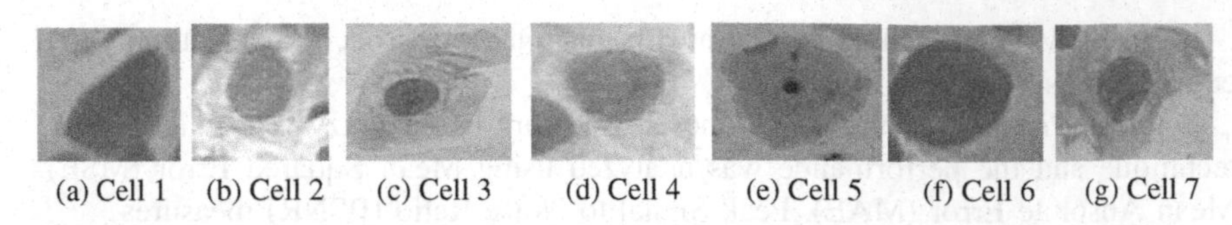

(a) Cell 1 (b) Cell 2 (c) Cell 3 (d) Cell 4 (e) Cell 5 (f) Cell 6 (g) Cell 7

**Fig. 3** Input cervical Pap smear images

proposed Fuzzy Edge Detection Method (FEDM). The Proposed method is implemented on seven Cervical Pap Smear Images using MATLABR2015a tool for the experiment. Seven images are related to different Cervical cell classes like Mild Dysplasia, Moderate Dysplasia, Severe Dysplasia and Carcinoma in situ, etc. (Fig. 3).

**Image Pre Processing**: In Image Processing there are different theories, methodologies and techniques to solve different problems occur in computer vision. Fuzzy Image Processing is the combination of different approaches that segment and represent features as fuzzy sets [16].

In the proposed method Image Pre Processing includes: Considered RGB Image is converted to gray Image as Edge Detection works on gray tones of the image. Median Filter is used to avoid salt and pepper noise occurred in the input image. Median Filter replaces each pixel with the median value of surrounding N-by-N neighborhood and removes the noise of the image without significantly reducing the sharpness of an image. Histogram Equalization is used to adjust the uneven intensity distribution of the image due to the microscope over expose or under expose of the image to enhance contrast.

Fuzzy Image Processing will be used when imprecise and incomplete information presented in the problem. It has the advantage of building an efficient model which is easily predictable for desired accuracy with a simple set of Fuzzy rules [17].

A new image I of size M × N with L gray levels will be defined as an array of fuzzy singletons with the membership value $\mu_{mn}$ of each image point $x_{mn}$ with respect to the image properties like brightness, noise and homogeneity [18].

$$\mathrm{I} = \bigcup_{m=1}^{M} \bigcup_{n=1}^{N} \frac{\mu_{mn}}{x_{mn}}. \tag{1}$$

The membership value $\mu_{mn}$ depends on the requirements of the application.

**Image Fuzzification**: Image Fuzzification is the process of generating suitable membership values to the image. Fuzzy Image Processing operates on membership values which should be defined in the intensity levels with a range 0 to 255. To apply fuzzy algorithm data should be in the range of 0 to 1 only [19, 20].

In proposed method fuzzification is carried out by the image gradient calculated along x-axis and y-axis. Image Gradient is considered to locate breaks in uniform regions. Hence, edges will be formed based on the gradient (Fig. 4).

**Edge Detection Fuzzy Inference System (EDFIS)**: A new Edge Detection Fuzzy Inference System (EDFIS) is created and the image gradients of x-axis, y-axis (Ix, Iy)

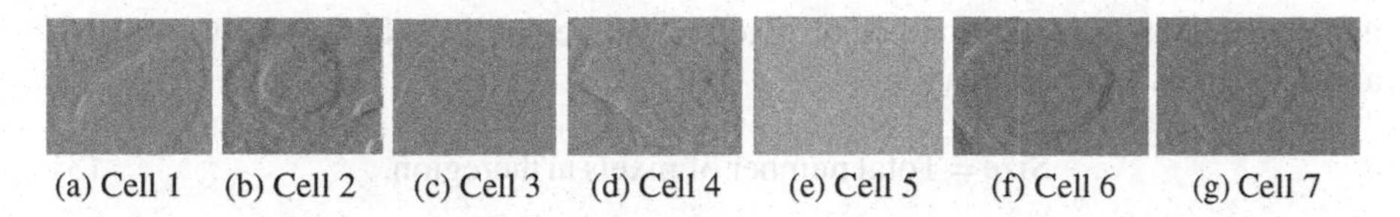

(a) Cell 1 (b) Cell 2 (c) Cell 3 (d) Cell 4 (e) Cell 5 (f) Cell 6 (g) Cell 7

**Fig. 4** Fuzzified cervical Pap smear images

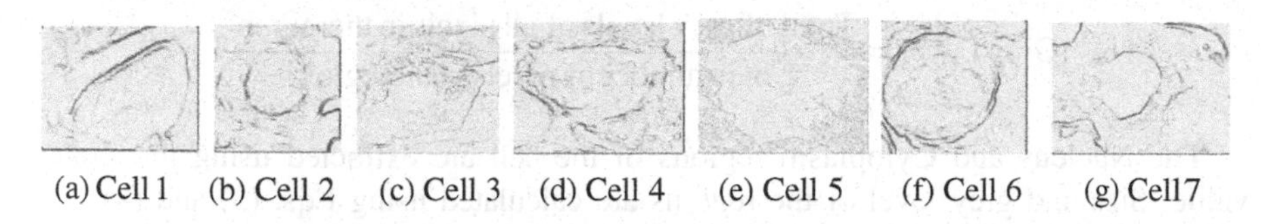

(a) Cell 1 (b) Cell 2 (c) Cell 3 (d) Cell 4 (e) Cell 5 (f) Cell 6 (g) Cell7

**Fig. 5** Output cervical Pap smear images

are given as inputs to the new EDFIS. A zero-mean Gaussian membership function is specified for each input.

A wide range of membership functions are available but as per the literature survey [21] most popular membership functions are Gaussian and Trapezoidal shape functions. Triangle and Square shape functions are the special cases of Trapezoidal [22].

In general, a Gaussian membership function is defined as:

$$\mu = \frac{I_{max}}{1 + \left(\frac{x-c}{a}\right)^{2b}}. \tag{2}$$

where $I_{max}$ is the maximum membership value, x is the input and a, b, c are the parameters which can be modified by the membership function. Based on the problem the three parameters vary. The Edge FIS specifies the intensity of the edge-detected image as an output.

**Defuzzification**: In Defuzzification a fuzzy quantity is converted into a precise quantity. In proposed method triangular membership function is used to defuzzify the fuzzy Pap smear images. A widespread assumption in Fuzzy Logic, Triangular membership function is used to specify the intensities to the output image. Triangular membership consists of three parameters which indicate start, peak and end of the triangles of the membership function. These parameters influence the intensity of the detected edges.

Fuzzy Inference System (FIS) rules are added to get the required image as an input to the next stage. The rules are added to make a pixel to white color if it belongs to uniform region otherwise to make it as black. The rules are specified by the gradients along x-axis and y-axis. If x-axis and y-axis gradients are zero then the pixel will appear as white.

After defuzzification the output images are given in Fig. 5.

**Feature Extraction**: The output image from the Defuzzification stage is used to extract the four features: Size of the Nucleus, Size of the Cytoplasm, Grey level of the Nucleus and Grey level of the Cytoplasm. The size of the Nucleus or Cytoplasm is calculated as the total number of pixels in the region i.e. Nucleus or Cytoplasm and it is given by the equation.

$$\text{Size} = \text{Total number of pixels in theregion.} \tag{3}$$

The grey level is calculated as mean of all pixels in the region by:

$$\text{Grey Level} = \frac{\text{Total of grey levels of all pixels in theregion}}{\text{Total number of pixels in theregion}} \tag{4}$$

The Nucleus and Cytoplasm regions of the cell are extracted using threshold value. Size and grey level of the regions are calculated using Eqs. (3) and (4).

# 4 Experimental Analysis

Features extracted from the seven images using proposed method are tabulated in the Table 1.

**Statistical Analysis**: The accuracy of the extracted features is analyzed by the Correlation Coefficient. A Correlation Coefficient >0.8 represents the strong relationship between two variables and the Correlation coefficient <0.5 represent that the relation is weak [23].

$$\text{Correlation Coefficient} = \frac{\sum (x-\bar{x})(y-\bar{y})}{\sqrt{\sum (x-\bar{x})^2 \sum (y-\bar{y})^2}}. \tag{5}$$

The Correlation Coefficient values of extracted features are compared with other. popular Image Segmentation methods using k-Means Clustering algorithm and Fuzzy C-Means algorithm. The Correlation Coefficient values are presented in the Table 2.

From the results in Table 2, it is found that the performance of the proposed Fuzzy Logic Based Edge Detection Method is best based on the Correlation Coefficient values of the method. The same is represented graphically in Fig. 6.

**Table 1** The features extracted from the cervical Pap smear images using proposed method

| Image | Nucleus size (Pixels) | Cytoplasm size (Pixels) | Nucleus grey level | Cytoplasm grey level |
|---|---|---|---|---|
| Cell1 | 3490 | 2499 | 173 | 171 |
| Cell2 | 2306 | 7546 | 170 | 170 |
| Cell3 | 5206 | 32823 | 176 | 168 |
| Cell4 | 2808 | 3878 | 173 | 166 |
| Cell5 | 2204 | 78176 | 179 | 174 |
| Cell6 | 4552 | 3378 | 171 | 163 |
| Cell7 | 3811 | 14769 | 175 | 179 |

**Table 2** Correlation coefficient values of the cervical cell images for four extracted features

| | Correlation coefficient value | | |
|---|---|---|---|
| Extracted features | FEDM | K-means clustering | Fuzzy C-means |
| Nucleus size | 0.9484 | 0.8367 | 0.8836 |
| Cytoplasm size | 0.9992 | 0.9277 | 0.902 |
| Nucleus grey level | 0.9384 | 0.4793 | 0.4151 |
| Cytoplasm grey level | 0.8721 | 0.5052 | 0.5139 |

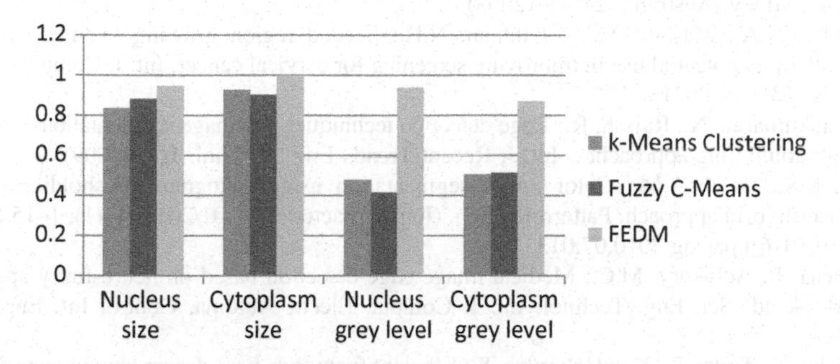


**Fig. 6** Comparison of correlation coefficient values of cervical Pap smear images

# 5 Conclusions and Future Directions

In this study, Fuzzy Edge Detection Method (FEDM) is proposed. The proposed method automatically extracted the Nucleus Size, Cytoplasm Size, Nucleus Grey Level and Cytoplasm Grey Level from the Cervical Pap Smear Images. The accuracy of extracted features is analyzed using Correlation Coefficient. The performance of proposed method is compared with other popular methods. The proposed method performed better than existing methods.

The extracted features may be used for Classification of Cervical Pap Smear Images in future using different techniques.

# References

1. Sulaimana, S.N., Mat-Isab, N.A., Othmanc, N.H., Ahmada, F.: Improvement of features extraction process and classification of cervical cancer for the neuralpap system. Procedia Comput. Sci. **60**, 750–759 (2015)
2. Sajeena, T.A., Jereesh, A.S.: Cervical cancer detection through automatic segmentation and classification of pap smear cells. Int. J. Appl. Eng. Res. **10**(18), 39078–39084 (2015). ISSN 0973-4562
3. Muthukrishnan, R., Radha, M.: Edge detection techniques for image segmentation. Int. J. Comput. Sci. Inf. Technol. (IJCSIT) **3**(6) (2011)

4. El Emary, I.M.M.: On the application of artificial neural networks in analyzing and classifying the human chromosomes. J. Comput. Sci. **2**(1), 72–75 (2006)
5. Mondal, K., Dutta, P., Bhattacharyya, S.: Efficient fuzzy rule base design using image features for image extraction and segmentation. In: Fourth International Conference on Computational Intelligence and Communication Networks (2012)
6. Mustafa, N., Mat Isa, N.A., Mashor, M.Y., Othman, N.H.: New Features of Cervical Cells for Cervical Cancer Diagnostic System Using Neural Network (2007)
7. Ghafar, R., Mat-Isa, N.A., Ngah, U.K., Mashor, M.Y., Othman, N.H.: Segmentation of stretched pap smear cytology images using clustering algorithm. In: CDROM Proceedings of World Congress on Medical Physics and Biomedical Engineering (WC2003). Paper no. 2356, vol. 4, Sydney, Australia. 24–29 (2003)
8. Mat-Isa, N.A., Mashor, M.Y., Othman, N.H.: Seeded region growing features extraction algorithm; its potential use in improving screening for cervical cancer. Int. J. Comput. Internet Manag. **13**(1) (2004)
9. Senthilkumaran, N., Rajesh, R.: Edge detection techniques for image segmentation—a survey of soft computing approaches. Int. J. Recent Trends Eng. Technol. **1**(2) (2009)
10. Tan, K.S., Isa, N.A.M.: Color image segmentation using histogram thresholding—Fuzzy C-meanshybrid approach. Pattern Recogn. (Impact Factor: 3.1). 01/2011; **44**(1), 1–15 (2011). doi:10.1016/j.patcog.2010.07.013
11. Mehena, J., Adhikary, M.C.: Medical image edge detection based on neuro-fuzzy approach. World Acad. Sci. Eng. Technol. Int. J. Comput. Electr. Automa. Control Inf. Eng. **10**(1) (2016)
12. Mondal, K., Dutta, P., Bhattacharyya, S.: Efficient fuzzy rule base design using image features for image extraction and segmentation. In: Fourth International Conference on Computational Intelligence and Communication Networks (2012)
13. Prasanna, M.K., Rai, S.C.: Fuzzy logic—a comprehensive study. Int. J. Adv. Found. Res. Comput. (IJAFRC) **1**(10) (2014). ISSN 2348 – 4853
14. Raj, A., Srivastava, A., Bhateja, V.: Computer aided detection of brain tumor in magnetic resonance images. Int. J. Eng. Technol. (IACSIT-IJET) **3**, 523–532 (2011)
15. Gupta, A., Ganguly, A., Bhateja, V.: A noise robust edge detector for color images using hilbert transform. In: Proceedings of (IEEE) 3rd International Advance Computing Conference, pp. 1207–1212, February (2013)
16. Dagar, N.S., Dahiya, P.K.: Soft computing techniques for edge detection problem: a state-of-the-art review. Int. J. Comput. Appl. (0975–8887) **136**(12) (2016)
17. Jahne, B., Haubecker, H., Geibler, P.: Handbook of Computer Vision and Applications, vol. 2. Academic Press Publishers (1999)
18. Divya, B., Shanthi, T.K., Sethuramalingam, T.K.: Edge detection technique by fuzzy logic CLA and canny edge detector using fuzzy image processing. Int. J. Recent Innov. Trends Comput. Commun. **2**(4), 954–957 (2014). ISSN 2321–8169
19. Tizhoosh, H.R.: Fuzzy Image Processing (in German). Springer, Berlin (1997)
20. Manikandan, G., Sairam, N., Harish, V., Saikumar, N.: Fuzzy logic—a comprehensive study. Int. J. Adv. Found. Res. Comput. (IJAFRC) **1**(10), (2014). ISSN 2348 – 4853
21. dos Santos Schwaab, A.A., Nassar, S.M., de Freitas Filho, P.J.: Automatic methods for generation of type-1 and interval type-2 fuzzy membership functions. J. Comput. Sci. **11**(9), 976–987 (2015)
22. Alikhani, A., Ahmadi, A., Alirezaee, S., Ahmadi, M., Erfani, S.: A CMOS implementation of programmable gaussian fuzzifier. Canadian Conference on Electrical and Computer Engineering (2015)
23. http://mathbits.com/MathBits/TISection/Statistics2/correlation.htm (2016)

# Minimizing Frequent Itemsets Using Hybrid ABCBAT Algorithm

**Sarabu Neelima, Nallamothu Satyanarayana and Pannala Krishna Murthy**

**Abstract** The expansion in information technology field leads to the increase in amount of data collected. Huge amount of data is stored in databases, data warehouses and repositories. Data mining is the process of analyzing the database and extract the required information and finding the relationships among the items of datasets using association rule mining. Apriori is a familiar algorithm for association rule mining which generates frequent itemsets. In this paper, we propose a new algorithm called hybrid ABCBAT which minimizes the generation of frequent itemsets and also reduces the time, space and memory. In the proposed algorithm, ABC is hybridized with random walk of BAT algorithm. Random walk is used in the place of onlooker bee phase in order to increase the exploration. Hybrid ABCBAT algorithm is applied over the frequent itemsets gathered from apriori algorithm, to minimize frequent itemsets. Different datasets from UCI repository are considered for experiment. The proposed algorithm has better optimization accuracy, convergence rate and robustness.

**Keywords** Apriori algorithm • Association rule mining • Artificial bee colony algorithm • Frequent itemsets • Random walk

S. Neelima (✉)
Department of CSE, JNTUH, Hyderabad, India
e-mail: sarabu.neelima@gmail.com

N. Satyanarayana
Department of CSE, Nagole Institute of Technology and Science, Hyderabad, India
e-mail: nsn1208@gmail.com

P. Krishna Murthy
Swarna Bharathi Institute of Science and Technology, Khammam, Andhra Pradesh, India
e-mail: krishnamurthy.pannala@gmail.com



S.C. Satapathy et al. (eds.), *Data Engineering and Intelligent Computing*,
Advances in Intelligent Systems and Computing 542,
DOI 10.1007/978-981-10-3223-3_9

## 1 Introduction

In recent years the size of the database is rapidly increasing. This led to a growing interest in the development of tools capable of extracting knowledge from data. Data mining is the process of analyzing large amounts of data and extracting useful information. Data mining functionalities include classification, clustering, link analysis (association) and prediction. Association rule mining is one of the important data mining techniques which find the relationship between the set of items in the database [1]. The association rules were first proposed by R.Agrawal in 1993 for mining frequent itemsets. Set of items which has minimum support are frequent itemsets. Association rule mining include two major tasks: Generating the frequent itemsets using minimum support that occur in the datasets and generation of strong association rules by using these frequent itemsets [2]. Hence, generation of frequent itemsets is the major task of association rule mining. There are number of techniques to generate the frequent itemsets. Apriori is an innovative algorithm for finding frequent itemsets from the huge amount of data.

## 2 Apriori Algorithm

For mining frequent itemsets for Boolean association rules R. Agrawal and R. Srikant in 1994 initiated the apriori algorithm. The name of the algorithm represents that this uses the previous knowledge of frequent itemset properties [2]. The apriori algorithm generates the frequent itemsets in a level-wise and apriori property can be used to improve the efficiency of level-wise generation of frequent itemsets [3]. Apriori property states that if all the subsets of the itemset are frequent then an itemset is frequent. This algorithm executes in two phases:

1. Join: Candidates are generated by joining frequent itemsets with itself.
2. Prune: Eliminate those itemsets for which the support value is less than minimum threshold value and also eliminate itemsets whose subsets are not frequent [4].

Apriori may generate more number of frequent itemsets. So, there is a need to minimize the frequent itemsets generated. Here we propose a new algorithm which minimizes frequent itemsets and also reduces the time taken for minimizing frequent itemsets.

## 3 Artificial Bee Colony (ABC) Algorithm

Swarm intelligence determines the measure of corporate behaviour of social insect colonies or some other animal societies to design an algorithm of solving various problems. Swarm intelligence algorithms are implemented to solve optimization problems [5]. ABC is one of the metaheuristic algorithms in swarm intelligence which simulates the intelligent foraging behaviour of honey bee. ABC is an optimization algorithm proposed by Karaboga in 2005. In ABC, colony consists of three groups of bees: employee bee, onlooker bee and scout bee. Each bee has its own task to improve the amount of nectar stored in the hive. The amount of nectar in food source is fitness value of food source. Each and every food source is kept in D-dimensional search space which exhibits the possible solution for optimization problems [6].

The employee bee search for the food sources and pass information to the onlooker bee in the dance area. The onlooker bee selects the best food source from the information shared by the employee bee to further proceed for searching the food. The food source is abandoned by employee bee when the fitness value is not improved in predefined cycles (limit). In this stage employee bee becomes a scout and search for new food source [6].

## 4 BAT Algorithm

In 2010, Xin-She Yang developed a metaheuristic optimization algorithm called Bat algorithm which is based on the echolocation or bio-sonar behavior of microbats with varying pulse rates of emission and loudness. Microbats use echolocation to hunt the food, avoid hurdles and to locate their roosting crevices in the dark. These bats emit a very loud and short sound pulse and receive the echo that returned back from the surrounding objects. Their pulses vary in properties and can be correlated with their hunting strategies, depending on the species [7]. Depending on the species their signal bandwidth varies, and often increased by using more harmonics. A random walk is an organization of the path that consists of a sequence of random steps. By using random walk a new solution for each bat is generated from the solution which is selected from the current best solutions [8].

## 5 Proposed Hybrid ABCBAT

Metaheuristic algorithms can also be used in data mining field for optimization [9–11]. To improve the efficiency of apriori algorithm, we propose hybrid ABCBAT algorithm which is a hybridized ABC algorithm with random walk of BAT algorithm. Hybrid ABCBAT performs global search and attribute interaction is better than greedy rule induction algorithms which are regularly used in data mining. Hybrid ABCBAT uses random walk of BAT algorithm in place of onlooker bee phase of ABC. In the onlooker bee phase, the random selection strategy will be used for looking the local optimization value in the neighborhood of food source and the higher probability solution will be chosen by onlooker bees. The standard ABC algorithm had disadvantages of easily prematurely falling into local optima and slow convergence rate in later stage. So that we are using random walk step in the place of onlooker bee phase. In order to use the ABC algorithm with random walk of BAT, the following phases must be addressed: initial population, fitness value, employed bee, random walk (BAT algorithm) and scout bees.

The steps of hybrid ABCBAT are given below:

Step 1: Generate the initial population
Step 2: Evaluate the fitness value of the population Repeat
Step 3: For each employed bee produce new solution (food source position)
Step 4: Select a solution among the best solutions
Step 5: Update the solution using random walk Rank the bats and find the current best solution
Step 6: If there is an abandoned solution then replace it with a new randomly produced solution for the scout. Memorize the best food source position

Until (Requirements not met)

## 6 Proposed Methodology

Proposed methodology minimizes frequent itemsets and decreases time, space and memory required to generate frequent itemsets. Hybrid ABCBAT algorithm is applied over the frequent itemsets gathered from apriori algorithm, to optimize frequent itemsets. Dataset is given as an input to the apriori algorithm. Frequent itemsets which are generated using apriori are given as an input to the hybrid ABCBAT.

The steps involved in the proposed methodology:

1. Start
2. Load Dataset from UCI repository.
3. Using Apriori algorithm to generate frequent itemsets.
4. Set the termination condition for hybrid ABCBAT.
5. Apply the hybrid ABCBAT on the frequent itemsets generated using apriori to minimize the frequent itemsets.
6. Evaluate the fitness value.
7. If the fitness value satisfies the termination criteria then add these itemsets to the output set.
8. Stop

The flow chart for the proposed methodology is given (Fig. 1):

## 7 Experimental Results

The proposed work is implemented in MATLAB so as to minimize the generation of frequent itemsets and to reduce the elapsed time for generating frequent itemsets. The datasets car, retail and student are taken from UCI repository. Data is converted

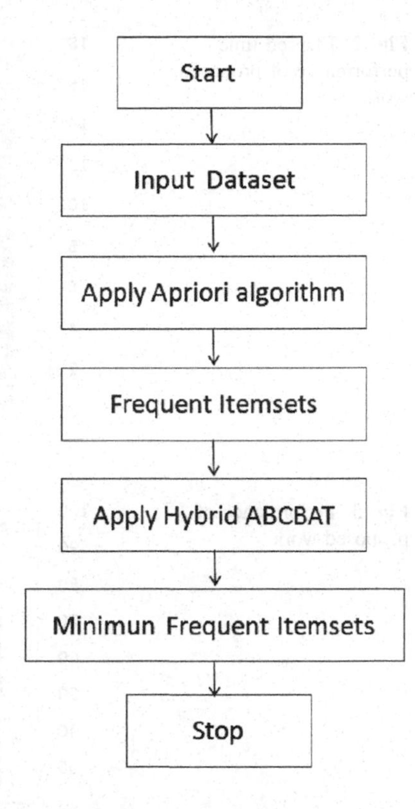


**Fig. 1** Flow chart for proposed methodology

**Table 1** Elapsed time to optimize frequent itemsets

| Dataset | Apriori-ABC | Proposed |
|---|---|---|
| Car | 15.8659 | 15.8279 |
| Student | 10.3022 | 10.2667 |
| Retail | 15.9927 | 15.8941 |

into binary format for implementation in MATLAB. Threshold value depends on the number of instances in the dataset.

The proposed methodology is compared with apriori with ABC. Table 1 shows the elapsed time taken to generate frequent itemsets for apriori with ABC and proposed methodology. From the Fig. 2 it is clear that the time taken by apriori with hybrid ABCBAT is less when compared with apriori with ABC.

Figure 3 shows the overall performance of proposed methodology when compared with apriori with ABC.

From the experimental results, time taken for proposed methodology for optimizing frequent itemsets is less when compared to apriori with ABC. It is also clear that the performance of proposed methodology is more when compared to apriori with ABC. Hence, proposed work minimizes the number of frequent itemsets and reduces the elapsed time taken for optimizing frequent itemsets. This also reduces the space and memory occupancy when compared to apriori with ABC.

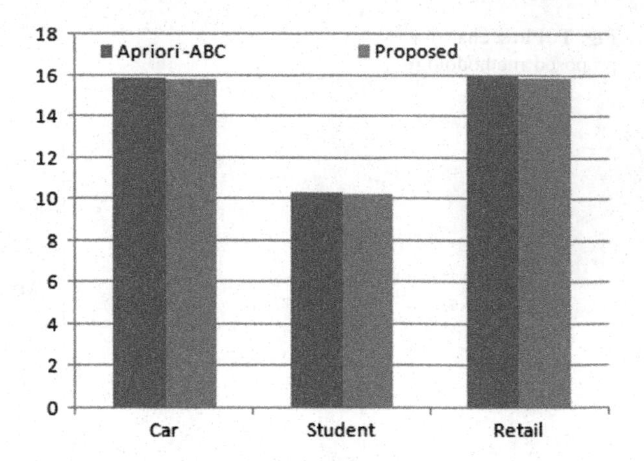


**Fig. 2** Elapsed time performance of proposed work

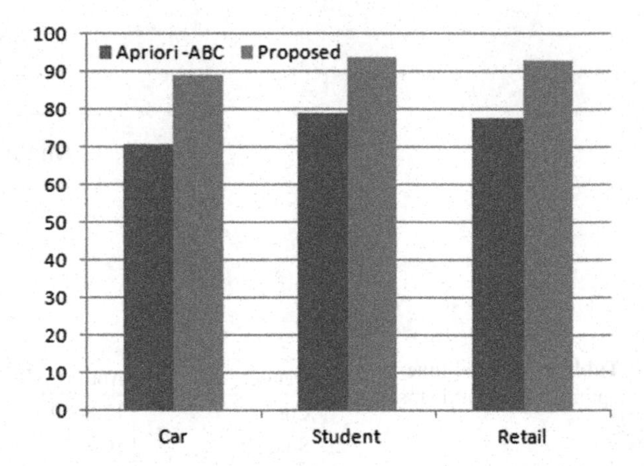


**Fig. 3** Performance of proposed work

# 8 Conclusion

In this paper, the proposed methodology is compared with apriori with ABC. The overall performance of proposed work is more and elapsed time taken for proposed work is less when compared to apriori with ABC. Hence, apriori with hybrid ABCBAT minimizes the frequent itemsets and also reduces the time, space and memory required for generation frequent itemsets.

# References

1. Dunham, M.H., Xiao, Y., Gruenwald, L., Hossain, Z.: A Survey of Association Rules
2. Han, J., Kamber, M.: Data Mining: Concepts and Techniques, 2nd edn

3. Gupta, B.: A Better Approach to Mine Frequent Itemset Using Apriori and FP Tree Approach (2011)
4. Raval, M.R., Rajput, I.J., Gupta, V.: Survey on several improved apriori algorithms. IOSR J. Comput. Eng. (IOSR-JCE) e- **9**(4), 57–61 (2013). e-ISSN: 2278-0661, p-ISSN: 2278-8727
5. Qin, Q., Cheng, S., Zhang, Q., Li, L., Shi, Y.: Artificial bee colony algorithm with time-varying strategy. Hindawi Publishing Corporation Discrete Dynamics in Nature and Society, vol. 2015, Article ID 674595 (2015)
6. Yurtkuran, A., Emel, E.: An Enhanced Artificial Bee Colony Algorithm with Solution Acceptance Rule and Probabilistic Multisearch. Hindawi Publishing Corporation Computational Intelligence and Neuroscience, vol. 2016, p. 13, Article ID 8085953 (2016)
7. Yang, X.-S.: A New Metaheuristic Bat-Inspired Algorithm (2010)
8. Yang, X.-S.: Bat Algorithm for Multi-objective Optimisation (2012)
9. Bhadoriya, V.S., Dutta, U.: Improved association rules optimization using modified ABC algorithm. Int. J. Comput. Appl. **122**(13), 23–26 (2015). (0975 – 8887)
10. Sahota, S., Verma, P.: Improved association rule mining based on ABC. Int. J. Comput. Appl. **135**(10), 6–10 (2016). (0975 – 8887)
11. Sharma, P., Tiwari, S., Gupta, M.: Optimize association rules using artificial bee colony algorithm with mutation. In: International Conference on Computing Communication Control and Automation, pp. 370–373. IEEE (2015)

# Hybrid Models for Offline Handwritten Character Recognition System Without Using any Prior Database Images

**Kamal Hotwani, Sanjeev Agarwal and Roshan Paswan**

**Abstract** In this paper a new method of classification is proposed by making hybrid models by using 3 different technique. One of them is correlation method, which use statistical template matching technique. Other one is principal component analysis in which for each image (character) there are some principal component, named Eigen value, and Eigen vectors. Third is Hough line detection technique, with the help of this we can find number of line segments in a character. Here with the experiments we can say that with the help of mixture of two or more different methods we can get better result. In this paper, we have implemented above techniques without using previous database of character images and getting 94.8% accuracy.

**Keywords** Offline characters recognition · Handwritten character recognition · Correlation · Hough line detection · Eigen value · Hybrid models

## 1 Introduction

Handwritten character recognition is one of the most emerging field of research. We can easily work on machine printed words/characters but handwritten characters are somewhat different. In handwritten, there are some challenges, like, uncertainty in writing styles, noise, two or more characters touching each others, some character are not connected, i.e., broken, improper scanning, lightning. It also depends on camera quality, angle at which image is taken, etc. It has number of applications

K. Hotwani (✉) · S. Agarwal
Malaviya National Institute of Technology, Jaipur, India
e-mail: kamalhotwani3@gmail.com

S. Agarwal
e-mail: sanagrawal@hotmail.com

R. Paswan
PEC University of Technology, Chandigarh, India
e-mail: roshanpaswan303@gmail.com



S.C. Satapathy et al. (eds.), *Data Engineering and Intelligent Computing*,
Advances in Intelligent Systems and Computing 542,
DOI 10.1007/978-981-10-3223-3_10

like recognizing formulas; perform analysis in banks, corporate, automatic form feeding, etc.

Character recognition is of two types, online character recognition, and offline character recognition. In online character recognition, there are two types of information, that is, spatial-luminance information. The luminance of character and the instance of time. Whereas in offline character recognition there is only one, factor that is, spatial information. Therefore offline characters recognition is little bit more complex than online character recognition. In this paper, we have taken some handwritten images of different persons that have upper case English alphabets and we try to recognize that characters via our technique. Handwritten recognition is now a day's one of the most challenging and emerging field. It can be used in various fields, number of areas. It can help to convert a handwritten data in image format to computer understandable characters. Here in this paper we have done experiments on uppercase English alphabets and converted in computer understandable format with 95% accuracy. There are some phases of our algorithm. They are as following:

1. Creating database by document itself
2. Preprocessing
3. Feature extraction or classification

These are explained in methodology section.

## 2 Previous Work

There are number of papers regarding character recognition but some of them are present in my following table, which acts as a torchbearer for us (Table 1).

**Table 1** Literature survey

| Author and year | Paper title | Contribution/work |
|---|---|---|
| Aravinda CV<br>Dr H.N. Prakash<br>@2014 IEEE | Template matching for Kannada handwritten recognition | In this paper author tried to recognize Kannad language via template matching (correlation) [1] |
| Nisha Sharma<br>Bhupendra Kumar<br>Vandita Singh<br>@2014 IEEE | Recognition of off-line handwritten characters | In this paper characters, numerals and special symbols are also recognized via NN and SVM [2] |
| Mahesh Jangid<br>Sumit Srivastava<br>@2014 IEEE | Gradient Local auto correlation for handwritten devanagari characters recognition | They used further extension of HOG and SIFT and give 90–95% accuracy with GLAC-SVM [3] |
| Quan-Sn Sun<br>Sheng-Gen Zeng<br>De-Sen Xia<br>@2004 IEEE | Feature Fusion Method based on canonical Correlation Analysis and HCR | In this paper they have used PCA, Eigen value and Eigen vectors to identify characters [4] |

## 3 Challenges

There are some challenges in offline handwritten character recognition. Following are some glimpse of challenges to make concept clearer (Figs. 1, 2, 3, 4 and 5).

These are some of the problems faced during image processing. Here in this paper we have tried to solve these issues.

## 4 Methodology

See Fig. 6.

**A. Creating database by document image itself**
We have applied training and testing from the same image. We want to reduce the burden of extra database and training. We tried of minimize the difference between input characters and database characters. Therefore, for this we give handwritten A, B, C, D… characters in image format as input or database.

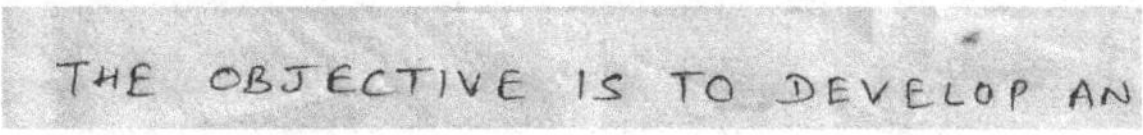

**Fig. 1** Scanning quality

**Fig. 2** Disconnected T and D

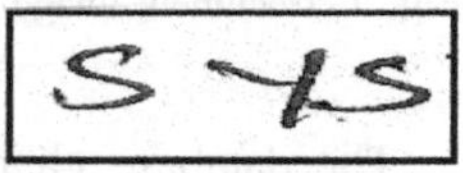

**Fig. 3** Connected Y and S

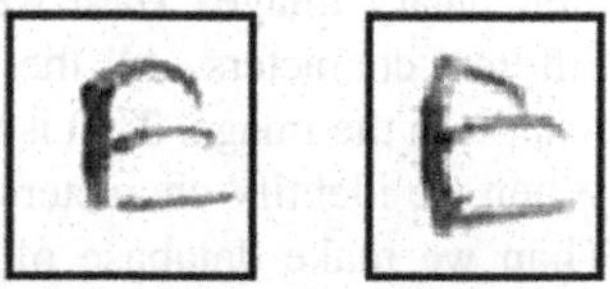

**Fig. 4** Different types of intra-personal writing skills

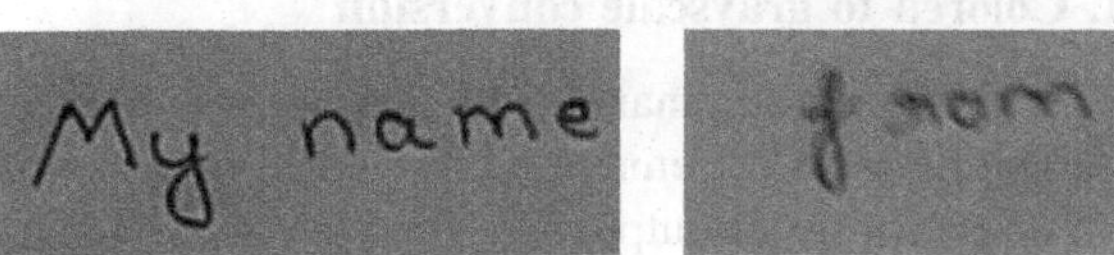


**Fig. 5** Brightness of image and blurred image

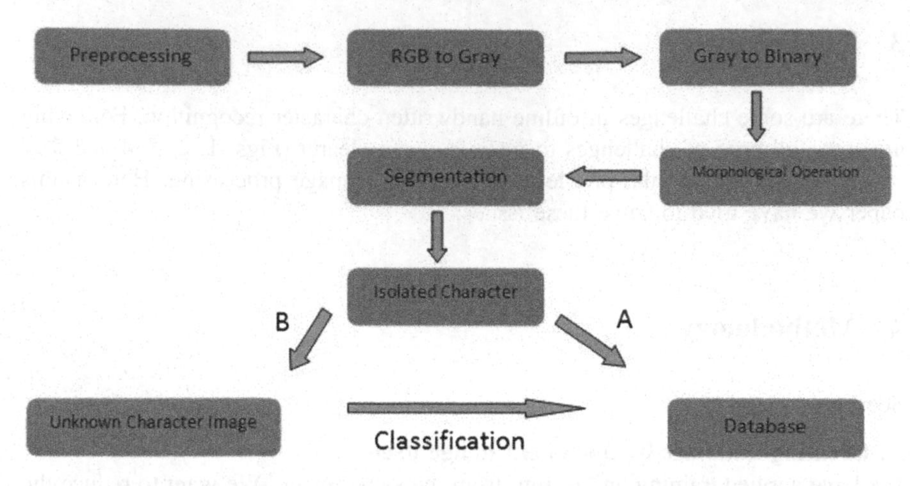


**Fig. 6** Complete methodology **a** creating database by document itself **b** taking an unknown character from document image and classification is done by the database created

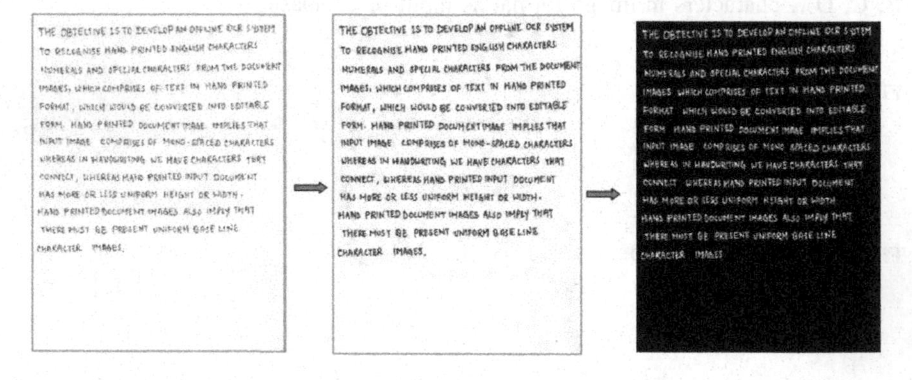


**Fig. 7** Document image is enhanced and then binary conversion

The database characters are processed same as document image itself. The database image is processed, noise removal, colored image to grayscale image and then binary image. Then we try to identify different lines in an image and then different characters. All the individual characters displayed and we have to enter what is in the image. That is we are teaching computer that this is A, or B character. When we identify characters and make computer understand what is in the image. Than we make database of that character. This database of character is used to identify document characters.

**B. Colored to grayscale conversion**

We to retain the luminance while eliminating hue and saturation information. This method helps us to remove all unwanted information. After this, we also enhance the image for better output. We tried to remove noise and increase contrast (Fig. 7).

**C. Binarization**

In grayscale image, the value of each pixel is between 0 and 255. Zero represent black pixel and 255 represent white pixel. Now with the help of threshold value we will convert this grayscale image into binary image. The range of binary image is 0 to 1. Zero represent black pixel and one represent white pixel. We will also perform negative binarization, that is, 0 to 1 and 1 to 0.This helps us to visualize better and computer can work efficiently on this type of image.

**D. Segmentation**

We have to extract object of our interest from document image. This is required as classifier can only classify or identify isolated characters not a line or word as a whole [5]. First, we crop line segment from document image and after a single isolated character from that segmented line (Fig. 8).

**E. Morphological operations**

In morphological operations, we have used some edge detections techniques that is, canny edge detection to find edges and with the help of these edges, we can connect the non-touching part and fill it so that it can give better result as compared to non-fill characters (Fig. 9).

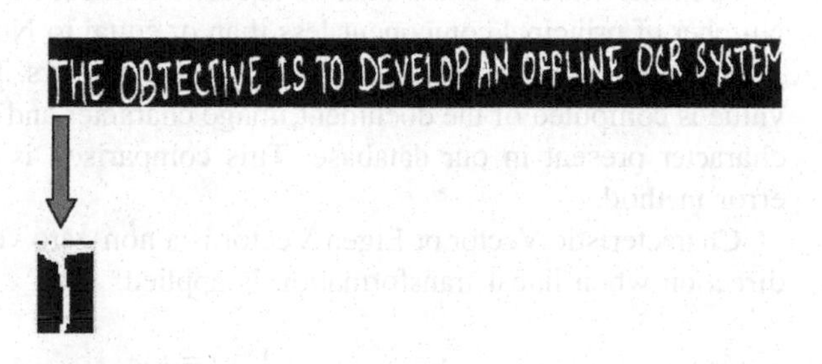


**Fig. 8** Segmentation of line and then single isolated character

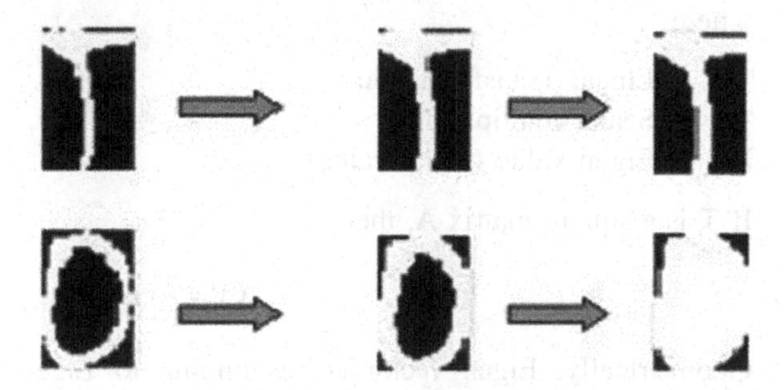

**Fig. 9** Morphological operations, filling regions of character

# 5 Classification

## A. Correlation Coefficient

Correlation coefficient is a method to find relation between two matrixes A and B. In our case, we have two images with same size and in binary format and we are finding correlation coefficient that is how much image A is related with image B [6]. It is a quantative method that can find dependency between two 2-D array.

$$r = \frac{1}{n-1} \sum \frac{(x_i - \overline{X})(y_i - \overline{Y})}{s_x s_y} \tag{1}$$

where,

r Relation between two matrix
n Total number of pixels
$x_i$, $y_i$ Value of ith pixel in x and y matrices

## B. PCA: Principal component analysis

PCA is another statistical method converts observations into linearly uncorrelated data variable called as principal component. In this method we find principal components which are less than or equal to number of original variables [7]. Number of principal component less than or equal to Number of original variables.

In this, we have used Eigen value and Eigen vectors. The Eigen vector and Eigen value is computed of the document image character and then compared with all the character present in our database. This comparison is done by minimum square error method.

Characteristic Vector or Eigen Vector is a non-zero vector. It does not change its direction when linear transformation is applied.

$$\mathrm{T(v)} = \lambda\, \mathrm{v} \tag{2}$$

where,

T Linear Transformation
T (v) Scalar multiple of v
λ Eigen value (scalar value)

If T is a square matrix A, then

$$\mathrm{A\,v} = \lambda\, \mathrm{v} \tag{3}$$

Geometrically, Eigen vector corresponding to Eigen value, points in stretched direction by transformation and Eigen Value is quantative unit of stretching. We have calculated Eigen values of various isolated characters and database images and then we find the minimum square error between them. The minimum error value is our expected character.

### C. Hough Transform

Hough transform is a feature extraction technique. Here this method is used to detect lines in an image. First, we use canny edge detector and find out the edges in the image and after that, we find out the number of lines in an image. This line detection will give the number of line in an image and number of line is compared with the database images whose number of line we already know during database creation process.

The equation of straight line is as following:

$$r = x\cos\theta + y\sin\theta \tag{4}$$

where,

x, y Coordinates
r distance from origin
θ Angle between x-axis and line connecting origin and line segment.

## 6 Experiments and Analysis

For better results, we have mixed two or more methods to make hybrid model for classification.

### A. Hybrid Model 1: Correlation + PCA

In this experiment, we have mixed two methods, Co-relation, and PCA. However, the problem is higher the co-relation better the relation and its value lies between −1 to 1. Moreover, for PCA we have used minimum square error technique that is minimum the error better the relation. Therefore, for making a hybrid method we have made a formula after doing some experiments and analysis [8], that is, co-relation has correct answer in second place or third place, if we arrange in decreasing order. Therefore, we have taken top five values and processed that value for PCA analysis and we get output with the following formula.

$$HC1 = r \times 100 \times K1 + (100 - pca\_cons) \times K2 \tag{5}$$

where,

HC1 Hybrid constant 1
r correlation coefficient
pca_cons minimum square error of PCA
K1 preference constant for correlation method
K2 preference constant for PCA method

The value of r, pca_cons is calculated and put here. The value of K1 is 2.5 and K2 is 1.

**B. Hybrid Model-2: Correlation + Hough transform**

In this experiment, we have made a hybrid model that comprises of correlation and Hough transform. We tried to detect lines of each character and compared with characters present in the database. This can be helpful to detect or differentiate between M and N or T and I. Again, we taken top five values and processed that value for line detection. We get output with the following formula.

$$HC2 = r \times 100 \times K1 + K2 + hough\_constant \tag{6}$$

where,

| | |
|---|---|
| HC2 | Hybrid constant 2 |
| r | correlation coefficient |
| K1 | preference constant for correlation method |
| K2 | preference constant for Hough transform method |
| hough_constant | if number of lines matches than some scalar value, else 0 |

The value of r is calculated and put here. The value of k2 is 1 here.

**C. Hybrid Model-3: Correlation + PCA + Hough transform**

In this experiment, we have mixed all three techniques and made a hybrid model that comprises of all three models.

We detected number lines of each character and compared with characters present in the database. Again, we taken top five values after correlation analysis and processed that value for PCA analysis and after that line detection. We get output with the following formula. The results obtained is better than other two models because numbers of lines and principal component have to find best match in top five expected character processed by correlation.

$$HC3 = r \times 100 \times K1 + (100 - pca\_cons) \times K2 + hough\_constant \tag{7}$$

where,

| | |
|---|---|
| HC3 | Hybrid constant 2 |
| r | correlation coefficient |
| pca_cons | minimum square error of PCA |
| K1 | preference constant for correlation method |
| K2 | preference constant for pca method |
| hough_constant | if line matches than some scalar value, else 0 |

The value of r, pca_cons is calculated and put here. The value of K1 is 2.5 and K2 is 1.

**Table 2** Results (accuracy) comparison

| Persons/methods | Hybrid model-1 Corr + pca (%) | Hybrid model-2 Corr + Hough (%) | Hybrid model-3 Corr + pca + Hough (%) |
|---|---|---|---|
| Person x | 93.5 | 93.3 | 94.5 |
| Person y | 96.23 | 93.9 | 95.8 |
| Person z | 93.1 | 92.67 | 94.14 |
| Average | **93.1** | **93.29** | **94.82** |

## 7 Results

In this paper, we have taken handwritten sample of some people and execute our hybrid models on different document images. I have taken three random images of three different persons, that is, three different handwriting and tabulated the results of experiment done (Table 2).

The hybrid model-3 gives us result up to 95%. Still there are some limitations and error but mainly error are due to handwriting error like two connected characters and disconnected single alphabet. The clear you write the better result you get.

## 8 Conclusion

In this paper, we tried to make a hybrid technology that can use multiple technique to get us a good result, that is, a better and reliable, handwritten character recognition technique. Hybrid method-1 gives result with 93.1% accuracy, hybrid method-2 gives 93.3% accuracy, and hybrid method-3 gives 94.8% accuracy.

Still there are some areas where future work can be done we can find out better algorithms or method which can give better results.

## References


1. Aravinda, C.V., Prakash, H.N.: Template matching method for Kannada handwritten recognition based on correlation analysis. In 2014 International Conference on Contemporary Computing and Informatics (IC3I), pp. 857–861. IEEE (2014)
2. Sharma, N., Kumar, B., Singh, V.: Recognition of off-line hand printed english characters, numerals and special symbols. In: 2014 5th International Conference on Confluence The Next Generation Information Technology Summit (Confluence), pp. 640–645. IEEE (2014)
3. Jangid, M., Srivastava, S.: Gradient local auto-correlation for handwritten Devanagari character recognition. In: 2014 International Conference on High Performance Computing and Applications (ICHPCA), pp. 1–5. IEEE (2014)

4. Sun, Q.S., Zeng, S.G., Heng, P.A., Xia, D.S.: Feature fusion method based on canonical correlation analysis and handwritten character recognition. In: Control, Automation, Robotics and Vision Conference, 2004. ICARCV 2004 8th, vol. 2, pp. 1547–1552. IEEE (2004)
5. Dunn, C.E., Wang, P.S.P.: Character segmentation techniques for handwritten text-a survey. In: 11th IAPR International Conference on, Pattern Recognition, 1992. vol. II. Conference B: Pattern Recognition Methodology and Systems, Proceedings, pp. 577–580. IEEE (1992)
6. Prasad, J.R., Kulkarni, U.V., Prasad, R.S.: Offline handwritten character recognition of Gujrati script using pattern matching. In: 3rd International Conference on, Anti-counterfeiting, Security, and Identification in Communication, 2009. ASID 2009, pp. 611–615. IEEE (2009)
7. Deepu, V., Madhvanath, S., Ramakrishnan, A.G.: Principal component analysis for online handwritten character recognition. In: Proceedings of the 17th International Conference on Pattern Recognition, 2004. ICPR 2004, vol. 2, pp. 327–330. IEEE (2004)
8. Kimura, F., Wakabayashi, T., Tsuruoka, S., Miyake, Y.: Improvement of handwritten Japanese character recognition using weighted direction code histogram. Pattern Recogn. **30**(8), 1329–1337 (1997)

# A Novel Cluster Algorithms of Analysis and Predict for Brain Derived Neurotrophic Factor (BDNF) Using Diabetes Patients

**Dharma Dharmaiah Devarapalli and Panigrahi Srikanth**

**Abstract** Brain Derived Neurotrophic Factor (BDNF) is involved Diabetes disease is associated with metabolic syndrome. Disease is mainly Type-2 Diabetes Mellitus (T2DM) parameters related to BDNF also. Today's most people suffered Diabetes Disease. Diabetes Mellitus is a metabolic disorder. Current research is Cluster analyses of T2DM of BDNF data based on predicting the diabetes and identify patients. In this paper, Evaluated as a clustering method for the cluster regarding T2DM of BDNF dataset classifies several clusters. Data Mining is one of the primary methods in clustering. This method examines measurements based on compute minimum, maximum and average values based predict of patients. These algorithms and mathematical problems applied into dataset, evaluate Normalize data and similarity measures based on identifying accurate results. Identification of the BDNF Korley et al. (J Neurotrauma, 33(2):215–225, 2015, [1]) gene these factors help the neurological affected, Change Behavior thing and Mind Depression.

**Keywords** T2DM of BDNF data · Clustering algorithms · Euclidean distance measure · Manhattan distance measure

---

D.D. Devarapalli
Department of Information Technology, Shri Vishnu Engineering College for Woman, Bhimavaram, India
e-mail: dharma@svecw.edu.in

P. Srikanth (✉)
Department of Information Technology, VNR Vignana Jyothi Institute of Engineering and Technology, Kukatpally, Hyderbad 500090, India
e-mail: srikanth.panigrahi@gmail.com



S.C. Satapathy et al. (eds.), *Data Engineering and Intelligent Computing*, Advances in Intelligent Systems and Computing 542,
DOI 10.1007/978-981-10-3223-3_11

# 1 Introduction

Brain derived neurotrophic factor (BDNF) is a protein. Human body is mainly behavior of brain and nervous extensions of factor. BDNF gene of human body related to pathological and protein related working system will be considered to damage of body affected as disease. Diseases are BDNF related central nervous system (CNS), Depression, epilepsy, and diabetes.

Prediction procedure is one of the process using machine learning and data mining. Machine learning concepts are mainly followed as Evolutionary algorithms, Fuzzy system, and rough set. Data mining algorithms of clustering, Classification, regression and rules using predict patients [1–3]. Other related work [4–8].

## *1.1 Problem Definition*

Diabetes diagnosis of BDNF concept of classifies several clusters. Introduce new model based construct identify of the patients positive or negative. This is achieved to one most research challenge to face and identify of the medical data and gene data variations.

T2DM of BDNF Convey of the body weight, height of weight depletion program and depressions of the fasting blood samples and calculated scores of each Attribute wise [9]. Banu et al., main aim studied as Relational ship between BDNF and metabolic parameters in patients with T2DM.In this aim to study of calculate data of error rate reduce Nosie calculate average value dataset [10]. Ram B. Singh et al., T2DM of BDNF important discuss depression, cardiovascular disease, BDNF,BDNF polymorphism, BDNF Insomnia, BDNF Alcoholism, obesity and insulin resistance [11]. Dharmaiah et al., main invention of Design and build intelligent expert system diagnosis of T2DM based on classify of the BDNF levels those type levels based on identify of behavior patients [12]. Multiple layer perceptron(MLP) neural network based on Diagnosis of the BDNF [13]. Marcel Dettling et al., clustering techniques applied expression levels of thousands of genes. Mainly it is using as predictive of genes clustering in microarray data presented and Most of the disease prediction problems are depending on clustering and classification [14]. Benhuai Xie et al., model using to clustering and proposed as high dimensional data but sample size data proposed as grouping multiple parameters of the same variable across cluster [15]. Genes related diseases diagnosis using microarray gene expression data demonstrated into cluster structure and predicted [16]. Other related works done [17, 18].

## 2 Dataset Detailed Description

**Diabetes Parameters** Diabetes Disease Parameters are age, gender, weight, and height, BMI (Body Mass Index), FBG (Fasting Blood Glucose), CRP (C—reactive protein), HDL (High Density Lipoprotein) and LDL (Density Lipoprotein).

**Diabetes Mellitus (DM) Cluster Representation**: Positive (Diabetic), Negative (Non-Diabetic).

| Diabetes parameter ranges: | | | |
|---|---|---|---|
| 1. | Age | | 30–60 |
| 2. | Gender | | Male (1), Female (0). |
| 3. | Weight | | (50–80 Kgs) |
| 4. | Height | Male | 1.5240–1.8268 Mts |
| | | Female | 1.4224–1.6460 Mts |
| 5. | BMI | Men | Low: 0–18.5 kg/mt/Medium: 18.6–24.9/high 25-29.9/Very high 30 above |
| | | Women | Low: 0–18.5 kg/mt/Medium: 18.6–22.9/high 23–24.9/Very high 25 above |
| 6. | FBG | | Low: 70–99/Medium: 100–125/high:126 and above |
| 7. | CRP | | Low: 0–0.9/Medium: 1.0–2.9/high: 3 and above |
| 8. | HDL | Men | Low: 0–40/Medium: 50–59/high:60 and above |
| | | Women | Low: 0.50/Medium: 50–59/high:60 and above |
| 9. | LDL | | Low: 2.6–3.3/Medium: 3.4–4.0/high:4.1–4.9 |
| 10. | BDNF | | Low: 0–950 pg/ml, Medium: 951–1599 pg/ml, High 1600–2400 pg/ml |
| 11. | DM | | Low: Healthy(Cluster1), |
| | | | Medium: Prediabetes (Cluster2), |
| | | | High: Diabetic(Cluster3). [12, 13, 19] |

## 3 Proposed Measure

Clustering algorithm using T2DM of BDNF patients dataset of diabetes patient predicting. Clustering algorithms of the our dataset is compute normalize and different types of similarities measures like as Euclidean and Manhattan using identify of the each attribute minimum value of the data. These processes based on identify patients [1, 3].

### *3.1 Decentralization*

Decentralization is method based on numerical attributes into nominal attributes or categorical attributes used in this context. We have considered to normalizing statistical techniques for decentralization. it is showing below.

In this equation $x = \{x_1, x_2 \ldots x_n\}$ dataset of Numerical attributes

With L classes of the class variable.it is showing as C = {C1...Cn}

$X_i$ is attributing of the data set [20]
$\overline{Xi}$ is the mean
$\sigma$ is standard deviation

Let us consider as the normalization or decentralization, such that

$$Z = \frac{Xi - \overline{Xi}}{\sigma} \tag{1}$$

### *3.2 Clustering Analysis*

This method is elementary as well as very useful access as long as partition a dataset within k distant clusters. Clustering procedure is defining as simple and mathematical problems notations $C_1 \ldots C_k$ indices observe of the specific cluster.

1. $C_1 \cup C_2 \cup C_3 \cup \ldots \cup C_K = \{1,\ldots,n\}$ each observation at least one k-cluster.
2. $C_K \cap C_{K'} = \varnothing$ for $k \neq k'$ in this clustering as non-overlapping of the no observations of the more than one cluster.

In specific dataset on one sample, ith observations in the number of elements of kth cluster of the data, then i € $C_k$.

$$\text{Minimize} = \left\{ \textstyle\sum_{(k=1)}^{K} W(Ck) \right\} \tag{2}$$

Clustering better simple choice to involve as balanced Euclidean distance

$$\mathrm{W}(\mathrm{C_k}) = \frac{1}{|\mathrm{Ck}|} \textstyle\sum_{i,i' \in \mathrm{Ck}} \sum_{(j=1)}^{p} (xij - xi'j) \wedge 2 \tag{3}$$

where |ck| denotes as number of observations in the *k*th cluster. It is considered Eqs. 2 and 3 gives optimization problem that defines clusters [2, 21]

$$\text{Minimize} = \frac{1}{|\mathrm{Ck}|} \textstyle\sum_{i,i' \in Ck} \sum_{(j=1)}^{p} (xij - xi'j) \wedge 2 \tag{4}$$

### 3.3 Eigen Vectors and Eigen Values

After compute similarity measures using to calculate Eigen vectors and Eigen value. Eigen values based on compute low value is predict gene.

λ is scalar value as the Eigen value same associated as the Eigen vector is v.

Eigen values and Eigen vector based on transformation formed as square matrix is performed below equation.

$$AV = \lambda V \tag{5}$$

Now transformation n-dimensional vectors defined as n × n matrix

$$A\nu = \omega \tag{6}$$

or

$$\begin{matrix} A_{11} & A_{12} & A_{13} & \cdots & \cdots & A_{1n} \\ A_{21} & A_{21} & A_{23} & \cdots & \cdots & A_{2n} \\ \vdots & \vdots & \vdots & & \ddots & \vdots \\ A_{n1} & A_{n2} & A_{n3} & \cdots & \cdots & A_{nn} \end{matrix}$$

5 and 6 equation based on Eigen values (A − λI) $\nu$ = 0. Where is I is n × n identify matrix.

## 4 Decentralization Possess of Clustering Algorithm

This section Decentralization existing similarity Euclidean and Manhattan, T2DM of BDNF dataset applied to identify of the patients.

**Input**: T2DM of BDNF patient of dataset most of the dataset patient behavior low, medium and high based on given clusters of two types of classes positive and negative.

**Output**: BDNF dataset using Applied to decentralization and similarities methods, predicted patients.

Step 1: Build pathological data T2DM of BDNF dataset. In this attributes of data in dataset numerical order and each record in the dataset ‘D’

Step 2: After getting BDNF data attributes classify ranges of each attribute wise like as low, medium and high.

Step 3: To Data numeric attributes in the dataset ‘D’ then apply z-score and Decentralization discuss Sect. 3.1

Step 4: After getting data Z is applied similarity measures Sect. 3.2

Step 5: Similarity measures applied then compute minimum each attribute wise and after getting minimum value is predicted value process discuss Sect. 3.2.

Step 6: Then compute similarities of Eigen values and Eigen vectors. After getting values minimum values is same attribute of the similarities procedure explained Sect. 3.3.

Step 7: compare similarities and Eigen values both are same minimum value of attribute is predicted.

Step 8: repeat Sects. 3.2 and 3.3 for each attribute wise and record wise based on classification rules.

Step 9: stop.

## 5 Case Studies of T2DM of BDNF

T2DM of BDNF data defined as new existing and new concept based on identify based those existing measurement using compute similarity.

**Step 1**: Initialize each cluster of the data (Table 1 and Fig. 1)

**Step 2**: Summarize dataset

Summarize of the each attribute of mean, standard deviation, minimum and maximum (Tables 2, 3 and 4), (Figs. 2, 3 and 4).

**Step 3**: Decentralized data To Data numeric attributes in the dataset then apply z-score and Decentralization discuss 3.1section (Table 5)

**Step 4**: Applied similarity Measures

After getting data Z is applied similarity measures section. Similarity measures applied then compute minimum each attribute wise and after getting minimum value is predicted value process discuss Sect. 3.2.

**Euclidean Distance Measure** (Table 6)

**Manhattan distance measure** (Table 7)

**Cluster of distance measure Averages**: (Table 8)

**Step 6**: Eigen Values based on Clusters (Table 9 and Fig. 5)

After compute distance measure values and Eigen values distance measure based on calculate this low true value is predicted. Comparison values is same attribute is predicted any distance measure.

## 6 Experimental Results

In this Research as, we followed complete dataset of Diabetes of BDNF of 259 patient's record. We use patient's records as followed as training and testing cases of predicted specific attribute based on maximum value computed as existing of proposed thing of predicted BDNF diagnosis data.

**Table 1** Cluster of initialized combination between two attributes of cluster like as classes

| S No | Age | Gender | Weight | Height | BMI | FBG | LDLL | HDLL | CRP | BDNF | Status | Class of cluster |
|---|---|---|---|---|---|---|---|---|---|---|---|---|
| 1 | 58 | 1 | 58 | 180 | 17.9 | 70 | 2.6 | 80 | 0.1 | 2384 | 1 | Cluster1 |
| 2 | 52 | 2 | 63 | 164 | 23.42 | 105 | 3.6 | 50 | 2.7 | 1297 | 2 | Cluster2 |
| 3 | 50 | 2 | 62 | 155 | 25.81 | 126 | 4.1 | 34 | 2.7 | 660 | 3 | Cluster3 |
| 4 | 59 | 2 | 51 | 164 | 18.96 | 70 | 2.6 | 80 | 0.1 | 2383 | 1 | Cluster1 |
| 5 | 59 | 2 | 55 | 160 | 21.48 | 88 | 3.1 | 63 | 0.8 | 1816 | 1 | Cluster2 |
| 6 | 57 | 2 | 59 | 158 | 23.63 | 105 | 3.6 | 50 | 2.5 | 1270 | 2 | Cluster2 |
| 7 | 58 | 1 | 75 | 169 | 26.26 | 106 | 3.6 | 50 | 2 | 1255 | 2 | Cluster2 |
| 8 | 57 | 2 | 51 | 161 | 19.68 | 88 | 3.1 | 63 | 0.8 | 1810 | 1 | Cluster1 |
| 9 | 50 | 1 | 60 | 167 | 21.51 | 89 | 3.1 | 63 | 0.8 | 1808 | 1 | Cluster1 |
| 10 | 60 | 1 | 76 | 173 | 25.39 | 125 | 4.1 | 34 | 2.8 | 1523 | 2 | Cluster2 |

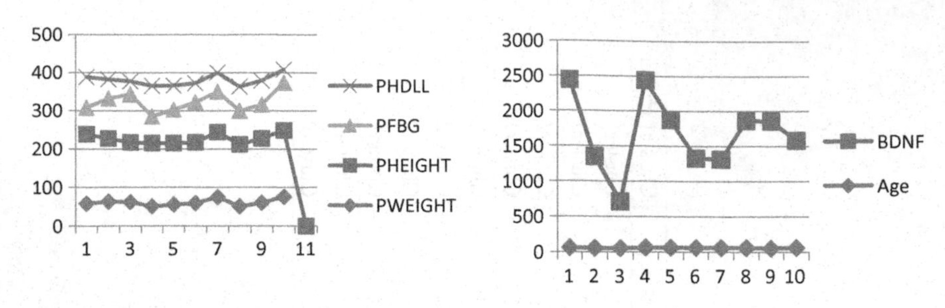


**Fig. 1** Combination between age, BDNF, BMI, weight and height

Case 1: Training Results

In this case training results of the completely 259 records checking as followed proposed of existing method of the predicted attribute as dependent of BDNF of the performed as either positive nor negative BDNF, HDL and BMI. In this data set of 259 records of 129 male records and 130 records.

BDNF of 259 records of the male records as 129 training as mostly verified into followed proposed of existing method of the predicted attribute as dependent of BDNF of the performed as either positive nor negative BMI and HDL.

BDNF of 259 records of the female records as 130 training as mostly verified into followed proposed of existing method of the predicted attribute as dependent of BDNF of the performed as either positive nor negative BMI and LDL.

Case 2: Testing Results

In this case, testing neither identifying number patients of the predicted as either positive nor negative. After identified cased on this phase will be predicted as 68 records. Those records based on predicted as specific attribute as BDNF of the FBG and CRP. Complete 259 record of in testing case is predicted as BMI and BDNF.

## 7 Conclusions and Future Work

In this paper, we studied Clustering is a significant method in emanate pasture of Data mining. Prominent Clustering Algorithms were considered to similarity measure of Clustering Algorithm using T2DM of BDNF of data distance measure using Eigenvalues based on is identified. Most of the computing of the similarity measure based on Identified of the BDNF these factors help the neurobiological

**Table 2** Each attribute of mean and SD combination of each attribute mean and SD

| | Age | Gender | Weight | Height | BMI | FBG | LDLL | HDLL | CRP | BDNF | Status |
|---|---|---|---|---|---|---|---|---|---|---|---|
| Mean | 56 | 1.6 | 61 | 165.100 | 22.404 | 97.200 | 3.350 | 56.7 | 1.53 | 1621 | 1.61 |
| SD | 3.829 | 0.51 | 8.66 | 7.46 | 2.96 | 19.76 | 0.54 | 16.22 | 1.11 | 534.03 | 0.699 |

**Table 3** Each attribute of minimum and maximum

| | Age | Gender | Weight | Height | BMI | FBG | LDLL | HDLL | CRP | BDNF | Status |
|---|---|---|---|---|---|---|---|---|---|---|---|
| Minimum | 50 | 1.0 | 51 | 155 | 17.90 | 70.0 | 2.60 | 34.0 | 0.10 | 600 | 1.0 |
| Maximum | 60 | 2.0 | 76 | 180 | 26.26 | 12.6 | 4.10 | 80.0 | 2.80 | 2384 | 3.0 |

**Table 4** Combination specific attribute

| | Age | Gender | Weight | Height | BMI |
|---|---|---|---|---|---|
| Mean | 56 | 1.6 | 61 | 165.100 | 22.404 |
| SD | 3.829 | 0.51 | 8.66 | 7.46 | 2.96 |
| Minimum | 50 | 1.0 | 51 | 155 | 17.90 |
| Maximum | 60 | 2.0 | 76 | 180 | 26.26 |

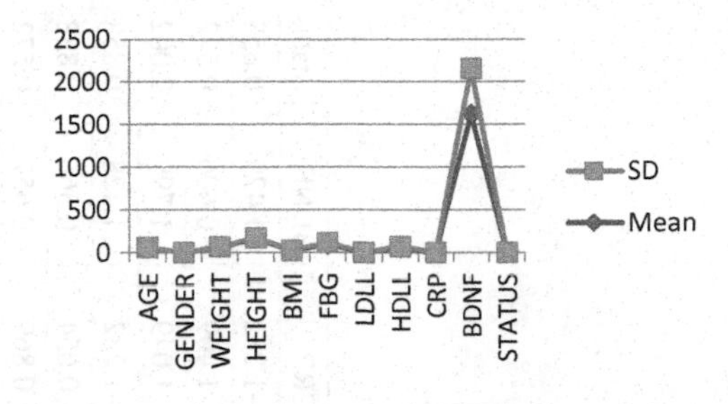


**Fig. 2** Combination of each attribute mean and SD

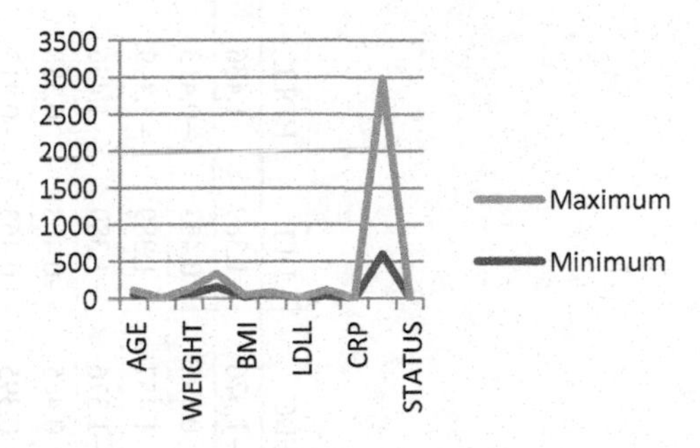


**Fig. 3** Combination of each attribute mean and SD

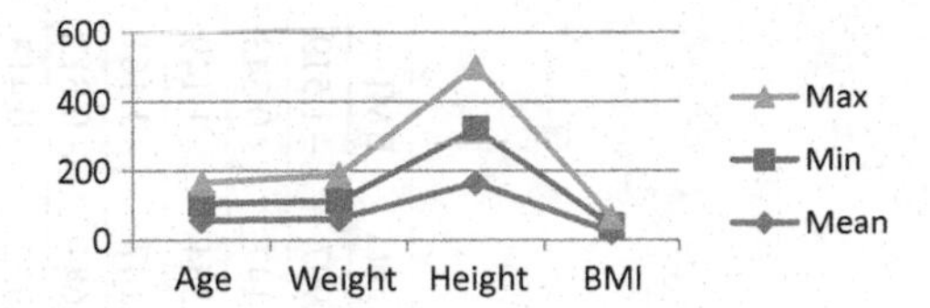


**Fig. 4** Relational ship between Age, weight, height and BMI

**Table 5** Decentralized data

| Age | Gender | Weight | Height | BMI | FBG | LDLL | HDLL | CRP | BDNF | Status | Class of cluster |
|---|---|---|---|---|---|---|---|---|---|---|---|
| 0.522 | −1.162 | −0.346 | 1.997 | −1.519 | −1.376 | −1.389 | 1.436 | −1.282 | 1.429 | 0.858 | Class1 |
| −1.044 | 0.775 | 0.231 | −0.147 | 0.343 | 0.395 | 0.463 | −0.413 | 1.409 | −0.606 | 0.572 | Class2 |
| −1.567 | 0.775 | 0.115 | −1.354 | 1.149 | 1.457 | 1.389 | −1.399 | 1.049 | −1.799 | 2.002 | Class3 |
| 0.783 | 0.775 | −1.154 | −0.147 | −1.161 | −1.376 | −1.389 | 1.436 | −1.282 | 1.428 | 0.858 | Class1 |
| 0.783 | 0.775 | −0.692 | −0.684 | −0.312 | −0.465 | −0.463 | 0.388 | −0.654 | 0.366 | 0.858 | Class2 |
| 0.261 | 0.775 | −0.231 | −0.952 | 0.413 | 0.395 | 0.463 | −0.413 | 0.869 | −0.657 | 0.572 | Class2 |
| 0.522 | −1.162 | 1.1615 | 0.523 | 1.3 | 0.445 | 0.463 | −0.413 | 0.421 | −0.685 | 0.572 | Class2 |
| 0.261 | 0.775 | −1.154 | −0.55 | −0.919 | −0.465 | −0.463 | 0.388 | −0.654 | 0.355 | 0.858 | Class1 |
| −1.567 | −1.162 | −0.115 | 0.255 | −0.301 | −0.415 | −0.463 | 0.388 | −0.654 | 0.351 | 0.858 | Class1 |
| 1.044 | −1.162 | 1.731 | 1.059 | 1.007 | 1.406 | 1.389 | −1.399 | 1.138 | −0.183 | 0.572 | Class2 |

**Table 6** Euclidean distance measure

| Euclidean distance measure | | | | | | | | | |
|---|---|---|---|---|---|---|---|---|---|
| | 1 | 2 | 3 | 4 | 5 | 6 | 7 | 8 | 9 |
| 2 | 6.016867 | 0 | 0 | 0 | 0 | 0 | 0 | 0 | 0 |
| 3 | 8.6301604 | 2.9693096 | 0 | 0 | 0 | 0 | 0 | 0 | 0 |
| 4 | 3.0328285 | 5.3956469 | 7.8562696 | 0 | 0 | 0 | 0 | 0 | 0 |
| 5 | 4.1098705 | 3.6107245 | 5.9026419 | 2.3509938 | 0 | 0 | 0 | 0 | 0 |
| 6 | 6.104667 | 1.6137255 | 3.2407972 | 5.024867 | 2.9554126 | 0 | 0 | 0 | 0 |
| 7 | 5.7919812 | 3.1449291 | 4.5212281 | 6.1290873 | 4.4532246 | 3.2235645 | 0 | 0 | 0 |
| 8 | 3.9559691 | 3.6716821 | 5.9923453 | 2.1958765 | 0.9339664 | 3.2274496 | 4.916384 | 0 | 0 |
| 9 | 3.6565939 | 3.5685666 | 5.858458 | 3.9602409 | 3.2386522 | 4.0751741 | 4.0632452 | 3.0330773 | 0 |
| 10 | 6.7703624 | 3.9115416 | 4.866897 | 7.2306031 | 5.4908866 | 3.969343 | 2.0441556 | 5.8866615 | 5.2978821 |

**Table 7** Manhattan distance measure

| Manhattan distance measure | | | | | | | | | |
|---|---|---|---|---|---|---|---|---|---|
| | 1 | 2 | 3 | 4 | 5 | 6 | 7 | 8 | 9 |
| 2 | 19.353192 | 0 | 0 | 0 | 0 | 0 | 0 | 0 | 0 |
| 3 | 27.369392 | 8.246969 | 0 | 0 | 0 | 0 | 0 | 0 | 0 |
| 4 | 5.509327 | 15.981483 | 23.997683 | 0 | 0 | 0 | 0 | 0 | 0 |
| 5 | 11.006834 | 10.633247 | 17.577099 | 6.420584 | 0 | 0 | 0 | 0 | 0 |
| 6 | 18.332431 | 2.872033 | 9.036961 | 14.960722 | 8.540138 | 0 | 0 | 0 | 0 |
| 7 | 17.023879 | 7.272436 | 13.649016 | 18.865588 | 13.517351 | 6.932002 | | 0 | 0 |
| 8 | 10.73856 | 11.034278 | 18.246217 | 5.751466 | 1.736054 | 9.209256 | 14.440616 | 0 | 0 |
| 9 | 9.920148 | 10.47751 | 17.449244 | 11.225683 | 5.877447 | 12.298683 | 11.817769 | 6.278478 | 0 |
| 10 | 19.920046 | 10.832397 | 11.903858 | 22.31187 | 16.963634 | 10.951631 | 5.558957 | 18.409131 | 15.786284 |

**Table 8** Cluster of distance measure averages

| Cluster of distance measure averages: | | |
|---|---|---|
| | Cluster of euclidean distance avg low value (true) | Cluster of Manhattan distance avg low value (true) |
| 1 | 3.62 | 15.46 |
| 2 | 2.96 | 8.59 |
| 3 | 2.514 | 12.42 |
| 4 | 2.08 | 8.837 |
| 5 | 1.682 | 5.181 |
| 6 | 1.543 | 4.376 |
| 7 | 1.324 | 3.977 |
| 8 | 0.8511 | 2.74 |
| 9 | 0.29 | 1.75 |

**Table 9** Eigen value based on each cluster of low true value

| S. No | Eigen values of Euclidean measure of cluster | Eigen values of Manhattan measure of cluster |
|---|---|---|
| 1 | 4.26 | 9.093 |
| 2 | 4.02 | 9.038 |
| 3 | 4.02 | 8.489 |
| 4 | 3.62 | 8.64 |
| 5 | 3.17 | 5.916 |
| 6 | 3.17 | 5.152 |
| 7 | 2.73 | 4.896 |
| 8 | 2.61 | 4.881 |
| 9 | 2.49 | 3.717 |

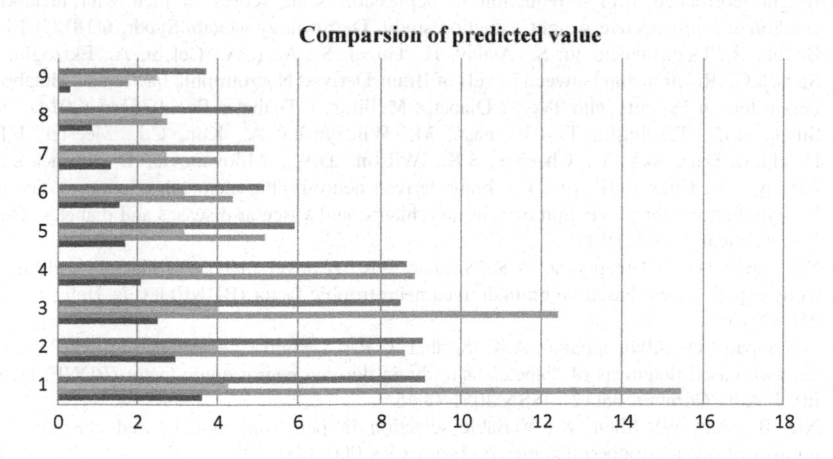


**Fig. 5** Comparison of predicted value

effected Change Behavior thing and Mind Depression. Future work developed as pathological and proteomics of T2DM of BDNF data dimensionality reduction based on reduce the data whether using clustering and classification.

## References


1. Korley, F.K., Diaz-Arrastia, R., Wu, A.H.B., Yue, J.K., Manley, G T., Sair, H.I., Van Eyk, J., Everett, A.D., Okonkwo, D.O., Valadka, A.B., Gordon, W.A., Maas, A.I.R., Mukherjee, P., Yuh, E.L., Lingsma, H.F., Puccio, A.M., Schnyer, D.M.: Circulating brain derived neurotrophic factor (BDNF) has diagnostic and prognostic value in traumatic brain injury. J. Neurotrauma **33**(2), 215–225 (2015)
2. Santhanam, T., Padmavathi, M.S.: Application of K-Means and Genetic Algorithms for Dimension Reduction by Integrating SVM for Diabetes Diagnosis. vol. 47, pp 76–83 (2015)
3. Srikanth, P., Anusha, C., Deveraplli, D.: A Computational Intelligence Techniques for Effective Medical Diagnosis using Decision Tree. i-Manager's J. Comput. Sci. 21–26 (2015)
4. Deverapalli, D., Anusha, C.H., Srikanth, P.: Identification of Deleterious SNPs in TACR1 Gene Using Genetic Algorithm, pp. 87–97 (2015)
5. Bhagat, A., Kshirsagar, N., Khodke, P., Dongre, K., Ali, S.: Penalty parameter selection for hierarchical data stream clustering **79**, 24–31 (2016)
6. Srikanth, P., Rajasekhar, N.: A Novel cluster analysis for Gene-miRNA interactions documents using improved similarity measure. In: MorocMorocco, 2016 IEEE International Conference on Engineering & MIS (ICEMIS-2016), pp. 1–7 (2016)
7. Srikanth, P.: Clustering algorithm of Novel distribution function for dimensionality reduction using big data of OMICS: Health, clinical and Biology Research Information. In: 2016 IEEE International Conference on Computational Intelligence and Computing Research (ICCIC) (2016)
8. Devarapalli, D., Srikanth, P., Narasinga Rao, M., Rao, V.: Identification of AIDS disease severity based on computational intelligence techniques using clonal selection algorithm. Int J Convergence Comput (IJCONVC), Inderscience Publications (2017)
9. Lee, T., Fu, C.P., Lee, W.J., Liang, K.W.: Brain-derived neurotrophic factor, but not body weight, correlated with a reduction in depression scale scores in men with metabolic syndrome: a prospective weight-reduction study. Diabetology Metab. Syndr. **6**(18) (2014)
10. Boyuk, B., Degirmencioglu, S., Atalay, H., Guzel, S., Acar, A., Celebi, A., Ekizoglu, I., Simsek,C.: Relationship between Levels of Brain-Derived Neurotrophic Factor and Metabolic Parameters in Patients with Type 2 Diabetes Mellitus. J. Diabetes Res. 978143 (2014)
11. Singh, R.B., Takahashi, T., Tokunaga, M., Wilczynska, A., Kim, C.J., Meester, F.D., Handjieva-Darlenska, T., Cheema, S.K., Wilson, D.W., Milovanovic, B., Fedacko, J., Hristova, K., Chaves, H.: Effect of brain derived neurotrophic factor, in relation to diet and lifestyle factors, for prevention of neuropsychiatric and vascular diseases and diabetes. Open Nutraceuticals J. **7**, (2014)
12. Devarapalli, D., Allamapparao, A.K., Sridhar, G.R.: A novel analysis of diabetes mellitus by using expert system based on brain derived neurotrophic factor (BDNF) levels. Helix **1**, 251–256 (2013)
13. Devarapalli, D., Allamapparao, A.K., Sridhar, G.R.: A Multi-layer perceptron (MLP) neural network based diagnosis of diabetes using brain derived neurotrophic factor (BDNF) levels. Int. J. Adv. Comput. **35**(12). ISSN:2051–0845
14. Xie, B., Pan, W., Shen, X.: Variable selection in penalized model-based clustering via regularization on grouped parameters. Biometrics 000, 000, (2007)
15. Ma, S., Song, X., Huang, J.: Supervised group lasso with applications to microarray data analysis. Res. Rep. (2003)

16. http://www.ncbi.nlm.nih.gov
17. Srikanth, P., Deverapalli, D.: A critical study of classification algorithms using diabetes diagnosis. In: IEEE 6th International Conference on Advanced Computing-2016 (IACC), Feb (2016)
18. Garcia, S., Luengo, J., Herrera, F.: Data pre-processing in data mining. In: Intelligent Systems References Library, Series vol. 72, Springer (2015)
19. Deverapalli, D., Srikanth, P.: Identification of AIDS disease using Genetic Algorithm, pp. 99–111. (2015)
20. Hazemi, F.A., Youn, C.H., Al-Rubeaan, K.A.: Grid-Based Interactive Diabetes System, pp. 258–263. IEEE (2011)
21. Suh, S.C., Vudumula, G.P.: The Role of Conceptual Hierarchies in the Diagnosis and Prevention of Diabetes, pp. 267–275. IEEE (2011)

# Image Frame Mining Using Indexing Technique

D. Saravanan

**Abstract** Data mining is a technique the bring out hidden information effectively from an available data set. Most of this extraction works well when performed for binary and character information. Mining information form images is a challenge today for many researchers. Creating of images and videos is easy as it does not require any domain knowledge, but extracting the required knowledge is difficult. For this reason, today video data mining is an interesting area for many researchers. To overcome these problems many researchers are motivated for finding an effective retrieval and indexing technique. This research paper brings a new technique for video content retrieval using hierarchical clustering technique. Objective of this work is to extract image key frames from the trained image set and use this as an image input query. The experiment proved that the proposed technique provided better results than existing video retrieval and indexing technique.

**Keywords** Data mining · Key frame selection · Clustering · Video data mining · Image mining · Histogram · Hierarchical clustering

## 1 Introduction

Increasing the demand of video content, video data retrieval is most challenge and complex process today. On the other hand, the quantity of video and audio data produced in earlier days is gradually growing due to the large quantity of digital devices such as surveillance cameras, digital cameras and camcorders etc. This vast amount of video data is not managed and mined effectively by the effective tools. Increasing this data's day by day need urgent attention in this particular field. Today number of technique and tolls available to extract the content from the web,

D. Saravanan (✉)
IFHE University, IBS-Hyderabad, Hyderabad, Telangana, India
e-mail: sa_roin@yahoo.com



S.C. Satapathy et al. (eds.), *Data Engineering and Intelligent Computing*,
Advances in Intelligent Systems and Computing 542,
DOI 10.1007/978-981-10-3223-3_12

but none of them never produce the effective result it force to develop a new technique or algorithm in this particular field. However, there exists valuable information behind the video data; the increasing data volume creates complexity to humans in extracting them without sufficient tools. Number of technique produce effective result in data mining for which group the data set reduces the burden and reduce the searching time of the user. Hence nowadays, the video database is the research area and its characteristic should be examined. The significant attributes in the videos are Motion, Texture and Color, which are not fully utilized by the retrieving and indexing. For the above discussed problems, the research work has the following tasks:

- Video clustering and indexing
- Video retrieval

The research work was proposed to perform video arrangement based on the content property and arranged based on their similarity. The techniques to classify the various types of videos, which make use of the result of the content based on the input frames. Most the existing technique perform effective grouping based on the content nature. clustering consists of dividing the data into homogenous groups or else granules, which depend on the same objective function which maximize the inter cluster distance. Consequently the video clustering is differing from the traditional clustering algorithms. However the videos are in the unstructured format. Thus preprocessing the video data is necessary to obtain the structured format of the video using the computer vision techniques or image processing. Most of the data sets are grouped because to reduce the searching time. It must be taken as a parameter when processing the video data.

### *1.1 Problems in the Existing System*

1. Video clustering differ from the traditional clustering techniques, existing technique allow the user one scan, it generally very difficult for video data mining.
2. Video data consist of various attributes such as motion, texture and color this attributes are dynamic in nature they are changed during time. Among this dynamic nature identify the particular character is difficult.
3. The growth of video data and insufficient of data storage it lead difficulty for data extraction.
4. Existing clustering algorithms are not support effective parameters for data extraction.
5. Existing system produce not eliminate the noise effectively, it increasing the searching time.

## *1.2 Advantage of Proposed System*

1. The propose technique offer less memory space and we want to minimize the time required for I/O.
2. Easy to handle any forms of distance.
3. Applicable to all kind of attributes.
4. Embedded flexibility about the granularity level.

# 2 Literature Survey

This paper discusses the current video data mining technique and indexing problem. This paper brings the video data retrieval using one of the hierarchical clustering algorithm CHEAMELEON clustering. This paper also made comparisons with existing clustering algorithm against chameleon in terms of various video files. Author concluded the truthfulness of groping the item set based on nature of the data set D. Saravanan [1]. This paper brings design of hierarchical clustering algorithm best suit for video data retrieval it brings the number of frame based on the input frame. The proposed technique based on the hierarchy model for classify the items. This model occupy less space compare to the existing techniques and it well suited for any type of image files The proposed technique also gives the similarity between hierarchical clustering algorithm which currently used for video data mining. D. Saravanan [6]. This paper brings the size of video files currently available due to the growth of media. It also brings the idea of cluster a video files based on the metadata, Algur et al. [4]. As per Abhilasha Yadav et al. [9] video security is more important but in the case of video data all cryptography algorithm are not suitable. They proposed new advanced encryption standard based algorithm to solve the above problem. This paper brings the image segmentation using fuzzy harmony search based algorithm using different image sources such as MR, CT and RGB images, Panda et al. [3] (Fig. 1).

# 3 Experimental Setup

The behavior of this work is the input video is converted into number of frames. Using frame extraction process, redundant frames are eliminated. Using clustering technique, frames are grouped. Finally, user retrieves relevant frame for given query image.

- Video frame progression
- Train the input image
- Image mining
- Video Indexing

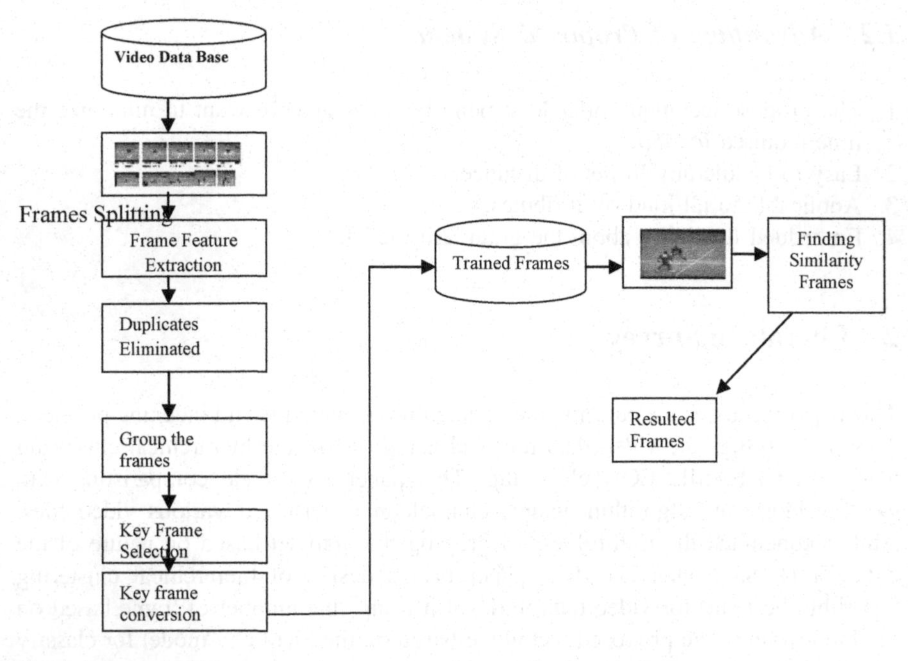


**Fig. 1** Proposed architecture

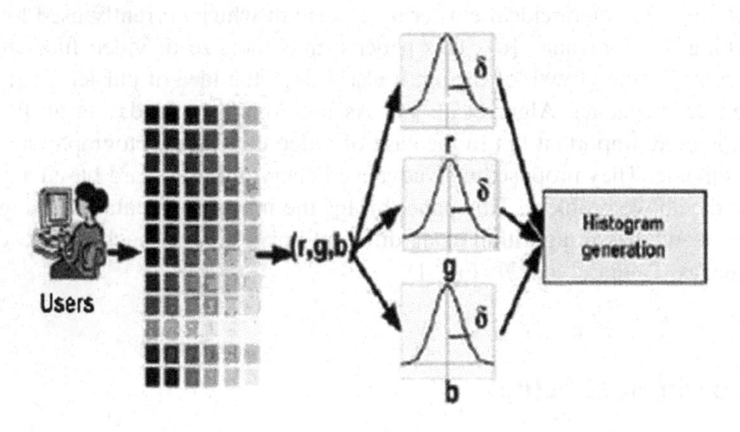


**Fig. 2** Histogram generation

## *3.1 Video Frame Progression*

A set of frames ordered by a time manner it is called video sequence. Initially the motion images are transformed into frames and that are stored into the database. This process shown in the Figs. 3, 4 and 5.

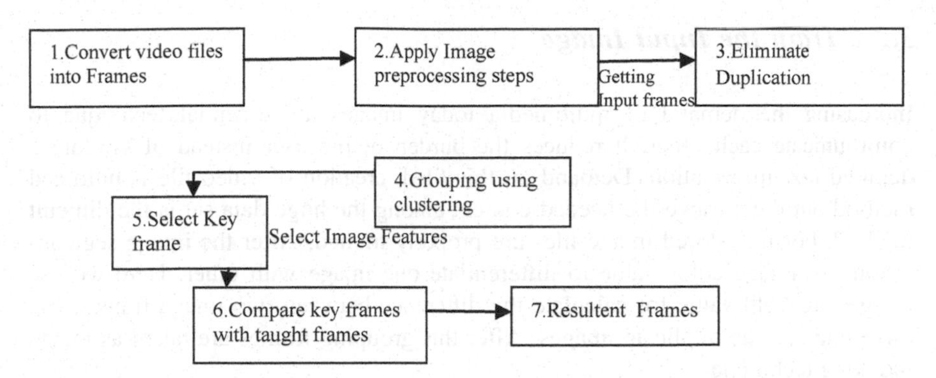


**Fig. 3** Overall process of video indexing

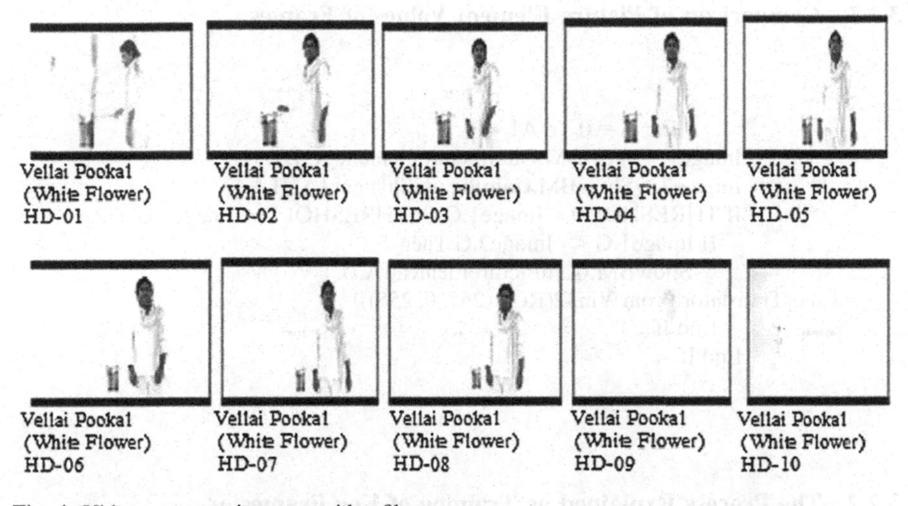


**Fig. 4** Video segmentation song video file

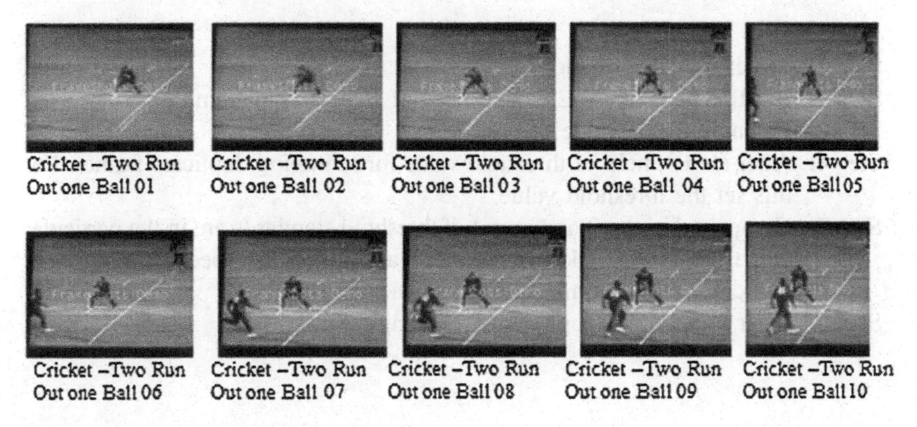


**Fig. 5** Video segmentation sport video file

## *3.2 Train the Input Image*

Increasing the demand of multimedia today images are essential technique to communicate each other. It reduces the burden of the user instead of sending a detailed text information. Demand on this filed, creation of video file is unforced method but the retrieve the needed content among the huge data set is the difficult method. For this stored image files are properly trained. After the image segmentation use image color value to differentiate one image with other. Here we use image threshold value for calculate the difference between two image frames, use this eliminate the duplicate images. After this grouping frames are taken as image indexing technique.

### 3.2.1 Comparison of Picture Element Values of Frames

```
            For A = 0 To A1 - 1
        Image1 = ShowBM.GainPictureElement(A, B)
        Image 2 = ShowBM.GainPictureElement (A, B + 1)
        If THRESHOLD < Image1.G Or THRESHOLD < Image2.G Then
          If Image1.G <> Image2.G Then
            ShowBM.GainPictureElement(A,B,
ColorTranslator.FromWin32(RGB(255, 0, 255)))
          End If
        End If
```

### 3.2.2 The Process Explained as Training of Key Frames as

Step 1: Select Image formation
Step 2: Identify the Picture Element value of step 1 input frame.
Step 3: Calculate RGB value of each frame.
Step 4: After step 3, help of threshold value for removing duplicate frames. For this set the threshold value.
Step 5: Compare the step 3 and step 4, if threshold standards are in the particular value the frames take, if the value are not in the specified range that particular frame eliminated as replica frame.
Step 6: After eliminating duplication, rearranging the frame.

## 3.3 *Image Mining*

Video data contain many image frames also called images. Every frames contain RGB picture element that used to differentiate one image frame with other frames. Toady most of the image extracting technique based on RGB color model. This RGB picture element used to compare the input query image and stored database image. After extracting the image property the pixel values are trained in the database by labeling the features of the image. The query image picture content compare with the proposed clustered like hierarchical clustering. The relevant result are then return to the user based on the input query image. Figure 2 here show the calculation of image threshold value used to compare two different frames.

## 3.4 *Video Indexing*

For effective retrieval images are arranged properly. Every images contain RGB pixels that used to arrange the frames effectively.

### 3.4.1 Suede Code for Comparing Frames

```
A =Picture width
    B = Picture Height
    A1 = A / 2,        B1 = B / 2
    bmptemp = DisplayBM
    For f= 1 To fl – 1,            For g = 1 To gl - 1
        colorpixel = DisplayBM.GetPixel(x, y)
        Ingcolorpixelvalue = (0.299 * shadcvalue.Red) + (0.587 * shadevalue
Green) + (0.1114 * shadevalue.Blue)
        GreyScale(x, y) = lngGrayScaleValue color
```

# 4 Advantage and Disadvantage of Video Indexing

There are a lot of advantages and disadvantages in video indexing techniques. For this here a table constructed compare the functions of video indexing such as Annotation based, Attribute oriented functions, and Area precise. It is shown in the Table 1.

**Table 1** Features and characteristics of video indexing techniques

| Functions | Annotation-based | Attribute oriented | Area precise |
|---|---|---|---|
| Essential method | Video OCR, CC analysis, speech recognition | Information's are processed through image action and point extraction. Image attributes such as sound identified through the attributes volume, pitch rate. Image analyzed with the attributes such as pixel, text and more | It works every functions. Property of sound and image used for entrainment, news, songs |
| Data set arrangements | semantic concepts are soaring by can be mined by language response | Image functions are extracted used image data set | Used general method for video type |
| Feature explanation | Certain function are regular since video OCR, language identification | Low level properties are unable to join with semantics. In certain type of videos attributes may unneeded example sport videos | Information's are completely shared via general attribute of the video files. Example in song video various information such as singer name, musician names can be shared |
| Mine the image low level feature | Low-level functions are eliminated | Based on image low level property the semantic annotations are identified | With help of re take part in the segment limits are identified |
| Information extraction | Information's are extracted based on keywords | Investigate are done with help of image | Use the extraction characteristics and functions |

# 5 Experimental Outcomes

See Figs. 6, 7, 8, 9, 10, 11.

# 6 Conclusion and Future Enhancement

Development of technology brings the enormous image application. Create and uploading of this image information's are easy, but retrieve the proper content is complex task for many user. There is no proper technique and algorithms are not available in this domain. Today image perform major role almost every application of users day to day life. The most challenge task for image retrieval, they never captured properly., most of the image files are capture reduced illumination, worst

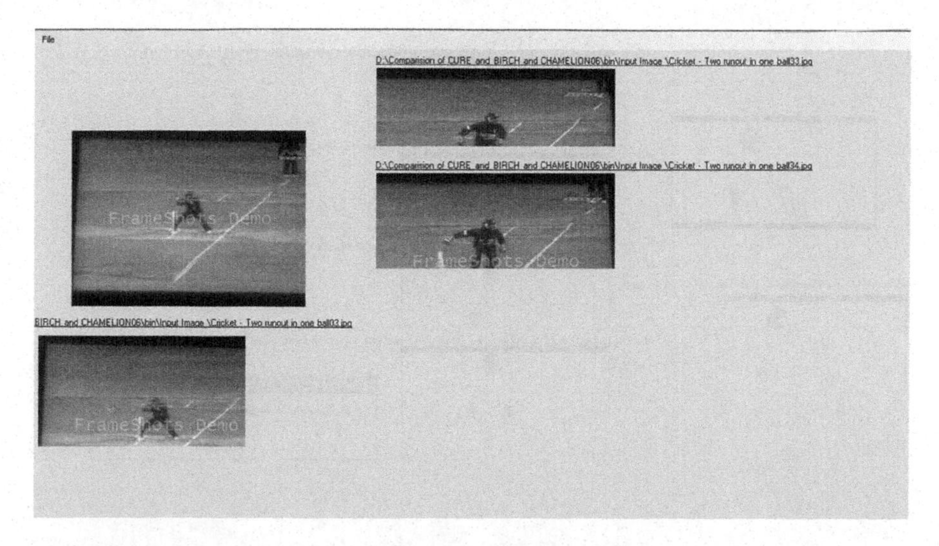


**Fig. 6** Based on Sport input key frame output Case1: 1-input file and 9 output files

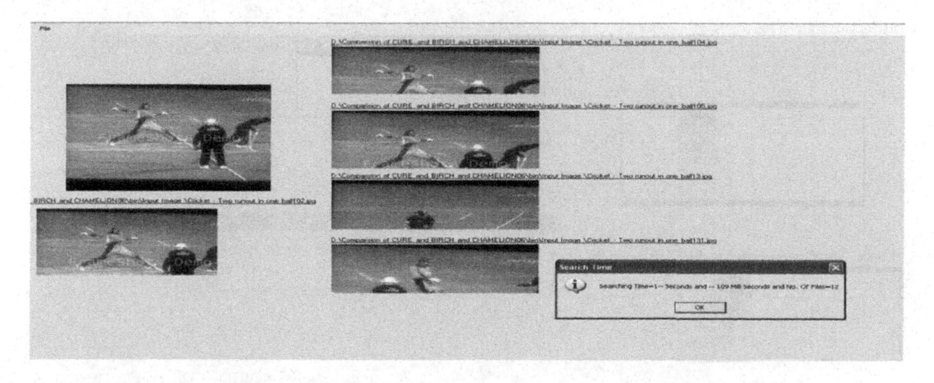


**Fig. 7** Based on Sport input key frame output Case3: 1 input and 13 output files

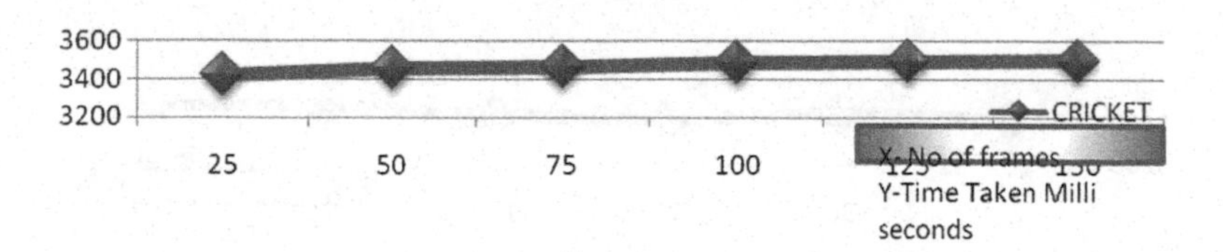


**Fig. 8** Performance in sports video

background and images are capture by un trained persons. Here the experiments prove that propose method produce good results, outcomes are verified that. In future the same technique applied to other hierarchical clustering algorithms.

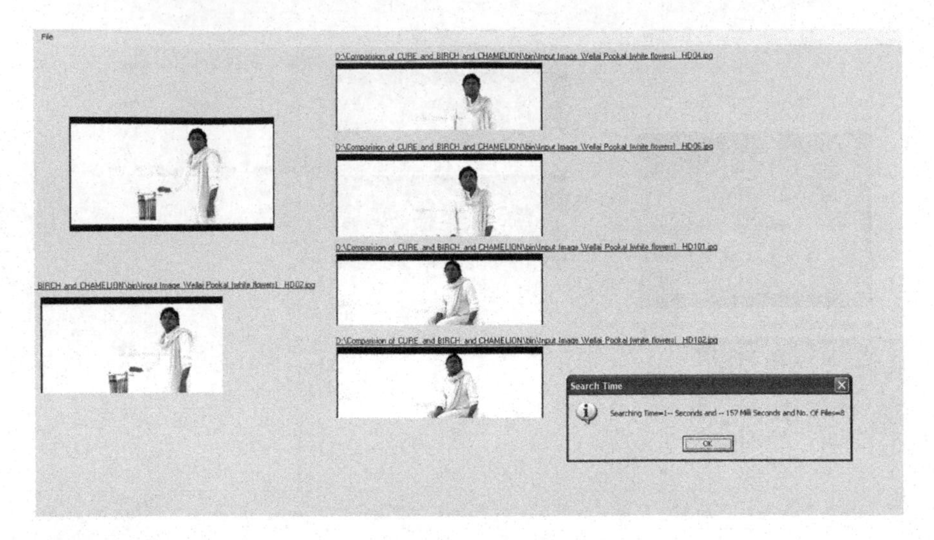


**Fig. 9** Based on Song input key frame output Case 1: 1 input and 4 output file

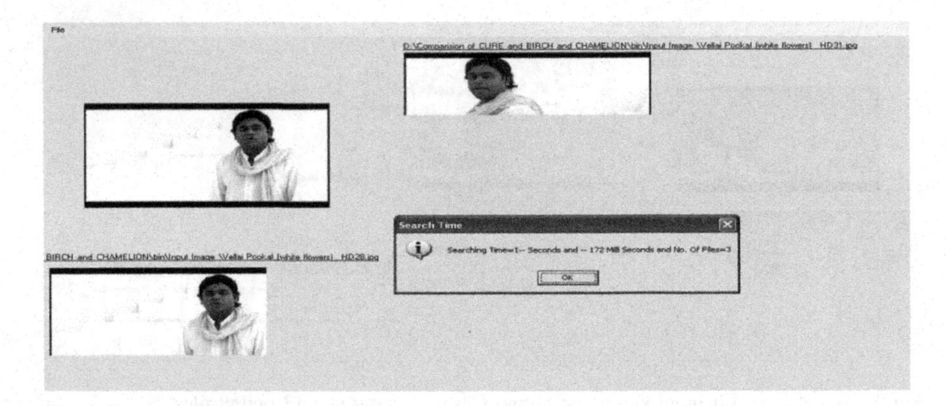


**Fig. 10** Based on Song input key frame output Case 2: 1 input and 1 output files

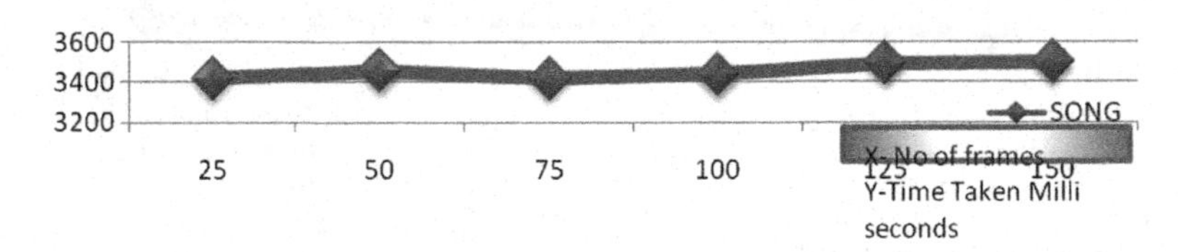


**Fig. 11** Performance in news video

## References

1. Saravanan, D.: Effective multimedia content retrieval. Int. J. Appl. Environ. Sci. **10**(5), (2015), 1771–17783 (2015)
2. Saravanan, D.: Performance Analysis of video data image using clustering technique. Indian J. Sci. Technol. **9**(10), 01–06 (2016)
3. Panda et al.: Hybrid data mining approach for image segmentation based classification. Int. J. Rough Sets And Data Anal. **3**(2), 65–81 (2016)
4. Algur et al.: Web video object mining: a novel approach for knowledge discovery. Int. J. Intell. Syst. Appl. **8**(4), 67–75 (2016)
5. Saravanan, D.: Text information retrieval using data mining clustering technique. Int. J. Appl. Eng. Res. **10**(3), 7865–7873 (2015)
6. Saravanan, D.: Segment based indexing technique for data file. Procedia comput. Sci. **87** (2016), 12–17 (2016)
7. Saravanan, D.: Video substance extraction using image future population based techniques. ARPN J. Eng. Appl. Sci. **11**(11), 7041–7045, (2016)
8. Bhojani, S.H., et al.: Data mining techniques and trends–a review. Glob. J. Res. Anal. **5**(5), 252–254 (2016)
9. Yadav, A., et al.: An efficient video data security mechanism based on RP_AES. Int. J. Adv. Technol. Eng. Explor.**3**(16), 36–42, (2016)
10. Yang, Y., Nie, F., Xu, D., Luo, J., Zhuang, Y., Pan, Y.: A multimedia retrieval framework based on semi-supervised ranking and relevance feedback. IEEE Trans. Pattern Anal. Mach. Intell. **34** (5), 723–742, (2012)
11. Saravanan, D., et al.: Data mining framework for video data. In: Recent Advances in Space Technology Services and Climate Change (RSTSCC), 2010, Chennai, pp. 167–170 (2010)
12. Saravanan, D., et al.: Matrix based sequential indexing technique for video data mining. J. Theor. Appl. Inf. Technol. **67**(3), 725–731 (2014)

# A Novel Adaptive Threshold and ISNT Rule Based Automatic Glaucoma Detection from Color Fundus Images

**Sharanagouda Nawaldgi, Y.S. Lalitha and Manjunath Reddy**

**Abstract** Glaucoma, an eye disease recognized to be the second most leading cause of blindness worldwide. Early detection and subsequent treatment of glaucoma is hence important as damage done by glaucoma is irreversible. Large scale manual screening of glaucoma is a challenging task as skilled manpower in ophthalmology is low. Hence many works have been done towards automated glaucoma detection system from the color fundus images (CFI). In this paper, we propose a novel method of automated glaucoma detection from CFI using color channel adaptive thresholding and ISNT rule. Structural features such as cup-to-disk ratio (CDR), neuro-retinal rim (NRR) area of the optic nerve head (ONH) are extracted from CFI using color channel adaptive thresholding and morphological processing in order to segment Optic Disk (OD) and Optic Cup (OC) required for calculating the CDR value. The results obtained by the proposed methodology are very promising yielding an overall efficiency of 99%.

**Keywords** Glaucoma · Color fundus images · Feature extraction · Cup-to-disk ratio · ISNT rule

## 1 Introduction

Glaucoma is a degenerative disease of the retina resulting in a gradual, progressive and irreversible degeneration of optic nerve fibers eventually leading to blindness. It is caused by increased Intraocular Pressure (IOP) [1]. Glaucoma has been recognized to be the second most common cause of blindness worldwide [2].

S. Nawaldgi (✉) · Y.S. Lalitha · M. Reddy
Appa Institute of Engineering and Technology, Kalaburagi, India
e-mail: sharan_yn@yahoo.co.in

Y.S. Lalitha
e-mail: patil.lalitha12@gmail.com

M. Reddy
e-mail: manjureddy.patil1@gmail.com



S.C. Satapathy et al. (eds.), *Data Engineering and Intelligent Computing*,
Advances in Intelligent Systems and Computing 542,
DOI 10.1007/978-981-10-3223-3_13

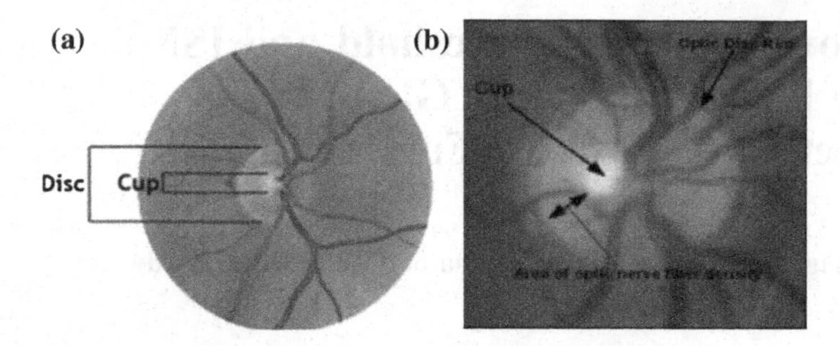


**Fig. 1** L to R: **a** Healthy CFI, **b** structures for glaucoma detection in CFI

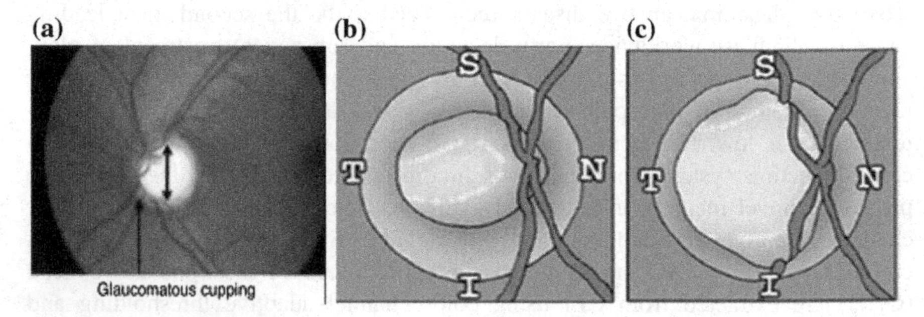


**Fig. 2** L to R: **a** CFI of a glaucomatous eye, **b** normal eye satisfying ISNT rule, **c** glaucomatous eye not satisfying ISNT rule

The asymptomatic nature and irreversible vision loss of glaucoma in its early and later stages respectively makes early detection a must for the prevention of the disease. Large scale manual screening programs requires skilled ophthalmologists which is scarce. CFI based screening procedures offer new screening alternative to manual screening. So efforts have been made toward automatic CFI analysis using Computer-aided diagnosis (CAD) for glaucoma detection [3–18]. This type of CAD—based pre-screening solution is more suitable for large scale screening. Automatic glaucoma detection using color fundus images includes detecting structural changes in the optic nerve head (ONH). Figure 1 shows healthy eye color fundus image, and structures of interest for glaucoma detection.

The major components of retina are Optic Disk (OD), Optic Cup (OC), and neuroretinal rim (NRR). Cup-to-disk ratio (CDR) is defined as ratio of OC to OD diameter. The region between the OD and OC is the neuroretinal rim. The CDR value in case of a healthy eye is $<0.3$. The cup size increases as IOP is elevated in case of glaucoma, the result of which is NRR thinning. The CDR value in case of glaucoma is high due to increased cup region. So CDR can be used as a fundamental measure for glaucoma detection. In addition to CDR, NRR thickness can also be used as an important feature for glaucoma detection. A healthy eye should satisfy Inferior, Superior, Nasal, Temporal (ISNT) rule i.e. the thickness of $I \geq S \geq N \geq T$. Figure 2 shows CFI of glaucomatous eye, normal eye satisfying the ISNT rule, and glaucomatous eye not satisfying the ISNT rule.

## 2 Related Work

Various past works have proposed CDR based glaucoma detection where Optic Disk and Optic Cup are segmented using fixed threshold [3–18]. Hatanaka [4] Proposed segmenting the Optic Disk region using the P-tile thresholding method and Canny edge detector to enhance edges. Around the center of Optic Disk, twenty profiles were obtained on blue channel of CFI in the vertical direction, and a smoothed profile was obtained by averaging these profiles. The Optic Cup edge was determined by the thresholding technique on the vertical profile. The vertical cup-to- disc ratio was calculated from segmented OD and OC. The proposed method was evaluated on seventy nine images which include twenty five glaucomatous images, a specificity of 85% and sensitivity of 80% was achieved. Kavitha [6] Proposed calculating the region of interest (ROI) in the original CFI using the mathematical morphology operations like dilation, erosion. To extract the Optic Disk and Optic Cup, component labeling and component analysis method was used. The Optic Cup was extracted from the green channel of the CFI. The method was tested against image data sets from Madurai Eye Care Centre, Coimbatore. Joshi [7] Proposed a novel method of Optic Cup detection based on vessel bend in the cup region. Anatomically vessel bends marks cup boundary which is considered relevant by Glaucoma experts. But vessel bends can occur at any place within the Optic Disk region. Hence, only a subset of these points (referred to as relevant vessel bend or r-bend) defines the Optic Cup boundary. The r-bends location within the Optic Disk region is derived and a local interpolating spline approximates the cup boundary in regions where r-bends are absent. The method was tested on 133 images comprising 32 normal and 101 glaucomatous images.

The drawback of the past literature is that it uses fixed thresholding for CDR calculation. Moreover CDR alone does not elaborate the inherent distribution of the abnormality in the retinal image. Therefore in this work we propose a novel method of glaucoma detection from CFI using color channel adaptive thresholding and morphological processing for Optic Disk and Optic Cup segmentation. Further we introduce ISNT rule to verify whether the cup progression in the NRR is in accordance with the ISNT rule. Thus the combination of CDR and the ISNT rule helps to differentiate between glaucomatous and normal eye better. The results obtained by this methodology are very promising yielding an overall efficiency of 99%.

## 3 Proposed Methodology

To calculate CDR value, extraction of Optic Disk and Optic Cup are required. Optic Disk and Optic Cup are extracted using color channel adaptive thresholding and morphological processing. Neuro-retinal rim area is the area between the OD and OC. Then we find the cup progression in the NRR area in inferior, superior, nasal,

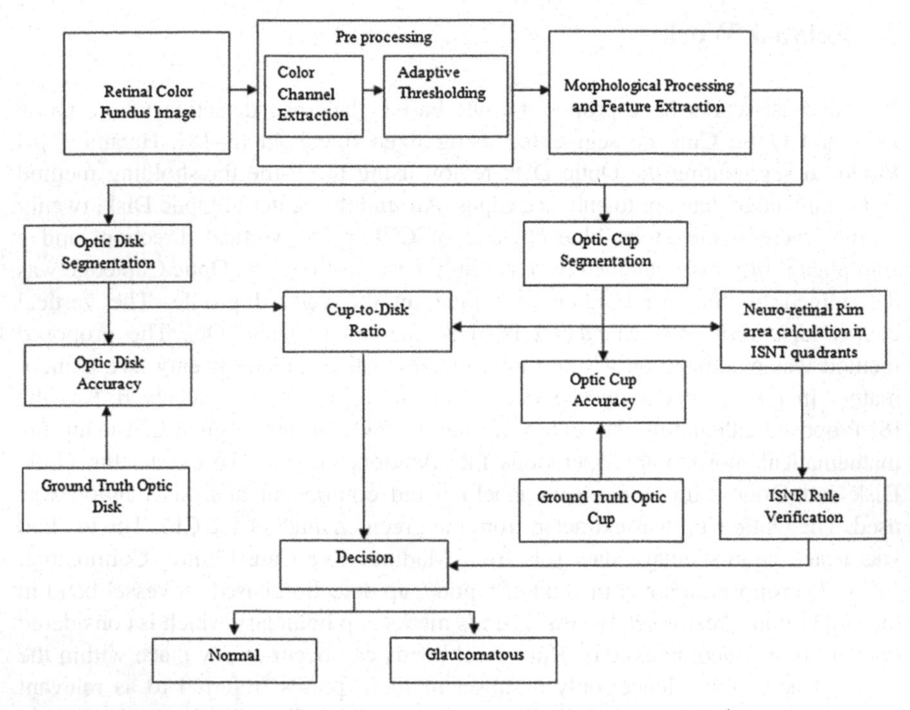


**Fig. 3** CDR calculation and ISNT rule

and temporal area to verify the ISNT rule. The proposed methodology is depicted in the Fig. 3 and is discussed in the following sub-sections.

A. Optic Disk and Optic Cup Extraction

For Optic Disk segmentation, the proposed method scans the red channel of the image and finds the average intensity value and based on the average intensity generates an automatic thresholding. This is followed by morphological erosion and dilation to compensate for loss of data in Optic Disk area. This provides a better Optic Disk segmentation. For Optic Cup segmentation, we use green channel of the image for automatic thresholding. CDR is calculated from the segmented Optic Disk and Optic Cup.

B. Neuroretinal Rim Extraction and ISNT rule Verification

Once the Optic Disk and Optic Cup is segmented, unlike the present system which uses only CDR as a measure to detect glaucoma, we propose a novel ISNT rule based glaucoma detection. The cup progression in the four regions of Neuroretinal Rim i.e. Inferior, Superior, Nasal, and Temporal is calculated and verified against the ISNT rule.

C. Decision Making

Based on the CDR value and ISNT rule, the proposed methodology evaluates the image to be glaucomatous or normal. Also the segmented OD and OC by the proposed methodology are verified against the ground truth OD and OC by superimposing the segmented OD, OC with ground truth OD, and OC respectively. The OD, OC accuracy is then calculated by the percentage of overlap between the segmented and ground truth OD, and OC respectively.

## 4 Dataset for Glaucoma Evaluation

Dataset plays a crucial role in determining the effectiveness and validating the accuracy of automated glaucoma detection system. The reference dataset DRISHTI-GS1 provided by Medical Image Processing (MIP) group, IIIT Hyderabad is used for evaluating our proposed work [19]. DRISHTI-GS1database consists of 50 training and 51 testing images. Manual segmentations were collected for both OD and OC from four different glaucoma experts and were marked with a dedicated marking tool. In addition to this, diagnostic opinion for each image being normal or glaucomatous was obtained from four glaucoma experts and a gold standard was derived based on the majority opinion (i.e. 3 out of the 4 expert's opinion).

Figure 4 demonstrates the results of our proposed methodology when applied to a sample CFI of DRISHTI-GS1database.

## 5 Performance Analysis

As the objective is to detect glaucoma from CFI, we opt for accuracy as the performance metric. DRISHTI-GS1dataset comes with a glaucoma expert marked truth images for both Optic Disk and Optic Cup areas. Both the OD and OC area detected by the proposed system is compared with the glaucoma expert marked OD and OC respectively and accuracy is measured by the percentage of overlap of detected OD and OC with that of glaucoma expert markings. The computed CDR by our proposed method is also compared against the glaucoma expert's value and the CDR accuracy is measured.

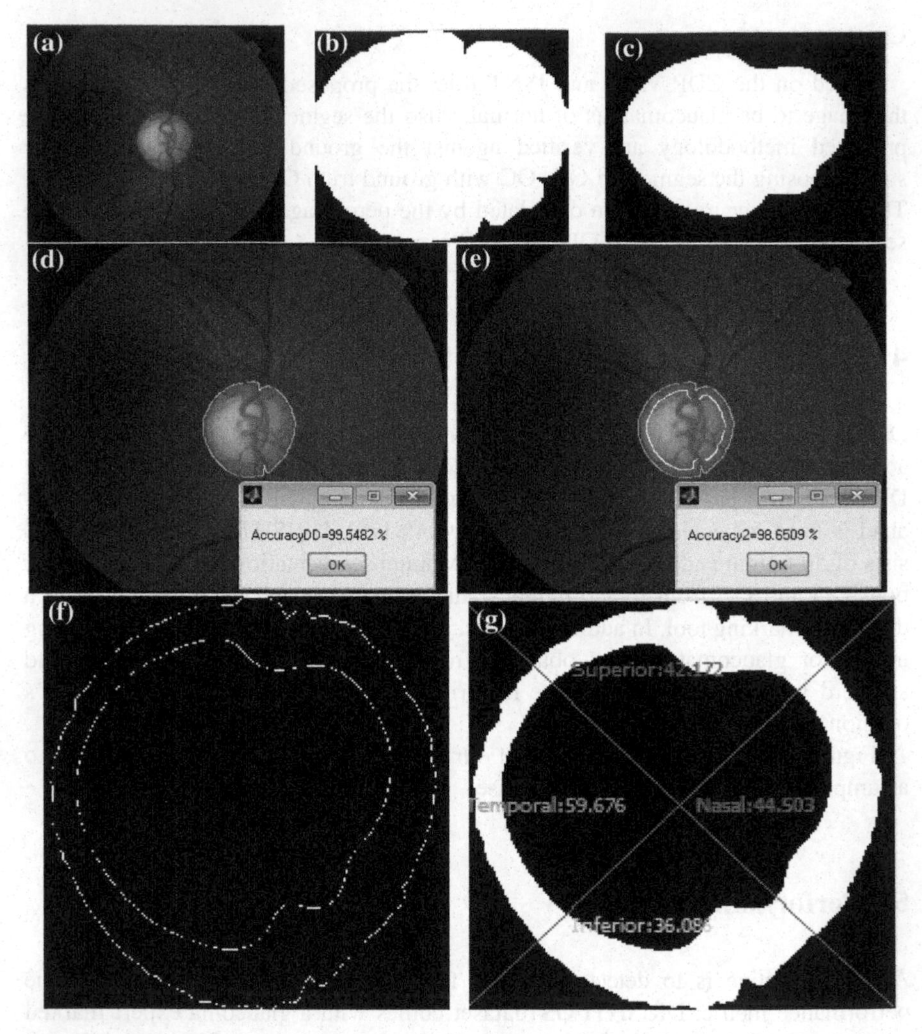


**Fig. 4** L to R: **a** CFI of eye, **b** detected optic disk, **c** detected Optic cup, **d** detected OD superimposed with actual optic disk, **e** detected OC superimposed with actual optic cup, **f** detected NRR area, **g** Optic Cup progression in NRR area (ISNT rule)

# 6 Experimental Results

We tested our proposed methodology against DRISHTI-GS1dataset. The proposed methodology's Optic Disk segmentation accuracy is more than 99% and Optic Cup segmentation is more than 96%. The OD and OC segmentation is followed by the calculation of cup-to-disk ratio. The CDR accuracy is more than 98% in comparison to the average computed value of CDR given by the 4-glacuoma experts. We also verified whether the cup progression in the neuro-retinal rim area is obeying the

**Table 1** Optic Disk, Optic Cup and CDR accuracy

| Image no. | Proposed system | | Proposed method CDR | Actual CDR | CDR accuracy |
|---|---|---|---|---|---|
| | Disk accuracy | Cup accuracy | | | |
| drishtiGS_010 | 90.39 | 90.25 | 0.82 | 0.89 | 92.13 |
| drishtiGS_012 | 96.54 | 92.08 | 0.80 | 0.87 | 91.95 |
| drishtiGS_017 | 100 | 94.98 | 0.58 | 0.53 | 91.37 |
| drishtiGS_024 | 94.61 | 99.65 | 0.67 | 0.74 | 90.54 |
| drishtiGS_026 | 95.66 | 92.23 | 0.81 | 0.88 | 92.04 |
| drishtiGS_032 | 100 | 94.73 | 0.63 | 0.70 | 90.00 |
| drishtiGS_058 | 95.84 | 98.20 | 0.78 | 0.84 | 92.85 |
| drishtiGS_051 | 99.67 | 94.68 | 0.68 | 0.74 | 91.89 |
| drishtiGS_069 | 97.50 | 94.95 | 0.73 | 0.78 | 93.58 |
| drishtiGS_094 | 98.73 | 92.75 | 0.52 | 0.49 | 94.23 |

**Table 2** Glaucoma detection based on CDR value and ISNT rule

| Image no. | CDR value | ISNT rule satisfied | Glaucoma detected |
|---|---|---|---|
| drishtiGS_010 | 0.82 | No | Yes |
| drishtiGS_012 | 0.80 | No | Yes |
| drishtiGS_017 | 0.58 | No | Yes |
| drishtiGS_024 | 0.67 | No | Yes |
| drishtiGS_026 | 0.81 | No | Yes |
| drishtiGS_032 | 0.63 | No | Yes |
| drishtiGS_058 | 0.78 | No | Yes |
| drishtiGS_051 | 0.68 | No | Yes |
| drishtiGS_069 | 0.73 | No | Yes |
| drishtiGS_094 | 0.52 | No | Yes |

ISNT rule (normal eye) or not obeying (glaucomatous eye) by calculating the rim area in the Inferior, Superior, Nasal, and Temporal quadrants. The proposed method classification accuracy between normal and glaucomatous eye based on both CDR and ISNT rule is more than 99%. The OD and OC accuracy for sample images of 10 are shown in Table 1 and glaucoma detection based on CDR and ISNT rule verification is depicted in Table 2.

## 7 Conclusion

Various past works have proposed different techniques for extraction of Optic Disk and Optic Cup area for automated glaucoma detection. Most of past works suffer from less accuracy. We have proposed a novel adaptive threshold based technique which adjusts itself according to the image data for better OD and OC segmentation. Better OD and OC segmentation results in accurate CDR value calculation. Our proposed method has better accuracy in comparison to previous proposed state of art methods like Hough transform, ellipse fitting etc. We also introduced a novel ISNT rule along with the calculated CDR for glaucoma assessment.

This work can be further improved by incorporating fuzzy logic into the final decision making of automated glaucoma detection.

**Acknowledgements** We would like to express our sincere gratitude and deep regard to Poojya Dr. Sharnbaswappa Appaji, President, Sharanabasveshwar Vidya Vardhaka Sangha, Kalaburagi, for his immense support and encouragement. We would also like to give our sincere gratitude to Dr. V.D Mytri, Principal, APPA IET, Dr. Anilkumar Bidve, Dean of Administration, APPA IET for their invaluable suggestions and support. We thank Dr. Pradeep Reddy, Dr. Rohit Patil for providing us the clinical insights of glaucoma.

## References

1. Glaucoma Research Foundation. http://www.glaucoma.org/glaucoma/typesofglaucoma.php
2. Quigley, H.A., Broman, A.T.: The number of people with glaucoma worldwide in 2010 and 2020. Brit. J. Ophthalmol. **90**(3), 262–267 (2006)
3. Inoue, N., Yanashima, K., Magatani, K., Kurihara, T.: Development of a simple diagnostic method for the glaucoma using ocular fundus pictures. In: Proceedings of 2005 IEEE, Engineering in Medicine and Biology 27th Annual Conference, pp. 3355–3358. Shanghai, China (2006)
4. Hatanaka, Y., Noudo, A., Muramatsu, C., Sawada, A., Hara, T., Yamamoto, T., Fujita, H.: Automatic measurement of cup to disc ratio based on line profile analysis in retinal images. Conf Proc IEEE Eng Med Bioi Soc. (2011)
5. Narasimhan, K., Vijayarekha, K.: An efficient automated system for glaucoma detection using fundus image. J. Theor. Appl. Inf. Technol. **33**, 104–110 (2011). E-ISSN: 1817- 3195
6. Kavitha, S., Karthikeyan, S., Duraiswamy, K.: Early detection of glaucoma in retinal image using cup to disc ratio. In: Second International Conference on Computing, Communication and Networking Technologies, vol 10, IEEE (2010)
7. Joshi, G., Sivaswamy, J., Karan, K., Prashanth, R., Krishnadas, S.R.: Vessel bend-based cup segmentation in retinal images. In: 20th International Conference on Pattern Recognition (ICPR), pp. 2536–2539. (2010)
8. Joshi, G.: Sivaswamy, Krishnadas S.R.: Optic disk and cup segmentation from monocular retinal images for glaucoma assessment. IEEE Trans. Med. Imaging **30**, 1192–1205 (2011)
9. Murthi, A., Madheswaran, M.: Enhancement of optic cup to disc ratio detection in glaucoma diagnosis, In: International Conference on Computer Communication and Informatics (ICCCI), pp. 1–5. IEEE (2012)
10. Joshi, G., Sivaswamy, J., Krishnadas, S.R.: Depth discontinuity-based cup segmentation from multi-view colour retinal images. IEEE Trans. Biomed. Eng. **59**, 1523–1531 (2012)
11. Ahmad, H., Yamin, A., Shakeel, A., Gillani, S.O., Ansari, U.: Detection of glaucoma using retinal fundus images. In: International Conference on Robotics and Emerging Allied Technologies in Engineering (iCREATE), pp. 321–324. IEEE (2014)
12. Alghmdi, H., Tang, H.L., Hansen, M., O'Shea, A., Al Turk, L., Peto, T.: Measurement of optical cup-to-disc ratio in fundus images for glaucoma screening. In: International Workshop on Computational Intelligence for Multimedia Understanding (IWCIM), pp. 1–5. IEEE (2015)

13. Vijapur, N.A., Kunte, R.S.R.: Glaucoma detection by using Pearson-R correlation filter. In: International Conference on Communications and Signal Processing (ICCSP), pp. 1194–1198. IEEE (2015)
14. Shekhar, S., Al-Nuaimy, W., Nandi, A.K.: Automated Localization of Retinal Optic Disk Using Hough Transform. IEEE, ISBI, pp. 1577–1580 (2008)
15. Yin, F., Liu, J., Wong, D.W.K., Tan, N.M., Cheung, C., Bhaskaran, M., Wong, T.Y.: Automated segmentation of optic disk and optic cup in fundus images for glaucoma diagnosis. In: 25th International Symposium on Computer Based Medical System, pp. 1–6 (2012)
16. Aquino, A., Gegundez-Arias, M.E., Marin, D.: Detecting the optic disk boundary in digital fundus images using morphological, edge detection and feature extraction techniques. IEEE Trans. Med. Imaging **29**, 1860–1869 (2010)
17. Liu, J., Wong, D., Lim, J., Li, H., Tan, N., Wong, T.: Argali-an automatic cup-to-disc ratio measurement system for glaucoma detection and analysis framework. In: Proceeding of SPIE Medical Imaging, vol. 7260, p. 72603 K. (2009)
18. Joshi, G.D., Sivaswamy, J., Karan, K., Krishnadas, R.: Optic disk and cup boundary detection using regional information. In: Proceeding of IEEE International Symposium on Biomedical Imaging (ISBI), pp. 948–951 (2010)
19. Sivaswamy, J., Krishnadas, S.R., Chakravarty, A., Joshi, G.D., Ujjwal, et al.: A Comprehensive retinal image dataset for the assessment of glaucoma from the optic nerve head analysis. JSM Biomed. Imaging Data Pap. **2**(1), 1004 (2015)

# Comparison of Classification Techniques for Feature Oriented Sentiment Analysis of Product Review Data

**Chetana Pujari, Aiswarya and Nisha P. Shetty**

**Abstract** With the rapid increase in popularity of e-commerce services over the years, all varieties of products are sold online today. Posting online reviews has become a common means for people to express their impressions on any product, while serving as a recommendation for others. To enhance customer satisfaction and buying experience, often the sellers provide a platform for the customers to express their views. Due to the explosion of these opinion rich sites where numerous opinions about a product are expressed, a potential customer finds it difficult to read all the reviews and form an intelligent opinion about the product. In this research, a new framework comprising of the inbuilt packages of python is designed which mines many customers' opinions about a product and groups them accordingly based on their sentiments, which aids the potential buyers to form a capitalized view on the product. Here classification of the reviews is done using three different classification algorithms i.e. Naïve Bayes Algorithm, Maximum Entropy Classifier and SVM (Support Vector Machine), and their performance is compared. The methodology showcased in this work can be extended easily in all domains.

**Keywords** Feature based opinion mining • Sentiment analysis • Tokenization • Bigram collocation • Classification • Nltk • Anaconda

C. Pujari (✉) · Aiswarya · N.P. Shetty
Manipal Institute of Technology, Manipal 576104, India
e-mail: chetana.pujari@manipal.edu

Aiswarya
e-mail: aiswarya.bhat@manipal.edu

N.P. Shetty
e-mail: nisha.pshetty@manipal.edu



S.C. Satapathy et al. (eds.), *Data Engineering and Intelligent Computing*,
Advances in Intelligent Systems and Computing 542,
DOI 10.1007/978-981-10-3223-3_14

## 1 Introduction

"What other's think??", [1] often influence the judgment of the people regarding any issue. In the recent years owing to the increased familiarity of social networking sites, blogs, feedback systems etc. there are hundreds of uninhibited reviews readily available for the interested customers to form an unbiased view about any product. With the increase in the usage of Internet and Web Services, more and more business use this platform to publicize and market their goods, discern new opportunities, and maintain their reputations. Browsing this immense collection of reviews to look for applicable information is a dreary and irksome task. The current scenario thus, emphasizes the need for an effective opinion mining system [2].

The opinion system has 3 components [3];

1. Opinion Holder: An individual or institution enunciating their point of view or experience with the product.
2. Commodity: The target entity or trait about which the user expresses his views.
3. Opinion Orientation: Categorizes the opinion as neutral, negative or positive. Also determines the intensity of the opinion (strong, weak, extreme, medium etc.).

For example "XYZ has good multimedia features." Here the opinion holder is the reviewer, object is multimedia features of XYZ and the orientation is positive.

Using appropriate packages or dictionaries a sentiment orientation system classifies the extracted opinion words (adjectives) into positive negative or neutral sentiments [4]. Thus, the polarity of the review for each feature of the product can be determined. Sentiments can be generally classified in 3 levels i.e. [5] document, sentence and aspect level. This summarized sentiments and opinions serve as a means for decision making in various aspects and enable the retailers to understand the wants and needs, fears and hitches [6] experienced by the customers while shopping online.

The sections in the paper are catalogued in the following manner. Section 1 showcases the need for and importance of opinion mining system. Section 2 illustrates the related work of prominent researchers in this area. Sections 3 and 4 discusses the proposed methodology and the results obtained. Conclusion and future work is elucidated in the last segment.

## 2 Related Work

Text mining, sentiment analysis and opinion identification is one among the common research streams in the field of natural language processing and information retrieval [7]. Many significant contributions are done to this domain on texts of various languages. Below some of the prominent works is discussed.

P.D. Turney [8] proposed an unsupervised learning algorithm which classifies the sentiment as good or bad. Phrases extracted using POS taggers were subjected to PMI-IR algorithm which determined their semantic association. The entire review was categorized accordingly on the basis of the mean semantic orientation of all phrases in the document. An average of 77% accuracy was attained with this method. Time required for processing was one of the main stringent factors in this method. To increase the accuracy the author suggested the incorporation of supervised learning algorithms.

Chien-Liang Liu et al. contrived a movie rating and review summarization system [9] in the mobile environment. The authors used latent semantic analysis (LSA) [9] to pinpoint various aspects or traits in the movie reviews while ignoring unsuitable sentences and then condensed the movie reviews highlighting the important features present. Based on the polarity of the words SVM was implemented to categorize the reviews. The summarized description of the review which contains only relevant aspects was then presented. Since this was implemented at sentence level the review had sentences obtained from different paragraphs causing a lack of fluency for the end users who read it. Even though this work could be extended to many languages, it suffered from limitations such as inability to deal with amalgamate data and inaptness to place the words in broader sense.

Jingbo Zhu et al. [10] fabricated an aspect-based opinion polling system where unstructured reviews were used as a dataset. For effective aspect identification, various terms related to a particular aspect were studied using multi-aspect bootstrapping method. An aspect-based segmentation model was developed which sliced the compound sentences in the user reviews into multiple single-aspect units which were classified in an opinion poll. Experiments conducted on real Chinese restaurant gave them around 75.5% accuracy.

Hu and Liu [11] employed NL Processor linguistic parser which used POS tagging to split the sentence into its constituents. They have employed algorithms of their own for frequent feature identification, opinion word extraction and orientation determination and presented with the straightforward summary of the reviews. Future work in this domain would be monitoring customer reviews for determining strength of the opinions and expanding this work to concoct views expressed via verbs, nouns and adverbs.

Ali Harb et al. [12] developed a web mining approach which classified the reviews based on the extracted adjectives in context with the specific domain rather than using a broad and vague dictionary. Tree Tagger tool was used to split the sentence into its constituent parts and they employed Apriori Algorithm to evaluate the list of adjectives collected and to rank the adjectives accordingly. The document was categorized as positive if it contained more number of positive adjectives. Invert polarities were also considered while classification. The accuracy of their method was reduced due to the presence of low quality and noisy text which affected the extraction process.

Xuanjing Huang et al. [13] retrieved most frequently used features using the concept of frequent item set and used the adjectives near them as opinion words of that feature. Using Word Net dictionary these features were classified as positive or negative sentiments. It was found that this approach worked well data in a particular domain but was difficult to train models while dealing with reviews intermixed from multiple domains like in Twitter.

In their endeavor [14] Xing Fang and Justin Zhan aimed to resolve the problem of accurately classifying the polarity of the sentences by incorporating techniques such as POS tagging and classification models namely Naïve Bayesian, Random Forest, and Support Vector Machine. They have tested their methodology on Amazon Data Set at sentence and review level. The major limitation of the process proposed by them was that it doesn't classify reviews that purely contain implicit sentiments.

Nathalie Camelin et al. designed a system [15] which analyzed the opinions in a spoken message. A strategy was proposed for detecting those segments of the speech which contains opinions and inferring their polarity using Ada Boost algorithm accurately.

U. Krcadinac et al. modeled a sentence based emotion recognition system called as Synesketch [16] which individual sentences in the text as input and labeled it with respect to the six emotional types defined by Ekman [17]. In future they aim to improve their algorithm by adding POS tagging and improved negation detection to their work.

In [18] Dilara Torunoğlu et al. improved the efficiency of Naïve Bayes algorithm by making use of Laplacian Smoothing approach. They found that their approach even exceeded the efficiency of SVM classifier. The authors would like to expand their work to categorize the sentiments in Turkish Twitter datasets.

## 3 System Design

In Fig. 1 the flow of data through various modules in the system is depicted.

### *3.1 Data Set*

Product review dataset of Amazon named Fine Foods downloaded from the Stanford Large Network Dataset Collection [19] is considered for this work. The dataset shows the review about various products from Amazon. In this work, a subset of the dataset is considered, which consists of around 214 reviews. Certain amount of preliminary assessment is done to identify only relevant details from the dataset.

### *3.2 Data Preprocessing*

Two major steps carried out in the process are tokenizing and stop word removal. Each sentence is broken down into its constituents (tokens) first. Then stop word removal is performed and also punctuations are removed.

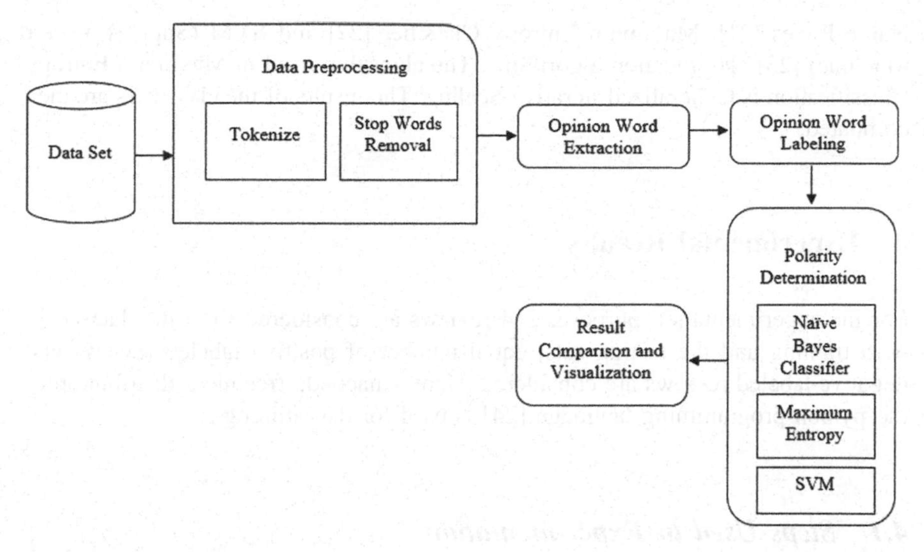


**Fig. 1** System architecture

## 3.3 *Opinion Word Extraction*

Employing POS Tagging [20] technique the tokens are classified into their respective genres (nouns, pronouns, adjectives etc.) to identify the words that depict the emotion of the reviewer (usually adjectives) regarding each feature of the product.

## 3.4 *Labeling of Opinion Word*

After the extraction of opinion words, each opinion word is labeled as positive or negative based on the sentiment it depicts, using bigram collocation. An example for bigram collocation is the phrase "*not a good product*" which shows negative sentiment even though it contains the word *good* which comes under positive list of sentiments.

This labeled list is used for the polarity determination of the entire review. BigramCollocationFinder and BigramAssocMeasure (Chi Square measure) which are inbuilt packages of python are used for the accurate classification of the sentiment expressed by the reviewer.

## 3.5 *Polarity Determination Using Classification Algorithms*

Appropriate categorization of end user's comments on the basis of the orientation of their opinion is done by the application of three classification algorithms namely,

Naïve Bayes [21], Maximum Entropy Classifier [22] and SVM (Support Vector Machine) [23] classification algorithms. The algorithm used in Maximum Entropy Classification is Generalized Iterative Scaling. The results of the classifiers are then compared.

## 4 Experimental Results

For the experimentation purpose, 214 reviews are considered from the dataset. In both training and the testing sets, equal number of positive labeled reviews and negative labeled reviews are considered. Here, Anaconda freemium distribution of the python programming language [24] is used for data mining.

### *4.1 Steps Used in Experimentation*

1. Prepare two data sets. One contains all positive reviews and other has all negative reviews.
2. Read the datasets separately.
3. Perform tokenizing and stop word removal.
4. Retain required stop words like *over, under, not* etc.

**Table 1** Without bigram collocation

| Classifier | Accuracy (%) | Precision (%) | Recall (%) | F-measure (%) |
|---|---|---|---|---|
| Naïve Bayes | 54.54 | 75.49 | 56.89 | 45.88 |
| SVM | 78.18 | 79.41 | 78.71 | 78.11 |
| Maximum Entropy | 56.36 | 68.89 | 58.42 | 50.45 |

**Table 2** Without bigram collocation but with N-Fold cross validation (where N = 5)

| Classifier | Accuracy (%) | Precision (%) | Recall (%) | F-measure (%) |
|---|---|---|---|---|
| Naïve Bayes | 75.23 | 80.12 | 75.98 | 73.7 |
| SVM | 80 | 79.11 | 79.16 | 78.9 |
| Maximum Entropy | 76.66 | 78.75 | 79.35 | 76.44 |

**Table 3** With bigram collocation

| Classifier | Accuracy (%) | Precision (%) | Recall (%) | F-measure (%) |
|---|---|---|---|---|
| Naïve Bayes | 50.9 | 74.52 | 53.44 | 39.36 |
| SVM | 76.36 | 77.15 | 76.79 | 76.33 |
| Maximum Entropy | 54.54 | 75.49 | 56.89 | 45.88 |

**Table 4** With bigram collocation and N-Fold cross validation (where N = 5)

| Classifier | Accuracy (%) | Precision (%) | Recall (%) | F-measure (%) |
|---|---|---|---|---|
| Naïve Bayes | 74.76 | 79.54 | 75.86 | 73.75 |
| SVM | 82.85 | 84.45 | 82.13 | 82.38 |
| Maximum Entropy | 79.04 | 81.75 | 79.99 | 78.59 |

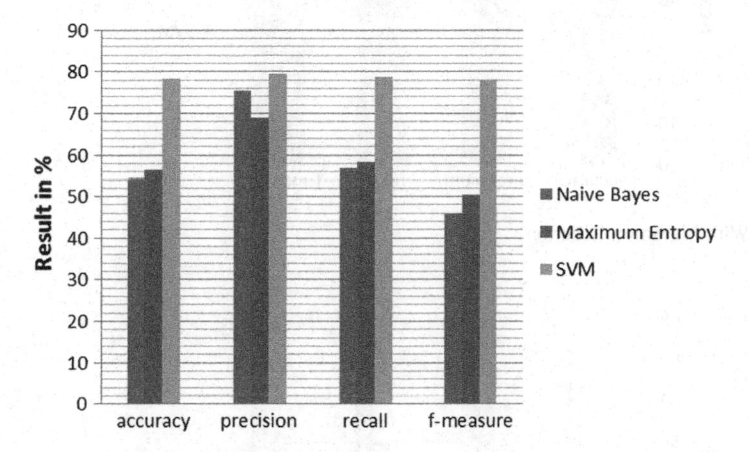


**Fig. 2** Without bigram collocation

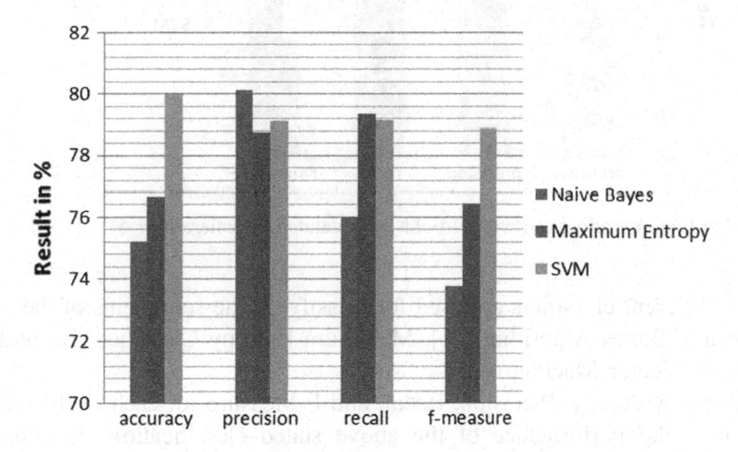


**Fig. 3** Without bigram collocation with N-Fold cross validation (where N = 5)

5. Fetch the required features (sentiment words) using BigramCollocationFinder and BigramAssocMeasure (Chi Square measure).
6. Depending on the features selected 75% of positive and 75% of negative sets are considered for training the model.

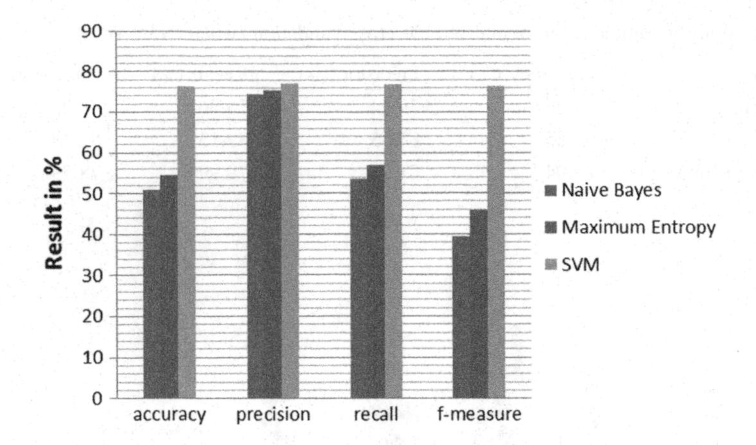


**Fig. 4** With bigram collocation

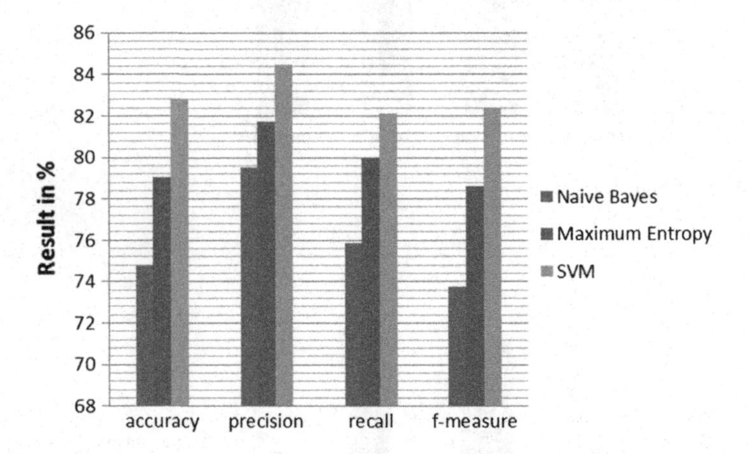


**Fig. 5** With bigram collocation with N-Fold cross validation (where N = 5)

7. Three different classifiers are used for classifying the sentiments of the reviews i.e. Naïve Bayes Algorithm [21], Maximum Entropy Classifier [22] and SVM (Support Vector Machine) [23].
8. Calculate Accuracy, Precision, Recall and F-Measure for each of the classifier.
9. Compare the performance of the above stated classification algorithms and visualize the result.

From the experimental results as shown in Tables 1, 2, 3 and 4, it was observed that performance of classifiers with bigram collocation were better than the performance of classifiers without bigram collocation. But, classifiers with bigram collocation along with the N-fold cross validation gave the best results. Here, the value of N was taken as 5. Thus, it was concluded that the performance of SVM with bigram collocation and N-Fold cross validation is comparatively better than Naïve Bayes and Maximum Entropy Classifier with bigram collocation and N-fold cross validation. Graphical representation of the same is shown is Figs. 2, 3, 4 and 5.

## 5 Conclusion and Future Work

The proposed methodology showcases a feature based opinion mining system constituting of Bigram Collocation (for extracting opinion words) and comparison of Naïve Bayes, SVM (Support Vector Machine) and Maximum Entropy Classifier's performance to classify product reviews based on the sentiments of the user. From the experiments carried out, it was found that SVM had better performance compared to Naïve Bayes and Maximum Entropy Classifier. The method presented in our paper can be extended to any domain and also manifests the implied emotion of the reviewer easily. In future our intention is to upgrade our concept to detect fake comments (online Spams) which promote or discredit a product. Our proposed methodology serves as an ideal recommendation system and also can contribute to areas such as marketing research, quality improvement etc.

## References


1. Pang, B., Lee, L.: Opinion mining and sentiment analysis. Found. Trends Inf. Retrieval **2**(1–2), 1–135 (2008)
2. Liu, B., Hu, M., Cheng, J.: Opinion observer: analyzing and comparing opinions on the Web. In: Proceedings of the 14th International Conference on World Wide Web, WWW'05. ACM, New York, USA, pp. 342–351 (2005)
3. Sharma, R., Nigam, S., Jain, R.: Polarity detection at sentence level. Int. J. Comput. Appl. **86**(11) (2014)
4. Cambria, E., Schuller, B., Xia, Y., Havasi, C.: New avenues in opinion mining and sentiment analysis. IEEE Intell. Syst. **28**(2), 15–21 (2013)
5. Medhat, W., Hassan, A., Korashy, H.: Sentiment analysis algorithms and applications: a survey. Ain Shams Eng. J. (Production and hosting by Elsevier B.V. on behalf of Ain Shams University.) **5**, 1093–1113 (2014)
6. Khan, S.A., Liang, Y., Shahzad, S.: An empirical study of perceived factors affecting customer satisfaction to re-purchase intention in online stores in China. J. Serv. Sci. Manag. **8**, 291–305 (2015)
7. Khairnar, J., Kinikar, M.: Machine learning algorithms for opinion mining and sentiment classification. Int. J. Sci. Res. Publ. **3**(6) (2013)
8. Turney, P.D.: Thumbs up or thumbs down? Semantic orientation applied to unsupervised classification of reviews. In: Proceedings of the 40th Annual Meeting of the Association for Computational Linguistics (ACL), Philadelphia, pp. 417–424 (2002)
9. Liu, C.L., Hsaio, W.H., Lee, C.H., Lu, G.C., Jou, E.: Movie rating and review summarization in mobile environment. IEEE Trans. Syst. Man Cybern. Part C (Appl. Rev.) **42**(3), 397–407 (2012)
10. Zhu, J., Wang, H., Zhu, M., Tsou, B.K., Ma, M.: Aspect-based opinion polling from customer reviews. IEEE Trans. Affect. Comput. **2**(1), 37–49 (2011)

11. Hu, M., Liu, B.: Mining and summarizing customer reviews. In: Proceeding KDD'04 Proceedings of the Tenth ACM SIGKDD International Conference on Knowledge Discovery and Data Mining, pp. 168–177. ACM, New York, (2004)
12. Harb, A., Plantie, M., Dray, G., Roche, M., Trousset, F., Poncelet, P.: Web opinion mining: how to extract opinions from blogs? In: CSTST'08: International Conference on Soft Computing as Transdisciplinary Science and Technology, p. 7 (2008)
13. Zhang, Q., Wu, Y., Li, T., Ogihara, M., Johnson, J., Huang, X.: Mining product reviews based on shallow dependency parsing. In: SIGIR'09, Proceedings of the 32nd International ACM SIGIR Conference on Research and Development in Information Retrieval (2009)
14. Fang, X., Zhan, J.: Sentiment analysis using product review data. J. Big Data (a Springer Open Journal) **2**(1), 1–14 (2015)
15. Camelin, N., Bechet, F., Damnati, G., De Mori, R.: Detection and interpretation of opinion expressions in spoken surveys. IEEE Trans. Audio Speech Lang. Process. **18**(2), 369–381 (2010)
16. Krcadinac, U., Pasquier, P., Jovanovic, J., Devedzic, V.: Synesketch: an open source library for sentence-based emotion recognition. IEEE Trans. Affect. Comput. **4**(3), 312–325 (2013)
17. Ekman, P.: Facial expression and emotion. Am. Psychol. **48**(4), 384–392 (1993)
18. Totunoglu, D., Telseren, G., Sagturk, O., Ganiz, M.C.: Wikipedia based semantic smoothing for Twitter sentiment classification. IEEE (2013)
19. Leskovec, J., Krevl, A.: SNAP Datasets: Stanford Large Network Dataset Collection. http://snap.stanford.edu/data, June 2014
20. Finch, A., Sumita, E.: Phrase-based part-of-speech tagging. In: International Conference on Natural Language Processing and Knowledge Engineering, 2007. NLP-KE 2007, Beijing. (Publisher IEEE), pp. 215–220 (2007)
21. Karim, M., Rahman, R.M.: Decision tree and Naïve Bayes algorithm for classification and generation of actionable knowledge for direct marketing. J. Softw. Eng. Appl. **6**, 196–206 (2013)
22. Nigam, K.: Using maximum entropy for text classification. In: IJCAI-99 Workshop on Machine Learning for Information Filtering, pp. 61–67 (1999)
23. Hsu, C.W., Chang, C.C., Lin, C.J.: A Practical guide to Support Vector Classification. www.csie.ntu.edu.tw/~cjlin/papers/guide/guide.pdf
24. Continuum Analytics Homepage. https://www.continuum.io/downloads

# Review of Texture Descriptors for Texture Classification

**Philomina Simon and V. Uma**

**Abstract** Texture classification is a process of distinguishing or classifying different textures into separate classes. Finding an efficient texture descriptor is a vital step for performing accurate texture classification. The research area of texture classification is widely investigated in several computer vision and pattern recognition problems. In this paper, texture descriptors applied for the texture classification in the literature are summarized. A general framework for texture classification and significance of texture descriptors are also presented in this paper.

**Keywords** Texture descriptor • Feature extraction • Texture classification • Texture • Local binary pattern

## 1 Introduction

Texture plays an important role as an image descriptor which is represented by the variations in the spatial arrangements of intensity of an image. Texture can easily be perceived by the humans and it is also generated as a result of variations of albedo of the imaged surface image [1]. Texture can be represented as a scale dependent feature. Textures are intricate visual patterns which are composed of spatially ordered entities that have brightness, color, shape or size. These local sub-patterns are distinguished by coarseness, fineness, regularity, smoothness and so on. Till now no satisfactory definitions for texture has been proposed in the literature. Texture can be defined as some property which we can feel through touch (roughness, depth, regularity, uniformity) or see (repeated and stochastic patterns, brightness variations). Texture classification is widely applied in automatic

P. Simon (✉)
Department of Computer Science, University of Kerala, Thiruvananthapuram, India
e-mail: philomina.simon@gmail.com

V. Uma
Department of Computer Science, Pondicherry University, Pondicherry, India
e-mail: uma.csc@pondiuni.edu.in



S.C. Satapathy et al. (eds.), *Data Engineering and Intelligent Computing*,
Advances in Intelligent Systems and Computing 542,
DOI 10.1007/978-981-10-3223-3_15

inspection, fabric inspection, medical image analysis and satellite imagery. There are different tasks of texture analysis such as texture classification, texture segmentation and texture recognition. According to Materka et al. [2], there are four approaches for solving texture classification, Statistical methods deal with the non deterministic properties of texture. These methods analyze the distribution and statistics between the pair of pixels in the image. Structural methods employ well defined parameters from the spatial arrangements of the pixels. Model based methods apply a stochastic model for computation. Fractal based methods are good for modeling natural textures; but it is not appropriate for representing the local image structure. Transform based methods performs texture analysis in the frequency domain and it consumes more time.

Texture descriptors represent the features of a texture in an image. Most of the researches on texture classification aim at finding an efficient texture descriptor that is powerful and discriminative. Textual descriptors extracts the features and efficiently represents them yielding higher classification accuracies irrespective of the classifier used. Most texture descriptors are simple and they are based on orientations, spatial arrangements of pixels, uniformity, histogram and gradients. In the literature, conventional texture descriptors such as Local Binary Pattern (LBP), Gray-Level Co-Occurrence Matrix(GLCM), Law's Textures, Statistical features, Autocorrelation models, Markov Random based, Weigner based, Gabor based, Wavelet based, Fourier domain based [1] are discussed. Section 1 discusses the introduction and a general framework of texture classification. Section 2 explains the significance of texture descriptors and discusses about recent texture descriptors from the literature. Section 3 concludes the paper.

## *1.1 General Framework of Texture Classification*

Application of image processing for Texture lies in the areas of texture classification, texture segmentation and texture recognition. Texture Classification is the process of identifying a surface by analyzing the texture characteristics of the image by texture descriptors. ie, classify textures and to distinguish them from one another such as bark, wood, fabric, sand and so on. Texture Segmentation deals with partitioning the image into different textured regions. Texture Recognition recognizes and analyses the textures for detecting faults or defects. Texture Classification can be performed using two approaches viz. Supervised and Unsupervised. A general framework for supervised and unsupervised texture classification is proposed in this paper. There are three main blocks in texture classification namely Feature Extraction, Feature Selection and Classification. Feature extraction is the initial step in the classification problem. Identifying the proper methods and texture descriptors for feature extraction is vital for texture classification. Classification can be performed either by training the classifier or by using any clustering or statistical methods. In Unsupervised classification, each pixel is compared to the clusters to find the nearest cluster that pixel belongs to based on a distance or similarity metric.

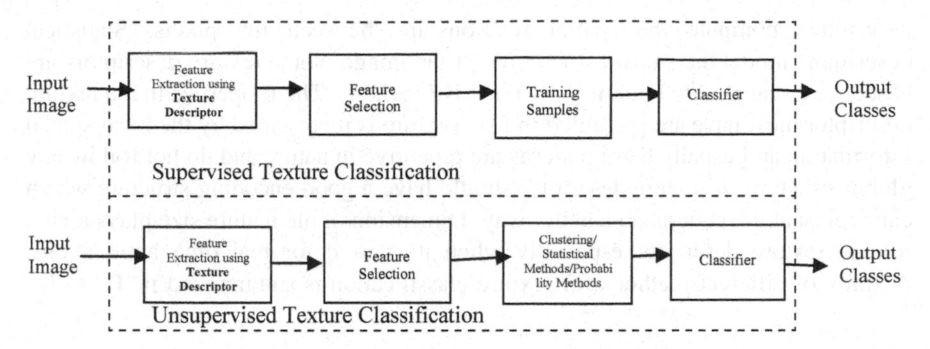


**Fig. 1** General block diagram of supervised and unsupervised texture classification

In Supervised classification, classes are predefined and classification is performed by appropriate training of the classifier to make it 'learn' the features. It assigns images in the test set to one of the texture classes learned from the training set (Fig. 1).

In this section, we also discuss a conventional descriptor Local Binary Pattern which is popular and widely used for texture classification.

Local Binary Pattern (LBP): It is a simple and efficient texture descriptor which summarizes the local structure in an image by comparing each pixel with its neighborhood and a binary code is obtained using Eq. 1.

$$LBP_{p,r}(N_c) = \sum_{p=0}^{P-1} g(N_p - N_c)2^p \tag{1}$$

where $N_p$ denotes neighborhood pixels in each block. $N_c$ is the central pixel value

$p$ is the sampling points for a $3 \times 3$ cell, where $P = 8$.

$r$ is the radius (for $3 \times 3$ cell, it is 1)

Binary Threshold function $g(x)$ is given using Eq. 2.

$$g(x) = \begin{cases} 0, & |x < 0 \\ 1, & |x \geq 0 \end{cases} \tag{2}$$

## 2 Significance of Texture Descriptors

Feature descriptors are the characteristics that uniquely identify the image. In the case of textures, it may be intensity, roughness, uniformity, regularity, repetitive patterns. Texture descriptors are generally extracted from gray scale images. In color images, descriptors are generated by computing the intensity map. There are three categories of texture descriptors viz. Spatial domain based texture descriptors, Statistical Texture descriptors, Frequency domain based texture descriptors. Spatial descriptors compute the spatial relationships between the pixels. Statistical descriptors model the statistical features of the image. Some texture descriptors are based on Local Binary Pattern (LBP) or LBP variants. The properties that a texture descriptor must have are presented in [3]. Texture is represented by the local spatial information and usually these patterns are repetitive in nature and do not follow any global structure. A texture descriptor should have a good encoding structure which can represent the patterns in a better way. Dimension of the feature size plays a vital role in texture descriptor especially when it is used for real time applications. A study of different methods for texture classification is summarized in Table 1.

### *2.1 Review of Texture Descriptors*

See Table 1.

## 3 Discussion

Texture descriptors or Texture feature extraction methods are responsible for efficient and accurate texture classification. Based on the above table, a new classification scheme for texture classification is discussed. From the literature survey discussed above, it is found that the texture classification methods can be grouped into Filter based [4–6], Histogram based [7–9], Statistical based [10–12], LBP based [13–15] and Fractal based [16–18] methods. Statistical methods are very popular as they provide better accuracy with less computational cost. Some methods may use combination of different efficient descriptors that model the texture properties such as pattern variations, roughness, edge patterns, stochastic model, Coarseness and so on. Several descriptors discussed are based on direction and spatial neighborhood. Different local or global texture features should be investigated before selecting an algorithm which can very well solve the problem of texture classification problem that can be illumination invariant, rotation invariant or scale invariant. The possibility of developing the texture classification methods that are robust to noise should be investigated. Proper tuning of parameters should be performed to find out the optimal values. From the above table, it is evident that most of the works in texture classification used Nearest Neighbor (NN) classifiers with Chi squared distance. Support Vector Machines and Nearest Neighbor (NN) based classifiers are commonly used for classification.

Based on the literature, LNIRP is a local descriptor which captures micro patterns and macro patterns. PTP is a good sampling technique to mimic the retinal sampling. In URIG, representation histogram is simple and LBP can be considered as the binary version of URIG. MTD model texture produced from the local gray level pixel intensity variations present in radiological images. HGM is low dimensional rotation invariant with orientation neutral descriptor. In BGP, feature

**Table 1** Summary of texture descriptors

| Author | Texture descriptor | Techniques based | Application | Classification techniques | Advantages | Databases used |
|---|---|---|---|---|---|---|
| Zhao et al. (2013) [19] | Uniform Rotation Invariant Gradient (URIG) | Gradient Vector | Texture classification | K-Means with Euclidean Distance | Variant rotation and Illumination condition | Outex and CUReT |
| Girija Chetty et al. (2015) [20] | Micro texture descriptors (MTD) | Fuzzy logic based | Early Diagnosis of Osteoarthritis | K-means clustering | Fuzzy logic based features allow better texture modeling in presence of uncertainties | OAI database |
| Monika Sharma et al. (2015) [7] | Histogram of Gradient Magnitudes (HGM) | Histogram of Magnitudes of Gradients. | Image Segmentation | Nearest-neighbor approach<br>Canopy clustering followed by K-means with Manhattan distance | Low-dimensional rotation invariant local descriptor<br>Low computational complexity<br>Orientation of the local gradients of pixel-intensities are ignored | Outex Database<br>Mirflickr 25000 image-collection<br>Satellite-imagery data |
| Lin Zhang et al. (2012) [4] | Binary Gabor Pattern (BGP) | Gabor filters and Local Binary Pattern | Texture classification | Even-symmetric Gabor filters<br>Odd-symmetric ones<br>Chi square measure is dissimilarity measure | Training free Rotation invariant texture representation | CUReT |
| Ziqi Zhu et al. (2014) [21] | Adaptive Hybrid Pattern(AHP) | Hybrid texture description model and Adaptive Quantization algorithm | Can be applied in object detection and Texture analysis | Nearest Neighborhood classifier | Noise-robust feature extraction discriminating textures in complex environments. | Outex texture database<br>CURet texture database |

(continued)

**Table 1** (continued)

| Author | Texture descriptor | Techniques based | Application | Classification techniques | Advantages | Databases used |
|---|---|---|---|---|---|---|
| Akoushideh et al. (2015) [22] | feature's value range (FR) | Local textural information.<br>FR act as a pre classifier | Texture classification | Nearest neighborhood classifier with Chi-square distance<br>SVM classifier | Pre-classifier and selects a few candidate categories for an input texture<br>Less time complexity | Scene-13, Outex and UIUC data sets |
| Costa et al. (2012) [16] | Segmentation-based Fractal Texture Analysis (SFTA) | Fractal Based<br>Two threshold Binary decomposition (TTBD) algorithm | Content based Image Retrieval<br>Medical Imaging | SVM built on a polynomial Kernel using Sequential Minimal Optimization (SMO) | Computed Fractal Dimension using Box counting method<br>Faster Computation Time | KTH-TIPS<br>Textured Surfaces (ROIs) of lung CT |
| Hayder Ayad (2012) [23] | Filter based Orientation Descriptor | Gabor filter based | Visual Object Categorization | SVM classifier<br>K-fold cross-validation | Improved the performance of the edge histogram descriptor | Caltech 101 dataset |
| Ben Ayed et al. (2014) [24] | Multi-scale Gray Level and Local Difference (MGLLD) | Local and Global information | Texture Classification | Nearest Neighborhood classifier<br>Chi-square distance | Better Classification accuracy | Outex database |
| Samuele Salti et al. (2014) [8] | Signature of Histograms of OrienTations (SHOT) | Based on Signatures and Histograms | Object recognition<br>Surface matching Applications | Nearest Neighborhood Classifier | rotation invariant and robust to noise as well as spurious shape variations | Stanford 3D Scanning Repository<br>RGB-D dataset |
| Faten Sandid et al. (2015) [9] | Local Combination Adaptive Ternary Pattern (LCATP)<br>Mean Histogram (MH) | Adaptive Thresholding techniques<br>Histogram based Color spaces | Material classification | Least squares Support Vector Machines classifier | Robustness to scale variation<br>Scale and Pose invariance | KTH-TIPS2b dataset |

(continued)

**Table 1** (continued)

| Author | Texture descriptor | Techniques based | Application | Classification techniques | Advantages | Databases used |
|---|---|---|---|---|---|---|
| Hussain Dawood et al. (2012) [10] | Weber Local Descriptor Variance (WLDV) | Based on Weber Local Descriptor Probability Weighted Moment for variance estimation | Texture classification | SVM classifier | Contrast information is used with WLD to improve classification accuracy | Brodatz textures dataset KTH-TIPS2 |
| Bashier et al. (2016) [25] | Extended Local Graph Structure (ELGS) | Graph structure and Histogram based | Texture classification | Nearest neighborhood classifier Chi-square, histogram intersection, | Spatial information is encoded in two directions Better classification rate with robustness and stability | UIUC and XU High Resolution texture databases |
| Garc´ıa-Olalla et al. (2012) [11] | Local oriented statistical information booster (LOSIB) LOSIB | Gray level differences along several orientations are extracted | Texture Classification | Support Vector Machine K-Nearest Neighbourhood with Chi Square distance | Local oriented information is extracted by calculating mean of the gray value differences of pixels and their neighbors along different orientations. | KTH-Tips-2a and Brodatz32 |
| Jun Zhang et al. (2013) [26] | Local Energy Pattern (LEP) | Local Feature Descriptor Represented by Statistical histogram | Texture classification Material Classification | Nearest neighbor (NN) classifier Chi-square distance | Robust to Rotation and Scale invariance Local features generated by normalized oriented energies. N-nary coding used for vector quantization. | KTH-TIPS a database. UCLA dynamic texture database |

(continued)

**Table 1** (continued)

| Author | Texture descriptor | Techniques based | Application | Classification techniques | Advantages | Databases used |
|---|---|---|---|---|---|---|
| Demir et al. (2016) [27] | Histogram-based Morphological Aps (HAPs) | Texture Descriptors created by Concatenating Local Histograms | Very High Resolution Remote Sensing Image Classification | Support vector machine classifier with Histogram intersection kernel | Capture the complex texture appearance | Trento-1 data set Trento-2 data set |
| Ali Hassan et al. (2013) [28] | Covariate shift based descriptor | Linear shift in the HT feature vector is reduced by odeling it as a covariate shift in the data. | Rotation and scale invariant Texture classification | Importance weighted support vector machines (IW-SVM) with Kullback–Leibler importance estimation | It can be applied to any good texture features that interpret image rotations and scaling as a shift in the resultant feature vector | Brodatz dataset |
| Jiangping He et al. (2013) [29] | Local Shearlet-based Energy Pattern (LSEP) | Shearlet-Based Texture Descriptor | Rotation Invariant Texture classification | Nearest neighbor classifier Dissimilarity measure used | Rotation invariant descriptor based on the Shearlet transform Robust to additive noise | Outex, Brodatz and CUReT texture databases |
| Riaz et al. (2013) [5] | Rotation- and Scale-Invariant Gabor Texture Features | Based on Gabor filters | Rotation and Scale invariant texture classification | Support Vector Machines with Polynomial Kernel (SVM) | Invariant to rotation and scale changes Gabor filters with 6 orientations and 4 scales. | Brodatz texture album |
| Heikki et al. (2010) [30] | Local Phase quantization (LPQ) | Based on decorrelation schemes | Texture Classification | Nearest Neighbor classifier with Chi squared distance | Blur tolerant decorrelation scheme work with anisotropic motion blur | Sharp as well as Blurred images of Outex datasets |
| Himeur et al. (2015) [31] | Binarized Statistical Image Features (BSIF) | BSIF binary code generated by learning basis vectors using ICA and scalar quantization | Robust Video Copy Detection | BSIF descriptor represents each pixel by a binary code. | Compact, non redundant and can be highly robust to rotation and flipping | TRECVID 2009 Large Video base |

(continued)

**Table 1** (continued)

| Author | Texture descriptor | Techniques based | Application | Classification techniques | Advantages | Databases used |
|---|---|---|---|---|---|---|
| Banerji et al. (2013) [32] | 3D Local Binary Pattern | Based on LBP descriptor and color cue | Scene and Object image Classification | EFM-NN classifier | Better Classification accuracy in different color spaces and in gray scale | Caltech 256, UIUC Sports dataset, MIT Scene dataset |
| Jabid et al. (2010) [13] | Local Directional Pattern (LDP) | Based on LBP and direction based edge response values | Gender Classification | Support vector machine (SVM) | Robust against non-monotonic changes in illumination and in random noise | FERET database |
| Larsen et al. (2014) [33] | Shape Index Histograms | Second order image structure is captured at multiple scales | Hep-2 Cell Classification | RBF kernel based SVM classifier. | Low Complexity of the method | ICIP 2013 Evaluation dataset<br>ICPR 2012 |
| Xiangping Sun et al. (2013) [14] | Local Pattern Co-occurrence Matrix LBPCM | Based on LBP and GLCM | Scale Invariant Texture Classification | Nearest neighbor classifier<br>Minimum distance used as similarity | Efficient extraction of scale invariant features<br>Large set of labeled training images with different scales is not required | KTH-TIPS2<br>CUReT |
| Backes et al. (2013) [34] | Complex Network based descritpor | Texture considered as a Pixel Network | Texture classification | Linear Discriminant Analysis (LDA) with Euclidean distance | Robustness against noise<br>Less computational complexity<br>Rotation invariance | Brodatz texture album<br>VisTex color textures<br>Outex texture database |
| Hao et al. (2016) [35] | Earth Mover's Distance (EMD) | Ground distances between Gaussian components and image features | Texture Classification | kernel-based SVM | Improved Gaussian embedding distance to compare Gaussians | KTH-TIPS-2b, FMD and UIUC |

(continued)

**Table 1** (continued)

| Author | Texture descriptor | Techniques based | Application | Classification techniques | Advantages | Databases used |
|---|---|---|---|---|---|---|
| Yong Xu et al. (2010) [17] | Wavelet based multifractal spectrum (WMFS) | Wavelet and Fractal based. Combines Spatial and frequency information | Texture Classification | SVM classifier using RBF kernel | Robustness to geometric transformation and photometric variations<br>Low dimensionality of texture descriptor Efficient computation | UIUC dataset<br>UMD data set |
| Nguyen et al. (2016) [12] | Statistical Binary Patterns (SBPs) | Based on as series of moment images, defined by local statistics | Rotational invariant texture classification | Nearest Neighbor (NN) classifier with Chi-square distance | Improves the discriminative power of LBP based on local statistical moments | KTH-TIPS 2b, CUReT, UIUC and DTD |
| MesquitaSa´ Junior et al. (2012) [18] | Gravitational Collapse model | Gravitational system and Fractal based | Analysis of texture roughness | Linear Discriminant Analysis (LDA) in a leave-one-out cross-validation scheme | Classify texture pattern through the variations of fractal dimension, computed using Bouligand–Minkowski method | Brodatz texture<br>Plant Leaf Textures |
| Zhang et al. (2013) [6] | Continuous Maximum Responses (CMR) based Local Descriptor | Based on Gaussian derivatives filters to construct the texton dictionary. | Texton dictionary-based texture classification | Nearest Neighbor (NN) classifier with Chi-square distance | CMR descriptor has good inter-class distinguish ability and PC descriptor has strong intra-class congregate ability | CUReT, KTH-TIPS and KTH-TIPS2-a datasets |
| Wang et al. (2012) [36] | Scale-Invariant Bag-of-Words Model (SIBW) | Based on SIFT and Bag of Words | Texture Classification in Compressed Domain | Sparse Modelling | Better classification accuracy<br>Less computational complexity<br>Robust Method | Caltech 256, UIUC Sports Event dataset<br>MIT Scene dataset |

(continued)

**Table 1** (continued)

| Author | Texture descriptor | Techniques based | Application | Classification techniques | Advantages | Databases used |
|---|---|---|---|---|---|---|
| Ferna´ndez et al.(2011) [37] | Binary Gradient Contours (BGC) | Based on binary gradients | Image Classification | Nearest neighbor rule with the L1 norm with Manhattan distance | BGC texture operator outperforms the well-known LBP Classification accuracy Improved | OuTeX USC-SIPI KTH-TIPS and Mondial Marmi databases |
| Chen et al. (2010) [15] | Weber Law Descriptor (WLD) | Based on SIFT and LBP | Face Detection Texture Classification | K-nearest neighbor Similarity between the WLD histogram features used as distance measure | Detecting edges elegantly, Robustness to noise and illumination change Powerful representation ability | Brodatz and KTH-TIPS2-a AR face data |
| Qi et al. (2016) [38] | Local Orientation Adaptive Descriptor (LOAD) | Based on point description on an Adaptive Coordinate System (ACS), | Illumination invariant Texture and Material Classification | one-vs-the-rest linear SVM classifier | Extract regional texture information Robustness to illumination and rotation variation Superior classification accuracy | Flickr Material Database Outex dataset UIUC data set. |
| Wang et al. (2016) [39] | Varied Local Edge Pattern (VLEP) | Based on LBP and Zernike moments features | texture image classification | nearest neighbor classifier. with chi-square distance | Descriptor has multi-scale, multi-direction (or multi-resolution) properties | CUReT database Outex database |
| Paci et al. (2013) [40] | Multi-threshold Local Quinary Patterns (MLQP) | Multi thresholding and Quinary coding | Texture classification | Support vector machines Gaussian process classifiers | obtains performance only slightly lower than SVM builds a high performance ensemble that works well on different datasets without parameters tuning | OUTEX TC 00000 dataset, KTH-TIPS Brodatz dataset |
| Mesquita Sá Junior et al. (2016) [41] | ELM based signature | Based on Extreme Learning Machine | Texture classification | Nearest neighbor classifier. with chi-square distance | To extract texture signature, based on weights of a single-hidden layer | Brodatz, Outex and Vistex |

(continued)

**Table 1** (continued)

| Author | Texture descriptor | Techniques based | Application | Classification techniques | Advantages | Databases used |
|---|---|---|---|---|---|---|
| | | | | | neural network. Extreme Learning Machine (ELM). | |
| Zhang et al. (2015) [42] | Combined Scale invariant wedge filter and gradient orientation | Frequency decomposition and gradient orientation | Scale invariant Texture representation | Simple Nearest neighbor(NN) classifier Chi-square distance | Obtained satisfactory results when compared with BSIF and LEP | KTH-TIPS CUReTdataset KTH-TIPS2-b dataset |
| Maani et al. (2013) [43] | Local Frequency descriptor LFD | Based on the local frequency components extracted from the LBP | Rotation invariant Texture Classification | Nearest neighborhood classifier with L1 distance | Uses frequency representation that is compact and more informative than LBP Robust to noise, less features, High accuracy | Outex database CUReT, and KTH-TIPS datasets |
| Linggvun Cai et al. (2015) [44] | Phased Congruency-based Binary Pattern (PCBP) | Phase-based texture feature descriptor | Classification of breast ultrasound images | Nonlinear SVM with radial basis function (RBF) kernel a | Robust descriptor against illumination changes of the breast ultrasonic images. ROI generation is not automatic; it is manual | BUS image database from Huashan hospital, Shangai Hospital |
| Guo et al. (2010) [45] | Completed Local Binary Pattern (CLBP) | Based on LBP | Texture Classification | Nearest Neighbor Classifier with $\chi^2$ distance | Very high classification accuracy Most popular | CUReT Outex |
| Xin Wang et al. (2014) [46] | Binary Rotation Invariant and Noise Tolerant (BRINT) | Based on Local Binary Pattern | Rotation Invariant and Noise Tolerant Texture Classification | Nearest Neighbor Classifier with Chi squared distance | robust to illumination variations, rotation changes, and noise | Brodatz CUReT Outex KTHTIPS2b |
| Li et al. (2010) [47] | Refined Histogram Model (RH) | Based on Bit-plane Probability & | Supervised Texture Segmentation | K-Nearest Neighbor Symmetrized Kullback–Leibler | No need to estimate the probability mass functions | Brodatz texture |

(continued)

**Table 1** (continued)

| Author | Texture descriptor | Techniques based | Application | Classification techniques | Advantages | Databases used |
|---|---|---|---|---|---|---|
| | | Generalized Gaussian Density (GGD) | | Divergence (SKLD) as similarity metric | Computational Cost Comparable | |
| Khellah et al. (2011) [48] | Dominant Neighborhood Structure (DNS) | Based on intensity similarity between any given image pixel and its surrounding neighbors | Texture Classification | K-nearest-neighbor with Euclidean Norm (L2 norm) | new global texture features that are not only rotation-invariant but also highly robust to noise by design | CUReT database |
| Qiu et al. (2016) [49] | Weyl Transform | Based on binary Heisenberg-Weyl group | Texture Classification | Nearest Neighbor classifiers | powerful tool for capturing image texture at the patch level<br>Transform invariance | Fabric texture samples<br>Natural textures |
| Mehta et al. (2016) [3] | Dense micro-block difference(DMD) | capturing the granularities at multiple scales and orientations. | Texture Classification | Linear SVM<br>Features are encoded using Fisher vector method to obtain an image descriptor, which considers high-order statistics | Fast computational scheme using integral images;<br>Low dimensionality;<br>Implementation simplicity<br>Free of tuning parameters setup | KTH-TIPS, UMD, KTH-TIPS-2a, Brodatz, and Curet |
| Wang et al. (2013) [50] | Local Neighboring Intensity Relationship Pattern (LNIRP)<br>Pixel to Patch (PTP) | Based on neighboring gray-scale properties and LBP | rotation invariant Texture Classification | 1-Nearest Neighbour Chi Squared distance | computational simplicity<br>Small feature dimensionality<br>No texton dictionary learning step &training-free | Outex database |
| Y. Quan. et al. (2014) [51] | Pattern Fractal Spectrum (PFS). | Fractal based | Texture Classification | Support vector machine (SVM) with the RBF kernel | Finding the self similarity of local pattern using fractal dimension | KTH-TIPS texture dataset<br>UMD<br>UIUC data set |

(continued)

**Table 1** (continued)

| Author | Texture descriptor | Techniques based | Application | Classification techniques | Advantages | Databases used |
|---|---|---|---|---|---|---|
| VijayaLakshmi et al (2016) [52] | Combination of Haralick texture features, Gabor features, shape features, and color features | Combination of Conventional methods | Leaf, Classification | Fuzzy Relevance Vector Machine (FRVM) | Leaf type is predicted accurately | ICL leaf dataset |
| Regniers O et al. [53] | Wavelet-Based Textural Features | Wavelet based | Supervised classification of VHR images | K-nearest neighbors K-NN<br>Maximum likelihood (ML)<br>SVM classifier | Highter Classification accuracies than GLCM<br>High Adaptability<br>Number of parameters are less | VisTex or Brodatz databases |

size is small and the classification speed is high which suits for real time applications. In AHP, feature histogram of the whole image is split into multiple binary patterns for reducing length of feature. SFTA performs better than Gabor and GLCM in image retrieval and image classification because of the TTBD algorithm. SHOT is a 3D match descriptor based on local histograms which is descriptive, robust to noise and clutter. LCATP encodes color and local structure information which is a combination of three local adaptive thresholding techniques. LOSIB is a texture descriptor enhancer. It is based on the additional local oriented statistical information and it considers the depth of neighborhood to calculate moments. In LEP, normalized local energies are used to generate features. LEP works well for dynamic textures also. In HAP, Marginal local distribution of attribute filter responses is calculated to obtain texture information. In LSEP, shearlet based energy features are computed and energy histograms are generated to describe texture. LSEP has comparable results with LBP and DNS. LPQ is a blur tolerant texture descriptor which shows a loss of minimal accuracy with sharp images. In BSIF, each pixel is represented by a binary code which is generated by learning basis vectors using ICA. 3D-LBP descriptor is based on color, texture, shape and wavelets which generates three new color images for embedding both texture and color information. LDP finds edge responses in several different directions and texture is encoded. Since LDP encode textures by the variations in directions, this descriptor is not sensitive towards illumination changes and presence of noise. LPCM is a descriptor which combines LBP and GLCM to obtain the co-occurrence matrix of local patterns. WMFS is a scale invariant descriptor obtains information from both spatial and frequency domain. It is based on the multi fractal analysis of wavelet pyramids. SBP is not sensitive towards small noise variations without compromising the discriminative power. SIBW is based on compressive sensing. Visual Words are extracted as local features in SIBW. BGC is a gradient based descriptor. WLD is a descriptor which based on the webers law and it depends on difference in the perceived human difference. VLEP is a multi scale and multi direction descriptor based on histogram spectrum features. LMP extends LBP to multiple patterns to preserve the structural information and to reduce the noise. LFD is a frequency based descriptor where feature size is small. It uses directional filters to multiple orientations. CLBP is popular and it performs better than variants of LBP which is based on the sign component and magnitude component. Multi Scale CLBP can construct the feature histogram fast based on sign and magnitude features. BRINT is a noise tolerant descriptor which does not require smoothing as a preprocessing step. BRINT is evaluated with different types of noise and with different noise levels. DNS is a global texture descriptor which is based on the similarity of dominant neighborhood. DNS map is constructed to generate intensity and illumination invariant features and it is robust to noise. DMD is a local descriptor which densely captures the granularities of the image patches at different scales and orientations. Dense features are compressed to reduce the dimensionality. In this section, a detailed discussion of the different features of the texture descriptor is presented.

## 4 Conclusion

Texture Classification groups the textures which belong to the same class. Researchers proposed several novel texture descriptors for extracting the features from the images. The algorithms showed promising results in extracting features that are robust to noise, rotation, scale and illumination variations. A general frame work of texture classification is presented in this paper. Over fifty, recently developed texture descriptors used for classification of textures, its significance, different classification techniques, texture databases have been reviewed in this paper.

## References

1. Petrou, M., García-Sevilla, P.: Image Processing: Dealing with Texture. Wiley Publishers (2006)
2. Materka, A., Strzelecki, M.: Texture Analysis Methods—A Review, Technical University of Lodz, Institute of Electronics, COST B11 report, Brussels (1998)
3. Mehta, R., Eguiazarian, K.E.: Member S. Micro-Block Difference **25**(4), 1604–1616 (2016)
4. Binary Gabor Pattern : An Efficient And Robust Descriptor For Texture Classification 81–84 (2012)
5. Riaz, F., Hassan, A., Rehman, S., Qamar, U.: Texture classification using rotation- and scale-invariant gabor texture features. IEEE Signal Process. Lett. **20**(6), 607–610 (2013)
6. Zhang, J., Zhao, H., Liang, J.: Continuous rotation invariant local descriptors for texton dictionary-based texture classification. Comput. Vis. Image Underst. **117**, 56–75 (2013)
7. Sharma, M., Ghosh, H.: Histogram Of Gradient Magnitudes: A Rotation Invariant Monika Sharma, Hiranmay Ghosh TCS Innovation Labs, Delhi TATA Consultancy Services Limited, 4614–4618 (2015)
8. Salti, S., Tombari, F., Di Stefano, L.: SHOT: unique signatures of histograms for surface and texture description. Comput. Vis. Image Underst. **125**, 251–264 (2014). Elsevier Inc.
9. Douik, A., Sandid, F.: Texture descriptor based on local combination adaptive ternary pattern. IET Image Process. **9**(8), 634–642 (2015)
10. Dawood, H.: Combining the Contrast Information with WLD for Texture Classification
11. Garc, O., Alegre, E., Fern, L., Etienne, D.S.: Local Oriented Statistics Information Booster (LOSIB) for texture classification (2014)
12. Nguyen, T.P., Vu, N.-S., Manzanera, A., Wang, S.-J.: Statistical binary patterns for rotational invariant texture classification. Neurocomputing **173**, 1565–1577 (2015)
13. Jabid, T., Kabir, M.H., Chae, O.: Facial expression recognition using Local Directional Pattern (LDP). In: 17th IEEE Int Conference on Image Processing ICIP 2010, 1605–1608 (2010)
14. Sun, X., Wang, J., She M.F.H., Kong, L.: Neurocomputing scale invariant texture classification via sparse representation. Neurocomputing **122**, 338–348 (2013). Elsevier
15. Chen, J., Shan, S., He, C., Zhao, G., Pietikainen, M., Chen, X., et al.: WLD: a robust local image descriptor. IEEE Trans. Pattern Anal. Mach. Intell. **32**(9), 1705–1720 (2010)
16. Costa, A.F., Humpire-Mamani, G., Traina, A.J.M.: An efficient algorithm for fractal analysis of textures. Braz. Symp. Comput. Graph Image Process. 39–46 (2012)
17. Xu, Y., Yang, X., Ling, H., Ji, H.: A new texture descriptor using multifractal analysis in multi-orientation wavelet pyramid. Proc. IEEE Comput. Soc. Conf. Comput. Vis. Pattern Recogn. **60603022**, 161–168 (2010)

18. De Mesquita, S., Junior, J.J.: Ricardo Backes A. A simplified gravitational model to analyze texture roughness. Pattern Recogn. **45**(2), 732–741 (2012)
19. 2015 IEEE International Conference on Texture Classification Using Uniform Rotation Invariant Gradient Image Processing (ICIP), China **1**, 3650–3654 (2015)
20. Chetty, G., Scarvell, J., Mitra, S.: Fuzzy texture descriptors for early diagnosis of osteoarthritis. IEEE Int. Conf. Fuzzy Syst. (2013)
21. Zhu, Z., You, X., Chen, C.L.P., Tao, D., Jiang, X., You, F., et al.: An noise-robust adaptive hybrid pattern for texture classification 0–5 (2014)
22. Texture K-. Keywords- Texture classification; features' value range; Haralick feature; local binary pattern (LBP); effective features., 65–9 (2015)
23. Ayad, H., Abdullah, S.N.H.S., Abdullah, A.: Visual object categorization based on orientation descriptor. In: Proceedings—6th Asia International Conference Math Model Computer Simulation, pp. 70–74. AMS (2012)
24. Gargouri, N., Ayed, B., Larousi, M.G.: Multi-scale Gray Level and Local Difference for texture classification (2014)
25. Bashier, H.K., Hoe, L.S., Hui, L.T., Azli, M.F., Han, Y., Kwee, W.K., et al.: Texture classification via extended local graph structure. Opt.—Int. J. Light Electron Opt. **127**, 638–643 (2016)
26. Thresholds USQ, Zhang, J., Liang, J., Zhao, H.: Local Energy Pattern for Texture Classification **22**(1), 31–42 (2013)
27. Demir, B., Bruzzone, L.: Histogram-based attribute profiles for classification of very high resolution remote sensing images **54**(4), 2096–2107 (2016)
28. Hassan, A., Shaukat, A.: International I, On W, Learning M, Signal for. Covariate shift approach for invariant texture classification (2013)
29. He, J., Ji, H., Yang, X.: Rotation invariant texture descriptor using local shearlet-based energy histograms. IEEE Signal Process. Lett. **20**(9), 905–908 (2013)
30. Heikkila, J., Ojansivu, V., Rahtu, E.: Improved blur insensitivity for decorrelated local phase quantization. Proc.—Int. Conf. Pattern Recogn. **2**, 818–821 (2010)
31. Himeur, Y., Sadi, K.A.: Joint Color and Texture Descriptor Using Ring Decomposition for Robust Video Copy Detection in Large Databases 495–500 (2016) (January)
32. Banerji, S., Sinha, A., Liu, C.: New image descriptors based on color, texture, shape, and wavelets for object and scene image classification. Neurocomputing, **117**, 173–85 (2013). Elsevier
33. Larsen, A.B.L., Vestergaard, J.S., Larsen, R.: Cell classification using shape index histograms with donutish pooling **33**(7), 1 (2014)
34. Backes, A.R., Casanova, D., Bruno, O.M.: Texture analysis and classification: a complex network-based approach. Inf. Sci. (Ny) **219**, 168–180 (2013)
35. Hao, H., Wang, Q., Li, P., Zhang, L.: Evaluation of ground distances and features in EMD-based GMM matching for texture classification. Pattern Recogn. (2016). Elsevier
36. Wang, X., Wang, Y., Yang, X., Zuo, H.: Background A. Texture Classification Based On SIFT Features and Bag-of-Words in Compressed Domain **41076120**, 941–945 (2012)
37. Fernndez, A., lvarez, M.X., Bianconi, F.: Image classification with binary gradient contours. Opt. Lasers Eng. **49**(9–10), 1177–1184 (2011)
38. Qi, X., Zhao, G., Shen, L., Li, Q., Pietikäinen, M.: Neurocomputing LOAD: Local orientation adaptive descriptor for texture and material classification **184**, 28–35 (2016)
39. Wang, Y., Zhao, Y., Cai, Q., Li, H., Yan, H.: A varied local edge pattern descriptor and its application to texture classification (2016)
40. Paci, M., Nanni, L., Severi, S.: An ensemble of classifiers based on different texture descriptors for texture classification. J King Saud Univ.—Sci. **25**, 235–244 (2013)
41. de Junior, J.J.M.S., Backes, A.R.: ELM based signature for texture classification. Pattern Recogn. (2015)
42. Zhang, J., Liang, J., Zhang, C., Zhao, H.: Scale invariant texture representation based on frequency decomposition and gradient orientation. Pattern Recogn. Lett. **51**, 57–62 (2014)

43. Maani, R., Kalra, S., Yang, Y.-H.: Noise robust rotation invariant features for texture classification. Pattern Recogn. **46**, 2103–2116 (2013)
44. Cai, L., Wang, X., Wang, Y., Guo, Y., Yu, J., Wang, Y.: Robust phase-based texture descriptor for classification of breast ultrasound images. BioMed. Eng. OnLine **14**, 26 (2015)
45. Zhenhua, G., Zhang, D.: A completed modeling of local binary pattern operator for texture classification. Image Process. IEEE Trans. **19**(6), 1657–1663 (2010)
46. Liu, L., Long, Y., Fieguth, P., Lao, S., Zhao, G.: BRINT : Binary Rotation Invariant and Noise Tolerant Texture Classification 1–13 (2014)
47. Li, L., Tong, C.S., Choy, S.K.: Texture classification using refined histogram. IEEE Trans. Image Process. **19**(5), 1371–1378 (2010)
48. Khellah, F.: Textured image denoising using dominant neighborhood structure. Arab. J. Sci. Eng. **39**(5), 3759–3770 (2014)
49. Qiu, Q., Thompson, A., Calderbank, R., Sapiro, G.: Data representation using the Weyl transform **64**(7), 1844–1853 (2016)
50. Wang, K., Bichot, C.-E., Zhu, C., Li, B.: Pixel to patch sampling structure and local neighboring intensity relationship patterns for texture classification. Signal Process. Lett. **20**(9), 853–856 (2013)
51. Quan, Y., Xu, Y., Sun, Y.: A distinct and compact texture descriptor. Image Vis. Comput. **32**(4), 250–259 (2014). Elsevier B.V.
52. VijayaLakshmi, B., Mohan, V.: Kernel-based PSO and FRVM: an automatic plant leaf type detection using texture, shape, and color features. Comput. Electron Agric. **125**, 99–112 (2016). Elsevier B.V.
53. Regniers, O., Bombrun, L., Lafon, V., Germain, C.: Supervised classification of very high resolution optical images using wavelet-based textural features. IEEE Trans. Geosci. Remote Sens. **54**(6), 3722–3735 (2016)

# Improvisation in HEVC Performance by Weighted Entropy Encoding Technique

**B.S. Sunil Kumar, A.S. Manjunath and S. Christopher**

**Abstract** Now a day multimedia applications are growing rapidly and at the same time the volume of video transactions is raising exponentially. This demands an efficient technique to encode the video and to reduce the congestion in the transmission channel. This paper presents an improvisation technique; weighted encoding for High Efficiency Video Coding (HEVC). This method optimizes the spatial and temporal redundancy during the motion compensation by the optimal choice of code block. The blocks are chosen on the basis of weights- assigned to it using the firefly algorithm. On encoding it reduces the size of the video with perceptually better quality video or Peak Signal to Noise Ratio (PSNR).

**Keywords** Compression · Encoding · Entropy HEVC · Scalable video coding · PSNR

## 1 Introduction

HEVC performance is high when compared to the conventional video coding standards but it has the drawback of computational complexity and storage problems while encoding [1]. It consists of multiple coding tools such as–Prediction unit, coding unit, transform unit in quadtree coding block partitioning tool. The

B.S. Sunil Kumar (✉)
Department of Computer Science & Engineering, GM Institute of Technology, Davangere 577006, India
e-mail: sunilkumarbs@gmit.ac.in

A.S. Manjunath
Department of Computer Science & Engineering, Siddaganga Institute of Technology, Tumkur 572102, India
e-mail: asmanju@gmail.com

S. Christopher
Department of Defence R & D, DRDO, DRDO Bhawan, New Delhi 110011, India
e-mail: s.christopher@hqr.drdo.in



S.C. Satapathy et al. (eds.), *Data Engineering and Intelligent Computing*,
Advances in Intelligent Systems and Computing 542,
DOI 10.1007/978-981-10-3223-3_16

quadtree is a phenomenon used for subdividing the picture into many blocks for coding and prediction [2]. In video coding, intra coding provides good quality videos but these too have a lot of draw backs. The challenges in new HEVC standard are solved and it is shown in the next section. The overview of the HEVC, particularly, focussing on the recent developments in the 3D and multi view video are addressed in 2016 by Tech et al. [3]. They studied the overview and various features of multi view (MV) video and depth-based 3D video formats in High efficiency video coding (HEVC). They used the multiview extension feature which has the ability to aid efficient coding of the multi-camera views and the associated auxiliary pictures and it is implemented along with single layer decoders. They modified the MV extension into advanced type of video extension called as 3D-HEVC, used in coded representation which has multiple views and the associated depth maps. Moreover, the bit rate reduction comparison is studied with MV-HEVC and found increased enhancement by using new video coding tools.

Increasing the coding efficiency by improving the bit allocation is experimentally studied in 2014 by Wang et al. [4]. They developed a gradient based R-lambda (GRL) for controlling the intra frame rate in HEVC to minimize the BER and enhancing the video quality. The algorithm has the capability to measure the frame-content complexity and it is an advanced method for improving the performance of the conventional R-lambda. Additionally, they also developed a coding tree unit level bit allocation method. Several works are focussed on the transform coding techniques in HEVC. In 2013, Nguyen et al. [5] have studied on various HEVC techniques which are used for transforming the codes and entropy coding. They used the quadtree-based partitioning dubbed as residual quadtree as one of the new transform coding technique that supports in increasing the size of transform blocks and dividing the residual blocks into multiple blocks and in 2012, Sole et al. [6] have worked on transform coefficient coding in HEVC that includes the scanning patterns and coding methods, sign data, coefficient levels and significant map. Our paper focuses on improving the entropy encoding of HEVC to accomplish better encoding performance.

## 2 Proposed Work

The key motive of the digital video coding standard is the optimization of coding efficiency. The ability of a coding standard to reduce the bit rate that represents a video content with desired video quality level is known as coding efficiency. It can also be defined as the ability of the coding standard to increase the quality of the video, with limited bit rate.

The first version of the High Efficiency Video HEVC standard [7], approved as ITU-T H.265 and ISO/IEC 23008-2, achieved the first motive and then they concentrated on the key extensions of its abilities to suit for the needs for a wide number of applications. The old version of the HEVC standard had a better scope. But, it haven't given any importance on the key features on the designing of the core elements In the development of the new HEVC standard encoding scheme, the extensions that are considered, while preparing this paper (pointing to the present status that the Vienna meetings of July/August 2013 holds), can be divided under three areas: the range extensions, the scalability extensions and the 3D video extensions. In range extension, the bit depth ranges and the colour sampling formats are enlarged. It also focuses on the performance of high-quality coding, screen-content coding and lossless coding. The scalability extensions allows the usage of the embedded bit stream subsets represented in the form of reduced-bit-rate The 3D video extensions increases both stereoscopic and multi-view representations in addition to the capability of the novel 3D which involves the use of depth maps and view-synthesis methodologies [8].

Our key contribution relies on proposing an improved version of standard entropy encoding method for HEVC. Since the standard entropy encoding has lack of competing performance, we improve it by introducing weightage factors for the rate of coding bits. The weighing range is depending upon the resolution and perceptivity of the video contents. Hence, a recently introduced optimization algorithm called as firefly algorithm is exploited.

## *2.1 Encoding Technique*

CABAC, which is the only entropy coding method of HEVC in modified form, has special features such as adaptive coefficient scanning, context modelling and coefficient coding. The context modelling is used for maximizing the efficiency of the CABAC and the indices for the context modelling are obtained from splitting depth of the transform tree. These indices include many syntax elements such as split_coding, unit_flag, cbf_luma, skip_flag, cbf_cr, cbf_cb, and split_transform flag. The split_transform_flag indicates the splitting of TB and spli_coding_unit_flag is used for indicating the further splitting of CB and it is coded depending upon the spatially neighbouring information. The skip_flag points out that the CB coded in the form of inter picture predictively skipped and the coding for cbf_cb, cbf_cr and cbf_luma are based on the splitting depth of the transform tree. To increase the throughput, the data is minimized with bypass mode of CABAC in HEVC.

In order to scan, various kinds of coefficient scanning methods such as horizontal, diagonal up-right and vertical scans are applied and depending upon the intra picture predicted regions directionality, the scanning orders are selected. The horizontal scan is used, when the prediction direction is nearer to vertical and the vertical scan is applied when the direction of prediction is nearer to horizontal regions. For other than the directions of vertical and horizontal, the diagonal up-right scan is used. Mostly, the scanning is performed in $4 \times 4$ and $8 \times 8$ sub blocks TB sizes. The $4 \times 4$ diagonal up-right scan is applied in sub blocks for coding the $16 \times 16$ or $32 \times 32$ intra picture predictions. In HEVC, the last non-zero transform coefficients position, sign bits, significant map and level of the transform coefficient can be transmitted. The sign bits are coded depending upon the number of coded coefficients and the new compression effects can be detected using the method of sign data hiding. From the parity of the total coefficient amplitudes, the first nonzero coefficients sign bit can be detected, if the value of changes between the first and last non-zero coefficients scanning positions and the sign data hiding having two non-zero coefficients in 4 × 4 sub block is greater than 3.

## 3 Weighted Entropy Encoding

A. *Weighted Entropy Model*

Let $X$ be the set containing $M$ video sequences $\{x_1, x_2, \ldots, x_M\}$, with individual $x_k$ for $1 \leq k \leq M$. It can have categorical attribute vector of $[y_1, y_2, \ldots, y_N]^T$, where $N$ represents the number of attributes and $y_j$ had a domain value which can be estimated by $[y_{1,j}, y_{2,j}, \ldots, y_{ni,j}]$ for $1 \leq j \leq N$, where, $y_j$ represents the number of distinct values in the attribute $y_j$. Let us consider the attribute $y_j$ as the random variable and so, the random vector $[y_1, y_2, \ldots, y_N]^T$ can be indicated as $Y$. The attribute $x_i$ is represented as $[x_{i,1}, x_{i,2}, \ldots, x_{i,m}]^T$.

The reverse sigmoid function of the entropy is used to weight the entropy of each attribute and so,

$$W_i = 1 - \log it^{-1}(w_i E_i) \tag{1}$$

$$W_i = 1 - \frac{1}{1 + e^{-w_i E_i}} \tag{2}$$

The weighted entropy model insists determining optimal $w_i$ based on the pixel wise correlation of the decoded video sequence with the original video sequence. Hence, the process of determining optimal $w_i$ can be formulated as a maximization problem given as follows.

$$w^* = \arg\max_{w_i} \sum_{l=1}^{M_s} \left( 2 \log x_l^{\max} - \left[ \log \frac{1}{|x_l|} \sum_u \sum_v \left( x_l(u,v) - \hat{x}_l(u,v) \right)^2 \right] \right) \tag{3}$$

where, $x_l(u,v)$ and $\hat{x}_l(u,v)$ refer to $(u,v)$th pixel element of a frame corresponds to $l$th video sequence and the decoded video sequence, respectively.

B. *Optimizing Entropy Weights*

This paper exploits firefly algorithm proposed by Xin-She Yang to optimize the entropy weights. In other words, the firefly algorithm is used to solve the objective model given in Eq. (3). This algorithm is closely related to the behaviour and the flashing of the fireflies. The algorithm works based on the three basic concepts such as,

(1) The fireflies are naturally unisex in character and so, one firefly will attract towards any other butterfly.
(2) In flashing butterflies, the brightness generated is directly proportional to the attractiveness and the lesser brightness emitting firefly will move near the brighter firefly. The attractiveness and brightness will decrease with respect to increasing distance. It takes a random movement if the brighter firefly is nil.
(3) The objective functions landscape identifies the fireflies brightness.

Since, the attractiveness and the brightness are directly proportional to each other, the changes in the brightness $\beta$ with respect to distance $r$ is given by $\beta = \beta_0 e^{-\gamma r^2}$ where, $\beta_0$ represents the attractiveness at distance $r = 0$. If one firefly emits high brightness, the movement of the firefly $i$ will be toward the brighter one $j$ and it is given as

$$w_i^{t+1} = w_i^t + \beta_0 e^{-\gamma r_{ij}^2}\left(w_j^t - w_i^t\right) + \alpha_t \in_i^t \tag{4}$$

where, the $\in_i^t$, $\alpha_t$ represents, the random number vector obtained from the Gaussian distribution or the uniform distribution at a time period $t$, randomization parameter having the third term randomization respectively and the generation of second term is because of the attraction. When $\beta_0 = 0$, the movement will be a random walk and if $\gamma = 0$, it converts to a variant of particle swarm optimization [9]. In addition to this, the $\in_i^t$ randomization will move towards the other distributions, known as Levy flights [9]. This algorithm is gives us a optimal weighted entropy value by moving from smaller weighted entropy to larger weighted entropy.

The pseudo code to optimize entropy weights is illustrated in the below algorithm

**Algorithm 1: Firefly algorithm to optimize entropy weights $W_i$**

```
1   Set t to zero
2   Generate initial fireflies w_i^0
3   t <- t+1
4   Determine the objective function
5   Update fireflies to determine w_i^t
6   If termination criteria is not met
7        Go to step 3
8   Else
9        Return w_i^t as w*
```

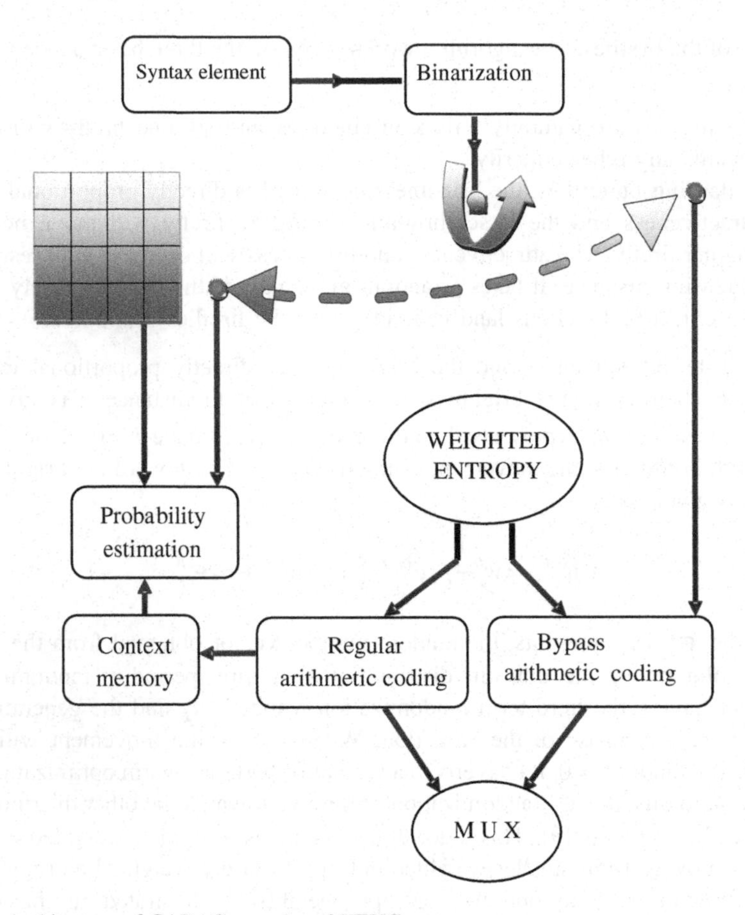


**Fig. 1** Architecture of CABAC encoder of HEVC

According to the algorithm, the initial fireflies are arbitrarily generated within the interval [0, 1] and they are subjected to determine the objective value given in Eq. (3). The fireflies are updated using Eq. (4) to obtain better fireflies, which will used to terminate the process when satisfactory improvement is exhibited by the updated fireflies. The improved compression performance can be experienced using the weighted entropy with the HEVC architecture as mentioned in Fig. 1.

## 4 Results and Discussion

The experimental study for the proposed and the standard entropy coding in HEVC standard has been done with the selected eight video sequences available in http://www.cipr.rpi.edu/resource/sequences/sif.html in YUV file format. The eight video sequences are diversified by its contents and number of frames such as coastguard, garden and football, respectively, with 300, 140, 112, 300, 300, 300, 115 and 125 sequences respectively. The PSNR of the decoded video sequences is analysed for understanding the performance of the encoded principles. The efficiency of the proposed method is evaluated by comparing with the standard encoding method.

A. *PSNR analysis*

The performance of the proposed method is analysed with PSNR metric and the results are compared with the standard coding method. Figure 2 represents the PSNR analysis for the video sequences; football, garden, mobile and tennis. PSNRs are plotted for four different numbers of transmitted bits, which are being determined using the block sizes: 2, 4, 8 and 16 respectively. For instance, in garden video sequence, the increase in transmitted bits increases the PSNR and reaches 76 dB and 73.5 dB for the proposed and the standard method, respectively. But in football video sequence, at block size 8, the PSNR values are 72 and 75 dB for the standard and proposed method. The performance deviation between the proposed and the standard method are noted as 2.9, 0.72, 4.03, 2.04, and 3.47% with respect to video sequences hall monitor, foreman, tennis, garden, and football.

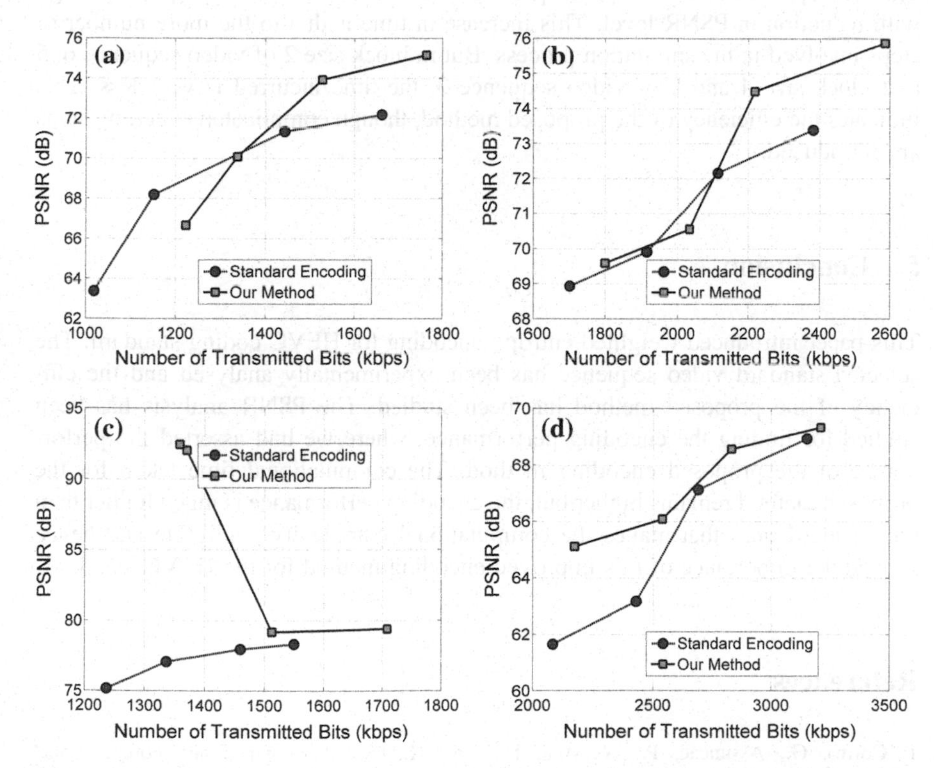


**Fig. 2** PSNR of decoded video sequences: **a** Football. **b** Garden. **c** Mobile. **d** Tennis of HEVC

**Table 1** Performance improvement over standard encoding principle and the computation cost incurred for the improvement

| Block size | 1 | | 2 | | 4 | | 8 | |
|---|---|---|---|---|---|---|---|---|
| Improvement metrics | PSNR | Time | PSNR | Time | PSNR | Time | PSNR | Time |
| Video sequence 1 | 4.12 | 0.25 | 3.63 | 0.54 | 2.77 | 0.24 | 5.19 | 0.88 |
| Video sequence 2 | 3.37 | 0.42 | 3.22 | 0.07 | 0.92 | 0.61 | 0.95 | 0.007 |
| Video sequence 3 | 1.37 | −0.31 | 1.56 | −0.91 | 19.41 | 0.22 | 23.0 | 0.74 |
| Video sequence 4 | 0.58 | 2.73 | 2.14 | 3.64 | 4.62 | 0.58 | 5.61 | 1.02 |
| Video sequence 5 | 0.79 | 1.03 | 1.58 | −1.02 | 1.56 | 0.42 | 1.1 | 0.98 |
| Video sequence 6 | 1.76 | 0.04 | 1.45 | −0.49 | 1.85 | 1.43 | 1.57 | 1.2 |
| Video sequence 7 | 2.58 | 0.14 | 2.33 | 0.42 | 1.59 | 3.72 | 1.26 | 1.44 |
| Video sequence 8 | 0.97 | 0.35 | 1.17 | 1.67 | 2.64 | 0.92 | 1.74 | 0.69 |

B. *Complexity analysis*

The performance of the proposed method with respect to the computational time and PSNR is given in Table 1 For video sequence 1, 2, 7, 8, 4, 6 (except block size 2), 5 (except block size 2), 3 (except block size 1 and 2), the time required is high with reduction in PSNR level. This increase in time is due to the more number of steps involved in the simulation process. But in block size 2 of video sequence 6, 5 and block size 4 and 8 of video sequence 3, the time incurred is very less. This indicates the efficiency of the proposed method, though complicated encoding steps are introduced.

# 5 Conclusion

This paper introduced weighted entropy encoding for HEVC coding standard. The selected standard video sequence has been experimentally analysed and the efficiency of the proposed method has been studied. The PSNR analysis has been studied for finding the encoding performance, where we had asserted the performance of the proposed encoding method. The computational time taken for the proposed method remains higher but, the encoding performance is much higher than the standard ones that makes the computational complexity negligible and hence proved the importance of this improved encoding method for the HEVC standard.

# References

1. Correa, G., Assuncao, P., Agostini, L., Cruz, L.D.S.: Performance and computational complexity assessment of high-efficiency video encoders. IEEE Trans. Circuits Syst. Video Technol. **22**(12), 1709–1720 (2012)

2. Bossen, F., Bross, B., Suhring, K., Flynn, D.: HEVC complexity and implementation analysis. IEEE Trans. Circuits Syst. Video Technol. **22**(12), 1685–1696 (2012)
3. Tech, G., Chen, Y., Muller, K., Ohm, J.R., Vetro, A., Wang, Y.K.: Overview of the multiview and 3D extensions of high efficiency video coding. IEEE Trans. Circuits Syst. Video Technol. **26**(1), 35–49 (2016)
4. Wang, M., Ngan, K.N., Li, H.: An efficient frame-content based intra frame rate control for high efficiency video coding. Signal Process. Lett. IEEE **22**(7), 896–900 (2014)
5. Nguyen, T., Helle, P., Winken, M., Bross, B., Marpe, D., Schwarz, H., Wiegand, T.: Transform coding techniques in HEVC. IEEE J. Sel. Top. Signal Process. **7**(6), 978–989 (2013)
6. Sole, J., Joshi, R., Nguyen, N., Ji, T., Karczewicz, M., Clare, G., Henry, F., Duenas, A.: Transform coefficient coding in HEVC. IEEE Trans. Circuits Syst. Video Technol. **22**(12), 1765–1777 (2012)
7. Sullivan, G., Ohm, J., Han, W.-J., Wiegand, T.: Overview of the high efficiency video coding (HEVC) standard. IEEE Trans. Circuits Syst. Video Technol. **22**(12), 1649–1668 (2012)
8. Sullivan, G.J., Boyce, J.M., Chen, Y., Ohm, J.R., Segall, C.A., Vetro, A.: Standardized extensions of High Efficiency Video Coding (HEVC). IEEE J. Sel. Top. Signal Process. **7**(6), 1001–1016 (2013)
9. Hakli, H., Uguz, H.: A novel particle swarm optimisation algorithm with levy flight. Appl. Soft Comput. **23**, 333–345 (2014)

# Improvement in PMEPR Reduction for OFDM Radar Signal Using PTS Algorithm

**C.G. Raghavendra, I.M. Vilas and N.N.S.S.R.K. Prasad**

**Abstract** In this paper we suggest a method to address the complication of variable amplitude in multicarrier signal. Multifrequency Complementary Phase-Coded (MCPC) signal has fluctuations in amplitude because it is the sum of carriers with different frequency. To avoid non linear working of power amplifiers at transmitter, it is desirable to reduce Peak to Mean Envelope Power Ratio (PMEPR) of the signal. We have tried two algorithms to reduce PMEPR and sidelobes. Namely, clipping technique which is a signal distortion technique and one from signal scrambling technique called as Partial Transmit Sequence (PTS) algorithm.

**Keywords** Multifrequency Complementary Phase-Coded radar (MCPC) · Peak to Mean Envelope Power Ratio (PMEPR) · Partial Transmit Sequence (PTS)

## 1 Introduction

The high range resolution in the radar system can be obtained by increasing the radar functions are personally identified with properties and attributes of electro-magnetic waves as they interface with physical items. Radar can perform various functions like resolution, detection and measurement. Resolution corresponds to radar's ability to separate one desired target signal from another and to separate desired from undesired target signals. The high range resolution in radar

C.G. Raghavendra (✉) · I.M. Vilas
Department of Electronics and Communication, MSRIT, Bengaluru, India
e-mail: cgraagu@msrit.edu

I.M. Vilas
e-mail: vilasmuchandi004@gmail.com

N.N.S.S.R.K. Prasad
Aeronautical Development Agency (ADA), Bengaluru, India
e-mail: nnssrkprasad2007@gmail.com



S.C. Satapathy et al. (eds.), *Data Engineering and Intelligent Computing*,
Advances in Intelligent Systems and Computing 542,
DOI 10.1007/978-981-10-3223-3_17

can be attained by expansion of the signal's bandwidth. Expansion of bandwidth in phase coded modulation signal is identical to decreasing in chip width. But to reduce chip width, it faces technological limitations. For excellent detection of target a radar needs a large signal power but transmitter are usually operate near their peak power limitations. These dissimilar needs of long pulses for detection and short pulses for range precision in measurements prohibited early radars from concurrently performing both capacities well. This made motivation for development of new technique called pulse compression to meet the both needs. There are two widely used methods of pulse compression. Namely, phase coded waveforms and Linear Frequency modulated waveform. In phase coded waveform method, the long pulse is divided into sub pulses of same duration and each duration of sub pulse is defined by a specific phase value. Linear frequency modulated waveform consists of rectangular amplitude modulation of pulse duration and linear frequency modulation.

Multicarrier signals for radar applications was first time presented in 1998 by Jankiraman. MCPC scheme was developed on basis of OFDM concept for radar applications and it was first proposed by Levanon in 2000 [1–4]. Every multifrequency phase coded signal is composed of N subcarriers which can be transmitted all together. Every subcarrier is modulated by phase sequences, where all phase sequences consists of M phase modulated bits. Each subcarrier is frequency separated by inverse duration of phase element. MATLAB software has been used to simulate the results.

## 2 Multifrequency Complementary Phase Coded Signals

Multifrequency Complementary Phase Coded (MCPC) signals has been proposed to attain good range resolution in radar systems. Several methods have been initiated to achieve this objective. In last few years, due to their well-known properties like good resolution, very simple realization and good bandwidth performance has been specially taken into consideration.

### *2.1 Polyphase Codes*

Polyphase codes [5, 6] adapt harmonically based phases which depend on a certain fundamental phase increment. If the pulse can take phase values more than two, then it is called as a polyphase code. In the polyphase codes, a pulse of time duration $\tau$ is divided into N number of parts of equal interval. Each sub divided duration is again partitioned into N number of sub-parts of width $\Delta\tau$.

Polyphase codes were proposed by Frank which has the properties of non-periodic correlation and they were named as Frank codes. Polyphase codes have lesser sidelobes when compared with bi-phase codes. Different types of Frank codes were proposed by Krestcher and Lewis which are better tolerant than Frank codes. The advantages of polyphase codes are, they are more doppler tolerant and less range of sidelobes can be observed compared to bi-phase codes.

A P × P MCPC signal consists of P carriers each of length P phase elements. Each sub carriers are phase modulated by different complementary phase sequences of length P. Each sequence consisting of P elements with each duration of $t_b$. Hence the complete pulse duration is given by $Pt_b$. The autocorrelation mainlobe width of these signals is $t_b/P$.

The complex envelope of the signal to be transmitted is therefore

$$x(t) = \left\{ \begin{array}{l} \left\{ \sum_{q=1}^{P} W_q \exp\{j[2\pi t f_s(\frac{P+1}{2} - q)]\} \right\} \\ \sum_{p=1}^{P} x_{q,p}[t-(p-1)t_b], \quad 0 \le t \le Pt_b \\ 0, \quad elsewhere \end{array} \right\} \tag{1}$$

where,

$$x_{q,p}(t) = \left\{ \begin{array}{ll} \exp(j\phi_{q,p}), & 0 \le t \le t_b \\ 0, & elsewhere \end{array} \right\} \tag{2}$$

$\phi_{q,p}$ is the *p*th phase element of *q*th sequence and $W_q$ is the amplitude scaling assigned to *q*th subcarrier. Equations 1 and 2 describe the complex envelope of a P × P MCPC signal.

# 3 PMEPR Reduction Techniques

## 3.1 Introduction and Classification

The immediate output of a MCPC signal has large amplitude variations when compared to conventional single carrier signal. The large fluctuations of amplitude are due to addition of multiple subcarriers of different frequencies. Higher values of PMEPR would drive power amplifiers at the transmitter into non-linear region. This results in interference of subcarriers and it may corrupt the spectrum of the signal. It also degrades performance of the system. Hence, it is advisable to reduce PMEPR value as low as possible to avoid non linearity of system. The various PAPR reduction techniques in OFDM can be classified as follows.

I. Signal Distortion Technique
II. Signal Scrambling Techniques
III. Coding Techniques

I. **Signal Distortion Technique**

In this technique, it tries to reduce the PAPR by distorting the signal before the power amplifier block. One of the simple methods is clipping and filtering the transmitter signal to eliminate fluctuations of amplitude. Due to clipping of the signal, there is a possibility of destruction of orthogonality to some extent which introduces in-band and out-band noises. Peak windowing and companding transforms are the examples of signal distortion technique.

II. **Signal Scrambling Techniques**

The basic principle in scrambling techniques is to scramble each signal with different scrambling sequence and decide one which has the minimum PAPR value. Evidently, this strategy does not ensure decrement of PAPR underneath a specific limit, but it can reduce the appearance probability of high PAPR to a incredible degree. The approaches in this method are selective mapping (SLM) and Partial Transmit Sequences (PTS).

III. **Coding Techniques**

The principle behind this technique is to introduce Forward Error Correction (FEC) methods to reduce the OFDM signals with larger PAPR value. The idea of forward error correction method here is to add redundant bit to reduce overall PAPR value. Forward Error Correction has two types of codes namely block codes and run length codes. Linear block codes, Golay complementary codes, Reed Mullar, Bose Chaudhari Hochquenghem (BCH), low density parity check (LDPC) are few block codes.

## 3.2 *Clipping Technique for Reduction of PMEPR*

This method employs modification of peak amplitudes of the signal. It limits the signal amplitude to predetermined threshold value. When the signal value is greater than the threshold value, then for that instance its value is replaced by threshold value else the clipper allows the signal without change. Clipping is easier to implement but has one major disadvantage. Since it is non-linear method, it introduces both in-band and out-band noises which degrades the performance. The clipping of the signal is defined by

$$T(x[n]) = \begin{cases} x[n] & if\ |x[n] \le T \\ TL & if\ |x[n] > T \end{cases} \tag{3}$$

where x[n] is the MCPC signal and TL = Threshold Level.

### 3.3 Partial Transmit Sequence (PTS) Algorithm for Reduction of PMEPR

In the PTS approach [7], the whole input data block are divided into disjoint sub-blocks. Every sub-block is multiplied by a phase weighting element, which can be achieved by the optimization algorithm to reduce the PMEPR factor. The phase scaling factors are decided in such a way that PMEPR is reduced. At the receiver end the original data is recovered by removing phase weighting factors. The schematic of PTS algorithm is shown in Fig. 1.

Consider input block of data $\{A_n, n = 0, 1, 2, \ldots\ldots\ldots, N\}$ defined as a vector. The input vector A is divided into V number of disjoint sets. These sub-joints are of same length without gap in between. The sub-joints are represented by vectors as

$$\{A_v, v = 1.2, \ldots\ldots, V\}$$

where

$$\begin{aligned} A_0 &= [A_1 \ldots\ldots A_{(N/V)} \ldots\ldots 0 \ldots\ldots 00 \ldots\ldots 0]^T \\ a_1 &= [0 \ldots A_{(n-1)(N/V)} \ldots 0 \ldots 00 \ldots 0]^T \\ A &= \sum_v^V A_v \end{aligned} \tag{4}$$

$A_v$ is the sub-blocks which are successively positioned and they are of same range.

Every divided sub-block is multiplied by corresponding complex phase vector. The phase vector is chosen such that PMEPR value is minimized. $b_v = [b_1, b_2, \ldots b_V]$. Choosing values for phase vectors $b_v \varepsilon \{\pm 1, \pm j\}$, is widely used in conventional PTS algorithm. The resultant PTS modified signal is given by

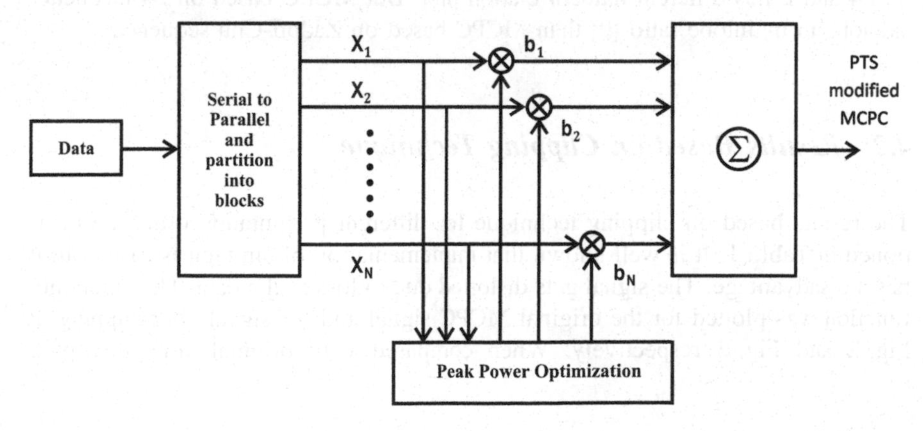


**Fig. 1** Schematic diagram of PTS algorithm for MCPC signal

$$a = \sum_{v-1}^{V} A_v b_v \tag{5}$$

The selection of phase vectors $\{b_v\}_{v-1}^{V}$ is restricted to a set of elements which actually decrease the search operation complexity. The search computational difficulty will accumulate exponentially with the number of sub-blocks. The PTS method based PMEPR performance depends on number of phase factors, the number of sub-blocks and even the sub-block segregation. The results based on PTS algorithm are discussed in the further section.

# 4 Results and Discussion

## 4.1 MCPC Based on Zadoff-Chu Sequence

A Zadoff-Chu [8] sequenced is also termed as Zadoff-Chu sequence (ZC) or Frank Zadoff-Chu (FZC) sequence, which is a complex valued mathematical sequence. Zadoff-Chu sequence is one of the polyphase codes and it has ideal periodic autocorrelation and constant magnitude. The Zadoff-Chu sequence of length L is defined as below, for q = 1 and r = 1.

$$z(k) = \left\{ \begin{array}{ll} e^{\frac{j2\pi r}{L}\left(\frac{k^2}{2} + qk\right)} & for\, L\, Even \\ e^{\frac{j2\pi r}{L}\left(\frac{k(k+1)}{2} + qk\right)} & for\, L\, Odd \end{array} \right\} \tag{6}$$

where k = 0, 1, 2… L − 1

MCPC signal based on Zadoff-Chu sequence was simulated and results were analyzed. An interesting result we observed in MCPC signal based on Zadoff-Chu, it has the same PMEPR values for all permutation orders as of MCPC signal based on P4 and it has different autocorrelation plot. But MCPC based on P4 has better sidelobe to mainlobe ratio [9] than MCPC based on Zadoff-Chu sequence.

## 4.2 Results Based on Clipping Technique

The results based on clipping technique for different permutation order are mentioned in Table 1. It is well known that implementation of clipping is easier but it has a disadvantage. The signal gets distorted due to loss in the data. The ambiguity function was plotted for the original MCPC signal and for signal after clipping in Fig. 2 and Fig. 3 respectively. When compared with original ambiguity plot,

**Table 1** PMEPR results after clipping technique

| Sequence order of phases | PMEPR (Original) | PMEPR (with PTS algorithm) |
|---|---|---|
| MCPC [1 2 3 4 5] | 1.73 | 1.42 |
| MCPC [1 3 4 5 2] | 3.48 | 1.55 |
| MCPC [1 4 5 2 3] | 2.97 | 1.63 |
| MCPC [1 5 2 3 4] | 4.39 | 1.75 |

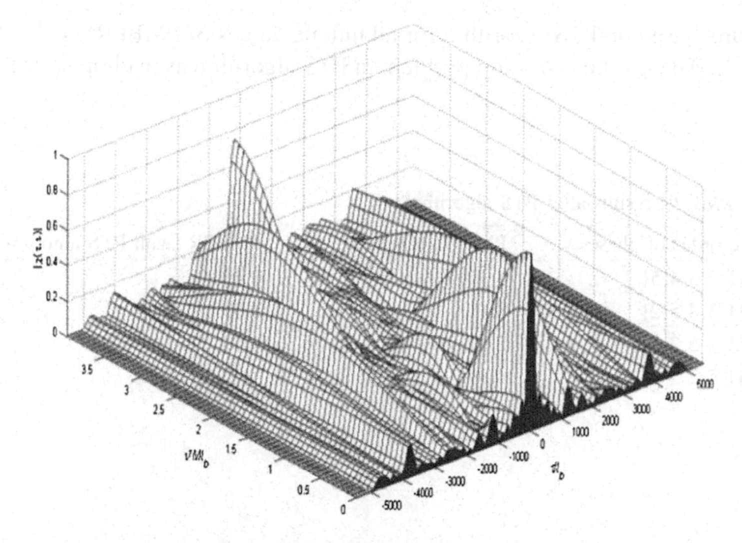


**Fig. 2** Ambiguity plot of original MCPC signal of order [1 2 3 4 5]

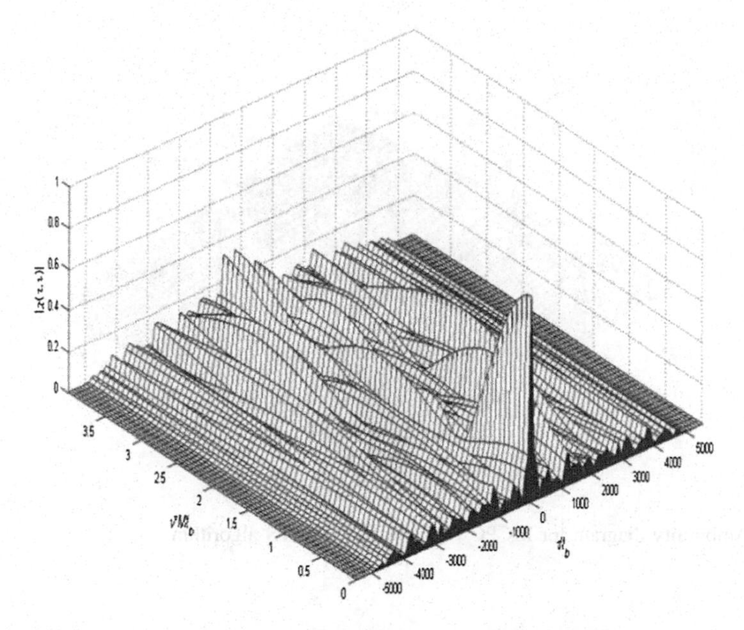


**Fig. 3** Ambiguity plot of MCPC signal after clipping of order [1 2 3 4 5]

the clipping technique plot looks similar to original one with reduced amplitude and sidelobe amplitude is reduced which is an advantage in radar applications.

## 4.3 *Results Based on Partial Transmit Sequence (PTS) Algorithm*

The results based on PTS algorithm for all unique values of PMEPR are mentioned in Table 2. There is no distortion problem in PTS algorithm as in clipping technique

**Table 2** PMEPR results after PTS algorithm

| Sequence order of phases | PMEPR (Original) | PMEPR (with PTS algorithm) |
|---|---|---|
| MCPC [1 2 3 4 5] | 1.73 | 1.42 |
| MCPC [1 3 4 5 2] | 3.48 | 1.55 |
| MCPC [1 4 5 2 3] | 2.97 | 1.63 |
| MCPC [1 5 2 3 4] | 4.39 | 1.75 |

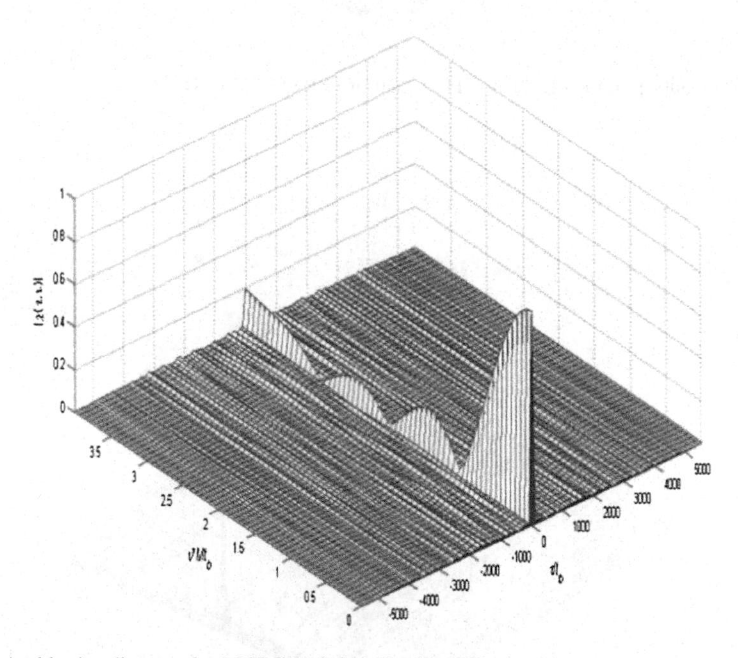


**Fig. 4** Ambiguity diagram for MCPC [1 2 3 4 5] with PTS algorithm

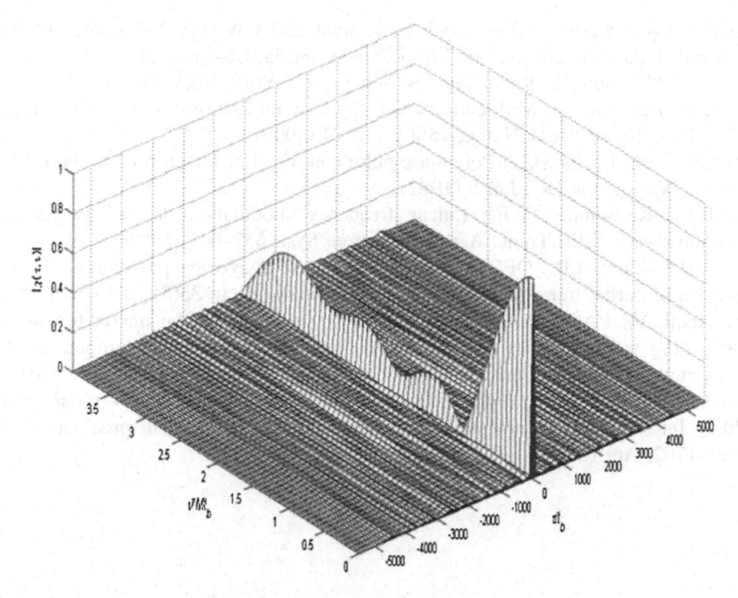

**Fig. 5** Ambiguity diagram for MCPC [3 5 2 1 4] with PTS algorithm

but the complexity increases with increase in number of sub-blocks. The ambiguity plot of the permutation order [1 2 3 4 5] and for the order [3 5 2 1 4] is shown in Fig. 4 and Fig. 5 respectively. Here also we can observe that there is almost suppression of sidelobes which makes it advantageous.

# 5 Conclusion

In this paper, the structure and generation of MCPC signal is explained. The results of MCPC based on Zadoff-Chu are exactly same as of P4. Two algorithms were applied to reduce PMEPR value. Using clipping technique we could successfully reduce PMEPR value and reduce sidelobes efficiently. Clipping technique has disadvantage of loss of information signal, introducing in-band and out-band losses. Hence, another algorithm called Partial Transmit sequence (PTS) was implemented. We could reduce PMEPR value efficiently and nullify the sidelobes in ambiguity plot.

# References

1. Levanon, N.: Multifrequency complementary phase-coded radar signal. IEEE Proc. Radar Sonar Navig. **147**(6), 276–284 (2000)

2. Levanon, N.: Multicarrier radar signals-pulse train and CW. In: Proceedings of the IEEE International Radar Conference, pp. 707–720, Alexandria, USA (2002)
3. Levanon, N., Mozeson, E.: Radar Signals. Wiley, New York, USA (2004)
4. Mozeson, E., Levanon, N.: Multicarrier radar signals with low peak-to-mean envelope power ratio. IEE Proc.-Radar Sonar Navig. **150**(2), 71–77 (2003)
5. Kretschmer, F. Jr., Lewis, B.L.: Polyphase pulse compression waveforms. In: Naval Research Laboratory, Washington DC, Jan 5 (1982)
6. Lewis, B.L., Kretschmer, F.F.: Linear frequency modulation derived polyphase pulse compression codes. IEEE Trans. Aerosp. Electron. Syst. 637–641 (1982)
7. Muller, S.H., Huber, J.B.: OFDM with reduced peak-to-average power ratio by optimum combination of partial transmit sequences. IEEE Electron. Lett. **33**(5), 368–369 (1997)
8. Baig, I., Jeoti, V.: PAPR reduction in OFDM systems: Zadoff-Chu matrix transform based pre/post-coding techniques. In: Second International Conference on Computational Intelligence, Communication Systems and Networks (CICSyN), pp. 373–377. IEEE (2010)
9. Muchandi, V.I., Raghavendra, C.G.: Analysis of multi-tone signal for radar application. In: IEEE International conference on communication and signal processing (ICCSP), pp. 1699–1702, Apr (2016)

# Design and Implementation of High Speed VLSI Architecture of Online Clustering Algorithm for Image Analysis

**M.G. Anuradha and L. Basavaraj**

**Abstract** A novel architecture for computing On-line clustering using moving average method for handling varied dimension data up to eight is proposed. The architecture proposed can perform clustering operation in a single clock cycle for any given dimension. A new method for division is proposed using parallel multiplier architecture and power of two which computes the division operation in single clock cycle. The architecture is tested for its working using Xilinx/ISim tool and the design is implemented using FPGA Spartan 3A.

**Keywords** Clustering · FPGA · K-Means · Fuzzy C-Means algorithm

## 1 Introduction

Clustering is used as data processing technique in many areas such as artificial intelligence, computer vision, data mining, image analysis, image segmentation, pattern recognition and object recognition to gain insight into the data and identify the underlying features. Many clustering algorithms have been developed for different applications. The Clustering of a high dimensional data is a time consuming task and the operation needs to be accelerated for real time environment [1]. In software implementation flexibility and low cost can be achieved but, it difficult to exploit parallelism for high performance applications. In order to accelerate the clustering, a hardware implementation could be used. The related works [2–6, 8–10] presents various architecture for K-Means clustering that can be applied for different applications. However, architectures are based on traditional iterative

M.G. Anuradha (✉) · L. Basavaraj
ATME College of Engineering, Mysore, India
e-mail: anuarun.19@gmail.com

L. Basavaraj
e-mail: principal@atme.in



S.C. Satapathy et al. (eds.), *Data Engineering and Intelligent Computing*,
Advances in Intelligent Systems and Computing 542,
DOI 10.1007/978-981-10-3223-3_18

partitional Forgy's K-means clustering algorithm. Fuzzy C-Means architecture developed in [7] works well for gray scale image but becomes prohibitive for large multi-dimensional data set. The architectures for different clustering proposed so far are built using iterative algorithm for which the input multi-dimensional data needs to be stored. Hence there is memory overhead in the design and due to iterative mechanism, the clustering speed is reduced. The above drawbacks can be reduced by using the Mac Queen's K-Means algorithm or the online clustering algorithm which is one of the oldest clustering techniques. The online clustering algorithm performs only one complete pass through data. Also the input data can be discarded after computation of centroid. The online clustering architecture developed in [10] also provide low power solution and the divider developed uses parallel subtractors and comparators and uses pipeline approach to obtain the throughput of one clock cycle per dimension. The proposed new architecture computes moving average of the new centroid where all the dimensions can be processed in a single clock cycle using the divider and an adaptable hardware.

## 2 On-Line Clustering Algorithm

The Mac-Queens K-Means algorithm or the On-line Clustering is very popular algorithm that updates the cluster center immediately after every assignment of the data. The cluster size K is to be specified by the user and it may vary from 1 to 16. The algorithm steps used to build the architecture is briefed in the following section considering N data points with each data point xi being a d-dimensional data vector with d varying from 1 to 8 represented as in (1).

$$\mathrm{xi} = \{\mathrm{x1}, \mathrm{x2}, \ldots\ldots\ldots\ldots\ldots \mathrm{xd}\}. \tag{1}$$

1. Begin with K clusters C1, C2, Ck, each consisting of one of the first K data points x1, x2,…xk.
2. For each of the (N-K) samples, find the centroid near it using Squared Euclidean distance. The distance between a data point xi and the centroid is computed as

$$D(xi, Ci) = \Sigma_{(i=1)}^{d} (xi - Ci)^2. \tag{2}$$

3. Put the sample in the cluster identified with the nearest centroid according to the nearest mean center.
4. For an input xi, if the nearest centroid is cj and the number of input nearer to centroid cj is nj then the new centroid using moving average method is given as

$$Cj(new) = \frac{1}{nj} xi + \left(1 - \frac{1}{nj}\right) Cj. \tag{3}$$

5. Go through the data for the second time and for each sample, find the centroid near it. Put the sample in the cluster identified with nearest centroid but do not recompute the centroid.

## 3 Proposed Hardware Architecture

The proposed hardware architecture works on the 64 bit input line with each dimension of the input occupying 8 bits and hence the 8'd data exactly fit the 64 bit input line. The architecture can handle up to 16 clusters. The blocks of the architecture are illustrated in Fig. 1 and the subsequent sections give the internal details of each of the block.

### *3.1 Temporary Register and Centroid Memory*

The input data Data-in is stored in a temporary register of 64 bits with each dimension occupying 8 bits so that exactly 8'D data each of 8 bits exactly fit the register. The number of dimensions in a data may vary from 1 to 8. Initial K data inputs are stored in the centroid memory from the temporary register. The data from the temporary register is stored in consecutive 8 memory locations of the centroid memory. Since the number of clusters may vary from 1 to 16, the centroid memory size is fixed to 128 bytes. After storing initial K centroids, the remaining (N-K) data must be compared with the K centroids. This is performed using square distance calculator.

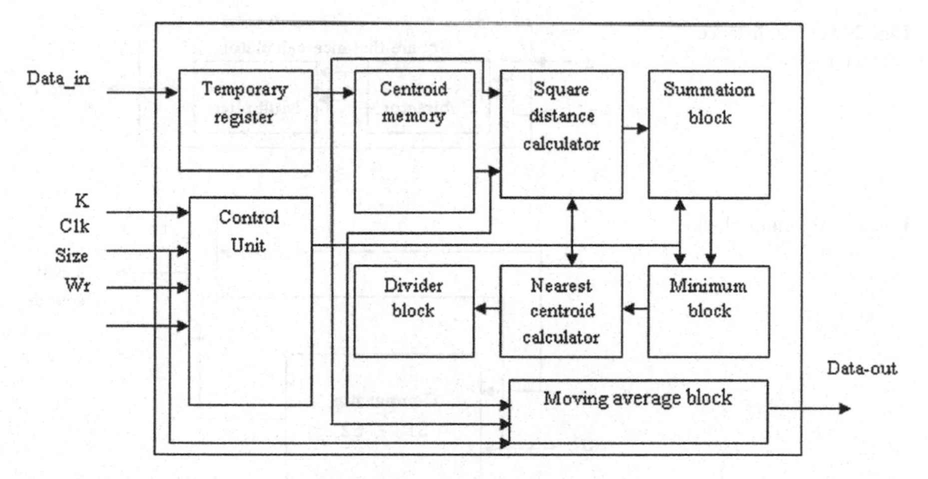


**Fig. 1** Proposed architecture of online clustering algorithm

### 3.2 Square Distance Calculator (SDC) and Summation Block

The distance between the input data should be compared with all the centroids. To accommodate all 16 centroids, 128 square distance calculator is required. An SDC used for computing square distance of one dimension is illustrated in Fig. 2. Depending on the dimensions and the number of centroids, totally size × K number of square distance calculators are utilized. The SDC computes the squared distance between 1'D of data with 1'D of centroid. Eight SDC are hence required for comparing the data with one centroid. If the data dimension is less than eight, then the SDC required is reduced.

The Summation block is used to obtain the distance between the data and the centroid. The distance of 1'D data obtained from the SDC must be added to find the overall distance between the data and the centroid. Hence dimension wise adding of the data is done using the summation block. Since there are maximum of 16 centroids, we require 16 summation unit to add eight 8 bits of data to produce 16 bit sum.

### 3.3 Minimum Block and Nearest Centroid Calculator

The data must be grouped to the nearest centroid. The comparison of the two sums using minimum block is illustrated in Fig. 3 where the data from summation block are compared to obtain the nearest centroid. The similar 14 minimum blocks are cascaded to compute the nearest centroid and place in the particular group.

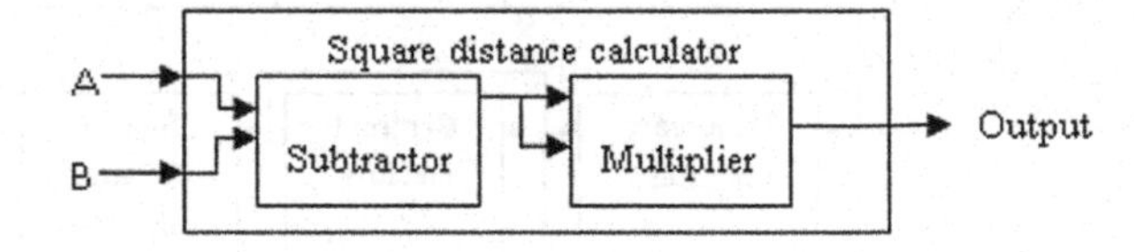


**Fig. 2** Square distance calculator

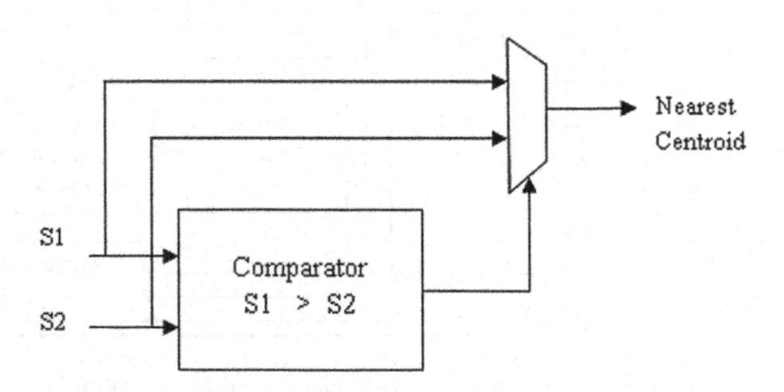


**Fig. 3** Minimum block

### 3.4 Divider Block

For computing new centroids using moving average method, it is required to calculate 1/N and 1-1/N. These computations are accomplished using floating point divider. To speed up the calculation, a new algorithm is developed to obtain 4 bit precision quotient in a single clock cycle. The algorithm for computing 4 bit precision quotient is given below.

1. Set the precision Pr to be initially 1.
2. Check the value of N. If $N < 10$, then set the Numerator $Nr = 10$, else if $N < 100$ set the Numerator $Nr = 100$. Similarly check for N less than 1000, 10000, 100000 and 1000000 and correspondingly set the Numerator value Nr to be 1000, 10000, 100000 and 1000000 respectively.
3. Set current iteration value i to be 1.
4. Perform $N \times i$
5. Check if $N \times i$ is less than Nr. If true then increment i and go to step 4.
6. Set the quotient $Q_{pr}$ to be $i - 1$ and remainder R to be Nr mod N.
7. Check if precision Pr is less than or equal to 3. If true then increment Pr and go to step 3.
8. Output the four bit precision quotient mantissa $Q_1Q_2Q_3Q_4$ and exponent $10^{-4}$.

The four bit quotient obtained is in the BCD form. Hence we need to convert the BCD quotient to binary for further processing. The process of conversion again involves multipliers and adders. The binary equivalent of the quotient is obtained by computing $Q_{fianl} = (Q_1 \times 1000) + (Q_2 \times 100) + (Q_3 \times 10) + Q_4$. The obtained quotient $Q_{fianl}$ is then used to compute 1-1/N.

N in 1/N computation denotes the number of pixel in a given cluster. Since the maximum image size we have considered is $1024 \times 1024$, the maximum pixels that need to be processed are $2^{20}$. Since the minimum number of clusters is two, the denominator in 1/N can never exceed the number of pixels. Hence we limit ourselves to check the value of N to be less than 10 lakh in step 2. The algorithm can be modified to set the higher numerator value if the image size is larger. The block diagram representation of the divider is as shown in Fig. 4.

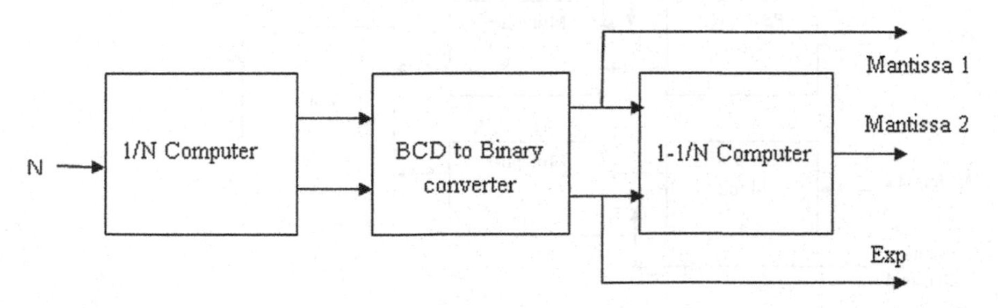


**Fig. 4** Divider block

### 3.5 Moving Average Block

The new centroid is calculated using moving average method given by (3) and the block diagram representation of moving average block is illustrated in Fig. 5. The factors 1/N and 1-1/N in (3) must be multiplied to each dimension. The dimensions may vary from 1 to 8. Each dimension is 8 bit wide. 1-1/N and 1/N are 4 bit precision floating point numbers represented as 16 bit mantissa and four bit exponent. The centroids and input is eight bit wide and hence there is incompatibility in multiplication. To accomplish this, we make use of sum of power of two which performs division using shift operation. Mantissa of 1/N and 1-1/N are divided by $2^{13}$ which approximates the 4 bit exponent value of the floating point number to fit the result in 8 bits after multiplication with data and previous centroid. Hence this division operation is also performed in single clock cycle.

## 4 The Experimental Results

The experiment contains four parts. The first part is the algorithm verification for varied dimension data and centroid using Xilinx ISim simulator. The second part is the visual results of image clustering.

The third part is the architecture verification of each module using FPGA Spartan 3A and the fourth part is the comparison of our work with the previous works.

The algorithm was verified by an exhaustive test bench which provided different dimension data and different number of centroids. The simulation result in Fig. 6 shows that the data is clustered to its nearest centroid as it arrives in a single clock cycle for 2 dimensional data with K = 2. The architecture was tested taking 256 samples of different dimensions and various values of K.

The simulation in Fig. 7 illustrates the assignment of data to the computed centroids in a single clock cycle for 2'D data with K = 2 in 2nd pass of the online

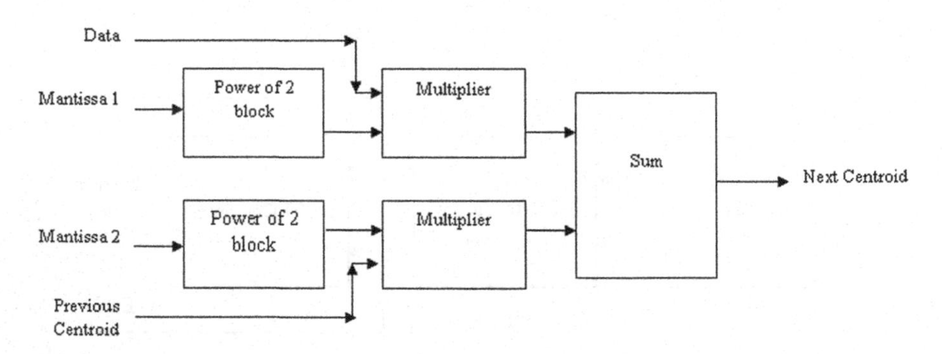


**Fig. 5** Moving average block

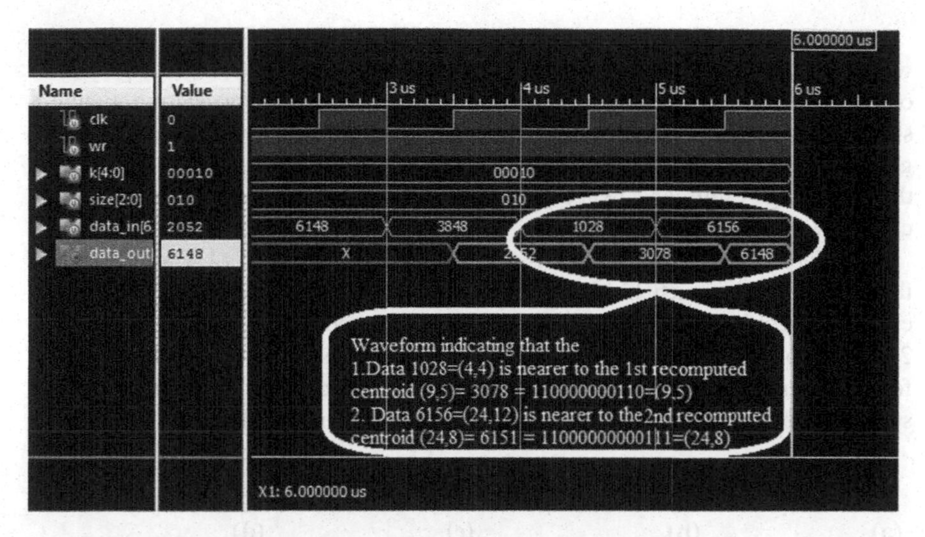


Fig. 6 Recomputation of nearest centroid in single clock cycle for the data

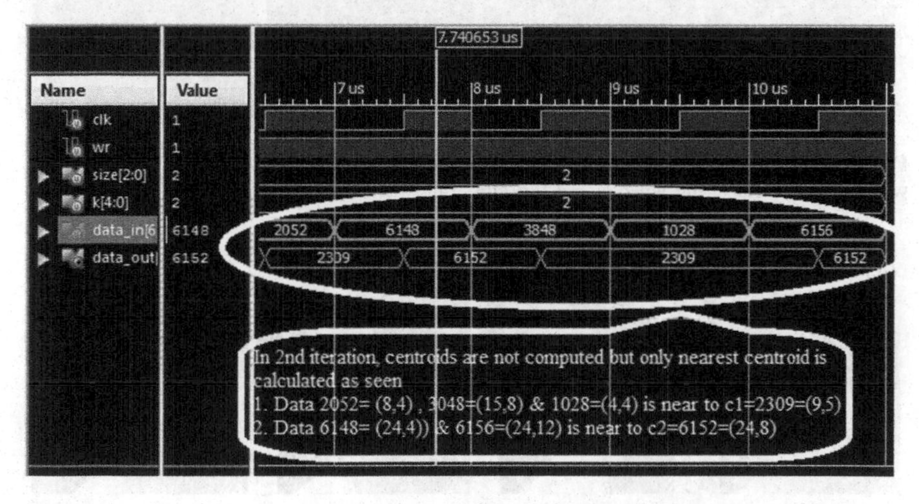


Fig. 7 Computation of the data to its nearest centroid in 2nd iteration of online clustering

clustering algorithm. The second part is the verification of the architecture developed for image analysis. The architecture was verified for both color and gray scale image.

The image data was stored in an external memory and one pixel per clock cycle was given to the proposed architecture for verification. The clustered data was again put back in an external memory for further analysis. For color image clustering, the data is taken as 3'D data with each dimension data being eight bits wide so as to form 24 bit full color pixel. After clustering, the color of each image pixel is represented by the color of its corresponding centroid.

For gray image clustering, the data is considered as single dimension of eight bit wide with each pixel having varied gray-scale levels. After clustering, the gray level of each pixel is represented by the gray level of its corresponding centroid. Figure 8 shows color clustering results for different values of K and Fig. 9 illustrate the gray-scale clustering results for varied values of K. It is seen for both the cases that the clustering results and hence the quality of image is better as the number of clusters K is increased.

The Table 1 shows the comparison result of the number of LUT's utilized and the time taken by each block of the architecture. As seen, one square distance calculator requires 13.8 ns to compute 1'D data and summation block requires 21 ns for computation of sum. Also the divider block used in the design takes 26 ns for computing 4 bit precision floating point quotient and hence can be used for high speed clustering or any DSP application where speed is the criteria.

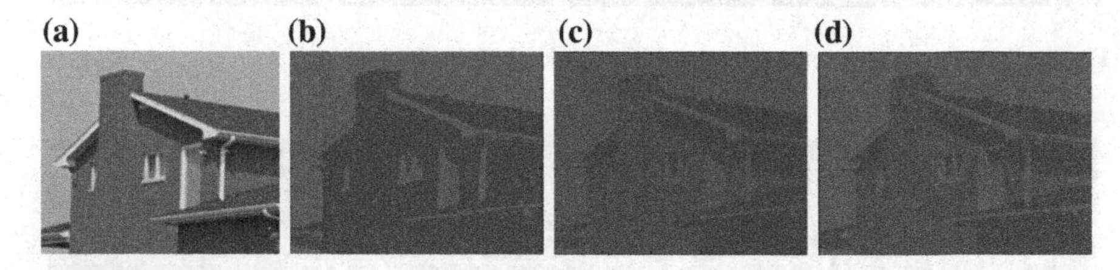


**Fig. 8** Example of color image clustering. **a** Original image "House" 256 × 256, **b** clustering when K = 8, **c** clustering when K = 12, **d** clustering when K = 16

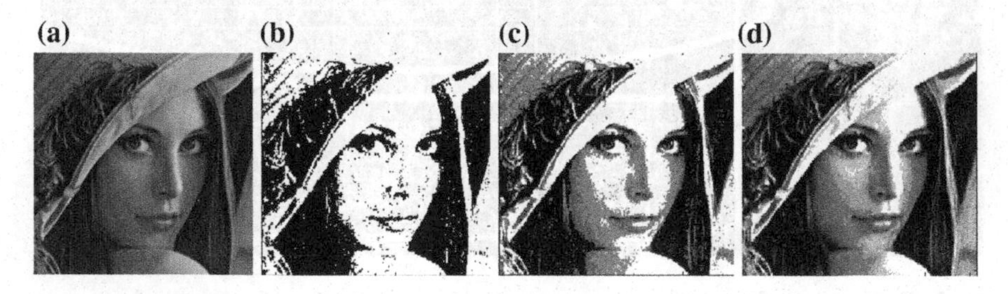


**Fig. 9** Example of image clustering. **a** Original image "Lena" 256 × 256, **b** clustering when K = 2, **c** clustering when K = 4, **d** clustering when K = 16

**Table 1** Comparison of timing and LUT's utilized for different blocks in the architecture

| Sl No. | Block | No. of LUTs utilized | Time (ns) |
|---|---|---|---|
| 1. | Square distance calculator | 8 out of 7168 | 13.82 |
| 2. | Summation block | 106 out of 7168 | 21.705 |
| 3. | Minimum block | 301 out of 7168 | 51.996 |
| 4. | Divider block | 129 out of 7168 | 26.16 |

**Table 2** Comparison of the proposed architecture with previous architecture

| | ICPACS 2011 [11] | ISCAS 2013 [10] | This work |
|---|---|---|---|
| Implementation | FPGA XC2VP100 | TSMC 90 nm | FPGA Spartan 3 |
| Algorithm | Self-organizing map | On-line clustering | On-line clustering |
| Distance measurement | Manhattan | Manhattan | Squared Euclidean |
| Number of clusters | 16 | 1–768 | 1–16 |
| Vector dimension | 3 | 1–768 | 1–8 |
| Module for division | Barallel shifter | Pipelined architecture | Parallel multiplier architecture |
| No. of clock cycles for division operation | Less than 10 cycles | 1 cycle | 1 cycle |
| Maximum throughput | Not specified | 1 dimension/cycle | 8 dimension/cycle |
| Target image size for color clustering | 640 × 480 pixels | 256 × 256 pixels | Support up to 1024 × 1024 pixels |

The fourth part is the comparison of our work with the previous literature. This work is compared with work of [10, 11]. The proposed work supports higher dimensions of input vectors and performs division using single clock cycle using parallel multiplier architecture and achieves higher throughput of 8 dimensions/cycle compared to the previous architectures as indicated in Table 2. Moreover, the proposed work has resource adaptive mechanism to handle vectors with different dimensions efficiently.

# 5 Conclusion

The architecture proposed is a solution for high speed clustering application used in varied multimedia application like pattern recognition, image segmentation, video analysis and data mining. A new method to compute floating point division is proposed using parallel multiplier and power of 2 blocks for computing new centroids which operates in single clock cycle and hence the each incoming data can be processed in a single clock cycle irrespective of the number of dimension. The experiment shows that the architecture formulated can handle up to 16 clusters with varied data dimension from 1 to 8. Hence the proposed architecture is a best solution for high speed image analysis.

**Acknowledgements** Author Anuradha M.G. working as Asst Prof, Department of ECE, JSSATE, Bengaluru and research scholar, ATME, Mysore, would like to thank the Management, Principal and the Department of ECE, JSSATE, Bengaluru for providing the technical and the moral support.

## References

1. Saegusa, T., Maruyama, T.: An FPGA implementation of real-time K-Means clustering for color images. J. Real-Time Image Process. **2**(4), 309–318 (2007). Clerk Maxwell, A.: Treatise on Electricity and Magnetism, 3rd edn., vol. 2, pp. 68–73. Oxford, Clarendon, (1892)
2. Filho, A.G.D.S., Frery, A.C., de Araújo, C.C., Alice, H., Cerqueira, J., Loureiro, J.A., de Lima, M.E., Oliveira, M.D.G.S., Horta, M.M.: Hyperspectral images clustering on reconfigurable hardware using the K-means algorithm. In: Proceedings of Symposium on Integrated Circuits and Systems Design, pp. 99–104 (2003)
3. Maruyama, T.: Real-time K-Means clustering for color images on reconfigurable hardware. In: Proceedings of International Conference on Pattern Recognition, pp. 816–819 (2006)
4. Chen, T.-W., Sun, C.-H., Bai, J.-Y., Chen, H.-R., Chien, S.-Y.: Architectural analyses of K-means silicon intellectual property for image segmentation. In: Proceedings of IEEE International Symposium on Circuits Systems, pp. 2578–2581 (2008)
5. Chen, T.-W., Chien, S.-Y.: Bandwidth adaptive hardware architecture of K-means clustering for video analysis. IEEE Trans. Very Large Scale Integr. (VLSI) Syst. **18**(6), 957–966 (2010)
6. Chen, T.-W., Chien, S.-Y. (Member, IEEE): Flexible hardware architecture of hierarchical K-means clustering for large cluster number. IEEE Trans. Very Large Scale Integr. (VLSI) Syst. **19**(8) (2011)
7. Li H.Y., Hwang W.J., Chang. C.Y.: Efficient fuzzy c-means architecture for image segmentation. Sensors **11**, 6697–6718 (2011)
8. Chen, T.-W., Sun, C.-H., Su, H.-H., Chien, S.-Y., Deguchi, D., Ide, I., Murase, H.: Power-efficient hardware architecture of K-means clustering with Bayesian-information-criterion processor for multimedia processing applications. IEEE J. Circuits Syst. **1**(3), 357–368 (2011)
9. Hernanez, O.J.: High performance VLSI architecture for the histogram peak climbing data clustering algorithm. IEEE Trans. Very Large Scale Integr. (VLSI) Syst. **14**(2), 111–121 (2006)
10. Chen, T.-W., Ikeda, M.: Design and implementation of low-power hardware architecture with single-cycle divider for on-line clustering algorithm. IEEE Trans. Circuits Syst. **60**(8), 2168–2175 (2013)
11. Oba, Y., Yamamoto, K., Nagai, T., and Hikawa, H.: Hardware design of a color quantization with self-organizing map. In: Proceeding of International Symposium on Intelligent Signal Processing and Communications Systems 1–6 (2011)

# Entity Level Contextual Sentiment Detection of Topic Sensitive Influential Twitterers Using SentiCircles

**Reshma Sheik, Sebin S. Philip, Arjun Sajeev, Sinith Sreenivasan and Georgin Jose**

**Abstract** Sentiment analysis, when combined with the vast amounts of data present in the social networking domain like Twitter data, becomes a powerful tool for opinion mining. In this paper we focus on identifying 'the most influential sentiment' for topics extracted from tweets using Latent Dirichlet Allocation (LDA) method. The most influential twitterers for various topics are identified using the TwitterRank algorithm. Then a SentiCircle based approach is used for capturing the dynamic context based entity level sentiment.

**Keywords** Sentiment analysis · LDA · Twitterers · Twitterrank · SentiCircle · Entity level sentiment

## 1 Introduction

The emergence of Microblogging services as a medium for rapid information exchange and as a marketing platform is the main advancement of this modern era. Twitter is a popular microblogging service which employs a social-networking model called "following". In one instance of "following" relationship, the twitterer

R. Sheik (✉) · S.S. Philip · A. Sajeev · S. Sreenivasan · G. Jose
Department of Computer Science and Engineering, TKM College of Engineering, Kollam, Kerala, India
e-mail: reshmasheik@tkmce.ac.in

S.S. Philip
e-mail: sebinsphilip@gmail.com

A. Sajeev
e-mail: arjunsajeev04@gmail.com

S. Sreenivasan
e-mail: siniths91@gmail.com

G. Jose
e-mail: georginjosemaliekal@gmail.com



S.C. Satapathy et al. (eds.), *Data Engineering and Intelligent Computing*, Advances in Intelligent Systems and Computing 542,
DOI 10.1007/978-981-10-3223-3_19

whose updates are being followed is called the “friend”, while the one who is following is called the “follower”.

Two common approaches used for sentiment analysis are supervised learning and lexical analysis. Supervised learning techniques [1] are highly dependent on training data and due to the lack of any order or pattern to the data present in the social networking domain, they tend to give poor results. Lexicon based approaches on the other hand blindly use predefined sentiment scores of words and ignore the contextual sentiment. SentiCircle is an improved version of the lexicon based approach which uses dynamic sentiment scores based on context and hence provides more accurate results.

Approaches to Twitter sentiment analysis tend to focus on the identification of sentiment of individual tweets (tweet-level sentiment detection). In this paper, we are interested in identifying sentiments of the influential twitterers using a context based sentiment analyzer. The benefit of solving this problem is multifold.

First, it brings rank to the real time web which allows the search results to be sorted by the influence of the contributing twitterers and their sentiments towards the search term giving a timely update of the thoughts of influential twitterers. Second, since Twitter is also a marketing platform, targeting those sentiments of influential users will increase the efficiency of the marketing campaign.

## 2 Proposed Work

The tweets of the top 100 twitterers are obtained by parsing twitaholic.com. The topics are extracted using LDA (Latent Dirichlet Allocation) [2] method. TwitterRank algorithm [3] uses a page rank based method to determine the most influential twitterers. SentiCircle [4] which is an improved lexical sentiment analysis method which considers contextual sentiment is used to perform entity level sentiment analysis on the tweets of the most influential twitterers. The workflow of the proposed work is shown in Fig. 1.

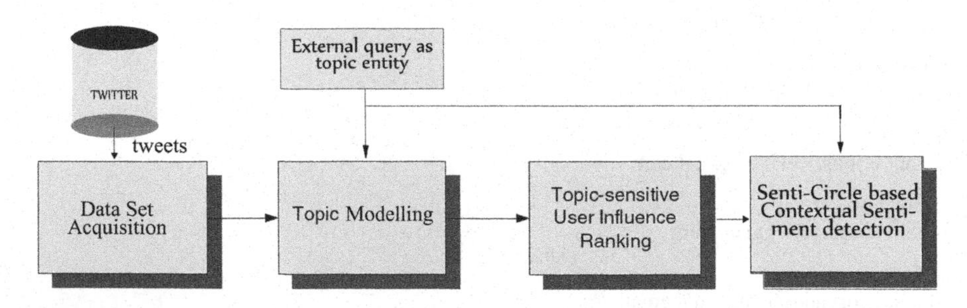


**Fig. 1** Workflow of the proposed work

## 2.1 Data Set Acquisition

Most recent 200 tweets from the top 100 twitterers were obtained based on their follower count from twitaholic.com. The information about their follower list and tweets count is stored in separate files.

## 2.2 Topic Modelling

Topic modelling uses a topic distillation technique that automatically identify the topics that twitterers are interested in based on the tweets they published. The Latent Dirichlet Allocation (LDA) [2] model is performed on the 100 documents, and each document is represented as a probability distribution over some topics, while each topic is represented as a probability distribution over a number of words.

Thus a document can be regenerated from the LDA model as each document is a probability distribution over a no. of topics. Sample a topic from the distribution. Similarly sample a word from each topic as it is a distribution over a no. of words. The repeated sampling creates a document. The main output of the above topic-distillation process is: DT, a D $\times$ T matrix, where D is the number of twitterers and T is the number of topics. $DT_{ij}$ contains the number of times a word in twitterer $s_i$'s tweets has been assigned to topic $t_j$.

Alternately the user can enter the query and construct the DT matrix based on the term frequency from the tweet collection.

## 2.3 Identification of Influential Twitterers

The topics extracted from tweets by LDA or a specific user given topic is used to model DT matrix. By using DT matrix we can pick up the most influential users for each topic.

**Metric for influence measure**. The concept of PageRank algorithm can be used in the sense that a twitterer has high influence if the sum of influence of her followers is high; at the same time, the influence on each follower is determined by the relative amount of content the follower received as shown in Fig. 2.

Here influence (A) = influence (B) + influence (C) + influence (D).

**Topic-specific influential Twitterers**. The topic-specific TwitterRank [3] for each topic t is computed using a random surfer model which makes use of a transition matrix $P_t$. Random surfer visits each *twitterer* with certain probability by following the appropriate edge in *D*. Thus performing a topic-specific random walk, which is essentially the construction of a topic-specific 'follower relationship' network. The transition matrix for topic *t*, denoted as $P_t$ is defined as:

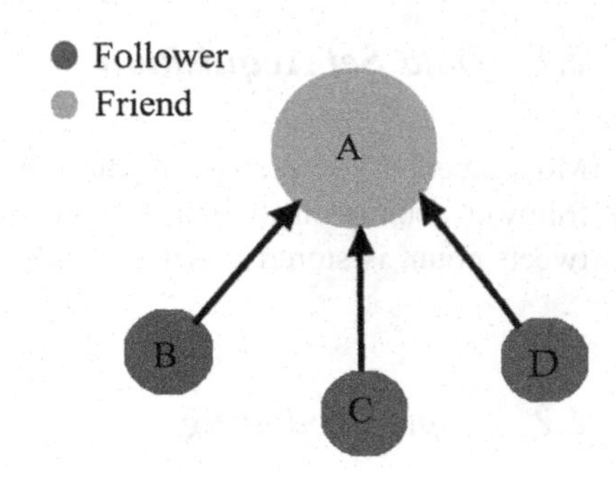


**Fig. 2** Example of a twitter graph D (V,E) with "following" relationship

$$P_t = \frac{|T_j|}{\sum\limits_{a:\, S_i follows S_a} |T_a|} * sim_t(i,j) \tag{1}$$

where each element in the $P_t$ is the transition probability for the surfer to move from follower $s_i$ to friend $s_j$. |Tj| is number of tweets published by $s_j$.

$sim_t(i, j)$ is the similarity between $s_i$ and $s_j$ in topic t, which is defined as:

$$sim_t(i,j) = 1 - \left|\overline{DT}_{it} - \overline{DT}_{jt}\right|. \tag{2}$$

To avoid loops, a teleportation vector $E_t$ is also introduced, which basically captures the probability that the random surfer would "jump" to some twitterers instead of following the edges of the graph D. $E_t$ is defined as follows.

$$E_t = \overline{\overline{DT}}_{.t}. \tag{3}$$

where $\overline{\overline{DT}}_{.t}$ is the tth column of the column normalized DT matrix, which records the number of times a word in the tweets has been assigned to topic t.

The TwitterRank for topic t can be calculated literately as:

$$\underset{TR}{\longrightarrow} = \gamma P_t \times \underset{TR_t}{\longrightarrow} + (1-\gamma)E_t. \tag{4}$$

$\gamma$ is a parameter between 0 and 1 to control the probability of teleportation.

**Most influential Twitterer for a user given topic**. TwitterRank vectors $TR_t$, basically measure the twitterers' influence in individual topics t. An aggregation of TwitterRank can also be obtained to measure twitterers' overall influence. So for a user given query with w words, we consider each word in the query as individual topic and find corresponding $TR_t$. Finally all $TR_t$ are combined together to get the influential twitterer on the given query. It can be calculated as:

$$\underset{TR}{\longrightarrow} = \sum_t r_t \times \underset{TR_t}{\longrightarrow} \tag{5}$$

$r_t$—weight assigned to each topic/word.

## 2.4 Sentiment Detection

The entity-level sentiment detection on twitter will give us the overall contextual semantics and sentiment of the given entity from the tweets of the influential twitterer, rather than individual tweet level sentiment classification. The input to this stage is the tweets of the most influential twitterer and the query given by the user in finding the influential twitterer as the entity.

**Tweets Preprocessing**. Tweet-NLP [5] developed by ARK Social Media Search tokenizes the tweet and returns the POS tags of the tweet along with the confidence score. It tags the twitter specific entries like Emoticons, Hashtag and Mentions etc. making it easy to work with tweets. The words that are non-english in nature are filtered out in this stage. Emoticons are replaced with its associated sentiment by means of a manually created emoticon dictionary. The acronyms in the tweets are replaced with their associated expansions from an online acronym list [6].

The URL's seen in the tweets usually do not convey any special sentiment, and hence they are removed. Tweets may mention another twitterer by means of special strings prepended with a '@' symbol and hashtags are topic words prepended with a '#' symbols, are removed from the rest of the text for sentiment detection. Tweets consist of various notions of negation. In general, words ending with 'nt' are appended with a not. Again 'not' is replaced by the word 'negation'. Negation plays a very important role in determining the sentiment of the tweet. When a negation is detected we reverse the sentiment of the remaining terms after the negation.

Twitter being a dynamic platform for expressing opinions and emotions, people often repeat the characters in certain 'highly emotional words' to stress its importance. For example the word 'cool' could be used as 'coooool' to show the importance of the word. We have taken this fact into consideration and removed the repeated characters from the word (if it repeats more than 3 times). We added a weight factor along with the original sentiment score of the word from AFINN-111 [7] dictionary, proportionate to the repeating sequence length. We filter out prepositions and pronouns, as it does not contribute any sentiment value. The stop-words from the tweet are not removed completely as such, as they might carry some important sentiment information with them. Instead we created another list from the standard stop word list, by filtering out the sentiment bearing terms and used this to filter out unwanted terms from tweets.

**Approach**. For sentiment detection of a term SentiCircle [4] based method is used to detect contextual semantic orientation of a term. The workflow of the sentiment detection is shown in Fig. 3.

**Data Structures Used**

*Term-Index.* A dictionary of terms in the whole tweets after preprocessing is generated such that value of each term is two tuple consisting of the index of the tweets in which the term appeared and the prior sentiment score of the term from AFINN-111 [7] dictionary.

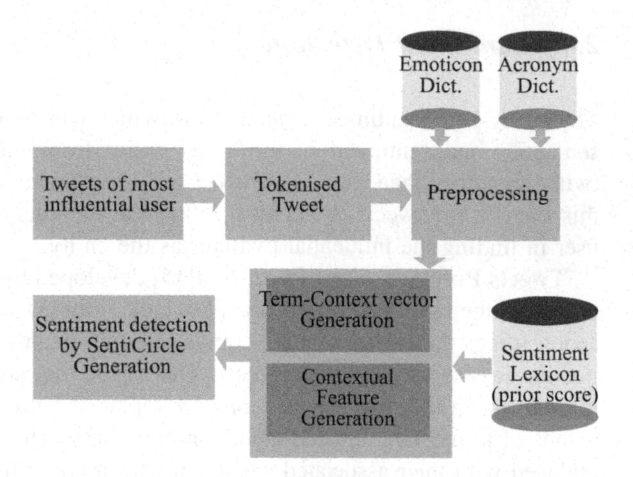


**Fig. 3** Entity-level sentiment detection system

*Term-Context Vector.* A Term-Context vector dictionary is created from Term-Index, for each term having value as list of all context terms appeared along with the key term in the tweets.

*Contextual-Feature.* A Contextual Feature is generated for each term from Term-Context vector and Term-Index vector. Each term has value as a 3 tuple list of all context terms along with their prior sentiment score (θ) and TDOC (Term Degree Of Correlation). TDOC gives the degree of relation between the term and the context term. The prior sentiment score for each term is obtained from the AFINN-111 dictionary and the TDOC is calculated as:

$$TDOC(m, c_i) = f(c_i, m) \times \log\left(\frac{N}{Nc_i}\right). \tag{6}$$

**SentiCircle for dynamic sentiments.** The context vector of a term m is converted into 2D geometric circle, which is composed of points denoting the context terms of m. Each context term is located in the circle based on its angle (defined by its prior sentiment), and its radius (defined by its degree of correlation with the term m).

$$\begin{aligned} r_i &= TDOC(m, c_i), \\ \theta_i &= Prior\_Sentiment(c_i) \times \pi. \end{aligned} \tag{7}$$

where (r,θ) is the polar coordinate of a context term on the circle. The SentiCircle in the polar coordinate system can be divided into four sentiment quadrants as shown in Fig. 4.

Y-axis in the Cartesian coordinate system defines the sentiment of the term, i.e., a positive y value denotes a positive sentiment and negative y for negative sentiment. The X-axis defines the sentiment strength of the term. The smaller the value

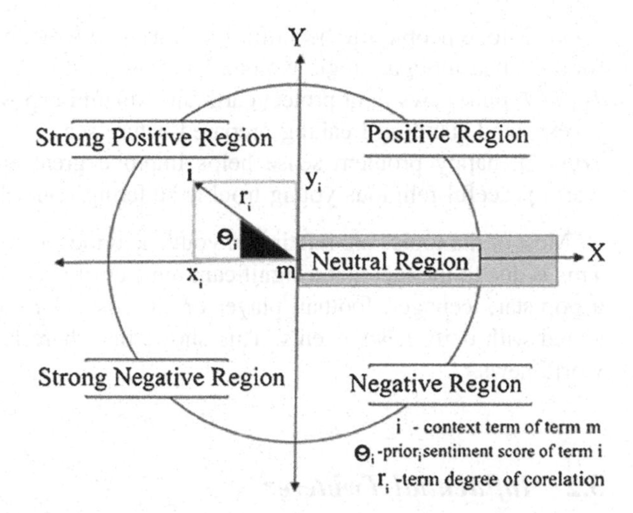


**Fig. 4** SentiCircle representation of term 'm'

of x, the stronger is the sentiment. Besides, a small region called the "Neutral Region" can be defined, near +ve X axis to represents a weaker sentiment.

**Entity-level Sentiment detection**. The overall sentiment of the SentiCircle is summarized into a single value by finding the geometric median of the Context terms in the sentiment circle which denote the sentiment of the term.

# 3 Results and Discussions

The dataset doesn't need to be very big but users in this database should have close relationship, because in this case we can ensure most of users have influence on others. The most recent 200 tweets from top 100 users were obtained by parsing the site 'http://twitaholic.com/' and storing it in a separate file. The users were classified into top 100 based on their follower count, which ensures high inter-relationship.

## *3.1 Topic Distribution*

After obtaining the feature matrix for 100 users the sample output of the LDA process (Topic modelling) as:

*Topic 1*: show live coming ready head full weeks run home set enter concert single case told people air catch bowl
*Topic 2*: love day time great good happy show special family friends big meet photo story friend true man listen long

*Topic 3*: love people life beautiful girl kids give women heart ready bring children forget called thought single woman bad body
*Topic 4*: pink news fight protect earth air explain deep woods problems peace river invest stand standing freaking prepared nature sea
*Topic 5*: happy problem sense helps improve great enjoy peace enjoying place warm peaceful religious young trouble suffering control teachings motivation.

Most of the topics are relating to youth activities and life events, expressions etc. This is due to the fact that a significant number of users in top 100 user list is either a pop star, teenager, football player or an artist. Here topic 4 denotes word associated with daily news events. This shows that there is a user account relating to world news.

## 3.2 *Influential Twitterer*

The most important part of this work is to find the influential twitterer for a user given query. Let the user query be 'trump'. The Fig. 5 shows the distribution of the word 'trump' in the entire tweet set for the set of influential twitterers.

The most influential twitterer for the topic 'trump' is 'CNN', which is the user account name of the world news channel CNN news. There is total of 48 occurrences of the word 'trump' in the entire 200 tweets of the user 'CNN'. This is the twitterer mentioning 'trump' most frequently.

## 3.3 *Sentiment Detection*

This stage detects the sentiment associated with the given topic words from the tweets of the identified influential twitterer. The tweets of CNN (only 5 out of 48 shown) associated with trump are (comma separated):

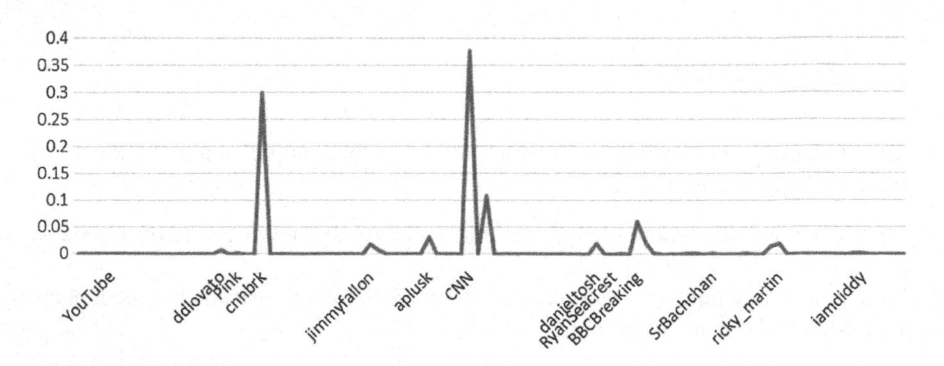


**Fig. 5** TwitterRank distribution for 'trump'

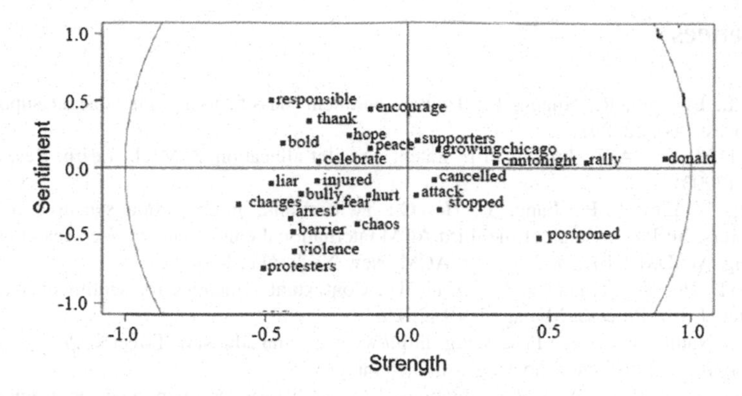


**Fig. 6** Example SentiCircle for "trump"

Donald Trump "is a bully," says man who tried to rush stage in Ohio over the weekend, Bernie Sanders hits Donald Trump's plan to build a border wall at the #DemTownHall, Hillary Clinton at the #DemTownHall: Donald Trump is "trafficking in hate and fear", @RealBenCarson warns @realDonaldTrump protesters: 'There could be an escalation', The confusion that's driving support for @realDonaldTrump via @CNNOpinion.

We have removed points near the origin for easy visualization as shown in Fig. 6. Dots in the upper half of the circle represent terms bearing a positive sentiment while dots in the lower half are terms bearing a negative sentiment.

The term 'donald', having high degree of correlation with 'trump' is observed (TDOC = 1).This is because 'Donald Trump' is the name of the US president candidate in 2016. Wherever the word 'trump' is used, it is prefixed by 'donald'. The overall sentiment detected by our system is 'weak negative' for the term 'trump'. Since there are a no. of 'neutral' context terms associated with 'trump' the sentiment strength is 'weak'.

## 4 Conclusions and Future Work

The proposed approach for sentiment analysis of Twitter data combines determination of influential twitterers and usage of a dynamic context based lexical sentiment analysis method. It gives more accurate results when compared to supervised learning or lexical methods. Social networking data enables marketing companies and analytics firms to gain market insights and make informed decisions. The sentiment based analyzer proposed in this paper can be used in such commercial applications. The efficiency of the sentiment analyzer can be improved by considering retweets, mentions reply's etc. in determining topical similarity ranking. Real time data analysis methods can also be integrated into the analyzer to provide real time social media analytics.

## References

1. Go, A., Bhayani, R., Huang, L.: Twitter sentiment classification using distant supervision. J. Process. 1–9 (2009)
2. Blei, D.M., Ng, A.Y., Jordan, M.I.: Latent dirichlet allocation. J. Mach. Learn. Res. **3**, 993–1022 (2003)
3. Weng, J., Lim, E.-P., Jiang, J., He, Q.: TwitterRank: finding topic-sensitive influential twitterers. In: Proceedings of the third ACM international conference on Web search and data mining (WSDM '10), pp. 261–270. ACM, New York, NY, USA (2010)
4. Saif, H., He, Y., Fernandez, M., Alani, H.: Contextual semantics for sentiment analysis of Twitter. J. Inf. Process. Manag. 5–19 (2016)
5. Twitter Natural Language Processing. http://www.cs.cmu.edu/~ark/TweetNLP/
6. Acronym list. http://www.noslang.com/dictionary/
7. Nielsen, F.A.: A new ANEW: evaluation of a word list for sentiment analysis in microblogs. In: Proceedings of the ESWC2011 Workshop on 'Making Sense of Microposts' (2011)

# Prediction of Human Ethnicity from Facial Images Using Neural Networks

**Sarfaraz Masood, Shubham Gupta, Abdul Wajid, Suhani Gupta and Musheer Ahmed**

**Abstract** This work attempts to solve the problem of ethnicity prediction of humans based on their facial features. Three major ethnicities were considered for this work: Mongolian, Caucasian and the Negro. A total of 447 image samples were collected from the FERET database. Several geometric features and color attributes were extracted from the image and used for classification problem. The accuracy of the model obtained using an MLP approach was 82.4% whereas the accuracy obtained by using a convolution neural network was a significant 98.6%.

**Keywords** Ethnicity identification · Artificial neural networks · Convolutional neural networks · FERET database

## 1 Introduction

Face analysis is one of the most studied research topic in the field of computer vision and pattern recognition for the past few decades. Although, face of a person provides a variety of demographic information like gender, age, ethnicity, etc yet ethnicity remains one of the invariant and fundamental attribute that cannot be easily masked like age and gender even in disguise. Therefore, grouping people

---

S. Masood (✉) · S. Gupta · A. Wajid · S. Gupta · M. Ahmed
Department of Computer Engineering, Jamia Millia Islamia, New Delhi 110025, India
e-mail: smasood@jmi.ac.in

S. Gupta
e-mail: shubham_0906@yahoo.com

A. Wajid
e-mail: abdulw976@gmail.com

S. Gupta
e-mail: suhanigupta139@gmail.com

M. Ahmed
e-mail: musheer.cse@gmail.com



S.C. Satapathy et al. (eds.), *Data Engineering and Intelligent Computing*,
Advances in Intelligent Systems and Computing 542,
DOI 10.1007/978-981-10-3223-3_20

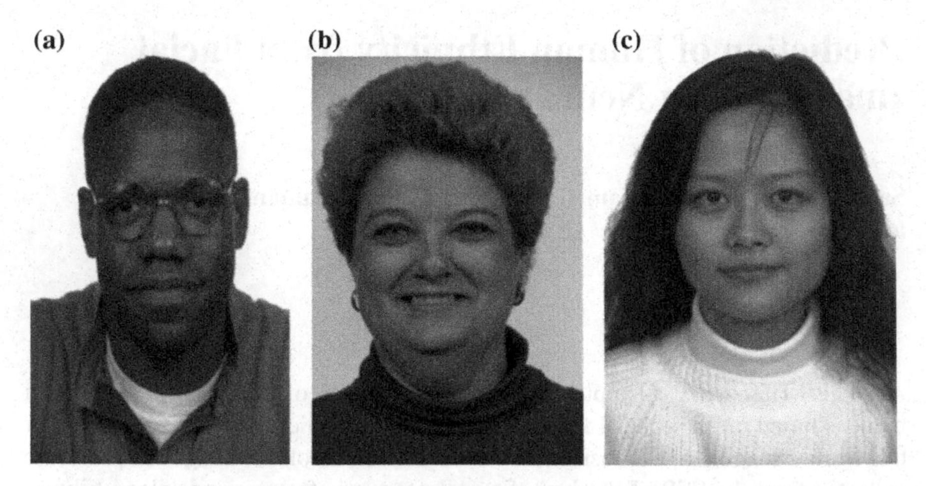


**Fig. 1** Images of sample from **a** Negroid, **b** Caucasia and **c** Mongolian ethnicity

based on age and gender would not only complex the problem but may also yield wrong results. For this reason, ethnicity classification is a key component that can be deployed in various video surveillance systems at security checkpoints. Furthermore, this classification statement has potential application in image search query where prior knowledge of ethnicity would narrow down the search space in the database, thus simplifying the process.

In this paper, three racial categories have been considered: The Mongolian, the Caucasian and the Negroid. As the people belonging to same category will have similar features. Similarly, people belonging to different racial category will have distinguishing features. This idea has helped us to extract and study fundamental features of each of these categories and classify them according to their distinguishing values (Fig. 1).

This paper is organized into five sections: Sect. 2 highlights the previous related research work and their contribution, Sect. 3 describes the design of our algorithm and the methodology adopted, Sect. 4 deals with the results and discussions and Sect. 5 finally concludes the paper.

## 2 Literature Review

Viola and Jones [1] have provided the efficient and rapid method for detecting face in input image. This is a novel approach which uses Adaboost classifier. This has high detection rate with very less computation time even on the dataset consisting of images under varying condition like illumination, pose, color, camera variation etc. This algorithm is used in our approach to detect face in the image which will be processed further.

Lu and Jain [2] has proposed ethnicity classification algorithm in which image of the faces were examined at multiple scales. The Linear Discriminant Analysis (LDA) scheme is used for input face images to improve the classification result. The accuracy of the performance of this approach is 96.3% on the database of 2,630 sample images of 263 subjects. However, the dataset considered in this work consisted only of two classes i.e. Asian and non-Asian.

Hosoi et al. [3] have integrated the Gabor wavelet features and retina sampling for their work. These features were then used with the Support Vector Machines (SVM) classifier. This approach has used three categories: Asian, African and European. And the accuracy achieved for each category is: 96%, 94% and 93% respectively. However their approach seemed to have issues when considered for other ethnicities.

In [4], ethnicity classification under the varying age and gender was performed on very large scale dataset for the first time. The dataset used was MORPHII which had 55,000 images. Guo and Mu had used Gabor features for classification problem of five ethnicities: Black, White, Hispanic, Asian and Indian. The prediction results for Black and White were good: 98.3% and 97.1% respectively. But due to insufficient dataset for other three races, prediction results deteriorated to 74.2% for Hispanic, 59.5% for Asian and 6.9% for Indian.

Roomi et al. [5] has used Viola and Jones [1] algorithm for face detection problem. After the detection of face, various features namely skin color; lip color and normalized forehead area were extracted from the image. This classification problem has used the Yale, FERET [6] dataset of Mongolian, Caucasian and Negroid images. The overall accuracy achieved in this work with these features was81%.

Tariq et al. [7] had used silhouetted face profiles for gender and ethnicity classification. The ethnicities considered were: Black, East and South East Asian, White and South Asian. The accuracy achieved is 71.66% with a standard error of 2.15%.

In [8], ethnicity classification was based on hierarchical fusion which involved fusion of frontal face and lateral gait. Although the correct classification rate was over 95%, but the database had only 22 subjects containing 12 East-Asian and 10 South-Americans.

It is evident from the above mentioned work that none had considered geometric features for their solution. Also the scope of pre-trained Convolution Neural Network has not yet been explored for this problem so far. Hence in this problem, we have considered geometric features for training the ANN, and have also tried to use convolutional neural networks for solving this problem.

## 3 Proposed Model

In an attempt to provide an efficient solution to the ethnicity classification problem, we have conducted two experiments. First experiment was done using Artificial Neural Network and the second one with Convolution Neural Network. In both the setups, the dataset consists of 447 FERET images. Out of them, 357 are used for training and 90 for testing the Neural Network.

## *3.1 Experiment 1 (Artificial Neural Network)*

In this experiment, the face area is detected from the target image using the Viola and Jones [1] algorithm. The various facial features are extracted from the sample images like geometric features, color of the skin and normalized forehead area. These features were then trained on an artificial feed forward Neural Network using a backpropagation based training algorithm. The various steps involved in this experiment are explained in Fig. 2.

### 3.1.1 Geometric Feature Calculation

Once the face has been detected using the cascade classifiers, other facial areas like Nose, mouth, left and right eye were marked. Then the distances and the ratios between these facial areas were calculated. The combinations of different geometric features of face were found to be different for different ethnicities (Fig. 3).

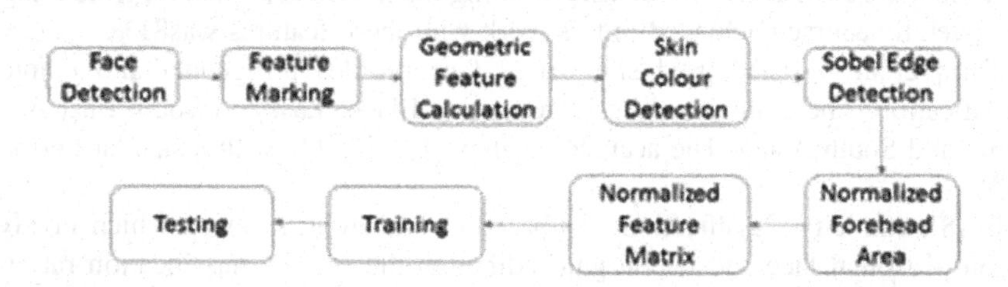


**Fig. 2** Block diagram for experiment 1

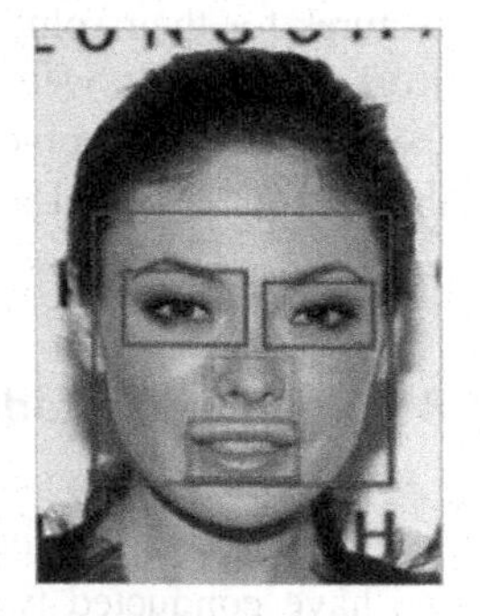

**Fig. 3** Face and facial feature detection using Viola and Jones [1]

**Table 1** Various geometric features ratio

| S. no | Features |
|---|---|
| 1 | Left eye-nose/right eye-nose |
| 2 | Left eye-mouth/right eye-mouth |
| 3 | Left eye-nose/left eye-mouth |
| 4 | Right eye-nose/right eye-mouth |
| 5 | Left eye-right eye/nose-mouth |
| 6 | Left eye-right eye/left eye-nose |
| 7 | Left eye-right eye/left eye-mouth |
| 8 | Right eye-nose/nose height |
| 9 | Nose height/nose-mouth |

Table 1 shows the various ratios of distances between various facial areas calculated earlier.

### 3.1.2 Extraction of Skin Color

Different ethnicities have varying skin colors. The dominant color of the face is used to classify the race because the distribution of skin color of different ethnicities is found to be grouped in a small area of color space. The training set contains both RGB and gray scale images. The RGB color space varies largely in intensity. In order to overcome illumination variation, the YCbCr color model is adopted and the input image is converted from RGB to YCbCr color space as follows:

$$Cb = -0.148R - 0.291G + 0.439B \tag{1}$$

$$Cr = 0.439R - 0.368G - 0.071B \tag{2}$$

The skin color classification is a good technique for classifying ethnicities but this often overlaps for Mongolian and Negroid, though it gives clear picture about Caucasian.

### 3.1.3 Normalized Forehead Area Calculation

The normalized forehead area is an important feature for ethnicity classification. The ratio of area of forehead and total face is a unique characteristic especially for Negros. Negros usually have large normalized forehead area as compared to other ethnicities. Therefore, with the help of this feature, we can easily distinguish Negroid ethnicity from others.

The forehead area is calculated with the help of Sobel Edge Detection [9] method. The intersection of smoothened vertical and horizontal edges gives the approximate position of eyes. The region above the eyes indicates the forehead area. With the help of coordinates of eyes and face, the normalized forehead area is calculated as:

$$Normalized\,forehead\,area = \frac{Forehead\,area}{total\,face\,area} \quad (3)$$

The forehead to face ratio is found to be more than 25% for Negros and less than 25% for Mongolians. In case of Caucasians, this ratio varies largely (Fig. 4).

#### 3.1.4 Training, Validation and Testing of Dataset

We implemented a multi-layered perceptrons for the purpose of training. Of the total 447 sample images, 320 images were used for training the network, 37 for validation and 90 were used for testing the trained network.

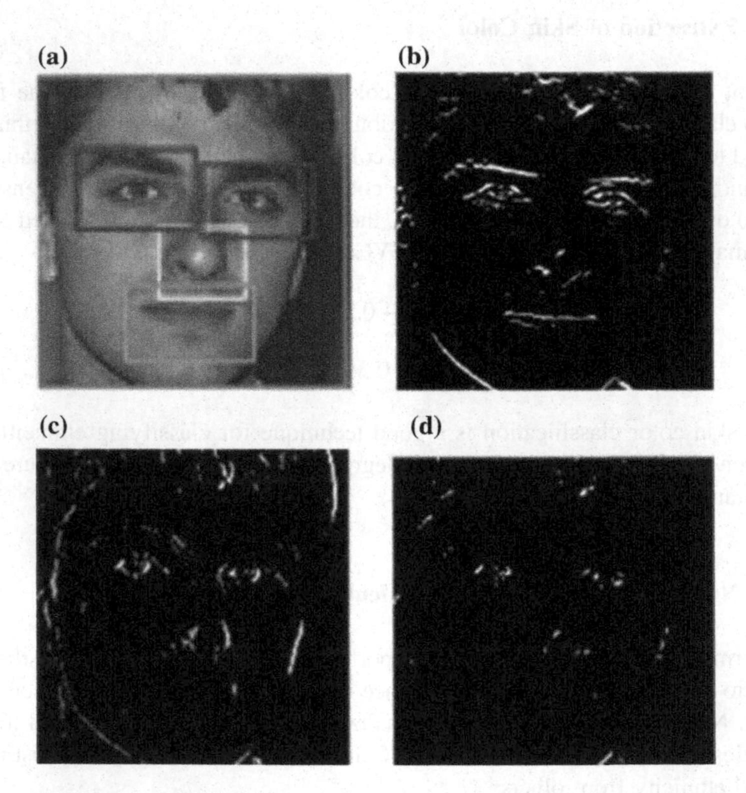


**Fig. 4** **a** Detected face, **b** horizontal Sobel edge (Gx), **c** vertical Sobel edge (Gy), **d** intersection of Gx and Gy

### *3.2 Experiment 2 (Convolutional Neural Network)*

Deep learning has been used as a tool for image classification in [10] which has yielded good results. In this experiment, we have used a pre-trained model of 16-layer architecture which has been pre-trained on various faces due to limited computation available with us. This pre-training model as described in [11] was used to extract features from the training and testing samples. During training, input image is passed through the stack of convolution layers which has a filter of size $3 \times 3$ and the activation function of “Relu”. The VGGNet used has 13 convolution layers. Pooling is done through max pool layer of $2 \times 2$ window size and a stride of 2. These stack of convolution layer is followed by three fully connected layer: first two has 4096 channels and the last has 3 channels, each representing an ethnic class.

The network was compiled with categorical cross entropy for loss function (we were using Softmax output) and nesterov momentum was used. Learning rate was set to 2.5e-4 and freezing all the layers except the final fully connected layers (train at about ~1/10th of the initial learning rate). With all these parameters initialized, this network is used for feature extraction. After this, training and testing was performed.

## 4 Results

The ethnicity identification model was developed by using 357 images for training, each image being $240 \times 360$ in pixels and 90 images for testing purpose. The face samples were taken almost in equal amount: Caucasians 120, Mongolians 120 and Negros 117. Result obtained through ANN and CNN varies largely.

### *4.1 Result of Experiment 1 (ANN)*

ANN is trained by changing the various attributes of the network. These networks were then saved and used to evaluate 90 testing samples. The results obtained in various experiments for different classes (C1 C2 and C3) are tabulated below. The best result is achieved with epochs 150, learning rate 0.17, Lavenberg Marquadt training algorithm and hidden neurons 17 (Table 2).

Testing was done using 31 Caucasian, 32 Mongolian and 30 Negroid sample images. The confusion matrix for the experiment 4 shows that out of 31 Caucasians, 25 were classified correctly. In case of Mongolian success rate was 25 out of 31 and in Negroid 25 out of 30. The overall classification accuracy in this approach was observed to be 82.4% (Fig. 5).

**Table 2** Result obtained in various experiments in ANN

| Attribute | Set1 | | | Set3 | | | Set2 | | |
|---|---|---|---|---|---|---|---|---|---|
| | C1 | C2 | C3 | C1 | C2 | C3 | C1 | C2 | C3 |
| True positive | 25 | 25 | 25 | 25 | 24 | 25 | 25 | 24 | 25 |
| True negative | 55 | 55 | 56 | 55 | 55 | 54 | 55 | 54 | 56 |
| False positive | 6 | 5 | 5 | 6 | 5 | 7 | 6 | 6 | 5 |
| False negative | 5 | 6 | 5 | 6 | 7 | 5 | 5 | 7 | 5 |
| Precision (%) | 80.6 | 83.3 | 83.3 | 80.0 | 82.8 | 78.1 | 80.6 | 80.0 | 83.3 |
| Recall (%) | 83.3 | 80.6 | 83.3 | 80.0 | 77.4 | 83.3 | 83.3 | 77.4 | 83.3 |
| F-measure (%) | 81.9 | 91.9 | 83.3 | 80.0 | 80.0 | 80.6 | 81.9 | 78.6 | 83.3 |
| Accuracy (%) | 80.6 | 83.3 | 83.3 | 77.4 | 80.0 | 83.3 | 80.6 | 80.0 | 83.3 |
| False positive rate (%) | 9.8 | 8.3 | 8.1 | 9.8 | 8.3 | 11.4 | 9.8 | 10 | 8.1 |

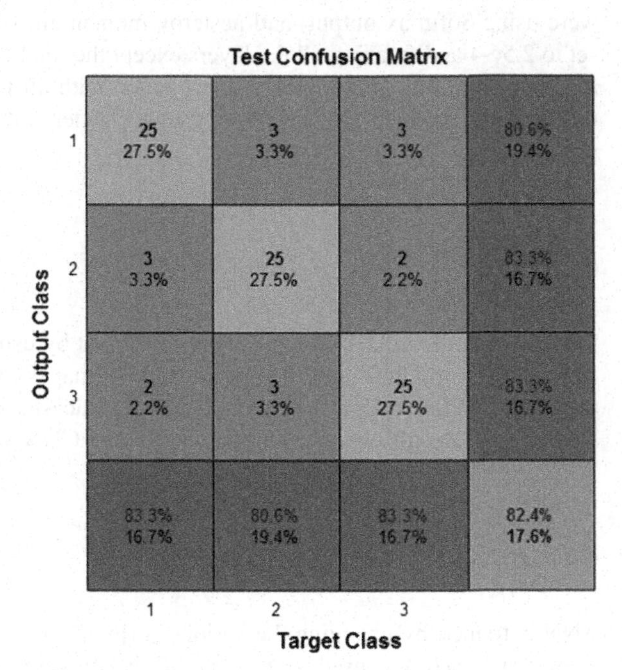


**Fig. 5** Confusion matrix obtained for set 1

## 4.2 *Result of Experiment 2 (CNN)*

The samples were tested on VGGNet by varying the number of epochs. Two experiments were conducted: one for 10 epochs and the other for 20 epochs. The testing and training accuracies are tabulated in Table 3.

**Table 3** Results obtained in various experiments in CNN

| Attribute | Set1 | Set2 |
|---|---|---|
| Epochs | 10 | 20 |
| Training time (s) | 413.074 | 824.03 |
| Training accuracy | 0.9942 | 0.9967 |
| Testing accuracy | 0.93301 | 0.986 |

## 5 Conclusion

Although both the experiments were able to provide the solution to the problem of ethnicity classification among three prominent races: Negroid, Mongolian and Caucasian. But on comparing the results obtained through ANN and CNN, it can be concluded that the performance of CNN approach is far superior than the ANN one. The accuracy achieved in ANN was 82.4% whereas in CNN, it was 98.6%.

The results obtained using an ANN in our work were better than the results achieved in [5]. However the accuracy obtained in both the experiments indicate that the Convolution Neural Network(CNN) approach has better yield rate. This can be attributed to the pre-trained model used in CNN and the 3-D arrangement of neurons.

But time required for feature extraction and training that network in case of CNN was much more than that in ANN. This work could be extended for other known ethnicities as well. It can play a major role in surveillance systems for security purposes.

## References

1. Viola, P., Jones, M.: Robust real-time face detection. Proc. Int. J. Comput. Vis. **57**(2), 137–154 (2004)
2. Lu, X., Jain, A.K.: Ethnicity identification from face images. In: Proceedings of SPIE 5404, Biometric Technology for Human Identification, Department of Computer Science & Engineering, Michigan State University (2004)
3. Hosoi, S., Takikawa, E., Kawade, M.: Ethnicity estimation with facial images. In: Sixth IEEE International Conference on Automatic Face and Gesture Recognition, 2004. Proceedings, pp. 195–200 (2004)
4. Guo, G., Mu, G.: A study of large-scale ethnicity estimation with gender and age variations. In: 2010 IEEE Computer Society Conference on Computer Vision and Pattern Recognition—Workshops, San Francisco, CA, 2010, pp. 79–86
5. Roomi, S.M.M., Virasundarii, S.L., Selvamegala, S., Jeevanandham, S., Hariharasudhan, D.: Race classification based on facial features. In: 2011 Third National Conference on Computer Vision, Pattern Recognition, Image Processing and Graphics (NCVPRIPG), Hubli, Karnataka, pp. 54–57 (2011)
6. Phillips, P.J., Wechsler, H., Huang, J., Rauss, P.: The FERET database and evaluation procedure for face recognition algorithms. Image Vis. Comput. J. **16**(5), 295–306 (1998)

7. Tariq, U., Hu, Y., Huang, T.S.: Gender and ethnicity identification from silhouetted face profiles. In: 2009 16th IEEE International Conference on Image Processing (ICIP), Cairo, pp. 2441–2444 (2009)
8. De Zhang, Wang, Y., Zhang, Z.: Ethnicity classification based on a hierarchical fusion. In: Biometric Recognition: 7th Chinese Conference, CCBR 2012, Guangzhou, China, December 4–5, 2012. Proceedings, pp. 300–307. Springer, Heidelberg (2012)
9. Jin-Yu, Z., Yan, C., Xian-Xiang, H.: Edge detection of images based on improved Sobel operator and genetic algorithms. In: 2009 International Conference on Image Analysis and Signal Processing, Taizhou, pp. 31–35 (2009)
10. Liu, S., Deng, W.: Very deep convolutional neural network based image classification using small training sample size. In: 2015 3rd IAPR Asian Conference on Pattern Recognition (ACPR), Kuala Lumpur, Malaysia, pp. 730–734 (2015)
11. Parkhi, O.M., Vedaldi, A., Zisserman, A.: Deep face recognition. In: Proceedings of the British Machine Vision, vol. 1.3 (2015)

# Glyph Segmentation for Offline Handwritten Telugu Characters

**C. Naga Manisha, Y.K. Sundara Krishna and E. Sreenivasa Reddy**

**Abstract** Segmentation plays a crucial role in the recognition of offline handwritten characters from the digitized document images. In this paper, the authors propose the glyph segmentation method for offline handwritten Telugu characters. This method efficiently segments the top vowel ligature glyph, main glyph, bottom vowel ligature glyph and consonant conjunct glyph from the offline handwritten Telugu character images. It efficiently identifies the small glyphs that are related to the unconnected main glyphs or consonant conjuncts and also efficiently segments the connected top vowel ligature from the main glyph. This approach of segmentation efficiently reduces the train data size for the purpose of offline handwritten Telugu characters recognition system. The result shows the efficiency of proposed method.

**Keywords** Telugu · Handwritten · Offline · Glyph · Segmentation

## 1 Introduction

Segmentation is a crucial step at the pre-processing stage for the recognition of offline handwritten characters from document images. It is directly influences the recognition success rate. This paper presents a glyph segmentation method for offline handwritten Telugu characters. "A glyph is a graphical representation of either a character, a part of a character, or a sequence of characters" [1]. In this paper glyph consider as a part of the offline handwritten Telugu character.

---

C. Naga Manisha (✉) · Y.K. Sundara Krishna
Krishna University, Machilipatnam, Andhra Pradesh, India
e-mail: ch.n.manisha@gmail.com

Y.K. Sundara Krishna
e-mail: yksk2010@gmail.com

E. Sreenivasa Reddy
Acharya Nagarjuna University, Guntur, Andhra Pradesh, India
e-mail: esreddy67@gmail.com



S.C. Satapathy et al. (eds.), *Data Engineering and Intelligent Computing*,
Advances in Intelligent Systems and Computing 542,
DOI 10.1007/978-981-10-3223-3_21

Various segmentation methods developed for document images of which the line segmentation, word segmentation, and character segmentation are the most common.

Line segmentation is segmenting the lines of the characters from the document image in [2]. Overlapping and touching lines segmentation methods developed in [3, 4]. Word segmentation is segmenting the words in the document images. Handwritten Arabic words segmentation method developed in [5]. Mathivanan et al. [6] proposed morphological based algorithm for handwritten text segmentation.

Srinivasa Rao et al. [7] Proposed modified drop fall algorithm for Telugu characters segmentation in noisy environment. Segmentation method for the touched Telugu consonant conjuncts and consonants developed in [8–10]. Telugu characters contain the largest set of glyphs and there is highest similarity between the glyphs, the offline handwritten Telugu characters recognition system is more complex. Therefore, the recognition system may encounter time and space complexity. The purpose of the proposed method is to decrease the train data size for the purpose of train the system. Thus, this method will decrease the time and space complexity of the recognition system and improve its efficiency.

The remaining paper is organized as follows. In Sect. 2, the pre-processing steps for applying to the offline handwritten Telugu characters are presented. In Sect. 3, the proposed method is described. In Sect. 4, the analysis of experimental results is provided, and in the final section the conclusions of this proposed method are presented.

## 2 Pre-processing Steps

Before applying the glyph segmentation method, it is necessary to apply the other pre-processing steps to the document image. To apply preprocessing techniques such as handwritten word skew correction found in [11]. The offline handwritten Telugu characters slant correction method found in [12]. The Otsu's method in [13] used to select the threshold value for convert the gray level image into binary image. For proper identification of the glyph, the background and foreground pixels can be converted into black and white pixels respectively. The character segmentation and thinning methods found in [4, 14].

## 3 Proposed Method

This section contains the proposed method for glyph segmentation for offline handwritten Telugu characters. The proposed method deals with connected and unconnected glyphs; the glyph is identified as 'G', 'm' and 'n' represent the width and height of 'G'. Let 'x' and 'y' are represented as vertical and horizontal points of the pixel. Section 3.1 contains propose method of identify the unconnected small glyphs. Section 3.2 contains the propose method of segment point identification for the connected vowel ligature and main glyph. Section 3.3 describes glyph segmentation for offline handwritten Telugu characters.

## 3.1 Unconnected Small Glyphs Identification

In Telugu, the unique nature of the main glyph can be identified by its unconnected small glyphs as shown in Fig. 1. The small glyphs cannot segment as vowel ligatures or consonant conjuncts. It is consider as a part of the main glyph or consonant conjunct glyph.

At Fig. 2a–c have a small glyph found at top, bottom and middle of the main glyph. And Fig. 2d has two small glyphs which is one is at top and another at middle of the main glyph. These small glyphs identified with their size. The small glyph identified by the Eq. 1.

Let 'b' and 'l' is the width and height of the offline handwritten Telugu character. Let 'S' is the small glyph, 'w' and 'h' are the width and height of 'S'. Let 'nG' is the number of glyphs. The authors applied trial-and-error method to identify '0.1' multiplicand for small glyphs identification.

$$S = \begin{cases} G & \text{if } m < (b \times 0.1) \text{ OR } n < (l \times 0.1) \\ \phi & \text{else} \end{cases} \tag{1}$$

Consider the following conditions

1. $S\{y_1, \ldots, y_h\} \cap G_i\{y_1, \ldots, y_n\} \neq \phi$
2. $\{V_1, \ldots, V_{nG}\} > 0$ where $V_i = S(y_1) - G_i(y_n)$

Consider the first condition, If $S\{y_1, \ldots, y_h\}$ and $G_i\{y_1, \ldots, y_n\}$ points intersected then '$G_i$' is related to 'S' and 'S' may be middle position of '$G_i$'.

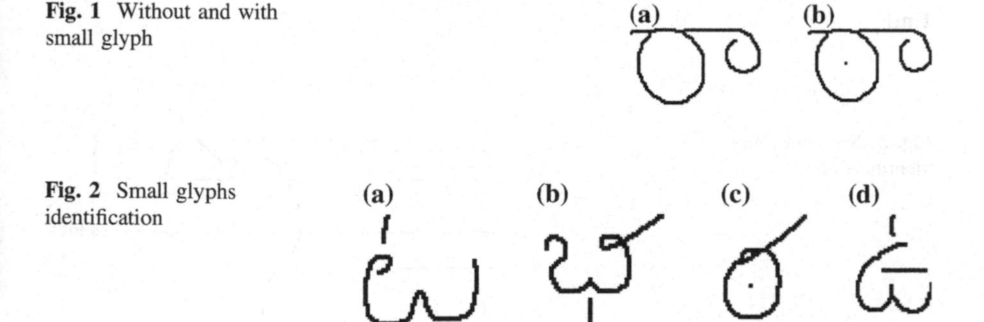


Fig. 1 Without and with small glyph

Fig. 2 Small glyphs identification

If first condition does not satisfied then consider second condition. It identifies that whether 'S' is bottom position of any 'G'. Identify minimum positive element from $\{V_1, \ldots, V_{nG}\}$ say '$V_k$' then '$G_k$' related to 'S'. If $\{V_1, \ldots, V_{nG}\}$ does not contains any positive elements then 'S' is the top position of 'G' therefore from $\{G_1(y_1), G_2(y_1), \ldots, G_{nG}(y_1)\}$ identify the minimum element '$G_j$' which is related to S.

## 3.2 Segmentation Point Identification

Generally, vowel ligature connected to main glyph. Therefore, for better recognition, it is necessary to segment vowel ligature from the main glyph. The methods for extraction of connected components from the image found in [15, 16]. The top vowel ligature segmentation will apply to glyph without its small glyphs. After ligature segmentation, the small glyphs will add to its main glyph. The segmentation point will identify with the help of number of vertical lines for each row as shown in Fig. 3.

Consider sample data 0011101100100001010

For the vertical line identification, the single '1' or sequence of 1's considering as single vertical line. Therefore, the sample data contains five vertical lines.

Let $\{L_1, \ldots, L_n\}$ is the set of number of vertical lines for 'n' rows. The algorithm developed for segmentation point identification shown in Algorithm Segmentation_Point_Identification. 'segpt' is represented as segmentation point.

**Algorithm:** Segmentation_Point_Identification

**Input:**
$\{L_1, \ldots, L_n\}$:Set of No. of Vertical Lines for 'n' rows
**Output:** segpt: Segmentation Point
**Begin**

1. Let segpt=0.
2. Let i=1.
3. Do Sequential Search for 'i' where $L_i > 1$
4. Do Sequential Search for 'i' where $L_i == 1$
5. Do Sequential Search for 'i' where $L_i > 1$ or $i > n-i$
6. if $i-1 < n-i$ and $i < n$ then assign 'i-1' to 'segpt'.

**End**

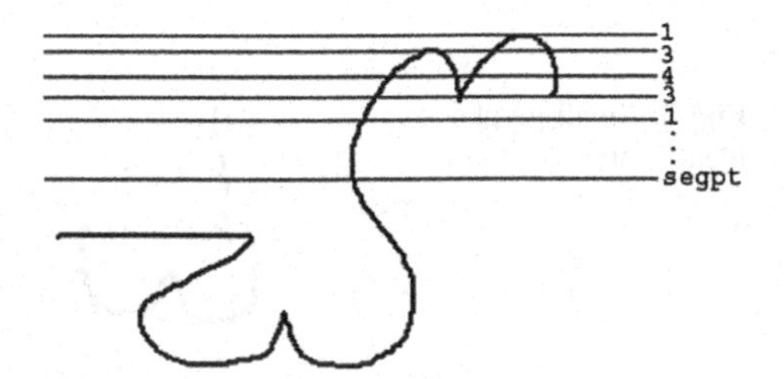


**Fig. 3** Segment point identification

Verify from '$L_1$'. The elements match with zero or more 1's followed by one or more elements which are greater than 1 and followed by 1. If $L_i$ is equal to 1 means 'i' may be vowel ligature index and $i - 1 < n - i$ verifies that vowel ligature height is always less than main glyph height.

### *3.3 Glyph Segmentation*

The characters in Telugu language are formed with one to four unconnected glyphs. Therefore, this section is divided into four cases. In this section; glyph 'G' index is identified based on the value of '$y_1$' for each G. The lowest '$y_1$' becomes the '$G_1$' and so on. Top vowel ligature glyphs identified as 'T'; main glyph is identified as 'M' and the consonant conjunct glyphs and bottom vowel ligature glyphs identified as 'B' with indices 1 and 2.

**Case-1**. In this case a character forms with single glyph as show in Fig. 4. The glyphs is may be single glyph or contains connected vowel ligature with main glyph shown in Fig. 4a and b respectively. The propose segment point identification for connected vowel ligature and main glyph algorithm found at Sect. 3.2.

If the segmentation point is zero or the small glyph at the top of main glyph then the entire glyph consider as main glyph shown in Eq. 2. If the segmentation point is more than zero then the glyph segmented into vowel ligature glyph and main glyph as shown in Eqs. 3 and 4 respectively.

$$M\{(x_1, y_1), \ldots, (x_m, y_n)\} \leftarrow G\{(x_1, y_1), \ldots, (x_m, y_n)\} \tag{2}$$

$$T\left\{(x_1, y_1), \ldots, \left(x_m, y_{segpt}\right)\right\} \leftarrow G\left\{(x_1, y_1), \ldots, \left(x_m, y_{segpt}\right)\right\} \tag{3}$$

$$M\left\{(x_1, y_1), \ldots, \left(x_m, y_{n-(segpt+1)}\right)\right\} \leftarrow G\left\{\left(x_1, y_{segpt+1}\right), \ldots, (x_m, y_n)\right\} \tag{4}$$

**Case-2**. This case deals with two unconnected glyphs. The unconnected vowel ligature-main glyph and main glyph-consonant conjunct glyph forms two combinations. The Telugu character forms with unconnected vowel ligature glyph and main glyph as shown in Fig. 5a and b. The Telugu character forms with main glyph and consonant conjunct as shown in Fig. 5c and d.

Let $G_1$ and $G_2$ identified as two unconnected glyphs. In this case it is needed to identify a point 'P' for main glyph identification as shown in Eq. 5.

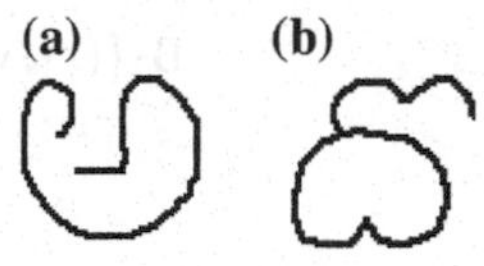


**Fig. 4** Telugu character forms with single glyph

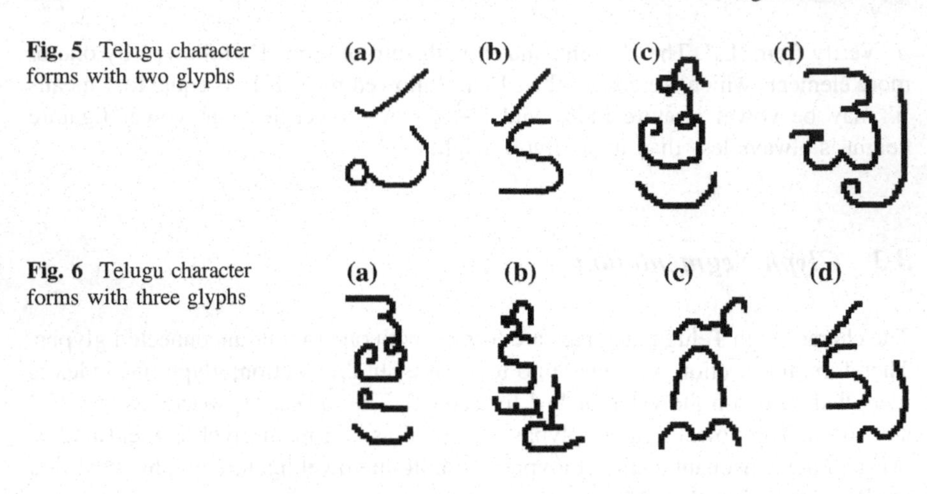


**Fig. 5** Telugu character forms with two glyphs

**Fig. 6** Telugu character forms with three glyphs

$$P \leftarrow G_1(y_1) + 0.75 \times \text{height}(G_1) \quad (5)$$

Consider the following conditions

1. if $G_1\{y_1, \ldots, y_{n1}\} \cap G_2\{y_1, \ldots, y_{n2}\} == 0$ and $\text{height}(G_1) > \text{height}(G_2)$
2. if $G_1\{y_1, \ldots, y_{n1}\} \cap G_2\{y_1, \ldots, y_{n2}\} \neq 0$ and $P > G_2(y_1)$

If any one of the above condition satisfied then $G_1$ contains main glyph and it may be segmented into vowel ligature and main glyph as described in case-1 at Sect. 3.3. $G_2$ identified as bottom glyph as shown in Eq. 6.

$$B_1\{(x_1, y_1), \ldots, (x_m, y_n)\} \leftarrow G_2\{(x_1, y_1), \ldots, (x_m, y_n)\} \quad (6)$$

If none of the above conditions does not satisfied then '$G_1$' consider as top vowel ligature glyph 'T' and '$G_2$' consider as main glyph 'M' as shown in Eqs. 7 and 8 respectively.

$$T\{(x_1, y_1), \ldots, (x_m, y_n)\} \leftarrow G_1\{(x_1, y_1), \ldots, (x_m, y_n)\} \quad (7)$$

$$M\{(x_1, y_1), \ldots, (x_m, y_n)\} \leftarrow G_2\{(x_1, y_1), \ldots, (x_m, y_n)\} \quad (8)$$

**Case-3**. This case deals with three unconnected glyphs.

If $\text{height}(G_1) > \text{height}(G_2)$ then the glyph $G_1$ contains the main glyph as shown in Fig. 6a and b. The method of connected vowel ligature and main glyph segmentation described in case-1 at Sect. 3.3. $G_2$ and $G_3$ segmented as consonant conjuncts and/or bottom vowel ligature glyphs '$B_1$' and '$B_2$' as shown in Eqs. 6 and 9.

$$B_2\{(x_1, y_1), \ldots, (x_m, y_n)\} \leftarrow G_3\{(x_1, y_1), \ldots, (x_m, y_n)\} \quad (9)$$

**Fig. 7** Telugu character forms with four glyphs

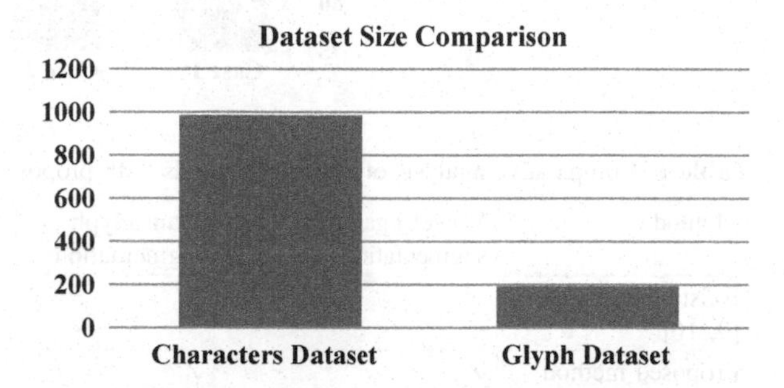


**Fig. 8** Telugu train dataset size comparison

If height($G_1$) $\leq$ height ($G_2$) then the $G_1$, $G_2$ and $G_3$ are unconnected glyphs shown in Fig. 6c and d. The glyphs $G_1$, $G_2$ and $G_3$ will segment to top vowel ligature glyph 'T', main glyph 'M' and consonant conjunct or bottom vowel ligature glyph '$B_1$' as shown in Eqs. 7, 8 and 10 respectively.

$$B_1\{(x_1, y_1), \ldots, (x_m, y_n)\} \leftarrow G_3\{(x_1, y_1), \ldots, (x_m, y_n)\} \quad (10)$$

**Case-4**. In this case the character contains four unconnected glyphs as shown in Fig. 7. In this case '$G_1$' segmented as top vowel ligature glyph 'T' and '$G_2$' segmented as a main glyph 'M' as shown in Eqs. 7 and 8 respectively. '$G_3$' and '$G_4$' are segmented as two bottom glyphs may be consonant conjunct and/or bottom vowel ligature glyph '$B_1$' and '$B_2$' as shown in Eqs. 10 and 11 respectively.

$$B_2\{(x_1, y_1), \ldots, (x_m, y_n)\} \leftarrow G_4\{(x_1, y_1), \ldots, (x_m, y_n)\} \quad (11)$$

## 4 Experiments Results

The proposed method very efficiently decreases the train dataset size of Telugu offline handwritten Telugu characters dataset. The train dataset size of 983 characters as described in [17] decreased to 191 glyphs. Therefore the proposed method very efficiently eliminates 80.57% glyphs from train dataset and identifies 19.43% glyphs for the recognition of offline handwritten Telugu characters for the purpose of train dataset. The Telugu train dataset size comparison as shown in Fig. 8. The proposed glyph segmentation method has been verified on 2,576 offline handwritten Telugu characters with different vowel ligatures and consonant conjuncts

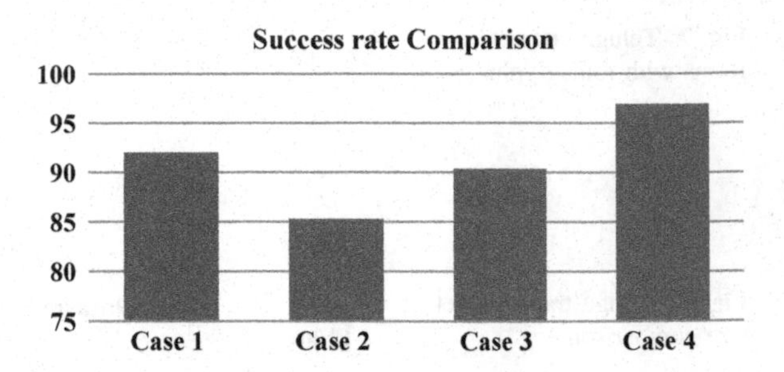


**Fig. 9** Success rate comparison

**Table 1** Comparative analysis of existing methods with proposed method

| Method | Vowel ligature segmentation | Main glyph segmentation | Consonant conjunct segmentation |
|---|---|---|---|
| Existing methods [9, 10] | × | × | ✓ |
| Proposed method | ✓ | ✓ | ✓ |

combinations. These characters were written by different people. The proposed method got overall success rate is 91.14%. The success rate comparison for four cases as shown in Fig. 9. The comparative analysis of existing Telugu based segmentations methods with proposed method as shown in Table 1.

## 5 Conclusion

In conclusion, the proposed method efficiently identifies the small glyphs that are related to the unconnected main glyphs or consonant conjuncts. The proposed method also segmented the connected vowel ligature from the main glyph. The proposed method efficiently segments the offline handwritten Telugu characters into the top vowel ligature glyphs, main glyphs, bottom vowel ligature glyphs and/or consonant conjuncts glyphs. The proposed method very efficiently eliminates 80.57% glyphs from Telugu train dataset and identifies 19.43% glyphs for the recognition of offline handwritten Telugu characters for the purpose of train dataset. Therefore, the proposed method will improve the recognition rate. The overall success rate of the proposed method is 91.14%.

## References

1. Glyph. https://glosbe.com/en/te/glyph
2. Dos Santos, R.P., Clemente, G.S., Ren, T.I., Cavalcanti, G.D.: Text line segmentation based on morphology and histogram projection. In: ICDAR'09. 10th International Conference, pp. 651–655. IEEE (2009)

3. Ouwayed, N., Belaïd, A.: Separation of overlapping and touching lines within handwritten Arabic documents. In Computer Analysis of Images and Patterns,, pp. 237–244. Springer, Berlin (2009)
4. Das, M.S., Reddy, C.R.K., Govardhan, A., Saikrishna, G.: Segmentation of overlapping text lines, characters in printed telugu text document images. Int. J. Eng. Sci. Technol. **2**(11) 6606–6610 (2010)
5. Elzobi, M., Al-Hamadi, A., Al Aghbari, Z.: Off-line handwritten arabic words segmentation based on structural features and connected components analysis. In: WSCG 2011 Communication Papers, pp. 135–142 (2011)
6. Mathivanan, P., Ganesamoorthy, B., Maran, P.: Watershed algorithm based segmentation for handwritten text identification. ICTACT J. Image Video Process. **4**(03), 767–772 (2014)
7. Srinivasa Rao, A.V., Mary Junitha, M., Shankara Bhaskara Rao, G., Subba Rao, A.V.: Segmentation of touching telugu characters under noisy environment. J. Emerg. Trends Comput. Inf. Sci. **5**(9), 698–702 (2014)
8. Bharathi, J., Reddy, P.C.: Improvement of telugu OCR by segmentation of Touching Characters. Int. J. Res. Eng. Technol. **03**(10), 333–341 (2014)
9. Bharathi, J., Chandrasekar Reddy, P.: Segmentation of touching conjunct consonants in telugu using minimum area bounding boxes. Int. J. Soft Comput. Eng. **3**(3), 260–264 (2013)
10. Bharathi, J., Chandrasekar Reddy, P.: Segmentation of telugu touching conjunct consonants using overlapping bounding boxes. Int. J. Comput. Sci. Eng. **5**(06), 538–546 (2013)
11. Blumenstein, M., Cheng, C.K., Liu, X.Y.: New preprocessing techniques for handwritten word recognition. In: Proceedings of the Second IASTED International Conference on Visualization, Imaging and Image Processing (VIIP 2002), Calgary, pp. 480–484. ACTA Press (2002)
12. Manisha, Ch.N., Sundara Krishna, Y.K., Sreenivasa Reddy, E.: Slant correction for offline handwritten telugu isolated characters and cursive words. Int. J. Appl. Eng. Res. **11**(4), 2755–2760 (2016)
13. Otsu, N.: A threshold selection method from gray-level histograms. IEEE Trans. Syst. Man Cybern. **SMC-9**(1), 62–66 (1979)
14. Lam, L., Lee, S.W., Suen, C.Y.: Thinning methodologies-a comprehensive survey. IEEE Trans. Pattern Anal. Mach. Intell. **14**(9), 869–885 (1992)
15. Di Stefano, L., Bulgarelli, A.: A simple and efficient connected components labeling algorithm. In: Proceedings of International Conference on Image Analysis and Processing, 1999, pp. 322–327. IEEE (1999)
16. Extraction of Connected Components. http://angeljohnsy.blogspot.com/2012/03/extraction-of-connected-components.html
17. Varalakshmi, A., Negi, A., Krishna, S.: dataset generation and feature extraction for telugu hand-written recognition. Int. J. Comput. Sci. Telecommun. **3**(3), 57–59 (2012)

# A Roadmap for Agility Estimation and Method Selection for Secure Agile Development Using AHP and ANN

**Amit Sharma and R.K. Bawa**

**Abstract** The modern software industry is expected to provide fast software delivery and because of dynamic environment the customer requirements changes very rapidly, which has lead to inclination towards agile development approaches over other traditional approaches. It has the advantages like fast release and simplified documents which eventually lead to maximizing profit and productivity. However, it is a mammoth task to make a calculative decision about whether to use an agile approach for a given project or not because of the lack of any empirical decision making process. This paper provides a roadmap for making decision using Analytic Hierarchy Process (AHP) and Artificial Neural Network (ANN) with Agility Indicator and if selected, it further suggests which Agile Development method is better suited for among popular methods like Feature-driven Development (FDD), Lean development, Scrum, Crystal Clear, Extreme Programming (XP) and Dynamic Software Development Method (DSDM). It also addresses the major concern about security requirements to enhance the security features by integrating security activities from security engineering processes without degrading the agility of the agile process.

**Keywords** Agile security · Agile development · Agility indicator · AHP · ANN · Scrum · XP · DSDM · FDD · Crystal clear

## 1 Introduction

In agile development a different approach is followed as compared to plan-driven development which relies on the formalization and control. Rather, agile development only requires formalizing processes only where it is very necessary and emphasizes

---

A. Sharma (✉) · R.K. Bawa
Punjabi University, Patiala, Punjab, India
e-mail: amitsharmapkl@gmail.com

R.K. Bawa
e-mail: rajesh.k.bawa@gmail.com



S.C. Satapathy et al. (eds.), *Data Engineering and Intelligent Computing*,
Advances in Intelligent Systems and Computing 542,
DOI 10.1007/978-981-10-3223-3_22

intensive and informal communication to develop projects with high business-value. The Agile Manifesto [1] describes the core values in the following terms:

- Responding to change over following a plan,
- Individuals and interactions over processes and tools,
- Customer collaboration over contract negotiation,
- Working software over comprehensive documentation.

The meaning and understanding of agile differs in practice. Defining agile methods itself is rather a difficult task, as it is a collection of well-defined methods, that also differ in practice.

### *1.1 Agile Development Methods*

Since 1980 several agile development methods have came into existence [2]. In order to systematically analyze the depth and breadth in agile development, we have selectively chosen popular and widely used methods like Scrum [3], Extreme Programming [4, 5], Crystal Clear [6], Lean development [7], Dynamic Software Development Method (DSDM) [8] and Feature-driven Development (FDD) [9]. These methods of agile approach have a broad range of characteristics [10] and they use adaptive approach rather than a predictive approach to develop software.

### *1.2 Security in Agile Development*

The main security concern with agile development is that the osmotic communication and tacit knowledge driven methods, self organizing team and trust on individuals, conflict with the quality and assurance activities as required by conventional secure software development methods. However, the studies also indicate that the agile approach improves quality. In addition, plan-driven development may also pose threat to secure software development that in case of agile development might be less critical. Dynamic and changing requirements may conflict with the early planning of security requirements, which agile development approach is intrinsically prepared for.

## 2 Proposed Roadmap

In the absence of any accepted and verified approach to analyze and select the appropriate development method for a software project, it becomes a complicated task for the project analyst; especially when every related aspect of the project has to be considered [11]. The purpose of this section is to suggest a roadmap for simplifying this task of selection of the most appropriate development method for the analyst.

## 2.1 Selection Procedure to Choose Agile or Plan Driven Approach

In order to help the project analyst to decide the best approach to develop a project, we have taken one famous method used for Multi Criteria Decision Making which is called Analytic Hierarchy Process (AHP). It takes care of both, subjectiveness of the human judgments as well as at the same time objective evaluation by using Eigen vector. Further with the help of Eigen Value, it also verifies the consistency of the evaluation.

Making decision is a process to select among different alternatives based on several criteria according a problem. In our work we have chosen these criteria from the agile manifesto, agile principles and different agile methodologies and considering every aspect of the project like customer, project manager, team etc. The input value for each criterion is given between 1 and 10, where 1 indicates the least weight age and 10 indicates the highest weight age. After thorough analysis and study the major criteria selected are the followings:

Project Complexity and Reliability (PCR)
Familiarity and Experience of Project (FEP)
Quality and Risk Management (QRM)
Experienced and Adaptive Team (EAT)
Team Communication and Democratic Culture (TCD)
Formalization and Documentation (FAD)
Customer Support and Collaboration (CSC)
Clarity of Requirements (COR)

As every project has different requirements thus the project analyst can add more criteria if needed and vice versa. To calculate the weights and ranks, comparison matrix has been got filled by five experts from industry working in different organizations and consolidated matrix is created using Weighted Geometric Mean of all participants as shown in Table 1.

Number of Criteria = 8, Number of participants = 5, Consistency Ratio = 2.8%, Lambda = 8.275, Consensus = 93.7%.

It has been proved by Prof. Saaty that if the value of Consistency Ratio is less than 10%, we can accept the judgment [12], otherwise we have to revise the subjective judgment. Thus in our case the approximation is quite good as Consistency Ratio is 2.8%. Based upon this matrix the corresponding ranks and weights are calculated as given in the Table 2 and the value of Global Agility Indicator is calculated as shown in Table 3. If the value of agility indicator is high it reflects the

**Table 1** Weighted geometric mean of all participants

| | PCR | FEP | QRM | EAT | TCD | FAD | CSC | COR |
|---|---|---|---|---|---|---|---|---|
| PCR | 1 | 0.251 | 1 | 0.237 | 0.333 | 1 | 0.281 | 2.080 |
| FEP | 3.979 | 1 | 2.080 | 1 | 1.442 | 4.217 | 1.709 | 3 |
| QRM | 1 | 0.480 | 1 | 0.281 | 0.480 | 1 | 0.333 | 0.693 |
| EAT | 4.217 | 1 | 3.556 | 1 | 1 | 2.268 | 1 | 1.442 |
| TCD | 3 | 0.693 | 2.080 | 1 | 1 | 3.556 | 1 | 2.080 |
| FAD | 1 | 0.237 | 1 | 0.440 | 0.281 | 1 | 0.333 | 1 |
| CSC | 3.556 | 0.584 | 3 | 1 | 1 | 3 | 1 | 3 |
| COR | 0.480 | 0.333 | 1.442 | 0.693 | 0.480 | 1 | 0.333 | 1 |

**Table 2** Weights and ranks of criteria

| Criteria | Weights (%) | Rank |
|---|---|---|
| Project complexity and reliability (PCR) | 6.50 | 6 |
| Familiarity and experience of project (FEP) | 21.70 | 1 |
| Quality and risk management (QRM) | 6.40 | 7 |
| Experienced and adaptive team (EAT) | 17.90 | 2 |
| Team communication and democratic culture (TCD) | 16.50 | 4 |
| Formalization and documentation (FAD) | 6.00 | 8 |
| Customer support and collaboration (CSC) | 17.70 | 3 |
| Clarity of requirements (COR) | 7.10 | 5 |

**Table 3** Calculation of global agility indicator

| Criteria | Weights | Input | |
|---|---|---|---|
| Project complexity and reliability (PCR) | 0.065 | 4 | 0.26 |
| Familiarity and experience of project (FEP) | 0.217 | 8 | 1.736 |
| Quality and risk management (QRM) | 0.064 | 5 | 0.32 |
| Experienced and adaptive team (EAT) | 0.179 | 8 | 1.432 |
| Team communication and democratic culture (TCD) | 0.165 | 8 | 1.32 |
| Formalization and documentation (FAD) | 0.06 | 3 | 0.18 |
| Customer support and collaboration (CSC) | 0.177 | 8 | 1.416 |
| Clarity of requirements (COR) | 0.071 | 5 | 0.355 |
| Global agility indicator | | | 7.019 |

high suitability of using agile development approach and vice versa. Subjectivity in defining the criteria and inconsistent input can be compensated using Artificial Neural Network as shown in next section.

## 2.2 Selection of Most Appropriate Agile Development Method

As we know that each and every agile development method is a collection of processes and practices that are supported by principles and values. Based upon the previous calculations and characteristics of different agile methods, which are discussed next, the most appropriate agile development method is selected.

**Criteria for the Selection of best agile method**. The decision of selecting a particular agile method which is best according to the project requirements cannot be taken by considering only one dimensional value, rather we have to consider different characteristics of the agile methods according to the context. In order to further analyze agile methods, we have chosen four major criteria which are deeply in tune with the agile values as mentioned in the Agile Manifesto [1]. The four criteria are:

Level of Formalization
Rigidity to Change
Process Cost
Project Complexity and Reliability

To calculate the weights and ranks, the same AHP process is repeated for these four criteria. Thus we have four Criteria, five numbers of participants and Consistency Ratio comes out to be 0.2%, with consensus of 94.7%. Thus the approximation is quite good. Based upon this, the corresponding weights and ranks are calculated and the value of Local Agility Indicator is calculated for each method as shown in Table 4.

The input taken can vary from person to person as there is a lot of subjectivity in defining the criteria so we have used Artificial Neural Network with Back Propagation Approach to train the network and producing the correct results even if the input data is inconsistent [13]. Figure 1 shows the trained network with modified weights and the output for a given input. Figure 2 shows the percentage of accuracy for a given selection of output.

**Table 4** Calculation of local agility indicator

| Criteria | Weights | Lean | Scrum | Crystal | XP | DSDM | FDD |
|---|---|---|---|---|---|---|---|
| Level of formalization | 0.225 | 2 | 4 | 5 | 6 | 7 | 8 |
| Rigidity to change | 0.323 | 4 | 2 | 3 | 1 | 8 | 9 |
| Process cost | 0.214 | 6 | 4 | 1 | 3 | 8 | 7 |
| Project complexity and reliability | 0.238 | 7 | 5 | 2 | 4 | 9 | 9 |
| Local agility indicator | | 4.69 | 3.59 | 2.78 | 3.26 | 8.01 | 8.34 |

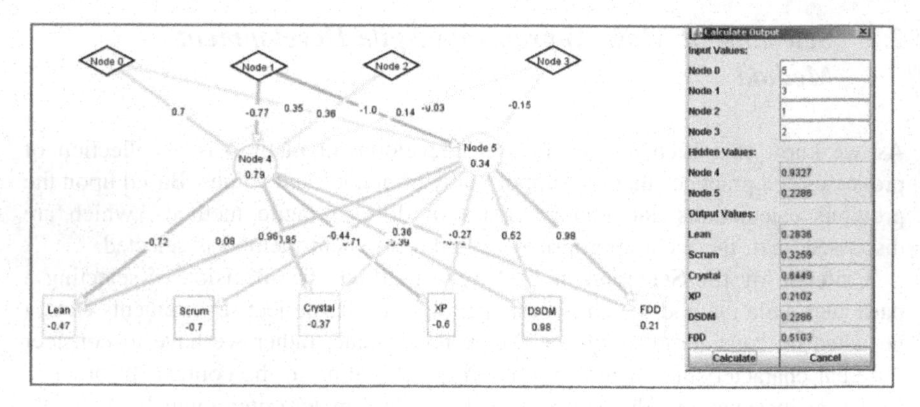


**Fig. 1** Trained network producing output for a given input

Test Results

Correctly Predicted Examples (10):

| Node 0 | Node 1 | Node 2 | Node 3 | Lean | Scrum | Crystal | XP | DSDM | FDD | Predicted Value |
|---|---|---|---|---|---|---|---|---|---|---|
| 2.0 | 4.0 | 6.0 | 7.0 | 1.0 | 0.0 | 0.0 | 0.0 | 0.0 | 0.0 | 0.2335 |
| 3.0 | 5.0 | 7.0 | 5.0 | 1.0 | 0.0 | 0.0 | 0.0 | 0.0 | 0.0 | 0.2727 |
| 4.0 | 2.0 | 4.0 | 5.0 | 0.0 | 1.0 | 0.0 | 0.0 | 0.0 | 0.0 | 0.2197 |
| 4.0 | 2.0 | 5.0 | 4.0 | 0.0 | 1.0 | 0.0 | 0.0 | 0.0 | 0.0 | 0.2432 |
| 5.0 | 3.0 | 1.0 | 2.0 | 0.0 | 0.0 | 1.0 | 0.0 | 0.0 | 0.0 | 0.2191 |
| 6.0 | 3.0 | 2.0 | 1.0 | 0.0 | 0.0 | 1.0 | 0.0 | 0.0 | 0.0 | 0.2355 |
| 6.0 | 1.0 | 3.0 | 4.0 | 0.0 | 0.0 | 0.0 | 1.0 | 0.0 | 0.0 | 0.2273 |
| 7.0 | 1.0 | 3.0 | 3.0 | 0.0 | 0.0 | 0.0 | 1.0 | 0.0 | 0.0 | 0.2403 |
| 8.0 | 9.0 | 7.0 | 9.0 | 0.0 | 0.0 | 0.0 | 0.0 | 0.0 | 1.0 | 0.2064 |
| 8.0 | 8.0 | 8.0 | 9.0 | 0.0 | 0.0 | 0.0 | 0.0 | 0.0 | 1.0 | 0.2067 |

Incorrectly Predicted Examples (2):

| Node 0 | Node 1 | Node 2 | Node 3 | Lean | Scrum | Crystal | XP | DSDM | FDD | Predicted Value |
|---|---|---|---|---|---|---|---|---|---|---|
| 7.0 | 8.0 | 8.0 | 9.0 | 0.0 | 0.0 | 0.0 | 0.0 | 1.0 | 0.0 | 0.2071 |
| 7.0 | 7.0 | 8.0 | 8.0 | 0.0 | 0.0 | 0.0 | 0.0 | 1.0 | 0.0 | 0.208 |

Predicted Correctly: 83%
Predicted Incorrectly: 17%
Select an output to analyze:
Lean Scrum Crystal XP DSDM FDD

**Fig. 2** Percentage of accuracy for a given selection of output

The following aspects about the given criterion are elaborated and at last these criteria are plotted against the different agile approaches as shown in Fig. 3.

**Level of Formalization**. It has been found that if the process formalization is high than it increases the rigidness to change as the formalized processes have fixed and specific operations, which also leads to higher process cost. But off course this may help in increasing the quality and reliability of processes.

**Rigidity to Change**. With dynamic environment the requirements are ever changing, which poses difficulties to larger teams as it hampers parallel development. High complexity projects with high formalization increases the rigidity to change.

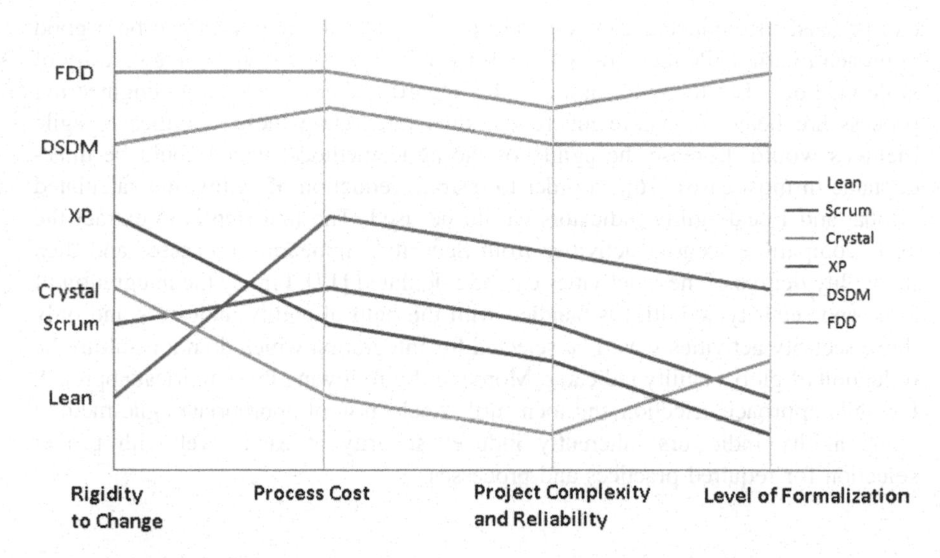


**Fig. 3** Characteristics of the compared agile methods

**Process Cost**. Process cost is directly proportional to the level of process formalization, as by defining specific practices and fixed roles may lead to more overhead.

**Project Complexity and Reliability**. More the project complexity and reliability more is the level of formalization, documentation and other overhead involved in implementing well defined and specific process.

**Comparison of Agile Methods**. We have plotted different agile methods like Lean, Scrum, Crystal clear, XP, DSDM and FDD against the four criteria as shown in Fig. 3. When the comparison of these agile methods is done against the four criteria, it has been clearly found that these agile methods can be classified into two broad categories. One category consists of more liberal and flexible methods like XP, Crystal, Scrum and Lean as shown in lower side of Fig. 3 and the other category consists of more rigid and heavyweight methods like DSDM and Scrum as shown in upper part of the Fig. 3. The lower category is best suited for flexible projects and the upper category is best for large and high reliability projects.

## *2.3 Integration of Security Activities for Selected Agile Development Method*

As the more and more organizations are adopting agile methods for developing projects and with increase of security issues it becomes very essential to use widely used Security engineering processes to provide security to agile processes [14]. Security engineering is based upon best practices, methods and other guidelines that can be used to produce secure software [15]. Theoretically it seems to be a good approach to arm agile methods with security activities for enhancing the security of agile methods. But the major issue in this regard is that these security engineering process are heavy in nature inherently, thus integrating these activities to agile methods would decrease the agility of the agile methods, which could be unacceptable in most cases [16]. In order to restrain reduction of agility, the calculated Global and Local agility indicators can to be used. The first step is to extract the agile compatible security activities from Security engineering processes and then the agility degree of these activities can be calculated [17]. Finally the integration of agile and security activities is handled with the help of agility indicators and only those security activities would be selected for integration which do not result in the reduction of global agility indicator. Moreover by following this empirical approach for agile approach selection and then further selection of appropriate agile method using agility indicators inherently induces security at each level with proper selection for required practices and processes.

## 3 Conclusion

This work provides an integrated approach to select agile development methodology as well as the best agile method suited according to the specific project requirements. As there was not much empirical work done in this field thus we tried to use globally accepted and tested methods like AHP and Artificial Neural Network so as to produce more authentic and reliable results. Global and local agility indicators have been used to measure the agility as there are no other metrics defined for this purpose so far. We hope that this roadmap will serve as a pivotal point for agile method selection and will generate better results in future for this field.

For future work fuzzy logic can be used for improving the inability of AHP to deal with human subjectiveness and imprecision in the comparison process. The process of security enhancement can be further improved by including factors according to industry settings.

## References

1. Beck, K., et al.: Manifesto for Agile Software Development (2001)
2. Schwaber, K., Beedle, M.: Agile Software Development With Scrum. Prentice-Hall, Upper Saddle River (2002)
3. Schwaber, K.: Scrum development process. In: Presented at OOPSLA'95 Workshop on Business Object Design and Implementation (1995)
4. Beck, K.: Extreme Programming Explained. Addison-Wesley, Reading (1999)
5. Beck, K.: Extreme Programming Explained: Embrace Change (2000)
6. Highsmith, J.: Agile Software Development Ecosystems. Pearson Education, Boston (2002)

7. Poppendieck, M., Poppendieck, T.: Lean Software Development An Agile Toolkit. Addison Wesley, Boston (2003)
8. Stapleton, J.: DSDM: The Method in Practice, 2nd edn. Addison Wesley, Longman (2003)
9. Palmer, S.R., Felsing, J.M.: A Practical Guide to Feature-Driven Development. Prentice Hall PTR, Upper Saddle River (2002)
10. Sharma, A., Sharma, R.: A systematic review of agile software development methodologies. In: Proceedings of the National Conference on Innovation and Developments in Engineering and Management (2015)
11. Nasr-Azadani, B., Mohammad Doost, R.: Estimation of agile functionality in software development. In: Proceedings of the International Multiconference of Engineers and Computer Scientists 2008, vol I IMECS (2008)
12. Saaty, T.L.: The Analytic Hierarchy Process. McGraw-Hill, New York (1980)
13. Sharma, A.: Automated design and implementation of ANN. In: Proceedings of the International Symposium, ISCET (2010)
14. Baca, D., Carlsson, B.: Agile development with security engineering activities. In: ACM International Conference on Software Engineering ICSE'11 (2011)
15. Keramati, H., Hassan, S., Hosseinabadi, M.: Integrating software development security activities with agile methodologies. In: IEEE/ACS International Conference on Computer Systems and Applications, pp. 749–754. AICCSA (2008)
16. Sharma, A., Bawa, R.K.: A comprehensive approach for agile development method selection and security enhancement. Proc. Int. J. Innovations Eng. Technol. **6**, 36–44 (2016)
17. Sharma, A., Bawa, R.K.: An integrated framework for security enhancement in agile development using fuzzy logic. Proc. Int. J. Comput. Sci. Technol. **7**, 150–153 (2016)

# Hadoop Framework for Entity Recognition Within High Velocity Streams Using Deep Learning

**S. Vasavi and S. Prabhakar Benny**

**Abstract** Social media such as twitter, Facebook are the sources for Stream data. They generate unstructured formal text on various topics containing, emotions expressed on persons, organizations, locations, movies etc. Characteristics of such stream data are velocity, volume, incomplete, often incorrect, cryptic and noisy. Hadoop framework is proposed in our earlier work for recognising and resolving entities within semi structured data such as e-catalogs. This paper extends the framework for recognising and resolving entities from unstructured data such as tweets. Such a system can be used in data integration, de-duplication, detecting events, sentiment analysis. The proposed framework will recognize pre-defined entities from streams using Natural Language Processing (NLP) for extracting local context features and uses Map Reduce for entity resolution. Test results proved that the proposed entity recognition system could identify predefined entities such as location, organization and person entities with an accuracy of 72%.

**Keywords** Entity recognition · Natural language processing · Stream data · Entity resolution · Hadoop framework · Tweets · Supervised learning

## 1 Introduction

Entity recognition involves identifying named entities in the given formal text and classifying them to one of the predefined entities. This process helps in data integration process of Extract, Transform and Load (ETL). Applications such as automated question answering system, entity duplication also require entity

S. Vasavi (✉)
Department of CSE, VR Siddhartha Engineering College, Vijayawada, India
e-mail: vasavi.movva@gmail.com

S. Prabhakar Benny
University College of Engineering for Women, Kakatiya University, Warangal, India
e-mail: prab_ku@yahoo.co.in



S.C. Satapathy et al. (eds.), *Data Engineering and Intelligent Computing*,
Advances in Intelligent Systems and Computing 542,
DOI 10.1007/978-981-10-3223-3_23

recognition. Entity represents real world objects such as persons, locations, organizations, objects etc. For instance, in the sentence.

"Jawaharlal Nehru is the first prime minister of India", "Jawaharlal Nehru" and "INDIA" are person, location entities respectively. Literature provides approaches, methods, techniques and tools to carry out entity recognition. Natural language processing plays vital role in recognizing and extracting entities. It performs various syntactic and semantic analysis on the text to recognize atomic elements (nouns, verbs, prepositions, quantifiers) of information such as person, organization, location, numeric value, date and time, unit of measurement. Following are the 5 categories of Named Entity Recognition (NER) methods that are used for automatic recognition and classification of named entities:

1. Rule-based NER
2. Machine learning-based NER
3. Hybrid NER
4. Statistical based
5. Deep learning

In rule based NER, names are extracted using human-made rules set. In machine learning based NER, supervised and unsupervised algorithms are used to recognize and classify various linguistic features into pre defined entities. Hybrid NER combines the advantages of both rule based and machine learning-based methods for entity recognition. Statistical based NER is based on training corpus that is manually labelled to train a model. Deep learning is based on a set of algorithms such as Artificial Neural Network (ANN), Deep Neural Network (DNN), Convolutional Neural Network (CNN), Deep Belief Network (DBN) at multiple processing layers. Section 2 presents literature survey. Proposed Hadoop framework for entity recognition using deep learning based NER is given in Sect. 3. Conclusions and future work are given in Sect. 4.

## 2 Literature Survey

A Tweet segmentation process is explained in [1]. Tweet is split into a sequence of consecutive words n-grams (n >= 1), each of which called a segment. Figure 1 presents sample segmentation process.

Framework called HybridSeg is explained in [1]. Tweets are split into segments. Their experiments showed that segment-based NER achieves better accuracy than the word-based alternative. Arabic NER system for social media data is given in [2]. Conditional Random field (CRF) is used by them. NER features such as lexical, contextual, gazetteers, morphological are used. 14 tags in Punjabi language such as Person, Organization, Location, Facility, Event, Relationship, Time, Date, Designation, Title-Person, Number, Measure, Abbreviation and Artifact has been used in [3]. Word window 7, 5 and 3 as context word features with different training and

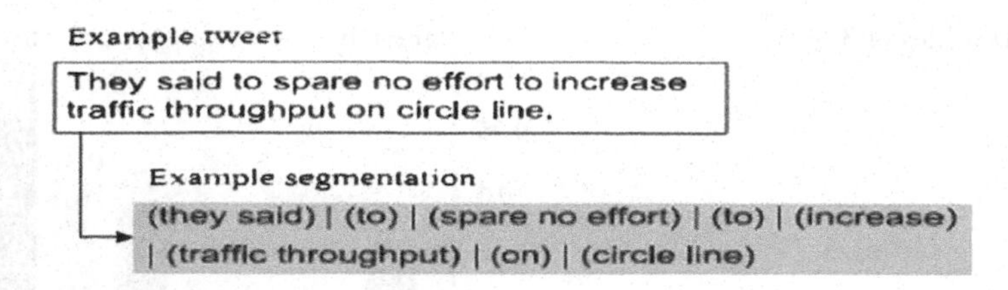


**Fig. 1** Segmentation process [1]

test sets have been discussed. Annotated corpus and gazetteers were manually created by them using online newspapers. C++ based OpenNLP CRF++ package5 is used in that work for segmenting sequential data. A named entity recognition method for concept extraction in micro posts is proposed in [4]. A detailed study on named entity recognition systems can be found in [5]. As per that article, the task of entity recognition started in 1995 for extracting company and defence related information in news papers. Entity recognition from Chinese and Japanese languages started in 1998. Later in 2002 and 2003, this task is performed for Spanish, Dutch and English, German languages respectively. During 2004–2008, entity recognition for Portuguese language is conducted. In 2008, this is extended to Hindi, Bengali, Oriya, Telugu and Urdu languages, in 2007–2011, for Italian and during 2009–2012 for French languages. Entity recognition tools were developed such as Stanford named entity recognizer, Lingpipe, Yamcha, Sanchay, CRF++, Mallet, MITIE. Deep Neural Network (DNN) for engineering Named Entity Recognizers (NERs) in Italian is given in [6]. They used recurrent context window network (RCWN) architecture to predict tags. This network uses recurrent feedback mechanism so that the dependencies are properly modelled. NER using deep learning is proposed in [7]. From a given input text their system could annotate block of text and to which category of entity each word belongs to. Existing algorithms for various 5 steps of the proposed system and the method chosen are explained in the following paragraphs.

1. Tokenization
2. Stop words removal
3. Stemming
4. POS tagging
5. Model building

Tokenization process: Affix removal algorithm, will remove suffixes or prefixes of a word. In Statistical method stemmers are based on statistical analysis techniques. N-gram (tri-gram, bi-gram etc.), HMM stemmer are the examples for this type of tokenizers. Stemmer uses state transition diagram in higher level lexical-Constrain Method. PTB Tokenizer is an efficient tokenizer that could decide when single quotes are parts of words, when periods do and don't imply sentence boundaries, etc. Proposed system uses PTB tokenizer because it has been used to label a wide variety of corpora, including the Brown corpus, the Wall Street Journal

Fig. 2 Densities of 3 entities

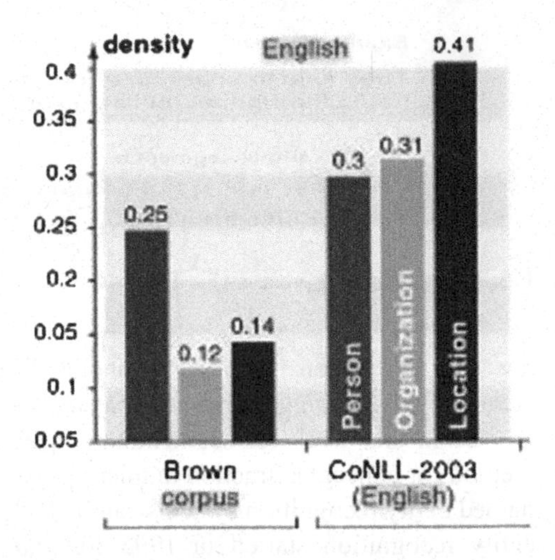


corpus, and the Switchboard corpus [8]. The following Fig. 2 presents densities of the entities person, location, and organization in English text [9].

"Location" ranging from 0.07 to 0.41, "person" as 0.25. Open American National Corpus (OANC) provides 15 million word corpus of American English, Linguistic Data Consortium (LDC) developed annotation guidelines, corpora and other linguistic resources to support the Automatic Content Extraction (ACE) Program, Manually annotated Sub-Corpus (MASC) contains 500 K words drawn (OANC), Ritter Twitter corpus with 10-class annotations.

Stemming: Existing algorithms for stemming are: A Hybrid Algorithm for Stemming of Nepali Text, Malay word stemming, Porter's Stemming Algorithm, Snowball Algorithm. Proposed system uses snowball algorithm.

Stop word removal: Existing algorithms for stop word removal are: Term based random sampling approach, Stop word removal algorithm for Arabic language, Stop list of Stanford (318 stop words), Onix text retrieval. Proposed system uses Stop list of Stanford NLP (318 stop words).

POS tagging: Existing algorithms for POS are: rule based POS tagging, Hidden Markov Model (HMM) POS tagging, transformation based POS tagging, Probabilistic models for POS tagging, HMM and Viterbi Algorithm, Stanford tagger, statistical POS tagging, Natural Language Tool kit (NLTK). The proposed system uses HMM and Viterbi algorithm.

Model building: Hidden Markov Model models are used for natural language processing based tasks like identify part of speech of words in a sentence. Conditional Random Field are Probabilistic models for segmenting and labelling sequence data. Maximum Entropy Model (MEM): The maximum entropy framework estimates probabilities based on the principle of making as few assumptions as possible, other than the constraints imposed. DNN is used in the proposed system.

## 3 Proposed System

Figure 3 presents the overall Hadoop framework for entity resolution. Hadoop framework performs entity recognition and resolution. Mapper phase matches two real world objects and reducer phase generates rules. Similarity between two entities is calculated using 13 different semantic measures and is implemented in Hive programming. This system is tested using e-catalogues of Amazon and Google [10]. Data from web such as product e-catalogs and from data suppliers such as tweets is stored in the data layer. Data consumers send and receive data using presentation layer. This presentation layer also helps in visualization, and reporting jobs on the platform in order to get back results and insights. Pre-processing layer helps in data acquisition, data cleaning, feature extraction and entity recognition. MapReduce does mapping of entities and rule generation in an iterative fashion. Generated rules and mapped entities are stored using storage layer. For resolving entities MapReduce layer uses Snowball stemming algorithm to perform pre-processing, then Token Blocking, blocking process graph techniques are used to remove the redundant comparisons and a total of 13 measures are used for measuring similarity. Detailed explanation about MapReduce functionality can be found in [10]. Figure 4 presents the detailed process of pre-processing layer in the context of stream data. First step in entity recognition is tokenization using PTB tokenizer. A total of 318 words list given by Cambridge University is used as a stop word dataset. Snowball algorithm is used for stemming. Duplicate words are removed from the resulting set of tokens and then given to the next phase. POS tagging is done using HMM and Viterbi algorithm. Reason for using PTB Tokinizer is, it could detect single quotes that are embedded in the text. Tags which are the output from POS tagger step forms bag-of-words. Penn TreeBank, a large annotated corpus of English tag set with 45 tags is used in the proposed system.

Reason for choosing HMM is that, it can calculate highest probability tag sequence for a given sequence of word forms. Since the corpus is labelled, training of HMM became easier. Ambiguity of words (words having more than one

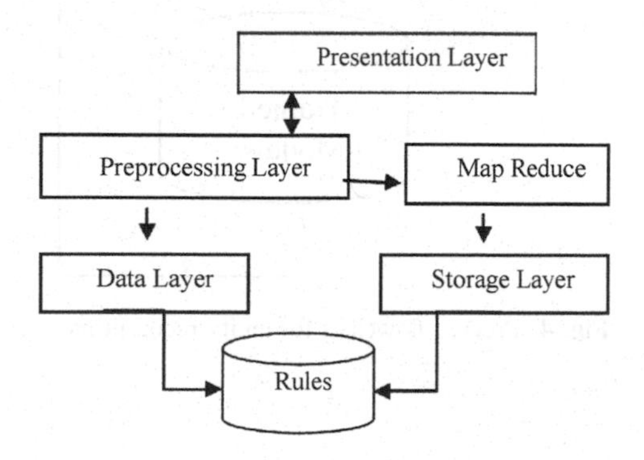


**Fig. 3** Hadoop framework for entity resolution

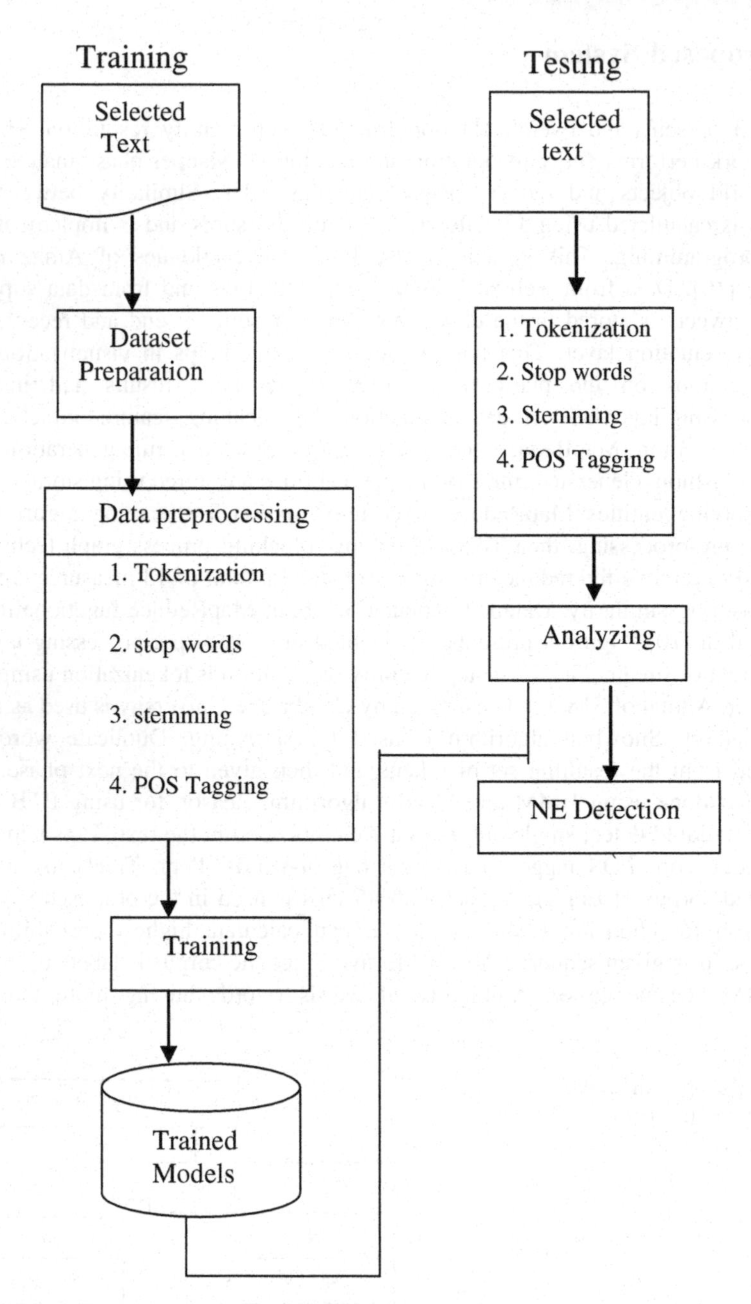


**Fig. 4** Process flowchart for entity recognition

interpretation) is resolved by considering the linguistic features of preceding and succeeding words.

In the experimentation 1: DNN is trained with back propagation algorithm. Stochastic gradient descent function is used for weight updation as shown in (1).

$$wij(t+1) = wij(t) + N\,\partial C/\partial\, wij \tag{1}$$

where N is the learning rate and C is the cost function and is chosen as Cross entropy function as given in (2)

$$C = -\sum dj \log\,(pj) \tag{2}$$

where dj represents the target probability for output unit j and pj is the probability output for j after applying the activation function. In experimentation 2, continuous skip gram model (1-skip tri gram) is modelled. Sample statements of about 125 from a class strength of 70 students is taken as input and given to both models. Table 1 presents tokens obtained by PTB Tokenizer. Table 2 presents the tokens after removing stop words and duplicate words.

For each token in the Table 2, using HMM and Viterbi algorithm, a dictionary is created that best maps each tag X to the probability of the best tag sequence of

**Table 1** Tokens for the given input text

| | | | | | | |
|---|---|---|---|---|---|---|
| Vijayawada | the | city | Had | state | Making | 2011 |
| is | and | is | an | The | it | Suburbs |
| a | the | one | population | City | the | Population |
| city | | of | of | is | in | Trading |
| on | headquarters | the | 1,491,202 | one | the | 1,034,358 |
| the | of | two | VIJAYAWADA | of | state | Business |
| banks | The | in | City | the | in | 1,491,202 |
| of | city | the | is | suburbs | terms | Proposal |
| the | is | state | declared | of | of | Declared |
| in | one | with | as | the | population | One |
| the | of | the | of | state | and | Complete |
| of | the | other | Andhra | capital | It | Two |
| It | major | being | Pradesh | Under | Vijayawada | Civic |
| is | trading | As | until | is | ( | Construction |
| a | and | of | the | also | VMC | Body |
| and | business | 2011 | complete | known | ) | Amaravathi |
| the | centers | the | Construction | as | put | Conversion |
| | of | city | of | The | Proposal | major |
| headquarters | the | had | Amaravathi | Business | for | Centers |
| of | state | a | The | Capital | conversion | capital |
| in | and | population | Civic | of | of | |
| of | hence | of | Body | Andhra | Municipal | |
| the | It | 1,034,358 | of | Pradesh | Corporation | |

**Table 2** Tokens after stop word and duplicates removal

| Vijayawada | Major | 2011 | Civic |
|---|---|---|---|
| City | Trading | Population | Body |
| Banks | Business | 1,034,358 | VMC |
| Headquarters | Centers | 1,491,202 | Proposal |
| State | Andhra | Declared | Conversion |
| One | Pradesh | Complete | Municipal |
| Suburbs | Two | Construction | Corporation |
| Capital | | Amaravathi | |

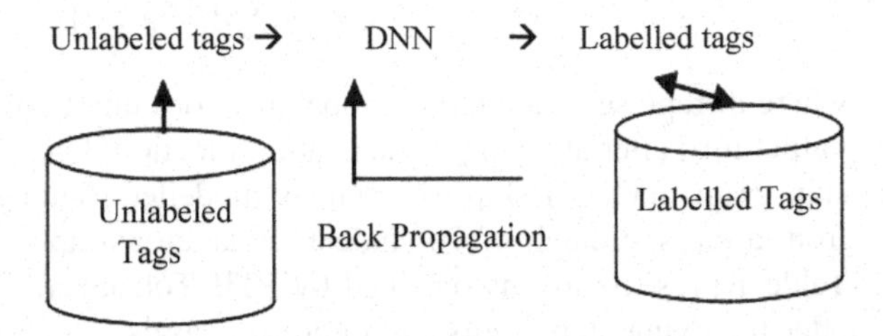


**Fig. 5** Iterative context window network

length i that ends in X. Viterbi algorithm, a dictionary is created that best maps each tag X to the probability of the best tag sequence of length i that ends in X. Viterbi algorithm works in two steps [11]: in the forward step, Find the best path to each node with the lowest negative log probability and in the Backward step, reproduce the path. For middle words calculate the minimum score for all possible previous POS tags. Simple DNN that uses linguistic features is operated on word level to predict entities. Context window network [12] states that, given an input sentence s = [w1, …, wn], for each word wi, the sequences of word contexts [wi − k/2 + 1, …, wi, …, wi + k/2] of size k around the target word wi (i = 1, …, n) are used as input to the network. This is done in iterative fashion as shown in Fig. 5.

At each iteration, recognizing the correct tag for each of the entity is done. Figure 6. presents the sample execution that highlights various identified entities. The following Fig. 7. presents output given by Stanford named entity tagger for the same input text. Table 3 presents the accuracy analysis of the proposed system in comparison to Stanford entity tagger.

In the given text persons are 0, locations are 6 and organizations are 1, the proposed system could correctly recognize that persons are 0, 2 locations are not recognized, one organization is correctly classified but 3 false positives are given. confusion matrix of the proposed system is shown in the following matrix M1.

$$M1 = \begin{matrix} 0 & 0 & 0 \\ 0 & 4 & 2 \\ 0 & 0 & 1 \end{matrix}$$

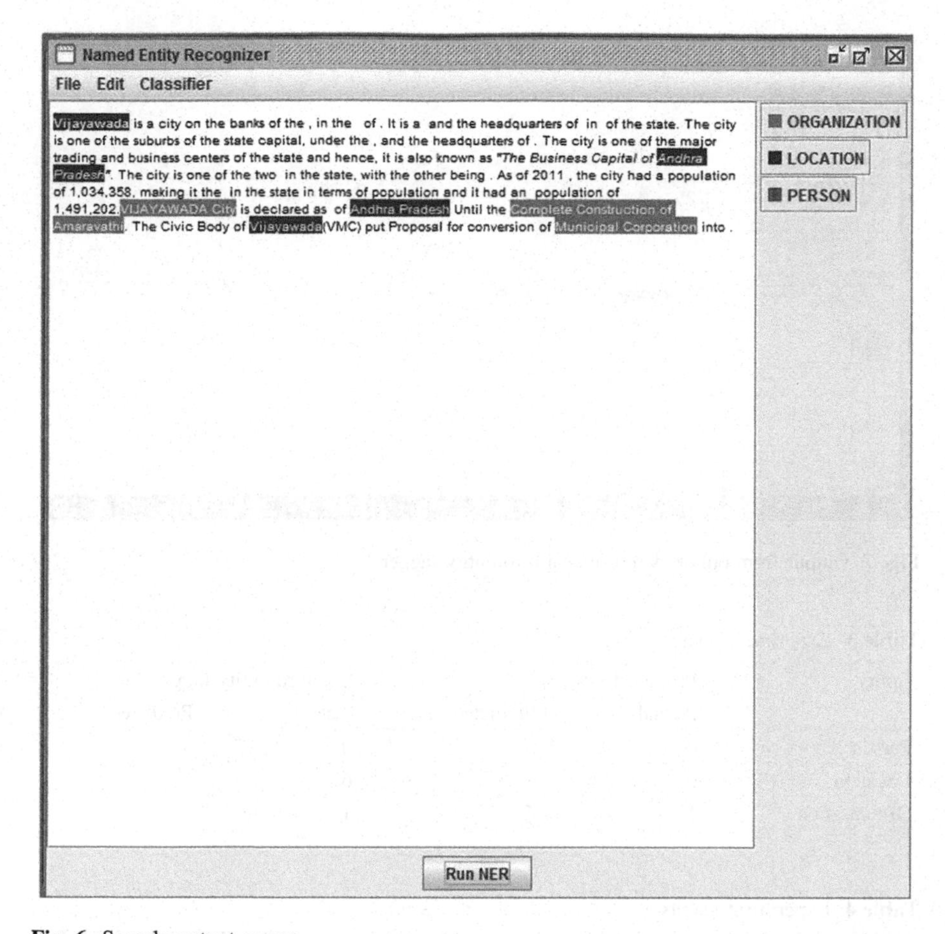


**Fig. 6** Sample output screen

Stanford entity tagger system recognized 3 persons (mis classification), only 2 locations are recognized, and, 2 locations are not recognized, one organization is correctly classified but 1 location is recognized as organization. Confusion matrix for Stanford entity tagger is shown in the following matrix M2.

$$M2 = \begin{matrix} 0 & 0 & 0 \\ 2 & 1 & 0 \\ 0 & 0 & 1 \end{matrix}$$

Even though experimentation 2 (DNN with continuous skip gram model) gave results in fast amount of time, experimentation 1 (DNN) gave 72% of accuracy. Table 4 presents the results of various entity types for both experiments.

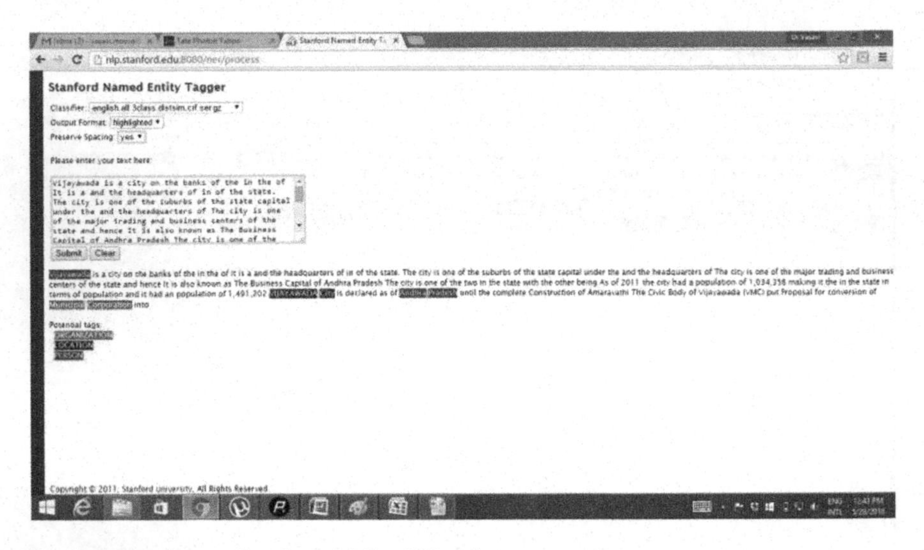


**Fig. 7** Output from online demo of Stanford entity tagger

**Table 3** Experiment results

| Entity | Proposed system | | Stanford entity tagger | |
|---|---|---|---|---|
| | Actual | Predicted | Actual | Predicted |
| Person | 0 | 0 | 0 | 3 |
| Location | 6 | 4 | 6 | 2 |
| Organization | 1 | 4 | 1 | 2 |

**Table 4** Experiment results

| Entity | Experimentation 1 | | | Experimentation 2 | | |
|---|---|---|---|---|---|---|
| | Actual | Predicted | % | Actual | Predicted | % |
| Person | **230** | **176** | **76.52** | **250** | **169** | **73.47** |
| Location | **158** | **117** | **74.05** | **158** | **125** | **79.11** |
| Organization | **54** | **36** | **66.66** | **54** | **28** | **5.85** |
| Average | | | **72.41** | | | **68.14** |

## 4 Conclusion and Future Work

This paper proposed an entity recognition system using NLP features and deep learning. Local context linguistic features are extracted using NLP. Nouns normally from the entities and are extracted. Two experiment setups were tested for recognizing person, location and organization entities. Even though second setup gave the result in a faster way, but the first setup could give better results. Future work is to test on more number of tweets and with more number of entity types. Proposed system is compared with Stanford entity tagger for sample text and proved to be producing better results. This work will be extended to extract Verbs that form relationships, numbers Geo-political and unit of measurement.

## References


1. Li, C., Sun, A., Weng, J., He, Q.: Tweet segmentation and its application to named entity recognition. IEEE Trans. Knowl. Data Eng. 558–570 (2015)
2. Zirikly, A., Diab, M.: Named entity recognition for arabic social media. In: Proceedings of NAACL-HLT 2015, pp. 176–185 (2015)
3. Kaur, A., Josan, G.S.: Evaluation of Punjabi named entity recognition using context word feature. In: IJCA, vol. 96, no 20, pp. 32–38 (2014)
4. Dlugolinsky, S., Krammer, P., Ciglan, M.: Combining named entity recognition methods for concept extraction in microposts. Microposts 1–41 (2014)
5. Patil, N., Patil, A.S., Pawar, B.V.: Survey of named entity recognition systems with respect to Indian and foreign languages. IJCA, vol. 134, no. 16, pp. 21–26 (2016)
6. Bonadiman, D., Severyn, A., Moschitti, A.: Deep neural networks for named entity recognition in Italian. In: QCRI (2016)
7. ERIC: Named-entity recognition using deep learning. http://eric-yuan.me/ner_1/. Accessed Apr 2015
8. Jurafsky, D., Martin, J.H.: Speech and Language Processing, Chapter 9 (2015)
9. Wachsmuth, H.: Text analysis pipelines: towards ad-hoc large-scale text mining, p. 139 (2015)
10. Prabhakar Benny, S., Vasavi, S., Anupriya, P.: International Conference on Computational Modeling and Security (CMS 2016). Elsevier Procedia Computer Science (2016)
11. Neubig, G.: NLP programming tutorial 5—part of speech tagging with hidden Markov models. http://www.phontron.com/slides/nlp-programming-en-04-hmm.pdf. Accessed Apr 2016
12. Collobert, R., Weston, J., Bottou, L., Karlen, M., Kavukcuoglu, K., Kuksa, P.: Natural language processing (almost) from scratch. J. Mach. Learn. Res. **1**(12), 2493–2537 (2011)

# A Novel Random Forest Approach Using Specific Under Sampling Strategy

**L. Surya Prasanthi, R. Kiran Kumar and Kudipudi Srinivas**

**Abstract** In Data Mining the knowledge is discovered from the existing real world data sets. In real time scenario, the category of datasets varies dynamically. One of the emerging categories of dataset is class imbalance data. In Class Imbalance data, the percentages of instances in one class are far greater than the other class. The traditional data mining algorithms are well applicable for knowledge discovery from balance datasets. Efficient knowledge discovery is hampered in the case of class imbalance datasets. In this paper, we propose a novel approach dubbed as Under Sampling using Random Forest (USRF) for efficient knowledge discovery from imbalance datasets. The proposed USRF approach is verified on the 11 benchmark datasets from UCI repository. The experimental observations show that an improved accuracy and AUC is achieved with the proposed USRF approach with a good reduction in RMS error.

**Keywords** Data mining · Knowledge discovery · Imbalance data · Random forest

## 1 Introduction

Data mining techniques can be broadly classified into classification and clustering. Classification is the process of classifying the labeled data into predefined classes. A general issue encountered in data mining is dealing with imbalance datasets, in which one class is predominantly outnumbers the other class. This issue results in

L. Surya Prasanthi (✉) · R. Kiran Kumar
Department of Computer Science, Krishna University, Machilipatnam, India
e-mail: prasanthi.latike@gmail.com

R. Kiran Kumar
e-mail: kirankreddi@gmail.com

K. Srinivas
Department of Computer Science & Engineering, V.R. Siddartha Engineering College, Vijayawada, India
e-mail: vrdrks@gmail.com



S.C. Satapathy et al. (eds.), *Data Engineering and Intelligent Computing*,
Advances in Intelligent Systems and Computing 542,
DOI 10.1007/978-981-10-3223-3_24

high accuracy for the instances of majority class i.e. instances belonging to the predominant class and less accuracy for the instances of minority class. Therefore when dealing with class imbalance datasets a specific strategy has to be implemented for efficient knowledge discovery from the datasets. There are different type of approaches exists in the literature to handle the problem of class imbalance nature, to name a few are oversampling, under sampling, subset approaches, cost sensitive learning, algorithm level implementations and hybrid techniques which combine more than one approaches.

In oversampling, the instances in the minority subset are oversampled by following different strategies. In under sampling, the instances in the majority subset are reduced by several techniques. In subset approaches, the dataset is split into different subsets to reduce the imbalance nature. In cost sensitive learning, the instances are assigned with cost values and the reshuffling of the dataset is performed by considering the cost values. In algorithmic level approaches, the base algorithm applied to the class imbalance data is modified to suit with the imbalance data learning. In hybrid level implementation, more than one above said approaches are applied to solve the problem of class imbalance learning. The existing approaches suffer from the one or more of the drawbacks; either they performed the excessive oversampling, or/and they performed the excessive under sampling etc. We addressed the above issue by following a specific strategy for efficient under sampling using nearest neighbor technique. The results of experimental simulation show a good improvement against the benchmark traditional methods. To overall contributions of our work are as follows,

i. We presented the framework which shows how to pickup only a few instances for performing specific under sampling, and justify this selection process both theoretically and empirically.
ii. The proposed approach will work as a prototype for elaborating experimental analysis; due to the open availability of datasets, compared algorithms and evaluation measures etc.
iii. Finally, our proposed USRF approach outperform almost all the compared benchmark algorithms in terms of accuracy, AUC and root mean square error.

The reminder of the manuscript is organized as follows. Section 2 presets the related work connecting to the class imbalance learning. Section 3 initially presents the formal description of the framework and in the later section the algorithmic approach is also presented. Section 4 presents the experimental methodology and datasets. Section 5 elaborates the results of the proposed approach with the benchmark algorithms. Finally, is Sect. 6 conclusion is presented.

## 2 Class Imbalance Learning Approaches

This section, presented the summarized view of the recent proposals in the domain of class imbalance learning.

In [1] Brown et al. have conducted several experiments on credit scoring imbalance datasets and they shown that random forest is one of the best performing algorithm on the imbalance credit scoring datasets. In [2] Lorena et al. have applied machine learning algorithms especially random forest classifier for modeling species potential distribution.

Molaei and Vadiatizadeh [3] have developed a safe distributed algorithm which is using improved secure sum algorithm and performed on classic ID3. In [4] Hu et al. have investigated on stock trading techniques using the combined approaches of trend discovery and extended classifier system. In [5] Lópezet al. have proposed imbalance domain learning technique which uses iterative instance adjustment approach for efficient knowledge discovery. Kumar et al. [6] have proposed an improved approach using ID3 as the base algorithm with Havrda and Charvatentrophy for building decision tree. Manohar et al. [7] have presented a classification approach for predicting future events.

In [8] Verbiest et al. have propose two prototype selection techniques both based on fuzzy rough set theory which removes noisy instances from the imbalanced dataset and generated synthetic instances. The above descried approaches are analyzed for discovering shortcomings and a novel algorithm know as under sampled random forest is proposed.

## 3 Framework of Under Sampled Radom Forest

This section presents the detail architecture of the proposed USRF approach which consists of four major modules. The detailed working principles of the USRF approach are explained below in the sub-sections.

In the initial phase of our proposed USRF the dataset is split into two subsets known as majority and minority subsets. In majority subset, the percentage of instances is more than other class. In minority subset, the percentage of instances is less than the other class. Since the proposed approach is a under sampling approach. We considered the majority subset for more elaborate analysis for reduction of instances.

The instances in the majority subset are reduced by following the below mentioned techniques; one of the technique is to eliminate the noise instances, the other technique is to find the outliers and the final technique is to find the range of weak instances for removal. The noisy and outlier instances can be easily identified by analyzing the intrinsic properties of the instances. The range of weak instances can be identified by first identifying the weak features in the majority subset. The correlation based feature selection [9] technique selects the important features by following the inter correlation between feature - feature and the inter correlation between feature and class. The features which have very less correlation are identified for elimination. The range of instances which belong to these weak features are identified for elimination from the majority subset. The number of features and instances eliminate by the correlation based feature selection technique will vary from dataset to dataset depending upon the unique properties of the dataset.

The proposed USRF algorithm is summarized as below.

**Algorithm: Under Sampled Radom Forest (USRF)**

**Algorithm:** New Predictive Model
**Input:** D – Data Partition, A – Attribute List, GR – Gain Ratio
**Output:** A Decision Tree

**Procedure:**
**Processing Phase:**
***Step 1.*** *Take the class imbalance data and divide it into majority and minority sub sets. Let the minority subset be P Є pi (i = 1,2,..., pnum) and majority subset be N Є ni(i = 1,2,..., nnum).*

*Let us consider*
*m' = the number of majority nearest neighbors*
*T= the whole training set*
*m= the number of nearest neighbors*

***Step 2.*** *Find mostly misclassified instances pi*
*pi = m'; where m'* $(0 \leq m' \leq m)$
*if m/ 2 ≤ m'<m then pi is a mostly misclassified instance. Then remove the instances m' from the minority set.*

*Let us consider*
*m' = the number of minority nearest neighbors*
***Step 3.*** *Find mostly misclassified instances ni*
*ni = m'; where m'* $(0 \leq m' \leq m)$
*if m/ 2 ≤ m'<m then pi is a mostly misclassified instance. Then remove the instances m' from the majority set.*

*Let us consider*
*m' = the number of majority nearest neighbors*
***Step 4.*** *Find noisy instances pi'*
*pi' = m'; where m'* $(0 \leq m' \leq m)$
*If m'= m, i.e. all the m nearest neighbors of pi are majority examples, pi' is considered to be noise or outliers or missing values and are to be removed.*

*Let us consider*
*m' = the number of minority nearest neighbors*
***Step 5.*** *Find noisy instances ni'*
*ni' = m'; where m'* $(0 \leq m' \leq m)$
*If m'= m, i.e. all the m nearest neighbors of pi are minority examples, ni' is considered to be noise or outliers or missing values and are to be removed.*

***Step 6.*** *For every pi' (i = 1,2,..., pnum') in the minority class P, we calculate its m nearest neighbors from the whole training set T. The number of majority examples among the m nearest neighbors is denoted by m'* $(0 \leq m' \leq m)$.
*If m'= m, i.e. all the m nearest neighbors of pi are majority examples, pi' is considered to be noise or outliers or missing values and are to be removed.*

***Step 7.*** *In this step, we generate s × dnum synthetic minority examples from the minority sub set, where s is an integer between 1 and k . One percentage of synthetic examples generated is replica of minority examples and other are the hybrid of minority examples.*

***Building Predictive Model:***
1. *Create a node N*
2. *If samples in N are of same class, C* ***then***
3. *return N as a leaf node and mark class C;*
4. *If A is empty* ***then***
5. ***return*** *N as a leaf node and mark with majority class;*
6. ***else***
7. *apply Radom Forest*
8. ***endif***
9. ***endif***
10. *Return N*

In the concluding phase of the algorithm, the subset in which irrelevant instances are removed is merged with the minority subset to form the strong dataset, which is further applied to the base algorithm for experimental simulation. In this context random forest [10] is used as the base algorithm for experimental simulation and results generation.

## 4 Methodology and Datasets

The methodology used for validation of generated experimental results is tenfold cross validation. The tenfold cross validation for 10 runs is considered as a decent validation set up in most of the benchmark empirical results simulation in the field of classification. Since, in the tenfold cross validation the mean of 10 runs of each and every measure is considered, the precision of the results can be agreed on any terms. The proposed approach is compared with benchmark set of algorithms C4.5 [11], Reduced Error Pruning (REP) Tree [11], Classification and Regression Trees (CART) [12] and NB Tree [13]. The experiments are implemented within the weka [14] environment on windows 7, i5-2410 M CPU running on 2.30 GHz unit with 4.0 GB of RAM.

***Datasets used in the experimental set up***
The datasets for the experiments are downloaded from the UCI [15] machine learning repository, which are described in Table 1.

The set of eleven UCI datasets: Breast-cancer, Breast-cancer-w, Horse-colic, German_credit, Pima diabetes, Hepatitis, Ionosphere, Labor, Sick, Sonar and Vote are used for experimental simulation. The Imbalance Ratio (IR) shown on the last column of the Table 1 gives the value of imbalance ratio. The value of IR can be calculated for the dataset by dividing the number of instances in the majority subset with number of instances in the minority subset.

**Table 1** UCI datasets and their properties

| S. no. | Dataset | Inst | Attributes | IR |
|---|---|---|---|---|
| 1. Breast | 286 | 9 | 2.37 | |
| 2. Breast-w | | 699 | 9 | 1.90 |
| 3. Colic_h | | 368 | 22 | 1.71 |
| 4. Credit_g | | 1,000 | 20 | 2.33 |
| 5. Diabetes | | 768 | 8 | 1.87 |
| 6. Hepatitis | | 155 | 20 | 3.85 |
| 7. Ionosphere | | 351 | 35 | 1.79 |
| 8. Labor | | 57 | 17 | 1.85 |
| 9. Sick | | 3772 | 30 | 15.32 |
| 10. Sonar | | 208 | 13 | 1.15 |
| 11. Vote | | 43517 | 17 | 1.58 |

**Table 2** Accuracy on all the datasets with summary of tenfold cross validation performance

| Dataset | C4.5 | REP Tree | REP Tree | NB Tree | ID3 | USRF |
|---|---|---|---|---|---|---|
| Breast | 74.28 ± 6.05○ | 69.35 ± 5.34● | 70.22 ± 5.19○ | 70.99 ± 7.94○ | 58.95 ± 9.22● | 69.86 ± 7.96 |
| Breast-w | 95.01 ± 2.73● | 94.79 ± 2.74● | 94.74 ± 2.60● | 96.38 ± 2.23● | 90.62 ± 3.20● | 98.95 ± 1.22 |
| Colic_h | 85.16 ± 5.91○ | 84.94 ± 5.73○ | 85.37 ± 5.41○ | 81.71 ± 6.39● | 52.58 ± 8.09● | 82.00 ± 7.71 |
| Credit_g | 71.25 ± 3.17○ | 72.02 ± 3.38○ | 73.43 ± 4.00○ | 74.27 ± 4.22○ | 8.94 ± 3.03● | 67.29 ± 4.54 |
| Diabetes | 74.49 ± 5.27● | 74.46 ± 4.39● | 74.56 ± 5.01● | 75.24 ± 5.23● | 26.15 ± 4.31● | 81.24 ± 4.57 |
| Hepatitis | 79.22 ± 9.57● | 78.62 ± 7.07● | 77.10 ± 7.12● | 80.93 ± 9.66● | 27.75 ± 10.18● | 82.54 ± 9.45 |
| Iono | 89.74 ± 4.38● | 89.46 ± 4.56● | 88.87 ± 4.84● | 89.15 ± 5.00● | 17.32 ± 4.79● | 94.49 ± 4.23 |
| Labor | 78.6 ± 16.58● | 78.2 ± 17.09● | 80.03 ± 16.67● | 92.27 ± 11.79○ | 59.33 ± 20.60● | 87.10 ± 14.47 |
| Sick | 98.72 ± 0.55○ | 98.68 ± 0.57○ | 98.85 ± 0.54○ | 97.82 ± 0.76 | 80.78 ± 1.88● | 97.82 ± 0.90 |
| Sonar | 73.61 ± 9.34● | 72.69 ± 10.19● | 70.72 ± 9.43● | 77.07 ± 9.65● | 70.96 ± 1.93● | 79.29 ± 10.42 |
| Vote | 96.57 ± 2.56 | 95.33 ± 3.10● | 95.81 ± 2.64● | 95.03 ± 3.29● | 93.15 ± 3.32● | 96.24 ± 3.0 |

○ Empty dot indicates the loss of USRF. ● Bold dot indicates the win of USRF

**Table 3** AUC on all the datasets with summary of tenfold cross validation performance

| Dataset | C4.5 | REP Tree | REP Tree | NB Tree | ID3 | USRF |
|---|---|---|---|---|---|---|
| Breast | 0.606 ± 0.087● | 0.580 ± 0.109● | 0.587 ± 0.110● | 0.663 ± 0.107● | 0.593 ± 0.097● | 0.696 ± 0.108 |
| Breast_w | 0.957 ± 0.034● | 0.959 ± 0.029● | 0.950 ± 0.032● | 0.986 ± 0.015● | 0.953 ± 0.024● | 0.998 ± 0.007 |
| Colic_h | 0.840 ± 0.070● | 0.847 ± 0.065● | 0.847 ± 0.070● | 0.859 ± 0.070● | 0.716 ± 0.060● | 0.905 ± 0.059 |
| Credit_g | 0.640 ± 0.062● | 0.712 ± 0.053○ | 0.716 ± 0.055○ | 0.760 ± 0.056○ | 0.513 ± 0.035● | 0.703 ± 0.060 |
| Diabetes | 0.751 ± 0.070● | 0.761 ± 0.057● | 0.743 ± 0.071● | 0.804 ± 0.055● | 0.539 ± 0.052● | 0.877 ± 0.042 |
| Hepatitis | 0.668 ± 0.184● | 0.620 ± 0.150● | 0.563 ± 0.126● | 0.826 ± 0.135● | 0.474 ± 0.043● | 0.867 ± 0.125 |
| Iono | 0.891 ± 0.060● | 0.899 ± 0.055● | 0.896 ± 0.059● | 0.920 ± 0.048● | 0.738 ± 0.064● | 0.982 ± 0.025 |
| Labor | 0.726 ± 0.224● | 0.768 ± 0.233● | 0.750 ± 0.248● | 0.964 ± 0.093 | 0.713 ± 0.193● | 0.952 ± 0.101 |
| Sick | 0.952 ± 0.040● | 0.968 ± 0.030● | 0.954 ± 0.043● | 0.938 ± 0.038● | 0.871 ± 0.033● | 0.992 ± 0.009 |
| Sonar | 0.753 ± 0.113● | 0.749 ± 0.105● | 0.721 ± 0.106● | 0.831 ± 0.099● | 0.498 ± 0.013● | 0.879 ± 0.080 |
| Vote | 0.979 ± 0.025● | 0.975 ± 0.024● | 0.973 ± 0.027● | 0.987 ± 0.017○ | 0.937 ± 0.036● | 0.984 ± 0.025 |

○ Empty dot indicates the loss of USRF. ● Bold dot indicates the win of USRF

**Table 4** RMS Error on all the datasets with summary of tenfold cross validation performance

| Dataset | **C4.5** | **REP Tree** | **REP Tree** | **NB Tree** | **ID3** | **USRF** |
|---|---|---|---|---|---|---|
| Breast | 0.444 ± 0.037○ | 0.466 ± 0.032● | 0.458 ± 0.039 | 0.473 ± 0.057● | 0.567 ± 0.072● | 0.458 ± 0.051 |
| Breast_w | 0.205 ± 0.060● | 0.209 ± 0.056● | 0.213 ± 0.058● | 0.169 ± 0.062● | 0.185 ± 0.070● | 0.087 ± 0.044 |
| Colic_h | 0.352 ± 0.060○ | 0.353 ± 0.058○ | 0.346 ± 0.059○ | 0.379 ± 0.072● | 0.391 ± 0.105● | 0.358 ± 0.048 |
| Credit_g | 0.476 ± 0.028● | 0.441 ± 0.025○ | 0.435 ± 0.026○ | 0.428 ± 0.034○ | 0.595 ± 0.114● | 0.458 ± 0.024 |
| Diabetes | 0.439 ± 0.042● | 0.430 ± 0.032● | 0.432 ± 0.036● | 0.417 ± 0.037● | 0.624 ± 0.059● | 0.366 ± 0.037 |
| Hepatitis | 0.404 ± 0.096● | 0.402 ± 0.057● | 0.419 ± 0.052● | 0.371 ± 0.099● | 0.510 ± 0.221● | 0.344 ± 0.076 |
| Iono | 0.299 ± 0.081● | 0.293 ± 0.065● | 0.302 ± 0.068● | 0.299 ± 0.078● | 0.050 ± 0.131○ | 0.205 ± 0.053 |
| Labor | 0.401 ± 0.170● | 0.387 ± 0.166● | 0.380 ± 0.183● | 0.200 ± 0.163○ | 0.425 ± 0.274● | 0.285 ± 0.117 |
| Sick | 0.105 ± 0.024○ | 0.106 ± 0.023○ | 0.099 ± 0.027○ | 0.136 ± 0.024● | 0.118 ± 0.025○ | 0.127 ± 0.018 |
| Sonar | 0.491 ± 0.093● | 0.452 ± 0.071● | 0.474 ± 0.078● | 0.434 ± 0.098● | 0.130 ± 0.344○ | 0.374 ± 0.054 |
| Vote | 0.157 ± 0.065○ | 0.186 ± 0.061 | 0.180 ± 0.060● | 0.185 ± 0.068● | 0.239 ± 0.076 ● | 0.172 ± 0.063 |

○ Empty dot indicates the loss of USRF. ● Bold dot indicates the win of USRF

The accuracy is the percentage of instances correctly classified by a classifier. The accuracy can also be defied in the terms of True positive (TP): Actual positive instances which are classified as positive by the classifier, True Negative (TN): Actual negative instances which are classified as negative by the classifier, False Positive (FP):Actual negative instances which are classified as positive instances by the classifier and False Negative (FN):Actual positive instances which are classified as negative instances by the classifier. The accuracy can be given below as Eq. (1),

$$Accuracy = \frac{TP + TN}{TP + FN + FP + TN} \tag{1}$$

AUC is the arithmetic mean of TP Rate and TN rate for only one run of the classifier. When there are multiple runs of the classifier, AUC is the captured area in the Receiver Operative Curve (ROC) of TP Rate and TN rate for multiple runs. Another important measure used in Root Mean Square (RMS) Error.

## 5 Experimental Results

In this section, we presented the completed set of experimental observations. The proposed approach (Under Sampled using Random Forest) USRF is predominant on all the evaluation metrics. The accuracy, AUC and RMS Error are generated using tenfold cross validation method.

From Table 2, we can observe that the proposed algorithms USRF accuracy value is improved on almost all the datasets. Table 3 compares the AUC value of the ID3 algorithm with the proposed USRF algorithm. The AUC value of the USRF

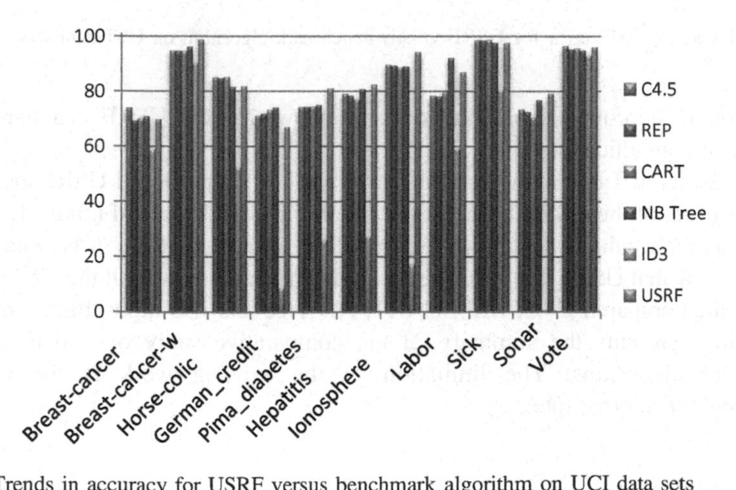


**Fig. 1** Trends in accuracy for USRF versus benchmark algorithm on UCI data sets

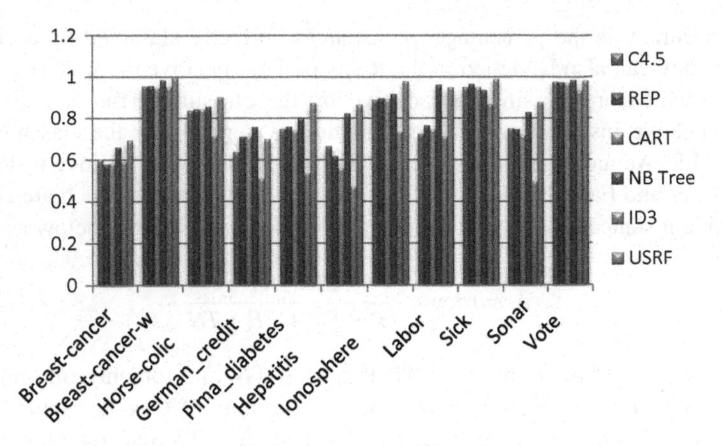


**Fig. 2** Trends in AUC for USRF versus benchmark algorithm on UCI data sets

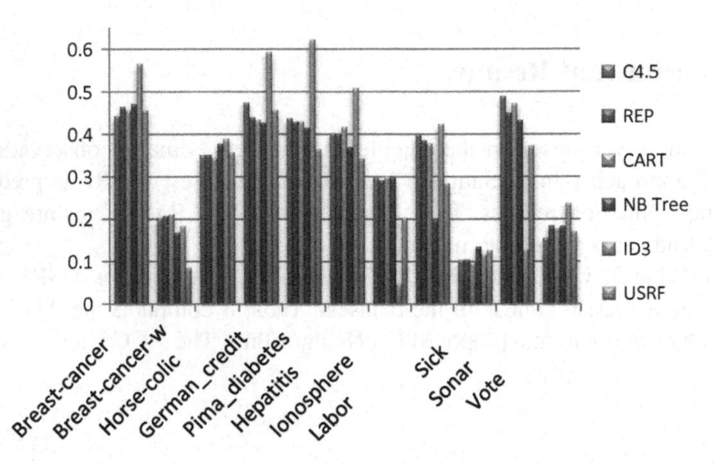


**Fig. 3** Trends in RMS error for USRF versus benchmark algorithm on UCI data sets

algorithm is improved on all the datasets show that the USRF can handle the imbalance data efficiently.

The results in Table 4 shows RMS error rate for the proposed USRF algorithm. The error rate of the USRF is decreased for all the UCI datasets. Figures 1, 2 and 3 represents the results of accuracy, AUC ad RMS error in the form of bar charts. The figures show that USRF has improved o all the metrics o almost all the UCI datasets against the compared C4.5, REP, CART, NB Tree and ID3 algorithms. The result in Table 5 presents the summary of the comparative study of USRF with the traditional algorithms. The limitation of the existing work is the need of improvement in error rate.

**Table 5** Summary of experimental results for USRF

| Results | Systems | Wins | Ties | Losses |
|---|---|---|---|---|
| Accuracy | USRF v/s C4.5 | 6 | 1 | 4 |
| | USRF v/s REP Tree | 8 | 0 | 3 |
| | USRF v/s CART | 7 | 0 | 4 |
| | USRF v/s NB Tree | 7 | 1 | 3 |
| | USRF v/s ID3 | 11 | 0 | 0 |
| AUC | USRF v/s C4.5 | 11 | 0 | 0 |
| | USRF v/s REP Tree | 10 | 1 | 1 |
| | USRF v/s CART | 10 | 0 | 1 |
| | USRF v/s NB Tree | 8 | 1 | 2 |
| | USRF v/s ID3 | 11 | 0 | 0 |
| RMS Error USRF v/s C4.5 | 7 | 0 | 4 | |
| | USRF v/s REP Tree | 8 | 0 | 3 |
| | USRF v/s CART | 7 | 1 | 3 |
| | USRF v/s NB Tree | 9 | 0 | 2 |
| | USRF v/s ID3 | 8 | 0 | 3 |

# 6 Conclusion

We have proposed an effective and simple classification algorithm for handling class imbalance problem. This method uses the under sampling strategy which uses a unique way for identifying the surplus instances from the majority subset and balance the dataset to some extend. Empirical results have shown that USRF reasonable improved the results for reducing the imbalance effect compared with traditional methods. The proposed method can be extended to better visualize the unique properties of each and every datasets.

# References


1. Brown, I., Mues, C.: An experimental comparison of classification algorithms for imbalanced credit scoring data sets. Expert Syst. Appl. **39**, 3446–3453 (2012)
2. Lorena, A.C., Jacintho, L.F.O., Siqueira, M.F., De Giovanni, R., Lohmann, L.G., de Carvalho, A.C.P.L.F., Yamamoto, M.: Comparing machine learning classifiers in potential distribution modeling. Expert Syst. Appl. **38**, 5268–5275 (2011)
3. Molaei, E., Vadiatizadeh, H., Amirmahdimohammadighavam, Rajabpour, N.: Fatemehziasistani Distributed algorithm for privacy preserving data mining based on ID3 and improved secure sum. Int. J. Adv. Stud. Comput. Sci. Eng. IJASCSE **3**(1), 28–34 (2014)
4. Hua, Y., Feng, B., Zhang, X., Ngai, E.W.T., Liu, M.: Stock trading rule discovery with an evolutionary trend following model. Expert Syst. Appl. **42**, 212–222 (2015)
5. López, V., Triguero, I., Carmona, C.J., García, S., Herrera, F.: Addressing imbalanced classification with instance generation techniques:IPADE-ID. Neurocomputing **126**, 15–28 (2014)

6. Kumar, S., Jain, S.: Intrusion detection and classification using improved ID3 algorithm of data mining. Int. J. Adv. Res. Comput. Eng. Technol. **1**(5), 352–356 (2012)
7. Manohar, S., Mittal, A., Naik, S., Ambre, A.: A dynamic classifier using decision tree algorithm. Int. J. Adv. Res. Comput. Sci. Softw. Eng. **5**(1), 628–631 (2015)
8. Verbiest, N., Ramentol, E., Cornelisa, C., Herrerac, F.: Preprocessing noisy imbalanced datasets using SMOTE enhanced with fuzzy rough prototype selection. Applied Soft Comput. **22**, 511–517 (2014)
9. Hall, M.A.: Correlation-based feature subset selection for machine learning. PhD thesis (1998)
10. Breiman, Leo: Random forests. Mach. Learn. **45**(1), 5–32 (2001)
11. Quinlan, J.: Induction of decision trees. Mach. Learn. **1**, 81C106 (1986)
12. Breiman, L., Friedman, J., Olshen, R., Stone, C.: Classification and Regression Trees. Wadsworth, Belmont, CA (1984)
13. Kohavi, R.: Scaling up the accuracy of Naive-Bayes classifiers: a decision-tree hybrid. In: Second International Conference on Knoledge Discovery and Data Mining, pp. 202–207 (1996)
14. Witten, I.H., Frank, E.: Data Mining: Practical Machine Learning Tools and Techniques, 2nd edn. Morgan Kaufmann, San Francisco (2005)
15. Asuncion, H.A., Newman, D.: UCI Repository of Machine Learning Database (School of Information and Computer Science). University of California, Irvine, CA (2007). http://www.ics.uci.edu/ ~ mlearn/MLRepository.html

# Improved Whale Optimization Algorithm Case Study: Clinical Data of Anaemic Pregnant Woman

**Ravi Kumar Saidala and Nagaraju Devarakonda**

**Abstract** WOA is a meta-heuristic algorithm possessing the proper potentiality in solving complex numerical function optimization problems. It works well, but poor in the convergence at exploration and exploitation phases. In order to enhance the convergence enforcement of WOA, a novel constitutional appraising strategy based WOA has been set forth in this paper. In this scenario, constituent states are fully utilized in each of the iterations to supervise the subsequent gazing process, and to counterbalance the local exploration with global exploitation. We fix up with the mechanism together with the convergence straight stuff of the enhanced algorithm. Comparable investigations are supervised on various mathematical benchmark function optimization problems. Simulation results confirm, with statistical significance, that the proposed scenario is more efficient in the convergence performance of WOA. In addition to this, we applied the same technique to a clinical dataset of an anaemic pregnant woman and obtained optimized clusters and cluster heads to secure a clear comprehension and meaningful insights in the clinical decision-making process.

**Keywords** Meta-heuristic optimization techniques · WOA · Clinical decision-making process · Clinical data analysis · Outlier detection

---

R.K. Saidala (✉)
University College of Engineering and Technology, Acharya Nagarjuna University, Guntur, India
e-mail: saidalaravikumar@gmail.com

N. Devarakonda
Lakireddy Bali Reddy College of Engineering, Mylavaram, Andhra Pradesh, India
e-mail: dnagaraj_dnr@yahoo.co.in



S.C. Satapathy et al. (eds.), *Data Engineering and Intelligent Computing*, Advances in Intelligent Systems and Computing 542,
DOI 10.1007/978-981-10-3223-3_25

# 1 Introduction

Nature inspired Meta-heuristic algorithms have been long used to solve the large, complex and dynamic numerical function optimization problems by mimicking biological or physical phenomena [1–3]. Authors [1] proposed WOA (Whale Optimization Algorithm) based on foraging etiquette of humpback whales which has been shown to be more competitive when compared with that of other conventional biological-Inspired optimization algorithms, such as DE (differential evolution), MBO, AFSA, ACO, ABC, Monkey Gaze, BCPA, FA, HS, BMO, CS, FOA, DE (Dolphin Echolocation), DPO, BA, KH, PSO, GWO [4], ALO [5] etc. The potency of these algorithms is that each one in the swarm always shares the information of the best solution that obtained individually is combined together. This process reduces the difficulty of getting an optimized solution over the course of generations. Authors [1] mentioned three operators to simulate in gaze of prey, encircling of prey, and bubble-net foraging etiquette of humpback whales. The simulated foraging etiquette of humpback whales with random or the best gaze agent to chase the prey is in two ways. They are upward-spiral and bubble-net attacking mechanism. The selection of attacking mechanism is based on a random number of $p$, where $p$ is in [0, 1].

As far as we studied many research articles regarding convergence problems in exploration and exploitation phases of nature inspired optimization algorithms and found that application of convergence rates in the preceding iteration as retaliation information to supervise the subsequent gazeing process. We confined that WOA suffers from low convergence rate. To obtain fast convergence rate, we tried various convergence formulas [6–8] and successfully secured by a convergence factor $\gamma$ in exploration and exploitation phases. In the modified process of obtaining a fast convergence rate, each gaze agent would also take a gait toward another topography selected from a set of qualified solutions. By performing so, these have more chances in finding fitter topographies and supervise the crossover as well as the mutation process in global exploration. Instead of the selection strategy that the obtained solutions for recombination as stated in standard WOA, there may be a chance to improve the convergence performances of whales by measuring the truths, whether they are better than the previous one but not by the corresponding fitness values. Based on this, we believe that the embryonic convergence can be efficiently avoided.

**Roadmap**: The remainder of this work is systematized as follows. In Sect. 2 we presented the standard WOA and in the next section we bestowed the proposed algorithm, followed by the convergence factor in Sect. 3.1 and the improved WOA in Sect. 3.2. In Sect. 4 detailed tabular and graphical experimental evaluation results has been furnished. The last section concludes the whole paper and points out the future scope.

## 2 Standard WOA

This algorithm is first coined by Mirjalili and Lewis [1] which is inspired by the special foraging etiquette of humpback whales. The whales perspicaciously create distinctive bubbles to make the school of small fishes move to the surface. It has been observed that these bubbles are in either '*circular*' manner or '9' shaped path. They found two maneuvers associated with bubble and named them '*upward-spirals*' and '*double- loops*'. This foraging process is divided into two phases' *exploration* and *exploitation*. In the first phase gaze agents (whales) randomly gaze for prey and then update the topography of a gaze agent. To percolate global randomly chosen a gaze agent instead of the best gaze agent obtained so far.

$$\vec{X}_{t+1} \leftarrow \vec{X}_{rand} - \vec{A}.\vec{D} \tag{1}$$

where, $\vec{D} = \left|\vec{A}.\vec{X}_{rand} - \vec{X}\right|$, $\vec{X}_{rand}$ is random topography vector, $\vec{A} \geq 1$ or $\vec{A} < -1$. If $\vec{A} \geq 1$ then new topography becomes the Eq. (1) else if $\vec{A} < -1$ new topography becomes the Eq. (2). WOA assumes that updated topography is optimal solution to target the prey. The updated topography is obtained by

$$\vec{X}_{t+1} \leftarrow \overrightarrow{X^*}_t - \vec{A}.\vec{D} \tag{2}$$

where, $\vec{D} = \left|\vec{C}.\overrightarrow{X^*}_t - \vec{X}_t\right|$ $\vec{A}$ and $\vec{C}$ are internal parameters, the subscript't 'specifies the current iteration, $\vec{X}$ is the topography vector, $\overrightarrow{X^*}$ is the new topography vector. '| |', '. ' are absolute value and element-by-element multiplication respectively. Both $\vec{A}$ and $\vec{C}$ are coefficient vectors where, $\vec{A} = 2\vec{a}.\vec{r} - \vec{a}$, $\vec{C} = 2.\vec{r}$. Here, $\vec{a}$ straightaway decreases from 2 to 0 during the course of maneuver (in both phases: exploration and exploitation) and $\vec{r} = (-1).*\mathrm{rand}(0, 1)$. In the second phase bubble-net attacking will happen. This consists of two approaches. WOA assumes probability of 50% to choose between either these two approaches. They are *shrinking encircling mechanism* and *Spiral updating topography*. The new updated topography becomes

$$\vec{X}_{t+1} = \begin{cases} \overrightarrow{X^*}_t - \vec{A}.\vec{D}, & p < 0.5 \quad (3) \\ \overrightarrow{D'}.e^{bl}.\cos(2\pi l) + \overrightarrow{X^*}_t & ,p \geq 0.5 \quad (4) \end{cases}$$

where, $\overrightarrow{D'} = \left|\overrightarrow{X^*}_t - X^*_t\right|$, $p = (-1).*\mathrm{rand}(0, 1)$. In first approach $\vec{a}$ is decreased to achieve to shrinking the encircling mechanism. Then $\vec{A} = (-2a).*\mathrm{rand}(1, 1) + a$ where, 'a' is an integer and coefficient $\vec{A}$ is decreased by $\vec{a}$. In the second approach, constant $b$ is used in defining the shape of the logarithmic spiral, $l = (-2).*\mathrm{rand}(1, 1) + 1$, and '. ' is a dot product operator.

# 3 Proposed Algorithm

## 3.1 Fast Convergence γ Factor

Optimization algorithms are tested based on number of gazeing agents and iterations. The fitness of each agent is calculated on each iteration. If the algorithm works good when it secure best fitness in the exploration phase. The performance of the algorithm may fall down when it gets inefficient gazeing iterations before better topography is derived. To overcome this problem gamma $\gamma$ factor also called convergence factor is used in many literatures.

$$\gamma_i = e^{\left\{-[trial(i)-1]\cdot\frac{\ln 10}{D-1}\right\}} \tag{5}$$

where, $D$ represents dimension, *trial* represents the number of inefficient gazeing iterations. *trial(i)* is a number and it should satisfy $i \leq trial(i) \geq D$. Instead of inefficient gazeing iterations before obtaining better topography, there may be a chance to improve the convergence performances of gaze agents by measuring the truths, whether they are better than the previous one but not by the corresponding fitness values using the above convergence factor.

**Table 1** Notations of uni and multimodal benchmark functions where **V_no** = 30, $f_{\mathbf{min}} = 0$

| Function name | Range |
|---|---|
| $F_1(x) = \sum_{i=1}^{n} x_i^2$ | [−100, 100] |
| $F_2(x) = \sum_{i=1}^{n} \lvert x_i \rvert + \prod_{i=1}^{n} \lvert x_i \rvert$ | [−10, 10] |
| $F_3(x) = \sum_{i=1}^{n} \left(\sum_{j-1}^{i} x_j\right)^2$ | [−100, 100] |
| $F_4(x) = \max_i\{\lvert x_i \rvert, 1 \leq i \leq n\}$ | [−100, 100] |
| $F_5(x) = \sum_{i=1}^{n-1} [100(x_{i+1} - x_i^2)^2 + (x_i - 1)^2]$ | [−30, 30] |
| $F_6(x) = \sum_{i=1}^{n} ([x_i - 0.5])^2$ | [−100, 100] |
| $F_7(x) = \sum_{i=1}^{n} ix_i^4 + \text{random}[0, 1)$ | [−1.28, 1.28] |
| $F_8(x) = \sum_{i=1}^{n} -x_i^2 \sin(\sqrt{\lvert x_i \rvert})$(Note: $f_{\mathbf{min}} = -418.9829*5$) | [−500, 500] |
| $F_9(x) = \sum_{i=1}^{n} [x_i^2 - 10\cos(2\pi x_i) + 10]$ | [−5.12, 5.12] |
| $F_{10}(x) = -20e^{\left(-0.2\sqrt{\frac{1}{n}\sum_{i=1}^{n} x_i^2}\right)} - e^{\left(\frac{1}{n}\sum_{i=1}^{n}\cos(2\pi x_i)\right)} + 20 + e$ | |
| $F_{11}(x) = \frac{1}{4000}\sum_{i=1}^{n} x_i^2 - \prod_{i=1}^{n}\cos\left(\frac{x_i}{\sqrt{i}}\right) + 1$ | [−32, 32] |
| $F_{12}(x) = \frac{\pi}{n}\{10\sin(\pi y_1) + \sum_{i=1}^{n} (y_i - 1)^2[1 + 10\sin\pi y_{i+1}^2] + (y_n - 1)^2\} + \sum_{i-1}^{n} u(x_i, 10, 100, 4)$ | [−600, 600] |
| $y_i = 1 + \frac{x_i + 1}{4} u(x_i, a, k, m) = \begin{cases} k(x_i - a)^m & x_i > a \\ 0 & -a < x_i < a \\ k(-x_i - a)^m & x_i < -a \end{cases}$ | [−50, 50] |
| $F_{13}(x) = 0.1\{\sin(3\pi x_1)^2 + \sum_{i=1}^{n} (x_i - 1)^2\left[1 + \sin(3\pi x_i + 1)^2\right] + (x_n - 1)^2\left[1 + \sin 2\pi x_n)^2\right]\} + \sum_{i-1}^{n} u(x_i, 5, 100, 4)$ | [−50, 50] |

**Table 2** Notations of fixed-dimension multimodal benchmark functions

| Function name | V_No | Range | $f_{min}$ |
|---|---|---|---|
| $F_{14}(x)=\left(\frac{1}{500}\sum_{j=1}^{25}\frac{1}{j+\sum_{i=1}^{2}(x_i-a_{ij})^6}\right)^{-1}$ | 2 | [−65, 65] | 1 |
| $F_{15}(x)=\sum_{i=1}^{11}\left[a_i-\frac{x_1(b_i^2+b_ix_2)}{b_i^2+b_ix_3+x_4}\right]^2$ | 4 | [−5, 5] | 0.0003 |
| $F_{16}(x)=4x_1^2-2.1x_1^4+\frac{1}{3}x_1^6+x_1x_2-4x_2^2+4x_2^4$ | 2 | [−5, 5] | 1.0316 |
| $F_{17}(x)=(x_2-\frac{5.1}{4\pi^2}x_1^2+\frac{5}{\pi}x_1-6)^2+10(1-\frac{1}{8\pi})\cos x_1+10$ | 2 | [−5, 5] | 0.398 |
| $F_{18}(x)=\left[1+(x_1+x_2+1)^2\begin{pmatrix}19-14x_1+\\3x_1^2-14x_2+6x_1x_2\end{pmatrix}+3x_2^2\right]*[30+(2x_1-3x_2)^2*(18-32x_1+12x_1^2+48x_2-36x_1x_2+27x_2^2)]$ | 2 | [−5, 5] | 3 |
| $F_{19}(x)=-\sum_{i=1}^{4}c_ie^{\left(-\sum_{j=1}^{3}a_{ij}(x_j-p_{ij})^2\right)}$ | 3 | [1, 3] | −3.86 |
| $F_{20}(x)=-\sum_{i=1}^{4}c_ie^{\left(-\sum_{j=1}^{6}a_{ij}(x_j-p_{ij})^2\right)}$ | 6 | [0, 1] | −3.32 |
| $F_{21}(x)=-\sum_{i=1}^{5}[(X-a_i)(X-a_i)^T+c_i]^{-1}$ | 4 | [0, 10] | −10.1532 |
| $F_{22}(x)=-\sum_{i=1}^{7}\left[(X-a_i)(X-a_i)^T+c_i\right]^{-1}$ | 4 | [0, 10] | −10.4028 |
| $F_{23}(x)=-\sum_{i=1}^{10}[(X-a_i)(X-a_i)^T+c_i]^{-1}$ | 4 | [0, 10] | −10.5363 |

## 3.2 Algorithm of Improved WOA

In the proposed method called Improved WOA, the objective is to include a set of qualified solutions found so far in the gaze process of whales. Improved WOA is based on a new convergence rate in exploration and exploitation phases. It adopts other operations as the standard WOA. It is described as follows.

**Step 1** Initialize the whale population $X_i\ (i=1,2,\ldots,n)$

**Step 2** Calculate the fitness of each gaze agent (whales)
$X^*$ is the best gaze agent obtained

**Step 3** Update a, A, C, l and p every gaze agent

**Step 4** If p< 0.5
If $|A|<1$
Update the topography using $\vec{X}_{t+1}\leftarrow\gamma_i.\vec{X^*}_t-\vec{A}.\vec{D}$ instead of Eq. (2)
else If $A\geq 1$
Select a random gaze agent $X_{rand}$ update the topography using $\vec{X}_{t+1}\leftarrow\gamma_i.\vec{X}_{rand}-\vec{A}.\vec{D}$ Instead of Eq. (1)
else If p $\geq$ 0.5
Update the topography using
$\vec{X}_{t+1}\leftarrow\gamma_i.\vec{D'}.e^{bl}.\cos(2\pi l)+\vec{X^*}_t$ instead of Eq. (4)

**Step5** Check is any gaze agent goes beyond the gaze space and amend it

**Step 6** Calculate the fitness of each gaze agent

**Step 7** Update $X^*$ if there is a better solution

**Step 8** Go to **Step 3** t times and return $X^*$

## 4 Experimental Results

In this first section, we examine the convergence performance of improved WOA on 23 mathematical benchmark functions are listed in Tables 1 and 2. Note that for all of the experiments, we have utilized 30 gaze agents and maximum 500 iterations. Second, discussed the case study; application of the clinical decision making process using clinical data of anaemic pregnant woman. To evaluate our proposed algorithm conducted several experiments using this data. As test beds, we chose 23 standard benchmark functions.

**Table 3** Statistical results of WOA for the 23 mathematical benchmark functions

| F_Name | Whale Optimization Algorithm (WOA) | | | | |
|---|---|---|---|---|---|
| | Average | Median | Std. Dev | Best | Worst |
| $F_1$ | 976.0925 | 1.4934e-29 | 6.0966e+03 | 1.0730e-81 | 6.2422e+04 |
| $F_2$ | 2.9207e+07 | 1.7031e-18 | 4.6134e+08 | 3.4225e-52 | 7.3017e+09 |
| $F_3$ | 5.7966e+04 | 5.2965e+04 | 3.4381e+04 | 1.6992e+04 | 1.2689e+05 |
| $F_4$ | 59.9157 | 57.1905 | 5.2690 | 57.1905 | 90.7160 |
| $F_5$ | 2.6080e+06 | 27.9527 | 1.9373e+07 | 27.9516 | 2.5396e+08 |
| $F_6$ | 629.8011 | 0.3204 | 4.7976e+03 | 0.3204 | 5.5280e+04 |
| $F_7$ | 1.3233 | 0.0057 | 11.1443 | 0.0052 | 139.6725 |
| $F_8$ | −1.1491e+04 | −1.2337e+04 | 1.7492e+03 | −1.2347e+04 | −2.4467e+03 |
| $F_9$ | 27.1213 | 0.0000 | 76.7544 | 0.0000 | 450.6246 |
| $F_{10}$ | 0.4300 | 6.2172e-15 | 2.5248 | 4.4409e-15 | 20.4593 |
| $F_{11}$ | 6.8496 | 0.0000 | 51.1150 | 0.0000 | 573.1051 |
| $F_{12}$ | 6.5367e+06 | 0.0472 | 5.6013e+07 | 0.0472 | 7.4401e+08 |
| $F_{13}$ | 9.6103e+06 | 0.2720 | 8.2630e+07 | 0.2720 | 9.5398e+08 |
| $F_{14}$ | 3.4340 | 2.9821 | 2.6263 | 2.9821 | 43.0259 |
| $F_{15}$ | 9.7449e-04 | 4.6921e-04 | 0.0038 | 4.6921e-04 | 0.0746 |
| $F_{16}$ | −1.0305 | −1.0316 | 0.0089 | −1.0316 | −0.8963 |
| $F_{17}$ | 0.4256 | 0.3982 | 0.1260 | 0.3979 | 1.2673 |
| $F_{18}$ | 3.5112 | 3.0000 | 3.5737 | 3.0000 | 34.9664 |
| $F_{19}$ | −3.8218 | −3.8560 | 0.0616 | −3.8565 | −3.4609 |
| $F_{20}$ | −2.9480 | −2.9899 | 0.0695 | −2.9900 | −2.3866 |
| $F_{21}$ | −4.9423 | −5.0547 | 0.4996 | −5.0547 | −0.5249 |
| $F_{22}$ | −3.6383 | −3.7213 | 0.3054 | −3.7213 | −0.5457 |
| $F_{23}$ | −2.3074 | −2.4192 | 0.3146 | −2.4192 | −0.6555 |

### 4.1 *Discussion of the Results*

Tables 3 and 4 provides the statistical results of the WOA and improved WOA. These two table shows that the improved WOA is able to provide the best fitness convergence results on all the statistical metrics for unimodal, multimodal, fixed-dimension multimodal benchmark functions respectively.

The significance of the results is graphically illustrated in Fig. 1a, b. These figures show that the plot of improved WOA is significantly secured fast

**Table 4** Statistical results of Improved WOA for the 23 mathematical benchmark functions

| F_Name | Improved Whale Optimization Algorithm (IWAO) | | | | |
|---|---|---|---|---|---|
| | Average | Median | Std. Dev | Best | Worst |
| $F_1$ | 141.0601 | 1.6987e-69 | 2.8017e+03 | 8.5331e-315 | 6.2234e+04 |
| $F_2$ | 2.6816e+08 | 1.1694e-32 | 5.9961e+09 | 2.1158e-159 | 1.3408e+11 |
| $F_3$ | 1.2377e+03 | 2.4741e-60 | 1.6451e+04 | 9.7553e-270 | 2.5731e+05 |
| $F_4$ | 0.2815 | 1.1628e-36 | 4.2173 | 4.6036e-149 | 87.9804 |
| $F_5$ | 4.0073e+05 | 27.8797 | 8.7116e+06 | 27.8797 | 1.9476e+08 |
| $F_6$ | 168.6582 | 0.1528 | 3.2574e+03 | 0.1528 | 7.2028e+04 |
| $F_7$ | 0.2960 | 2.3889e-04 | 6.2970 | 4.0297e-05 | 140.7692 |
| $F_8$ | −1.2393e+04 | −1.2560e+04 | 897.0959 | −1.2569e+04 | −1.7542e+03 |
| $F_9$ | 2.8204 | 0.0000 | 28.9688 | 0.0000 | 467.2299 |
| $F_{10}$ | 0.1190 | 8.8818e-16 | 1.2921 | 8.8818e-16 | 20.8727 |
| $F_{11}$ | 2.5364 | 0.0000 | 37.8199 | 0.0000 | 605.3266 |
| $F_{12}$ | 1.7024e+06 | 0.0101 | 3.1591e+07 | 0.0101 | 6.8750e+08 |
| $F_{13}$ | 2.5621e+06 | 0.3602 | 5.6061e+07 | 0.3602 | 1.2533e+09 |
| $F_{14}$ | 1.8742 | 0.9980 | 18.6189 | 0.9980 | 416.9956 |
| $F_{15}$ | 0.0015 | 3.2965e-04 | 0.0205 | 3.2964e-04 | 0.4502 |
| $F_{16}$ | −1.0198 | −1.0316 | 0.1012 | −1.0316 | 0.1630 |
| $F_{17}$ | 0.4115 | 0.3979 | 0.1003 | 0.3979 | 1.2713 |
| $F_{18}$ | 3.3809 | 3.0000 | 3.0850 | 3.0000 | 33.3362 |
| $F_{19}$ | −3.8585 | −3.8613 | 0.0411 | −3.8614 | −2.9634 |
| $F_{20}$ | −3.2169 | −3.3014 | 0.1696 | −3.3015 | −1.7692 |
| $F_{21}$ | −9.9569 | −10.0979 | 0.8079 | −10.1128 | −2.0944 |
| $F_{22}$ | −6.3471 | −6.8573 | 0.8925 | −6.8695 | −1.0269 |
| $F_{23}$ | −5.0517 | −5.1273 | 0.3852 | −5.1273 | −1.3220 |

convergence and have lower and narrower convergence curve when compared to WOA for the some of the problems. The performance is measured in terms of statistical metrics average, median, standard deviation, best and worst. These statistical results show the accuracy and convergence of improved WOA.

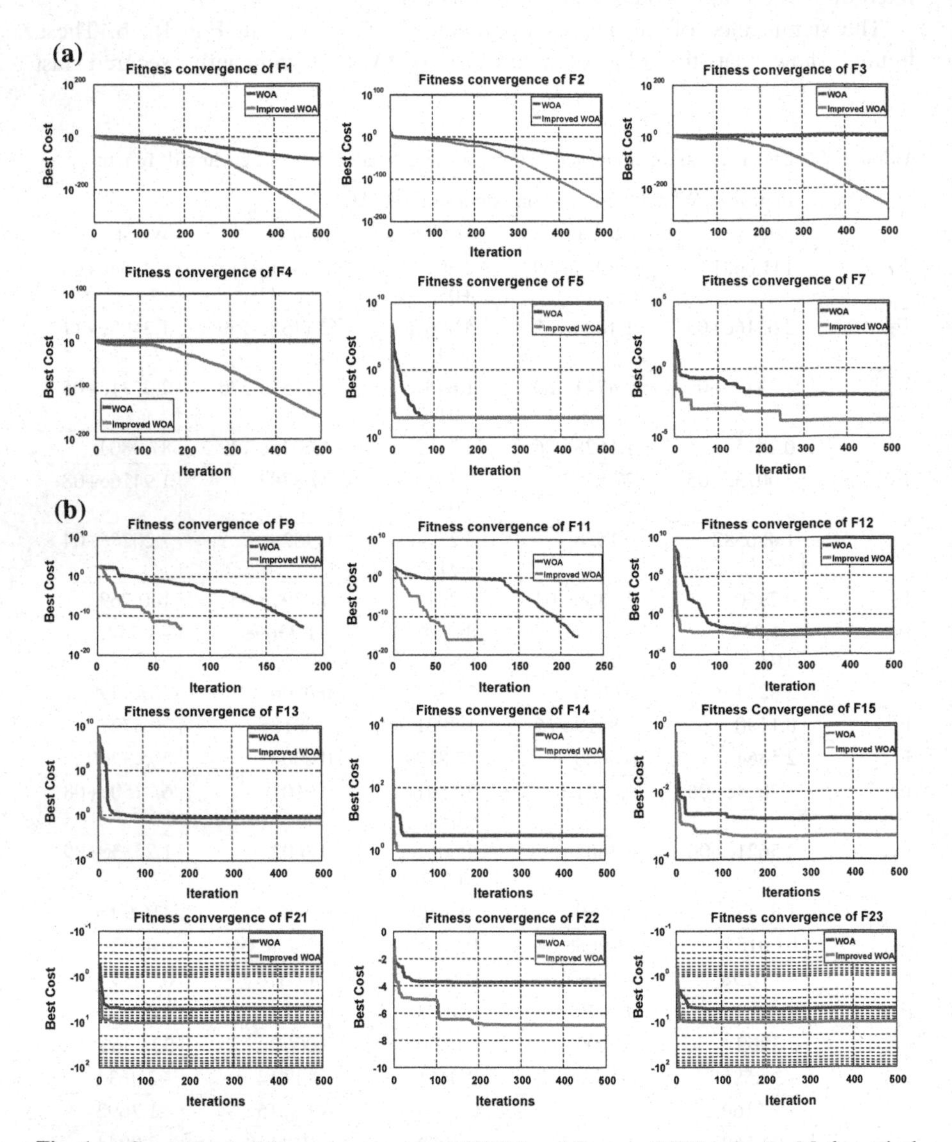


**Fig. 1** **a** Comparison of fitness convergence of WOA and Improved WOA for the Mathematical benchmark functions from $F_1$ to $F_4$, $F_7$, $F_9$, **b** Comparison of fitness convergence of WOA and Improved WOA for the mathematical benchmark functions from $F_{11}$ to $F_{15}$ and $F_{20}$ to $F_{23}$

So, it can be stated that the improved version of WOA algorithm is able to provide superior convergence than WOA in all mathematical benchmark functions. The quality of data is an important component in certifying the validity of the end-results of pragmatic evaluations. Moreover, it is required for a clear decipherment of data characteristics for numerous application performances. Among the distinctive attributes of data, such as dimensionity, heterogeneity, complexity, data-types, it is true that the complexity and sensitivity integrated mining algorithm is demanding for its scalable and adaptable mining process for clinical data sets. The test case taken is of clinical decision making process which is based on optimized clustering and outlier unmasking. Improved WOA is used for distance function optimization. First, centroids for natural clusters are gazed and the initial clusters and unmasking of outliers are done. Finally, clusters and outliers are iteratively modified and refined according to the relationship among them. This approach aims to unmask abnormal entries, which allows medicator to know essential information to improve clinical decision making processes. Based on this patient care process; in this test case clinical data are clustered into k = 5. Table 5 represents the unmasked outliers and the summary of outliers in individual patient population. Figure 2 represents the detected outliers that shows us the importance

**Table 5** Attribute wise unmasked outliers

| Attribute | No. of outliers |
|---|---|
| Hb | 4 |
| HCT | 6 |
| WBCs | 21 |
| RBCs | 10 |
| Platelets | 11 |
| Trimester | 4 |

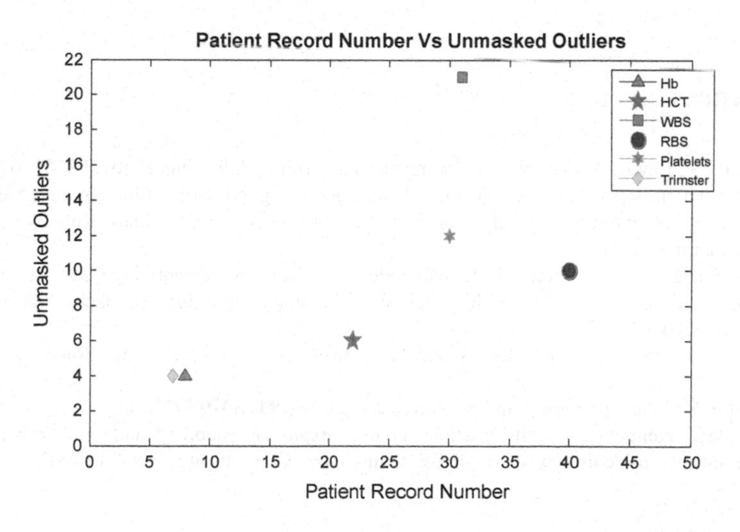


**Fig. 2** Patient record number versus Outliers unmasked

of the attributes, which can be more influential parameters than other factors. We can observe that WBC is vitally essential to investigate the anaemic pregnant woman patient safety care as their health conditions. The notations of the clinical dataset used in this case are presented in Table 5.

## 5 Conclusion and Future Scope

The contribution of this research work is twofold. Firstly, improved version of WOA is presented. Secondly, the improved WOA is used in clinical decision making process by applying it to a clinical dataset. An experimental investigation of the improved WOA was done on 23 mathematical benchmark functions and secured improved performance. Then we applied this algorithm for improving the faultless Clinical decision-making process. We practical this model to a clinical dataset of anaemic pregnant woman. Firstly, the dataset is biloculate into several partitions titled clusters and cluster heads are identified within each cluster. Then, optimized partitions and cluster heads are secured. Extensive simulation results obtained in this investigate achieved that the improved WOA significantly improves the clinical decision-making process for predicting health risks and disease prevalence. In further research, we are planning to enhance the proposed algorithm and we try to improve the outlier unmasking process using optimized supervised algorithms by increasing the number of clinical attributes and instances. Also, we try to investigate the same for other clinical data sets.

**Acknowledgements** The authors would like to thank Prof. E. Sreenivasa Reddy (The Dean, University college of Engineering and Technology, Acharya Nagarjuna University, A.P, India) for his advices on writing this paper.

## References

1. Mirjalili, S., Lewis, A.: The Whale optimization algorithm. Adv. Eng. Softw. **95**, 51–67 (2016)
2. Ghaffarizadeh, A., Ahmadi, K., Flann, N.S.: Sorting unsigned permutations by reversals using multi-objective evolutionary algorithms with variable size individuals. In: Evolutionary Computation (CEC)
3. Ghaffarizadeh, A., Eftekhari, M., Esmailizadeh, A., Flann, N.: Quantitative trait loci mapping problem: an extinction-based multiobjective evolutionary algorithm approach. Algorithms **6**, 546–564 (2013)
4. Mirjalili, S., Mirjalili, S.M., Lewis, A.: Grey wolf optimizer. Adv. Eng. Softw. **69**, 46–61 (2014)
5. Mirjalili, S.: The ant lion optimizer. Adv. Eng. Softw. **83**, 80–98 (2015)
6. Clerc, M., Kennedy, J.: The particle swarm: explosion stability and convergence in a multi-dimensional complex space. IEEE Trans. Evol. Comput. **6**(1), 58–73 (2002)

7. Andre, J., Siarry, P., Dognon, T.: An improvement of the standard genetic algorithm Figurehting premature convergence in continuous optimization Adv. Eng. Soft. **32**, 49–60 (2001)
8. Low, S.H., Lapsley, D.E.: Optimization flow control—I: basic algorithm and convergence. IEEE/ACM Trans. Netw. (TON) 7.6, 861–874 (1999)

# A Novel Edge Detection Algorithm for Fast and Efficient Image Segmentation

**P. Prathusha and S. Jyothi**

**Abstract** Edge detection determines the boundaries of objects in an Image. Edge detection is a vital concept in object recognition and Image analysis. This paper evaluates the existing edge detection methods and proposes a new edge detection algorithm which uses the morphological operations, sobel operator, Gaussian Smoothing and masking. The novelty of the proposed algorithm is extracting continuous edges in the Image and removing spurious edges using m-connectivity. The paper introduces performance parameters for edge detection to determine which method gives good results. A parameter named Human Perception Clarity (HPC) is mathematically modeled and experimentally proves the efficacy of proposed algorithm.

**Keywords** Image segmentation methods · Otsu segmentation · Canny edge detection · Edge detection operators · Human perception clarity

## 1 Introduction

An Image processing application development involves various steps like Image acquisition, Pre-processing, Image Enhancement, Image Segmentation and Image Classification. Pre-Processing involves the removal of noise from Images. Conversion of color Images to gray Images and resizing the Images is part of Image Pre-processing. Image Enhancement increases the brightness and reducing the blurring in the images. Real time Images have poor contrast and induced noise during Image acquisition because of flaws in scanning devices and illumination. Segmentation and Edge detection methods extract important portions of images.

P. Prathusha · S. Jyothi (✉)
Department of Computer Science, Sri Padmavati Mahila University,
Tirupati 517502, Andhra Pradesh, India
e-mail: jyothi.spmvv@gmail.com

P. Prathusha
e-mail: prathusha.p@rediffmail.com



S.C. Satapathy et al. (eds.), *Data Engineering and Intelligent Computing*,
Advances in Intelligent Systems and Computing 542,
DOI 10.1007/978-981-10-3223-3_26

The results obtained at the end in Image classification depend on preprocessing and segmentation and edge detection. The success of the classification procedure is eventually decided by pre-processing and Segmentation. Hence this paper proposes a new edge detection algorithm which uses sobel operator, morphological operations, masking and Gaussian Smoothing. New performance parameter namely Human Perception Clarity is mathematically introduced in the paper. The performance of the edge detection algorithm is enhanced by adapting morphological operations and Gaussian Smoothing. The experimental results show that the proposed algorithm integrated with morphological operations yields better results than classical edge detection methods.

## 1.1 Literature Review

Image Segmentation has been a challenging research area studied in [1–6]. Bhargavi [7] had developed a detailed survey on threshold based segmentation techniques in Image processing and refers that segmentation techniques were classified as contextual or non-contextual. Khan [8] had made a comparative study on Image segmentation methods and classifies Image segmentation as semi-interactive and fully automatic approaches. Nain [9] gave a dynamic thresholding based edge detection and emphasized the adaptive efficient peak detection of the image histogram and usage of morphological operations. Al-Kubati [10] had evaluated canny and Otsu image segmentation methods. Sucharita [11] made a comparative study on various edge detection techniques like sobel, Robert, prewitt and canny edge detection operators. Khalil [12] classified fish based on color histogram using back propagation classifier and he used color histogram and gray level co-occurrence matrix to classify fish species. Nagalakshmi [13] used canny edge detection for identification of prawn species. Das [14] made a study on Image segmentation techniques and classified segmentation methods. Thakare [15] classified segmentation methods into edge based, region based and Hybrid methods. Sainia [16] made a comparative study on edge detection algorithms. Narendra [17] gave a study of edge detection techniques in quality inspection of food products.

## 1.2 Image Segmentation

Image Segmentation Methods are broadly classified into Pixel Based methods, Edge Based Methods and Region Based Methods. Pixel based Segmentation is a process of segmenting an image based on pixels or group of pixels [18].

Image Segmentation methods are classified on two basic properties of intensity values: discontinuity and similarity. Edge detection operators are the part of discontinuity based approach [19]. They divide the image based on abrupt changes in intensity. Region growing, Region splitting & merging, Thresholding are similarity based approaches.

## 2 Proposed Algorithm

The existing first derivative operators like sobel, Robert and prewitt give edges on applying corresponding masks on the image. However the edges are not continuous and some edge information is lost in this process. To overcome this a unique algorithm has been proposed.

The proposed algorithm is

Step 1: Apply sobel vertical edge operator on the input image.
Step 2: Apply sobel horizontal edge operator on the image obtained from step 1.
Step 3: Use Masking and add the horizontal and vertical edges to obtain the thick edges of the edge detected image obtained from step 2.
Step 4: Apply Gaussian Smoothing filter to remove noise.
Step 5: The boundary extracted image is one or more pixel thick image. So apply the following masks to get one pixel thick image and using Hit-miss transform [20] to get one pixel thick image.

The masks are $\text{Mask } 1 = \begin{bmatrix} 1 & 1 & 1 \\ 0 & 1 & 1 \\ 0 & -1 & 1 \end{bmatrix}$ $\text{Mask } 2 = \begin{bmatrix} 1 & 1 & 1 \\ 1 & 1 & 0 \\ 0 & -1 & 1 \end{bmatrix}$.

Step 6: The operation of Hit-miss transform creates breaks in the pixel boundary. So the gaps are to be filled. To thin and fill gaps the following masks are used.

$$\text{Mask } 3 = \begin{bmatrix} -1 & -1 & -1 \\ -1 & -1 & -1 \\ -1 & -1 & -1 \end{bmatrix}$$

Step 7: To avoid the ambiguity in paths of 4-connected and 8-connected neighbors, m-connectivity [20] is used. The mask used for this purpose is

$$\text{Mask } 4 = \begin{bmatrix} 1 & 1 & 0 \\ 1 & 1 & 0 \\ 0 & 0 & 0 \end{bmatrix}$$

Step 8: At the end Boundary is calculated as

Boundry = Boundry − (Boundary Ø Mask 4) where Ø is Hit-Miss transform defined in [20].

# 3 Experimental Setup and Results

Mat Lab 2013 is used to develop the GUI for Image segmentation methods (Fig. 1).

# 4 Performance Criteria of Edge Detection

To assess the efficiency of the performance of proposed approach we made use of the classical images of Image processing. Many years of research gave numerous edge detection operators and algorithms, but the performance evaluation of each edge detection algorithm is still an ambiguity. Every researcher claims that his algorithm is ultimate and outperforms the existing methods. But defining parameters to measure the performance of edge detection method is still a question mark [21]. Gives some performance evaluation of edge detection methods. Performance Ratio (PR) and PSNR are the parameters and the edge detected images are checked against ground truth images available in BSD database. But in real time applications ground truth images are not available. Here we define some parameters to check the performance of edge detection.

| Original Image | Sobel operator | Prewitt operator | Robert operator | Canny Edge detection | Proposed algorithm |
|---|---|---|---|---|---|
| Cameraman | | | | | |
| Lena Noise | | | | | |
| Crab | | | | | |
| Lobester | | | | | |
| Ultra sound | | | | | |

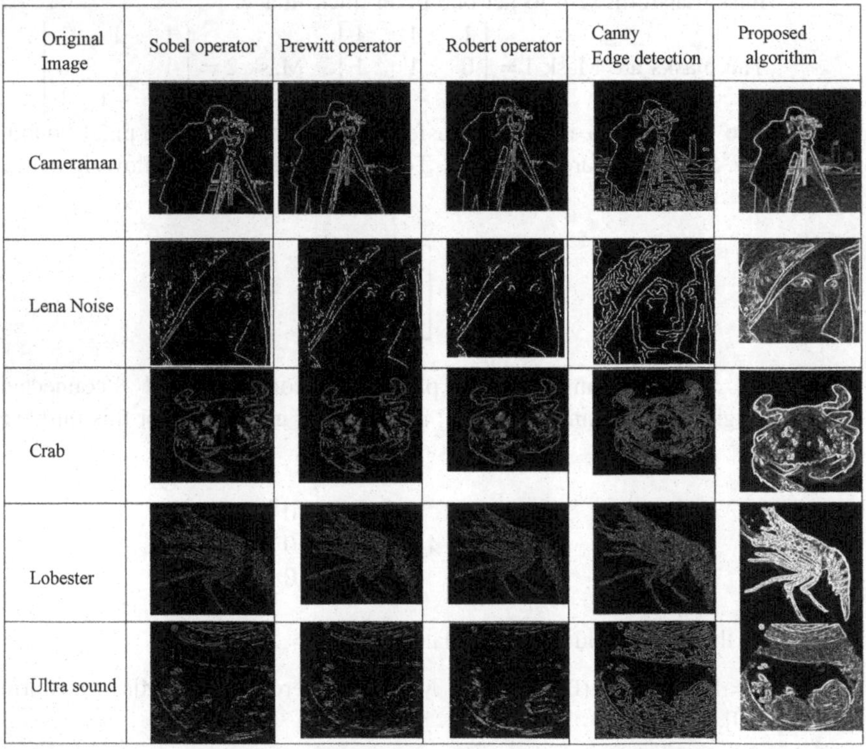

**Fig. 1** Comparision of results of applying various edge detection operators and proposed algorithms

***Pixel thickness***: Pixel thickness gives clarity to the image. Some edge detection algorithms identify 1-pixel thick edge or 2-pixel thick edges.
***Connectivity***: The 4-connectivity or 8-connectivity of neighbors in the edge lines gives spurious edges and lead to false edges. The existence of m-connectivity gives true edges and this is ensured in our algorithm.
***Human Perception Clarity (HPC)***: Here we are defining a new parameter and we have experimented this parameter with our algorithm and compared with existing algorithms.

Let Xi define a values which denotes the number of votes for supporting the algorithm as good edge detection

$$\mathrm{Xi} = \begin{cases} 1 & \text{accepted algorithm} \\ 0 & \text{otherwise} \end{cases}$$

Let Yi define a value which denotes the number of votes against the algorithm
Let N denote the number of humans participating in the evaluation.
We can define the Human Perception Clarity (HPC) as $\mathrm{HPC_V}$ and $\mathrm{HPC_A}$
$\mathrm{HPC_V}$ gives the percentage of humans giving vote for perfect edge detection.
$\mathrm{HPC_A}$ gives the percentage of humans giving negative vote of improper edge detection for an algorithm (Fig. 2, Tables 1, 2, 3, 4 and 5).

$$\mathrm{HPC_A} = \frac{N - Xi}{N} \tag{1}$$

$$\mathrm{HPC_V} = \frac{N - Yi}{N} \tag{2}$$

Finally

$$\mathrm{HPC_A} + \mathrm{HPC_V} = 1. \tag{3}$$

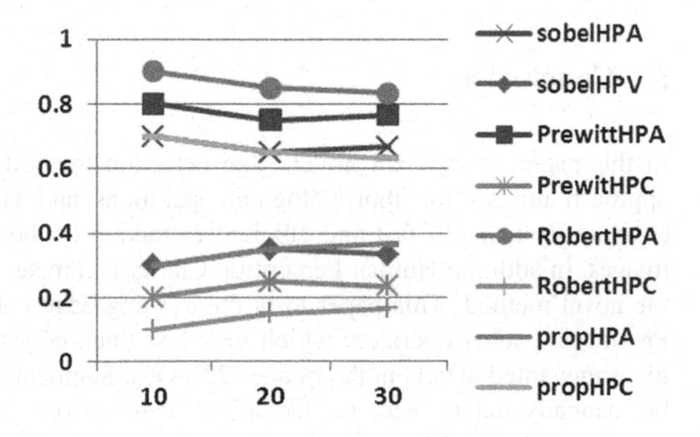


**Fig. 2** Graph showing the performance of all existing methods and comparision with proposed algorithm

**Table 1** Values of HPCA and HPCv for sobel operator

| Sobel operator | Xi | Yi | $HPC_A$ | $HPC_v$ |
|---|---|---|---|---|
| N = 10 | 3 | 7 | 0.7 | 0.3 |
| N = 20 | 7 | 13 | 0.65 | 0.35 |
| N = 30 | 10 | 20 | 0.6667 | 0.3334 |

**Table 2** Values of HPCA and HPCv for Robert operator

| Robert operator | Xi | Yi | $HPC_A$ | $HPC_v$ |
|---|---|---|---|---|
| N = 10 | *1* | *9* | *0.9* | *0.1* |
| N = 20 | *3* | *17* | *0.85* | *0.15* |
| N = 30 | *5* | *25* | *0.8333* | *0.1666* |

**Table 3** Values of HPCA and HPCv for Prewitt operator

| Prewitt operator | Xi | Yi | $HPC_A$ | $HPC_v$ |
|---|---|---|---|---|
| N = 10 | 2 | 8 | *0.8* | *0.2* |
| N = 20 | 5 | 15 | *0.75* | *0.25* |
| N = 30 | 7 | 23 | *0.7666* | *0.234* |

**Table 4** Values of HPCA and HPCv for Canny edge detector

| Canny edge detector | Xi | Yi | $HPC_A$ | $HPC_v$ |
|---|---|---|---|---|
| N = 10 | *6* | *4* | *0.4* | *0.6* |
| N = 20 | *12* | *8* | *0.4* | *0.6* |
| N = 30 | *17* | *13* | *0.433* | *0.5667* |

**Table 5** Values of HPCA and HPCv for Proposed Algorithm

| Proposed algorithm | Xi | Yi | $HPC_A$ | $HPC_v$ |
|---|---|---|---|---|
| N = 10 | *7* | *3* | *0.3* | *0.7* |
| N = 20 | *13* | *7* | *0.35* | *0.65* |
| N = 30 | *19* | *11* | *0.3666* | *0.6333* |

## 5 Conclusion

In this paper an efficient novel edge detection method has been presented. This approach utilizes the morphological operations and Gaussian smoothing which enormously helps in fast and efficient extraction of the continuous thick edges in images. In addition Human Perception Clarity parameter has been integrated to test the novel method. This paper tests the existing edge detection techniques Robert, Prewitt and sobel operators which give less thick edges and Canny edge detector gives unwanted details in the image [22]. Otsu Segmentation needs the threshold to be manually determined. Laplacian of Gaussian operator also gives gray image which has fewer details. Compared to existing Edge detectors, the proposed algorithm extracts precisely thick continuous image boundary which is very important to extract prominent and significant corners [23] in images and also in computing image semantics [24].

The GUI designed helps to successfully check all the edge detection methods at one place and choose appropriate method for your particular application. Performance evaluation measures give comparative results of all the existing methods. The proposed algorithm has highest $HPC_V$. This shows the efficiency of the proposed algorithm.

**Acknowledgements** Authors would like to thank DBT, New Delhi for sanctioning the Project. Currently this work is carried out under DBT Project.

## References

1. Dr. Rama Bai, M.: A new approach for edge extraction using various preprocessing methods on Skeletonization technique. Int. J. Appl. Innov. Eng. Manag. (IJAIEM) **2**(6) (2013). www.ijaiem.org. ISSN:2319–4847
2. Kezia, S., Shanti Prabha, I., Vijay Kumar, V.: Innovative segmentation approach based on LRTM. Int. J. Soft Comput. Eng. (IJSCE) **2**(5) (2012). ISSN:2231–2307
3. Han, F., Tu, Z., Zhu, S.-C.: Range image segmentation by an effective jump-diffusion method. IEEE Trans. Pattern Anal. Mach. Intell. **26**(9) (2004)
4. Gupta, A., et al.: An edge detection approach for images contaminated with Gaussian and impulse noises. In: Proceedings of (Springer) 4th International Conference on Signal and Image Processing (ICSIP 2012), vol. 2, pp. 523–533 (2012)
5. Bhateja, V., Devi, S.: A reconstruction based measure for evaluation of mammogram edge-maps. In: Proceedings of (Springer) International Conference on Frontiers in Intelligent Computing Theory and Applications AISC vol. 199, pp. 741–746 (2012)
6. Bhateja, V., Misra, M., Urooj, S.: Non-linear polynomial filters for edge enhancement of mammogram lesions. Elsevier-Computer Methods and Programs in Bio-medicne, vol. 129C, pp. 125–134 (2016)
7. Bhargavi, K., Jyothi, S.: A survey on threshold based segmentation technique in image processing. Int. J. Innov. Res. Dev. **3**(12) (2014)
8. Khan, A.M., Ravi, S.: Image segmentaion methods: a comparative study. IJSCE **3**(4) (2013). ISSN:2231–2307
9. Nain, N., Jindal, G., Garg, A., Jain, A.: Dynamic threshoding based edge detection. In: Proceedings of the World Congress on Engineering 2009, vol. I WCE 2008, 2–4 July 2008, London, UK
10. Al-Kubati, A.A.M., Saif, J.A.M., Taher, A.A.: Evaluation of Canny and Otsu image segmentation. In: International Conference on Emerging Trends in Computer and Electronics Engineering, 24–25 March 2012, Dubai
11. Sucharita, V., Jyothi, S., Mamatha, D.M.: A comparative study on various edge detection Techniques used for the identification of Penaeid Prawn species. Int. J. Comput. Appl. (0975–8887) **78**(6) (2013)
12. Khalil, M., Omar, K.B., Noah, S.A.M.: Fish classification based on robust features extraction from color signature using back-propagation classifier. J. Comput. Sci. **7**(1), 52–58 (2011). ISSN 1549-3636
13. Nagalakshmi, G., Jyothi, S.: Image acquisition, noise removal, edge detection methods in image processing using Matlab for prawn species identification. In: Proceedings of

International Conference on Emerging Trends in Electronics & Telecommunications, 29th-31st May 2015, Kualalumpur, Malaysia
14. Dass, R., Priyanka, Swapna Devi.: Image sementation techniques. IJECT **3**(1) (2012). ISSN:2230-7109 (online)
15. Thakare, P.: A study of image segmentation and edge detection techniques. IJCSE **3**(2) (2011)
16. Saini, S., Kasliwal, B., Bhatia, S.: Comparative study of image edge detection algorithms
17. Narendra, V.G., Hareesh, K.S.: Study and comparision of various edge detection techniques used in quality inspection and evaluation of agricultural and food products by computer vision. Int. J. Agric. Biol. Eng. (2011)
18. Evangeline, D.: Image Segmentation: From the Beginning to Current Trends. CSI Communications (2015)
19. Gonzalez, R.C., Woods, R.E.: Digital Image Processing, 3rd edn. (2008)
20. Gonzalez, R.C., Woods, R.E.: Digital Image Processing, 2nd edn. Addison Wesley Longman (2000)
21. Khaire, P.A., Dr. Thakur, N.V.: A fuzzy set approach for edge detection . Int. J. Image Process. (IJIP) **6**(6) (2012)
22. Canny, J.: A computational approach to edge detection. IEEE Trans. Pattern Anal. Mach. Intell. PAMI-**8**(6), 679–698 (1986)
23. Nain, N., Laxmi, V., Bhadviya, B., Gopal, A.: Corner detection using difference chain code as curvature. In: The Third IEEE International Conference on Signal Image Technology and Internet Based Systems, SITIS'07, 2007, Track III, pp. 766–770
24. Nain, N., Laxmi, V., Agarwal, D., Khandelwal, M.: Transformation Invariant Shape Descriptors. International Conference on Image Processing and Computer Vision, IPCV'07, June 25–28, Nevada, USA, 2007, vol. I, pp. 545–550

## Author Biographies

**P. Prathusha** is working as Asst. Professor in the Department of Computer Science in RGMCET, Nandyal. Presently she is persuing Ph.D under guidance of Dr. S. Jyothi in Sri Padmavathi Mahila Viswavidyalayam (SPMVV), Tirupathi. She completed her M.Tech from JNTU, Anantapur in 2012. She has over 10 years of teaching experience. She is a life member of CSI. Her areas of interest are Image Processing and Pattern Recognition.

**S. Jyothi** is working as Professor in the Department of Computer Science and Director of IQAC, Sri Padmavathi Mahila Visvavidyalayam (SPMVV), Tirupati. She did her Ph.D in Theoritical Computer Science from S.V. University, Tirupati. She has 22 years teaching experience & 26 years research experience and she guided 11 Ph.D.s and 7 M.Phil.s. She has written and edited 10 books and proceedings. More than 150 research articles published and presented in International Journals and Conferences. She is senior member of IEEE & IACSIT, fellow of RSS, ISCA, IETE & SSARSC, member of IEEE CS, GSRS & WS, ACM, IET, IAENG & *Gyancity Research Labs* and life member of CSI, ISTE, ISCA, IFERP, IUPRAI and ISRS. She is honored and awarded AP State Best Teacher Award, Rashtriya Gaurav Award, inclusion of her Biography in 2000 Outstanding Intellectuals of the 21st Century, Great Britain, nominated for Top 100 Professional by International Biographical Centre, Cambridge and Best Citizens of India Award by International Publishing House, New Delhi. Her areas of interest in Image Processing, Soft Computing, Data Mining, Big Data analytics, Bioinformatics and Hyperspectra Image Analysis.

# Real Time Audio Steganographic Countermeasure

**T.M. Srinivas and P.P. Amritha**

**Abstract** Steganographic techniques are used to embed data into a cover file using different algorithms. In this paper audio steganography countermeasure is discussed which uses a technique called double stegging or steganographic jamming where variations of LSB embedding algorithms are used for audio steganography prevention which can be used in real time under acceptable information loss. We then proceed to show that this method renders hidden embedded data unrecoverable. The resulting audio quality after Steganographic Jamming is evaluated. Mean Opinion Score and Signal to Noise Ratio are used to calculate the quality of output audio file which shows the effectiveness of the technique described.

**Keywords** Audio steganographic countermeasure · Double stegging · Steganographic jamming · Least significant bit

## 1 Introduction

The science of hiding information within media is called Steganography. While cryptography protects by randomizing the content of secret data, Steganography keeps hidden the existence of secret data itself [5]. Steganography uses digital media like images, audio and video as cover. In this experiment the focus is only on audio steganography where audio files are used as cover media. Application of Steganography in audio is challenging as Human Auditory System (HAS) is more sensitive to small changes in audio data than Human Visual System (HVS) [6, 14]. Motivation for this experiment is to prevent use of this technology in organized crime [7] and also more recently increasing insider threats at organizations. Further

T.M. Srinivas (✉)
TIFAC CORE in Cyber Security, Amrita School of Engineering, Coimbatore, India
e-mail: srinivas31.meharwade@gmail.com

P.P. Amritha
Amrita Vishwa Vidyapeetham, Amrita University, Coimbatore, India
e-mail: pp_amritha@cb.amrita.edu



S.C. Satapathy et al. (eds.), *Data Engineering and Intelligent Computing*, Advances in Intelligent Systems and Computing 542,
DOI 10.1007/978-981-10-3223-3_27

new ideas and approaches emanate for robust design of steganographic algorithms by looking at steganographic prevention methods [1].

When steganography is applied to audio data there are three important requirements to be taken care of Robustness, Inaudibility or Undetectability and Capacity [8]. Hence audio quality is not to be degraded while defending against steganography to make the defense technique undetectable. This work separates itself work from steganalysis which focuses on detecting and finding hidden data present in cover medium whereas our goal is to interfere with the steganographic receiver whether or not there exists a steganographic communication. Hence this method can be applied to all audio material as this method does not introduce any noticeable disturbances to the audio signal.

In this paper the use of Least Significant Bit (LSB) embedding and it's variant for steganographic jamming are discussed to make the data, hidden using Steganographic techniques irrespective of underlying algorithm, irrecoverable without substantial reduction in audio quality of the steganographic jammed audio media. Audio quality measurement techniques are then used to grade the quality of the output audio data in terms of Mean Opinion Score (MOS), Waveform Amplitude Distribution Analysis-Signal to Noise Ratio (WADA-SNR) [2], National Institute of Standards and Technology-Signal to Noise Ratio (NIST-STNR) and Bit Error Rate (BER).

## 2 Related Work

Steganographic Jamming by way of "Double-Stegging" as an idea is presented by [9]. Here they have discussed about the same application doing the second embedding process which embedded the secret data the first time. Also they are discussing about image files whereas the focus here is on audio as cover file and also no evaluation of output audio quality or efficiency of the proposed algorithm is discussed.

In [1] although their aim is to prevent audio steganography, it is completely different from what is experimented here as their work is about combining basic signal processing techniques in a novel way to prevent audio steganography by interfering with the steganographic receiver.

In [3] they discuss about audio steganographic prevention in cloud storage systems where they propose two algorithms, the enhanced-RS algorithm and the SADI algorithm. Here they first detect whether there is steganography present in the stored stationary files using the enhanced-RS algorithm and then try to destroy the hidden data using the SADI algorithm which works by interchanging the bits based on the minimum Manhattan distance. As the author initially find the media with stego content and then try to destroy the hidden data, it cannot be used in real time whereas the system suggested here can work in real time.

In [11] examination on steganographic tools for hiding information is presented which also discusses the different approaches these tools used to hide data and the supported types of cover media is presented of which our interest is only on tools supporting audio as cover media.

In 2006, Floriano De Rango [4] discusses about subjective and objective measurement methods available for evaluating quality of audio data. It also discusses the drawbacks like more time consumption, very slow and expensiveness of subjective measurement methods over objective measurement methods. Perceptual Evaluation of Speech Quality (PESQ) methodology is used to evaluate quality of the output audio data as it is more accurate than others and also reference original file is available for comparison.

# 3 Procedure and Implementation

In this section we describe how steganographic jamming is implemented and also discuss the various tools that were implemented for audio steganography and the effectiveness of the jamming technique in destroying the data hidden by various tools. A tool was developed based on the below methodology to perform steganographic jamming of cover audio file. The data used for steganographic jamming can be any file or text which is encrypted before embedding into the cover media. The encryption key is derived from the key that is used for double stegging. Password based encryption scheme "PBEWithMD5AndDES" from java library is used to encrypt the message bits to be hidden so that the data is randomized, also a random salt is introduced in the code for generation of key so as to prevent dictionary attacks. The encryption algorithm used is Data Encryption Standard (DES) and the hashing algorithm used is Message Digest (MD5). The decoding process is not explained here as it is not mandatory or necessary to decode the hidden data from the double stegged file as the sole purpose here is the destruction of the data hidden in the cover file.

## 3.1 Methodology

**Method 1: LSB Embedding**

1. Read the cover audio file and then a copy of file is generated which is used to hide data.
2. Read the data to be used to for steganographic jamming, if the data size is less than size of cover audio divided by sample size of audio, convert it into binary sequence of message bits.
3. The above message bits are then encrypted using password based encryption and the LSB of each sample of cover audio is replaced with the encrypted message bits.

4. The modified cover samples are then written to a file forming output audio signal.

**Method 2: Variable LSB Embedding**

1. Read the cover audio file and then a copy of file is generated which is used to hide data.
2. Read the data to be used to for steganographic jamming, if the data size is less than size of cover audio divided by sample size of audio, convert it into binary sequence of message bits.
3. The above message bits are then encrypted using password based encryption and the first and third LSB of alternate samples of cover audio is replaced with the encrypted message bits.
4. The modified cover samples are then written to a file forming output audio signal.

## 3.2 Tools Implemented

A number of steganographic tools are available in the market today. Here the discussion is only on the tools that can work with audio as cover media and are either open source or freeware. Following are the tools of the many tools that were able to successfully hide and extract the secret data into and from the audio cover file. There are other tools available like Steganofile, Xiao, S_Tools, Silenteye but we will not be experimenting on them as they either failed to hide data successfully in cover file or could not successfully extract hidden data from the cover file.

1. **Openpuff**: Openpuff supports the following audio formats wav, aiff, next/sun, and mp3. It also supports various encryption algorithms for security along with scrambling and whitening. The highest capacity of data that can be embedded is one by sixteenth of the file size of cover audio data. Openpuff uses LSB embedding algorithm along with data whitening (addition of random noise) for hiding data in cover media.
2. **DeepSound**: DeepSound supports AES encryption for security and wav and flac audio file formats for cover media. It uses either LSB embedding algorithm or variable LSB embedding algorithm depending on the size of the secret data that is to be hidden. The maximum embedding capacity offered by this tool is half the size of cover media file.
3. **Steghide**: Steghide is a command line interface (CLI) tool which uses graph theoretic approach to hide secret data into the audio cover file. The maximum capacity offered by this tool is one by sixteenth the size of cover file. It supports. au and wav audio file formats and also implements a checksum to check for integrity of extracted data.

4. **DeEgger**: This tool takes binary of given secret file and merges the binary code of cover file with the secret file in turn increasing the total size of the cover file and hence not making it a good steganographic tool. This tool hides the secret data in the cover file based on unique secret tags assigned to the data that is to be hidden. Hence deleting, or adding noise to the secret tag will render the hidden secret data corrupted and unable to recover. Since file size of cover media increases as the size of the secret data increases, the maximum capacity is not dependent on cover media.
5. **OurSecret**: This tool works similar to DeEgger in the way that it also increases the size of cover file and hence making it hard not to be suspicious as it hides binary data of secret file in between samples of audio cover data and also has same drawbacks as DeEgger.

## 4 Evaluation and Results

The following performance measures have been incorporated in this experiment after steganographic jamming has been implemented on the cover media using the tool developed on Java platform. None of the tools discussed above were able to recover the hidden files after performing steganographic jamming on the cover file which had data hidden in it.

The quality of output audio signal after steganographic jamming was rated using PESQ [13] which automatically rates audio signal quality objectively through software based on human perception of speech quality. It grades the resulting output based on MOS scale as shown in Table 1. which varies from 1 (Very Annoying) to 5 (Imperceptible). For this test three audio signals as described in Table 2 were composed. These signals were used as cover files and secret data was first hidden into it by the above tools and then steganographic jamming technique was performed. The resulting quality of audio signal was measured using PESQ reference implementation software P.862 [12] recommendation and values obtained are listed in Table 3. As the PESQ method is a full reference method of objectively testing quality of audio signal, the steganographic cover files were used as reference for measuring output with double stegged audio signal. Results shows that the output audio quality is well maintained for both the methods of embedding.

**Table 1** ITU-T conversation opinion scale for MOS

| Perceived distortion level | Quality | Grade |
|---|---|---|
| Imperceptible | Excellent | 5 |
| Perceptible but not annoying | Good | 4 |
| Slightly annoying | Fair | 3 |
| Annoying | Poor | 2 |
| Very annoying | Bad | 1 |

**Table 2** Description of audio signal used for test

| | Music file 1 | Music file 2 | Music file 3 |
|---|---|---|---|
| Number of channels | 2 | 2 | 2 |
| Sample rate | 8000 | 16,000 | 16,000 |
| Length of audio signal (s) | 54.3 | 54.3 | 300 |
| Bits per sample | 16 | 16 | 16 |

**Table 3** PESQ obtained after steganographic jamming of cover signal

| Steganography tools | | LSB embedding | Variable LSB embedding |
|---|---|---|---|
| OpenPuff | Music file 1 | 4.499 | 4.201 |
| | Music file 2 | 4.389 | 4.1139 |
| | Speech signal | 4.201 | 4.031 |
| DeepSound | Music file 1 | 3.61 | 3.32 |
| | Music file 2 | 3.621 | 3.292 |
| | Speech signal | 3.502 | 3.04 |
| StegHide | Music file 1 | 4.4 | 4.15 |
| | Music file 2 | 4.36 | 4.13 |
| | Speech signal | 4.23 | 4.06 |
| DeEgger | Music file 1 | 3.90 | 3.65 |
| | Music file 2 | 3.84 | 3.6 |
| | Speech signal | 3.6 | 3.48 |
| OurSecret | Music file 1 | 4.0005 | 3.625 |
| | Music file 2 | 4.102 | 3.675 |
| | Speech signal | 3.85 | 3.312 |

Table 4 shows the bit error rate obtained when stego cover media was compared with double stegged cover media indicating corruption/partial destruction of hidden data and hence rendering the secret data unrecoverable by the stego tools and also maintains good output audio quality. As seen in Table 4 bit error rate is varying from 6.3 to 10.03% for steganographic jamming using LSB embedding and from 12.23 to 23.63% using Variable LSB embedding indicating good results steganographic jamming using variable LSB embedding technique because higher BER indicates the effectiveness of the algorithm without substantial reduction in MOS score.

Table 5 gives the NIST Signal to Noise ratio values computed using [10] of Original File, Cover File and the Double Stegged file which shows that there is very little degradation in audio quality or very less noise was introduced using steganographic jamming. And hence maintain good audio quality output.

Table 6 shows Waveform Amplitude Distribution Analysis—Signal to Noise Ratio [2] of original, steg and double stegged files. It can be seen that there is very little reduction in quality of audio signal after double stegging hence making it a good technique to effectively curb illegal and malicious use of steganography.

**Table 4** BER comparison table

| Steganography tools | | LSB embedding | Variable LSB embedding |
|---|---|---|---|
| OpenPuff | Music file 1 | 0.091 | 0.1337 |
| | Music file 2 | 0.095 | 0.1402 |
| | Speech signal | 0.08 | 0.1562 |
| DeepSound | Music file 1 | 0.102 | 0.212 |
| | Music file 2 | 0.956 | 0.2232 |
| | Speech signal | 0.1003 | 0.2363 |
| StegHide | Music file 1 | 0.063 | 0.1223 |
| | Music file 2 | 0.0593 | 0.1245 |
| | Speech signal | 0.0821 | 0.1335 |

**Table 5** NIST SNR values (dB)

| Steganography tools | | Original file | Cover file | Double stegged file |
|---|---|---|---|---|
| OpenPuff | Music file 1 | 7.8 | 7.8 | 7.8 |
| | Music file 2 | 7.5 | 7.5 | 7.5 |
| | Speech signal | 7.3 | 7.3 | 7.2 |
| DeepSound | Music file 1 | 7.8 | 7.8 | 7.7 |
| | Music file 2 | 7.5 | 7.3 | 7.2 |
| | Speech Signal | 7.3 | 7.3 | 7.2 |
| StegHide | Music file 1 | 7.8 | 7.8 | 7.7 |
| | Music file 2 | 7.5 | 7.5 | 7.5 |
| | Speech signal | 7.3 | 7.2 | 7.1 |

**Table 6** WADA-SNR Values (dB)

| Steganography tools | | Original file | Cover file | Double Stegged file |
|---|---|---|---|---|
| OpenPuff | Music file 1 | 8.8 | 8.8 | 8.6 |
| | Music file 2 | 8.9 | 9.1 | 8.7 |
| | Speech signal | 8.9 | 8.8 | 8.7 |
| DeepSound | Music file 1 | 8.8 | 8.8 | 8.7 |
| | Music file 2 | 8.9 | 8.5 | 8.4 |
| | Speech signal | 8.9 | 8.4 | 8.3 |
| StegHide | Music file 1 | 8.8 | 8.8 | 8.6 |
| | Music Ffe 2 | 8.9 | 9.1 | 8.7 |
| | Speech signal | 8.9 | 8.8 | 8.8 |

## 5 Conclusion

Double Stegging or Steganographic Jamming was implemented as technique to destroy embedded data in a given audio signal irrespective of the underlying steganographic algorithm used. It was proved that our method was effective as the steganographic tools failed to recover the hidden data and also there was very less noise/distortion introduced in the cover media after steganographic jamming. Quality of the output signal was measured using Mean Opinion Score, Bit Error Rates and Signal to Noise Ratio. The above results also suggest that double stegging using variable LSB embedding algorithm was more successful in destroying secret data which was indicated by higher bit error rates.

## References

1. Nutzinger, M.: Real time attacks on audio steganography. J. Inf Hiding Multimed. Signal Process. **3**(1), (2012) ISSN 2073-4212
2. Kim, C., Stern, R.M.: Robust Signal-to-Noise Ratio Estimation Based on Waveform Amplitude Distribution Analysis. In: Interspeech pp. 2598–2601 (2008)
3. Thwarting Audio Steganography Attacks in Cloud Storage Systems. In: International Conference on Cloud and Service Computing, pp. 259–265
4. De Rango, F., Tropea, M., Fazio, P., Marano, S.: Overview on VoIP: subjective and objective measurement methods, IJCSNS Int. J. Comput. Sci. Netw. Secur. **6**(1B), (2006)
5. Katzenbeisser, S., Petitcolas, F.A.P.: Information Hiding Techniques for Steganography and Digital Watermarking, pp. 121–148. Artech House, Boston, London (2000)
6. Basu, P.N., Bhowmik, T.: on embedding of text in audio-a case of steganography. In: Proceedings. Of International Conference on Recent Trends in Information, Telecommunication and Computing, pp. 203–206 (2010)
7. Shelley, L., Picarelli, J.: Methods not motives: Implications of the convergence of international organized crime and terrorism. Int. J. Police Pract.Res. **3**, 305–318 (2002)
8. Al-Ani, Z.K., Zaidan, A.A., Zaidan, B.B., Alanazi, H.O.: Overview: Main fundamentals for steganography. J. Comp. **2**(3), 158–165 (2010)
9. Adee, S.: Spy vs. spy. http://spectrum.ieee.org/computing/software/spy-vs-spy/
10. Ellis, D.: Lab for Recognition and Organization of Speech and Audio (LabROSA.) http://labrosa.ee.columbia.edu/projects/snreval/
11. Karadogan, I., Das, R.: An examination on information hiding tools for steganography. Int. J. Inf. Secur. Sci. **33**, 200–208
12. PESQ reference implementation software P.862 recommendation. http://www.itu.int/rec/T-REC-P.862-200511-I!Amd2/en
13. Recommendation ITU-T P. 862, Perceptual evaluation of speech quality: an objective method for end-to-end speech quality assessment of narrow-band telephone networks and speech codecs, Technical report (2001)
14. Premalatha, P., Amritha, P.P.: Optimally locating for hiding information in audio signal. Int. J. Comput. Appl. (0975–8887) **65** (14), (2013)

# A Hybrid Method for Extraction of Events from Natural Language Text

**Vanitha Guda and S.K. Sanampudi**

**Abstract** Events extraction is a significant and interesting task in the field of Natural Language Processing (NLP). Basically events are the dynamic occurrences, specific happenings, causes or things. An event plays a vital role in narrative of text and also important for many NLP applications. This paper presents a Hybrid/Composite way of events extraction from natural language text. Earlier work of events extractions were developed with rule based approach or machine learning methods. The Proposed hybrid makes use of both machine learning approaches and hand coded rules to extract the events. Experiments were conducted on SemEval-2010 data set, the results obtained shown better precision and recall when compared with the existing methods.

**Keywords** Events · Natural Language Processing (NLP) · Events extraction · Rules based approach · Machine learning techniques

## 1 Introduction

Nowadays, information storage on web is being increased on an exponential manner. With the reason web has become a major source to obtain the required information, the information is available in an unstructured form. Several tools exist to extract the information in structured form but different methods of information retrieval need to be used to extract relevant information from unstructured data.

"Events" place a prominent role in this context of processing natural language text. Events may be the causes, one or more changes in the real world happenings,

V. Guda (✉)
Research Scholar of JNTUH, CSE Department, Chaitanya Bharathi Institute of Technology, Gandipet, Hyderabad 500075, India
e-mail: vanithaguda@gmail.com

S.K. Sanampudi
Department of IT, JNTUH, Nachupally, Karimnagar, India
e-mail: sureshsanampudi@jntuh.ac.in



S.C. Satapathy et al. (eds.), *Data Engineering and Intelligent Computing*,
Advances in Intelligent Systems and Computing 542,
DOI 10.1007/978-981-10-3223-3_28

(e.g., one industry buys another industry here "buy" is an event). The basic meaning of the verb in English language provides a dynamic occurrence or action that shows the activity or status. Events are the primary terms in our work these are the dynamic entities which happens or occurs in a context. The present framework defines the events as the entities which can be represented in text as verbal form.

The state-of-art methods available in the literature for event extraction were classified as domain specific [1], annotation based [2, 3], ACE [4], ERE [5], context based and some of them are using machine learning approaches [6–8] and TempEval-2007 participants [8]. Most of the related works tried to improve the classification through feature engineering and machine learning approaches but their system's performance is limited by the available training data. This paper provides an approach that can detect relevant events and models the events information. The task of Event extraction is utilized in a number of natural language applications like temporal question answering, machine translation, question answering, information extraction, and document summarization [9].

The available event extraction methods developed so far were addressing the issue in a domain specific aspect. In this paper open domain aspect is used for event extraction task so that it can be utilized generically to all domain related applications. A Hybrid method defined as method-3 has been derived by merging different techniques of event extraction projected in method 1 and method 2.

- *Method 1*: Event Extraction using rules, framed by using lexical features.
- *Method 2*: Event Extraction with machine learning techniques.
- *Method 3*: Hybrid way or Composite way (Events Extraction with features + machine learning techniques).

Events are identified with hand coded rules by taking text or a text document is the input source and firstly performs tokenization to recognize valid tokens. After recognizing the tokens the generated tokens are compared with the rules from rules based event extraction algorithm [10]. This algorithm checks the tokens and maps the tokens with rules to declare events.

The major constraint of this rule based method is that the nonverbal events are not recognized by this model (e.g.: word "Park" can be identified as noun by POS but it can be verb also) these sort of nonverbal events are not recognized and the gain of this rule based method is that it is easy to implement for small class of grammars or less volume texts.

Widespread techniques of machine learning such as Conditional Random Field (CRF), Semantic Role Label (SRL) and WordNet were used for event extraction. Comparing with rules based techniques CRF obtains better results in extraction of the events and also identifies the nonverbal events, because the limited trained features of CRF can helps to label and recognize the events. The limitation of CRF is training part of the data, handling the different senses of token and nonverbal events. By using hand coded rules and CRF techniques most of the times the nonverbal events, named entities (NE) are not recognized. To overcome these limitations, SRL and WordNet which are two popular methods of machine learning were used to recognize the events.

**Semantic Role Label (SRL)** is the method based on event nominalization which focuses on the events or target class. SRL tries to extract high level information that is more independent from the word, and other side it is verbal SRL. If a word is treated as noun and also a verb two tasks needs to be performed one is *identifying the arguments related to the word and other is Argument Labelling.*

The major feature of SRL is it will extract all constituents of a word by determining their arguments and its adjuncts. The two important tasks of SRL are:

(1) Event and Result Nominalization is the process to get the constitutes and nominals of a word in bulk of nonverbal nouns.
(2) Nominals identification with the help of its suffixes (-ed, or, ee, er).

By using SRL mostly the constituents and context of the tokens were derived and nominals are detected. SRL parses the sentences and decides the words as Events. WordNet categories the features like lexical such as parts of speech (POS), Stem, hypernym, meronym etc. WordNet is used to identify nonverbal events.

Eg: "Ship" is word by using CRF and SRL can identify as nonevent because its POS can tagged as NN means noun so it will not treated as an event but the word may be verb also in other context, these type of nonverbal events can be identified by using WordNet. WordNet performs the task in two steps:

i. Tokens parsed by WordNet if the token appears as noun and verb that will be declared as events.
ii. The stem of noun words are checked by WordNet if any one of the word senses is verb then it will be announced as Event. (e.g.: "Declared" is a word the stem word is declare).

The remaining part of the paper is organized as follows, Sect. 2 describes a Hybrid model Framework for events extraction, Sect. 3 explains results while conclusion and future work were discussed in Sect. 4.

## 2 Hybrid Model for Event Extraction

Event Extraction includes various aspects to be addressed. The detailed steps were explained in a diagrammatic form depicted in Fig. 1. The framework of event extraction accepts the set of documents as input. It undergoes various preprocessing steps that involve tokenization, parts of speech identification and Chunking [11].

The next step of event extraction includes **feature identification** it considers the preprocessed text and identifies the set of features like lexical, syntactic and WordNet features.

- *Morph-Lexical*: lexical feature involves basic identification of the token it considers the lexical character pattern.

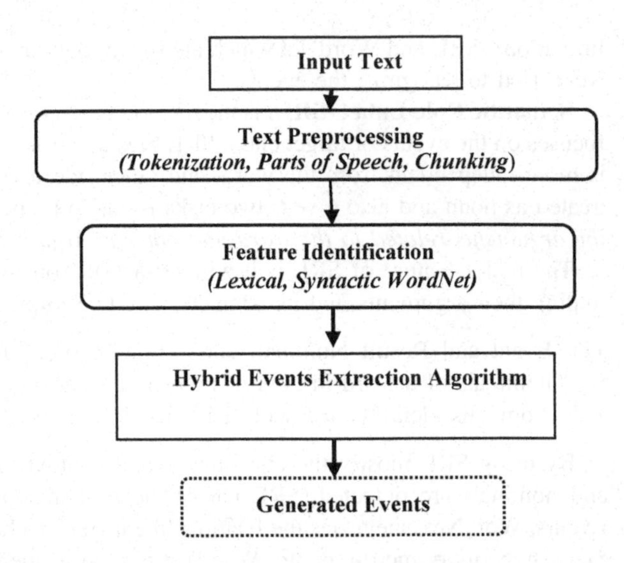


**Fig. 1** Hybrid model framework for event extraction

- *Syntactic*: A Part of Speech is basic syntax identification we use penTree Tagger [11]. Further, boolean features are included this features indicates if the word is lower-case, alphabetic, alphanumeric, titled, capitalized, an acronym (capitalized with dots), number, decimal number, stop-word elimination can be synthesized.
- *WordNet*: Syntactic token may have different senses that can be recognized by WordNet. For each token we considered the number of senses associated to that word. Firstly the most common senses of the lemmas, the next entailments for verbs, antonyms, hypernyms and hyponyms each of them is are considered using WordNet. A total of 23 WordNet-based features have been considered in this work.

**A Hybrid or Composite way of events extraction** is proposed to overcome the limitations of hand coded rules and machine learning techniques. Composite means combining the best features of the both the techniques such as rule based approach mentioned in method 1 and machine learning technique mentioned in method 2 and deriving new rules for events extraction the following points explains the derived method:

1. Consider Plain Text T
2. Perform Pre-processing and tokenize the text (involves elimination of the Stop words, lexical task, morphological, syntactical features)
3. Perform Event Extraction using following steps:
   a. Performs lexical task by PosTagger this will tag the verbal words as Events.
   b. Perform syntax and morphological Analysis using WordNet to identify different senses of a word to extract the

token that appears as both nouns and verbs which can be defined as event using composite rules.

c. Run CRF based standford Named Entity (NE) tagger this will tag remaining unidentified events.

The above stated steps are using lexical information of the token and machine learning techniques of CRF and WordNet. By using the above procedure rules are framed to extract events the rules are called composite rules because of the rules formation with combination of machine learning techniques.

***Composite Rules for Event Extraction***:

**Rule 1**: Morphologically derived nouns from verbs are distinguished as nominalizations (or nonverbal nouns). The nonverbal nouns are identified by its suffixes(-tion, ion, ing, ed) these are not NEs but may end with these suffixes are considered as *Event* words.
**Rule 2**: The token if it verb and also noun then the combinations are searched in the sentence of the test set, non-NE noun words are considered as the Events.
**Rule 3**: Nominal and nonverbal event nouns can be identified by the complements of its aspectual PPS headed by prepositions like (during, after, before, at the end of, at the beginning etc., these are clues) the next words after these clues are also Events.
**Rule 4**: Event noun can also appears like objects of aspectual and time related verbs, such as (e. g., began a hotel or carried out the test etc.) the non-NE that appears after these expressions are also Events.
**Rule 5**: Any of the token if it fails of the above rules that are treated like non Events as to how to shorten it would be most welcome.

## 3 Experimental Results

The data set used for evaluation of proposed wok by using SemEval-2010 [12] from TempVal and MUC [13] to get the evaluation results. SemEval and MUC datasets are normal text documents but not specific to any domain. First hand coded rules were applied on preprocessed input, next CRF based system along with WordNet were applied. Hand-coded rules are developed by using lexical information of the word, CRF technique is implemented with various features of the word, CRF and WordNet combined to extract more nonverbal words.

**SemEval dataset** [14] contain 323 verbal and 125 non-verbal event nouns. Table 1 presents the results of the proposed framework. The Evaluation metrics

**Table 1** Results of event extraction for semval data (In Percentages)

| Method | Precision | Recall | F-measure |
|---|---|---|---|
| Handcoded rules | 75.32 | 77.14 | 76.21 |
| CRF | 77.21 | 78.91 | 78.05 |
| CRF + WordNet | 79.20 | 81.22 | 80.19 |
| Hybrid (or) Compositerules | 81.11 | 84.23 | 82.64 |

**Table 2** Results of event extraction for MUC Data (In Percentages)

| Method | Precision | Recall | F-measure |
|---|---|---|---|
| Handcoded rules | 63.32 | 65.15 | 64.18 |
| CRF | 67.21 | 68.12 | 67.64 |
| CRF + WordNet | 70.20 | 72.21 | 68.18 |
| Hybrid (or) Compositerule | 72.21 | 73.38 | 73.74 |

such as e precision, recall and F-measure were used to gauge the performance of the proposed work. Performance of F-Measure increases by at least 1.84% from base hand-coded rules to hybrid method. Table 1 shows very high performance improvement (i.e., 3.98%) with the use of CRF + wordNet. The Inclusion of the hybrid/composite method rules obtained an increased F-measure of 6.43% when compared with earlier methods. The precision, recall and F-measure were calculated for the proposed hybrid model for event extraction and the values were found to be 81.11%, 84.23% and 82.64%, respectively. An improvement of approximately 6% F-measure is obtained when compared with other base methods.

**MUC dataset** [13] contain 200 sentences over the 30 documents and 253 verbal and 110 non-deverbal event nouns, in Table 2 results were presented. Performance of F-Measure increases by at least 2.54% from base hand-coded rules to hybrid method. Table 2 shows very high performance improvement (i.e., 3.99%) with the use of CRF + wordNet. The Inclusion of the hybrid/composite method rules obtained an increased F-measure of 7% when compared with earlier methods. The precision, recall and F-measure were calculated for the proposed hybrid model for event extraction and the values were found to be 73.38%, 72.21% and 73.74%, respectively. An improvement of approximately 7% F-measure is obtained when compared with other base methods.

## 4 Conclusion, Future Work

A Hybrid/Composite framework was presented for event extraction. I make use of both the gains of rule based and machine learning methods. Composite rules were derived that utilize both the word net and CRF and SRL techniques. The experiments were conducted on SemEval-2010 [12] and MUC [13] data sets. The results were evaluated using the Precision, Recall and F-Measure. Initially. Evaluation results yield the better precision value of 81.11% for Sem-Eval dataset and 72.21% for MUC data set. There is an improvement of approximately 6–7% of F-measure for the Hybrid method of event extraction for SemEval-2010 and MUC when compared with base methods.

A future direction of works includes the identification of more precise rules for event identification and multiword events. Investigating machine learning techniques like maximum entropy and support vector machine suitability for event extraction.

## References

1. Sohn, S., Wagholikar, K.B., Li, D., Jonnalagadda, S.R., Tao, C., Elayavilli, R.K., Liu, H.: Comprehensive temporal information detection from clinical text: medical events, time, and tlink identification. J. Am. Med. Inform. Assoc. (2013). doi:10.1136/amiajnl-2013-001622
2. Pustejovsky, J., Castano, J., Ingria, R., Saurí, R., Gaizauskas, R., Setzer, A., Katz, G., Radev, D.: TimeML: robust specification of event and temporal expressions in text. In: IWCS-5 Fifth Int. Workshop on Computational Semantics (2003)
3. Boguraev, B., Ando, R.K.: TimeBank-Driven TimeML analysis. In: Annotating, Extracting and Reasoning about Time and Events, Dagstuhl, Germany (2005)
4. Poveda, J., Surdeanu, M., Turmo, J.: A comparison of statistical and rule-induction learners for automatic tagging of time expressions in english. In: Proceedings of the International Symposium on Temporal Representation and Reasoning (2007)
5. Boguraev, B., Ando, R.K.: TimeMLCompliant text analysis for temporal reasoning. In: Proceedings of Nineteenth International Joint Conference on Artificial Intelligence (IJCAI-05), Edinburgh, Scotland, pp. 997–1003 (2005)
6. Mani, I., Wellner, B., Verhagen, M., Lee, C.M., Pustejovsky, J.: Machine learning of temporal relation. In: Proceedings of the 44th Annual meeting of the Association for Computational Linguistics, Sydney, Australia (2006)
7. Chambers, N., Wang, S., Jurafsky, D.: Classifying temporal relations between events. In: Proceedings of the ACL Demo and Poster Sessions, Prague, Czech Republic, pp. 173–176 (2007)
8. Verhagen, M., Gaizauskas, R., Schilder, F., Hepple, M., Katz, G., Pustejovsky, J.: Task 15: TempEval temporal relation identification. In: Proceedings of the 4th International Workshop on Semantic Evaluations, pp. 75–80, SemEval (2007)
9. Shou, L., Wang, Z., Chen, K., Chen, G.: Sumblr: Continuous summarization of evolving tweet streams. In: Proceedings of SIGIR, pp. 533–542. ACM (2013)
10. Guda, V., Sanampudi, S.K.: Rule based event extraction in natural language text. In: Proceedings of IEEE International Conference of Recent Trends in IEEE, pp. 47–51, (May 2016). ISBN 978-1-5090-0773-5/16/$31.00 © 2016
11. http://nlp.standford.edu/software/lex-parser.shtml
12. http://semeval2.fbk.eu/semeval2.php?location=download&task_id=5&datatype=trial
13. Miller, S., Fox, H., Crystal, M., Ramshaw, L., Stone, R., Schwartz, R., Weischedel, R.: Message Understanding Conference Proceedings (MUC-7) (1997)
14. Strotgen, J., Gertz, M.: Heideltime: high quality rule-based extraction and normalization of temporal expressions. In Proceedings of the 5th International Workshop on Semantic Evaluation, pp. 321–324. Association for Computational Linguistics (2010)

# A Proposal on Application of Nature Inspired Optimization Techniques on Hyper Spectral Images

**M. Venkata dasu, P. VeeraNarayana Reddy and S. Chandra Mohan Reddy**

**Abstract** Hyper spectral image are used in various applications such as geological systems, geo sciences and astronomy. These images are acquired using remote sensing. Remote sensing is the process of getting information about an object without making any physical contact with the object. Satellite Images referred as hyper spectral images are the most used images in remote sensing and are of more interest to find out the classification of objects in those images. The classification can give us the important factors like vegetation, buildings, roads and more. Satellite images can be of assistance in supervision of effects due to natural disasters, to recognize mining areas which are hidden from human view, biodiversity examination, rural and urban environment detection for analysis, etc. However, occasionally the Satellite images acquired can be affected by unforeseen distortions, artificial unwanted structures called artifacts that are formed by the tool itself or sometimes due to the diverse pre-processing procedures involved. Optimization algorithms in combination with Image processing methods are used to classify the objects in satellite images for easy perception and analysis. In this paper, various optimization techniques like particle swarm optimization (PSO), DPSO, HSO, and Proposed MFA optimization algorithms are compared to obtain optimal classification of objects in a satellite image.

**Keywords** Enhancement · Segmentation · Particle swarm optimization (PSO) · Darwinian PSO · HSO · MFA

---

M. Venkata dasu (✉)
Research Scholar-ECE, Jawaharlal Nehru Technological University,
Anantapuramu, Andhra Pradesh, India
e-mail: dassmarri@gmail.com

P. VeeraNarayana Reddy
Sri Venkateswara College of Engineering, Tirupati, Andhra Pradesh, India
e-mail: principal_svew@svcolleges.edu.in

S. Chandra Mohan Reddy
Jawaharlal Nehru Technological University, Pulivendula, Andhra Pradesh, India
e-mail: email2cmr@gmail.com



S.C. Satapathy et al. (eds.), *Data Engineering and Intelligent Computing*,
Advances in Intelligent Systems and Computing 542,
DOI 10.1007/978-981-10-3223-3_29

# 1 Introduction

In the last six decades, satellite imagery plays a important role in our day today life and has been used successfully in diverse fields. The images sent by the satellites endow with a lot of geographical information present on the earth's surface. Today, satellite images are used in various scientific applications, in particular measuring the changes which influence the environment. The images captured by the satellites have a massive volume of data in it and good number of information are generated from them has increased tremendously.

A proverb says, ***one picture is worth more than thousand words***. An image consist lot of information and can be retrieved through image processing. Data in various forms play a vital role in all walks of life. To analyze this huge amount of data embedded on the images, an interdisciplinary research field and digital image processing has emerged. Interpretation and analysis of satellite imagery is conducted using particular remote sensing methods. Digital images play a vital role in various image processing applications [1] which include oceanography, meteorology, fishing, biodiversity conservation, agriculture, forestry, geology, landscape, cartography, regional planning, intelligence, education and warfare. Satellite images are helpful in tracking of earth resources, geographical mapping, and prediction of agricultural crops, urban population, weather forecasting, flood and fire control [2]. In this research work, digital image processing is done on satellite images which involve image enhancement and denoising, segmentation, clustering, feature extraction and classification of objects in the scene. Better understanding of the image processing results.

The paper is structured as follows: In Sect. 1 Introduction. This section explores overview and background general information related to the satellite images and optimization techniques. In Sect. 2 the model of enhancement problem is discussed. In Sect. 3 theory of proposed MFA is discussed. Results and discussion are given in Sect. 4, and finally the conclusion.

# 2 Optimization Techniques

In the applications of image processing, image enhancement is to get better visibility or perception in images for human spectators. The enhanced image is

$$\text{represented as } q(x,y) = T[p(x,y)]. \tag{1}$$

where p(x,y) is an input image, q(x,y) is the enhanced output image and T is an operator. In this paper optimization techniques and histogram methods are used to classify the satellite images. Various optimization techniques are discussed below.

The main objective of the optimization methods is to inquire about best optimum attributes to certain restrictions for the purpose of minimizing or maximizing objective functions (Rardin 1998; Van den Bergh 2002). The selection of values for attributes always satisfies all restric which vigorously alters the velocitytions is termed as *feasible result.* optimal solutions is known to be the promising solution with the objective function values compared with other realistic solutions (Rardin 1998). These techniques are widely used in decision making systems, industrial planning, resource allocation scheduling, business, engineering and computer science to attain optimal solutions. Nowadays the research work in the domain of optimization is very lively.

## *2.1 Particle Swarm Optimization*

PSO principle operation is based on communal behavior of colony of insects like bees, ants, school of fish and a flock of birds. The entity swarm performs its own and group intelligence of the swarm [3]. In this optimization if one particle finds a fine path the remaining swarm follow the optimum path immediately even if their position is far-off in the swarm. PSO produces a preliminary population of particles randomly, each one signifies a possible solution of system, and each one is characterized by 3 indexes namely position, velocity, fitness. Firstly each particle allots a random velocity which vigorously alters the velocity & position of particles through their own voyage experience as well as their neighbor's. Hence, with superior fitness through incessant learning and updating the entire cluster will fly to the search region [4]. This operation is repeated until the highest iterations or the predetermined minimum fitness is reached. So PSO is known as a group or fitness based optimization algorithm, whose merits are quick convergence, minimalism, easy implementation and fewer parameters. The whole PSO performs a powerful convergence, and it may easily get fascinated in confined minimum points, which creates the swarm mislay diversity.

***Drawbacks:*** To discover an overall minimum the number of particles travels in space and may be attentive in incorrect local best possible points.

## *2.2 Darwinian Particle Swarm Optimization (DPSO)*

The basic principle of DPSO is that it considers the group of input points as discrete entities and base points that are stored in the position vector. To indicate its correspondence it has a related fitness value. The fitness value that is connected with its earlier best vector has a preceding finest value to entity particle vector. Each preceding finest vector of a particle undergo for optimal match of those related input and base points.

### *2.3 Harmony Search Optimization*

In basic Harmony Search Optimization each result is known a harmony and is characterized by real vectors with N dimensions. Early inhabitants of harmony vectors which are produced arbitrarily are accumulated inside a harmony memory. Moreover from the harmony memory either by a pitch regulation or arbitrary re-initialization operation a fresh harmony aspirant is generated [5]. The fresh candidate harmony were updated in harmony memory by evaluating the fresh and bad harmony aspirants In harmony memory the bad harmony vector is swapped by fresh candidate vector when second one delivers an improved solution [6]. This operation continues until a definite termination criterion is reached.

## 3 Proposed Model

The proposed implemented method in this research work is shown in Fig. 1 is meant to extract the fine details (Enhancement and segmentation) in hyper spectral images.

Hyper spectral images which are acquired from the Public database can be utilized as inputs for the proposed model. The acquired images are usually RGB images [7]. Hence MATLAB software with several algorithms is used to process the images to get an improved output.

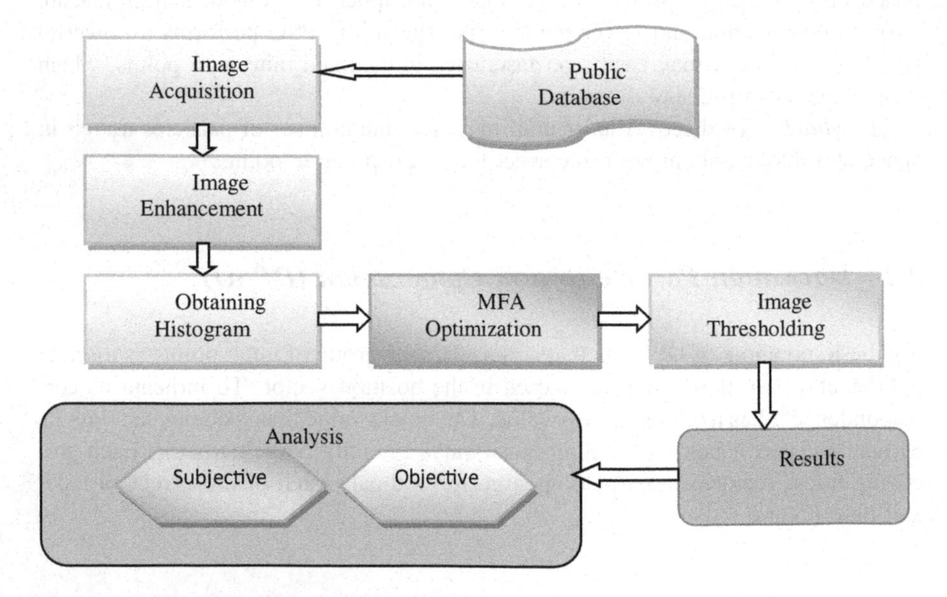


**Fig. 1** Model diagram of proposed method

### 3.1 MFA Optimization

Magnetic Force Attraction (MFA) is an inclusive algorithm which states the act of electromagnetism.MFA is an inhabitant based method which has a pull and revulsion process to progress the constituents of the inhabit ants directed by their objective function values [8, 9]. This algorithm is improved version of Electro Magnetism Like Optimization method. MFA mainly alters a particle through the space tracking and force which is generated from remaining inhabitants. By using the charge of individual particle and based on its objective function value force is determined [10, 11]. So for finding a global solution of a nonlinear optimization problem the MFA method is designed.

**Algorithm**

1. Input factors: $Gen_{max}$ (Maximum generations), $Gen_{local}$ & δ (local search parameter), and N (Population Size).
2. Initialization: Initially set the generation counter t = 1, Set the amount of $P_t$ homogenously in X and recognize the finest point in $P_{t.}$
3. while $t < Gen_{max}$ do
4. $K_i^t \leftarrow CalcK(P_t)$
5. $y_{i,t} \leftarrow Shift(x_{i,t,}\ K_i^t)$
6. $z_{i,t} \leftarrow local(Gen_{local,}\ \delta, y_{i.t})$
7. $x_{i,t+1} \leftarrow select(P_{t+1}, y_{i.t}, z_{i,t})$
8. end while

Some features of MFA dispense PSO and ACO. In recent works by considering the optimal attributes the better accuracy had confirmed and is used to decipher numerous kinds of engineering complications like image processing, neural network training, array pattern optimization, control systems, vehicle routing, flow-shop scheduling and communications [12]. In terms of precision and computation time MFA contributes many features to other evolutionary optimization techniques like HSO, PSO and DPSO.

### 3.2 Kapur's Thresholding

In image segmentation process especially for multilevel thresholding, Kapur's entropy is the finest threshold selection method for hyper spectral images with respect to shape and regularity procedures [2, 13–15]. It is computationally immaterial because of ineffective creation between the class variance.

Consider q(x,y) is a processed version of input image p(x,y) at some optimal threshold T,then q(x,y) is written as,

$$q(x,y) = \begin{cases} 1, & \text{for} \quad p(x,y) > T \\ 0, & \text{for} \quad p(x,y) < T \end{cases} \tag{2}$$

The kapur method distributes the isolated region more unified by maximize the segmented histogram [16, 17]. In the gray-level histogram if the threshold value is at lowest value then Foreground (object) is isolated from its background. Among all multilevel image thresholding techniques Kapur's entropy-based thresholding technique is more efficient, noteworthy and frequently used segmentation technique where it can acquires enhanced threshold function in terms of probability values.

In this proposed model MFA Optimization with Kapur Thresholding is utilized to achieve the optimum results over existing PSO, DPSO and HSO Methods. The images considered for experimental investigations are Aerial images of Tsunami effected Kalutara Beach (Courtesy: Digital globe) shore before and after Tsunami in Sri lanka during December of 2004. As a part of Experimental investigations, partial results obtained in first attempt are discussed below (Fig. 2).

The image shown in Figs. 3, 4 and 5 are output images of PSO, DPSO, and HSO which gives the threshold image with minimum visible edges when compared to Fig. 4 which is the output of proposed method. MFA optimization method gives the best minimal time over remaining methods which are discussed above (Figs. 6 and 7).

**Fig. 2** Original Image

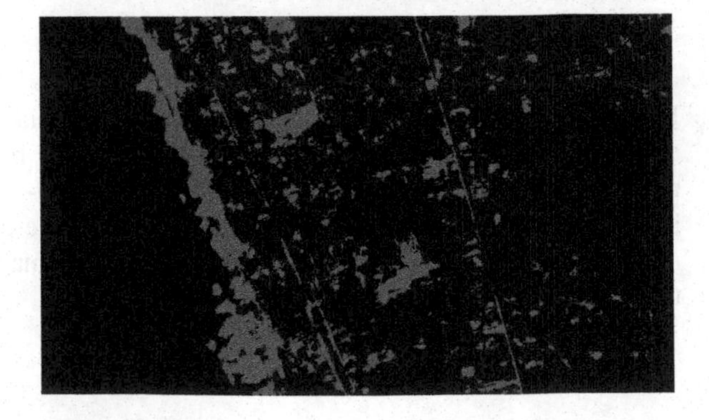

**Fig. 3** PSO Image

**Fig. 4** DPSO Image

**Fig. 5** HSO Image

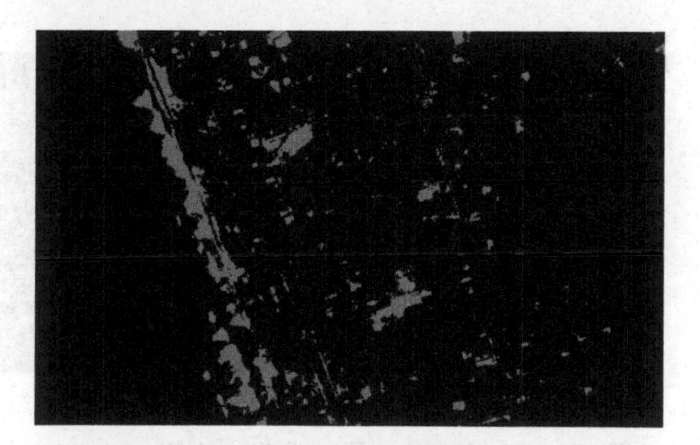

**Fig. 6** MFA Image

**Fig. 7** Original image After Tsunami

**Fig. 8** PSO

**Fig. 9** DPSO Image

**Fig. 10** HSO Image

The images shown in Figs. 8, 9 and 10 are output images of PSO, DPSO and HSO with Multilevel Image Thresholding which gives the threshold image with minimum visible edges when compared to Fig. 11 which is the output of proposed method. MFA optimization method gives the best minimal time over remaining methods which are discussed above. The results of proposed method for both after and before Tsunami images suggest that the unwanted edges present in results of existing procedures are removed and an improvement is observed from analysis in terms of water and land.

**Fig. 11** MFA Image

**Table 1** Result after apply the PSO, DPSO, HSO and, MFA to the set of benchmark images

| Before Tsunami (Kalutara Beach) | | | | | After Tsunami (Kalutara Beach) | | | |
|---|---|---|---|---|---|---|---|---|
| | Std | Mean | PSNR | CPU TIME | Std | Mean | PSNR | CPU TIME |
| PSO | 0.19 | 28.54 | 13.16 | 16.08 | 0.17 | 27.26 | 13.03 | 15.02 |
| DSO | 0.12 | 26.13 | 16.34 | 16.02 | 0.15 | 25.12 | 15.06 | 14.73 |
| HSO | 0.10 | 22.68 | 18.16 | 15.03 | 0.10 | 23.24 | 17.64 | 13.87 |
| MFA | 0 | 13.1 | 20.13 | 10.02 | 0 | 9.36 | 20.10 | 11.2 |

### 3.3 *Performance Measures*

The Experimental results after applying PSO, DPSO, HSO and, MFA to the set of benchmark images is shown in Table 1.

The parameters listed above are Standard deviation, Mean, Peak Signal to Noise Ratio and CPU time(s). It is concluded that the proposed method gives the best results over existing methods.

## 4 Conclusions and Future Scope

In this work the multilevel thresholding with Magnetic Force Attraction (MFA) is investigated on images of seashore in Srilanka before and after Tsunami. The results provided in this paper are of first attempt by setting the value of level '2'. It is clear from the results that unwanted edges in the images are removed for easy perception of the shore and land, which will be helpful in analyzing the effect of the calamity pictorially. Further in this work, it is slated to change the values of levels and include some enhancement algorithms for better results. The processing of images is carried only after the artifacts added to images during Image acquisition have been removed by enhancement methods with filters.

**Acknowledgements** The authors are thankful to JNTUCEA, Anantapuramu and Annamacharya Institute of Technology & Sciences, Rajampet, A.P. for their extensive support in carrying our research work by providing research facilities.

## References

1. Gonzalez, R.C., Woods, R.E.: Digital Image Processing Addison Wesley, Reading, Mass, USA (1992)
2. Naderi, B., Tavakkoli-Moghaddam, R., Khalili, M.: Electromagnetism-like mechanism and simulated annealing algorithms for flowshop scheduling problems minimizing the total weighted tardiness and makespan. Knowl.-Based Syst. **23**, 77–85 (2010)
3. Snyder, W., Bilbro, G., Logenthiran, A., Rajala, S.: Optimal thresholding: A new approach, pattern recognition letters, **11**(11) (1990)
4. Kennedy, J., Eberhart, R.: Particle swarm optimization. In: Proceedings of the IEEE International Conference on Neural Networks **4**, 1942–1948 (1995)
5. Geem, Z.W., Kim, J.H., Loganathan, G.V.: A new heuristic optimization algorithm: harmony search. Simulation **76**(2), 60–68 (2001)
6. İlkerBirbil, S., Fang, Shu-Cherng: An Electromagnetism-like Mechanism for Global Optimization. J. Global Optim. **25**, 263–282 (2003)
7. Rocha, A., Fernandes, E.: Hybridizing the electromagnetism-like algorithm with descent search for solving engineering design problems. Int. J. Comp. Math. **86**, 1932–1946 (2009)
8. Rocha, A., Fernandes, E.: Modified movement force vector in an Software, electromagnetism-like mechanism for global optimization. Optim. Methods & Softw. **24**, 253–270 (2009)
9. Tsou, C.S., Kao, C.H.: Multi-objective inventory control using electromagnetism-like metaheuristic. Int. J. Prod. Res. **46**, 3859–3874 (2008)
10. Wu, P., Yang, W.H., Wei, N.C.: An electromagnetism algorithm of neural network analysis an application to textile retail operation. J. Chin. Inst. Ind. Engineers **21**(1), 59–67 (2004)
11. Birbil, S.I., Fang S.C., Sheu, R.L.: On the convergence of a population-based global optimization algorithm, J. Glob. Optim., **30**(2), 301–318, (2004)
12. Kumar A, Shaik F, Image processing in diabetic related causes. Springer, Berlin (2015) ISBN: 978-981-287-623-2,
13. Cowan, E.W.: Basic Electromagnetism. Academic Press, New York (1968)
14. Hung, H.L., Huang, Y.F.: Peak to average power ratio reduction of multicarrier transmission systems using electromagnetism-like method Int. J. Innovat. Comput., Information and Control, **7**(5A) 2037–2050 (2011)
15. Akay, B. (2013). A study on particle swarm optimization and artificial bee colony algorithms for multilevel thresholding. Applied Soft Computing, 13(6),3066–3091.
16. Sathya, P.D., Kayalvizhi, R.: A new multilevel thresholding method using swarm intelligence algorithm for image segmentation. J. Intel. Learn. Syst. Appl. **2**, 126–138 (2010)
17. Kapur, J.N., Sahoo, P.K., Wong, A.K.C.: A new method for gray-level picture thresholding using the entropy of the histogram. Comp. Vis. Gr. Image Process. **29**, 273–285 (1985)

# Dual Notched Band Compact UWB Band Pass Filter Using Resonator Loaded SIR for Wireless Applications

**Arvind Kumar Pandey and R.K. Chauhan**

**Abstract** In this paper a compact and simple Ultra Wide Band (UWB) Band Pass Filter(BPF) with dual notch band is proposed. Structure of the proposed filter comprises a Stepped Impedance Resonator (SIR) loaded with Interdigital resonator. SIR is coupled with input and output ports using interdigital coupling to create UWB pass band range of the filter. Intedgital resonators are loaded on the SIR to create notch band to reject frequency in the pass band of the filter. Pass band of the filter covers 3.1 to 10.6 GHz with two notch bands at 5.2 and 7.8 GHz to avoid interfernce from the WiFi and satellite comunication. Filter is designed on Rogers RT/ Duriod 6010, with the dielectric substrate of dielectric constant εr = 10.8 and thickness h = 1.27 mm. There is no via or defected ground structure is used in the filter which makes its fabrication easier and cost effective.

**Keywords** UWB · BPF · SIR · Interdigital

## 1 Introduction

In 2002, Federal Communication Commission (FCC), madc available frequency band between 3.1 to 10.6 GHz for commercial use in wireless communication [1]. Due to this notification, the importance of filter design in the UWB range rapidly increased. Initially many filter structures based on SIR and multimode resonator were reported [2–5]. These designs were with good insertion loss and selectivity. However, in the pass band range of the filter interference of WiFi and satellite communication was big challenge for communication community. To get rid of these challenges structure of filter were reported which were able to pass UWB

A.K. Pandey (✉) · R.K. Chauhan
Department of Electronics and Communication Engineering,
Madan Mohan Malaviya University of Technology, Gorakhpur, India
e-mail: arvindmknk@gmail.com

R.K. Chauhan
e-mail: rkchauhan27@gmail.com



S.C. Satapathy et al. (eds.), *Data Engineering and Intelligent Computing*,
Advances in Intelligent Systems and Computing 542,
DOI 10.1007/978-981-10-3223-3_30

range frequency with notch at different interfering signals [6–9]. However, all reported structures suffer from poor insertion loss, poor selectivity and with wide rejection at interfering signals.

In this paper a simple, planar and compact structure of filter proposed which shows good insertion loss, sharp notch and interfering signals without via and defected ground structure in the structure. This filter is composed of SIR loaded with interdigital resonator. Different notch frequencies can be tuned by changing resonator length. Proposed filter is designed to pass UWB frequency between 3.1 to 10.6 GHz with notch at 5.2 and 7.8 GHz.

## 2 Design of Proposed Filter

The structure of proposed filter comprises of SIR based Multimode Mode Resonator (MMR) loaded with interdigital resonators as shown in the Fig. 1. To create a UWB pass band range a well-known MMR structure of SIR is designed for the centre frequency of 6.85 GHz. SIR structure is made up of a half-wavelength line ($L_1 = \lambda g/2$) of low impedance (Width, $W_1$) in the centre and two identical quarter-wavelength line ($L_2 = \lambda g/4$) of higher impedance(Width, $W_2$) on both sides, as shown in Fig. 1 [3]. The three resonating frequencies of SIR are used to create UWB pass band by coupling it through interdigital structure (IDS) to input and output ports. The coupling length of the IDS is used as around of length $\lambda g/4$ at 6.85 GHz. The coupling gap used is very small to reduce the insertion loss of the filter [3]. By varying length of coupling, $L_C$ the resonating frequencies can be adjusted. As we can see, on increasing $L_C$, the insertion loss of the filter getting lower and pass band becomes flat, as shown in Fig. 2a. To achieve UWB range in pass band for the value of $L_C = 3.95$ mm is used as shown in Fig. 2b.

The UWB filter designed by using SIR is loaded with resonator to create notch at required frequency. An interdigital resonator of $\lambda g/2$ length is loaded on SIR to create notch at 7.8 GHz frequency, as shown in Fig. 3a. The current distribution shown in Fig. 3b shows that the resonator is fetching out 7.8 GHz from the applied

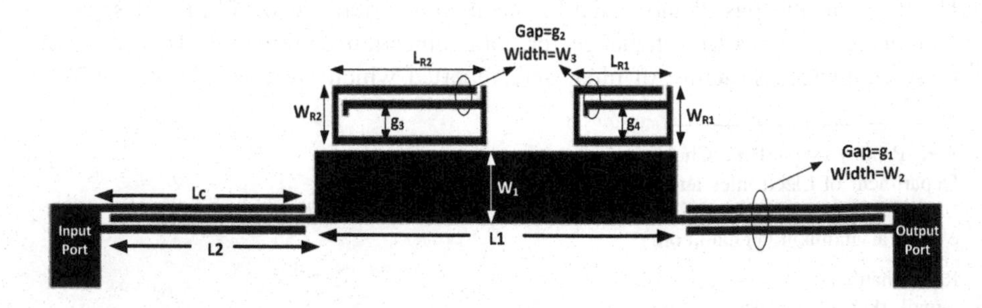


**Fig. 1** Structure of proposed filter

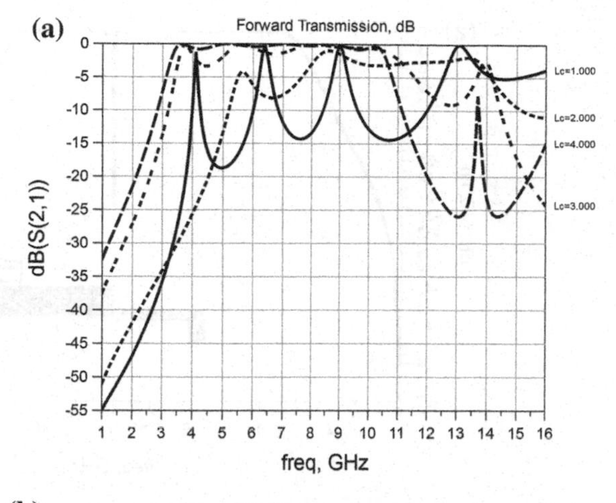


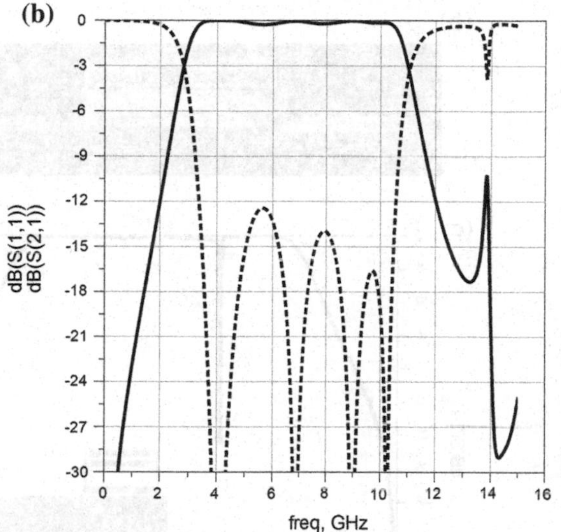


**Fig. 2** **a** Effect of coupling length, $L_C$ on the frequency response of the proposed. **b** UWB BPF flat response of IDS coupled SIR structure

signal and because of this a notch is created at 7.8 GHz in the frequency response shown in Fig. 3a.

A interdigital resonator with λg/2 length at 5.2 GHz frequency is loaded on SIR filter to create notch at 5.2 GHz. The frequency response of SIR loaded with resonator of λg/2 length at 5.2 GHz is shown in Fig. 3c. The current distribution of this structure also verifies the notch at 5.2 GHz, as shown in Fig. 3d.

To achieve the required pass band in UWB range between 3.1 to 10.6 GHz and notch band frequencies at 5.2 and 7.8 GHz the final structure is shown in Fig. 1. The dimensions of the proposed filter structure are such as—$L_1 = 7.34$ mm, $L_2 = 4.15$ mm, $W_1 = 1.08$ mm, $W_2 = 0.1$ mm, $Lc = 3.95$ mm, $L_{R1} = 2$ mm, $L_{R2} = 3.1$ mm, $W_{R1} = 0.8$ mm, $W_{R2} = 0.8$ mm, $g_1 = 0.05$ mm, $g_2 = 0.1$ mm, $g_3 = 0.4$ mm, $g_4 = 0.4$ mm.

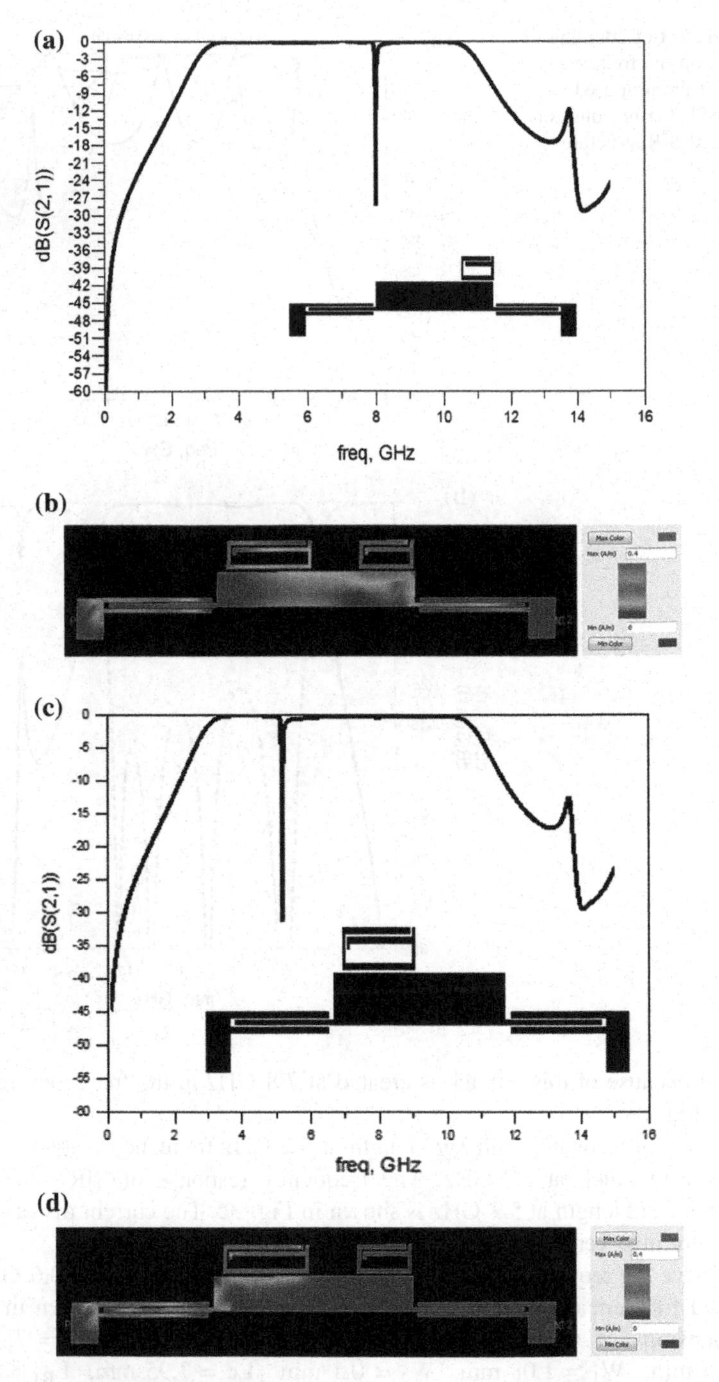


**Fig. 3** Notch creation by loading resonator **a** Notch creation at 7.8 GHz by loading. **b** Current distribution shows resonance by resonator at 7.8 GHz. **c** Notch creation at 5.2 GHz by loading resonator. **d** Current distribution shows resonance by resonator at 5.2 GHz

## 3 Results and Discussions

The frequency response of the filter shows good insertion loss of less than 0.1 dB in the pass band range between 3.1 to 10.6 GHz and notch at 5.2 and 7.8 GHz frequencies are created in the pass band range. The stop band range of the proposed filter is wide with loss greater than 20 dB. The response of the proposed filter is shown in Fig. 4.

(a)

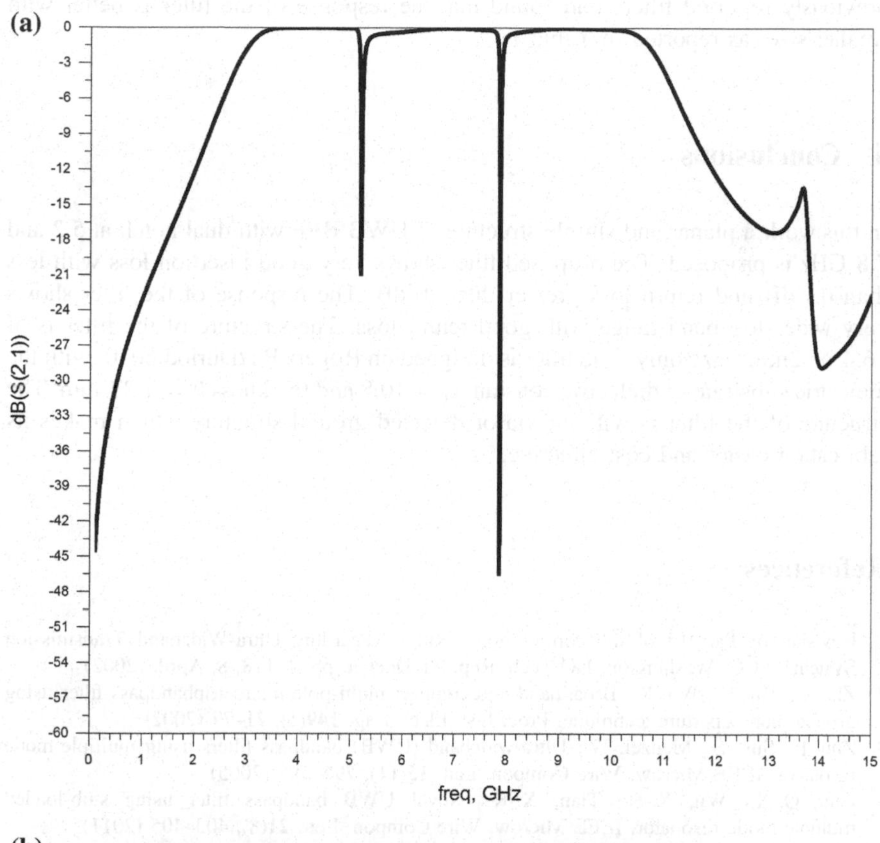


(b)

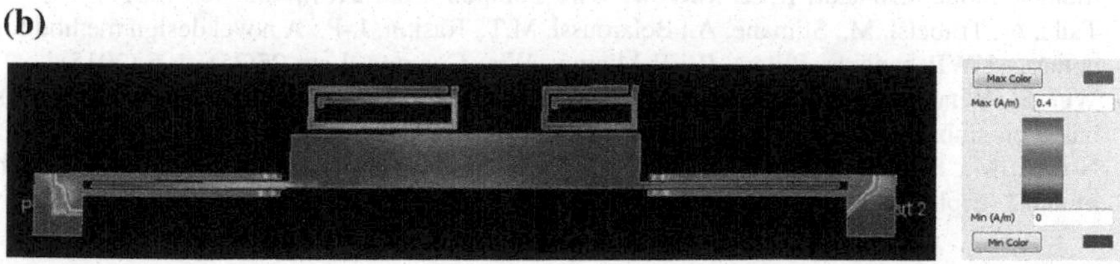


**Fig. 4** **a** Frequency response of the proposed filter **b** Current distribution of the filter other than 5.2 and 7.8 GHz frequencies

**Table 1** Comparison of proposed filter with different previously reported filters

| References | $\varepsilon_r$/h(mm) | IL(dB) | RL(dB) | Size($\lambda \times \lambda$)(mm$^2$) |
|---|---|---|---|---|
| [6] | 2.2/1 | <2 | >16 | 0.71 × 0.45 |
| [7] | 2.2/0.79 | <2.1 | >18 | 0.84 × 0.19 |
| [8] | 2.2/0.78 | <0.8 | >15 | 0.54 × 0.06 |
| [9] | 2.33/0.79 | <1.25 | >12 | 0.74 × 0.61 |
| **Proposed work** | **10.8/1.27** | **<0.1** | **>20** | **0.37 × 0.04** |

The response and structure of the proposed filter is compared with different previously reported filters and found that the response of the filter is better with smaller size, as reported in Table 1.

## 4 Conclusions

In this work a planar and simple structure of UWB BPF with dual notch at 5.2 and 7.8 GHz is proposed. The proposed filter shows very good insertion loss with less than 0.1 dB and return loss greater than 20 dB. The response of the filter shows very wide stop band range with good return loss. The structure of the filter is of 16.4 × 2mm$^2$ size only. The filter is designed on Rogers RT/Duriod 6010, with the dielectric substrate of dielectric constant $\varepsilon_r$ = 10.8 and thickness h = 1.27 mm. The structure of the filter is without via or defected ground structure which makes its fabrication easier and cost effective.

## References


1. Revision of Part 15 of the commission's Rules Regarding Ultra-Wideband Transmission Systems, FCC, Washington, DC Tech. Rep. ET-Docket, pp. 1–118, 8, April, (2002)
2. Zhu, L., Bu, H., Wu, K.: Broadband and compact multi-pole microstripbandpass filter using ground plane aperture technique. Proc. Inst. Elect. Eng. **149**(1), 71–77 (2002)
3. Zhu, L., Sun, S., Menzel, W.: Ultra-wideband (UWB) bandpass filters using multiple-mode resonator. IEEE Microw. Wire Compon. Lett. **15**(11), 796–798 (2005)
4. Chu, Q.-X., Wu, X.-H., Tian, X.-K.: Novel UWB bandpass filter using stub-loaded multiple-mode resonator. IEEE Microw. Wire Compon. Lett. **21**(8), 403–405 (2011)
5. Taibi, A., Trabelsi, M., Slimane, A., Belaroussi, M.T., Raskin, J.-P.: A novel design method for compact UWB bandpass filters. IEEE Microw. Wire Compon. Lett. **25**(25), 4–6 (2015)
6. Wei, F., Wang, Z.D., Yang, F., Shi, X.W.: Compact UWB BPF with triple-notched bands based on stub loaded resonator. Electron. Lett. **49**(2), 124–126 (2013)
7. Nosrati, M., Daneshmand, M.: Developing single-layer ultra-wideband band-pass filter with multiple (triple and quadruple) notches. IET Micro. Antennas Propag. **7**(8), 612–620 (2013)

8. Song, K., Xue, Q.: Compact ultra-wideband (UWB) bandpass filter with multiple notched bands. IEEE Microw. Wire. Compon. Lett. **20**(8), 447–449 (2010)
9. Kumar, S., Gupta, R.D., Parihar, M.S.: Multiple Band Notched Filter Using C-Shaped and E-Shaped Resonator for UWB Applications. IEEE Microw. Wire. Compon. Lett. **26**(5), 340–342, (2016)

# Frequent SubGraph Mining Algorithms: Framework, Classification, Analysis, Comparisons

**Sirisha Velampalli and V.R. Murthy Jonnalagedda**

**Abstract** Graphs and Trees are non-linear data structures used to organise, model and solve many real world problems and becoming more popular both in scientific as well as commercial domains. They have wide number of applications ranging from Telephone networks, Internet, Social Networks, Program flow, Chemical Compounds, BioInformatics, XML data, Terrorist networks etc. Graph Mining is used for finding useful and significant patterns. Frequent subgraph Mining mines for frequent patterns and subgraphs and they form the basis for Graph clustering, Graph classification, Graph Based Anomaly Detection. In this paper, classification of FSM algorithms is done and popular frequent subgraph mining algorithms are discussed. Comparative study of algorithms is done by taking chemical compounds dataset. Further, this paper provides a framework which acts as strong foundation in understanding any frequent subgraph mining algorithm.

**Keywords** Apriori · Chemical compounds · Frequent subgraph mining · Graphs · Graph mining · Pattern growth · Trees

## 1 Introduction

Now-a-days vast amount of data is being generated from various sources with rapid speed. Upon analysis we can mine knowledge from such huge data. Graph mining has become popular and active for data analysis. From a large collection of graphs, the very basic patterns called frequent substructures can be discovered. Frequent Subgraph Mining (FSM) algorithms [1] are useful for characterizing graph sets, discriminating different groups of graphs, classifying and clustering graphs, building graph indices and facilitating similarity search in graph databases. By Graph

S. Velampalli (✉) · V.R.M. Jonnalagedda
Department of CSE, University College of Engineering, JNTUK, Kakinada 533003, India
e-mail: sirisha.velampalli@gmail.com

V.R.M. Jonnalagedda
e-mail: mjonnalagedda@gmail.com



S.C. Satapathy et al. (eds.), *Data Engineering and Intelligent Computing*,
Advances in Intelligent Systems and Computing 542,
DOI 10.1007/978-981-10-3223-3_31

modeling we can solve a wide variety of problems. For example chemical compounds [5] have many recurrent substructures. We can mine chemical compounds using various graph-based pattern discovery algorithms. We can create a graph for each compound where vertices are atoms and edges are bonds between the vertices. Charge of each atom can be vertex label and type of bond can be edge label. Similarly, in a social network nodes represents persons and edges may be the type of relationship (friend, father, spouse etc.) between persons.

In this paper, chemical compounds dataset is taken and analysis is done by applying various algorithms. The rest of the paper is organized as follows: Basic definitions and concepts are provided in Sect. 2. In Sect. 3, framework for all frequent subgraph mining algorithms is given. In Sect. 4, we classified in-memory and parallel FSM algorithms, FSM algorithms that works on a single large graph and a set of graphs and FSM algorithms that uses MapReduce paradigm. Popular in-memory FSM algorithms are discussed in Sect. 5. In Sect. 6, we discussed dataset, results and comparisons. Conclusion of the paper with future scope is given in Sect. 7.

## 2 Background

**Graph** A graph G = (V, E) consists of a set of objects called vertices denoted as V = {v1, v2, v3....} and edges denoted as E = {e1, e2, e3....}

**Labelled Graph** If a label is associated with each vertex or edge then it can be called as Labelled graph.

**Subgraph** A subgraph S of a graph G is a graph whose set of vertices and set of edges are all subsets of G.

**Frequent subgraph** A subgraph is said to be frequent if its support count is not less than minimum support threshold.

**InducedSubgraph** A vertex-induced subgraph is one that consists of some of the vertices of the original graphs and all of the edges that connect them in the original. An edge-induced subgraph consists of some of the edges of the original graph and all vertices that are at their endpoints.

**Vertex Growing** Vertex growing [3] inserts a new vertex to an existing frequent subgraph during candidate generation.

**Edge Growing** Edge growing [3] inserts a new edge to an existing frequent subgraph during candidate generation.

**Isomorphism,Automorphism and Canonical Labelling** If we take two objects which consist of some finite sets and some relations on them. An isomorphism [4] between two objects is a bijection between their sets so that the relations are preserved. Automorphism can be defined as Isomorphisms from an object to itself. We can choose one member of each isomorphism class and call it the canonical member of the class. The process of finding the canonical member of the isomorphism class containing a given object is called canonical labeling.

## 3 Basic Framework for All Algorithms

Basic framework of all algorithms is summarized in Algorithm 1. Each step in the algorithm is explained in later part of the paper.

**Algorithm 1** Basic Framework of Frequent Subgraph Mining Algorithms

**1:** Frequent_Subgraph_Mining (G,minsup)
**2:** //G is database containing graphs
**3:** Mining starts with frequent pattern of size 1 i.e. **F_1**
**4:** **while** Generating patterns of size 2, 3,...**F_k**
**5:** C_(k+1)=**Candidate_generation**(F_k,G)
**6:** forall c in C_(k+1)
**7:** **if isomorphism_checking(c)=true**
**8:** **support_counting(c,G)**
**9:** **if** c.sup >= minsup
**10:** F_k+1=F_k+1
**11:** k=k+1
**12:** **return** set of frequent patterns

### *3.1 Candidate Generation*

Candidate generation is the process of generating candidates. Basic approaches include Apriori based, pattern-growth based and Approximate method.

**1. Apriori based method:** Apriori-based frequent substructure mining algorithms share similar characteristics with Apriori-based frequent itemset mining algorithms In frequent subgraph mining, the search for frequent graphs starts with graphs of small "size", and proceeds in a bottom-up manner by generating candidates having an extra vertex, edge or path. Apriori-based algorithms for frequent substructure mining include AGM [3], FSG [5], and a path-join method.

**2. Pattern Growth method:** In pattern growth-based frequent substructure mining for each discovered graph g, it performs extensions recursively until all the frequent graphs with g embeddings are discovered. The recursion stops once no frequent graph can be generated. Pattern Growth algorithms for frequent substructure mining include MoFa [10], gSpan [6], GASTON [7]

**3. Approximate method:** Approximate method does not focus on mining complete frequent subgraphs. SUBDUE [8] is based on approximate frequent subgraph mining method. It uses Minimum Description Length (MDL) as a evaluation metric and based on compression it discovers frequent subgraphs.

**Fig. 1** Depth first search versus breadth first search

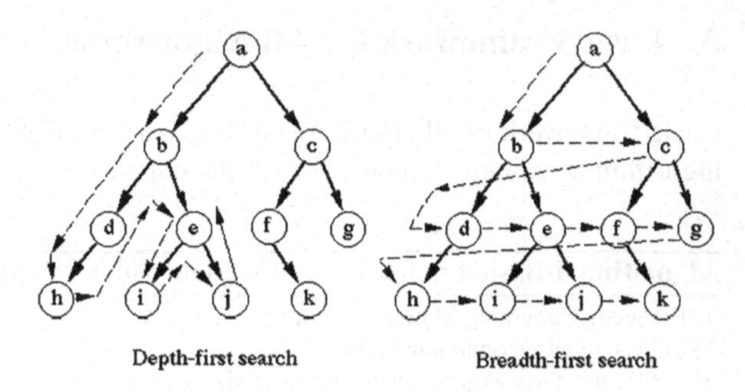


### *3.2 Search Order*

Each of the algorithms follows either Breadth First Search (BFS), Depth First Search (DFS) or Beam Search. Because of level-wise candidate generation, the apriori-based approach use the breadth-first search (BFS). In contrast, the pattern-growth approach can use BFS as well as DFS. DFS consumes less memory than BFS. The substructure discovery algorithm SUBDUE [8] uses computationally constrained beam search. Breadth first search explores vertices across the breadth of the frontier where as Depth First Search explores deeper into the graph whenever possible. Figure 1 shows depth first as well as breadth first search order.

### *3.3 Discovery Order*

Each algorithm uses either free extension or right most extension to discover various patterns. MoFa [10] uses free extension, whereas gSpan [6] uses right most path extension to discover patterns.

### *3.4 Duplicate Elimination*

Duplicate elimination can be done in 2 ways.

**1. Graph Isomorphism Checking:** Graph Isomorphism is discussed under Sect. 2. Isomorphism is one of the major challenges faced by many Frequent subgraph Mining algorithms.

**2. Canonical Labelling:** FSG algorithm [5] has powerful canonical labelling system. Isomorphism can be known easily using canonical labelling. A graph can be uniquely identified by assigning codes. Two graphs are isomorphic if they have same code.

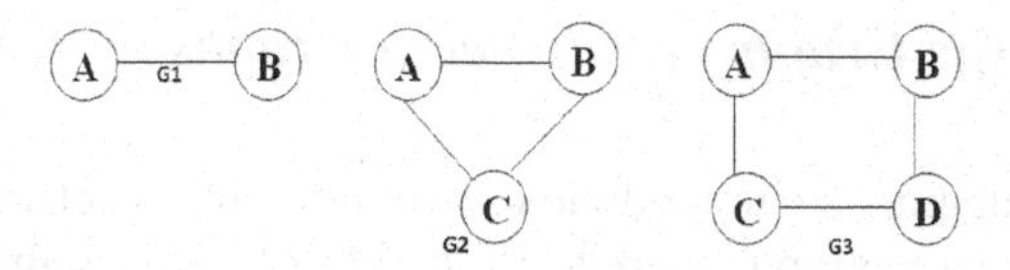


**Fig. 2** Sample Graphs

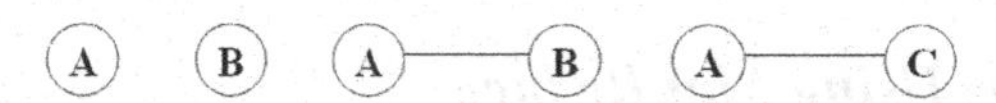


**Fig. 3** Frequent subgraphs with support count = 2

## *3.5 Support Calculation*

Support counting is the process of counting how many times each of the final candidates appear in the graph database. The most general naive Frequent Subgraph Mining algorithm will consider every possible subgraph during candidate generation, will prune none of them, and will then perform a support count on each candidate. Discovery of Frequent subgraphs is shown by taking three sample graphs G1, G2 and G3 shown in Fig. 2. Frequent subgraphs discovered from graphs shown in Fig. 2 with support count of two is shown in Fig. 3.

# 4 Classification of Algorithms

## *4.1 In-memory Versus Parallel Storage*

In this section we classify algorithms based on in-memory and parallel storage. For small-scale graph mining tasks, in-memory algorithms are sufficient. But for processing large data-sets we need shared memory parallel algorithms. Popular in-memory and shared memory Parallel algorithms are shown in Table 1.

**Table 1** FSM Algorithms

| In-memory algorithms | Parallel algorithms |
|---|---|
| AGM [3] | DB-SUBDUE [12] |
| FSG [5] | DB-FSG [13] |
| gSpan [6] | OO-FSG [14] |
| gaston [7] | MotifMiner [15] |
| DMTL [11] | Parmol [16] |
| SUBDUE [8] | ParSeMis [17] |

### *4.2 Single Large Graph Versus Set of Graphs*

Mining frequent subgraphs can be done in two different aspects. Papers [8, 18–20] deals with a single large graph where as [3, 6, 21, 22] deals with a set of graphs.

### *4.3 Algorithms Using MapReduce*

Now-a-days data is growing at immense speed. Computing Big Data has become a challenge. Mapreduce is a distributed programming model which is used for computing massive data. Algorithms which can mine large graphs using Mapreduce has become very popular. Papers [23–25] used Mapreduce to mine patterns in large graphs.

## 5 Algorithms

This section discusses popular in-memory Frequent Subgraph Mining Algorithms along with their advantages and limitations.

### *5.1 Apriori Based Graph Mining (AGM Algorithm)*

The AGM algorithm [3] uses a vertex-based candidate generation method that increases the substructure size by one vertex. Two size-k frequent graphs are joined only if they have the same size-(k − 1) subgraph. Here, graph size is the number of vertices in the graph. The newly formed candidate includes the size-(k − 1) subgraph in common and the additional two vertices from the two size-k patterns.

### *5.2 Frequent SubGraph Mining (FSG Algorithm)*

The FSG algorithm [5] adopts an edge-based candidate generation strategy that increases the substructure size by one edge.Two size-k patterns are merged if and only if they share the same subgraph having k − 1 edges, which is called the core. Here, graph size is taken to be the number of edges in the graph. The newly formed candidate includes the core and the additional two edges from the size-k patterns.

The limitations of AGM and FSG algorithms are they generate huge number of candidates, perform multiple database scans and it is very difficult to mine long patterns.

### 5.3 *Substructure Discovery Using Examples (SUBDUE Algorithm)*

SUBDUE is a heuristic algorithm [8] and does not discover complete set of frequent patterns. SUBDUE performs approximate substructure pattern discovery. It uses Minimum Description Length (MDL) heuristic and Background knowledge for pattern discovery.

### 5.4 *GSpan: Graph-Based SUbstructure Pattern Mining (gSpan Algorithm)*

gSpan [6] is one of the most popular pattern growth algorithm which restricts the number of redundant subgraph candidates that must be examined. It uses Depth-first search as its mining strategy. The following are the concepts introduced in gSpan:

(1) gSpan Encoding.
(2) Rightmost expansion.
(3) Lexicographic order.

### 5.5 *CloseGraph Algorithm*

The limitation of gSpan algorithm is it just shares the common frequency and generates so many candidates which are difficult for analysis. CloseGraph [9] uses a pattern growth approach which is built on top of gSpan. Given a graph dataset CloseGraph mines all closed frequent patterns. The generated patterns are easy to interpret.

### 5.6 *Graph/Sequence/Tree ExtractiON (GASTON)*

GASTON [7] follows a level-wise approach for finding frequent subgraphs. It first considers simple paths then complex trees and finally complex cyclic graphs. Gaston uses an occurrence list based approach, in which all occurrences of a small set of graphs are stored in main memory.

### 5.7 *Cost Analysis*

Total cost required to generate Graph patterns basically involves three major components i.e. number of candidates, Database and Isomorphism checking as shown in Fig. 4.

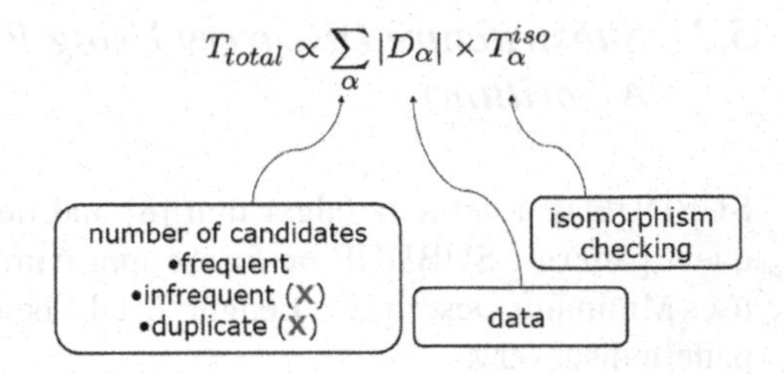


**Fig. 4** Cost analysis

# 6 Results

Dataset [5] contains totally 340 chemical compounds. In that there are 24 different atoms, 66 atom types and 4 types of bonds. On an average the dataset has 27 vertices and 28 edges per graph. The largest one has 214 edges and 214 vertices.

The number of patterns discovered by each algorithm with varying thresholds is shown in Fig. 5. For each of the algorithms Min. Support Thresholds are shown on X-axis and number of patterns discovered on applying the algorithm is shown on Y-axis. Comparison of all algorithms is given in Fig. 5e. From Fig. 5e we can clearly infer that “gspan produces more number of patterns” where as “closegraph produces less number of patterns”.

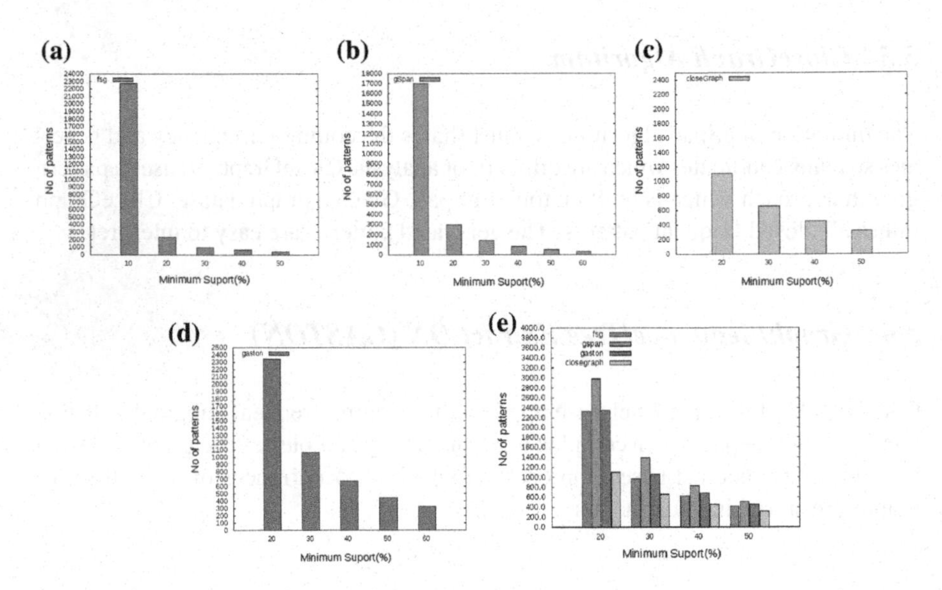


**Fig. 5** **a** fsg, **b** gspan, **c** closegraph, **d** gaston algorithms, **e** Comparisons

## 7 Conclusion

In this paper, basic framework of Frequent Subgraph Mining algorithms is given. Classification of FSM algorithms based on storage as well as number of graphs is given. Further, the purpose and limitations of popular FSM algorithms are discussed. For mining large datasets distributed processing FSM algorithms are needed. As big data is becoming a challenge, Graph Analytics will emerge drastically in near future for modelling and extracting intelligence from huge data.

## References

1. Han, J., Kamber, M.: Data Mining Concepts and Techniques, 2nd edn. Morgan Kaufmann Publishers (2006)
2. McKay., B.D., Piperno., A.: Nauty and Traces. Graph Canonical Labeling and Automorphism Group Computation. http://pallini.di.uniroma1.it/Introduction.html
3. Inokuchi, A., Washio, T., Motoda, H.: An apriori-based algorithm for mining frequent substructures from graph data. In: Proceedings. 4th European Conference Principles Data Mining Knowledge Discovery, pp. 13–23 (2000)
4. Tan, P.N., Steinbach, M., Kumar, V.: Introduction to Data Mining, Addison-Wesley (2005)
5. Kuramochi, M., Karypis, G.: Frequent subgraph discovery. In: Proceedings International Conference Data Mining, pp. 313–320 (2001)
6. Yan, X., Han, J.: gSpan: Graph-based substructure pattern mining. In: Proceedings International Conference Data Mining, pp. 721–724 (2002)
7. Nijssenm, S., Kok, J.: A quickstart in frequent structure mining can make a difference. In: Proceedings 10th ACM SIGKDD International Confernce Knowledge Discovery Data Mining, pp. 647–652 (2004)
8. Cook, D.J., Holder, L.B., Cook, D.J., Djoko, S.: Substructure discovery in the SUBDUE system. In: Proceedings of the AI Workshop on Knowledge Discovery in Databases, pp. 169–180 (1994)
9. Yan, X., Cheng, H.: CloseGraph: Mining closed frequent graph patterns. In: Proceedings of the ninth ACM SIGKDD International Conference on Knowledge Discovery and Data Mining, pp. 286–295 (2003)
10. Worlein, M., Meinl, T., Fisher, I., Philippsen, M.,: A quantitative comparison of the subgraph miners MoFa, gSpan, FFSM, and Gaston. Knowledge Discovery in Databases: PKDD pp. 392–403, (2005)
11. Chaoji, V., Hasan, M., Salem, S., Zaki, M.: An integrated generic apprach to pattern mining: data mining template library. Data Min. Knowl. Discov. J. **17**(3), 457–495 (2008)
12. Chakravarthy, S., Beera, R., Balachandran, R.: Db-subdue: database approach to graph mining. In: Proceedings Advances Knowledge Discovery and Data Mining, pp. 341–350 (2004)
13. Chakravarthy, S., Pradhan, S.: Db-FSG: An SQL-based approach for frequent subgraph mining. In: Proceedings 19th International Conference Database Expert System Applications, pp. 684–692 (2008)
14. Srichandan, B., Sunderraman, R.: Oo-FSG: An Object-oriented Approach to Mine Frequent Subgraphs. In: Proceedings Australasian Data Mining Conference, pp. 221–228
15. Parthasarathy, S., Coatney, M.: Efficient Discovery of common substructures in macromolecues. In: Proceefdings IEEE International Confernce Data Mining, pp. 362–369 (2002)
16. Meinl, T., Worlein, M., Urzova, O., Fischer, I., Philippsen, M.: The parmol package for frequent subgraph mining electronic communications of the EASST vol. 1, pp. 1–12, (2006)

17. Philippsen, M., Worlein, M., Dreweke, A., Werth. T.: Parmesis: The parallel and sequential mining suite (2011). https://www2.cs.fau.de/EN/research/ParSeMiS/index.html
18. Cook, D.J., Holder, L.B.: Graph-based data mining. IEEE Intell. Syst. **15**(2), 32–41, (2000)
19. Jiang, X., Xiong, H., Wang, C., Tan, A.H.: Mining globally distributed frequent subgraphs in a single labeled graph. Data Knowl. Eng. **68**(10), 1034–1058 (2009)
20. Kuramochi, M., Karypis, G.: Finding frequent patterns in a large sparse graph. Data Min. Knowl. Discov. **11**(3), 795–825 (2005)
21. Borgelt, C., Berthold, M.R.: Mining molecular fragments: Finding relevant substructures of molecules. In: Proceedings IEEE International Conference on Data Mining (ICDM). In: . pp. 51–58, (2002)
22. Huan, J., Wang, W., Prins, J.: Efficient Mining of Frequent Subgraphs in the Presence of Isomorphism. In: Third IEEE International Conference on Data Mining (ICDM).In: Proceedings IEEE, pp. 549–552 November (2003)
23. Hill, S., Srichandan, B., Sunderraman, R.: An iterative mapreduce approach to frequent subgraph mining in biological datasets. In: Proceedings of the ACM Conference on Bioinformatics, Computational Biology and Biomedicine (2012)
24. Xiao, X., Lin, W., Ghinita, G.: Large-scale frequent subgraph mining in mapreduce. In: Proceedings International Conference Data Engineering. pp. 844–855 (2014)
25. Bhuiyan, M.A.: An iterative mapreduce based frequent subgraph mining algorithm. IEEE Tans. Knowl. Data Eng. pp. 608–620 (2015)

# Facial Expression Classification Using Machine Learning Approach: A Review

A. Baskar and T. Gireesh Kumar

**Abstract** Automatic Facial Expression analysis has enthralled increasing attention in the research community in excess of two decades and its expedient in many application like, face animation, customer satisfaction studies, human-computer interaction and video conferencing. The precisely classifying different emotion is an essential problem in facial expression recognition research. There are several machine learning algorithms applied to facial expression recognition expedition. In this paper, we surveyed three different machine learning algorithms such as Bayesian Network, Hidden Markov Model and Support Vector machine and we attempt to answer following questions: How classification algorithm used its characteristics for emotion recognition? How various parameters in learning algorithm is devoted for better classification? What are the robust features used for training? Finally, we examined how advances in machine learning technique used for facial expression recognition?

**Keywords** Facial expression · Machine learning · Bayesian network · Hidden markov model · Support vector machine · Deep belief network

## 1 Introduction

Facial Expression (FE) [1] analysis paying attention to many research community and its increasing interest in this field mainly owing to its useful application like, customer satisfaction studies, face animation, human-computer interaction and

A. Baskar (✉)
Department of Computer Science and Engineering, Amrita School of Engineering,
Amrita Vishwa Vidyapeetham, Amrita University, Coimbatore, India
e-mail: a_baskar@cb.amrita.edu

T. Gireesh Kumar
TIFAC-CORE in Cyber Security, Amrita School of Engineering,
Amrita Vishwa Vidyapeetham, Amrita University, Coimbatore, India
e-mail: t_gireeshkumar@cb.amrita.edu



S.C. Satapathy et al. (eds.), *Data Engineering and Intelligent Computing*,
Advances in Intelligent Systems and Computing 542,
DOI 10.1007/978-981-10-3223-3_32

video conferencing. In extensive, automatic different emotional classification [2] is integration of three major module, called face detection [3], facial feature extraction and finally facial expression classification used to recognize correct Emotions. In most of the work, authors [2–5] used pre-processed face image or image sequence feeds to the facial feature extraction module. Facial feature extraction module plays the pivotal role for better emotion categorizations. In general, two types of facial features can have in human faces such as permanent facial features called eye brows, mouth, eyes and transparent facial features called deepening of facial furrows. Geometric-based methods and appearance-based method are the two most commonly available approaches for feature extractions. Extracted facial feature given inputs to classification module, finally it categorizes different emotions. The precisely classifying different emotion types are the covet problem in facial expression recognition research and advances in machine learning algorithms attempt to attains the better classification.

In this paper, we surveyed three different machine learning algorithms. First, we investigated how different Bayesian Networks used to classify facial expression with following accepts like Structure of Network of FE, Belief variable, associated event to belief variables, Likelihoods and Initial priors, Hidden Markov Model and Support Vector Machine. Next, we investigates how choices of five HMM models used to classify emotion recognition by analyzing following characteristics, What Image features used for design HMM? Structure of HMM, parameter $\lambda$, Initial probability, probability distribution and classifier used. Followed by we surveyed binary-class SVM, Multi-class SVM with different kernels functions. We further examined, influences of parameter for their better classification, the robust features to conquer the classification rate. Finally, we surveyed Deep Belief Network for facial expression recognition in advances in machine learning techniques.

## 2 Bayesian Network (BN)

BN is a graphical model, capable of exhibiting relationships clearly and intuitively using probabilistic approach among a set of attributes. This section we investigated, how human emotions are classified using four distinct Bayesian network classifier using following characteristics: Structure of Network, Belief variable, associated event to belief variables, Likelihoods and Initial priors.

First three approach [6, 7] uses two layer and three layer structure of network. In two layer structure, top layer or first level designed with only one belief variable called emotional or Facial_Expression (F_E) node. Second level have parent node F_E and belief variable in this level may varies depends on facial features, the more nodes used for independent Action Units and less number of nodes used for combined features of face. In three layer structure of network, bottom layer or third level [1] extracted twenty two angular information using geometric model and Appearance based model uses Gabor filter followed by moments applied around texture information present in skins region of face (it provides important evidence like skin deformation or wrinkle pattern) used as a belief variable in this level. There is no change in the middle and top layer.

Fourth approach authors [8] proposed how to learn the structure of Bayesian networks and uses for classifications. The structure of network in BN learns in two phase, called internal and external phase. Internal phase learns casual structure among facial feature nodes. The greedy approach K2 learns the network in this phase. The external phase learns casual structure of network between emotional node and facial feature node. The Table 1 summarizes different characteristics associated with different Bayesian networks.

## 3 Hidden Markov Model (HMM)

Hidden Markov Model describes the statistical demeanor of a signal using set of statistical model. In this paper, we investigates five choices of HMM models, called Left-Right HMM, Ergodic HMM, Emotion-Specific HMMs, Multilevel HMM and Mixture of HMM and neural network with following characteristics: Image features used for design HMM, structure of HMM, parameter $\lambda$, Initial probability, probability distribution and classifier used.

**Left-to-Right HMM**: Otsuka [9] obtained the feature of image in two steps. First, Velocity vector obtained using optical flow followed by 2D Fourier transform estimated to velocity vector around eye and mouth region. To improve the performance, they proposed Gaussian mixture density to approximate the output probability. This model reduce the degree of freedom.

**Ergodic HMM**: It allowed the transition of every state of the model can be reached in a predetermined number of time steps to any other state [10].

**Emotion-Specific HMMs**: Ira Cohen [10] proposed, AU for Pre-segmented sequences of the expressions from an image sequences feeds to the HMM architecture. The parameter $\lambda$ of model (Emotion Expression) learned using Baum-Welch algorithm and followed by maximum Likelihood (ML) classifier maximizes the likelihood of correct emotion using forward-backward procedure. The ML classifier for emotion-specific HMM is shown in the following equation.

$$c = \arg \max_{1 \le i \le 6} [P(O/\lambda i)]$$

**Multilevel HMM**: Ira Cohen [10], uses left-right HMM to train six different expression independently. Obtained state sequence of each HMM decoded using vitterbi algorithm and feeds input to the high-level HMM. The high-level HMM consists of seven states. The state transition between expressions are enacted to pass through neutral state.

**Mixture of HMM and neural network**: Mahmoud Khademi [11], modeled each single AU using two times, First, Wincandide-3 software used manually on first frame of the video and geometric feature points were extracted on the face

**Table 1** Summarizes different characteristics associated with different bayesian networks

| BBN approach | Structure of network | Belief variable | Associated event to belief variables | Likelihood | Initial priors |
|---|---|---|---|---|---|
| Simlicio et al. [6] | 2 Level | **Level 1**: one variable called, Facial_Expression (F_E)<br>**Level 2**: twelve variable (AU) 1, 4, (1 + 4), 6, 7, 12, 15, 17, 20, 23, 24, 25 | **Level 1**: {anger, fear, happy, sad, neutral}<br>**Level 2**: {Yes(Action Unit Present), No(Action Unit not present)} | Yes = 0.99<br>No = 0.01 | Uniform distribution<br>P(F_E = anger) = p<br>(F_E = fear) = p<br>(F_E = happy) = p<br>(F_E = sad) = P<br>(F_E = neutral) = 0.2 |
| Simlicio et al. [6] | 2 level | **Level 1**: one variable called, Facial_Expression (F_E)<br>**Level 2**: six variables called, EB, Ch, LE, LC, CB, MF, MA | **Level 1**: {anger, fear, happy, sad, neutral}<br>**Level 2**:<br>(EB) ∈{none, AU1, 4, 1 + 4}<br>(Ch) ∈{none, AU6}<br>(LE) ∈{none, AU7}<br>(LC) ∈{none, AU12, 15}<br>(CB) ∈{none, AU17}<br>(MF) ∈{none, Au20, Au23}<br>(MA) ∈{none, AU24, Au25} | Events varies from 0.01 to 0.99. It depends on How strongly Level 2 variable associates with Level 1 variable | Uniform distribution<br>P(F_E = anger) = p<br>(F_E = fear) = p<br>(F_E = happy) = p<br>(F_E = sad) = P<br>(F_E = neutral) = 0.2 |
| Singh Maninderjit et al. [1] | 3 Level | **Level 1**: one Emotion node<br>**Level 2**: fifteen Facial Action Unit<br>**Level 3**: Landmark measurement in face with angular feature | **Level 1**: {anger, disgust, fear, happy, sad, surprise}<br>**Level 2**: {Yes(Action Unit Present), No(Action Unit not present)}<br>**Level 3**: 22 points and each landmark is distributed into ten state | Events values are varies from 0.01 to 0.99. Level 2 variable associates with Level 1 variable | Uniform distribution<br>P(F_E = anger) = p<br>(F_E = fear) = p<br>(F_E = happy) = p<br>(F_E = sad) = P<br>(F_E = neutral) = 0.166 |

**Table 2** Summarization of choices of HMM and its characteristics

| HMM approach | Image features | Structure of HMM | Initial probability | The HMM parameter λ algorithm | Observation sequences (Ot) | Classifier |
|---|---|---|---|---|---|---|
| Left-Right HMM [9] | Velocity vector: optical flow 2D Fourier transform for eye and mouth region | Left-to-right HMM: 3 state expression | Clustering and its statistical parameter | Baum-Welch Algorithm | Gaussian Mixture density | – |
| Emotion-specific HMMs [10] | The AU for Pre-segmented sequences of the expressions | Left-to-right HMM 5 state expression with one return state | – | Baum-Welch Algorithm | Gaussian Mixture density | Maximum Likelihood (ML) classifier |
| Multi-level HMM [10] | Automatic segmentation and recognition enhance the discrimination between class | **First level**: independent HMMs for each emotions **Higher level**: seven states. | Neutral-1 other expression-0 | Decoding state sequence: vitterbi algorithm | Gaussian Mixture density | vitterbi algorithm |
| Mixture of HMM and neural network [11] | Each AU modeled two times, using Geometric and Appearance based feature separately | Left-to-right HMM | Neutral-1 other expression-0 | Neural network trained using back-propagation | Gaussian mixture density | Threshold |

followed by optical flow algorithm used to estimates the velocity in subsequent frames. Secondly, Gabor wavelets were used in appearance based model separately to extracts facial feature. Author's is applied Principal component analysis (PCA) on feature vector for dimensionality reduction. Choices of HMM and its characteristics are summarized in Table 2.

# 4 Support Vector Machine (SVM)

Support Vector Machines (SVMs) are supervised learning methods, It make use of different kernel function to transforms the data in input space to feature space better classifications [12]. In this section, we surveyed binary-class SVM, Multi-class SVM with different kernels namely, Linear, Polynomial, Sigmoid and Radial basis function. We further examined, which parameter inclined most to their better classification? What are the robust features, it will helps to conquers classification rate? and finally we explored simple and efficient SVM classification tool called LIBSVM. The Table 3 summarized Different type SVM and its characteristics.

Philipp Michel [4], trained series of binary SVMs for between all possible basic expressions (e.g. anger vs. disgust, anger vs. fear, anger vs. happy, anger vs. sad, anger vs. surprise and it repeated for all other expressions). In this work multi-class classification was modeled using cascade of binary SVM and a voting scheme.

Melanie Dumas [13], trained feature vectors using binary classification, it uses C-SVC formulations and multi-class classification, it uses C-SVC and nu-SVC formulation. The coefficient parameter c along x axis adjust center of the tanh function, value for c was set zero, because varying values of c not effect on scaling the sigmoid or adjusting y axis elevation and gamma parameter set. 1667 or 1 over no of classes makes, sigmoid kernel to outperforms in multi-class classification.

Hung-HSU TSAI [14], extracts shape and texture features from facial image using angular radial transform (ART), Gabor filter (GF) and discrete cosine transforms (DCT). In this work, classification attained using multi-class classification with radial basis function and observed that Shape and Texture various greatly for happy and surprise and it gives high accuracy rate. Shape and Texture feature almost similar for fear and disgust and it gives lower accuracy rate.

Shuaishi Liu [15], proposed least squares support vector machine (LS-SVM) model uses improved particle swarm optimization (PSO) to classify emotions in an image. In general, LS-SVM has advantages like global optimal solution, fast learning and strong generalization ability, but it finds difficult in optimal key parameter selection. The improved PSO was used to supports automatic selection of optimal parameter to LS-SVM. First it improves the global search capability followed by it improves key parameter like balance parameter C and kernel parameter $\sigma$. Feature vector designed using Local Gabor filter followed by fractional power polynomial kernel PCA used for dimensionality reduction.

**Table 3** Summarization of SVM classifiers and its characteristics

| SVM approach | Image features | SVM Type | Parameter | Classifier | Remarks |
|---|---|---|---|---|---|
| Michel [4] | Image: 22 facial landmark features<br>Video: Facial landmark motion for sequence of images | C-SVC<br>Nu-SVC | Kernel Type: Linear, Polynomial. degree = 2, RBF width = 1 and Sigmoid scale = 1, offset = 0 | Cascade of binary SVM and a voting scheme | Tool: LIBSVM [17, 18]<br>The linear kernel outperforms compares with polynomial kernel<br>To increase the width in RBF, outperforms over the linear |
| Dumas [13] | Gabor filter applied over selected region in face and feature vector generated for each image | Binary Classification: C-SVC<br>Multi-class classification | Kernel Type: Linear, Polynomial and Sigmoid. | Binary SVM | The Linear kernel outperforms for binary classification<br>Sigmoid kernel generates high accuracy for multi-class classifications |
| Tsai et al. [14] | Facial image was normalized using SQI and followed by ART, DCT, GF used to extracted shape and texture feature | Multi-class classification | Kernel Type: Radial basis function with c = 8 and g = 0.125 | Multi-class SVM | Shape and Texture various greatly for happy and surprise and it gives high accuracy rate<br>Shape and Texture feature almost similar for fear and disgust and it gives lower accuracy rate |
| Liu et al. [15] | Local Gabor filter followed by fractional power polynomial kernel PCA used for dimensionality reduction | LS-SVM with improved PSO | Kernel Parameter:<br>$c \in (0, 1000)$ *and* $\sigma \in (0, 10)$<br>Improved PSO parameter:<br>m = 25, Tmax = 100, Dmax = 0.24, Dmin = 0.001, Hmax = 2.4, Hmin = 0.2, Pm = 0.01, Pc = 0.5, c1 = c2 = 2.05 | LS-SVM | Improved PSO algorithm solves the reaching of local optimal solution problems |

# 5 Recent Techniques: A Review

In this section we reviewed what are advanced development in the facial expression recognition? And how this techniques are used to recognized the different type emotion. We examined how deep belief network was used for facial expression recognition and its characteristics.

## 5.1 *Facial Expression Recognition Using Deep Belief Network*

DBN [16] is a statistical model or deep neural network [16, 17]. It trained in the layer based algorithm. The authors [5] used Multi-Layer Restricted Boltzmann Machine (RBM) network, it endeavors to model RBM network based probability distribution of its input in every layer of the network in training and abstraction of image feature and followed by achieves higher level feature classification using the softmax probability model through Back Propagation (BP) network. The structure of DBN in this work as follows: first, pre-processed facial image as inputs to recognition model it includes pre-training, fine tuning and class prediction. The pre-training steps uses RBM network, which consists of multiple layer and obtain abstract higher level feature vector. In general, RBM used Gibbs sampling was used to attain the maximum likelihood estimation in second step. Third step, fine tuning model used single layered BP network: through supervised sample iterative, it achieves optimal network weight and hidden layer contains probabilistic classification model. Class prediction model finally recognizes facial expression through softmax multi-classification network.

# 6 Conclusions

We have discussed three different machine learning algorithms such as Bayesian Network, Hidden Markov Model, Support Vector machine and advances in machine learning techniques called Deep Belief Network used for facial expression recognition. In Bayesian Networks, the lower levels in belief variables used single or combination of facial features and top levels used global variable for classification. The Hidden Markov Model, attains better classification either 3 or 5 state model used independent expression or combination of expression using multilevel state models. In Support Vector machine, the series of binary SVMs for between all possible basic expressions helps for better classification. The layer based approach used in Deep Belief Network, improves facial expression classification.

## References

1. Singh, M., Majumder, A., Behera, L.: Facial expressions recognition system using Bayesian inference. In: International Joint Conference on Neural Networks (IJCNN), pp. 1502–1509. IEEE (2014)
2. Pantic, M., Rothkrantz, L.J.M.: Automatic analysis of facial expressions: the state of the art. IEEE Trans. Pattern Anal. Mach. Intell. **12**, 1424–1445 (2000)
3. Hjelmas, E., Low, B.K.: Face detection: a survey. Comput. Vis. Image Underst. **3**, 236–274 (2001)
4. Michel, P., El Kaliouby, R.: Real time facial expression recognition in video using support vector machines. In: Proceedings of the 5th International Conference on Multimodal Interfaces, pp. 258–264. ACM (2003)
5. Yang, Y., Fang, D., Zhu, D.: Facial expression recognition using deep belief network. Rev. Tec. Ing. Univ. Zulia. **39**(2), 384–392 (2016)
6. Simplicio, C., Prado, J., Dias, J.: Comparing bayesian networks to classify facial expressions. In: Proceedings of RA-IASTED, the 15th IASTED International Conference on Robotics and Applications, Cambridge, Massachusetts, USA (2010)
7. Datcu, D., Rothkrantz, L.J.M.: Automatic recognition of facial expressions using bayesian belief networks. In: IEEE International Conference on Systems, Man and Cybernetics, vol. 3, pp. 2209–2214. IEEE (2004)
8. Miyakoshi, Y., Kato, S.: Facial emotion detection considering partial occlusion of face using Bayesian network. In: IEEE Symposium on Computers & Informatics (ISCI), pp. 96–101. IEEE (2011)
9. Otsuka, T., Ohya, J: Recognizing multiple persons' facial expressions using HMM based on automatic extraction of significant frames from image sequences. In: International Conference on Image Processing Proceedings, vol. 2, pp. 546–549. IEEE (1997)
10. Cohen, I., Garg, A., Huang, T.S.: Emotion recognition from facial expressions using multilevel HMM. In: Neural Information Processing Systems, vol. 2 (2000)
11. Khademi, M., Manzuri-Shalmani, M.T., Kiapour, M.H., Kiaei, A.A.: Recognizing combinations of facial action units with different intensity using a mixture of hidden markov models and neural network. In: International Workshop on Multiple Classifier Systems, pp. 304–313. Springer, Berlin (2010)
12. Abdulrahman, M., Eleyan, A.: Facial expression recognition using support vector machines. In: 23nd Signal Processing and Communications Applications Conference (SIU), pp. 276–279. IEEE (2015)
13. Dumas, M.: Emotional expression recognition using support vector machines. In: Proceedings of International Conference on Multimodal Interfaces (2001)
14. Tsai, H.-H., Lai, Y.-S., Zhang, Y.-C.: Using SVM to design facial expression recognition for shape and texture features. In: International Conference on Machine Learning and Cybernetics, vol. 5, pp. 2697–2704. IEEE (2010)
15. Liu, S., Tian, Y., Peng, C., Li, J.: Facial expression recognition approach based on least squares support vector machine with improved particle swarm optimization algorithm. In: IEEE International Conference on Robotics and Biomimetics (ROBIO), pp. 399–404. IEEE (2010)
16. Ramachandran, R., Rajeev, D.C., Krishnan, S.G., Subathra, P.: Deep learning—an overview. Int. J. Appl. Eng. Res. **10**(10), 25433–25448 (2015). Research India Publications
17. Haridas, N., Sowmya, V., Soman, K.P.: GURLS vs LIBSVM: performance comparison of kernel methods for hyperspectral image classification. Indian J. Sci. Technol. **8**, 24 (2015)
18. Chang, C.-C., Lin, C.-J.: LIBSVM: a library for support vector machines. ACM Trans. Intell. Syst. Technol. (TIST), vol. 3 (2011)

# Heuristic Approach for Nonlinear $n \times n$ $(3 \leq n \leq 7)$ Substitution-Boxes

**Musheer Ahmad, M. Alauddin and Hamed D. AlSharari**

**Abstract** Substitution boxes are meant to enact nonlinear transformations of $n$-bit input streams to $n$-bit output streams. A highly nonlinear essence of them is imperative to induce obligatory confusion of data and to mitigate the potential linear cryptanalysis as well. It has been known that cryptographically potent S-boxes are creditworthy for the success of modern block encryption systems. This paper proposes to suggest an approach to frame a generic design that has the efficacy of synthesizing highly nonlinear balanced $n \times n$ S-boxes for $3 \leq n \leq 7$. The proposed approach is based on the heuristic optimization that seeks for local and global best S-box candidates on each iteration. The resultant optimized S-boxes are provided and tested for nonlinearity soundness. The performance outcomes and assessment analysis justify that the generic approach is consistent for contriving highly nonlinear key-dependent S-boxes.

**Keywords** Heuristic optimization • Nonlinearity • Substitution-box • Block encryption • Linear cryptanalysis

## 1 Introduction

Instinctively, the block ciphers are meant to action encryption and decryption of its input data one block at a time rather than bit-by-bit using a shared secret key. The size of block is predefined and fixed. They are symmetric in nature, meaning both

M. Ahmad (✉)
Faculty of Engineering and Technology, Department of Computer Engineering, Jamia Millia Islamia, New Delhi 110025, India

M. Alauddin
Department of Petroleum Studies, ZH College of Engineering and Technology, Aligarh Muslim University, Aligarh 202002, India

H.D. AlSharari
Department of Electrical Engineering, College of Engineering, AlJouf University, Al-Jouf, Kingdom of Saudi Arabia



S.C. Satapathy et al. (eds.), *Data Engineering and Intelligent Computing*, Advances in Intelligent Systems and Computing 542,
DOI 10.1007/978-981-10-3223-3_33

sender and receiver make use of same secret key. The working rule of block cryptosystems can essentially be derived from the structure of substitution-permutation (S-P) networks posed by Fiestel [1]. The S-P network is a prominent architecture which is opted by most of the modern block cipher systems. The network involves utilization of substitution-boxes during substitution phase in their rounds that improves substantially the essence of confusion and nonlinearity as an output of the cryptosystem. It takes a block of plaintext and the secret keys as inputs and applies them against several layers substitution and permutation phase operations to give out the cipher text block [2]. A strong S-box with all features concerning a cryptographic system forms the very base of secure encryption systems. The nonlinear nature of S-boxes has become key characteristic in making an effective security system. Factually, the cryptographic lineaments of S-boxes represent the main strength of the corresponding block based encryption system [3, 4]. Hence, the development of formidable S-boxes is of utmost significance for cryptologists in designing strong cryptosystems.

Often implemented as a lookup table, an $n \times n$ S-box takes $n$ input bits and transforms into $n$ output bits. An $n \times n$ substitution-box is mapping, from GF $(2^n) \rightarrow GF(2^n)$, that nonlinearly transforms $n$-bit input data to $n$-bit output data, where $GF(2^n)$ is the Galois field having $2^n$ elements [5]. It is also conceived as a multi-input and multi-output Boolean function. Meaning, a $5 \times 5$ S-box consists of five Boolean component functions where each Boolean function takes 5-bit stream as input and generates 1-bit as output, all five functions collectively yields 5-bit output stream. As a result, the performance features and characteristics meant for Boolean functions can be easily considered and extended to quantify the strength of S-boxes. To date, most of the S-box work carried out is dedicated to the design of $8 \times 8$ S-boxes which is due to the success of AES block cipher and its S-box introduced by NIST [6]. Almost all of them are balanced and whose design primarily based on concepts such as affine transformations, gray coding, power mapping, optimization, chaotic systems, etc., [7–21]. Substitution-boxes are merely nonlinear component and have a central role to play to decide the security strength of most block ciphers. The cryptographic features of S-boxes are of immense significance for the security of cipher systems. Thus, ample research has been dedicated to enhance the quality of S-boxes in order to restrict cryptanalysis assaults that endeavor imperfect designs. In [22], Matsui has suggested the method of conducting linear cryptanalysis to attacks block ciphers having low nonlinear essence. It is shown that the S-boxes having low nonlinear nature are susceptible to this kind of assault. Moreover, according to Massey [23], '*linearity is the crux of cryptography*', hence, some nonlinearity needs to be induced in the encryption system to meet the Shannon's confusion and diffusion requirements for a strong cryptosystem. Therefore, the cryptographically good nature and strength of S-boxes is of utmost grandness and needed to thwart linear cryptanalysis assaults. To fulfill the aforementioned need, a generic approach is design to yield highly nonlinear S-boxes. The approach is based on the heuristic search which is executed with the help of the chaotic system used as pseudo-random generation source.

The remaining of the paper is structured as: Sect. 2 prepared to give brief of linear cryptanalysis assaults and chaotic skew-tent map. The proposed heuristic approach is described in Sect. 3. The experimental results and analyses have been discussed in Sect. 4. Section 5 concludes the work carried out in this communication.

# 2 Related Concepts

## 2.1 Linear Cryptanalysis

The substitution-boxes are the main informant of nonlinearity and thereby the confusion in block ciphers. It is critical to comprehend the extent to which they can be proximated as linear equations [23]. With regards to cryptanalysis, for every input variable $x_i$ having n-dimension of an S-box and there is output variable $y_i$ of n-dimension of S-box. These variables $x_i$ and $y_i$ are not particularly independent from each other, because the likelihood of the output relies upon the input is always exists. The aim of linear cryptanalysis is to discover the linear combination of $x_i$ which is exactly as linear combination of $y_i$ that is fulfilled by finite probability. For a perfect S-P network, such connections will be fulfilled precisely 50% of the ideal opportunity for any choice of $x_i$ and $y_i$ variables [3]. It should estimate and find that there exists some determination of linear combinations such that the chance of fulfilling the relation is not 0.5, if so then this bias from the relationship can be utilized in the attack.

## 2.2 Chaotic Skew Tent Map

The chaotic skew-tent map is a widely used one-dimensional piece-wise dynamical system which has the governing equation as [24]:

$$y(i+1) = \begin{cases} \frac{y(i)}{p} & 0 < y(i) \leq p \\ \frac{1-y(i)}{1-p} & p < y(i) < 1 \end{cases} \quad (1)$$

where, $y$ is state of the map, $y(i) \in (0, 1)$ for all $i$ and $p \in (0, 1)$ is system parameter. Skew-tent map exhibits chaotic phenomenon for all $p \in (0, 1)$. The trajectory of map covers the entire space of (0, 1) for allowed values of $p$. It has simple iterative equation, in which the next value of chaotic variable $y$ is easily obtained on iteration over previous value of $y$ variable, where $p$ is usually kept fix. The map depicts better characteristics than famous 1D Logistic chaotic map as far as pseudo-random number generation is concern. The initial values assigned to $y(0)$

and $p$ constitute the secret key. The generation of optimized $n \times n$ generic S-Boxes, using proposed heuristic search approach, is dependent on this secret key.

# 3 Proposed Approach

The proposed heuristic optimization based approach for S-Boxes generation has following algorithmic procedure.

Step 1. Properly select $y_0$, $p$, $T$, *iterations*

Step 2. Read size of S-Box as $n$ ($3 \leq n \leq 7$), set $len = 2^n$

Step 3. Find $r = \lfloor n/2 \rfloor$, take local best *LSBox* and global best *GSBox* as empty tables

Step 4. Execute chaotic map (1) for $T$ times and dispose the values but the last

Step 5. Take $A_1, A_2, A_3 = \{\ \}$ as empty sets, where $|A_i| = 0$

Step 6. Further execute the map (1) and collect outputted $y$

Step 7. Extract random number $q \in [0, 2^n - 1]$ as:

$q = \lfloor y \times 10^{13} \rfloor \bmod (2^n)$ where, $\lfloor\ \rfloor$ is the *floor* operator

IF ($q \notin A_1$)

Append $q$ to $A_1$

ELSE IF ($q \notin A_2$)

Append $q$ to $A_2$

ELSE IF ($q \notin A_3$)

Append $q$ to $A_3$

Step 8. Check completion of $A_i$ arrays:

IF ($|A_1|= len$ & $|A_2|= len$ & $|A_3|= len$)

*move to* Step 9

ELSE

*move to* Step 6

Step 9. Reshape $A_1$, $A_2$ and $A_3$ to $2^r \times 2^{n-r}$ S-Box look-up tables

Step 10. Find average nonlinearities $nl_i$ of S-Boxes $A_i$ as:

$[nl_1, nl_2, nl_3] = gen\_nonlinearity(A_1, A_2, A_3)$

Step 11. Find local best among $A_1$, $A_2$ and $A_3$ and nominate as *LSBox*

Step 12. IF $NL(LSBox) > NL(GSBox)$, THEN update *GSBox* as current global best

Step 13. Repeat Steps 5 to 12 if desired iterations are not yet finished

# 4 Results and Analysis

The substitution boxes are the extension of cryptographic Boolean functions. Therefore, the performance measures used to assess the security strength of Boolean functions are opted for S-boxes too by the researchers worldwide. Here, we are concerned about the nonlinearity strength of generated S-boxes. Therefore, the heuristic optimization is carried out on the basis of nonlinearity as the objective function. To mitigate the linear cryptanalysis assaults and to achieve desired confusion, the designed S-boxes should have high scores of nonlinearity. The experimental set-up to conduct the simulation assumed arbitrarily the initial values of keys as: $y(0) = 0.56789$, $p = 0.567$, $T = 100$ and $iterations = 2000$. The optimized $3 \times 3$, $4 \times 4$, $5 \times 5$, $6 \times 6$, and $7 \times 7$ S-boxes obtained by proposed heuristic approach are listed in Table 1. The balanced Boolean functions and substitution boxes are principally desired for their usage while designing any block cryptosystem. The cryptographic Boolean functions which are not balanced reckoned as weak and insecure. A Boolean function is said to be balanced if it has same number of 0's and 1's in all its output vectors. All the generated S-boxes are verified for requisite condition of balancedness property for cryptographic sense of applicability. Since, all the S-boxes in Table 1 has unique values in the range of $[0, 2^n - 1]$, this further justifies their balancedness or bijectiveness.

## 4.1 Nonlinearity of Balanced Boolean Functions

The degree of nonlinearity directly corresponds to the amount of confusion induced by the cryptosystem to the output cipher content. Right back from the history, the linearity is always succumbs to serious assaults and considered as design weakness. The straight idea to build a strong cryptosystem is to build and employ cryptographically strong Boolean functions or substitution boxes. A Boolean function is arrogated as weak if it tends to have low nonlinearity score. The score of nonlinearity of a Boolean function $f$ is determined by finding the minimum distance of $f$ to the set of all affine functions [25]. Thus, the component functions of an S-box should have upright nonlinearities scores. The nonlinearity $NL_f$ of any Boolean function $f$ is accounted as:

$$NL_f = \frac{1}{2}\left(2^n - WH_{\max}(f)\right)$$

where, $WH_{max}(f)$ is the Walsh-Hadamard transform of Boolean function $f$ [26]. It has been examined that the maximum nonlinearity of a balanced Boolean function for odd $n$ (= 3, 5, 7) equals $2^{n-1} - 2^{(n-1)/2}$ in [27], and it is tightly higher than $2^{n-1} - 2^{(n-1)/2}$ as proved by Patterson in [28]. Whereas, for a balanced Boolean function in even $n$ (= 4, 6, 8) number of variables, it is possible to touch the nonlinearity

**Table 1** Proposed $n \times n$ substitution-boxes

3 × 3 S-Box

| | 0 | 1 | 2 | 3 |
|---|---|---|---|---|
| 0 | 2 | 3 | 5 | 1 |
| 1 | 6 | 4 | 7 | 0 |

4 × 4 S-Box

| | 0 | 1 | 2 | 3 |
|---|---|---|---|---|
| 0 | 2 | 6 | 15 | 8 |
| 1 | 14 | 13 | 7 | 3 |
| 2 | 11 | 5 | 1 | 10 |
| 3 | 4 | 12 | 9 | 0 |

5 × 5 S-Box

| | 0 | 1 | 2 | 3 | 4 | 5 | 6 | 7 |
|---|---|---|---|---|---|---|---|---|
| 0 | 29 | 31 | 19 | 2 | 3 | 26 | 12 | 11 |
| 1 | 27 | 1 | 10 | 28 | 18 | 21 | 15 | 14 |
| 2 | 20 | 0 | 4 | 7 | 9 | 13 | 22 | 8 |
| 3 | 25 | 6 | 23 | 5 | 30 | 16 | 17 | 24 |

6 × 6 S-Box

| | 0 | 1 | 2 | 3 | 4 | 5 | 6 | 7 |
|---|---|---|---|---|---|---|---|---|
| 0 | 55 | 58 | 5 | 47 | 9 | 39 | 1 | 61 |
| 1 | 11 | 32 | 52 | 26 | 25 | 12 | 21 | 18 |
| 2 | 31 | 23 | 8 | 19 | 29 | 44 | 56 | 3 |
| 3 | 46 | 38 | 24 | 0 | 57 | 43 | 16 | 30 |
| 4 | 41 | 27 | 36 | 7 | 42 | 33 | 51 | 34 |
| 5 | 10 | 13 | 15 | 6 | 22 | 53 | 17 | 48 |
| 6 | 49 | 2 | 50 | 45 | 20 | 60 | 59 | 62 |
| 7 | 35 | 54 | 28 | 4 | 40 | 14 | 37 | 63 |

7 × 7 S-Box

| | 0 | 1 | 2 | 3 | 4 | 5 | 6 | 7 | 8 | 9 | 10 | 11 | 12 | 13 | 14 | 15 |
|---|---|---|---|---|---|---|---|---|---|---|---|---|---|---|---|---|
| 0 | 34 | 88 | 65 | 42 | 105 | 124 | 104 | 117 | 113 | 58 | 86 | 37 | 72 | 14 | 101 | 100 |
| 1 | 23 | 15 | 7 | 12 | 90 | 60 | 66 | 57 | 67 | 18 | 118 | 30 | 51 | 40 | 70 | 75 |
| 2 | 32 | 22 | 1 | 73 | 54 | 106 | 115 | 80 | 53 | 17 | 25 | 125 | 50 | 97 | 0 | 33 |
| 3 | 91 | 69 | 94 | 55 | 35 | 4 | 116 | 95 | 63 | 98 | 92 | 126 | 21 | 79 | 46 | 61 |
| 4 | 93 | 5 | 102 | 6 | 64 | 41 | 109 | 31 | 121 | 16 | 44 | 8 | 122 | 2 | 56 | 36 |
| 5 | 52 | 110 | 62 | 10 | 38 | 96 | 27 | 48 | 59 | 81 | 123 | 83 | 82 | 68 | 84 | 43 |
| 6 | 112 | 85 | 13 | 119 | 9 | 29 | 120 | 47 | 99 | 3 | 71 | 26 | 45 | 11 | 87 | 19 |
| 7 | 49 | 39 | 114 | 24 | 20 | 78 | 108 | 103 | 111 | 28 | 76 | 89 | 74 | 107 | 77 | 127 |

bound of $2^{n-1} - 2^{n/2}$. The nonlinearity scores of the $n$ component function of proposed optimized $n \times n$ S-boxes are listed in Table 2. It is evident from the Table 2 that each of the component functions almost achieved the best value. The average nonlinearity scores of S-Boxes are enumerated and compared in Table 3. In [13], Millan have designed the $n \times n$ ($5 \le n \le 8$) S-Boxes by applying the concept of Hill Climbing to perform the optimization. Fuller et al. in [14] applied the concept of row-based technique to generate the bijective $n \times n$ S-Boxes for

**Table 2** Nonlinearities of $n$ Boolean functions $f_i$ of proposed $n \times n$ S-boxes

| S-Box | Nonlinearities | | | | | | |
|---|---|---|---|---|---|---|---|
| | $f_1$ | $f_2$ | $f_3$ | $f_4$ | $f_5$ | $f_6$ | $f_7$ |
| 3 × 3 | 2 | 2 | 2 | | | | |
| 4 × 4 | 4 | 4 | 4 | 4 | | | |
| 5 × 5 | 12 | 12 | 10 | 10 | 10 | | |
| 6 × 6 | 24 | 24 | 22 | 24 | 24 | 24 | |
| 7 × 7 | 50 | 52 | 50 | 50 | 52 | 48 | 50 |

**Table 3** Average nonlinearity scores of some $n \times n$ S-boxes

| S-Box | Millan [13] | Fuller et al. [14] | Laskari et al. [15] | Proposed |
|---|---|---|---|---|
| 3 × 3 | NR | NR | NR | 2 |
| 4 × 4 | NR | NR | NR | 4 |
| 5 × 5 | 10 | 6 | 10 | 10.8 |
| 6 × 6 | 20 | 18 | 20 | 23.66 |
| 7 × 7 | 46 | 42 | 46 | 50.3 |

$5 \leq n \leq 8$. In Ref. [15], Laskari et al. used the particle swarm optimization and differential evolution approaches for optimizing the S-Boxes. The results provided in Table 3 for Laskari et al., reports their best scores among a number of nonlinearity scores for $5 \leq n \leq 8$ provided by them in their article using PSO's and DE's individually. The comparison made in Table 3 evinced that proposed S-boxes provide significantly better scores of average nonlinearities than their counterparts investigated in [13–15]. The simulation and comparison outcome justify that the proposed approach is efficient enough to yield highly nonlinear S-Boxes.

## 5 Conclusion

In this communication, a heuristic optimization based generic approach is designed to generate $n \times n$ substitution boxes. A chaotic skew-tent map is used as source of random number generator to produce local generation consisting of three S-Boxes in each iteration. The local best of current generation and global best achieved so far are accredited. The whole process of optimized S-Box generation is under the control of key; hence the S-Boxes have dynamic nature. Meaning, with slight variations in any of the key component, the different set of efficient S-Boxes can be developed with great ease. The S-Boxes are assessed against nonlinearity, and it has been found that the proposed optimized S-Boxes have excellent performance strength. The proposed approach has the features of simplicity, low computations, credibility to produce efficient substitution-boxes. Furthermore, the comparison drawn with some optimized S-Boxes indicates that the anticipated approach is respectable in yielding highly nonlinear S-boxes of size $n \times n$.

## References


1. Feistel, H.: Cryptography and computer privacy. Sci. Am. **228**(5), 15–23 (1973)
2. Stinson, D.R.: Cryptography: Theory and Practice. CRC Press (2005)
3. Wood, C.A.: Large substitution boxes with efficient combinational implementations, M.S. thesis, Rochester Institute of Technology (2013)
4. Burnett, L.: Heuristic optimization of Boolean functions and substitution boxes for cryptography. Ph.D. dissertation, Queensland University of Technology (2005)
5. Nedjah, N., Mourelle, L.D.M.: Designing substitution boxes for secure ciphers. Int. J. Innov. Comput. Appl. **1**(1), 86–91 (2007)
6. Daemen, J., Rijmen, V.: The Design of Rijndael: AES—The Advanced Encryption Standard. Springer (2002)
7. Ahmad, M., Bhatia, D., Hassan, Y.: A novel ant colony optimization based scheme for substitution box design. Proc. Comput. Sci. **57**, 572–580 (2015)
8. Lambić, D.: A novel method of S-box design based on chaotic map and composition method. Chaos, Solitons Fract. **58**, 16–21 (2014)
9. Ahmad, M., Chugh, H., Goel, A., Singla, P.: A chaos based method for efficient cryptographic S-box design. In: Thampi, S.M., Atrey, P.K., Fan, C.-I., Pérez, G.M. (eds.) SSCC 2013, CCIS 377, pp. 130–137 (2013)
10. Cui, L., Cao, Y.: A new S-box structure named affine-power-affine. Int. J. Innov. Comput. Inf. Control **3**(3), 751–759 (2007)
11. Tran, M.T., Bui, D.K. Duong, A.D.: Gray S-box for advanced encryption standard. In: International Conference on Computational Intelligence and Security, pp. 253–258 (2008)
12. Wang, Y., Wong, K.W., Li, C., Li, Y.: A novel method to design S-box based on chaotic map and genetic algorithm. Phys. Lett. A **376**(6), 827–833 (2012)
13. Millan, W.: How to improve the nonlinearity of bijective S-Boxes. In: Australasian Conference on Information Security and Privacy. Lecture Notes in Computer Science, vol. 1438, pp. 181–192 (1998)
14. Fuller, J., Millan, W., Dawson, E.: Multi-objective optimisation of bijective S-boxes. New Gener. Comput. **23**(3), 201–218 (2005)
15. Laskari, E.C., Meletiou, G.C., Vrahatis, M.N.: Utilizing evolutionary computation methods for the design of S-boxes, In: International Conference on Computational Intelligence and Security, pp. 1299–1302 (2006)
16. Alkhaldi, A.H., Hussain, I., Gondal, M.A.: A novel design for the construction of safe S-boxes based on TDERC sequence. Alexandria Eng. J. **54**(1), 65–69 (2015)
17. Ahmad, M., Rizvi, D.R., Ahmad, Z.: PWLCM-based random search for strong substitution-box design. In: International Conference on Computer and Communication Technologies, pp. 471–478 (2015)
18. Ahmad, M., Ahmad, F., Nasim, Z., Bano, Z., Zafar, S.: Designing chaos based strong substitution box. In: International Conference on Contemporary Computing, pp. 97–100 (2015)
19. Ahmad, M., Khan, P.M., Ansari, M.Z.: A simple and efficient key-dependent S-box design using fisher-yates shuffle technique. In: International Conference on Security in Computer Networks and Distributed Systems, pp. 540–550 (2014)
20. Ahmad, M., Malik, M.: Design of chaotic neural network based method for cryptographic substitution box. In: International Conference on Electrical, Electronics, and Optimization Techniques, pp. 864–868 (2016)
21. Ahmad, M., Mittal, N., Garg, P., Khan, M.M.: Efficient cryptographic substitution box design using travelling salesman problem and chaos. Perspective in Science (2016). doi:10.1016/j.pisc.2016.06.001
22. Matsui, M.: Linear cryptanalysis method of DES cipher. In: Advances in Cryptology: EuroCrypt'1993 Proceedings. Lecture Notes in Computer Science, vol. 765, pp. 386–397 (1994)

23. Zeng, K., Yang, C.H., Rao, T.R.N.: On the linear consistency test in cryptanalysis with applications. In: Crypto 1989 Proceedings. Lecture Notes in Computer Science, vol. 435, pp. 167–174 (1990)
24. Li, S., Li, Q., Li, W., Mou, X. and Cai, Y., Statistical properties of digital piecewise linear chaotic maps and their roles in cryptography and pseudo-random coding. In: IMA International Conference on Cryptography and Coding, 205–221 (2001)
25. Cusick, T.W., Stanica, P.: Cryptographic Boolean Functions and Applications. Elsevier, Amsterdam (2009)
26. Hussain, I., Shah, T.: Literature survey on nonlinear components and chaotic nonlinear components of block ciphers. Nonlinear Dyn. **74**(4), 869–904 (2013)
27. Helleseth, T., Klve, T., Mykkelveit, J.: On the covering radius of binary codes. IEEE Trans. Inf. Theor. **24**(5), 627–628 (1978)
28. Patterson, N.J., Wiedemann, D.H.: The covering radius of the $[2^{15}, 16]$ Reed-Muller code is at least 16276. IEEE Trans. Inf. Theor. **29**(3), 354–356 (1983)

# Bi-Temporal Versioning of Schema in Temporal Data Warehouses

**Anjana Gosain and Kriti Saroha**

**Abstract** The temporal design of data warehouse (DW), which is an extension to multidimensional model gives a provision to implement the solution to handle time-varying info in dimensions. The dimension data is time-stamped with valid time (VT) to maintain a complete data history in temporal data warehouses (TDWs). Thus, TDWs manage evolvement of schema over a period of time by using versioning of schemas as well as evolution of data described under various versions of schema. But schema versioning in TDWs has not been covered in full detail. Mainly, the approaches to handle schema versions using valid time were proposed so far. This paper proposes an approach for bitemporal versions of schema in temporal DW model that allows for retroactive and proactive schema modifications and in addition also helps in tracking them.

**Keywords** Data warehouse · Temporal data warehouse · Versioning · Schema · Transaction time · Valid time · Bitemporal

## 1 Introduction

Data Warehouses (DWs) collect and store historical data from different heterogeneous and may be, distributed sources. They are designed to provide support for multidimensional analysis and decision making process. A DW schema as well as its data can undergo revisions to satisfy the changing and ever-increasing demands of the user. To manage these revisions in DWs, various approaches have been proposed in the literature namely, schema and data evolution, schema versioning and temporal extensions. Schema and data evolution [5–7, 17–19, 22, 25]

---

A. Gosain (✉)
USICT, GGSIPU, Dwarka, India
e-mail: anjana_gosain@hotmail.com

K. Saroha
SOIT, CDAC, Noida, India
e-mail: kritisaroha@gmail.com



S.C. Satapathy et al. (eds.), *Data Engineering and Intelligent Computing*,
Advances in Intelligent Systems and Computing 542,
DOI 10.1007/978-981-10-3223-3_34

approach provides a limited solution, since it maintains only one DW schema and delete the previous schema causing information loss. On the other hand, schema versioning [2–4, 8, 9, 16, 21] approach maintains complete history of the DW evolution defined by a series of schema revisions. But, it is a well-known fact that not only the conceptual schema, but its underlying data also evolve with time and thus, need to maintain multiple versions of data along with schema. Temporal extensions [11, 13, 15, 20, 23], timestamping the multi-dimensional data fulfill the requirement. This results in TDWs with schema versioning support. TDWs keep the evolution of data by using the VT as timestamp for the dimension data. To record and trace the revisions, VT (valid time: indicates time while a data is correct), and TT (transaction time: indicates time while a data is entered into the records) are generally employed. Sometimes, a combination of both referred as bitemporal (i.e. including VT as well as TT for timestamp) may also be used [24].

In the field of TDWs, lot of work is reported in the direction of temporal versioning of dimensional data and schema but none of the works focus on bitemporal versioning of schema. Most of these proposals target revisions in the structure of dimension instances [11, 13, 15, 20, 23] but do not track versions of a schema. The work proposed in [1] used a temporal star schema that uses valid time to timestamp the dimension and transaction data. The approach presented by Chamoni and Stock [11] also timestamped the dimension data using valid time. The temporal design for multidimensional OLAP was recommended by Mendelzon and Vaisman [22] together with temporal query language (TOLAP). The model stores information about the members and schema of the dimensions, and provides support for schema evolution. The model, however did not register versioning of data. Authors in [8, 9] presented a unique temporal model to support the evolution of multidimensional data. The proposed approach is based on time-stamping of level instances together with their hierarchical relations as well as transaction data. However, only modifications applied to the structure and members of the dimensions have been discussed. The COMET model presented by Eder and Koncilia [12–15] handles transaction data and structure data by time stamping the data elements with valid time. The model focuses on the evolvement of dimension members and does not consider the evolution of schema and cubes. They suggested mapping operations that would allow to convert from one version of schema to other using valid times.

The approaches for TDWs so far practice VT alone for managing versions of schema, thus do not allow for more than one version to exist with the same VT. Further, valid time versioning is essential for implementing retro-active (delayed updates) and pro-active (what-if analysis) modifications to schema. However, it fails to maintain the record of updates (i.e. it does not record when the update was originally proposed). Bitemporal versioning, on the other hand, is able to handle as well as keep record of retro-active and pro-active schema modifications, (i.e. records when an update was originally suggested and finally implemented).

This paper proposes a theoretical model for bitemporal schema versioning of TDW, where each temporal version of a TDW is timestamped with bitemporal time (the combination of VT and TT), thus allowing for a single or multiple versions to co-exist having the identical VT intervals, but distinct TT intervals. We also propose evolution functions adapted to suit the requirements of bitemporal versioning of schema using the temporal model of TDW proposed by Eder [14, 15]. Also, the structure of metaschema has been proposed for the support and management of bitemporal schema versioning.

The rest of this paper is structured as follows. Section 2 has an overview of the key concepts for the bitemporal schema versioning. Section 3 presents the proposed structural modification operations and suggests meta-schema for recording intentional data; and finally, Sect. 4 presents the conclusions.

## 2 Bi-Temporal Versioning of Schema

In bitemporal versioning of schema, a new version of schema is created when the desired revisions are applied to the latest version of schema, provided it satisfies the validity condition. This form of versioning uses bitemporal time (VT and TT) to timestamp every version of schema. The proposed schema revision can affect only the present version or the versions that have some overlap with the time interval specified for the revisions. It allows retro- and pro-active schema modifications and helps in keeping record of such changes. Thus, in a system which demands complete traceability, only bitemporal schema versioning can verify that a new version was generated as a result of retro- or a pro-active schema revision [10]. Figure 1a shows TDW structure with bitemporal schema versioning.

It may be noted that the older versions of schema that have complete overlap with VT interval of the revisions are also maintained in addition to the revised version as shown in Fig. 1b, where the schema versions SV1 and SV2 have the same validity interval but different TT. We suggest bitemporal versioning of schema in TDW to define, track and store modifications of TDW schema in the form of bitemporal versions of schema.

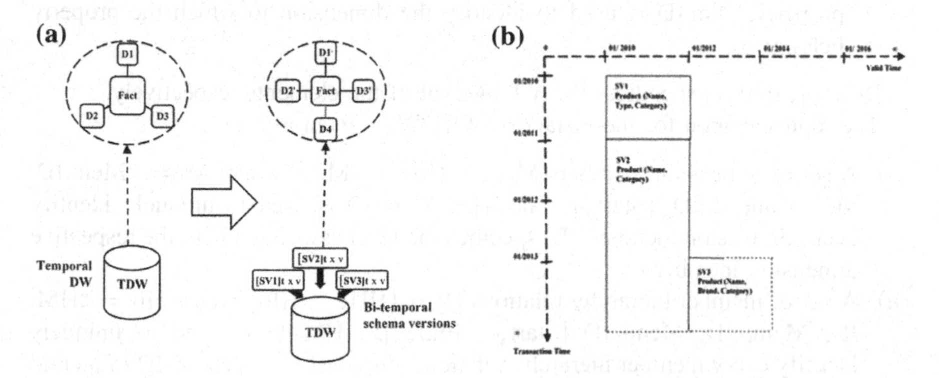


**Fig. 1** **a** Model of bitemporal schema versioning. **b** Representation of bitemporal schema versions in TDW

## 2.1 Symbols Used

In the following discussion, [$Start_{VT}$, $End_{VT}$] denotes the VT of a schema version where $Start_{VT}$, $End_{VT}$ describes 'begin and end' time of the VT respectively. Also, $End_{VT} \geq Start_{VT}$. [$Start_{TT}$, $End_{TT}$] defines the TT of a schema version where $Start_{TT}$, $End_{TT}$ gives 'begin and end' of the TT respectively. Also, $End_{TT} \geq Start_{TT}$.

Notation "∞" is used to denote the value 'Until-Changed' in TT (used for timestamping the latest version), and 'Forever' (maximum time) in VT. The notation "Current Time" defines the present TT and VT. The complete domain of TT and VT and the bitemporal are defined as:

$Ut = \{0\ldots \infty\}t$ $Uv = \{0\ldots \infty\}v$ $Ub = \{0\ldots \infty\}t \times \{0\ldots \infty\}v$, where, bitemporal pertinence is given by the Cartesian product of domains of TT and VT.

The representation for the TDW schema based on the model proposed in [14, 15], is as given below:

(i) A set of dimensions Dim = {D1, …,Dj} where Di = <Dim.ID, DimName, [$Start_{VT}$, $End_{VT}$]>. Dim.ID is the dimension ID, which uniquely identifies a dimension, DimName specifies the dimension name.
(ii) A set of levels L = {L1, …,LK} where Li = <L.ID, LName, [$Start_{VT}$, $End_{VT}$]>. L.ID is the level ID, which uniquely identifies a level, LName specifies the level name.
(iii) A set of hierarchy relations among the levels HR = {H1, …,Hm}, where Hi = <HR.ID, $L_C$.ID, $L_P$.ID, [$Start_{VT}$, $End_{VT}$]>. HR.ID is a hierarchy identifier which uniquely identifies a hierarchy relation. $L_C$.ID and $L_P$.ID defines ID for the child and parent level respectively. $L_P$.ID may be 'NULL' in case the given level is at the topmost position in the hierarchy. $L_C.ID \rightarrow L_P.ID$ denotes roll up function.
(iv) A set of properties P = {P1, …,Pv} where Pi = <P.ID, PName, PType, Dim. ID, [$Start_{VT}$, $End_{VT}$]>. P.ID is the property ID, which uniquely identifies each property, PType is used to specify the data type of the respective property, Dim.ID is used to identify the dimension to which the property belongs to.

[$Start_{VT}$, $End_{VT}$] specifiess the VT interval of the elements respectively.

The representation for the instances of TDW is given as:

(i) A set of dimension members Mem = {M1, …,MP } where Mi = <Mem.ID, MemName, L.ID, [$Start_{VT}$, $End_{VT}$]>. Mem.ID is used to uniquely identify every dimension member. 'L' specifies the level associated with the respective dimension member.
(ii) A set of member hierarchy relations HI = {HI1, …,HIb} where HIi = <HM. ID, $Mem_C$.ID, $Mem_P$.ID [$Start_{VT}$, $End_{VT}$]>. HM.ID is used to uniquely identify every member hierarchy relation. $Mem_C$.ID specifies the ID of a child member, $Mem_P$.ID gives the ID of the parent dimension member of $Mem_C$.ID

or it may be 'NULL' if the given member is the topmost dimension member in the hierarchy.

[$Start_{VT}$, $End_{VT}$] specifies the VT interval of the elements respectively.

Thus a bitemporal schema version can be expressed as a tuple $S_V$ = <S.ID, Dim, L, P, HR, [$Start_{VT}$, $End_{VT}$], [$Start_{TT}$, $End_{TT}$]> and the instance as I = <$S.ID_i$, $L_i$, $Dim_i$, $HR_i$, $P_i$, [$Start_{VT}$, $End_{VT}$]>, where S.ID is used to uniquely identify a schema version and thus allows to maintain and keep record of various versions of schema, L is a set of levels, Dim is the set of dimensions, HR is a set of hierarchy relations, P gives the set of properties. [$Start_{VT}$, $End_{VT}$], [$Start_{TT}$, $End_{TT}$] gives the VT and TT of the schema versions respectively.

Figure 1b gives the temporal representation of three bitemporal schema versions. The schema version SV1 includes a dimension 'Product' with attributes (Name, Type, Category). This schema version is updated to SV2 with the deletion of attribute 'Type' from the dimension 'Product'. SV3 is created with the addition of attribute 'Brand'. The timestamps of the three bitemporal schema versions is given as:

SV1 [1/2010–12/2010]t × [l/2010–12/2011]v
SV2 [1/2011–∞]t × [l/2010–12/2011]v
SV3 [1/2013–∞]t × [l/2012–∞]v.

## 3 Proposed Structural Modification Functions

In this section, we present and discuss a set of evolution functions for the support of bitemporal schema versioning. The functions are derived from the previous proposals in the field but neither examines the option of using bitemporal timestamps on schema versions. The paper describes the following 13 structural modification functions. Table 1 gives the details of all the functions.

- InsertDim: A new dimension can be inserted
- EraseDim: An old dimension can be removed
- RenameDim: A new name may be given to an old dimension
- InsertL: A new level can be inserted to a given dimension
- RearrangeL: A given level is repositioned among the levels of a dimension
- EraseL: An old level can be removed from a dimension
- RenameL: A new name may be given to an old level of a dimension
- InsertProp: A new property can be inserted to a dimension
- EraseProp: An old property can be removed from a dimension
- RenameProp: A new name may be given to an old property of a dimension
- ReorgL: A level instance can be reorganized
- MergeL: The instances of a level can be merged together
- DivideL: A level instance can be divided into multiple instances

**Table 1** Structural modification operations

| Operations | Definition |
|---|---|
| InsertDim <S.ID$_i$, DimName, L, P, HR, [Start$_{VT}$, End$_{VT}$], [Start$_{TT}$, End$_{TT}$]> | A new dimension is inserted to the schema version S.ID$_i$, giving a revised version S.Id$_{i+1}$. DimName specifies the name of the dimension, L, P, HR specifies the set of levels, set of properties, set of hierarchy relations contained in the dimension respectively. [Start$_{VT}$, End$_{VT}$] denotes the VT interval of the new dimension. The VT of S.Id$_{i+1}$ is [Start$_{VT}$, End$_{VT}$] and TT is [CurrentTime, ∞] |
| EraseDim <S.ID$_i$, Dim.ID, [Start$_{VT}$, End$_{VT}$], [Start$_{TT}$, End$_{TT}$]> | An old dimension with given ID (Dim.ID) is removed from the selected version of schema S.ID$_i$, giving a revised version S.ID$_{i+1}$. The end time of (Dim.ID) is set to End$_{VT}$. The VT of S.ID$_{i+1}$ is [Start$_{VT}$, End$_{VT}$] and its TT is [CurrentTime, ∞] |
| RenameDim <S.ID$_i$, Dim.ID, DimName, [Start$_{VT}$, End$_{VT}$], [Start$_{TT}$, End$_{TT}$]> | A new name (DimName) is assigned to an existing dimension (Dim.ID) in the selected version of schema S.ID$_i$, giving a revised version S.ID$_{i+1}$. The end time of (Dim.ID) is set to End$_{VT}$. [Start$_{VT}$, End$_{VT}$] denotes the VT interval of the renamed dimension. The VT of S.ID$_{i+1}$ is [Start$_{VT}$, End$_{VT}$] and TT is [CurrentTime, ∞] |
| InsertL <S.ID$_i$, Dim.ID, l.ID$_1$, ln, l.ID$_2$, [Start$_{VT}$, End$_{VT}$], [Start$_{TT}$, End$_{TT}$]> | A new level (ln) is inserted to an existing dimension (Dim.ID) in the selected version S. ID$_i$, giving a revised version S.ID$_{i+1}$. The newly inserted level (ln) have a hierarchy assignment defined as (l.ID$_1$ → ln; ln → l. ID$_2$) and l.ID$_1$ may be 'NULL' if ln is the new lowest level inserted in Dim.ID and l.ID$_2$ may be 'NULL' if ln is inserted at the top level in the dimension hierarchy. [Start$_{VT}$, End$_{VT}$] denotes the VT interval of the new level (ln). The VT of S.ID$_{i+1}$ is [Start$_{VT}$, End$_{VT}$] and TT is [CurrentTime, ∞] |
| RearrangeL <S.ID$_i$, Dim.ID, l0, ln, l1, [Start$_{VT}$, End$_{VT}$], [Start$_{TT}$, End$_{TT}$]> | Rearranges the location of level (ln) in the dimension hierarchy of (Dim.ID). It redefines the parent relation and therefore, the hierarchy, giving a revised version S.ID$_{i+1}$. l1 defines the set of Parent$_{New}$ of ln and a new hierarchy relation is set up between l1 and ln. l0 defines the set of *Parent$_{Old}$* of ln and the hierarchy relation between l0 and ln is removed. The end time of all old hierarchy roll up assignments is changed to End$_{VT}$. The VT interval of all new hierarchy roll up assignments is given by [Start$_{VT}$, End$_{VT}$]. The VT of S.ID$_{i+1}$ is [Start$_{VT}$, End$_{VT}$] and TT is [CurrentTime, ∞] |

(continued)

**Table 1** (continued)

| Operations | Definition |
|---|---|
| | It may be noted that the set of $Parent_{Old}$ or $Parent_{New}$ may be NULL |
| EraseL <S.ID$_i$, Dim.ID, l.ID, [Start$_{VT}$, End$_{VT}$], [Start$_{TT}$, End$_{TT}$]> | An old level (l.ID) belonging to the dimension (Dim.ID) is removed from the selected version S.ID$_i$, giving a revised version S.ID$_{i+1}$. The end time of (l.ID) is changed to End$_{VT}$. The VT of S.ID$_{i+1}$ is [Start$_{VT}$, End$_{VT}$] and its TT is [CurrentTime, ∞] |
| RenameL <S.ID$_i$, Dim.ID, l.ID, lname, [Start$_{VT}$, End$_{VT}$], [Start$_{TT}$, End$_{TT}$]> | A new name (lname) is assigned to an existing level (l.ID) belonging to (Dim.ID) of selected version S.ID$_i$, giving a revised version S.ID$_{i+1}$. The end time of (l.ID) is set to End$_{VT}$. [Start$_{VT}$, End$_{VT}$] denotes the VT interval of the renamed level. The VT of S.ID$_{i+1}$ is [Start$_{VT}$, End$_{VT}$] and its TT is [CurrentTime, ∞] |
| InsertProp <S.ID$_i$, Dim.ID, P, [Start$_{VT}$, End$_{VT}$], [Start$_{TT}$, End$_{TT}$]> | A new property (P) is inserted into an existing dimension (Dim.ID) of selected version S.ID$_i$, giving a revised version S.ID$_{i+1}$. [Start$_{VT}$, End$_{VT}$] gives the VT interval of the new property (P). The VT of S.ID$_{i+1}$ is [Start$_{VT}$, End$_{VT}$] and TT is [CurrentTime, ∞] |
| EraseProp <S.ID$_i$, Dim.ID, P.ID, [Start$_{VT}$, End$_{VT}$], [Start$_{TT}$, End$_{TT}$]> | An old property (P.ID) belonging to the dimension (Dim.ID) is removed in the selected version S.ID$_{i+1}$, giving a revised version S. ID$_{i+1}$. The end time of (P.ID) is set to End$_{VT}$. The VT of S.ID$_{i+1}$ is [Start$_{VT}$, End$_{VT}$] and its TT is [CurrentTime, ∞] |
| RenameProp <S.ID$_i$, Dim.ID, P.ID, Pname, [Start$_{VT}$, End$_{VT}$], [Start$_{TT}$, End$_{TT}$]> | A new name (Pname) is assigned to an existing property belonging to (Dim.ID) of selected version S.ID$_i$, giving a revised version S.ID$_{i+1}$. The end time of (P.ID) is set to End$_{VT}$. [Start$_{VT}$, End$_{VT}$] denotes the VT interval of the renamed property. The VT of S. ID$_{i+1}$ is [Start$_{VT}$, End$_{VT}$] and TT is [CurrentTime, ∞] |
| Reorg a level instance <S.IDi, Dim.ID, l.ID, lmem.ID, lmem.ID0, lmem.ID1, [Start$_{VT}$, End$_{VT}$], [Start$_{TT}$, End$_{TT}$]> | Redefines the parent relation for the given instance (lmem.ID) of level (l.ID) belonging to Dim.ID of selected version S.ID$_i$ giving a revised version S.ID$_{i+1}$. The $Parent_{Old}$ relation (lmem.ID0) is removed and a $Parent_{New}$ relation (lmem.ID1) is assigned. The end time of old hierarchy is set to End$_{VT}$. [Start$_{VT}$, End$_{VT}$] denotes the VT interval of the new hierarchy assignment. The VT of S.ID$_{i+1}$ is [Start$_{VT}$, End$_{VT}$] and its TT is [CurrentTime, ∞]<br>It may be noted that $Parent_{Old}$ and $Parent_{New}$ belong to the same level in the hierarchy |

(continued)

**Table 1** (continued)

| Operations | Definition |
|---|---|
| Merge level instances <S.IDi, l.ID, lmem.ID(1… n), lnew, lp.ID, [Start$_{VT}$, End$_{VT}$], [Start$_{TT}$, End$_{TT}$]> | Combines the 'n' given instances (lmem. ID1… n) of a given level (l.ID) to give a new instance (lnew) belonging to the same level (l. ID) of selected version S.ID$_i$ giving a revised version S.ID$_{i+1}$. The parent (lp.ID), which was the parent of instances (lmem.ID1… n) is now the parent of newly created instance (lnew). The old hierarchy relation is removed and new parent relation is assigned. Similarly, all the child instances of (lmem.ID1… n) are now the children of new instance (lnew). The end time of (lmem.ID1… n), old hierarchy from (lmem. ID1… n) to lmem.ID and old hierarchy assignment of children of (lmem.ID1… n) is set to End$_{VT}$. [Start$_{VT}$, End$_{VT}$] gives the VT interval of new instance (lnew), new parent hierarchy from (lnew) to lp.ID and new hierarchy assignment of children of (lmem. ID1… n) to (lnew). The VT of S.ID$_{i+1}$ is [Start$_{VT}$, End$_{VT}$] and its TT is [CurrentTime, $\infty$]<br>It may be noted that all instances (lmem.ID1… n) have the same parent |
| Divide a level instance into many instances <S.IDi, Dim.ID, lmem.ID, lnew(1… n), lp. ID, l.ID, [Vs, Ve], [Ts, Te]> | A given instance (lmem.ID) of level (l.ID) belonging to dimension (Dim.ID) is divided into 'n' new instances (lnew1… n) of the same given level (l.ID) of selected version S.ID$_i$ giving a revised version S.ID$_{i+1}$. The parent (lp.ID), which was the parent of the given instance (lmem.ID) is now the parent of newly created instances (lnew1… n). The old hierarchy relation is removed and new parent relation is assigned. Similarly, all the child instances of (lmem.ID) are now the children of (lnew1… n). The end time of (lmem.ID), old hierarchy from (lmem.ID) to parent (lp.ID) and old hierarchy assignment of children of (lmem.ID) is set to End$_{VT}$. [Start$_{VT}$, End$_{VT}$] gives the VT interval of new instances (lnew1… n), new parent hierarchy from (lnew1… n) to (lp.ID) and new hierarchy assignment of children of (lmem.ID) to (lnew1… n). The VT of S.ID$_{i+1}$ is [Start$_{VT}$, End$_{VT}$] and its TT is [CurrentTime, $\infty$] |

The structural revisions in schema versions (SV) are shown in example in Fig. 2 for the dimension 'Product'.

For e.g., addition of a level 'Brand' creates a new schema version SV2, merging of Category 'C1' and 'C2' creates a new version from SV2 to SV3, a split

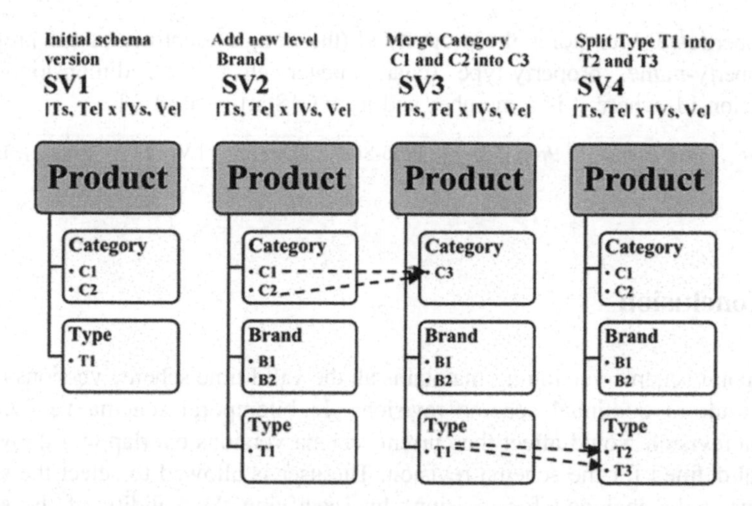


**Fig. 2** Example of structural revisions in schema versions

performed on Type 'T1' into 'T2' and 'T3' results in the version SV4. The resulted versions include valid time [$Start_{VT}$, $End_{VT}$] and transaction time [$Start_{TT}$, $End_{TT}$] for bitemporal schema versioning.

## *3.1 Meta-Schema Structure*

In this section, the structure of meta-schema based on [10] is presented. The structure is proposed to maintain a log of structural modifications. To provide support for bitemporal versions of schema, the meta-schema are described as well as maintained as temporal relations by time-stamping all the data records with VT, TT or both. The meta-schema discussed below can be used by any of the TDWs supporting schema evolution.

- Schema Versions Table: stores the version-number of the schema as unique identifier of the schema version and its VT [$Start_{VT}$, $End_{VT}$] and TT [$Start_{TT}$, $End_{TT}$].

| Ver-Id | [$Start_{TT}$, $End_{TT}$] | [$Start_{VT}$, $End_{VT}$] |
|---|---|---|

- Dimensions Table: stores the dimension-id (the unique identifier of the dimension), dimension-name, dimension-type, the version-id of the schema, where it is a member and its VT [$Start_{VT}$, $End_{VT}$].

| Dim-Id | Dim-Name | Dim-Type | Ver-Id | [$Start_{VT}$, $End_{VT}$] |
|---|---|---|---|---|

- Properties Table: stores the property-id (the unique identifier of the property), property-name, property-type (char, integer etc.), size, dimension-id and version-id, where it is a member and its VT [$Start_{VT}$, $End_{VT}$].

| Prop-Id | Prop-Name | Prop-Type | Prop-Size | Dim-Id | Ver-Id | [$Start_{VT}$, $End_{VT}$] |
|---|---|---|---|---|---|---|

## 4 Conclusion

Bitemporal schema versioning maintains all the valid time schema versions created as a result of continual schema revisions. In bitemporal schema versioning, a schema revision would affect the current schema versions overlapping the validity interval defined for the schema revision. The user is allowed to select the schema versions to be included for revisions by specifying the validity of the schema revision. Moreover, bitemporal versioning of schema contributes to support and manage retro- and pro-active schema changes and also keep record of the changes. The paper explains the different evolution functions for ever growing and emerging bitemporal schema versions for TDW and further can be extended to manage the modifications of fact data with time. A representation for meta-schema has been discussed to provide support for schema versioning at intensional level. The storage of multiple schema versions is another important issue which would also be addressed in future.

## References


1. Agrawal, R., Gupta, A., Sarawagi, S.: Modeling multidimensional databases. IBM Research Report, IBM Almaden Research Center (1995)
2. Bębel, B., Eder, J., Konicilia, C., Morzy, T., Wrembel, R.: Creation and management of versions in multiversion data warehouse. In: Proceedings of ACM Symposium on Applied Computing (SAC), pp. 717–723 (2004)
3. Bębel, B., Królikowski, Z., Wrembel, R.: Managing multiple real and simulation business scenarios by means of a multiversion data warehouse. In: Proceedings of International Conference on Business Information Systems (BIS). Lecture Notes in Informatics, pp. 102–113 (2006)
4. Bębel, B., Wrembel, R., Czejdo, B.: Storage structures for sharing data in multi-version data warehouse. In: Proceedings of Baltic Conference on Databases and Information Systems, pp. 218–231 (2004)
5. Benítez-Guerrero, E., Collet, C., Adiba, M.: The WHES approach to data warehouse evolution. Digit. J. e-Gnosis (2003). http://www.e-gnosis.udg.mx, ISSN: 1665-5745
6. Blaschka, M., Sapia, C., Hofling, G.: On schema evolution in multidimensional databases. In: Proceedings of International Conference on Data Warehousing and Knowledge Discovery (DaWaK). Lecture Notes in Computer Science, vol. 1676, pp. 153–164 (1999)
7. Blaschka, M.: FIESTA: a framework for schema evolution in multidimensional information systems. In: 6th CAiSE Doctoral Consortium, Heidelberg (1999)

8. Body, M., Miquel, M., Bédard, Y., Tchounikine, A.: A multidimensional and multiversion structure for OLAP applications. In: Proceedings of ACM International Workshop on Data Warehousing and OLAP (DOLAP), pp. 1–6 (2002)
9. Body, M., Miquel, M., Bédard, Y., Tchounikine, A.: Handling evolutions in multidimensional structures. In: Proceedings of International Conference on Data Engineering (ICDE), p. 581 (2003)
10. De Castro, C., Grandi, F., Scalas, M.R.: On schema versioning in temporal databases. In: Clifford, S., Tuzhilin, A. (eds.) Recent Advances in Temporal Databases, pp. 272–294. Springer, Zurich Switzerland (1995)
11. Chamoni, P., Stock, S.: Temporal structures in data warehousing. In: Proceedings of International Conference on Data Warehousing and Knowledge Discovery (DaWaK). Lecture Notes in Computer Science, vol. 1676, pp. 353–358 (1997)
12. Eder, J.: Evolution of dimension data in temporal data warehouses. Technical report 11, Univ. of Klagenfurt, Department of Informatics-Systems (2000)
13. Eder, J., Koncilia, C.: Changes of dimension data in temporal data warehouses. In: Proceedings of International Conference on Data Warehousing and Knowledge Discovery (DaWaK). Lecture Notes in Computer Science, vol. 2114, pp. 284–293 (2001)
14. Eder, J., Koncilia, C., Morzy, T.: A model for a temporal data warehouse. In: Proceedings of the International OESSEO Conference, Rome, Italy (2001)
15. Eder, J., Koncilia, C., Morzy, T.: The COMET metamodel for temporal data warehouses. In: Proceedings of Conference on Advanced Information Systems Engineering (CAiSE). Lecture Notes in Computer Science, vol. 2348, pp. 83–99 (2002)
16. Golfarelli, M., Lechtenbörger, J., Rizzi, S., Vossen, G.: Schema versioning in data warehouses. In: Proceedings of ER Workshops. Lecture Notes in Computer Science, vol. 3289, pp. 415–428 (2004)
17. Hurtado, C.A., Mendelzon, A.O., Vaisman, A.A.: Maintaining data cubes under dimension updates. In: Proceedings of International Conference on Data Engineering (ICDE), pp. 346–355 (1999)
18. Hurtado, C.A., Mendelzon, A.O., Vaisman, A.A.: Updating OLAP dimensions. In: Proceedings of ACM International Workshop on Data Warehousing and OLAP (DOLAP), pp. 60–66 (1999)
19. Kaas, Ch.K., Pedersen, T.B., Rasmussen, B.D.: Schema evolution for stars and snowflakes. In: Proceedings of International Conference on Enterprise Information Systems (ICEIS), pp. 425–433 (2004)
20. Letz, C., Henn, E.T., Vossen, G.: Consistency in data warehouse dimensions. In: Proceedings of International Database Engineering and Applications Symposium (IDEAS), pp. 224–232 (2002)
21. Malinowski, E., Zimanyi, E.: A conceptual solution for representing time in data warehouse dimensions. In: 3rd Asia-Pacific Conference on Conceptual Modelling, Hobart Australia, pp. 45–54 (2006)
22. Mendelzon, A.O., Vaisman, A.A.: Temporal queries in OLAP. In: Proceedings of International Conference on Very Large Data Bases (VLDB), pp. 242–253 (2000)
23. Schlesinger, L., Bauer, A., Lehner, W., Ediberidze, G., Gutzman, M.: Efficiently synchronizing multidimensional schema data. In: Proceedings of ACM International Workshop on Data Warehousing and OLAP (DOLAP), pp. 69–76 (2001)
24. SOO, M.: Bibliography on Temporal Databases. ACM SIGMOD Record (1991)
25. Vaisman, A., Mendelzon, A.: A temporal query language for OLAP: implementation and case study. In: Proceedings of Workshop on Data Bases and Programming Languages (DBPL). Lecture Notes in Computer Science, vol. 2397, pp. 78–96. Springer (2001)

# An Improved Mammogram Classification Approach Using Back Propagation Neural Network

**Aman Gautam, Vikrant Bhateja, Ananya Tiwari and Suresh Chandra Satapathy**

**Abstract** Mammograms are generally contaminated by quantum noise, degrading their visual quality and thereby the performance of the classifier in Computer-Aided Diagnosis (CAD). Hence, enhancement of mammograms is necessary to improve the visual quality and detectability of the anomalies present in the breasts. In this paper, a sigmoid based non-linear function has been applied for contrast enhancement of mammograms. The enhanced mammograms are used to define the texture of the detected anomaly using Gray Level Co-occurrence Matrix (GLCM) features. Later, a Back Propagation Artificial Neural Network (BP-ANN) is used as a classification tool for segregating the mammogram into abnormal or normal. The proposed classifier approach has reported to be the one with considerably better accuracy in comparison to other existing approaches.

**Keywords** Contrast enhancement · BP-ANN · GLCM · Region of interest (ROI)

A. Gautam (✉) · V. Bhateja · A. Tiwari
Department of Electronics and Communication Engineering,
Shri Ramswaroop Memorial Group of Professional Colleges (SRMGPC),
Lucknow 226028, Uttar Pradesh, India
e-mail: gautamaman543@gmail.com

V. Bhateja
e-mail: bhateja.vikrant@gmail.com

A. Tiwari
e-mail: absoluteananya@gmail.com

S.C. Satapathy
ANTIS, Visakhapatnam, India
e-mail: sureshsathapathy@gmail.com



S.C. Satapathy et al. (eds.), *Data Engineering and Intelligent Computing*,
Advances in Intelligent Systems and Computing 542,
DOI 10.1007/978-981-10-3223-3_35

## 1 Introduction

Cancer categorizes those set of diseases which includes uncontrolled growth of cells. Breast Cancer is the one in which an abnormal multiplication of breast cells takes place. The unknown root of this illness leads to a high mortality rate amongst middle aged women. The analysis of mammograms (X-ray image of the breast) is mostly done by experienced radiologists, who play a keen role in determining the features and shape(s) of the lesions present. However, the human factor results in a low precision outcome [1]. To improve the analysis of the radiologists a correlation with computer-based techniques is needed, which can be used for analysis, detection or the overall diagnosis of breast cancer. Computer-based diagnosis systems, however, are not self-generating in nature but they can be actively used by the radiologists so that the final decision is made faster and with a greater degree of accuracy. Another problem faced by the radiologist is that during the screening of mammograms, the image obtained from the X-ray machine having low contrast is extremely marginal. In poor quality images, the slight dissimilarity between the normal and malignant tissue is not distinguishable and thus it becomes difficult on the radiologists' part to give accurate results. Thus, enhancement of the mammogram becomes very essential after the screening of mammograms has been done [2]. Some related works and studies have been done in the past for the contrast-improvement of the mammograms. Tang et al. [3] applied Discrete Wavelet Transform (DWT) which is a multi-resolution enhancement technique over the mammograms. The enhancement method was same for every wavelet coefficient at every level due to which the enhancement response was not prominent. Anitha et al. [4] used morphological operations to extract the breast profile from the background. The method was capable to intensify the mass but the background was not suppressed appropriately. They reported an accuracy of 95% with a false negative measure of 6% and true negative measure of 36%. A. Abubaker performed a thresholding technique [5] followed by applying morphological operations to obtain the enhanced image. The method used by authors' distorted the shape of the ROI which affects the performance of the classifier yielding an accuracy of 93%. Wang et al. [6] performed CAD based visual enhancement of mammograms by removing major background tissues by applying image matting. Another approach applied Law's Texture Energy Measure (LAWS TEM) [7] as extraction of textural features and used ANN for final classification. The Law's textures approach improved the accuracy of the classifier in comparison to GLCM features but due to the use of single layer neural network the accuracy of their work was 93%. Recently Non-Linear Polynomial Filters [8, 9] and Non-Linear Unsharp Masking [10] approaches are also deployed for pre-processing of mammograms. In this paper, a non-linear enhancement algorithm has been used to improve the overall contrast of the image without affecting the information present in the original mammographic image. To define the textural features of the breast anomaly four GLCM features are computed. For final classification module a neural network based on backpropagation algorithm has been used. The detailed explanation of the methodology, analysis and discussions of results along with conclusions are mentioned in the sections to follow.

# 2 Proposed CAD Methodology for Mammogram Classification

The proposed CAD methodology comprises of modules such as pre-processing, features extraction and classification as shown in Fig. 1. Firstly, the input mammogram is fed for pre-processing where RGB to gray conversion and enhancement process takes place. This module uses a non-linear enhancement based pre-processing technique which suppresses background and refines the contrast of the foreground containing the tumor present in the breast.

Secondly, the detected ROI is sent to a features extraction module where a structure of quantitative data is obtained by calculating various attributes such as Energy, Entropy, and Correlation etc. for final classification. Lastly, the selected features are provided to BP-ANN classifier which categorizes the detected ROI into Normal and Abnormal.

## *2.1 Pre-processing Using Non-linear Enhancement*

For noise suppressed enhancement of mammograms, a logistic function is used for non-linear processing. It is given by Eq. (1).

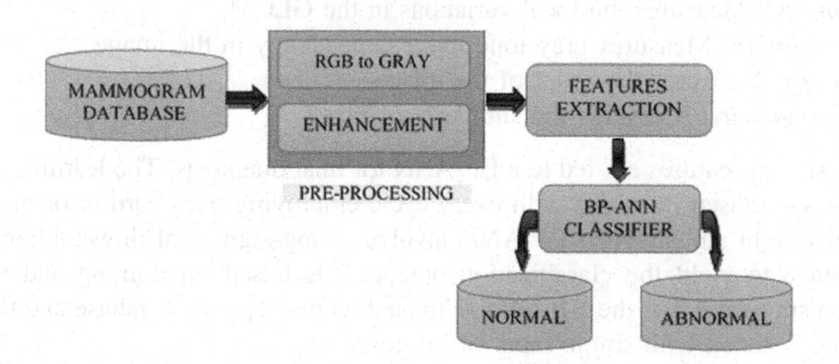


**Fig. 1** Block diagram of proposed methodology to classify normal and abnormal and mammograms

$$\log istic(x) = \frac{1}{1 + e^{-x}} \tag{1}$$

The function modifies the mammogram by defining a certain threshold; pixels having magnitude above this threshold are enhanced while the rest are suppressed. The enhancement using the aforesaid function has been carried out by combining the function of Eq. (1) in a linear fashion along with a gain parameter as given in Eq. (2).

$$y(x) = a[\log istic\{k(x-b)\} - \log istic\{-k(x+b)\}] \tag{2}$$

where: $x$ indicates the intensity of gray level for original image at coordinate $(i, j)$ and parameter $a$ is given by Eq. (3).

$$a = \frac{1}{\log istic\{k(1-b)\} - \log istic\{-k(1+b)\}} \tag{3}$$

where: $b \in R$, $k \in N$ parameters manages the threshold control and governs the contrast enhancement of the mammograms [1].

## 2.2 Features Extraction and Classification

The abnormal breast tissues reveal distinct characteristics as compared normal tissues. Therefore, before the classification procedure, features of detected ROI(s) should be extracted. GLCM is an analytical method of examining the texture of detected anomalies by considering the structural correspondence of pixels. It calculates how frequently a pair of pixel occurs in an image within a specified structural network to categorize the texture [11]. For each GLCM, four Haralick features are calculated which provide textural information of the ROI. These four features are:

- ***Contrast***: Measures the local variations in the GLCM.
- ***Correlation***: Measures gray tone linear dependency in the image.
- ***Energy***: Measures the order of the image.
- ***Homogeneity***: Measures the similarity of pixels.

These four features are fed to a BP-ANN for final diagnosis. The learning model used is supervised in nature with every cycle employing backward error propagation of weight adjustments. BP-ANN involves a log-sigmoidal thresholding at its final stage to yield the classification output. It is based on training and testing mechanism where first the BP-ANN is trained with a separate database and then its learning is tested with single input of images.

## 3 Results and Discussions

In the proposed classification methodology, the digital mammograms are taken from The Cancer Imaging Archive (TCIA) [12] which provides a classified set of images for cancer anomalies. During simulations, these mammograms are normalized by performing RGB to Gray conversion followed by enhancement process using non-linear transformation as discussed in the previous section.

The results of enhanced mammograms using the proposed non-linear enhancement function are shown in Fig. 2. Herein, Fig. 2a shows an original mammogram (Mam#1) which is containing a malignant mass with targeted ROI shown in Fig. 2b (ROI#1). After enhancement, as shown in Fig. 2c, it is observed that the background is suppressed while the contrast is improved and the targeted ROI is very clearly visible in Fig. 2d (EnROI#1); here, the distorted boundaries indicate the presence of malignancy in the mass. Similarly Fig. 2e, g shows original and enhanced mammograms with their ROI(s) shown in Fig. 2f and Fig. 2h respectively. Response of non-linear transformation on other test mammograms have been shown in Fig. 3. Now the enhanced and segmented ROI is used for features extraction where four GLCM features are extracted namely Contrast, Homogeneity,

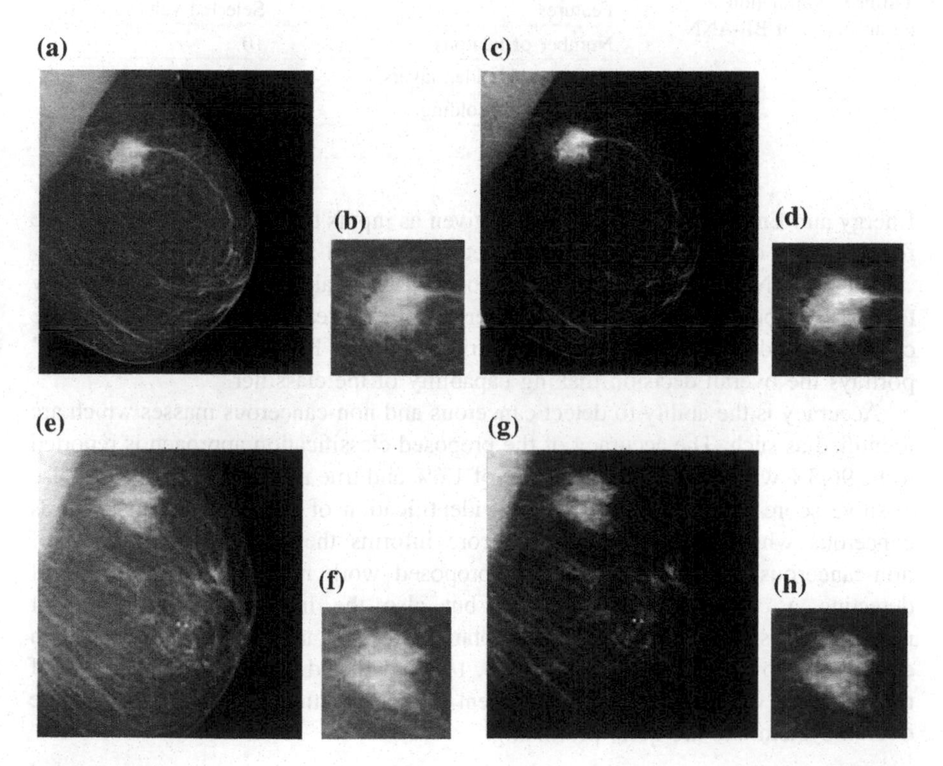


**Fig. 2** The original and enhanced mammograms with their subsequent ROI. **a** Mam#1, **b** ROI#1, **c** EnMam#1, **d** EnROI#1, **e** Mam#2, **f** ROI#2, **g** EnMam#2, **h** EnROI#2

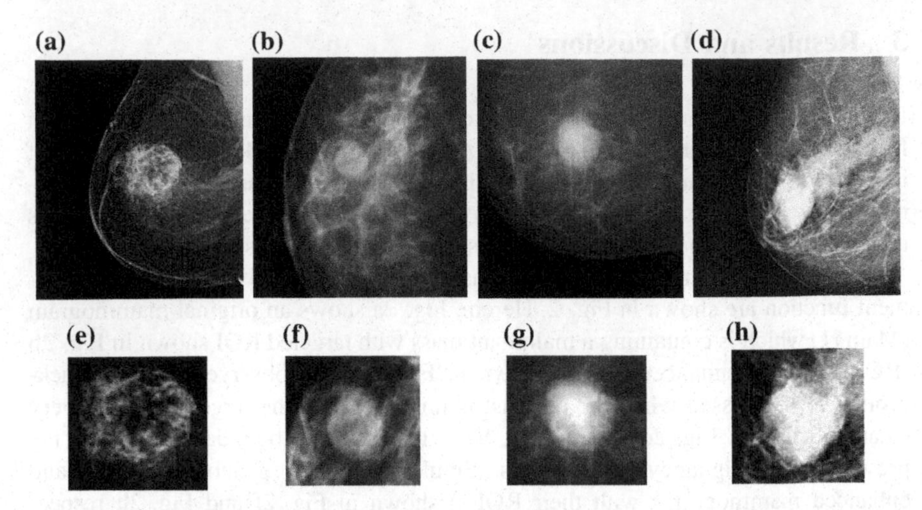


**Fig. 3** **a–d** Original mammograms: Mam#3, Mam#4, Mam#5, Mam#6, **e–f** Enhanced ROI: ROI#3, ROI#4, ROI#5, ROI#6

**Table 1** Simulation Parameter(s) of BP-ANN

| Features | Selected values |
|---|---|
| Number of neurons | 10 |
| Number of hidden layers | 50 |
| Type of thresholding | Log-sigmoidal |
| Epoch | 28 iterations |

Energy and Entropy. These features are given as inputs to BP-ANN comprising the following set of input simulation parameters as shown in Table 1.

The BP-ANN has been simulated to be used for training validation and finally testing. The performance of the classifier obtained at each of these stages can be communicated using the confusion matrix shown in Fig. 4. 'Confusion Matrix' portrays the overall decision making capability of the classifier.

Accuracy is the ability to detect cancerous and non-cancerous masses which are identified as such. The accuracy of the proposed classification approach is reported to be 96.3% with a false positive score of 1.6% and true negative of 32.3%. A false positive score determines the incorrect identification of a non-cancerous mass as cancerous whereas, a true negative score informs the correct identification of non-cancerous mass as such. In the proposed work not only the accuracy of detecting a cancerous mass is high but also the incorrect classification of non-cancerous mass is low. These obtained results are considerably better in comparison to the works reported in [13, 14] as depicted in Table 2. The results of the proposed work affirm an improvement in the classification performance of the overall system and are quite promising.

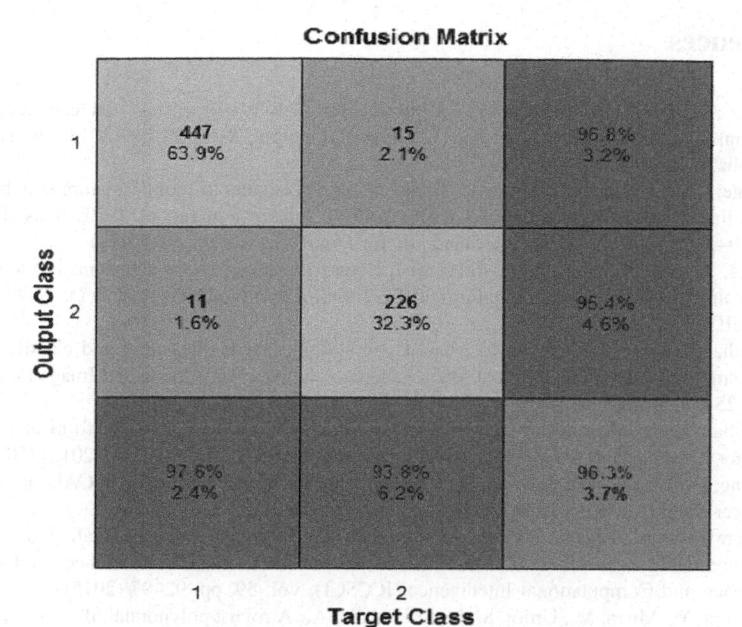


**Fig. 4** Confusion matrix of enhanced mammograms using BP-ANN classifier

**Table 2** Comparison of accuracy of proposed classification methodology with other existing approaches

| References | Classification accuracy (%) |
|---|---|
| Saini et al. [13] | 87.5 |
| Al-Najdawi et al. [14] | 90.7 |
| Proposed methodology | 96.3 |

# 4 Conclusion

In the proposed work, a non-linear enhancement approach for contrast improvement of mammogram has been deployed. The prime advantage of this method with respect to the other approaches mentioned in the literature is its ability to improve the pixel intensity in only the anomalous region. Also, it reduces the necessity for further segmentation of mammogram as the ROI is well defined after the enhancement procedure only. The simulation results of the classifier show that with enhancement of mammograms the classification accuracy is optimized and is proved to be better in comparison to other state of art approaches. Future development of the presented research concerns with the improvement of the adaptability of the proposed algorithm. The mentioned technique can be incorporated in the conventional CAD system under the pre-processing module to achieve precise diagnosis of breast cancer.

## References


1. Jain, A., Singh, S., Bhateja, V.: A robust approach for denoising and enhancement of mammographic breast masses. Int. J. Converg. Comput. **1**(1), 38–49 (2013) (Inderscience Publishers)
2. Bhateja, V., Urooj, S., Misra, M.: Technical advancements to mobile mammography using non-linear polynomial filters and IEEE 21451-1 information model. IEEE Sens. J. **15**(5), 2559–2566 (2015) (Advancing Standards for Smart Transducer Interfaces)
3. Tang, J., Liu, X., Sun, Q.: A direct image contrast enhancement algorithm in the wavelet domain for screening mammograms. IEEE J. Sel. Top. Signal Process. **3**(1), 74–80 (2009) (IEEE)
4. Anitha, J., Peter, J.D.: A wavelet based morphological mass detection and classification in mammograms. In: IEEE International Conference on Machine Vision and Image Processing, pp. 25–28. IEEE (2012)
5. Abubaker, A.: Mass lesion detection using wavelet decomposition transform and support vector machine. Int. J. Comput. Sci. Inf. Technol. (IJCSIT), **4**(2), 33–46 (2012) (IJCSIT)
6. Wang, H., Li, J.B., Wu, L., Gao, H.: Mammography visual enhancement in CAD-based breast cancer diagnosis. Clin. Imaging **37**(2), 273–282 (2013)
7. Setiawan, A.S., Elysia, Wesley, J., Purnama. Y.: Mammogram classification using law's texture energy measure and neural networks. In: International Conference on Computer Science and Computational Intelligence (ICCSCI), vol. 59, pp. 92–97 (2015)
8. Bhateja, V., Misra, M., Urooj, S., Lay-Ekuakille, A.: A robust polynomial filtering framework for mammographic image enhancement from biomedical sensors. IEEE Sens. J., **13**(11), 4147–4156 (2013) (IEEE)
9. Bhateja, V., Misra, M., Urooj, S.: Non-linear polynomial filters for edge enhancement of mammogram lesions. Comput. Methods Prog. Biomed. **129C**, 125–134 (2016) (Elsevier)
10. Bhateja, V., Misra, M., Urooj, S.: Human visual system based unsharp masking for enhancement of mammographic images. J. Comput. Sci. (2016)
11. Mohamed, H., Mabroukb, M.S., Sharawy, A.: Computer aided detection system for micro calcifications in digital mammograms. Comput. Methods Programs Biomed. **116**(3), 226–235 (2014)
12. The Cancer Imaging Archive (TCIA) (2016). http://www.cancerimagingarchive.net/. Accessed 31st Aug 2016
13. Saini, S., Vijay, R.: Mammogram analysis using feed-forward back propagation and cascade-forward back propagation artificial neural network. In: 5th IEEE International Conference on Communication Systems and Network Technologies, pp. 1177–1180. IEEE, Gwalior (2015)
14. Al-Najdawia, N., Biltawib, M., Tedmorib, S.: Mammogram image visual enhancement, mass segmentation and classification. Appl. Soft Comput. **35**, 175—185 (2015) (Elsevier)

# Performance Comparison of Pattern Search, Simulated Annealing, Genetic Algorithm and Jaya Algorithm

**Hari Mohan Pandey, Manjul Rajput and Varun Mishra**

**Abstract** In this paper, we have shown the performance comparison of four powerful global optimization algorithms, namely Pattern Search, Simulated Annealing, Genetic Algorithm and Jaya Algorithm. All of these algorithms are used to find an optimum solution. The standard benchmark functions are utilized for the implementation. The results are collected and analyzed that helps to classify the algorithms according to their computational capability to solve the optimization problems.

**Keywords** Genetic algorithm · Simulated annealing · Jaya algorithm · Pattern search

## 1 Introduction

Optimization is a technique that allows us to get the best out of a given system. It has a vital role in various disciplinary systems such as single or multi-disciplinary systems. In engineering, science and production optimization play a significant role because costs, resources all are crucial aspects. There exists several global optimization algorithms are proposed that are mainly falling into 2 categories: Evolutionary Algorithms (EAs) [1, 2], Swarm Intelligence (SI) [3]. The EA and SI both are applied to solve science and engineering problems. There is no doubt about the computational capabilities of these algorithms. The success of these algorithms are mainly depends on the parameters they used. For example, the GA's success

H.M. Pandey (✉) · M. Rajput · V. Mishra
Department of Computer Science and Engineering, Amity University,
Sector-125, Noida, Uttar Pradesh, India
e-mail: profharimohanpandey@gmail.com

M. Rajput
e-mail: mkigraj3@gmail.com

V. Mishra
e-mail: varunmishra321@gmail.com



S.C. Satapathy et al. (eds.), *Data Engineering and Intelligent Computing*,
Advances in Intelligent Systems and Computing 542,
DOI 10.1007/978-981-10-3223-3_36

depends on the crossover and mutation rate, selection method used [4–6]. Similarly, cooling rate is the key for the success of the Simulated Annealing (SA) [7].

The key challenge with any global optimization algorithm is the premature convergence and slow finishing, which occurs due to lack of diversity [8, 9]. The effective parameters setting is one of the ways to alleviate the premature convergence [10]. Also, there exist several techniques were proposed for the quantification of the parameters [10]. But, parameters tuning for any algorithms are still a matter of concern as it consumes most of the development time. In the past years, few optimization algorithms such as Teaching Learning Based Optimization (TLBO), Jaya algorithm has been developed that doesn't require any algorithm specific parameters/variables. The experimental results of these algorithms are found very effective. Jaya algorithm has been developed recently and shown as a powerful algorithm. The primary objective of this paper is to classify the computational capability of four global optimization algorithms. The authors have selected GA, SA, Pattern Search (PS) and Jaya algorithms for the comparison. The standard benchmark functions are implemented for each algorithm and results are collected for the analysis. The %success rate is considered as a quality measure that helps to categorize the computational ability of an individual algorithm.

The rest of the paper is organized as follows: Sect. 2 discusses the algorithms selected for the comparison. The standard benchmark functions that have been taken for the implementation are represented in Sect. 3. Discussion of computational experiments, analysis and observations are given in a comprehensive manner in Sect. 4. Lastly, the conclusion is drawn in Sect. 5.

## 2 Algorithms Selected: PS, SA, GA and Jaya Algorithm

This section sheds light on the basics of the algorithms that are selected for the comparison. The authors have selected 4 global optimization algorithms: PS, SA, GA and Jaya algorithm. PS, SA and GA are well know algorithm and has been utilized to solve different optimization problems. On the other hand, the Jaya algorithm is new optimization algorithm and it doesn't require any algorithm specific parameters.

PS is a numerical method based approach, which is used for optimization of functions (non-continuous and non-differentiable) [11, 12]. The working of the PS can be summarized as follows: "*They varied one theoretical parameter at a time by steps of the same magnitude, and when no such increase or decrease in any one parameter further improved the fit to the experimental data, they halved the step size and repeated the process until the steps were deemed sufficiently small.*"

SA is a meta-heuristic approach of finding the global optimum of a function [7, 13]. It is mostly used, when finding an acceptable local optimal result is more important than finding a precise global optimal solution in a given time frame.

GA is a stochastic approach for solving heuristic problems. It is based on Darwin's theory "*survival of the fittest*". It uses biological evolutionary mechanisms

**Table 1** Standard benchmark function adapted

| S. N | Function | Formulation | D | SR | C |
|---|---|---|---|---|---|
| F01 | Ackley function | $f(x,y) = -20\exp\left(-0.2\sqrt{0.5(x^2+y^2)}\right) - \exp(0.5(\cos(2\pi x) + \cos(2\pi y))) + e + 20$ | 30 | [–32, 32] | MN |
| F02 | Beale function | $f(x,y) = (1.5 - x + xy)^2 + (2.25 - x + xy^2)^2 + (2.625 - x + xy^3)^2$ | 5 | [–4.5, 4.5] | UN |
| F03 | Booth function | $f(\mathrm{x}) = (x_1 + 2x_2 - 7)^2 + (2x_1 + x_2 - 5)^2$ | 2 | [–10, 10] | MS |
| F04 | Goldstein-Price function | $f(x,y) = (1 + (x+y+1)^2(19 - 14x + 3x^2 - 14y + 6xy + 3y^2)) (30 + (2x-3y)^2(18 - 32x + 12x^2 + 48y - 36xy + 27y^2))$ | 2 | [–2, 2] | MN |
| F05 | Griewank function | $f(\mathrm{x}) = \sum_{i=1}^{d} \frac{x_i^2}{4000} - \prod_{i=1}^{d} \cos\left(\frac{x_i}{\sqrt{i}}\right) + 1$ | 30 | [–600, 600] | MN |
| F06 | Rastrigin function | $f(\mathrm{x}) = An + \sum_{i=1}^{n} \left[x_i^2 - A\cos(2\pi x_i)\right]$ | 30 | [–5.12, 5.12] | MS |
| F07 | Rosenbrock function | $f(\mathrm{x}) = \sum_{i=1}^{n-1} \left[100(x_{i+1} - x_i^2)^2 + (x_i - 1)^2\right]$ | 30 | [–30, 30] | UN |
| F08 | Schwefel function | $f(\mathrm{x}) = 418.9829d - \sum_{i=1}^{d} x_i \sin\left(\sqrt{\lvert x_i \rvert}\right)$ | 30 | [–100, 100] | UN |
| F09 | Sphere function | $f(x) = \sum_{i=1}^{d} x_i^2$ | 30 | [–100, 100] | US |
| F10 | Trid function | $f(\mathrm{x}) = \sum_{i=1}^{d} (x_i - 1)^2 - \sum_{i=2}^{d} x_i x_{i-1}$ | 10 | [–$d^2$, $d^2$] | UN |

like mutation, crossover, selection as a key operation. The GA learns with its experiences and adapt to change according to the environment. It is useful for all those problems in which there is not enough information to build a differentiable function or where the problem has such a complex structure.

Jaya algorithm is a novel approach developed by Ventaka Rao. It is a Sanskrit word, which means Victory. Jaya is a powerful algorithm is applied to optimization

**Table 2** The best, average and worst results obtained by PS, SA, GA and Jaya algorithm

| Function | Factor | PS | SA | GA | JAYA |
|---|---|---|---|---|---|
| F01 | B | 8.88E-16 | 8.88178E-16 | 1.339741874 | 0.018932 |
| | M | 8.88E-16 | 7.448557816 | 6.95397 | 0.1936591 |
| | W | 8.88E-16 | 17.29329434 | | 0.975111 |
| F02 | B | 8.80E-09 | 1.98245E-05 | 0.45013898 | 0.5406757 |
| | M | 1.46 | 0.285830097 | 2062351463 | 5369952106 |
| | W | 2.508757794 | 0.61493473 | | 1.04577E + 11 |
| F03 | B | 0 | 6.07238E-07 | 0.315841106 | 0.2215424 |
| | M | 6.11E-11 | 0.000110246 | 523.2587633 | 621.8396674 |
| | W | 4.66E-10 | 6.03E-04 | | 6232.3199 |
| FO4 | B | 30 | 3.000000002 | 96.81774047 | 5.0577859 |
| | M | 389196.6698 | 11.1000022 | 8.61506E+14 | 6.33642E+11 |
| | W | 3844424.658 | 84.00002049 | | 1.79891E+13 |
| F05 | B | 0 | 0 | 0.011581455 | 0.0062252 |
| | M | 0.086551408 | 0.095959437 | 0.46744774 | 0.0321846 |
| | W | 0.421653114 | 0.233033947 | | 0.0861527 |
| F06 | B | 0 | 0 | 2.376470556 | 0.0008155 |
| | M | 0 | 0.577971116 | 134.4679 | 0.0624826 |
| | W | 0 | 1.989920381 | | 0.5022965 |
| F07 | B | 0.06804122 | 0.000887016 | 8.296886837 | 0.0067647 |
| | M | 17.48571242 | 8.314071034 | 5275997.691 | 1.617736 |
| | W | 80.942906 | 66.79150398 | | 15.5644245 |
| F08 | B | –592.1827523 | 513.2464588 | 726.5836864 | 789.8026336 |
| | M | 687.8494018 | 590.247288 | 737.8034 | 789.9504101 |
| | W | 830.0751967 | 651.4546808 | | 790.3974054 |
| F09 | B | 0 | 0 | 0.040728283 | 0.0000069 |
| | M | 0 | 2.47862E-05 | 67.81484456 | 0.0002203 |
| | W | 0 | 2.12E-04 | | 0.0014298 |
| F10 | B | –2 | –1.999961706 | –1.856691669 | –1.9839064 |
| | M | –2 | –1.599995728 | 103.5373549 | 17.8760654 |
| | W | –2 | –1.999999947 | | 227.4628934 |

B: Best result, A: Average result, W: Worst result

of constrained and unconstrained problems. It is similar to TLBO algorithm, unlike TLBO, the Jaya only has one phase. Comparing to the other algorithm, it is easier to apply. Unlike the other algorithms, the Jaya is a parameter less algorithm, which requires only some common parameters for its implementation. When one applies the Jaya algorithm for any problem, then the solution that we obtained is always to move towards the best solution and moving away from the worst solution.

## 3 The Standard Benchmark Function Used

This section highlights the standard benchmark functions that are selected for the computational experiments [14–16]. Table 1 represents the benchmark functions with their formulation, dimension (D), search range (SR) and characteristic (U: Unimodal, M: Multimodal, S: Separable, N: Non-separable).

## 4 Computational Experiment, Analysis and Observations

Extensive experiments are conducted to evaluate the performance of the algorithms. The standard benchmark functions shown in Table 1 are implemented applying each algorithm.

The following system configurations have been utilized MATLAB 2015a, Intel ® Core™ i7-3632 QM, 2.20 GHz, x64 based processor, RAM-8 GB. Each algorithm is implemented 10 times for the benchmark function shown in Table 1. The results are collected successfully and presented in an organized manner. Table 2 represents the best, average and worst results obtained experimentally for each algorithm over 10 runs. Table 2 demonstrates that the performance of the PS is worse since it doesn't produce the minimized results rather it gives zero (F06 and F09). Also, the

**Table 3** The percentage success rate for PS, SA, GA and Jaya algorithm over 100 iterations

| Functions | Algorithms (% success rate) | | | |
|---|---|---|---|---|
| | PS | SA | GA | JAYA |
| F01 | 100 | 90 | 100 | 100 |
| F02 | 0 | 100 | 100 | 100 |
| F03 | 100 | 100 | 90 | 100 |
| F04 | 40 | 100 | 80 | 100 |
| F05 | 100 | 100 | 90 | 100 |
| F06 | 100 | 100 | 100 | 100 |
| F07 | 0 | 100 | 90 | 100 |
| F08 | 100 | 100 | 100 | 100 |
| F09 | 100 | 100 | 90 | 100 |
| F10 | 100 | 100 | 100 | 100 |

**Table 4** Mean (M) and Standard Deviation (S.D.) for PS, SA, GA and Jaya algorithm

| Functions | Factors | Algorithms | | | |
|---|---|---|---|---|---|
| | | PS | SA | GA | JAYA |
| F01 | M | 8.88E-16 | 7.44855782 | 6.95397 | 0.1936591 |
| | S.D | 0 | 5.82311076 | 1.737279108 | 0.210541 |
| F02 | M | 1.46 | 0.2858301 | 2062351463 | 5369952106 |
| | S.D | 1.6532251 | 0.25253311 | 3152914609 | 21095473728 |
| F03 | M | 6.1118E-11 | 0.00011025 | 523.2587633 | 621.8396674 |
| | S.D | 1.4288E-10 | 0.00020797 | 487.6083258 | 1250.108898 |
| FO4 | M | 389196.67 | 11.1000022 | 8.61506E+14 | 6.33642E+11 |
| | S.D | 1214082.76 | 25.6144555 | 2.69483E+15 | 3.29046E+12 |
| F05 | M | 0.08655141 | 0.09595944 | 0.46744774 | 0.0321846 |
| | S.D | 0.14607016 | 0.07408978 | 0.168564575 | 0.0191562 |
| F06 | M | 0 | 0.57797112 | 134.4679 | 0.0624826 |
| | S.D | 0 | 0.83112218 | 104.445407 | 0.1056152 |
| F07 | M | 17.4857124 | 8.31407103 | 5275997.691 | 1.617736 |
| | S.D | 27.2150038 | 21.00808 | 5188043.133 | 3.4173247 |
| F08 | M | 687.849402 | 590.247288 | 737.8034 | 789.9504101 |
| | S.D | 449.757454 | 60.6408758 | 60.29352953 | 0.1492436 |
| F09 | M | 0 | 2.4786E-05 | 67.81484456 | 0.0002203 |
| | S.D | 0 | 6.6503E-05 | 71.91079794 | 0.0003148 |
| F10 | M | −2 | −1.59999573 | 103.5373549 | 17.8760654 |
| | S.D | 0 | 1.26490712 | 105.7701684 | 50.2512172 |

important thing to note here is: SA also produces zero for some of the functions (F05, F06 and F09), which indicates that the SA is also found incapable in finding an optimum value for F05, F06 and F09. On the other hand, it can be seen from Table 2 that GA and Jaya algorithms are able to obtain the global optimum. The percent success rate (number of times the algorithm has successfully been achieved the global optimum) is calculated for each algorithm. Table 3 presents the percent success rate for each algorithm over 100 iterations.

Table 3 indicates that for some of the functions F02 and F07 the PS algorithm is found incapable to find the global solution. The SA is showing effective performance (100%) for all the function except the F01. The GA's performance is also good, but it showed the 100% success rate only for F01, F02, F06, F08 and F10. On the other hand, the Jaya algorithm is found very effective as it produces 100% success rate for each function. The mean and standard deviation for each algorithm is determined that demonstrate the superiority of the Jaya algorithm. Table 4 represents the mean and standard deviation of the algorithms. Again, it can be seen that the performance of the Jaya is superior that SA, whereas the PS is found worse.

## 5 Conclusions

This paper presents the comparative analysis of the 4 global optimization algorithms: PS, SA, GA and Jaya algorithm. The authors have used the standard benchmark functions for the implementation. The experimental results conclude that the Jaya algorithm is a powerful global optimization algorithm and has a greater computational ability to solve the complex optimization problems. The results also conclude that GA and SA both are also an effective approach to solve an optimization problem. It has been observed from the experimental results reported in Tables 2, 3 and 4 that PS algorithm is a worse algorithm and has very poor computational ability. The results shown in Table 2 indicate that the PS algorithm has shown the tendency of not reaching to the best solution for many benchmark functions. The results reported in this paper might be useful for those who are keen to identify a suitable algorithm for their own research. In addition, the results reported in this paper will motivate more research in this direction.

## References


1. Holland, J.H.: Genetic algorithms. Sci. Am. **267**(1), 66–72 (1992)
2. Goldberg, D.E., Richardson, J.: Genetic algorithms with sharing for multimodal function optimization. Genetic algorithms and their applications. In: Proceedings of the Second International Conference on Genetic Algorithms. Hillsdale. Lawrence Erlbaum, NJ (1987)
3. Krause, J. et al.: A survey of swarm algorithms applied to discrete optimization problems. In: Swarm Intelligence and Bio-inspired Computation: Theory and Applications. Elsevier Science & Technology Books, pp. 169–191 (2013)
4. Shukla, A., Pandey, H.M., Mehrotra. D.: Comparative review of selection techniques in genetic algorithm. In: 2015 International Conference on Futuristic Trends on Computational Analysis and Knowledge Management (ABLAZE), IEEE (2015)
5. Pandey, H.M.: Performance evaluation of selection methods of genetic algorithm and network security concerns. Proc. Comput. Sci. **78**, 13–18(2016)
6. Pandey, H.M. et al.: Evaluation of genetic algorithm's selection methods. In: Information Systems Design and Intelligent Applications. Springer India, pp. 731–738 (2016)
7. Aarts, E., Korst, J.: Simulated Annealing and Boltzmann Machines (1988)
8. Pandey, H.M., Chaudhary, A., Mehrotra, D.: A comparative review of approaches to prevent premature convergence in GA. Appl. Soft Comput. **24**, 1047–1077 (2014)
9. Pandey, H.M., Dixit, A., Mehrotra, D.: Genetic algorithms: concepts, issues and a case study of grammar induction. In: Proceedings of the CUBE International Information Technology Conference. ACM (2012)
10. Pandey, H.M.: Parameters quantification of genetic algorithm. In: Information Systems Design and Intelligent Applications. Springer India, pp. 711–719 (2016)
11. Lewis, R.M., Torczon, V.: A globally convergent augmented Lagrangian pattern search algorithm for optimization with general constraints and simple bounds. SIAM J. Optim. **12**(4), 1075–1089 (2002)
12. Yin, S., Cagan, J.: An extended pattern search algorithm for three-dimensional component layout. J. Mech. Des. **122**(1), 102–108 (2000)
13. Hwang, C.-R.: Simulated annealing: theory and applications. Acta Applicandae Mathematicae **12**(1), 108–111 (1988)

14. Rao, R.: Jaya: A simple and new optimization algorithm for solving constrained and unconstrained optimization problems. Int. J. Ind. Eng. Comput. **7**(1), 19–34 (2016)
15. Karaboga, D., Basturk, B.: A powerful and efficient algorithm for numerical function optimization: artificial bee colony (ABC) algorithm. J. Global Optim. **39**(3), 459–471 (2007)
16. Houck, C.R., Joines, J., Kay, M.G.: A genetic algorithm for function optimization: a Matlab implementation. NCSU-IE TR 95.09 (1995)

# Maven Video Repository: A Visually Classified and Tagged Video Repository

**Prashast Sahay, Ijya Chugh, Ridhima Gupta and Rishi Kumar**

**Abstract** WEB 2.0's accelerated growth has paved a way for the emergence of social video sharing platforms. These video sharing communities produce videos at an exponential rate. Unfortunately, these videos are incongruously tagged, leading to minimal amount of metadata to retrieve them. Categorizing and indexing these videos has become a pressing problem for these communities. Videos generated by these communities depend on users to tag them, thus they end up being loosely tagged. An innovative and novel application has been presented to classify and tag these large volumes of user-generated videos. The above proposed content-based automatic tagging application tags the videos, which further help in indexing and classifying them. This application first recognizes the person in the video and then discerns their emotions and then creating a MPEG-7 xml file to store the metadata. This application will drastically reduce human effort and radically increase the efficiency of video searching.

**Keywords** Video classification · Automatic video tagging · Face recognition in videos · Emotion recognition

P. Sahay (✉) · I. Chugh · R. Gupta · R. Kumar
Computer Science Department, Amity University, Noida, Uttar Pradesh, India
e-mail: pras1211@gmail.com

I. Chugh
e-mail: ijyac.06@gmail.com

R. Gupta
e-mail: rdhmgupta@gmail.com

R. Kumar
e-mail: rishikumar182000@gmail.com



S.C. Satapathy et al. (eds.), *Data Engineering and Intelligent Computing*,
Advances in Intelligent Systems and Computing 542,
DOI 10.1007/978-981-10-3223-3_37

# 1 Introduction

Unprecedented growth of multimedia-sharing communities has led to exponential boom of videos. An analysis conducted by Google states that 300 h of videos are uploaded online every minute.

Videos presently are being tagged by users. This process of tagging or adding meta-data (data about data) is cumbersome and time consuming. People are generally indolent to add desirable information about the video thus leading to poor information about the video. Video searches are dependent upon this information provided about the videos, to retrieve them when appropriate keywords are used to search them. Henceforth when this information provided is poor or less than adequate, classification or retrieval of videos becomes difficult [1].

'Folksonomy' is a term coined for tagging or associating online documents with specific keywords to help finding and re-finding these documents later on. This keyword-document association is called tagging. It is a kind of media-knowledge extraction. These tags associated with the documents such as images and videos provide cumulative information about the document.

Henceforth tagging videos with appropriate metadata has become need of hour to ease the process of video retrieval. Through this paper we combat this problem by use of automatic tag generating application. This application first detects the key-frames, followed by face recognition and emotion recognition. Resulting data from this process is compiled to form an MPEG-7 Meta-data file which is saved in an XML database like eXist DB. Thus, tags formed after these processes are like name of a recognized person, or a recognized emotion and so the videos can be tagged using these keywords. Hence, keywords can be used to classify videos and help in easy retrieval of videos [5].

# 2 Contemporary Works

## *2.1 Key Frame Detection*

Videos are segmented into temporal units called shots. A video comprises of scenes and scenes are made of shots. Each shot consists of 'n' number of frames. To detect key frames from each shot most widely used method is:-

Shot Boundary Detection: It's a Histogram based technique which uses Histogram Differencing to detect shot boundaries. Thereafter a threshold value is calculated using mean and standard deviation of n successive frames. This threshold value is used to obtain the Key Frame in a shot.

## 2.2 Face Recognition in Videos (FRiV)

Face detection and face recognition are parallel processes. FRiV involves a group of following steps:

(1) **Pre-processing**: The key framed extracted is accepted and processed into binary image by a process Binarization. After this, key frame is filtered using the techniques explained in [8].

(2) **Feature Extraction and Dimensionality reduction:** Feature Extraction is process of extracting features from the Binary image that will best help in solving the problem. Ex: In case of face recognition Facial features like skin colour, distance between eyes, and width of nose are useful. 80 such face nodal points can be extracted as features.

Some of the most widely used techniques for Feature Extraction and Dimensionality Reduction are:-

1. *Principal Component Analysis (PCA)*: It was proposed by Turk and Pentland [4] and makes use of Eigen values to choose principal component. The largest Eigen Value is chosen to re-orient features in the direction of maximum variance. This helps in recognizing people. It is robust and works accurately in unsupervised environment.
2. *Linear Discriminant Analysis (LDA)*: It generally works better than PCA as it never neglects important features as in case of PCA. It tries to decrease intra-class separability and increases inter-class separability which in turn helps in easy classification when training with large data set and size of class is large [10].
3. *Classifiers and Cross-Validation*: Various classifiers like Support vector Machines (SVM) and K-nearest neighbour are used in conjunction with Euclidean Distance. Leave out One cross is used to cross validate the trained models [3].

## 2.3 Video Annotation

Appropriate tagging of labels to a video helps improve its indexing, classification, and searching capabilities.

There are two sets of Tagging, one of them is "Open-set Tagging", which involves recognition of keywords, phrases, sentences and assigning them to the content. The other set has pool of tags and tags to be associated must be chosen from the pool of tags [2]. There is another term called "Geo-tagging" it basically tags documents like images and videos to the place where they were originally taken from.

To store video annotation as data, there is a requirement of a hierarchical structure because fundamental Wavelet Based Techniques work layer-wise. VideoML is an XML based Mark-up Language which helps in storing video annotation data in a hierarchical order [9]. MPEG-7 uses Description Definition language which follows a Description Scheme with smallest unit called "Descriptor".

The Descriptors have a schematic structure called Description Scheme and a field which is known as Descriptor Values that contains the Data/Tag to be annotated describing the particular Frame/Scene. Description Definition Language is developed on similar concepts as of XML language. It provides us the freedom to create and modify description schemas.

## 3 Existing Systems

Contemporary systems involves humungous amount of manual labour to label videos appropriately. The other applications and system which try to tag these videos have a large number of drawbacks which were recognized and the proposed system tries to address these drawbacks and makes an effort to combat these drawbacks.

*Drawbacks:-*

(1) Cumbersome process of tagging each and every video.
(2) Inefficient Video retrieval process.
(3) Unable to retrieve desired result.
(4) Videos are often classified under wrong titles.

## 4 Implemented System

Intensive research and analysis was done on the existing systems and to combat the drawbacks and with a motive to automate the process of Video Tagging, an innovative and novel method has been proposed in the paper [13]. Maven Video Repository is the implementation of the method proposed in this paper.

Maven Video Repository has three core modules namely

(1) Face Recognition
(2) Emotion recognition
(3) MPEG-7 XML File

M.V.R also has several sub-modules that require in-depth knowledge regarding alternative disciplines like classifiers, cross validations, training data, XML database, feasibility, performance and standardisation.

## 4.1 Implementation of System

(1) **Building Face Recognition Model**

(a) *Training Data*

Training Data is one of the most important parts of any supervised learning task. For the purpose of this project, we created our own Dataset, inspired by the AT&T image Dataset. It contains a main folder, which contains sub-folders with each person's name. Each sub-folder contains 10 images of that person in gray scale with equal dimensions. This becomes the dataset for the system [7].

(b) *Predicting Model*

It compromises of feature extraction algorithms like PCA, LDA, Local binary Patterns and classifiers like K-Nearest Neighbour, SVM. In our Prediction model we use LDA which is better than PCA and other feature extraction algorithms at low cost of computation. For classification, K-NN is used with K = 5 and distance metric as Euclidean distance.

(c) *Cross Validation*

To estimate the Precision level and accuracy we have used K-Cross validation (Fig 1).

(2) **Emotion recognition Model**

To recognize emotions we have used precompiled and trained dataset of 1000 images which contain 68 different facial poses or face Landmarks [3]. In addition to it, for feature extraction we have used PCA. It first extracts features from image and then performs PCA transformation. Then using the trained dataset it predicts the appropriate emotion. It is capable to detect 7 emotions namely: Anger, Contempt, Disgust, Fear, Happiness, Sadness, and Surprise [6].

```
Loading dataset...
Validating model with 10 folds...
2016-05-16 23:29:29,358 - facerec.validation.KFoldCrossValidation - INFO - Processing fold 1/9.
2016-05-16 23:29:29,638 - facerec.validation.KFoldCrossValidation - INFO - Processing fold 2/9.
2016-05-16 23:29:29,904 - facerec.validation.KFoldCrossValidation - INFO - Processing fold 3/9.
2016-05-16 23:29:30,170 - facerec.validation.KFoldCrossValidation - INFO - Processing fold 4/9.
2016-05-16 23:29:30,372 - facerec.validation.KFoldCrossValidation - INFO - Processing fold 5/9.
2016-05-16 23:29:30,575 - facerec.validation.KFoldCrossValidation - INFO - Processing fold 6/9.
2016-05-16 23:29:30,746 - facerec.validation.KFoldCrossValidation - INFO - Processing fold 7/9.
2016-05-16 23:29:30,903 - facerec.validation.KFoldCrossValidation - INFO - Processing fold 8/9.
2016-05-16 23:29:31,059 - facerec.validation.KFoldCrossValidation - INFO - Processing fold 9/9.
PredictableModel (feature=Fisherfaces (num_components=4), classifier=NearestNeighbor (k=5, dist_metr
ic=EuclideanDistance))
ValidationResult (Description=ExperimentName, Precision=82.22%, Accuracy=82.22%)
```

**Fig. 1** Cross validation to estimate accuracy of trained model

(3) **MPEG-7 XML Builder**

MPEG-7 XML is a metadata descriptor standard which describes about the video frames. Output from face recognition model and emotion recognition model are fed into an algorithm which gives output a XML based.

# 5 Maven Video Repository Application

## *5.1 Upload Video*

This sub-module is responsible for uploading video either from client's computer to server or provides the YouTube link to download video (Fig. 2).

## *5.2 Video Processing*

This sub-module carries out all the video processing. First of all face recognition module is carried out and the resulting output is displayed on screen (Fig. 3).

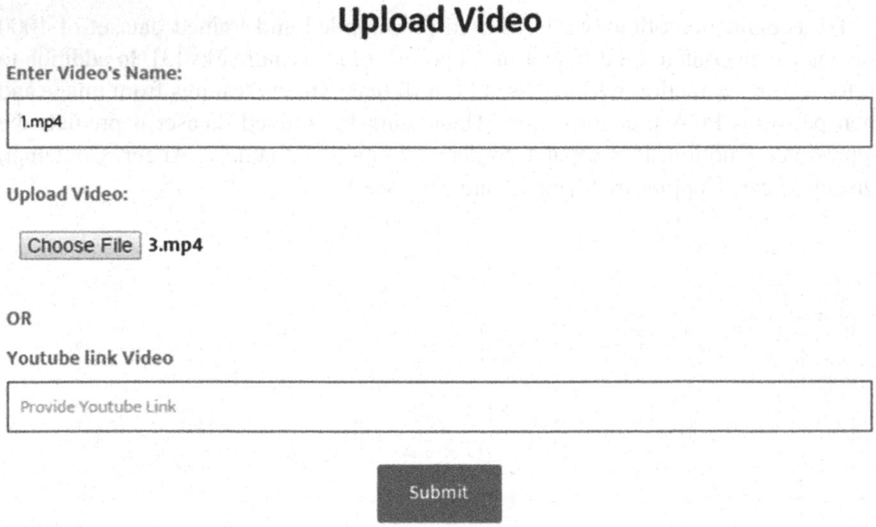

**Fig. 2** Uploading video

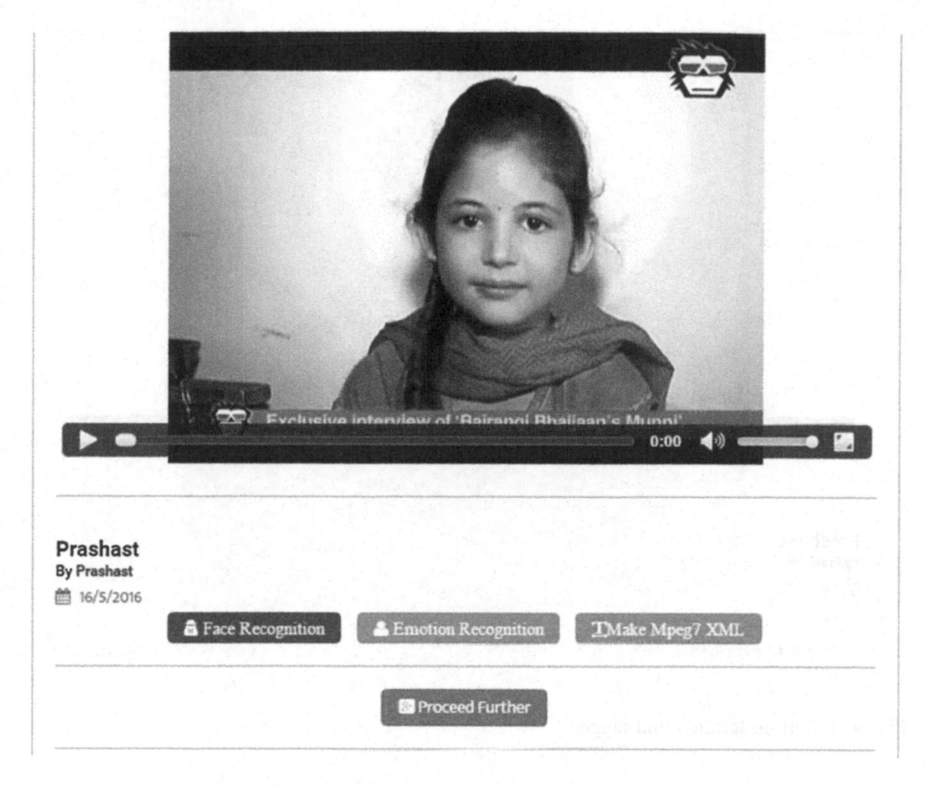


**Fig. 3** The respective video uploaded

Next emotion recognition module is run and followed by MPEG-7 XML generator. At server side XML file is generated and combined with html file to create tags and give information about the particular scene. This also creates tags that are appropriate for the video (Figs. 4 and 5).

## *5.3 Video Search*

It offers to search video by name or by keywords. Searching from keywords result in retrieval of all videos tagged with that keyword (Fig. 6).

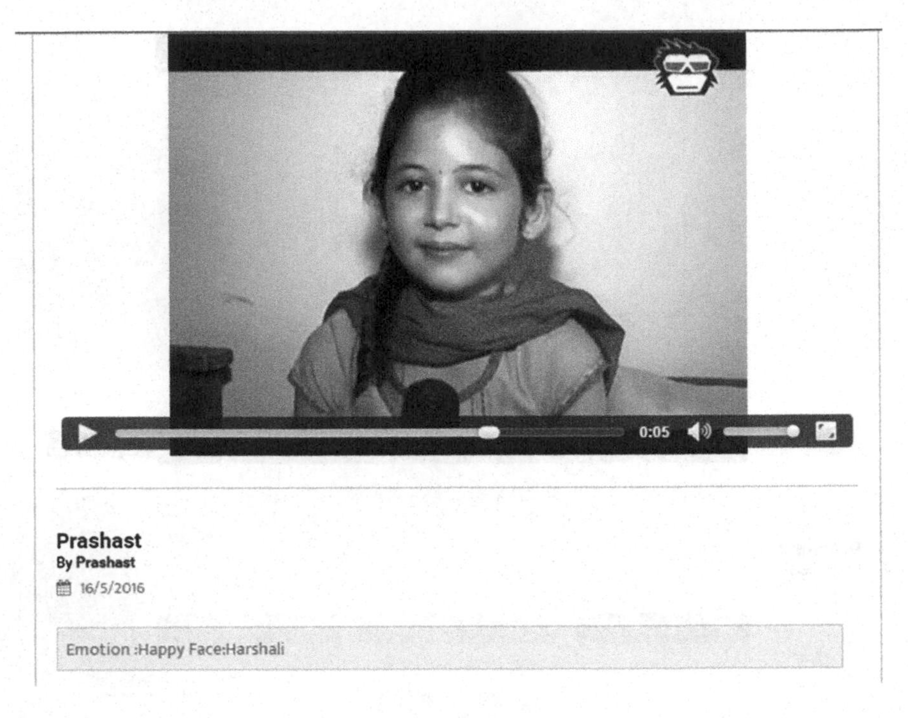


**Fig. 4** Emotion detected and tagged

**Fig. 5** Tags/metadata generated from video

## *5.4 Search Result*

This is the result of the query made. It shows all the video that are associated with the keyword. Ex: happy is queried, so all the video which are tagged happy are retrieved (Fig. 7).

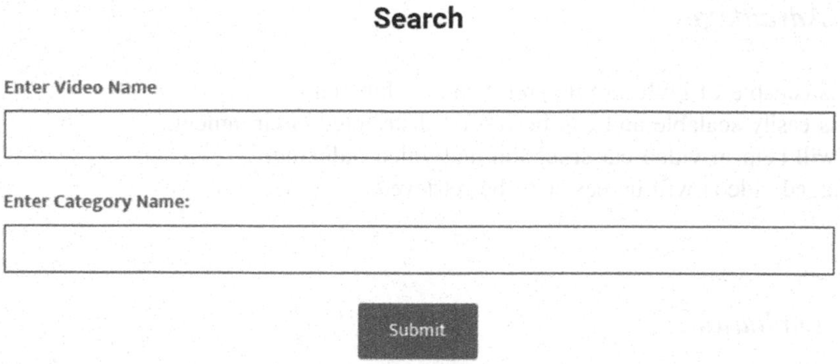

Search

Enter Video Name

Enter Category Name:

Submit

**Fig. 6** Searching through video names

Harshali
By Prashast
16/05/2016
Rate - 
Happy
Harshali is an Indian Child Actress . Dwells in Delhi
Harshali

**Fig. 7** Video retrieval result

# 6 Conclusion

The Maven Video Repository application deals with one of the biggest problem in the field of automation of video tagging. This application will drastically increase the video retrieval capabilities of video sharing communities. It will also help in better video classification making video search more user friendly. Hence The Maven Video Repository will pave the way for an efficacious video tagging system.

## 6.1 Advantages

1. It is capable of both mobile and web development.
2. It is easily scalable and can be run on distributed environment.
3. It will help in video classification and video indexing.
4. Tagged videos will be easier to be retrieved.

## 6.2 Limitations

1. Application has 70% precision and is prone to misjudge correct person or emotion.
2. Large videos will be slow in processing as face recognition and emotion recognition consumes a lot of computer resources.

# 7 Future Scope

The system proposed above has wide scope for future application, which will additionally reduce human labour and increase efficiency in domain of video indexing.

Application can be made more powerful by adding following modules:-

A. Recognition of Demo graphs: Recognize gender; age of a person so that metadata of the video increases in case there information is not in the training dataset.
B. Video Surveillance: Face recognition on live feed can be used in surveillance to track criminals. For an example a thief tries to rob a bank but, in mean while surveillance cameras identify the robber and hence he/she is caught easily.
C. Video tagging on Facebook: As Facebook provides image tagging similarly on uploading a video, tags related to video can be prompted.

# References

1. Gill, P., Arlitt, M., Li, Z., Mahanti, A.: Youtube traffic characterization: a view from the edge. In: Proceedings of the 7th ACM SIGCOMM Conference on Internet Measurement, pp. 15–28. ACM (2007)
2. Bloehdorn, S., Petridis, K., Saathoff, C., Simou, N., Tzouvaras, V., Avrithis, Y., Handschuh, S., Kompatsiaris, Y., Staab, S., Strintzis, M.G.: Semantic annotation of images and videos for

multimedia analysis. In: The semantic web: research and applications, pp. 592–607. Springer, Berlin (2005)
3. Zhu, X., Ramanan, D.: Face detection, pose estimation, and landmark localization in the wild. In: 2012 IEEE Conference on Computer Vision and Pattern Recognition (CVPR), pp. 2879–2886. IEEE (2012)
4. Zhang, D., Islam, Md.M., Lu, G.: A review on automatic image annotation techniques. Pattern Recognit. **45**(1), 346–362 (2012)
5. Baştan, M., Cam, H., Güdükbay, U., Ulusoy, Ö.: Bilvideo-7: an MPEG-7-compatible video indexing and retrieval system. IEEE MultiMed. **17**(3), 62–73 (2010)
6. Kliper-Gross, O., Gurovich, Y., Hassner, T., Wolf, L.: Motion interchange patterns for action recognition in unconstrained videos. In: Computer Vision–ECCV 2012, pp. 256–269. Springer, Berlin (2012)
7. Huang, G.B., Jain, V., Learned-Miller, E.: Unsupervised joint alignment of complex images. In: International Conference on Computer Vision (ICCV) (2007)
8. Valstar, M.F., Mehu, M., Jiang, B., Pantic, M., Scherer, K.: Meta-analysis of the irst facial expression recognition challenge. IEEE Trans. Syst. Man Cybern. Part B Cybern. **42**(4), 966–979 (2012)
9. Sahay, P., Kumar, R., Chugh, I., Gupta, R.: Visually classified & tagged video repository. In: 2016 6th International Conference-Cloud System and Big Data Engineering (Confluence), pp. 467–472. IEEE (2016)
10. Martínez, A.M., Kak, A.C.: Pca versus lda. IEEE Trans. Pattern Anal. Mach. Intell. **23**(2), 228–233 (2001)

# A Comprehensive Comparison of Ant Colony and Hybrid Particle Swarm Optimization Algorithms Through Test Case Selection

**Arun Prakash Agrawal and Arvinder Kaur**

**Abstract** The focus of this paper is towards comparing the performance of two metaheuristic algorithms, namely Ant Colony and Hybrid Particle Swarm Optimization. The domain of enquiry in this paper is Test Case Selection, which has a great relevance in software engineering and requires a good treatment for the effective utilization of the software. Extensive experiments are performed using the standard flex object from SIR repository. Experiments are conducted using Matlab, where Execution time and Fault Coverage are considered as quality measure, is reported in this paper which is utilized for the analysis. The underlying motivation of this paper is to create awareness in two aspects: Comparing the performance of metaheuristic algorithms and demonstrating the significance of test case selection in software engineering.

**Keywords** Optimization · Meta-heuristics · Ant colony optimization · Particle swarm optimization · Regression testing

## 1 Introduction

Every software system enters into the maintenance phase after release and keeps evolving continuously to provide the functionality required and to incorporate changing customer needs. Regression testing is an activity that tries to find any bugs introduced during the maintenance phase. One approach is to re-run all the test

A.P. Agrawal (✉)
Department of Computer Science and Engineering, Amity University,
Noida, Uttar Pradesh, India
e-mail: apagrawal@amity.edu

A. Kaur
University School of Information and Communication Technology, GGSIP University,
New Delhi, India
e-mail: arvinderkaurtakkar@yahoo.com



S.C. Satapathy et al. (eds.), *Data Engineering and Intelligent Computing*,
Advances in Intelligent Systems and Computing 542,
DOI 10.1007/978-981-10-3223-3_38

cases available from an earlier version of the software system [1]. But it is often too costly in terms of time and effort required to rerun all the test cases and is practically infeasible [2]. Two approaches have therefore emerged to optimize the regression test suite—Regression Test Case Selection and Prioritization [3]. The focus of this paper is on Regression Test Case Selection for re-execution to reduce the overall effort. Selecting the test cases would give the opportunity to optimize some performance goals like minimize execution time and maximize fault coverage.

The objective of this paper is to use two nature inspired metaheuristic techniques—Ant Colony Optimization and Hybrid Particle Swarm Optimization in testing for the purpose of minimizing the time required in regression testing by selecting the test cases and compare their performance. These two algorithms were empirically evaluated on flex object from the SIR repository and results of both the algorithms are compared on the basis of Fault Coverage and Execution Time. Results indicate that hybrid Particle Swarm Optimization outperforms the Ant Colony Optimization algorithm.

Rest of the paper is organized as follows: Sect. 2 briefly discusses the Regression Testing and states the Regression Test Case selection Problem. Sections 3 and 4 give an overview of the Ant Colony Optimization and Hybrid Particle Swarm Optimization algorithms respectively. Section 5 presents experimental design and results obtained results. Section 6 discusses the results. Finally Conclusion and Future work conclude the paper.

## 2 Problem Definition

### 2.1 Regression Testing

Primary purpose of Regression Testing is to increase the performance of software systems in terms of productivity and efficiency and to ensure quality assurance of applications [4]. The main objective of regression testing is to assure that the modification in one part of the software does not adversely affect the other parts of the software [5].

### 2.2 Regression Test Case Selection

Regression Test Case Selection is choosing a subset of test cases from the existing set of test cases that are necessary to validate the modified software. It reduces the testing cost by reducing the number of test cases to test the modified program [6, 7].

We consider following two criteria for the selection of test cases:

### 2.2.1 Fault Coverage

The aim is to maximize the Fault coverage. Let T = {$T_I$, $T_2$, …, Tn} indicates a test suite and F = {$F_1$,$F_2$, …,Fm} indicates faults covered by a test case [8]. F(Ti) indicates a function which returns the subset of faults covered by a test case. Then, the fault coverage of a test suite is given by

$$\text{Fault Coverage} = 100 * \sum_{t=1}^{s} \bigcup_{t=1}^{s} \{F(Ti)\}/k \tag{1}$$

where s reflects the selected test cases and k is the total number of selected test cases.

### 2.2.2 Execution Time

The execution time should be minimized. Execution time is defined as the total time required to execute a test [9]. The total Execution time of selected test cases is given by

$$\sum_{t=1}^{s} Time \tag{2}$$

where s = selected test cases.

However test case selection could reduce the rate of detection of faults because the effectiveness of a test suite could decrease with the decrease in the size of a test suite.

## 3 Ant Colony Optimization for Optimizing Test Suites

Ants demonstrate complexity in their social behavior which has been attracting the attention of humans since a long time. Apparently the most marked behavior is the formation of ant streets. The most astonishing behavioral pattern demonstrated by ants is that they are able to find the shortest path from its nest to food source. It is proven that this behavior is the result of communication by a hormone called pheromone, an odorous chemical substance ants deposit and smell [10]. Computer scientists are simulating real ants in artificial ants and one such example is ant colony optimization.

**Description of Algorithm**

1. Test Cases are treated as a graph.
2. Each node on the graph has a specific score.
3. All ants start from the same place i.e. first node.
4. Ants deposit pheromone on the paths i.e. edges between two nodes.
5. Ants move on any node within 20 nodes from node number.
6. Ant follows route with $[\tau]*[\eta]$ probability.
7. $\eta$ = (Number of faults)/(execution_time).
8. $\tau = \tau + \tau_0 * \rho$.
9. where $\rho = 0.9$.
10. $\tau_0 = 1/Cnn$; where Cnn is distance between current and next node.
11. Ants can only move forward.
12. Pheromone is appended on the paths.

## 4 Hybrid Particle Swarm Optimization for Test Case Selection

Particle Swarm Optimization is a population based memory using optimization technique. In PSO, each particle has its own position vector and velocity vector [11]. Position vector represents a possible solution to a problem and also a velocity vector. In testing, position means rank assigned to a test case in a test suite, whereas the velocity is the coverage or the execution time taken by a test case. Every particle stores its best position seen so far and also the best global position obtained through interaction with its neighbor particles. Particle Swarm Optimization algorithm guides its search by adjusting the velocity vector and Position of particles. Objective function is responsible for moving the particles. Particles which are very far from the optimal solution have higher velocity as compared to the particles that are very near to the optimal solution. Many variants of PSO with Genetic Algorithm, Gravitational Search Algorithm, Krill Herd and Cuckoo Search Algorithm etc. has been used by previous researchers to solve many optimization problems like frame selection in image processing, aircraft landing, feature selection in software defect prediction, quadratic assignment problem etc. [11–19].

This paper presents Hybrid PSO with Genetic Algorithm for test case selection. The algorithm uses Crossover and Mutation operator of genetic Algorithm and uses the Velocity Vector of PSO to guide the particles to Global Best Solution.

**Description of Algorithm**

1. Read Test cases from the text file and choose them randomly from the test suite.
2. Read execution time and statements covered by each test case.
3. Set MAX_VEL, MIN_VEL, MAX_LEN, MIN_LEN, TOTAL_PARTICLES and MAX_ITER.

4. Initialize each particle with zero Velocity and random Length. Each Particle contains set of Test Cases.
5. for i = 0 to MAX_ITER.

    i. Evaluate Fitness of Each Particle.
    ii. Use Bubble sort to sort Particles in the ascending order of their fitness level.
    iii. Set Velocity of Each Particle.
    iv. Update Each Particle on the basis of the changes required (Particle with bad fitness level have higher changes).
    v. Apply Mutation with 1, 2 and 5% rate on each particle.
    vi. Determine Global and Local Optimal Values.

# 5 Experimental Design and Results

Our Empirical study addresses the following research questions:

**Table 1** Results of ant colony optimization

| Selected test cases | Execution time of selected test cases | Exccution time for program | Fault coverage |
|---|---|---|---|
| T1, T3, T12, T17, T19, T21, T37, T54, T69, T77, T78, T85, T89, T106, T125, T129, T146, T153, T168, T178, T192, T199, T217, T236, T237, T243, T247, T266, TT267, T271, T289, T304, T314, T325, T343, T358, T368, T375, T379, T398, T416, T433, T438, T441, T459, T467, T477, T487, T488, T495, T502, T519, T523, T539, T554, T559, T563, T564, T566, T567 | 14.48 | 0.35 | 16 |

**Table 2** Result of Hybrid PSO at mutation probability 1, 2 and 5%

| Selected test cases | Execution time of selected test cases | Execution time for program | Fault coverage |
|---|---|---|---|
| Mutation probability 1% | | | |
| T303, T258, T164 | 0.872175 | 30.81 | 16 |
| Mutation probability 2% | | | |
| T553, T308, T258, T414 | 1.07353 | 33.455 | 16 |
| Mutation probability 5% | | | |
| T12, T412, T118, T308 | 1.06982 | 33.66 | 16 |

**Table 3** Combined Results of ACO and Hybrid PSO

| Algorithm | Mutation probability | No. of test cases selected | Total execution time of selected test cases | Total no. of faults covered | Total no of test cases | Total execution time of all test cases | Total no. of faults | %age of test cases selected | %age of faults covered | Execution time of Algo. | %age of time required |
|---|---|---|---|---|---|---|---|---|---|---|---|
| ACO | N/A | 60 | 14.48 | 16 | 567 | 170.4 | 19 | 10.58 | 84.21 | 0.35 | 8.7 |
| Hybrid PSO | 1% | 3 | 0.87 | 16 | 567 | 170.4 | 19 | 0.52 | 84.21 | 30.81 | 18.5 |
| Hybrid PSO | 2% | 4 | 1.07 | 16 | 567 | 170.4 | 19 | 0.70 | 84.21 | 33.45 | 20.2 |
| Hybrid PSO | 5% | 4 | 1.06 | 16 | 567 | 170.4 | 19 | 0.70 | 84.21 | 33.66 | 20.3 |

- *RQ1: Which algorithm is optimal for Regression Test Case selection?*
- *RQ2: How much saving can we achieve by Regression Test Case selection for the benchmark problem?*

We have used Flex—Fast Lexical Analyzer as the object under test which is available in open source from Software Artifact Infrastructure Repository [20]. It consists of 567 test cases and 19 seeded faults and execution time of each test case. Test cases have been numbered from T1 to T567 and faults have been numbered from f1 to f19. We ran both the algorithm on the input data set and collected results. We have considered bi-objective optimization as we want to maximize the fault coverage and minimize the execution time.

Table 1 above displays the results obtained upon execution of Ant Colony Optimization Algorithm on the input data set.

Table 2 above display the best results obtained upon execution of Hybrid Particle Swarm Optimization algorithm for particle size 30 and mutation probabilities of 1%, 2% and 5% respectively. Particle size was kept constant at 30 in each run and algorithm was executed 30 times for 500 numbers of iterations to avoid any biases.

## 6 Discussion on Results

Table 3 above displays the combined results of both the algorithms. It can easily be seen from the statistics below that the results are quite satisfactory. In case of ACO less than 11% of test cases have been able to detect 84.2% of the faults and requires only 8.7% of total time including the running time of the algorithm. Hybrid PSO algorithm has been run for 30 particles and for 500 iterations in each run and the best of 30 runs has been taken as the observed result as follows for Mutation Probability 1%, 2% and 5% respectively. Hybrid PSO is able to detect 84.21% faults with only 0.7% of the total test cases which is a very cost efficient in comparison of ACO, though the running time of Hybrid PSO is much more than ACO. Although hundred percent of the faults have not been identified this is still a great achievement since the tradeoff is quite high.

It is evident from the results that both the algorithms are optimal for regression test case selection which answers our research question 1. Hybrid PSO outperforms ACO in number of test cases selected but requires more time to run itself than ACO. We can save around 90% of the execution time through test case selection which answers our research question 2.

## 7 Conclusion and Future Scope

Primary focus of this paper was to present a comprehensive comparison of two meta-heuristic algorithms in the domain of software testing. Extensive experiments were conducted on the benchmark object flex and results obtained answered the research questions in study. This allows future researchers to conduct this kind of study for number of metaheuristic algorithms on number of benchmark problems.

## References

1. Mirarab, S., Akhlaghi, S., Tahvildari, L.: Size-constrained regression test case selection using multi-criteria optimization. IEEE Trans. Softw. Eng. **38**(4), 936–956 (2012)
2. Rothermel, G., Harrold, M.J., Dedhia, J.: Regression test selection for C++ software. Softw. Test. Verif. Reliab. **10**(2), 77–109 (2000)
3. Yoo, S., Harman, M.: Regression testing minimization, selection and prioritization: a survey. Softw. Test. Verif. Reliab. **22**(2), 67–120 (2012)
4. Mao, C.: Built-in regression testing for component-based software systems. In: 31st Annual International on Computer Software and Applications Conference, 2007. COMPSAC 2007, vol. 2, pp. 723–728. IEEE (2007)
5. Ali, A., Nadeem, A., Iqbal, M.Z.Z., Usman, M.: Regression testing based on UML design models. In: 13th Pacific Rim International Symposium on Dependable Computing, 2007. PRDC 2007, pp. 85–88. IEEE (2007)
6. Nagar, R., Kumar, A., Singh, G.P., Kumar, S.: Test case selection and prioritization using cuckoos search algorithm. In: International Conference on Futuristic Trends on Computational Analysis and Knowledge Management (ABLAZE), pp. 283–288, IEEE (2015)
7. Jeffrey, D., Gupta, N.: Experiments with test case prioritization using relevant slices. J. Syst. Softw. **81**(2), 196–221 (2008)
8. Kaur, A., Goyal, S.: A bee colony optimization algorithm for fault coverage based regression test suite prioritization. Int. J. Adv. Sci. Technol. **29**, 17–30 (2011)
9. Kumar, M., Sharma, A., Kumar, R.: An empirical evaluation of a three-tier conduit framework for multifaceted test case classification and selection using fuzzy-ant colony optimisation approach. Softw. Pract. Exp. **45**(7), 949–971 (2015)
10. Dorigo, M., Birattari, M., Stutzle, T.: Ant colony optimization. IEEE Comput. Intell. Mag. **1**(4), 28–39 (2006)
11. Liu, B., Wang, L., Jin, Y.H.: An effective hybrid pso-based algorithm for flow shop scheduling with limited buffers. Comput. Oper. Res. **35**(9), 2791–2806 (2008)
12. Apostolopoulos, T., Vlachos, A.: Application of the firefly algorithm for solving the economic emissions load dispatch problem. Int. J. Comb. (2011)
13. Chang, X., Yi, P., Zhang, Q.: Key frames extraction from human motion capture data based on hybrid particle swarm optimization algorithm. In: Recent Developments in Intelligent Information and Database Systems, pp. 335–342. Springer International Publishing (2016)
14. Wang, G.G., Gandomi, A.H., Alavi, A.H., Deb, S.: A hybrid method based on krill herd and quantum-behaved particle swarm optimization. Neural Comput. Appl. **27**(4), 989–1006 (2016)
15. Girish, B.S.: An efficient hybrid particle swarm optimization algorithm in a rolling horizon framework for the aircraft landing problem. Appl. Soft. Comput. **44**, 200–221 (2016)
16. Cui, G., Qin, L., Liu, S., Wang, Y., Zhang, X., Cao, X.: Modified PSO algorithm for solving planar graph colouring problem. Prog. Nat. Sci. **18**(3), 353–357 (2008)
17. Kakkar, M., Jain, S.: Feature selection in software defect prediction: a comparative study. In: 2016 6th International Conference-Cloud System and Big Data Engineering (Confluence), pp. 658–663. IEEE (2016)
18. Tayarani, N.M.H., Yao, X., Xu, H.: Meta-heuristic algorithms in car engine design: a literature survey. IEEE Trans. Evol. Comput. **19**(5), 609–629 (2015)

19. Agrawal, A.P., Kaur, A.: A comparative analysis of memory using and memory less algorithms for quadratic assignment problem. In: 2014 5th International Conference on Confluence the Next Generation Information Technology Summit (Confluence), pp. 815–820. IEEE (2014)
20. Do, H., Elbaum, S., Rothermel, G.: Supporting controlled experimentation with testing techniques: an infrastructure and its potential impact. Empir. Softw. Eng. **10**(4), 405–435 (2005)

# Parameter Estimation of Software Reliability Model Using Firefly Optimization

**Ankur Choudhary, Anurag Singh Baghel and Om Prakash Sangwan**

**Abstract** This paper, presents an effective parameter estimation technique for software reliability growth models using firefly algorithm. Software failure rate with respect to time has always been a foremost concern in the software industry. Every second organization aims to achieve defect free software products, which makes software reliability prediction a burning research area. Software reliability prediction techniques generally use numerical estimation method for parameter estimation, which is certainly not the best. Local optimization, biasness and model's parameter initialization are some foremost limitation, which eventually suffers the finding of optimal model parameters. Firefly optimization overcomes these limitations and provides optimal solution for parameter estimation of software reliability growth models. Goel Okumoto model and Vtub based fault detection rate model is selected to validate the results. Seven real world datasets were used to compare the proposed technique against Cuckoo search technique and CASRE tool. The results indicate the superiority of proposed approach over existing numerical estimation techniques.

**Keywords** Firefly optimization · Parameter estimation · Software reliability growth model · Metaheuristics

---

A. Choudhary (✉)
Department of Computer Science & Engineering, Amity University,
Noida, Uttar Pradesh, India
e-mail: ankur.tomer@gmail.com

A.S. Baghel
Department of Computer Science & Engineering, School of ICT,
Gautam Buddha University, Greater Noida, India

O.P. Sangwan
Department of Computer Science & Engineering, Guru Jambheshwar University,
Hisar, India



S.C. Satapathy et al. (eds.), *Data Engineering and Intelligent Computing*,
Advances in Intelligent Systems and Computing 542,
DOI 10.1007/978-981-10-3223-3_39

# 1 Introduction

Failure free software development brings various challenges for software reliability engineering. It opens door to optimize already developed models or to develop new reliability prediction models that can predict faults prior to its occurrence. Because the software failures may cause inconvenience all over the place especially in the human involved areas, such as economic loss, damage in home products or may be the loss of life in case of medically related software and so on. This makes software reliability a very important research area. Hence, to achieve failure free software, optimization and development of reliable software products are very essential. Software reliability techniques may help software developers to manage requirements, schedule, and budget.

Software Reliability analysis can be done from at various stages of software development life cycles [1]. Out of many modeling techniques, software testing based software reliability growth model (SRGM) is explored the most [2]. SRGMs further classified into Non-homogenous passion process based models, which is explored a lot [3–6]. Parameter estimation always plays a important role for any mathematical model. The SRGM also requires a good parameter estimation technique for better mapping of unknown model parameters for validation on real failure data sets. The most generalized approach of parameter estimation for SRGM is least square estimation (LSE) and Maximum likelihood estimation (MLE). But due to failure non-linearity in software these approach is not the panacea. These techniques suffers from various limitations such as for likelihood function maximization and error minimization and the existence of derivatives of the evaluation method are required [7, 8]. To overcome these limitations, some authors have proposed nature inspired optimization approach for parameter estimation of SRGM. Literature says that Genetic Algorithm, Particle Swarm Optimization, Simulated Annealing, Cuckoo Search Optimization techniques are used in parameter estimation of SRGM [7–12].

In this paper, firefly algorithm based parameter estimation approach for SRGM is proposed. Firefly algorithm is an efficient nature-inspired optimization algorithm, which is enthused by the flashing patterns and behavior of fireflies. To validate the proposed approach, we performed experiments using seven real-world datasets on Goel–Okumoto (GO model-1979) and Vtub based fault detection rate model (Phem model-2014) using firefly algorithm. The rest of the paper is structured as follows. Basic concepts of GO, Pham Model and existing parameter estimation techniques are discussed in Sect. 2. Section 3 explains the basic concept of Firefly optimization and proposed parameter estimation approach, Sect. 4 validates the experimental results and Sect. 5 concludes of this paper.

# 2 Software Reliability Model and Parameter Estimation Techniques

Software reliability of every software product is a measurable characteristic, which describes the number of failure at any point of time t. As already discussed software reliability can be predict from various phases. Software reliability growth model is one of them which are based on testing phase data [1]. Which can be further categories into Non Homogenous Poisson Process (NHPP) based models. An NHPP model maps the failures distribution of software at any point of time t. We have selected Goel and Okumoto model and Vtub-shaped fault-detection rate model to analyze parameter estimation of software reliability models for this study.

## *2.1 Goel and Okumoto Model*

GO model is well proven NHPP based model. This model makes the following assumptions [1, 13]: Such as, the total failure count observed during infinite time is N. All observed failures are autonomous to each other. The software failure count in Δt time is the proportional to undetected failure $-m(t)$, where $m(t)$ represents the mean value function. The numbers of failures observed between different failure-intervals are not related to each other. The hazard rate is constant and debugging and patching is perfect.

The Mathematical representation of above taken assumptions can be represented as following equation.

$$\frac{d}{dt}m(t) = b(a - m(t)) \tag{2.1}$$

By solving (2.1) at initial condition $m(0) = 0$, the mean value function and failure intensity function at time *t*.

$$m(t) = a(1 - e^{-bt})a > 0, b > 0 \tag{2.2}$$

$$\lambda(t) = abe^{-bt}a > 0, b > 0 \tag{2.3}$$

where the parameter 'a' represent predictable total number of failure in the software in infinite time, where as 'b' is failure detection/failure removal rate. The parameters a, b can be estimated from the historical failure data by using the numerical estimation techniques.

## 2.2 Vtub-Shaped Fault-Detection Rate Model [14]

Vtub based fault detection rate model also follows the category of NHPP based model. Most of the NHPP based models follows the common underlying assumptions of developing and operating environment conditions are same. But it cannot be correct all the time. So vtub based fault detection rate model incorporate that the uncertainty of system fault detection rate per unit time based on operating environment.

$$m(t) = N\left(1 - \left(\frac{\beta}{\beta + a^{t^b} - 1}\right)^{\alpha}\right) \tag{2.4}$$

where N represents the expected faults count present in the software after development and before testing, ‘a’ represent expected total number of failure in the software in infinite time, whereas ‘b’ is time dependent fault-detection rate per unit of time. Parameter $\alpha$ and $\beta$ is generalized probability density function. The parameters a, b, $\alpha$, $\beta$ and N can also be estimated from the experienced historical failure data by using the numerical estimation techniques.

## 2.3 Parameter Estimation Techniques

Any proposed mathematical model is said to be incomplete without parameter estimation. The nonlinearity of software failure makes it more crucial. The aim is to identify a set of parameter which is best fitted to the function and can map the failure data perfectly. But the parameter estimation of nonlinear models turns into an optimization problem. To address this problem numerical estimation such as least square estimation and maximum likelihood estimation techniques are used. Least square estimation (LSE) technique is used for determining the set of parameters having the highest likelihood of being accurate for a given experimental dataset [7]. Moreover LSE technique uses basics of curve fitting to the experimental dataset in order to estimate the unknown parameters [8]. LSE is very easy to apply [11, 13–15]. LSE offers steady outcomes in case of broader data sets and is a most preferred method by the organizational practitioners.

To overcome limitations of numerical estimation and better accuracy, nature inspired optimizations has been applied. In 1995, Minohara and Tohma [12] proposed Genetic algorithm based parameter estimation for SRGMs and found that it is a more suitable technique. In 2008, Zhang et al. [8] used Particle Swarm Optimization (PSO) technique as a new parameter estimation approach for SRGM, but observation says that this approach require high search range and low convergence speed. In 2009, Aljahdali et al. [17] used multi objective genetic algorithm for SRGM. In 2013 AL-Saati et al. [18] proposed a cookoo search based parameter estimation technique. In this proposed technique firefly algorithm is proposed for superior estimation, which is associated with the parameter values of SRGMs and its efficiency is compared to CS based approach and CASRE tool.

## 3 Parameter Estimation Using Firefly Optimization

Firefly Algorithm was proposed by Xin-She Yang in 2007, which is based on the fireflies' behavior and flashing patterns [19]. The FA follows following three ideal rules:

1. Fireflies attracted to other fireflies in spite of their sex because they are unisexual.
2. The attractiveness and brightness is proportional to each other, and they both depend on the distance in between. According to their brightness they follow their movement.
3. The brightness of a firefly is maps with objective function.

The proposed approach is as follows:

- First step is to formulate the objective function of the proposed approach Minimize obj_fun(p) subject to $p_i \in P_i = 1, 2, 3, 4 \ldots N$ Where obj_fun(p) is an objective function, p is the set of each unknown parameter $p_i$, n is the number of unknown parameters $P_i$ is the set of the possible values for each unknown parameter, where $L^{P_i} \le P_i \le U^{P_i}$ and $L^{P_i}$ and $U^{P_i}$ are the lower and upper bounds for each unknown parameter. In our problem a, b are the unknown parameter and n = 2 in case of GO model and a, b, α, β and N are unknown parameter and n = 5 for Vtub based fault detection rate model. The FA parameters are also initialized such as population size N, the number of maximum iteration (MI), or termination criterion, light absorbent coefficient etc.
- Repeat while t < max iteration
  - For i=1 to N
  - For j=1 to N
  - calculate the light intensity $I_i$ at distance $X_i$ by using obj_fun($X_i$)
    - o check if $(I_j > I_i)$
      - ▪ Then move i^th firefly towards the j^th firefly in all d dimensions
    - o end of if statement
  - Attractiveness of firefly varies with distance r via $e^{-\gamma r^2}$
  - Evaluate new solutions and update the light intensity
  - end of j loop
  - end of i loop
  - Rank the fireflies based on their fitness and find the current best solution.

- End of while

– Where as we compute for light variations

$$I(r) = I_0 e^{-\gamma r} \tag{1}$$

– For Attractiveness formulation

$$\beta(r) = \beta_0 e^{-\gamma r^m} \text{ where m} >= 1 \tag{2}$$

– For location moving

$$x_i(t+1) = x_i(t) + \beta\left(x_j(t) - x_i(t)\right) + \alpha\varepsilon_i \tag{3}$$

# 4 Experimental Design and Result Discussions

To analyze the accuracy of the firefly algorithm employed for parameter estimation, testing is conducted. The results are compared with previous proposed technique CS [18] using the same related datasets and CASRE tool. Starting with experiment design first of all set the firefly algorithm parameters. The selection of parameters is done by performing multiple runs of experiment. The iterations were varied from (100–2000) and found that 1000 iteration was quite enough to get the best results. As recommended [16] the range of α is 0.25. α is used to balance the local exploitation without jumping too far. The value of β is chosen 0.20 as suggested in the literature and $\gamma$ is Absorption coefficient which is selected as 1. To compare the proposed techniques parameter of cuckoo search has also been set. The iterations is selected as 1000, no of nests is 50, discovery rate is.25. Table 1 shows seven real

**Table 1** Software reliability failure data set

| Dataset | # of failures | # of weeks/months |
|---|---|---|
| DACS Datasets [13] | | |
| FC2 | 54 | 11 |
| FC6 | 73 | 9 |
| FC40 | 101 | 50 |
| Tandem computers software failure [20] | | |
| DS1 | 100 | 20 |
| DS2 | 176 | 18 |
| Misra's space shuttle software failure data [13] | | |
| STS | 231 | 38 |
| Tohma's software failure data [13] | | |
| TS | 86 | 22 |

**Table 2** Comparison Criteria

| Evaluation parameter | Formula |
|---|---|
| Bias | $\text{Bias} = \frac{\sum_{i=1}^{k} m(t_i)' - m(t_i)}{k}$ |
| Mean square error | $\text{MSE} = \frac{\sum_{i=1}^{k} (m(t_i)' - m(t_i))^2}{k-p}$ |
| Predictive-ratio risk | $\text{PPR} = \sum_{i=1}^{k} \left[\frac{m(t_i)' - m(t_i)}{m(t_i)'}\right]^2$ |
| Theil statistic | $\text{TS} = \frac{\sum_{i=1}^{k} (m(t_i)' - m(t_i))^2}{\sum_{i=1}^{k} m(t_i)'^2}$ |

**Table 3** Comparison results of CS based approach and proposed FA based approach

| Model | Dataset | Comparison | FA | CS |
|---|---|---|---|---|
| GO model | FC2 | MSE | 19.45261 | 29.00014 |
| | | TS | 12.01207 | 14.12895 |
| | FC6 | MSE | 8.779727 | 14.15892 |
| | | TS | 5.596757 | 6.98499 |
| | FC40 | MSE | 29.7217 | 94.75789 |
| | | TS | 7.975733 | 13.41843 |
| | DS1 | MSE | 25.66399 | 95.27767 |
| | | TS | 6.690663 | 12.34884 |
| | DS2 | MSE | 347.0833 | 352.9038 |
| | | TS | 15.53647 | 15.59411 |
| | Tohma | MSE | 5.511223 | 35.96293 |
| | | TS | 3.872172 | 9.806347 |
| | Mishra | MSE | 3407.218 | 139.8348 |
| | | TS | 42.22632 | 9.482933 |

world dataset for results validation. For the comparison criteria are shown in Table 2 The parameter estimation using FA is performed on tool Matlab 2009B. According to the literature optimization techniques does not give guarantee for exact solution so we perform 30 runs with maximum 1000 iteration. The selection of best techniques is done using MSE, TS comparison criteria's shown in the Table 3 given below. Table 3 shows the comparison between CS and FA based approach on GO model. Figure 1 shows the comparison of actual experienced failures and predicted no of cumulative failures using V stub model.

The effectiveness of proposed FA based approach is further compared against CASRE tool after comparing with existing numerical estimation methods. We have chosen CASRE tool for comparison. CASRE tool is one of the famous reliability tools. For comparison, we have selected the LSE based approach in the CASRE tool for the parameter estimation of SRGMs. The results are shown in Table 4.

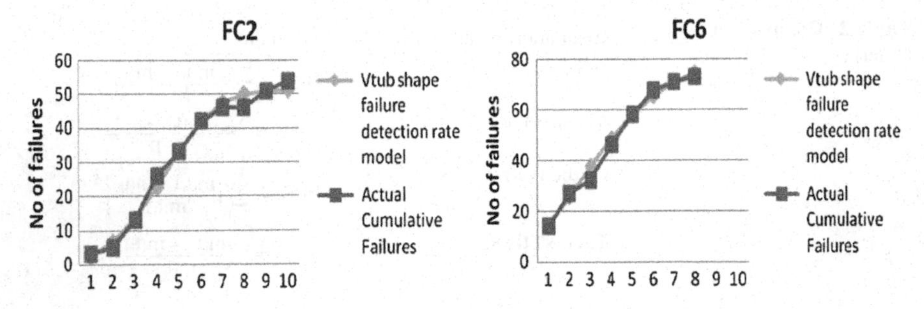


**Fig. 1** Comparison results of actual cumulative failures, FA based Vtub shape failure detection rate model

**Table 4** MSE values comparison of CASRE tool and FA

| | GO model | |
|---|---|---|
| Dataset | CASRE | FA |
| FC2 | 27.5608 | 19.4526 |
| FC6 | 12.8113 | 8.7797 |
| DS1 | 207.8749 | 25.6639 |

# 5 Conclusions and Future Scope

In this paper, an effective approach for parameter estimation for SRGM using FA is proposed. The experiment results save proved the effectiveness of FA based parameter estimation of SRGMs. To evaluate the proposed FA based technique, we performed extensive experiments on seven real world datasets using GO model and Vtub shape failure detection rate model. The MSE and TS values with comparison to existing cuckoo search based parameter estimation on GO model and FA based approach validated the effectiveness and accuracy of FA based approach. The experimental results with CASRE tool showed that the FA based approach can get the more accurate and effective optimal solution. For future work, we will carry out a detailed experimental study of other SRGMs using FA, its variants and other meta-heuristics to estimate more accurate parameter estimation of SRGMs.

# References

1. Kapoor P.K., Pham, H., Gupta, A., Jha, P.C.: Software reliability assessment with OR applications. Springer, London (2011)
2. Wood, A.: Predicting software reliability. Computer **29**(11), 69–77 (1996)
3. Goel, A.L., Okumoto, K.: Time-dependent error-detection rate model for software reliability and other performance measures. IEEE Trans. Reliab. **3**, 206–211 (1979)
4. Choudhary, A., Baghel, A.S., Sangwan, O.P.: Software reliability prediction modeling: a comparison of parametric and non-parametric modeling. In: 2016 6th International

Conference-Cloud System and Big Data Engineering (Confluence), pp. 649–653. IEEE (2016)
5. Xie, M.: Software reliability modelling. vol. 1. World Scientific (1991)
6. Goševa-Popstojanova, Katerina, Trivedi, Kishor S.: Architecture-based approach to reliability assessment of software systems. Perform. Eval. **45**(2), 179–204 (2001)
7. Hsu, C.-J., Huang, C.-Y.: A study on the applicability of modified genetic algorithms for the parameter estimation of software reliability modeling. In: 2010 IEEE 34th Annual COMPSAC. IEEE (2010)
8. Minohara, T., Tohma, Y.: Parameter estimation of hyper-geometric distribution software reliability growth model by genetic algorithms. In: 6th International Symposium on Software Reliability Engineering, 1995. Proceedings. IEEE. (1995)
9. Zhang, K.H., Li, A.G., Song, B.W.: Estimating parameters of software reliability models using pso. Comput. Eng. Appl. **44**(11), 47–49 (2008)
10. AL-Saati, D., Akram, N., Abd-AlKareem, M.: The use of cuckoo search in estimating the parameters of software reliability growth models. arXiv preprint arXiv:1307.6023 (2013)
11. Kim, T., Lee, K., Baik, J.: An effective approach to estimating the parameters of software reliability growth models using a real-valued genetic algorithm. J. Syst. Softw. (2015)
12. Bidhan, K., Awasthi, A.: Estimation of reliability parameters of software growth models using a variation of Particle Swarm Optimization. Confluence 2014 5th International Conference, vol. 800 no. 805, pp. 25–26 Sept. (2014)
13. Tohma, Y., Jacoby, R., Murata, Y., Yamamoto, M.: Hyper-geometric distribution model to estimate the number of residual software faults. In: Proceedings of COMPsac-89, pp. 610–617, Orlando, Sept. (1989)
14. Pham, H.: A new software reliability model with vtub-shaped fault-detection rate and the uncertainty of operating environments. Optimization **63**(10), 1481–1490 (2014)
15. Schneidewind, Norman F.: Software reliability model with optimal selection of failure data. In: IEEE Transactions on 19.11 Software Engineering, pp. 1095–1104 (1993)
16. Song, X., Tang, L., Zhao, S., Zhang, X., Li, L., Huang, J., Cai, W.: Grey wolf optimizer for parameter estimation in surface waves. Soil Dyn. Earthquake Eng. **75**, 147–157 (2015)
17. Aljahdali, S.H., El-Telbany, M.E.: Software reliability prediction using multi-objective genetic algorithm. In: International Conference on Computer Systems and Applications, pp. 293–300. IEEE (2009)
18. Geem, Z.W., Kim, J.H., Loganathan, G.V.: A new heuristic optimization algorithm: harmony search. Simulation **76**(2), 60–68 (2001)
19. Yang, X.S.: Firefly algorithm, stochastic test functions and design optimisation. Int. J. Bio-Inspir. Comput. **2**(2), 78–84 (2010)
20. Wood, A.: Software reliability growth models. Tandem Technical Report 96.130056 (1996)

# Prediction of Crime Trends Using Mk-MC Technique

**B.M. Vidyavathi and D. Neha**

**Abstract** Day by day the quantum of data has been increasing not only in terms of user generated content in social media but also outside the social media, due to which the data has gone from scarce to superabundant that conveys new advantages to users. This explosion of data has made it difficult to handle and analyze huge datasets. Therefore, the techniques of Data Mining assist in exploring and analyzing enormous datasets and helps in discovering meaningful patterns. Clustering is one such task of Data Mining that gathers all the data and partitions it into various groups taking into account their similarity or closeness measure. Clustering in the field of Social Science is used in identification, analysis and detection of various crime patterns. This paper proposes the Modified k-means clustering technique which is applied on the fictitious crime data in order to identify various crime patterns or trends and make a variety of predictions from the analysis of different crime patterns.

**Keywords** Pre-processing · Data cleaning · k-means clustering · Modified k-means clustering

**Acronym** Mk-MC · Modified k-means clustering

## 1 Introduction

The days when an ordinary pen and paper were used to report a crime and in the coming days when the computers are used to record such crimes and used as a database are now gone. Essentially crime is an act which is considered illegal as per the law of the land concerning the countries and it is an act against the society.

B.M. Vidyavathi (✉) · D. Neha
Ballari Institute of Technology and Management, Ballari, India
e-mail: vidyahm1@gmail.com

D. Neha
e-mail: neha.mahendranath@gmail.com



S.C. Satapathy et al. (eds.), *Data Engineering and Intelligent Computing*,
Advances in Intelligent Systems and Computing 542,
DOI 10.1007/978-981-10-3223-3_40

Crimes are committed due to various reasons such as poverty or political vindictiveness or due to religious bigotry or due to other complex reasons which include depression, alcoholism, drug induced, mental disorder, family conditions etc. Crimes committed now differ from country to country and very largely international crimes across all borders have affected every human life. Since crime is predictable that follows a pattern, for instance the Peshawar attack, Pathankot attack, Paris attack that took place recently, Times square attack and other growing concerns like theft, carrying arms, abuse of drugs and human trafficking, murders, burglary etc. there should be a need to identify such patterns and predict such crimes obtained from intelligent agencies. Therefore, crime pattern analysis or crime detection is a significant area for the intelligence agencies [1]. Identifying the crime characteristics is the initial step performed by the agencies. The role of a crime analyst differs from one agency to other as the information or knowledge collected by them is huge and also there is a complexity of relationship between these kinds of data. While some information is kept confidential, few becomes public data. Therefore storing and managing such information has to be done accurately and efficiently. With the rapid technological advances, the above pattern of the mind of a criminal can now be the subject matter or data by the recent innovative trends of storing the same in Data Mining. Clustering in Data Mining helps in performing crime analysis. In this work, the goal is to facilitate crime predictions and its patterns by applying the Modified k-means clustering (Mk-MC) technique. Using this technique, a variety of predictions can be made, which thus helps to reduce future crime incidents.

This paper is organized into the following sections: Sect. 2 contains the Literature Review. Section 3 contains the design and process explaining about the general working of the venture. Section 4 describes the pre-processing technique. Sections 5 and 6 describes the procedure to perform k-means and Modified k-means clustering technique. Section 7 describes the experimental results of the clustering techniques and their performances followed by the conclusion.

## 2 Literature Review

Zhang Haiyung [2] introduced the concepts of Data Mining and its applications and explained that transforming the data into valuable information and knowledge can be done by application techniques of Data Mining. Saurabh Arora, Inderveer Chana [3] and B.M. Vidyavathi, Neha. D [1] introduced about clustering and explained various clustering techniques which can help in the analysis of large quantities of data. Malathi and Dr. S. Santhosh Baboo [4, 5] presented a prediction model that searches for missing values and uses a clustering algorithm for the crime data. MV algorithm and Apriori algorithm were used to fill the missing values and for recognizing crime patterns that may help in giving accurate predictions. Using different Data Mining techniques, they developed a crime analysis tool that helps the criminal investigation departments to efficiently and effectively handle crime data and patterns. M. Ramzan Begaum et al. [6] proposed techniques for developing a crime analysis tool and explained the types of crime and how it is handled by suitable data mining techniques. Jyothi Agarwal et al. [7] introduced a rapid miner tool that performs the k-means clustering technique to analyze the crime data and the analysis was done by plotting a graph considering various attributes for the identification of various crime trends. Shi Na et al. [8] proposed an improved k-means clustering approach to solve a data structure to store information during every iteration. Some of the results show that the improved method can effectively enhance the speed of clustering and accuracy thereby reducing the computational complex nature of the method. Zhang Chen, Xia Shixiong [9] and Yugal Kumar, G. Sahoo [10] proposed a new clustering method based on k-means that avoid randomness of the initial center thereby overcoming the limitation of the k-means clustering technique.

# 3 Design and Process

The General working of the venture are as follows:

- To analyse the crime activities using clustering techniques, a given description of crime such as location, time (day, month and year), type and physical description of the suspects are to be used as a record for creating a database [1]. The crime records are categorized into three entities and they are:
- Accuser's Personal Information (API): The attributes of API are as follows: Accuser's id, Name, Gender, Identification, Date of birth/Age, Height, Weight, Marital Status, Skin tone, Address, Phone number, Nationality, Background and other information.
- Crime Committed information (Crime Status): The attributes of CStatus are: Crime type, Crime Subtype, Weapons, Date, Time, Location, Status, Involvement in other activities, Number of times/repeat offenders and other information.
- In addition, the Crime types and subtypes are categorized into the following: Violent crime (Murder, rape, sexual assault, kidnap, stalking), Property Crime (Robbery, burglary, theft-electronic crime theft, identity theft), Traffic Violations (Reckless driving, speeding, property damage, driving under the influence of drugs and alcohol, hit and run), Sex Crime (Rape, sexual abuse, prostitution, child molestation, trafficking in women and children) and Fraud (Money laundering, Insurance fraud, corruption, trafficking in movies, music and other intellectual property).
- Criminal's Family Background: The attributes are: Father's name, Mother's name, Siblings, Family income and other information.
- The data collected by the user from various sources are susceptible to noisy, missing and inconsistent data (also called as dirty data) and may lead to low quality results. Therefore, transformation of the above data has to be done by

filtering the dataset according to the requirements called as the pre-processing phase.

- Pre-processing of the data consists of the following stages: Data Cleaning, Data Integration, Data Transformation, and Data Reduction.
- Data Cleaning technique is employed in this work to produce clean and error-free data.
- Clustering techniques are applied on the results that are obtained from pre-processing phase.
- k-means clustering and Modified k-means clustering techniques for this purpose are employed that groups the crime data into different clusters based on the similarity measure.
- Modified k-means clustering technique is employed to improvise the limitations of the k-means clustering technique.
- A statistical data is depicted by plotting a graph based on the results of the clustering process, for example, the percentage of crime occurrences.
- A user will be able to facilitate multiple predictions from the graph thereby monitoring various crime patterns.

## 4 Pre-processing Phase

Pre-processing is a technique of transforming the raw data obtained from multiple sources into an understandable format and preparing the data for further processing. In this work, Data Cleaning technique is employed to pre-process the fictitious crime data. Data Cleaning is a process of determining and detecting inaccurate, incomplete or unreasonable data and improving the quality of data by correcting the errors. To correct the invalid records, special algorithms like Brute Force Pattern Matching and k-nearest neighbor are used. Both the algorithms are used for detecting and correcting the errors. The Data Cleaning algorithm helps to do the following:

1. Detecting of Missing Values.
2. Detection of uniqueness.
3. Referential Integrity.
4. Duplication Detection.
5. Detection of Mis-spellings.
6. Detection of invalid and inconsistent data.

## 5 Clustering Phase (k-means Clustering)

k-means is a partitional clustering technique that partitions ‘n’ different types of data or observations into ‘k’ clusters. ‘k’ is the number of clusters into which the datasets have to be grouped or partitioned into and the value of ‘k’ has to be specified by the user in advance. Once the ‘k’ value is specified or initialised, identification of seeds or centroids from the datasets by random observations and then assigning all the other remaining datasets to one of the seeds based on proximity to the seeds takes place. Euclidian Distance (Perpendicular Bisector) will be used as the distance measure to calculate the distance from the datasets to the seeds. Once the first set of clusters are obtained, the centroid for the clusters have to be calculated by adding all the data points present in the cluster(s) and assigning them as the seeds for the next iteration. In this way, different sets of clusters are obtained at every iteration and stops when none of the cluster assignment changes thereby producing final sets of clusters.

Some of the limitations of the k-means clustering technique are: The technique is applicable only when the numbers of clusters are defined by the user(s) in advance. Determination of the k-value is difficult. Randomly picking the cluster centers do not lead to a good result due to which calculating the distance from every data item to each cluster center during each cycle is a time-consuming process. When clusters are formed of different dimensions, the data elements present in clusters of different dimensions are not equal. The efficiency of clustering is affected since the execution time to perform the clustering is long.

## 6 Modified Clustering Phase (Modified k-means Clustering)

Figure 1 is a flow chart representation of the Modified k-means Clustering technique. In this technique, ‘k’ value i.e., the number of clusters are generated randomly without the user having to specify or initialize the value of ‘k’. In order to generate an equal number of elements in the clusters, the size of the number of elements and the number of clusters have to be taken into consideration in order to decide how many numbers of elements need to be present in cluster(s). Selection of cluster centers or seeds is done by initialising a range of cluster centers rather than identifying the seeds by random observations. This will speed up the process of assigning the remaining observations to the seeds. Euclidian Distance will be used as the distance measure to calculate the distance from the data sets to the seeds in order to obtain different sets of clusters during every iteration and stops when none of the cluster assignment changes. This technique reduces the time taken taken to form different clusters thereby increasing its efficiency.

## 7 Experimental Results

The following figure and tables provided are the results obtained by applying k-means and Modified k-means Clustering techniques. The experiment was conducted by taking into consideration fictitious crime data. The collected data was first

**Fig. 1** Flow chart representation of Modified k-means Clustering (Mk-MC) technique

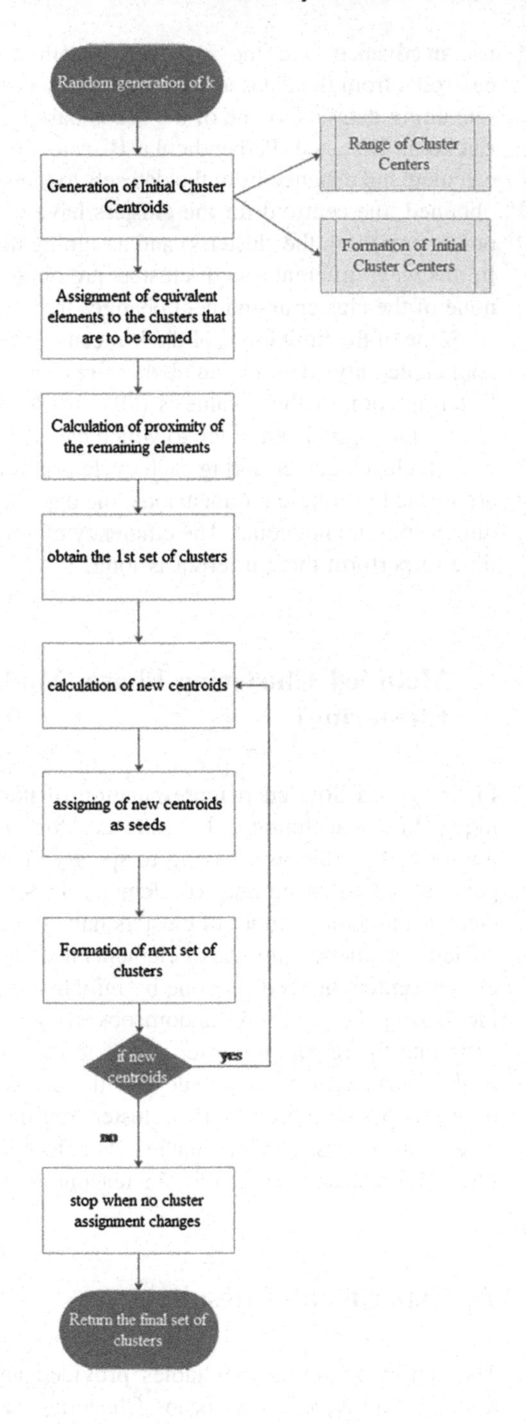

transformed into an understandable format by applying the Data Cleaning technique in order to subject it to further processing. The clustering techniques were applied on selected attributes of the pre-processed data.

- Result of the Modified k-means Clustering Technique

The graph plotted in Fig. 2 is based on the calculations of the Modified k-means clustering technique. The total numbers of crime records are 6 and numbers of clusters chosen randomly by the program are 2. The graph is plotted taking attribute Years (Date of Crime) on the x-axis and attribute Age on the y-axis. The final result is based on the assumption of the values assigned to both the axis is/are:

(1) Cluster 1 (represented by color orange) = Accusers of age group (15–20 and 21–25 years) committing different crime(s) in the year (1990–2000).
(2) Cluster 2 (represented by color red) = Accusers of age group (31–35, 41–45 and 56–60 years) committing different crime(s) in the year (2001–2015).
(3) The overall time taken by the Modified k-means clustering technique to form 2 clusters is 40 ms.

- Performance (Graph)

The datasets used for conducting pre-processing and clustering techniques correspond to fictitious crime data that are categorized into three different entities:

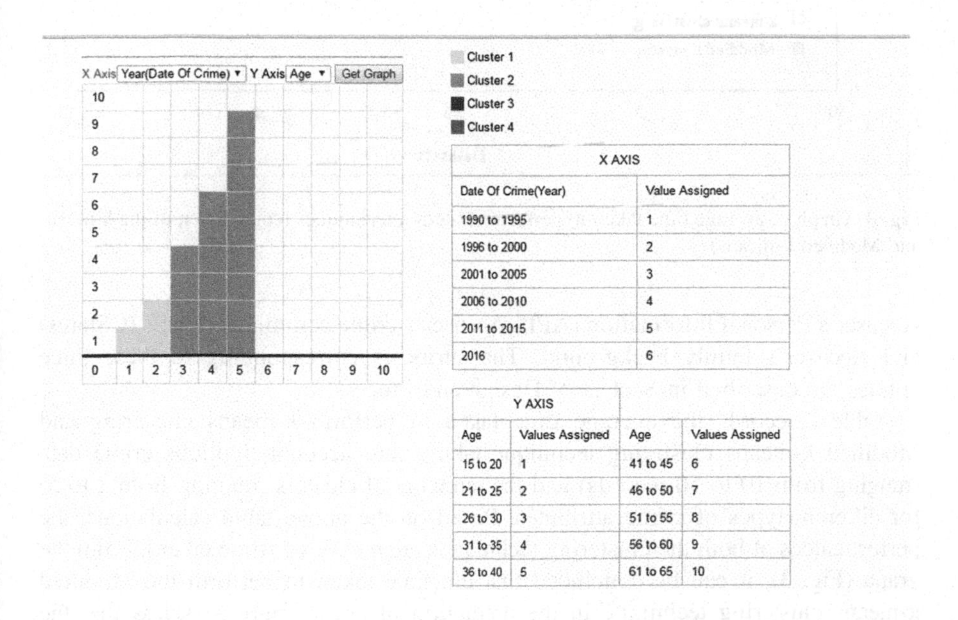


**Fig. 2** Graph plotted based on calculations of Modified k-means clustering technique (x-axis = year (date of crime), y-Axis = Age)

**Table 1** Average time taken to perform clustering techniques for the different attributes

| Datasets(n) | Number of Clusters(k) | Time taken (milliseconds) | |
|---|---|---|---|
| | | k-means | Modified k-means |
| n = 10 | k = 2–6 | 40 | 32 |
| n = 20 | k = 2–6 | 35 | 27 |
| n = 30 | k = 2–6 | 37 | 30 |
| n = 40 | k = 2–6 | 38 | 32 |
| n = 50 | k = 2–6 | 39 | 32 |

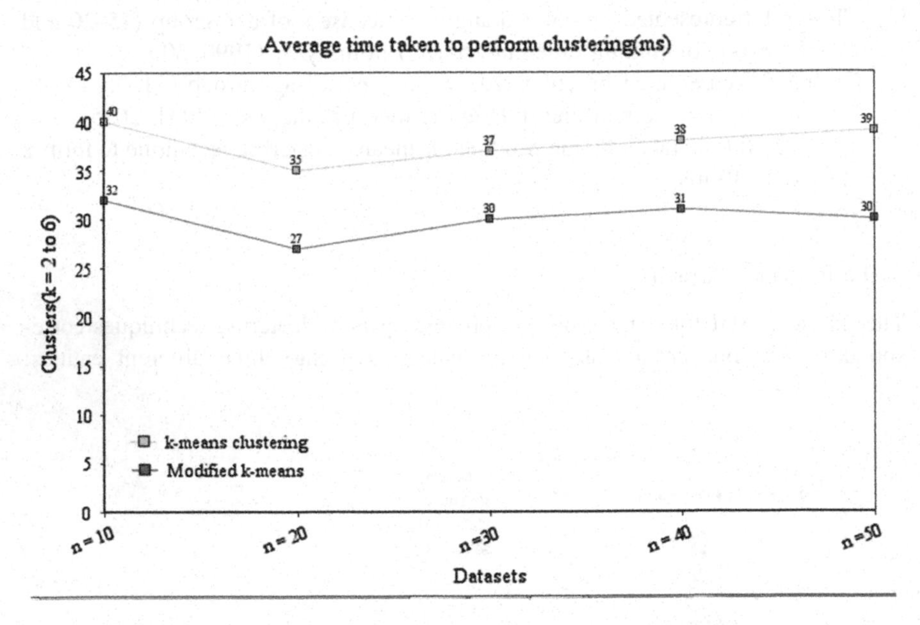


**Fig. 3** Graph of average time taken to perform clustering techniques (comparison of the k-means and Modified k-means)

Accuser's Personal Information (API), Accuser's crime committed details (CStatus) and Accuser's family background. The attributes corresponding to these three entities are described in Sect. 3 of Design and Process

Table 1 records the average time taken to perform k-means clustering and Modified k-means clustering technique taking into account fictitious crime data (ranging from 10 to 50 records) and the numbers of clusters (ranging from 2 to 5) for different types of crime attributes. Based on the above table calculations, the performances of both the clustering techniques are measured (time taken). From the graph (Fig. 3), it can be concluded that the time taken to perform the Modified k-means clustering technique in the prediction of crime analysis is less that the k-means clustering technique.

Since this work is specifically taking into consideration fictitious crime data, the results are based on the assumptions of that data. Good results can be obtained and good predictions can be made if real crime data is used.

## 8 Conclusion and Future Work

Crime pattern analysis is an essential task where efficient clustering techniques can be applied. From the clustered results, it is easy to identify different patterns of crime, thereby making it easier for the user to make a variety of predictions. A comparison of different clustered results of our approach can be made based on their analysis i.e., time taken to form clusters. The results from the comparisons can be used for predicting future crime trends. Prediction methods do not predict when and where the next crime will take place; they only reveal a common pattern associated with time, place and risk that enable predictions to be made. Therefore, the intricate nature of the crime related data and its existing unseen relations within itself have made Data Mining a progressing field aiding the criminologists, crime investigation departments, police departments and other crime related departments. In view of advancement of technology, in future, there is every possibility of improvements to be made in the Modified k-means clustering technique. There is also a scope for better visual representation of graphic patterns of crime data which can help in making the analysis faster and easier suiting the requirements.

## References

1. Vidyavathi, B.M., Neha, D.: A survey on applications of data mining using clustering techniques. Int. J. Comput. Appl. **126**(2), 7–12 (2015). (0975-8887)
2. Haiyang, Z.: A Short Introduction to Data Mining and Its Applications. IEEE (2011)
3. Arora, S., Chana, I.: A survey of clustering techniques for big data analysis, IEEE. In: 5th International Conference Confluence The Next Generation Information Technology Summit (Confluence), pp. 59–65 (2014)
4. Malathi, A., Baboo, S.S.: Evolving data mining algorithms on the prevailing crime Trend—An intelligent crime prediction model. Int. J. Sci. Eng. Res. **2**(6), 1–6 (2011)
5. Malathi, A., Baboo, S.S.: An enhanced algorithm to predict future crime using data mining. Int. J. Comput. Appl. **21**(1), 1–6 (2011)
6. Ramzan Begam, M., Sengottuvelan, P., Ramani, T.: Survey: tools and techniques implemented in crime data sets. IJISET—Int. J. Innov. Sci. Eng. Technol. **2**(6), 707–710 (2015)
7. Agarwal, J., Nagpal, R., Sehgal, R.: Crime analysis using k-means clustering. Int. J. Comput. Appl. **83**(4), 1–4 (2013)
8. Na, S., Xumin, L., yong, G.: Research on k-means Clustering Algorithm-An improved k-means Clustering Algorithm. In: IEEE Third International Symposium on Intelligent Information Technology and Security Informatics, pp. 63–67 (2010)

9. Zhang, C., Shixiong, X.: k-means clustering algorithm with improved initial center. In: IEEE, Second International Workshop on Knowledge Discovery and Data Mining, pp. 790–792 (2009)
10. Kumar, Y., Sahoo, G.: A new initialization method to originate initial cluster centres for k-means algorithm. Int. J Adv. Sci. Technol. **62**, 43–54 (2014)

# A Novel Map-Reduce Based Augmented Clustering Algorithm for Big Text Datasets

**K.V. Kanimozhi and M. Venkatesan**

**Abstract** Text clustering is a well known technique for improving quality in information retrieval, In Today's real world data is not organized in the essential manner for a precise mining, given a large unstructured text document collection it is essential to organize into clusters of related documents. It is a contemporary challenge to explore compact and meaning insights from large collections of the unstructured text documents. Although many frequent item mining algorithms have been discovered yet most do not scale for "Big Data" and also takes more processing time. This paper presents a high scalable speedy and efficient map reduce based augmented clustering algorithm based on bivariate n-gram frequent item to reduce high dimensionality and derive high quality clusters for Big Text documents and also the comparative analysis is shown for the sample text datasets with stop word removal the proposed algorithm performs better than without stop word removal.

**Keywords** Text documents · Frequent item · Similarity · Clustering

K.V. Kanimozhi (✉) · M. Venkatesan
School of Computing Science and Engineering, VIT University,
Vellore, Tamilnadu, India
e-mail: kanimozhi.kv2014@vit.ac.in

M. Venkatesan
e-mail: mvenkatesan@vit.ac.in



S.C. Satapathy et al. (eds.), *Data Engineering and Intelligent Computing*,
Advances in Intelligent Systems and Computing 542,
DOI 10.1007/978-981-10-3223-3_41

## 1 Introduction

With the advance of fast internet growth the World Wide Web text database is loaded with millions and billions of unstructured text document collection. Hence to retrieve information web search engines is used. In the area of text mining and information retrieval to explore the exact underlying patterns is a most important challenge. To increase the efficiency of precision in retrieval rate from the web search engine lots of techniques have been implemented like clustering. So text clustering plays a vital role in grouping of similar text documents into clusters such that the objects in a cluster are similar to each other and the objects which are not same are different from one another. Hence it is evaluated based on the similarity and dissimilarity of objects across clusters. Recently frequent item set have been implemented in the area of text clustering. The idea of clustering is broadly used and turned out as an area of typical attention for the researchers. In Sect. 2 the outline of some existing works regarding frequent items generations of interest is given. In Sect. 3, includes the problem statement and the proposed solution for efficient clustering of text documents in big text data sets using map-reduce framework. In Sect. 4 the experiments and result analysis is shown. Finally, the conclusion is given in Sect. 5.

## 2 Back Ground and Literature Review

In order to categorize the huge amounts of information into knowledgeable clusters, efficient document clustering algorithms plays a main role. Clustering Frequent item based method [1, 2, 3] is one of the recent research topic and essential challenge for researchers. The various problem [4–6] like spam detection, clustering customer behaviors, document clustering [7, 8], and sentiment analysis, concept and entity extraction, text categorization has been analyzed. Some of the recent literature survey related to frequent item set-based text clustering is reviewed in brief. Document clustering based on maximum frequent sequence is proposed in [9]. Frequent Term-Based Clustering (FTC) to generate the clusters with overlaps as few as possible is proposed in [10]. But it leads to large number of clusters. Frequent Item set-based Hierarchal clustering (FIHC) to generate clusters using tree based method and not able to handle the Big Data and leads to complexity is proposed in [11]. Two frequent item set mining algorithms for big data is proposed in [12]. Parallel FP-Growth algorithm using map reduce framework on distributed system which reduces the computational dependencies and achieve virtually linear speedup on query recommendation for search engine [13]. New parallel frequent item algorithm using spark and showed faster execution than apriori algorithm implemented in map reduce framework [14]. Maximum Capturing based on three different similarity measures for document clustering using frequent patterns [15] proved better than other methods like CFWS, CFWMS, FTC, FIHC but it do not scale for big text datasets.

## 3 Problem Statement and Proposed Solution

The main challenges faced by text clustering process are namely the huge volume of documents and the large size of text documents features. A Novel based map-reduce augmented clustering methodology is proposed to enhance the efficiency of clustering. The proposed system is divided into two phases: the objective of first phase is to decrease the very large number of document terms through our newly proposed bivariate n gram algorithm in order to address high dimensionality. Secondly a new fast augmented clustering method has been proposed to cluster the documents in efficient manner.

### *3.1 Map-Reduce Framework*

To handle the massive text datasets the Map-Reduce programming has been implemented. This algorithm works in different levels and clusters many document in parallel manner which leads to reduced computation speed and huge dimensionality reduction.

### *3.2 Text Preprocessing*

Using the map-reduce programming the algorithm does the preprocessing step for tokenization and stop words removal for our data set collection. The first step is called tokenization which divides the every line of text document into individual words called tokens. Next is stop words removal where the words which do not contain important significant information or occur so often that in text collection are removed.

### *3.3 Proposed Methodology*

**Algorithm for Text Clustering based on frequent items**

---

**Input:** Input Large text document D collection from local system to hadoop file system.
**Output:** Compact high quality clusters.
**Method:**

Begin of the Algorithm
**Step 1:** Convert the paragraph from each document into separate sentences.
**Step 2:** Remove all stop words from those converted documents
**Step 3:** For each document D in document set do
Begin
3.1 Count the number of words in each line present in each document
3.2 Obtain the minimum count.
End
End for
**Step 4:** For each document D in document set do
Begin
4.1 Generate the item sets using bivariate n-gram mechanism by pairing minimum two words for generating item set.
4.2 Continue pairing the words till minimum count.
4.3 Calculate the occurrence of each word pair.
4.4 Find out the item set which have occurred maximum times from the generated item sets
End
End for
4.5 Output the frequent item sets.
**Step 5:** Calculate similarity between the documents by number of item set they have in common and construct document matrix as per occurrence of frequent item set by generating the frequent item set and their corresponding documents.
**Step 6:** Calculate document occurrence by finding number of times each document occurs.
**Step 7:** Select pivot points based on maximum document occurrence and perform clustering as per pivot.
**Step 8:** Output the set of clusters obtained.
End of the algorithm.

## 3.4 *Case Study: Clustering the Sample Text Documents Using Proposed Algorithm*

### 3.4.1 Input Text Files

The input Datasets used in this paper is 20Newsgroup; all these datasets are converted to original text format. Once it is converted it is fed to hadoop file system from local file system and preprocessing like tokenization and stop word removal are done.

### 3.4.2 Calculate Minimum Count

To obtain the minimum count we need to count the number of words in each line for each document. In Table 1. The minimum count is 3.

### 3.4.3 Generate Item Sets

Frequent item sets are generated using bivariate n-gram mechanism. 'bi' means minimum two words will be taken for generating frequent item and 'variate' means it depends on two values n = 2 and n = minimum count where n is the number of words to be paired together. Example: The boy is playing football.

- 1st pairing The boy; boy is; is playing; playing football
- 2nd pairing The boy is; boy is playing; is playing football
- 3rd pairing The boy is playing; boy is playing football
- 4th pairing The boy is playing football.

Note: These pairing are the item sets and not frequent item sets. After generating the item sets, we need to calculate how many times each pair occurs (Table 2).

**Table 1** Obtain minimum count by counting the number of words

| Content of document D1 | No. of words |
|---|---|
| The boy is playing football | 5 |
| How are you? | 3 |
| The baby is sleeping | 4 |

**Table 2** Obtain the occurrence of each item pair

| Item set | Occurrence |
|---|---|
| How are | 50 |
| Where are | 55 |
| Going to | 20 |
| From school | 15 |
| Hello sir | 45 |
| Good morning | 65 |

From the generated item sets we need to extract those item sets which have occurred the maximum times. Those item sets are the frequent item sets (Table 3).

#### 3.4.4 Calculate Similarity as Per Item Set and Construct Document Matrix as Per Item Set Occurrence

The similarity measure between two documents is calculated by number of frequent item set they have in common. Then the document matrix by generating frequent item sets and their corresponding documents are found (Table 4).

#### 3.4.5 Calculate Document Occurrence

After obtaining the document matrix next to find out number of times each document occurs (Table 5).

**Table 3** Find the maximum occurred frequent item sets from Table 2

| Item set | Occurrence |
|---|---|
| How are | 50 |
| Where are | 55 |
| Hello sir | 45 |
| Good morning | 65 |

**Table 4** The reduced document similarity matrix obtained from Table 3

| Item set | Text documents |
|---|---|
| How are | 1.txt, 2.txt, 5.txt |
| Where are | 1.txt, 2.txt, 3.txt, 4.txt |
| Going to | 1.txt, 2.txt, 6.txt |
| Hello sir | 1.txt, 3.txt, 5.txt |
| Good morning | 1.txt, 2.txt, 5.txt |

**Table 5** Document occurrence obtained from Table 4

| Document | Occurrence |
|---|---|
| 1.txt | 5 |
| 2.txt | 4 |
| 3.txt | 2 |
| 4.txt | 1 |
| 5.txt | 3 |
| 6.txt | 1 |

**Table 6** Final Clusters obtained from Table 5

| Cluster | Pivot | Documents |
|---|---|---|
| 1 | 1.txt | 1, 5, 6, 8, 11, 12 |
| 2 | 2.txt | 2, 4, 14 |
| 3 | 5.txt | 3, 9, 10, 13, 14 |

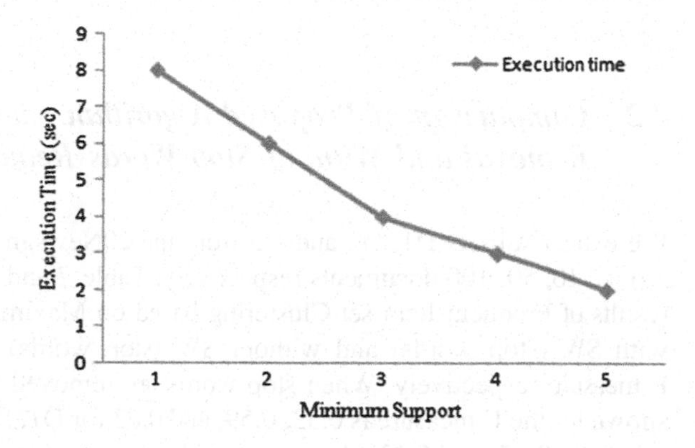


**Fig. 1** Scalability of finding frequent item sets with respect to the minimum support

### 3.4.6 Select Pivot Points Randomly and Perform Clustering as Per Pivot

After finding document occurrence, the next thing is to cluster the documents. This can be done by selecting pivot points. Selection of the pivot points are calculated based on maximum occurrence of documents. All the documents associated with a pivot will form a cluster (Table 6).

# 4 Experiments and Results Analysis

In this section, Implementation of algorithms and the performance evaluation are performed in map-reduce parallel programming on a Ubuntu 14.04 PC with 4 GB RAM and Intel Core i3 processor in hadoop 2.6.0 for sample example and the accuracy of clustering is shown.

## *4.1 Evaluation for Computing Frequent Item Sets*

The algorithm computes frequent item sets to deal with the scalability and dimension reduction. The experiment shows, for sample data the bivariate n-gram mechanism to find only frequent 2-pair item sets. From the 20_Newsgroup dataset sample 100 documents are taken. The bivariate n-gram algorithm generates 36,996

2-pair item sets and from these item sets extracts 293frequent 2-pair item sets when the minimum support is 2.

The algorithm is computed for different minimum support count on the 20_Newsgroup dataset, and the result is shown in Fig. 1. Whenever minimum support count increases, the execution time decreases since less frequent 2-word sets are found. Hence the output clearly shows the proposed algorithm is scalable.

## *4.2 Comparison of Proposed Algorithm with Stop Words Removal and Without Stop Words Removal*

We extract subsets D1, D2 and D3 from the 20Newsgroup dataset which contains around 10, 50, 100 documents respectively. Table 7 and Figs. 2, 3 and 4 shows the results of Frequent Item set Clustering based on Maximum Document Occurrence with SW (stop words) and without SW (stop words) for Precision, Recall and F-measure respectively. When stop words are removed, the difference observed is shown for the F-measure is 0.53, 0.59, and 0.62 for D1, D2 and D3 respectively and it is 0.42, 0.27, and 0.13 when stop words are not removed. Since the F-measure is

**Table 7** Result of Precision, Recall, and F-measure for 20Newsgroups dataset

| Dataset | Precision | | Recall | | F-measure | |
|---|---|---|---|---|---|---|
| | With SW | Without SW | With SW | Without SW | With SW | Without SW |
| D1 | 0.27 | 0.8 | 1 | 0.4 | 0.42 | 0.53 |
| D2 | 0.41 | 0.528 | 0.21 | 0.6667 | 0.27 | 0.59 |
| D3 | 0.4 | 0.75 | 0.08 | 0.52 | 0.13 | 0.62 |

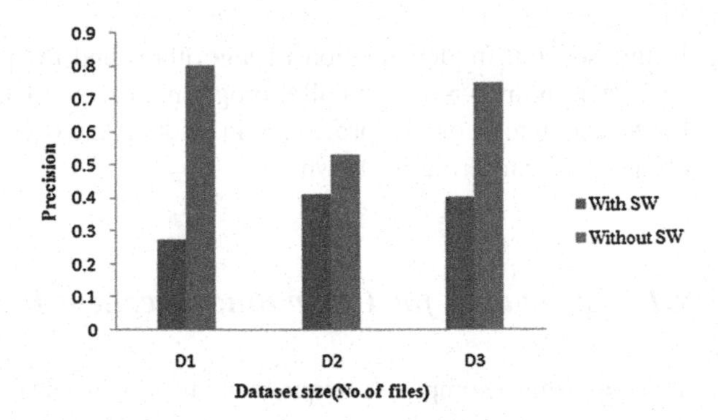


**Fig. 2** Precision with SW (*With Stop Word*) and Without SW (*Without Stop Words*)

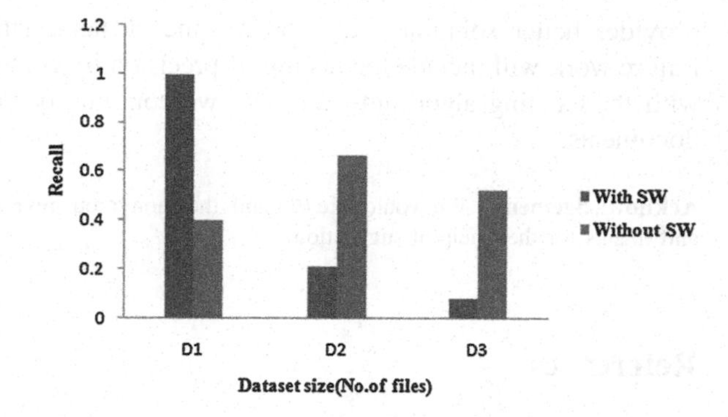


**Fig. 3** Recall with SW (*With Stop Word*) and Without SW (*Without Stop Words*)

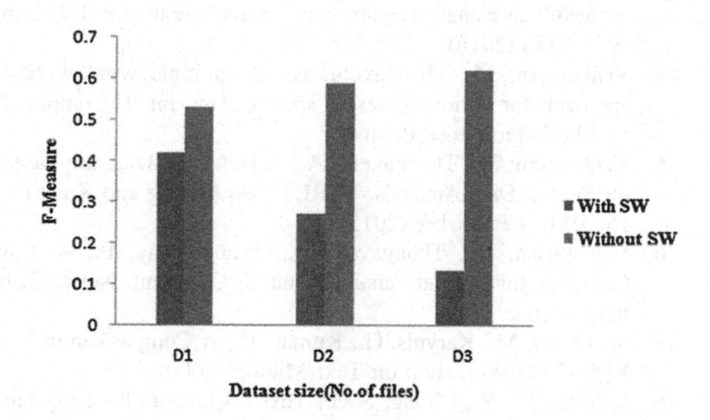


**Fig. 4** F-measure with SW (*With Stop Word*) and Without SW (*Without Stop Words*)

higher the clustering quality is better. Hence clustering with stop words removal is superior to the clustering when stop words are not removed.

## 5 Conclusion

The experimental results proves the proposed map reduce programming using novel bivariate n-gram mechanism for generating frequent items, solves the scalable problem using its parallel approach, which requires relatively low computational requirements and also the document with maximum similarity is selected as a pivot point and clustering is done as per pivot and generates compact and high quality clusters and hence it is faster and efficient and improves the performance on hadoop file system. And also comparison result shows without stop word the algorithm provides better solution and improves the cluster quality than with stop word. Future work will include the testing of precision by comparing proposed algorithm with the existing algorithms and also we continue our process in ranking of the documents.

**Acknowledgements** We would like to thank the anonymous reviewers for their useful comments and thanks for their helpful suggestion.

## References

1. Kanimozhi, K.V., Venkatesan, M.: Survey on text clustering techniques. Adv. Res. Electr. Electron. Eng. **2**(12), 55–58 (2015)
2. Kanimozhi, K.V., Venkatesan, M.: Big text datasets Clustering based on frequent item sets—a survey. Int. J. Innovat. Res. Sci. Eng. **2**(5). ISSN: 2454– 9665 (2016)
3. Naaz, E., Sharma, D., Sirisha, D., Venkatesan, M.: Enhanced k-means Clustering approach for health care analysis using clinical documents. Int. J. Pharm. Clin. Res. **8**(1), 60–64. ISSN-0975 1556 (2016)
4. Venkatesan, M., Thangavelu, A.: A multiple window based Co-location pattern mining approach for various types of spatial Data. Int. J. Comput. Appl. Technol. **48**(2), 144–154 (2013). Inderscience Publisher
5. Venkatesan, M., Thangavelu, A.: A Delaunay Diagram-based Min–Max CP-Tree Algorithm for Spatial Data Analysis, WIREs Data Mining and Knowledge Discovery, vol. 5, pp. 142–154. Wiley Publisher (2015)
6. Venkatesan, M., Thangavelu, A., Prabhavathy, P.: A Novel Cp-Tree based Co-located Classifier for big data analysis. Int. J. Commun. Netw. Distrib. Syst. **15**, 191–211 (2015). Inderscience
7. Steinbach, M., Karypis, G., Kumar, V.: A Comparison of Document Clustering Techniques: KDD-2000 Workshop on Text Mining (2000)
8. Luo, C., Li, Y., Chung, S.M.: Text document clustering based on neighbors. Data Knowl. Eng. **68**, 1271–1288 (2009). Elsevier
9. Edith, H., Rene, A.G., Carrasco-Ochoa, J.A., Martinez-Trinidad, J.F.: Document clustering based on maximal frequent sequences. In: Proceedings of FinTAL 2006, LNAI, vol. 4139, pp. 257–67 (2006)
10. Beil, F., Ester, M., Xu, X.: Frequent term based text clustering. In: Proceedings of ACM SIGKDD International Conference on knowledge Discovery and Data Mining. pp. 436–442 (2002)
11. Fung, B., Wang, K., Ester, M.: Hierarchal document clustering using frequent item sets. In: Proceedings of the 3rd SIAM International Conference on Data Mining (2003)
12. Moens, S., Aksehirli, E., Goethals, B.: Frequent Item set Mining for Big data (2014)
13. Li, H., Wang, Y., Zhang, D., Zhang, M., Chang, E.Y.: Parallel FP-Growth for query recommendation. In: Proceedings of ACM Conference on Recommender systems, pp 107–114 (2008)
14. Qiu, H., Gu, R., Yuan, C., Huang, Y.: YAFIM: a parallel frequent item set mining algorithm with spark. In: 28th International Parallel & Distributed Processing Symposium Workshops. IEEE (2014)
15. Zhang, W., Yoshida, T., Tang, X., Wang, Q.: Text Clustering using frequent item sets. Knowl.-based Syst. **23**, 379–388 (2010). Elsevier

# Scrutiny of Data Sets Through Procedural Algorithms for Categorization

**Prasad Vadamodula, M. Purnachandra Rao, V. Hemanth Kumar, S. Radhika, K. Vahini, Ch. Vineela, Ch. Sravani and Srinivasa Rao Tamada**

**Abstract** This paper evaluates the selected classification algorithms for classifying thyroid datasets. The classification algorithms considered here are principle component analysis method and partial least square regression method of machine learning algorithms. After successful prediction of disease levels, these algorithms resultant output levels are compared. The analysis suggests the best classifiers for predicting the exact levels of thyroid disease. This work is a comparative study of above said algorithms on thyroid dataset firmly collected from UCI Machine Learning Repository.

**Keywords** Thyroid disease · Ridge regression · PCA · PLS · Machine learning algorithm

## 1 Introduction

In this paper, thyroid data sets (TDS) [1] classification is done in the earliest stage and anon the calculations of decisive factors (DF) are considered to single out the most prominent classifiers which are required to classify the TD levels. Firstly, a string matching system (SMS) is introduced which further predicts thyroid disease accordingly with the available database. If the developed system fails, then the data sets of Andhra Pradesh thyroid (APTD) are passed as the input to the classification algorithms (CA's). The CA's undergoes the routine process of identifying, as well as predicting the actual levels of TD with which the patient is suffering. An SMS

M.P. Rao · V. Hemanth Kumar · S. Radhika · K. Vahini · Ch. Vineela · Ch. Sravani
Department of CSE, Raghu Institute of Technology, Visakhapatnam, India
e-mail: purnachandrarao.m@gmail.com

S.R. Tamada
Department of CSE, Gitam Institute of Technology, Visakhapatnam, India

P. Vadamodula (✉)
Department of CSE, GMR Institute of Technology, Visakhapatnam, India
e-mail: prasad_v542@yahoo.co.in; prasadv542@gmail.com



S.C. Satapathy et al. (eds.), *Data Engineering and Intelligent Computing*,
Advances in Intelligent Systems and Computing 542,
DOI 10.1007/978-981-10-3223-3_42

[2] would check whether the entered data string is available in the database or not. If data is available, then there is no need of a classifier, otherwise a classifier is required to predict the TD. The datasets which are passed and queried by the CA's are evaluated by considering the DF's i.e., meticulousness (mT), exactitude (eX), compassion (cP) and rigour (rI). The paramount values obtained by the DF's gives us the preeminent classifier among the selected classifiers in this work of thyroid disease diagnose (TDD) [3].

Similarly, UCITD datasets are passed through SMS & CA's and the same procedure is on track. The outcome obtained is recorded and further, a comparative study between APTD and UCITD is done. This comparative study is useful in identifying the apex data set and the preeminent classifier [4].

## *1.1 Role of Principal Component Analysis (PCA)*

The PCA [5] uses variance as a metric of interestingness and finds the content required i.e., orthogonal vectors from the feature space available in the data. The data patterns are identified by PCA and these data patterns are used for finding and predicting the anomalies from APTD & UCITD data. This algorithm is a measure or a module of dimensionality reduction [6] algorithms.

***Pseudo Code of the Principal Component Analysis (PCA)*** [7]
***Procedure implemented***

***Step 1***: Recover basis:

$$\text{Calculate } \mathbf{XX^T} = \Sigma_{i=1,2\ldots T}\mathbf{X_i X_i^T} \text{ and let } \mathbf{U} = \textbf{Eigen vectors of } \mathbf{XX^T}$$

corresponding to the top "d" Eigen values.
***Step 2***: Training data is encoded as:
$\mathbf{Z} = \mathbf{U^T X}$ where Z is a $\mathbf{d} \times \mathbf{t}$ matrix.
***Step 3***: Training data is reconstructed as:

$$\mathbf{X} \sim = \mathbf{UZ} = \mathbf{UU^T X}.$$

***Step 4***: Encode test example:
$\mathbf{Z} = \mathbf{U^T} > \mathbf{X}$, where Z is a d-dimensional encoding of x.
***Step 5***: Test example is reconstructed as:

$$\mathbf{X} \sim = \mathbf{U_y} = \mathbf{UU^T X}.$$

### 1.2 *Role of Partial Least Squares Regression (PLS) [8]*

Prediction of Y (Output) considering X (Symptoms Array Vector Input) to finalize the structure. Y is an output vector and likewise X is an attribute, the output is traced using normal regression of multiple data. When predictions count are large in number then, X is singular and the regression exists no more [9]. Many methods are urbanized to survive with this task. A simple approach is to eliminate predictor's data values, called PC regression, which performs PCA of the X matrix. Similarly m the X components are used on Y regression for tracing the data count,

Orthogonality PCA multi co-linearity. Subset of optimum prediction set remains constant. From this the detailed X and Y vales are traced [10].

PLS, PCA performance as mentioned in [11]. PLS traces the components set and PCA goes with the generalization of the components set. The following discussed pusedo code is considered for implementation of the PLS

***Pseudo Code of the* Partial Least Squares Regression (PLS)**
***Procedure implemented***:

***Step 1***: To create major two matrices: **M = X** and **N = Y.** These matrices are normalized The ss matrices are denoted as SSX and SSY.
***Step 2***: $\mathbf{w} \propto \mathbf{M}^T\mathbf{u}$ (Estimate score of X weights).
***Step 3***: $\mathbf{t} \propto \mathbf{Mw}$ (Estimate score of X factor).
***Step 4***: $\mathbf{c} \propto \mathbf{N}^T\mathbf{t}$ (Estimate score of X factor).
***Step 5***: $\mathbf{u} = \mathbf{Nc}$ (Estimate score of Y).

The elementary vectors t, u, w, c, and p are stored in corresponding matrices. The sum of squares is explained by vectors as $\mathbf{p}^T\mathbf{p}$, and the proportion of variance is obtained by division of sum of squares and corresponding total sum of squares [12].

If (M == null matrix) then Latent Vectors are traced else Re-do Step 1 [13].

## 2 Database Used for Evaluation [14]

Database consists of approximately 57 rules of 390 patients obtained from UCI repository [15, 16].

Boolean representation of Datasets
1-True-Yes 0-False-No

**Ruleset 1**
If the featured attributes are entered as Rule set-1 The diagnose is considered as Thyroid is Negative and the prevention methodology is regular checkups on thyroxin.

**Ruleset 2**
If the featured attributes are entered as Rule set-2

The diagnose is considered as Hyperthyroid and the curing methodology is a referral Source is SVI.

## 2.1 Decisive Factors

These factors help in identifying the best classifier for classification and also give the accuracy and correctness for predicting the TD (Table 1).

The term **Cumulative** gives the total count of existing elements in the Entered Input String which are successful Hits/Miss to the database, whereas term **individual** gives the occurrence, as of read at particular iteration.

***Meticulousness***: ***(mT)***: The meticulousness gives the correctness percentage of identified classifier.

$$mT = (CTP + CTN)/(CTP + CFP + CFN + CTN).$$

***Exactitude***: ***(eX)***: Exactitude is the percentage obtained through the positive count.

$$eX = CTP/(CTP + CFN).$$

***Compassion***: ***(cP)***: **Compassion** is true positive rate on positive classifiers.

$$cP = CTP/(CTP + CFP).$$

***Rigour***: ***(rI)***: Rigour is to calculate the true rate of negative elements in proportion with tuples which are negative or if wrongly identified.

$$rI = CTN/(CTN + CFP)$$

***PROCEDURE FOR CALCULATING DECISIVE FACTORS:***

***Example***: Consider a single tuple database for calculating Decisive factors (Table 2)

- *"*" indicates the TRUE POSITIVES and "~" indicates the TRUE NEGATIVES.*
- *The remaining 0's in the Rule Sets are False Negatives and 1's are False Positives.*

**Table 1** This table represents the terminology which is used to calculate the Decisive Factors for obtaining the preeminent Classifier and apex datasets

| Term (Individual) | Abbreviation | Term (Cumulative) | Abbreviation |
|---|---|---|---|
| TP | *True Positives* | ***CTP*** | *Count of True Positives* |
| FP | *False Positives* | ***CFP*** | *Count of False Positives* |
| TN | *True Negatives* | ***CTN*** | *Count of True Negatives* |
| FN | *False Negatives* | ***CFN*** | *Count of False Negatives* |

**Table 2** This is a sample table which is picked up from the original database of APTD, to display the diagnose method along with its prevention and referral source as a suggestion

| Entered input string [17]: | | | | | | | |
|---|---|---|---|---|---|---|---|
| SYMPTOMS | S3 | S6 | S7 | S9 | S10 | S11 | S12 |
| Rule Set-1 (Already Existing Knowledge Base) | *1** | *1** | *0~* | *0* | *0* | *1* | *0* |
| Rule Set-2 Entered Data String | *1** | *1** | *0~* | *1* | *1* | *0* | *1* |

The Entered Data String is the choice given by the User, if the Entered Data String directly hits the knowledge base, then the SMS system produces 100% Output, otherwise the system modifies itself by sending the same data to the classifiers, whether the HITS of the Data is successful or not, the values which are required by the Decisive factors are updated and maintained as shown below

$$Count\ of\ True\ Positives\ (*) = CTP = 02$$
$$Count\ of\ True\ Negatives(\sim) = CTN = 01$$
$$Count\ of\ False\ Positives = CFP = 03$$
$$Count\ of\ False\ Negatives = CFN = 06$$

*Therefore, the obtained Decisive Factors are*

$$mT = (02 + 01)/(02 + 01 + 03 + 06) = 03/12 = \mathbf{0.25}$$
$$eX = 02/(02 + 06) = 02/08 = \mathbf{0.25}$$
$$cP = 02/(02 + 03) = 02/05 = \mathbf{0.4}$$
$$rI = 01/(01 + 03) = 01/04 = \mathbf{0.25}$$

The procedure discussed above is calculated for a single tuple knowledge base, but in the beginning we already informed that the APTD and UCITD are using 390 and 549 tuples of data respectively, hence the matching and neighbouring elements trace would give more apex values for the decisive factors.

## 3 Results

When the input string is given to the SMS [18], it processes the entered string with the knowledge base automatically and then the diagnose is suggested, otherwise the ML Algorithms will predict the TD. After predicting the TD levels, the DF calculations are made. As per the apex value obtained by the DF's, the best classifier would be traced out, and the result obtained by the classifier is finally displayed as a suggestion.

**CASE 1**: SMS obtained 100% prediction levels [19].

**CASE 2**: Classification Algorithms generates 89% output [20, 23].

If Symptoms as per Rule set-2, then the **SMS** shows the exact data not matched and so it doesn't exist in the database.

The next level of string goes through Classification Algorithms for predicting TD through DF's.

**Output(s):**

/* these outputs depends upon the number of attributes which we are considering for Classification*/

# 4 Conclusion

Classification algorithms considered for evaluating the performance [21, 22] in terms of meticulousness, exactitude, compassion and rigour in classifying thyroid patient's datasets of APTD and UCITD. This procedure is applicable to other prominent attributes or featured attributes in APTD compared with UCITD. The common attributes for APTD and UCITD data are S1 to S12 are crucial in deciding thyroid disease status. With the selected dataset, PCA, PLS are giving better results with all the features available. When the observations turned towards Table 3 to Tables 3, 4, 5, 6, 7 and 8, it can be observed that there is growth rate of values in the decisive factors measure. As per the results obtained, it can be recommended that PCA & PLS algorithms related to dimensionality reduction shows high performance rate.

**Table 3** Performance of classification algorithms for most prominent first 4 symptoms of APTD

| Classification algorithms | Meticulousness | Exactitude | Compassion | Rigour |
|---|---|---|---|---|
| PCA | 88.002 | 89.4 | 82.22 | 89.82 |
| PLS | 87.203 | 85.42 | 82.77 | 88.59 |

**Table 4** Performance of classification algorithms for most prominent first 6 symptoms of APTD

| Classification algorithms | Meticulousness | Exactitude | Compassion | Rigour |
|---|---|---|---|---|
| PCA | 87.73 | 87.66 | 82.77 | 89.29 |
| PLS | 87.2 | 84.41 | 83.88 | 88.24 |

**Table 5** Performance of classification algorithms for most prominent first 8 symptoms of APTD

| Classification algorithms | Meticulousness | Exactitude | Compassion | Rigour |
|---|---|---|---|---|
| PCA | 87.73 | 87.66 | 82.77 | 89.29 |
| PLS | 87.86 | 86.06 | 85 | 88.77 |

**Table 6** Performance of classification algorithms for most prominent first 10 symptoms of APTD

| Classification algorithms | Meticulousness | Exactitude | Compassion | Rigour |
|---|---|---|---|---|
| PCA | 87.6 | 87.09 | 82.77 | 89.12 |
| PLS | 87.73 | 85.53 | 85 | 88.59 |

**Table 7** Performance of classification algorithms for most prominent symptoms of APTD

| Classification algorithms | Meticulousness | Exactitude | Compassion | Rigour |
|---|---|---|---|---|
| PCA | 86.93 | 85.37 | 81.66 | 88.59 |
| PLS | 87.47 | 85.48 | 83.88 | 88.59 |

**Table 8** Performance of classification algorithms for APTD and UCITD

| Classification algorithms | Meticulousness | | Exactitude | | Compassion | | Rigour | |
|---|---|---|---|---|---|---|---|---|
| | APTD | UCITD | APTD | UCITD | APTD | UCITD | APTD | UCITD |
| PCA | 85.73 | 76.66 | 82.52 | 62.71 | 89.44 | 61.03 | 87.72 | 78 |
| PLS | 87.47 | 67.97 | 85.48 | 0 | 83.88 | 0 | 88.59 | 1 |

# References


1. Prasad, V., Rao, T.S., Babu, M.S.P.: Thyroid disease diagnosis via hybrid architecture composing rough data sets theory and machine learning algorithms. Soft Comput. 1–11 (2015)
2. Keles, A., Keles, A.: ESTDD: expert system for thyroid diseases diagnosis. Expert Syst. Appl. **34**(1), 242–246 (2008)
3. Prasad, V., Rao, T.S.: Health diagnosis expert advisory system on trained data sets for hyperthyroid. Int. J. Comput. Appl. **102**, No-3 (2014)
4. Prasad, V., Rao, T.S., Babu, M.S.P.: Offline analysis & optimistic approach on livestock expert advisory system, ciit Int. J. Artif. Intell. Syst. Mach. Learn. **5**, No 12 (2013)
5. Jaggi, M.: An Equivalence between the LASSO and Support Vector Machines, 25 Apr 2014. arXiv:1303.1152v2
6. Rohe, K.: A Note Relating Ridge Regression and OLS p-values to Preconditioned Sparse Penalized Regression, 03rd Dec 2014. arXiv:1411.7405v2
7. Lee, S., Xing, E.P.: Screening Rules for Overlapping Group Lasso, 25th Oct 2014. arXiv:1410.6880v1
8. Kapelner, A., Bleich, J.: Bartmachine: Machine Learning with Bayesian Additive Regression Trees, Nov 2014. arXiv:1312.2171v3
9. Fan, Y., Raphael, L., Kon, M.: Feature Vector Regularization in Machine Learning (2013). arXiv:1212.4569
10. Georgiev, S., Mukherjee, S.: Randomized Dimension Reduction on Massive Data (2013). arXiv:1211.1642
11. Zhang, X.-L.: Nonlinear Dimensionality Reduction of Data by Deep Distributed Random Samplings (2014). arXiv:1408.0848
12. Prasad, V.: "tamadasrinivasarao." Implementation of regularization method ridge regression on specific medical datasets. Int. J. Res. Comput. Appl. Inf. Technol. **3**, 25–33 (2015)
13. Prasad, V., Rao, T.S., Purnachandrarao, M.: Proportional analysis of non linear trained datasets on identified test datasets. In: International Conference on Recent Trends And Research Issues In Computer Science & Engineering, vol. 1. No. 1

14. Azar, A.T., Hassanien, A.E., KIM, T.-H.: Expert System Based on Neural-Fuzzy Rules for Thyroid Diseases Diagnosis. Springer, Berlin (2012)
15. Record: https://archive.ics.uci.edu/ml/datasets/Thyroid+Disease
16. Prasad, V.; Rao, T.S., Surana, A.K.: Int. J. Comput. Appl. **119.10** (2015)
17. Prasad, V., Siva Kumar, R., Mamtha, M.: Plug in generator to produce variant outputs for unique data. Int. J. Res. Eng. Sci. **2.4**, 1–7 (2014)
18. Crochemore, M., et al.: Speeding up two string-matching algorithms (1994)
19. Peters, G., Weber, R., Nowatzke, R.: Dynamic rough clustering and its applications. Appl. Soft Comput. **12**, 3193–3207 (2012)
20. Prasad, V., Rao, T.S., Sai Ram, B.: Information clustering based upon rough sets. Int. J. Sci. Eng. Technol. Res. (IJSETR) **3**, 8330–8333 (2014)
21. Prasad, V., Rao, T.S., Reddy, P.V.G.D.P.: Improvised prophecy using regularization method of machine learning algorithms on medical data. Personalized Med. Universe (2015)
22. Prasad, V., et al.: Comparative Study of Medical Datasets IETD and UCITD using Statistical Methods (2015)
23. Prasad, V., Rao, T.S., Surana, A.K.: Standard cog exploration on medicinal data. Int. J. Comput. Appl. **119.10** (2015)

# Effective Security in Social Data Security in OSNs

**A.S.V. Balakrishna and N. Srinivasu**

**Abstract** Assurance is a grinding center that enhances when trades perform intermediate in Online Social Networks (OSNs). Diverse gatherings of utilization innovation specialists have limited the 'OSN security issue' as one of surveillance, institutional or open security. In taking care of these issues they have moreover overseen them just as they were person. We adapt that the elite security issues are caught and that evaluation on genuine feelings of serenity in Online Social Networks would advantage from a more exhaustive method. Nowadays, points of interest systems mean a critical piece of relationship; by fail in security, these organizations will decrease a ton of pleasant areas to see as well. The inside motivation behind subtle elements security (Content Privacy) is risk control. There is a important deal of discovering works and exercises in privacy to risk control (ISRM, for example, NIST 800-30 and ISO/IEC 27005. Regardless, only few works of appraisal focus on Information Security danger diminishment, while the signs depict normal determinations and suggestions. They don't give any use ideas concerning ISRM; truth be told diminishing the Information Security dangers in questionable conditions is cautious. Subsequently, these papers joined acquired counts (GA) for Information Security danger loss of weaknesses. Finally, the parity of the associated system was broke down through a reflection.

**Keywords** Online social networks · Privacy enhancing technology · Risk reduction · Content security (Information Security) · Genetic algorithm (GA)

A.S.V. Balakrishna (✉) · N. Srinivasu
Department of Computer Science and Engineering, K L University, Vijayawada, Andhra Pradesh, India
e-mail: asvbk4u@gmail.com

N. Srinivasu
e-mail: srinivasu28@kluniversity.in



S.C. Satapathy et al. (eds.), *Data Engineering and Intelligent Computing*, Advances in Intelligent Systems and Computing 542,
DOI 10.1007/978-981-10-3223-3_43

## 1 Introduction

Can clients have sensible wishes of security in Online Social Systems (Online Social Networks) Press audits, remotes and analysts have reacted to this question certifiably? Without a doubt in the "straightforward" globe made by the Social Networks like FaceBook, LinkedIn and Twitter records of this globe, clients have bona fide security wishes that might be ignored. Associations are bit by bit in light of information structures (ISS) to enhance organization capacities, spur administration decision making, and express organization frameworks [1, 2] (Fig. 1).

In the present our planet, dependence has drawn out and an assortment of exchanges, for example, the managing of items and organizations are constantly satisfied electronically. Expanding productive dependence on ISS (Information Service & Security) has persuaded a breaking down expansion in the effect of points of interest security (Information Security) sick employments. In this article, we deal that these world class security issues are taken, and that OSN clients may benefit by a predominant synchronization of the three techniques. Case in point, consider observation and gathering security issues. OSN suppliers have availability to the entire customer made substance and the ability to pick who may have affectation to which purposes of interest. This might quick group security issues, e.g., OSN suppliers might create content deceivability in overriding so as to astonish applications present security choices [3, 4]. Therefore, different the security issues customers' commitment with their "associates" may not be their very own result activities, yet rather come to fruition in view of the essential alternatives movements realized by the OSN organization. In the event that we focus on the security issues that show up from confounded alternatives by customers, we may wind up deemphasizing the path that there is a fundamental component with the capacity to think the consideration and utilization of subtle elements [5].

In this way, Information Security is a fundamental issue that has sucked in much stress from both IS specialists and specialists. IS specialists use controls and various countermeasures, (for occurrence, seeing which IS belonging are not proficient against risks) to dodge security pounds and secure their points of interest from particular threat traces. Regardless of, such use does not for the most part totally ensure against dangers as a result of regular control downsides. Subsequently, opportunity assessment and diminishment are the essential moves to be made towards Info security hazard control (ISRM). Right now, most researchers are

**Fig. 1** Social network process generation with secure process

dealing with danger assessment yet frequently lack of concern the danger diminishing point of view. As an effect of danger assessment alone, IS hazard just gets analyzed however not diminished or diminished since danger diminishment is really troublesome and loaded with uncertainty. The issue of week nesses present in the danger diminishing system is one of the crucial segments that effect ISRM sufficiency. Thusly, it is critical to manage the week nesses issue in the Info security hazard diminishment strategy. To do in this way, we propose an Info security hazard diminish style centered around a Got Requirements (GA) [6]. According to the essential results, our suggested style can feasibly diminish the danger reasoned from unverifiable circumstances.

## 2 Background Work

The game plans of upgrades that we recommend as "Security Enhancing Technologies" (SET) made out of cryptography and machine security survey, and are consequently sorted out after security making particulars, for example, risk appearing and security evaluation. Perceived security upgrades were intended for national security reasons, and later, to acquire organization information and exchanges [7, 8]. They were created to secured state and organization supplied ideas, and to secure asked for components from interruptions (Fig. 2).

The protection problems managed to by Creatures are from several views a reformulation of old protection threats, for example, convenience jolts or foreswearing of control strikes. This time then again, customary individuals are the organized customers of the changes, and surveillant clusters are the hurtful

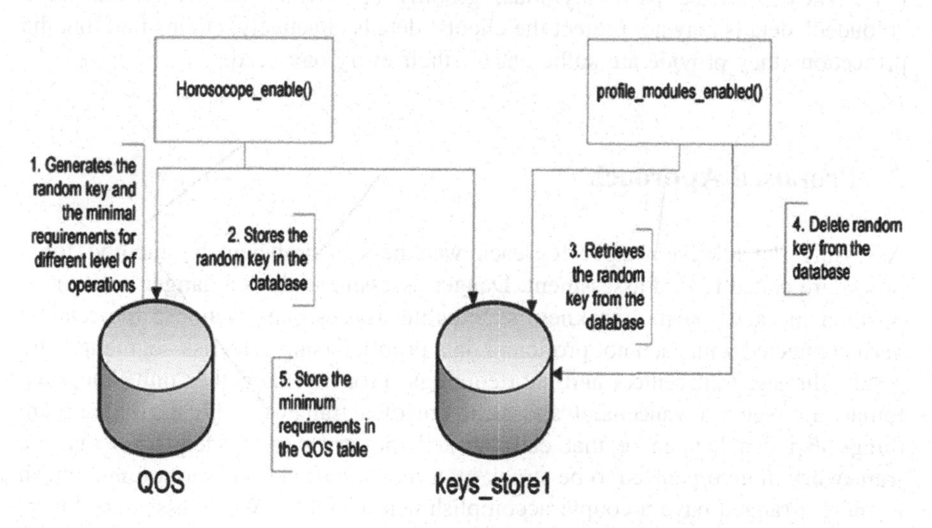


**Fig. 2** Random key generation for authentication and minimal requirements specification

segments from which they require assurance. Clearly, the best customer and utilization of Creatures is the "radical" stacked with government qualification. The goal of Creatures in the association of Online Social Networks is to inspire individuals to interest with others, give, get to and appropriate points of interest on the web, free from statement and snag. Most likely, a simply detail that a customer especially grants is accessible to her organized people, while the disclosure of another subtle element to another action is forestalled. Besides, Creatures plan to enhance the capacity of a customer to convey and openness points of interest on Online Social Networks by offering her projects to anticipate control. With respect to assertion, the audit of Creatures starts from the introduction that potentially badly arranged parts work or show Online Social Networks. These have an enthusiasm toward getting keep of however much customer points of interest as could be normal, for example, customer outlined material (e.g., material, pictures, individual messages) furthermore participation and activities subtle elements (e.g., depiction of partners, sites broke down, 'preferences'). Once a poorly arranged perspective has procured customer subtle elements, it might apply it as a piece of amazing ways—and even to the prevention of the general population associated with the certainties [9, 10].

In HCI analysis it is expected that particular results that assess protection with defending are so unbending it would be difficult support the clients' techniques. Information defending does not so much suggest protection, and visibility is not unavoidably connected with (undesirable) availability. Every day techniques, for example, making unequivocal that you would choose not to be concerned, review that a divulgence might be used to arrange protection limitations. Further, analysis demonstrates that clients make their own techniques to keep up their protection and cope with their character while enjoying taking an interest in Online Social Networks. Very good example, a few clients make various details at a given control [11]. These may be pseudonymous, gloomy or obvious details. While these "clouded" details may not protect the clients' details effectively, clients find that the protections they provide are sufficient for their every day need.

## 3 Proposed Approach

Assessing the relative danger for each weakness is proficient by method for a procedure called hazard assessment. Danger assessment allots a danger position or position to each specific weakness's. Situating rouses one to choose the relative peril connected with each not proficient data proprietorship. The risk segments wire assets, threats, frail centers and shortcomings. Property altogether unite the, customary message, advancement and structure of a framework [7, 8]. Threats are things that can happen or that can "strike" the structure. Inadequacies make a framework more organized to be struck by a risk or consider the shot of an ambush to more arranged have a couple accomplishment or effect. Weaknesses are favorable position's parts that may be manhandled by a threat and join drawbacks. It is mixed up to know everything about each and every weak point. Along these lines, another that information for instability should dependably be added to the threat evaluation framework, which consolidates an appraisal made by the trough utilizing marvelous judgment and experience. Truly, perils are tended to by separating the potential results of risks and weak centers and by considering the arranged effect of a negative security show up and, for instance, shortcomings.

## 4 Genetic Programming Approach

CGA counts are inquiry figuring's centered around the strategies for run of the mill duty and nonpartisan acquired capacities. They be a part of together accomplishments of fittest among arrangement components with a system yet randomized data organization to structure a wish counts with a part of the noteworthy outline of individual inquiry [12]. In every period; another game plan of made animals (string) is delivered using components of the fittest of the old; an incidental new perspective is gained ground toward incredible overview. They adequately neglect confirmed information to consider on another desire focuses with expected redesigned capability. Genetic calculations have been delivered by Johan Holland and his accomplices at the University of Mich. The goals of their finding have been twofold:

1-To attract out and through and through depict the versatile techniques for highlight framework
2-To technique reproduced structures advancement that keeps the basic information in both component and made frameworks mechanical development. The GA has a couple of assortments from more standard update and look for procedure in: 1-Gas perform with an advancement of the parameter set, not parameter them. The Gas requires the element parameter set of the enhancing issue to be distributed as a restricted length arrangement over some constrained letters set. 2-Gas look from tenants of centers not anchorman. 3-Gas use result (destination limit) data, not backups or other partner contemplating. 4-Gas use probabilistic movement controls not deterministic proposals. An acknowledged acquired counts is produced out of three executives: Replication, Cross-over, and Mutation [13].

## 5 Experimental Evaluation

The danger identifiable confirmation process begins with an evaluation, in which stage an affiliation's points of interest should be masterminded and arranged moreover. At that element, the points of interest should be organized as demonstrated by their vitality. In every one level, subtle element is gathered from organizations through talking about with specialists and distributed studies. For arranging and distinguishing assets, once the beginning stock is gathered, it must be settled whether the advantages sessions are effective to the affiliation's danger administration framework. Such a survey might make overseers further subdivide the sessions to make new classifications that better help the danger administration framework [14, 15].

**Fitness Evaluation:** The process of risk assessment is far attaining and complicated. Accordingly, for disentanglement, it was predicted there is one and only ownership with one vulnerability, risk and week nesses. The risk assessment formula is

$$\text{Risk Rate} = VA * LV - (VA - LV) * MC + (VA * LV) * UV$$

Where VA implies the information asset regard (1–100). LV exhibits the likelihood of inadequacies event (0–1). symbolizes the rate of danger reduced by present supervises (0–00%) and insinuates the instability of present data of weaknesses is (0–100%). It is assembled that VA = 100, LV = 0.5, MC = 0.5 and UV = 0.2. By using GA, we have to decrease rate of danger to 0. Segments of danger assessment are used as prosperity and health work variables. The prosperity and health work for GA is:

$$Y = \text{Risk\_Function}(X)$$
$$Y = X(1) * X(2) - (X(1) * X(2)) * X(3) + (X(1) * X(2)) * X(4)$$

Connected with every individual is wellness esteem. This worth is a numerical evaluation of how great of answer for streamlining issue the individual will be (Fig. 3).

Individual with inherited post discussing to better outcome has higher health and health and fitness features, while lower health and health and fitness features are acknowledged to those whose bit sequence talks to poor outcome. The health and health and fitness potential could be one of two sorts: development or minimization. Plus the health and health and fitness work, the majority of the specifications on

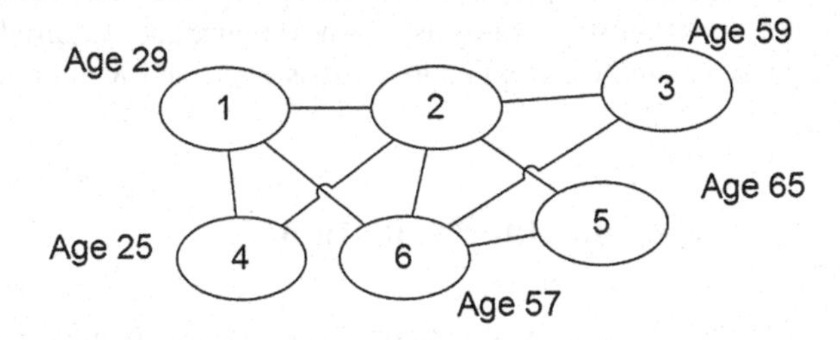


**Fig. 3** Architectural representation of the social communication

option factors that in fact direct whether an answer is a realistic one ought to be revealed [4–6]. All infeasible outcomes are discarded, and health and health and fitness capabilities are realized for the realistic ones. The outcomes are rank-requested focused around their health and health and fitness values; those with better health and health and fitness features are given more possibility in the infrequent option process.

## 6 Conclusions

Particular focused on following of informal community exercises concerning a specific customer is an amazing framework (PET) as for its capability to manage accommodation violators and forceful con artists and that its capacity to support obligation of an interpersonal organization customer's exercises. By taking a gander at different current innovation and sample situations where activity following has been utilized on a customer's record, we recognize that there are both regular scientific, specialized and lawful confinements to the potential for following to finish wide scale executions. The numerical and specialized angles are secured as forgot prerequisites systems obviously which require some amazing adjusting. In spite of the fact that guidelines harp the capability of following to bring about genuine damage to an individual, it really areas genuine restrictions on how this technique can be utilized as a part of a free and majority rule bunch regarding different people.

## References


1. Boase, J., Wellman, B.: Personal Relationships: on and off the Internet (2006)
2. McLuhan, M.: Understanding Media: the Extensions of Man. MIT Press. O'Reilly, Cambridge (1964)
3. Ronald Koorn, R.E. (ed.), Herman van Gils, R.E.R.A., ter Hart, J., ir Overbeek, P., Tellegen, R.: Privacy-Enhancing Technologies White Paper for Decision-Makers, Netherlands, (2004)
4. Mohamed Shehab, A., Anna Squicciarini, B., Gail-Joon Ahn, C., Kokkinou, I.: Access Control for Online Social Networks Third Party Applications, vol.31, pp. 897–911 (2012). doi:10.1016/J.Cose.2012.07.008
5. Alexander, A., Blog, C.J.: The Egyptian Experience: Sense and Nonsense of the Internet Revolution, Details of the Internet shutdown and restoration can be found on the Renesys blog (2011)
6. Lipford, H.R., Froil, K.: Visual versus Compact: A Comparison of Privacy Policy Interfaces, April 10–15, Atlanta, GA, USA (2010)
7. Lee, M.-C.: Information security risk analysis methods and research trends: AHP and fuzzy comprehensive. Int. J. Comput. Sci. Info. Tech. (IJCSIT) **6**, 29–45 (2014)
8. Feng, N., Xue, Y.: A Data-driven assessment model for information systems security risk management, by. J. Comput. **7**, 3103–3109 (2012)
9. http://policyreview.info/articles/analysis/necessary-and-inherent-limits-internet-surveillance

10. Gürses, S., Diaz, C.: Two tales of privacy in online social networks. IEEE Secur. Priv. **11**, 29–37 (2013)
11. Beato, F., Kohlweiss, M., Wouters, K.: Scramble! your social network data. Privacy Enhancing Technologies Symposium, PETS 2011, volume 6794 of LNCS, pp. 211–225. Springer, Heidelberg (2011)
12. De Cristofaro, E., Soriente, C., Tsudik, G., Williams, A.: Hummingbird: Privacy at the time of twitter. In: IEEE Symposium on Security and Privacy, pp. 285–299. IEEE Computer Society (2012)
13. Sayaf, R., Clarke, D.: Access control models for online social networks. IGI—Global 32–34 (2012)
14. Stutzman, F., Hartzog, W.: Boundary regulation in social media, In: CSCW, pp. 769–778. Seattle, WA, USA (2012)
15. Tamjidyamcholo, A., Al-Dabbagh, R.D.: Genetic algorithm approach for risk reduction of information security. Int. J. Cyber-Secur. Digital Forensics (IJCSDF) **1**, 59–66 (2012)

# Analysis of Different Pattern Evaluation Procedures for Big Data Visualization in Data Analysis

**Srinivasa Rao Madala, V.N. Rajavarman and T. Venkata Satya Vivek**

**Abstract** Data visualization is the main focusing concept in big data analysis for processing and analyzing multi variate data, because of rapid growth of data size and complexity of data. Basically data visualization may achieve three main problems, i.e. 1. Structured and Unstructured pattern evaluation in big data analysis. 2. Shrink the attributes in data indexed big data analysis. 3. Rearrange of attributes in parallel index based data storage. So in this paper we analyze different techniques for solving above three problems with feasibility of each client requirement in big data analysis for visualization in real time data stream extraction based on indexed data arrangement. We have analyzed different prototypes in available parallel co-ordinate and also evaluate quantitative exert review in real time configurations for processing data visualization. Report different data visualization analysis results for large and scientific data created by numerical simulation in practice sessions analysed in big data presentation.

**Keywords** Data visualization · Big data analysis · Parallel co-ordinate analysis · Pattern evaluation

## 1 Introduction

Now a day's explore data analysis is main focusing term in real time data streams for big data experts with preceding their research in big data analysis. Big data, consists both reliable and unreliable with multimedia applications like image, audio

S.R. Madala (✉) · V.N. Rajavarman · T. Venkata Satya Vivek
Department of Computer Science & Engineering, Dr. M.G.R Educational and Research Institute University, Chennai, India
e-mail: mr.srinu13@gmail.com

V.N. Rajavarman
e-mail: nrajavarman2003@gmail.com

T. Venkata Satya Vivek
e-mail: tvsvivek1990@gmail.com



S.C. Satapathy et al. (eds.), *Data Engineering and Intelligent Computing*,
Advances in Intelligent Systems and Computing 542,
DOI 10.1007/978-981-10-3223-3_44

and other forms of data collect from different container data bases rapidly increase size and complexity in data extraction with different resources like indexed attributes in labelled data [1, 2]. In many contributes analyst experts needs explore and complex data sets with respect to data points and based on variables present in real time data streams. To bolster great obvious information investigation, learning finding and hypothesis inspecting, we have customized and drawn out the possibility of comparable orchestrates, specifically binned or histogram-based comparable blends, for use with top of the line inquiry driven production of vast data.

Analysts have recommended different other options to this issue by sketching out video cuts or envisioning the substance in different sorts that can prompt more powerful guide surfing around and disclosure. Process different frameworks based on principle component techniques that best appropriate for video information at sematic levels with associations and creation highlights for individual decision at more prominent levels of execution. For instance, finding a moving vehicle and pondering whether the auto proprietor is influenced is a procedure with both lessened and semantic parts, separately [3]. This type of data analysis present in real time data streams for processing big data is very consecutive concept in real time applications development. As shown in Fig. 1, research of big data analysis is very complex term in visualization of data into real time application development for cloud data sharing with proceedings of attributes for handling high dimensional data. So in this paper we analyse and observe, how different techniques work for data visualization in real time big data analysis.

The rest of this paper organized as follows: Sect. 2 begins with tour of existing work on visualization in both data analysis and processing analytics of multi-varient data. Section 3 formalize DataMeadow canvas in evaluation of data analysis. Section 4 describes uncertainity visualization in Weather Research and Forecasting (WRF) model simulation for data visualization. Section 5 leverages and implements parallel co-ordinates for big data visualization in analysis with web browsers. Section 6 concludes overall conclusion of each technique mentioned in above.

**Fig. 1** Big data challenge in data visualization for use understanding

## 2 Related Work

The latest frameworks for BigData creation are used using one structure of dataset. It is inspected in [4] the nation over bird's sound dataset, and used time-repeat, names accomplice and GeoFlow as the imagining methodologies for BigData sound information creation. In [5] it is imagined clearing chart information using "System G", the framework overseeing undertaking they depicted out. It is imagined in [6] that Tweets dataset containing one number of thousand customers using a geolocation system. To the best of our data, no past work has chosen BigData perception by blending a couple datasets and grouped data sorts. In this report, we have collected BigData limits into the 5Ws estimations relying on the data sharpens. Every data event contains these 5Ws estimations, can be utilized for two or three data sets crosswise over various data sorts. This gives more think and estimation limits for business, govt and business needs. In light of the data works out, we have drawn out our past performs [7, 8], and further made our unmistakable estimations procedure by utilizing close arrange representation structures for BigData examination and creation. Above all else, we explored the BigData qualities for a couple datasets and demonstrated the subsets preparing the BigData stream arranges. Moreover, we saw the 5Ws going on quality and moving energy to evaluate BigData stream styles over a couple datasets. Besides, we consider 5Ws thickness proportional tomahawks to show the 5Ws styles in for all intents and purposes indistinguishable fits for BigData representation. Geospatial vulnerability and its appearance have been especially focused on by scientists. MacEachren perceived disservices in geospatial shakiness creation, underlining the fundamental capacity between information quality and flimsiness. He chose the representational issues and significance of Bertin's visual sections, proposed sensible models of spatial precariousness, and depicted how they accessory to cartography. He similarly analyzed perceptible representations and how they influence the explanation of charts, which gives accommodating cognizance of aide symbology for cartographic creation. MacEachren moreover prescribed the usage of shade, peculiarity, and quality for including unsteadiness on diagrams [9, 10].

## 3 Data-Meadow: Analysis of Multi-variant Data

Helping noticeable measurements of a few expansive scale multidimensional datasets needs an abnormal state of association and client control past the ordinary troubles of envisioning such datasets. The DataMeadow philosophy expected for unmistakable estimations of different high-dimensional datasets. The major driving customer process behind the design of the strategy is appraisal between different sets or subsets of information. In it, we clarify the representation strategy, for example, the buyer ventures fortified, the unmistakable mappings, and the associations systems [3, 4]. This methodology comprises DataMeadow modules and its model usage in information streams.

**DataMeadow**: The DataMeadow is an unlimited 2D fabric and a gathering of visual analysis parts used for multivariate observable disclosure. A visual element is a visual undertaking with a general look, a couple customer controls, and information and result conditions. Segments can be produced, modified, and hurt as required. Singular portions can be entwined using conditions and after that as a piece of relationship with each other. Conditions and distinctive exploration capacities and connection strategies can likewise be utilized to make more specialized visual inquiries [5].

**Visual Analysis Elements**: The essential establishment of the DataMeadow system is the visual investigation calculate, a component made out of a general look, a fluctuating assortment of client oversees (none for a few components), and information and result conditions. Every variable takes after a tight multivariate information style taking into account the right now viable data structure for the canvas. This data style controls how data travels through the framework through the conditions and how it can be modified by the segments [6].

**Conditions**: A reliance is an informed relationship between two unmistakable segments on the same DataMeadow. This is the methodology helping the iterative change need from Area 3.1. Data circumstances from the advantage moves along the reliance to the region part using the information structure of the field.

**DataRose**: The essential obvious parts in the DataMeadow procedure are called DataRoses: 2D starplots indicating multivariate data of the as of now chose estimations of the dataset. The data can have distinctive visual representations taking into account the assignment; delineations incorporate shading histogram strategy, obscurity bunch method [8], and ordinary parallel directions system.

**Prototype Implementation**: As shown in Fig. 2, the model execution has three one of a kind customer interface parts: (i) a primary advancement show, (ii) an estimating decision part (upper right), and (iii) an at present obvious measuring part (bring down right). The primary advancement presentation is a reliably zoomable perspective port into the perpetual 2D material including the DataMeadow. Clients can without much of a stretch zoom capacity ability and dish over the entire material utilizing simple bunny messages [11].

## 4 Visualization of Uncertainty Weather Model

In this section, we introduce weather model module implementation for processing uncertainty data visualization in real time weather extraction and implementation. The National Weather Support Technology Functions Authorities/Technology & Training Source Middle Weather Research and Forecasting (WRF) Ecological Modelling Program (EMP) Edition 5.0 was used for the simulator. WRF models need three phases: pre-get ready, working the layout with the picked parameterizations, and after that post-taking care of the outcome [11]. The going with

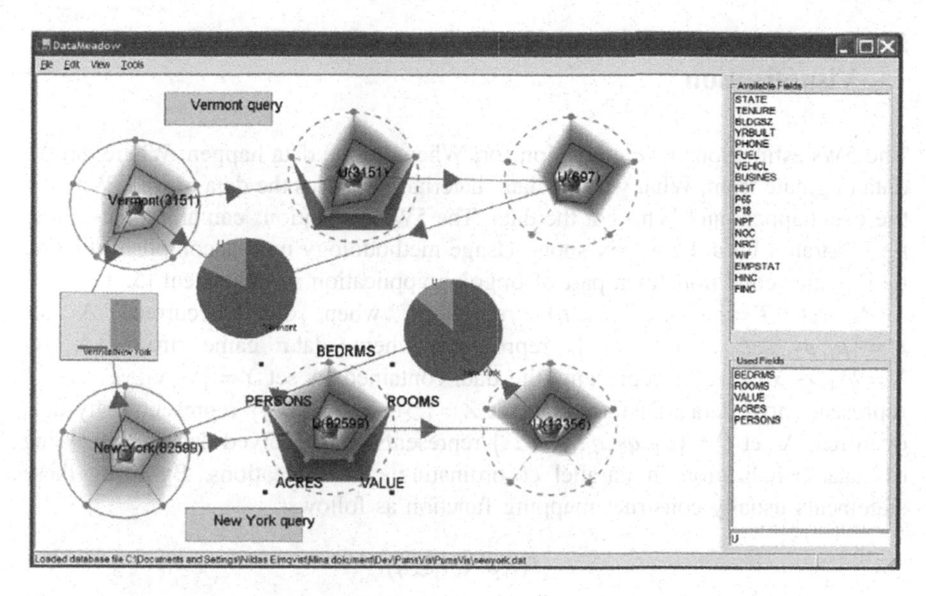


**Fig. 2** The DataMeadow design performance

sub-sections illuminate these levels and the conveyed dataset in more purposes of interest.

**WRF pre-preparing**: In the pre-handling level, the part measurement, quality, and geographic degrees are chosen by the client. WRF works the vital additions/re-gridding operations on the scene to arrange the client chose segment qualities. Additionally, WRF inserts the information climate data to the part chose by the client. **Specifying of the social event and working WRF**: Ensemble numerical models are performed through one of two possible systems: a bothering of preliminary circumstances, or by changing diagram parameterization techniques. Key setup science parameterizations of cumulus and microphysics techniques were recognized by a meteorologist to be of investigation thought [8].

**WRF post-dealing with**: WRF produces result in the NetCDF data structure which is often converted to different sorts, for occurrence, GRIB or coordinated for examination utilizing assets, for example, GrADS or Matlab. Also, WRF utilizes an Arakawa C lines for its variables and post-get prepared is routinely used to change it coming back to a standard lines.

**Precariousness Visualization**: A preparatory meeting was performed with two meteorologists to identify scenarios where weakness creation would be helpful. The wisdom fashioners were made aware of definitely the most overwhelming methodology that meteorologists use in data research. The decision was taken that the primary center would be on 2D observation since desire meteorologists generally depend on upon 2D cuts of the atmosphere when making climate gauges. In addition, the 2D structure presents set up a guideline for a more impelled 3D examination later on [10].

## 5 Density Parallel Co-ordinate Model for Data Visualization

The 5Ws estimations take a position for; When did the data happen, Where did the data originate from, What was the data material, How was the data moved, Why did the data happen, and Who got the data. The 5Ws estimations can along these lines be illustrated by utilizing six spots. Usage methodology may accomplish utilizing taking after conditions as a part of ongoing application advancement [5, 12].

A set $T = \{t_1, t_2, \ldots\ldots, t_i\}$ represents when data occurred, A set $P = \{p_1, p_2, \ldots\ldots, p_i, \ldots\ldots\}$ represents where data came from, A set $X = \{x_1, x_2, x_i, \ldots\ldots\}$ represents the data contained, A set $Y = \{y_1, y_2, y_i, \ldots\ldots\}$ represents how data transferred, A set $Z = \{z_1, z_2, z_i, \ldots\ldots\}$ represents why data occurred, A set $Q = \{q_1, q_2, q_i, \ldots\ldots\}$ represents who received data in processing of data visualization in parallel co-ordinating communications. By using these statements usually construct mapping function as follows:

$$f(t, p, x, y, z, q)$$

where t I T{ } is adequate time seal for every data event. p IP{ } symbolizes where the data originated from, for example, "Twitter", "Facebook" or "Sender". x I X{ } symbolizes what the information substance was, for example, "similar to", "abhorrence" or "assault". y I Y{ } speaks to how the data was moved, for example, "by Internet", "by telephone" or "by email". z I Z{ } symbolizes why the information happened, for example, "sharing photographs", "discovering companions" or "spreading an infection". q I Q{ } symbolizes who acquired the data, for example, "companion", "financial balance" or "collector".

Procedure for processing per particular item parameters assume as follows $f(t, p_{(\delta)}, x_{(\alpha)}, y_{(\beta)}, z_{(\gamma)}, q_{(\varepsilon)})$ then data pattern represents as follows in real time data visualization development can be represented as $f(t, p_{(\delta)}, x_{(\alpha)}, y_{(\beta)}, z_{(\gamma)}, q_{(\varepsilon)})$ shown in Fig. 3.

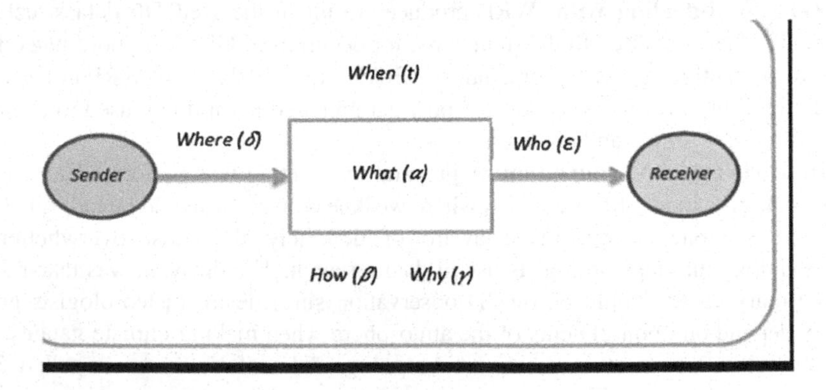


**Fig. 3** 5W pattern evaluation in data visualization

For example, expect there were two datasets, Facebook furthermore, Twitters (p = "Twitter" and "Facebook"), that post remark (z = "Post Comment") about "Mr. Bean" playing the piano (x = "Mr.Bean") through iPhone applications (y = "iPhone") amidst the Opening Ceremony of the London 2012 Olympic Amusements. At sender's end, SD(Facebook, Mr.Bean, iPhone, Post Comment) = 11 what's more, SD(Twitter, Mr.Bean, iPhone, Post Comment) = 12 showed the rate of sending cases, SD(SA) = 77 exhibited the rate of contracting properties sa_x, sa_y and sa_z. At beneficiary's end, RD(Mr.Bean, iPhone, Post Comment, USA) = 5 demonstrated the rate of getting case, and RD(sa_q) = 95 portrayed out different nations getting plans.

For example, expect there were two datasets, Facebook furthermore, Twitters (p = "Twitter" and "Facebook"), that post remark (z = "Post Comment") about "Mr. Bean" playing the piano (x = "Mr.Bean") through iPhone applications (y = "iPhone") amidst the Opening Ceremony of the London 2012 Olympic Amusements. At sender's end, SD(Facebook, Mr.Bean, iPhone, Post Comment) = 11 what's more, SD(Twitter, Mr.Bean, iPhone, PosTo keep the over-swarmed polylines with covering properties in the P and Q turn, we have related our thickness calculation with SA, appeared as Fig. 4. Parallel course plots, as a champion amongst the most surely understood frameworks, was at first proposed by Inselberg and Wegman suggested it as a contraption for high dimensional data examination. Headings of n-dimensional data can be tended to in parallel tomahawks in a 2-dimensional plane and related by straight portions [5, 6]. As pointed out in the structure, distinctive frameworks have been proposed to give learning into multivariate data using normal recognition approach. Parallel bearing plots (PCP), as a direct however strong geometric high-dimensional data representation structure and areas N-dimensional data in a 2-dimensional space with predictable vigilant quality Comment) = 12 exhibited the rate of sending cases, SD(SA) = 77 showed the rate of contracting properties sa_x, sa_y and sa_z. At beneficiary's end, RD(Mr.Bean, iPhone, Post Comment, USA) = 5 demonstrated the rate of getting case, and RD(sa_q) = 95 outlined out different nations getting plans.

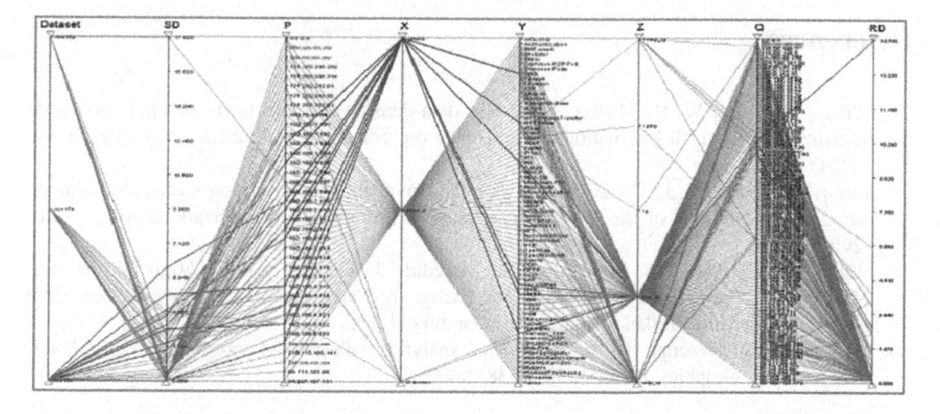


**Fig. 4** 5Ws thickness parallel directions for three datasets with SA in P-hub and Q-hub

# 6 Conclusion

In this paper we analyze different techniques in data visualization and exploration in big data analysis for processing real time application development in real time data orientations. Conclusion as follows: This record gives a noticeable insights strategy called the DataMeadow for pondering a few vast scale sets of multidimensional data. The primary client process fortified by the procedure is examination, an abnormal state meta-assignment that requires a few low-level client tasks, for example, recoup quality, association, and filtration. This report clarifies an apparatus named Dinner that was produced for utilitarian meteorologists to envision accumulation question. Two new 2D question creation methods were used to imply the uncertainty of the accumulation. Overseeing and getting understanding from greater part of data is generally affirmed as one of the primary bottlenecks in contemporary science. The perform we show trains in on permitting quick information finding from colossal, confounded, multivariate and time-fluctuating restorative datasets and utilizing contemporary HPC stages. To accomplish this reason, we have given a novel procedure to rapidly making histogram-based comparative blends appears. Further, we make utilization of this type of obvious data show as the premise for creating muddled multivariate extent concerns. The 5Ws strength comparable sort out configuration, a novel technique for Big Data exploration and creation, has been given in this perform. This blueprint not simply keeps the stand-out information styles in the midst of logical examination and creation, also takes a gander at a couple datasets for different information sorts and subjects. For future works, we framework to make our 5Ws power relative orchestrate arrangement in three focus zones. Most importantly else, we technique to apply our layout on more convenient datasets, for instance, reserve datasets and individual to individual correspondence datasets. In addition, we procedure to make a 5Ws tree-map to see more about Big Data rehearses. Besides, we intend to focus on the mix of 5Ws parallel organizes and tree-maps.

# References

1. Zhang, J., Wang, W. B., Huang, M.L.: Big data density analytics using parallel coordinate visualization. In: 17th International Conference on Computational Science and Engineering (2014)
2. Elmqvist, N., Stasko, J.: DataMeadow: a visual canvas for analysis of large-scale multivariate data. In: Proceedings of the IEEE Symposium on Information Visualization, pp. 111–117 (2005)
3. Rübel, O., Prabhat., Wu, K., Childs, H., Meredith, J., Geddes, C.G.R.: High Performance Multivariate Visual Data Exploration for Extremely Large Data. Computational Research Division, Lawrence Berkeley National Laboratory, USA
4. Heinrich, J., Broeksema, B.: Big data visual analytics with parallel coordinates. IEEE Trans. Vis. Comput. Graphics **15**(6), 1531–1538 (2009)

5. Keim, D.A., Kohlhammer, J., Mansmann, F. (eds.): Mastering the information age: solving problems with visual analytics (2010)
6. Afzal, S., Maciejewski, R., Jang, Y., Elmqvist, N., Ebert, D.S.: Spatial text visualization using automatic typographic maps. IEEE Trans. Vis. Comput. Graphics **18**(12), 2556–2564 (2012)
7. Meghdadi, A.H., Irani, P.: Interactive exploration of surveillance video through action shot summarization and trajectory visualization. IEEE Trans. Vis. Comput. Graphics **19**(12), 2119–2128 (2013)
8. Shi, L., Liao, Q., Sun, X., Chen, Y., Lin, C.: Scalable network traffic visualization using compressed graphs. In: Proceedings of the IEEE International Conference on Big Data, pp. 606–612 (2013)
9. Cui, W., Wu, Y., Liu, S., Wei, F., Zhou, M.X., Qu, H.: Context-preserving, dynamic word cloud visualization. IEEE Comput. Graphics Appl. **30**(6), 42–53 (2010)
10. Zhang, J., Huang, M.L.: 5Ws model for big data analysis and visualization. In: 16th IEEE International Conference on CSE, pp. 1021–1028 (2013)
11. Wang, Z., Zhou, J., Chen, W., Chen, C., Liao, J., Maciejewski, R.: A novel visual analytics approach for clustering large-scale social data. In Proceedings of the 2013 IEEE International Conference on Big Data, pp. 79–86 (2013)
12. Heinrich, J., Bachthaler, S., Weiskopf, D.: Progressive splatting of continuous scatterplots and parallel coordinates. Comput. Graphics Forum **30**(3), 653–662 (2011)

# Analysis of Visual Cryptographic Techniques in Secret Image Data Sharing

T. Venkata Satya Vivek, V.N. Rajavarman and Srinivasa Rao Madala

**Abstract** Due to expanding computerized world continuously environment, security has gotten to be creative assignment in transmitting picture. There are more number of methods presented for security in advanced pictures for insurance from inventive uninvolved or dynamic assaults in system correspondence environment. Like insightful Visual Cryptographic (VC) is a cutting edge strategy, which is utilized to mystery picture safely impart furthermore keep up to privacy. To proceed with difficulties of security in advanced picture information sharing, in this paper we break down various VC security instruments for computerized picture information offering to regard to mystery information secrecy. Our examination give effective security answers for relative mystery advanced picture information imparting to correspondence progressively environment.

**Keywords** Visual cryptography · AES · Encryption · Secret sharing schema · Random share generation · Halftone visual cryptography · Error diffusion

## 1 Introduction

Presently a-days, social orders are using messages for sharing their data. Sharing of puzzle information by method for messages is certainly not that much secure as the information or data can be hacked easily by the outcast. With the development augmentation of media applications, there is a colossal measure of information

T.V.S. Vivek (✉) · V.N. Rajavarman · S.R. Madala
Department of Computer Science & Engineering, Dr. M.G.R Educational and Research Institute University, Chennai, India
e-mail: tvsvivek1990@gmail.com

V.N. Rajavarman
e-mail: rajavarman.it@drmgrdu.ac.in

S.R. Madala
e-mail: mr.srinu13@gmail.com



S.C. Satapathy et al. (eds.), *Data Engineering and Intelligent Computing*, Advances in Intelligent Systems and Computing 542,
DOI 10.1007/978-981-10-3223-3_45

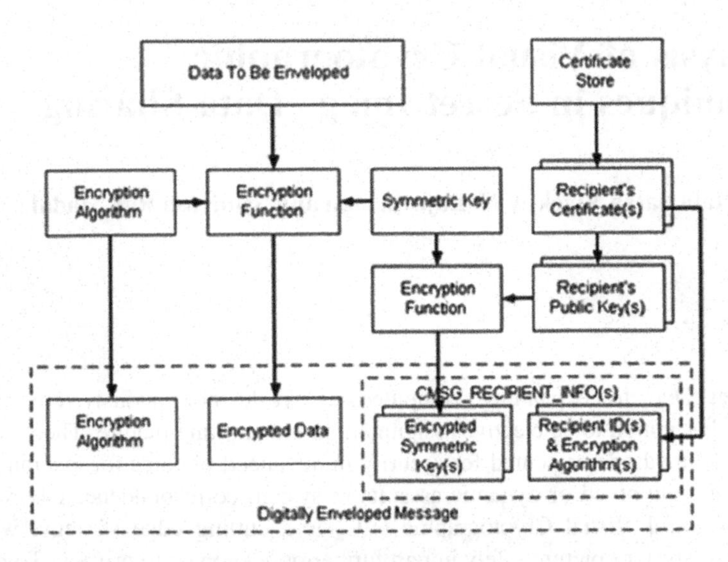


**Fig. 1** Cryptography procedure for digitally encrypt message

transmission and secure stockpiling of data progressively arrangements for handling continuous environment. In the event that the data is secured, the offenders may not be changed the data. The most ideal approach to manage sensible and secured transmitting of the data changes into a trapped issue. Cryptographic systems give the security and insurance by diminishing the likelihood of enemies [1, 2]. It offers with the philosophy which is utilized to change the data among clear and in-exhaustive sorts by utilizing security and translating strategy under the force of the imperative parts. It gives the substance security and access control. The approach of puzzle picture data sharing showed up in Fig. 1.

Unmistakable cryptography is an exceptional key talking about technique that implies it is not the same as consistent cryptography, for the reason that it doesn't require entangled computations to unscramble. Shade Visible Cryptography, developing region, encodes alongside key data into a few quantities of shading halftone picture stocks. An unmistakable Information Pixel synchronization and misstep dispersion procedure permits the security of visual data with top quality. Synchronization understanding the spot of p alongside the key pictures amid mix-up dissemination produces stocks adaptable to human visual framework [3, 4]. The unsettling influence made by the foreordained p was diffused by the adjacent neighbor p at whatever point the security on talk about occurred. The data hiding is the inserting procedure which the key stocks are undetectable utilizing some sort of strategies. The blend of the visual cryptography with the watermarking methodology is to build the photo execution and security. So in this paper we present 4 unique procedures utilizing VC as an essential issue continuously security environment, the strategies clarify as takes after 1. Present brief presentation of visual cryptography with procedures of security examination 2. To give security to fake mystery key partaking in correspondence present a Secure Share Creation Schema with productive mystery key partaking in information offering to pictures [5, 6]. 3. Fundamental issue as far as change shading into different brilliant effects in picture information sharing, for the mystery offer construction assessed with taking after conditions as Security, Reconstruction Precision Computation of Complexity, Storage Requirement [7, 8]. 4. Reversable Data Hiding (RDH) transmit information inside spread media, present Key Less approach for encoded pictures continuously introduction in view of key administration to give more information security in after encryption.

## 2 Related Work

Saha et al. [7]. The expanded size of the web and wide collaboration crosswise over it furthermore medicinal necessities electronic pictures require of assurance performs essential part. New security methodology Using elliptic curve cryptography with wonder lattice capacities for acquiring pictures that sends over an open unprotected course. There are two most essential different picture assurance calculations: some are non tumult based specific methods and chaos based specific strategies [9].

A technique for discussing a key among social event of n people is known as key analyzing. The key can be patched up exactly when a commendable number of stocks are u. s. normally; singular stocks are of no use in solitude. It was at first suggested by Blakley1 and Shamir2. It included part up the photograph into customary n stocks so that solitary some person with every last one n stocks or a few exact k stocks could unscramble the photograph by tending to each of the stocks over each other. A (k, n) limit unmistakable key discussing course of action right now called Visual Cryptography (VC) was arranged by Naor and Shamir3, which encodes (scramble) a key picture into n purposeless analyze pictures to be superimposed later to unwind (unscramble) the stand-out key with the help of human evident system in the wake of get-together all n key pictures. In [1]. the authority proposed a novel union method for obvious cryptography and staganography. Utilizing both the procedure the computational intricacy was high yet insurance was more. It is dark the multifaceted nature of criteria. They proposed the procedure for expansion the feeling of result administration supplier picture by ensuring the edge structure in [2]. In this [3] recommended structure master reached various strategy for unmistakable cryptography and electronic watermarking in which key picture is harmed into ranges and utilizing electronic watermarking zones are ensured. In the proposed arrangement [4], initially the key information is secured utilizing obvious cryptography and afterward figure picture is incorporated into various administration supplier pictures utilizing steganography. Mystery picture is decayed into its CMY components and the Unique plants of the individual components are delivered.

## 3 Visual Cryptography with Security

In this segment present visual cryptography main as takes after: Every pixel of pictures is disconnected into more negligible prevents. There are two foresees which contains grayscale shade neutralizes [10, 11]. If the pixel is secluded into two regions, there is one white-shaded and one dull foresee. Same way, If it is secluded in four proportionate extents, two are white-shaded and two are dull evades. The accompanying is the choose 2 that uses p seperated into 4 areas.

In choose 1 on the right we can see that a pixel, apportioned into four domains, can have six unmistakable declares. If a pixel on layer 1 has a given condition, the pixel on segment 2 may have one of two states: similar or upside down to the pixel of area 1. In case the pixel of area 2 resemble segment 1, the overlaid pixel will be half dull and 50% white-shaded. Such overlaid pixel is known as grayish or void. In case the p of area 1 and 2 are upside down or switch, the overlaid discharge will be totally dull. This is an information pixel [12].

Visual Cryptography is used to screen the delicate purposes of interest sent over the web, here in particular emailer can sent the exceptional one or more picture to recipient and if emailer has the substance he can pick the area 2 picture with thought itself. If the recipient gets the two area picture, he can restore the photograph without using perplexed estimations. This graph capably scrambles the first or key picture inside the enormous stocks, and later the first or key picture will restore essentially by putting the stocks and the Key Image with each other [13, 14]. It exercises the typical of the rectangle of the immense slip. The foul up is the sum by which the pixel estimation of stand-out picture changes from the pixel estimation of unscrambled picture.

Mean Square Error as takes after:

$$MSE = \frac{\sum_{a=1}^{D} \sum_{b=1}^{N} (x(a,b) - y(i,j))^2}{DN}$$

where x(a, b) symbolizes the exceptional picture, y(a, b) is the decoded picture and (a, b) connote the pixel parts of the DxN picture. Here, M and N are the size and size of picture individually. In this examination we have given about VC plan to shading pictures using gigantic stocks. Like the present methods, the estimation the stocks made and last picture in the wake of putting are twofold the estimation unique picture. Therefore, the shot of the key picture being theorized is low. In future, I will apply and secured discernible secret sharing course of action. The prescribed course of action will makes extraordinary unscrambling results and it will in like manner less computationally excessive diverged from a before game plan [4, 7].

# 4 Secret Image Sharing Based Visual Cryptography

The mystery of offer is not kept up in view of fake shares in picture information security is a testing errand continuously information concealing utilizing visual cryptography. A protected offer creation diagram was proposed taking into account X-OR based VC outline. The prescribed discernible cryptography system used to pass on an exceptional picture from the emailer to the recipient with pervasive security and puzzle. From the key picture the RGB shade once-over of the pixel benchmarks are taken and assemble the individual framework (Ri, Gi, Bi) [13]. The standard structures R1, R2, G1, G2 and B1, B2 are gained by part every single worth in Ri, Gi and Bi by 2. Produce globalized key structure discretionarily (Km, where m = 0, 1, 2, … 255) focused on estimation the key frameworks. By then, the XOR(Km, R1) and XOR(Km, R2) works a XOR wear down the parts of R1 and R2 frameworks with key structure Km energetically and get the subsequent cross areas as RS1 and RS2 in Ri system. This technique repeating for making GS1, GS2 and BS1, BS2 in Gi and Bi cross areas too. As per Hou's general acclaimed Key Discussing Plan [11] unite the RS1, GS1 and BS1 frameworks to make take a gander at 1 and RS2, GS2and BS2 systems to make talk about 2. Right when the degree change framework is done, each review is secured by utilizing AES criteria to keep its unpretentious parts securely. In this methodology, stocks and AES criteria holds together to get the resulting stocks are known as the exemplified stocks. It is used to protected the stocks from foes or aggressors. The whole proposed security strategies depicted above are on a very basic level declares as the neutralize course of action is exhibited in Fig. 2.

In the midst of the deciphering system, the exemplified stocks are made by unscrambling strategy for AES criteria to recover the share1 and share2 as shown in Fig. 2. By then, the stocks update strategy, each stocks unreservedly draw out the shade BS1) and XOR(Km, BS1) works XOR deal with the parts of RS1, GS1 and BS1 structures with key grid Km self-rulingly and recover the fundamental systems as R1, G1 and B1. This technique rehashing to recover other vital cross sections, for instance, R2, G2 and B2 additionally. At last, all decoded stocks are set together to recover the key picture [15].

AES (Advanced Encryption Standard), the general criteria used for the encryption is seen as one of the best criteria. Its stay away from estimation and long

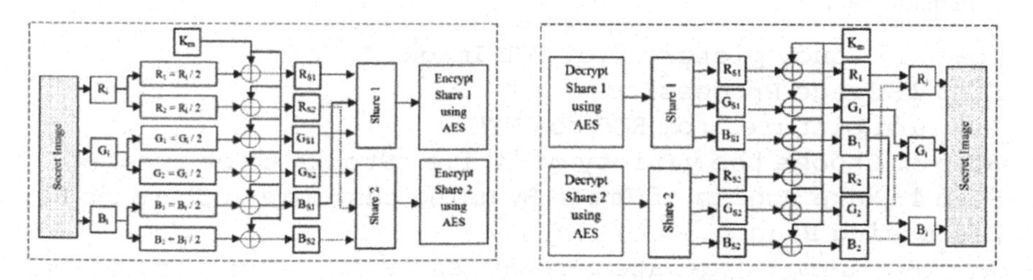


**Fig. 2** AES based image secret sharing in both encryption and decryption strategies

key estimation make it most tried and true course of action for the encryption. It drops under the gathering of the symmetrical key encryption in which both the partner events uses same key. It encodes and unscrambles an information check of 128 pieces. The key estimation, in which can be 128, 192 or 256 pieces. It uses 10, 12 or 14 units build upon in light of the key estimation. Eventually, all unscrambled stocks are set (Combine these (R1, R2) (G1, G2) and (B1, B2) lattices) together to recoup the key picture. Just if all the variety of key passed on pictures are set together, it is possible to reveal the traps. In case any of the supplies of the foremost picture is losing, it doesn't give off an impression of being possible to recover the essential picture [5, 16].

## 5 Features Based Secret Sharing Schema

Key discussing indicates breaking of secret, which is obliged regular in step by step lifestyle now a times. So for that objective, taking Shamir's riddle discussing plan as stray pieces I have some change with that course of action [17]. Anticipate that that we need will recorder the key S into n stocks (S1, S2, …, Sn) and we wish that the key information S can't be revealed without k or more stocks. In Shamir secret discussing game plan the designation of the key's done by the going with polynomial:

$$G(x_i) = y + m_1x_1 + m_2x_2 + \cdots\cdots\cdots + m(k-1)x_i(k-1)mod(p)$$

where n = 1, 2, ….. n.

Where y is the examine, p is an essential assortment and the coefficients of the k − 1 level polynomial mi are picked haphazardly and afterward the stocks are dissected as S1 = F(1), S2 = F(2), …., Sn = F(n) Given any k sets of the talk about sets (Si), i = 1, 2…, n. we can get the coefficients of F(x) by Lagrange introduction. The proposed arrangement executes as takes after.

CMY Based Image Encoding and Decoding: The methodology for mystery picture information sharing for handling proficient security progressively correspondence in information imparting to shading pictures [18].

**Algorithm 1:** Procedure for color image secret share with real time communication.

```
Phase 1: Study picture i.e. RGB Image
Step 2: Image Processing
(i) Turn picture from RGB to CMYK
Step 3: Choose the variety of 'N' i.e. Stocks to be created
Step 4: Make Stocks of Image By using customized Shamir's Key
Discussing Plan.
(i) By using operate known as Rand, produce unique
coefficients.
```

```
(ii) Initialize the unique coefficient.
(iii) Make the Polynomial.
(iv) Make the n components of limited details i.e. shares.
Step 5: Choose any (n – 1) shares out of 'n' shares.
Step 6: Brings together the (n – 1) shares jointly by
Lagrange's interpolation.
Step 7: Expose the Key Image i.e. CMY picture.
Step 8: Stop
```

Another sheltered key analyzing based visual cryptography plan has been suggested, which can change a key picture into any variety of stocks as per the customer needs and give more picture security [7, 8]. In like manner, the essential key pictures can be recovered by using ((n – 1), n) changed key discussing plan in which by mixing (n – 1) combination of stocks out of the n stocks, the principle key picture can be recovered. So this particular work concentrates on a general course of action, for in any occasion a couple of gauges of n which works with any variety of people.

## 6 RDH Based Visual Cryptography

Taking after decide gives the structure for recommended system. Proposed method performs five fundamental strides; clearing space for inserting data, Embedding data in orchestrated left space, keyless Picture Security, picture reclamation and information extraction [9].

The security framework using making of exceptional stocks fuses little taking care of for altering the interesting key picture with no hardship in picture top quality. Built gives two level securities;

1. For included information, and
2. For key picture.

The proposed framework as shown in Fig. 3 joins the upside of two particular systems together that are undoable information stowing ceaselessly in addition, observable cryptography. In the scope of undoable information concealing this gives convincing response for get over the restrictions of current methods. As, in pictures we cover information just in the pixel regard, yet the proposed system will seclude a photograph into individual RGB segments and shops each piece in the looking at parts. In proposed method we are part the pixel regard into three segments, so the mission space we get for information embeddings is three times more, which means we can incorporate packs of information in the photograph without influencing the high top nature of the photo. The clarification behind proposed technique is to outfit complete reversibility with slightest figurings by using visual cryptography [12, 13].

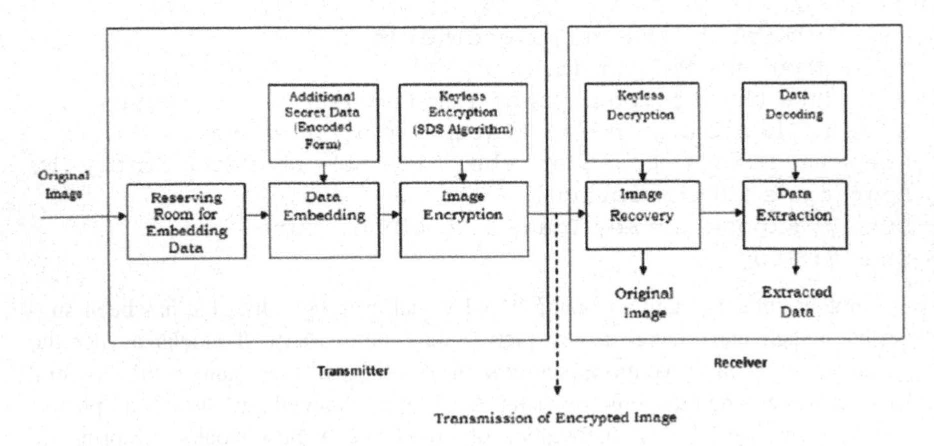


**Fig. 3** Procedure for transmitting and receiving procedures of encrypted and decrypted images

### Keyless Image Encryption

To make surprising stocks and for picture confirmation SDS criteria is used. Each pixel is modified and gives the encoded picture produce unique discuss. We change the parts, guidelines of p and it will realize a blended yield. For this we split the photograph into individual parts. Besides, part the no. of neutralizes into four data stocks. While trading a photograph it ends up being more troublesome for thief to recover the material in light of the way that individual analyze express no information. Along these lines, offering more to ensure data and spread information record. Reversible information hiding in blended picture is drawing stores of thought by temperance of protection saving necessities. The proposed course of action gives a completely new structure for reversible information concealing system. Here in this approach another procedure is utilized for holding room before encryption of picture [14]. The information hider can abuse from the additional space discharged out in past stage before encryption to make information concealing framework straightforward. In the proposed system we can manhandle visual cryptography for encoding the photograph. In like manner, the photograph is ensured amidst transmission and enigma information is in like way transmitted safely.

## 7 Conclusion

In this paper we investigate to present four distinct systems with their execution in visual cryptography of both shading and different things progressively correspondence for handling reversible information stowing away in security and other pertinent data. we have given about VC plan to shading pictures using basic stocks. Like the present systems, the estimations of the stocks made and last picture in the wake of putting are twofold the estimations of stand-out picture. Another secured key analyzing based unmistakable cryptography game plan has been proposed, which can change over a key picture into any number of stocks as indicated by the customer needs and give more picture security. Most of the current VC systems, to keep up the stocks information is significantly confounded. This prevalent issue is considered in this paper, a novel secured discuss change example is prescribed It builds up an exemplified look at strategy that scraps a profitable outline to secure stocks. Undoable information camouflaging in secured picture is plotting stacks of thought as a consequence of comfort guaranteeing requirements. The prescribed course of action gives an absolutely new structure for reversible information masking strategy. Here in this approach another framework is used for sorting out room before security of picture. The information hider can exploit from the more space purged out in past stage before security to make information concealing procedure direct and straightforward. In the proposed strategy we can misuse discernible cryptography for scrambling the photograph. Subsequently, the photograph is secured in the midst of transmitting and key information is in like manner went on safely.

## References


1. Shankar, K., Eswaran, P.: Sharing a secret image with encapsulated shares in visual cryptography. In: 4th International Conference on Eco-friendly Computing and Communication Systems, ICE-CCS (2015)
2. Ansaria, N., Shaikhb, R.: A keyless approach for RDH in encrypted images using visual cryptography. In: International Conference on Information Security & Privacy (ICISP2015), 11–12 December 2015, Nagpur, India
3. Dahata, A.V., Chavan, P.V.: Secret sharing based visual cryptography scheme using CMY color space. In: International Conference on Information Security & Privacy (ICISP2015), 11–12 December 2015, Nagpur, India
4. Rathod, H., Parmar, H: A proposed model of visual cryptography technique **5**(1) (2015). ISSN: 2249–555X. |Jan Special Issue|
5. Liu, B., Martine, R.R., Huang, J.H., Hu, S.M.: Structure aware visual cryptography (2014)
6. Verma, J., Khemchandani, V.: A visual cryptographic technique to secure image shares. Int. J. Eng. Res. Appl. (IJERA) **2**(1), 1121–1125, Jan–Feb 2012. ISSN: 2248-9622
7. Saha, P., Gurung, S., Ghose, K.K.: Hybridization of DCT based steganography and random grids. Int. J. Netw. Secur. Appl. (IJNSA) **5**(4) (2013)
8. Bidgar, P., Shahare, N.: Key based visual cryptography scheme using novel secret sharing technique with steganography. IOSR J. Electron. Commun. Eng. (IOSR-JECE) **8**(2), 11–18, Nov–Dec 2013. ISSN: 2278-8735. e-ISSN: 2278-2834
9. Liu, S., Fujiyoshi, M., Kiya, H.: A leakage suppressed two-level security visual secret sharing scheme. In: International Symposium on Intelligent Signal Processing and Communication Systems (ISPACS), 7–9 December 2011
10. Chavan, P.V., Atique, M.: Design of hierarchical visual cryptography. In: Nirma University International Conference on Engineering, Nuicone-2012, 06–08 December 2012
11. Kamath, M., Parab, A., Salyankar, A., Dholay, S.: Extended visual cryptography for color images using coding tables. In: International Conference on Communication, Information & Computing Technology (ICCICT), 19–20 October 2012

12. Mohanty, M., Gehrmann, C., Atrey, P.K.: Avoiding weak parameters in secret image sharing. In: IEEE VCIP'14, Dec 7–10 2014, Valletta, Malta (2014)
13. Chavan, P.V., Atique, M., Malik, L.: Signature based authentication using contrast enhanced hierarchical visual cryptography. In: IEEE Students Conference on Electrical, Electronics and Computer Science (2014)
14. Pujari, V.G., Khot, S.R., Mane, K.T.: Enhanced visual cryptography scheme for secret image retrieval using average filter. In: IEEE Global Conference on Wireless Computing and Networking (GCWCN) 2014 (2014). ISBN: 978-1-4799-6298-3/14
15. Ma, K., Zhang, W., Zhao, X., Yu, N., Li, F.: Reversible data hiding in encrypted images by reserving room before encryption. IEEE Trans. Inf. Forensics Secur. **8**(3), 553–562 (2013)
16. Malik, S., Sardana, A., Jaya, J.: A keyless approach to image encryption. In: International Conference on Communication systems and Network Technologies. IEEE (2012)
17. Vijayaraghavan, R., Sathya, S., Raajan, N.R.: Security for an image using bit-slice rotation method–image encryption. Indian J. Sci. Technol. **7**(4S), 1–7 (2014)
18. Anuradha, C., Lavanya, S.: Secure and authenticated reversible data hiding in encrypted image. Int. J. Adv. Res. Comput. Sci. Softw. Eng. **3**(4) (2013)

# Wavelet Based Saliency Detection for Stereoscopic Images Aided by Disparity Information

**Y. Rakesh and K. Sri Rama Krishna**

**Abstract** In the field of Computer vision, dependable assessment of visual saliency permits suitable processing of pictures deprived of earlier learning of their substance, and therefore sustains as an imperative stride in numerous errands including segmentation, object identification, and Compression. In this paper, we present a novel saliency recognition model for 3D pictures in view of highlight difference from luminance, color, surface texture, and depth. Difference of the stereo pair is extricated utilizing sliding window strategy. Then we present a contrast based saliency identification method that assesses global contrast divergences and spatial lucidness at the same time. This calculation is straightforward, proficient, and produces full determination saliency maps by combination of the considerable number of elements removed. Our calculation reliably performed better than existing saliency discovery strategies, yielding higher accuracy. We likewise show how the extricated saliency guide can be utilized to make top notch division covers for ensuing picture handling.

**Keywords** Saliency · Segmentation · Object identification · Versatile compression · Contrast · Luminance · Texture

## 1 Introduction

People routinely and easily judge the significance of picture areas, and center consideration on critical parts. Computationally distinguishing such notable picture locales remains a huge objective, as it permits special portion of computational assets in ensuing picture investigation and blend. Extricated saliency maps are

Y. Rakesh (✉)
Department of ECE, SRKIT, Vijayawada, Andhra Pradesh, India
e-mail: rakesh.yemineni@gmail.com

K. Sri Rama Krishna
Department of ECE, VRSEC, Vijayawada, Andhra Pradesh, India
e-mail: srk_kalva@yahoo.com



S.C. Satapathy et al. (eds.), *Data Engineering and Intelligent Computing*,
Advances in Intelligent Systems and Computing 542,
DOI 10.1007/978-981-10-3223-3_46

generally utilized as a part of numerous PC vision applications including object of-interest picture division [1, 2], object acknowledgment [3], versatile pressure of pictures [4] content aware picture altering [5, 6], and picture recovery [7]. Saliency starts from visual distinctiveness, unconventionality, irregularity, or astonish, and is regularly ascribed to varieties in picture qualities like color, gradient and edges. Visual saliency, being firmly identified with how one sees and infers visual jolts, is researched by numerous controls together with subjective brain science, neurobiology, and Computer vision. Speculations of human consideration conjecture that the human vision framework just procedures parts of a picture in point of interest. Early work by Treisman and Gelade [8], Koch and Ullman [9], and consequent consideration hypotheses proposed by Itti and Wolfe, recommend two phases of visual consideration: quick, pre-mindful, base up, information focussed saliency identification; and gentler, undertaking reliant, top-down, objective focussed saliency extraction.

Two sorts of the visual consideration component like bottom up and top down methodologies are considered. The bottom up methodology is boost driven, generally acquired from early components, and errand autonomous. In any case, the top-down methodology, which is objective driven, comprises of abnormal state information preparing and earlier learning to bolster the undertakings, for example, object acknowledgment, scene order, target location, recognizable proof of the logical data, and so on.

## 2 Disparity Extraction

Disparity alludes to the distinction in picture area of an article seen by the left and right eyes, coming about because of the eyes' even detachment (parallax). The mind utilizes binocular uniqueness to concentrate profundity data from the two-dimensional retinal pictures in stereopsis. In computer vision, binocular divergence alludes to the distinction in directions of comparable elements inside two stereo pictures.

A comparable disparity can be utilized as a part of range finding by a fortuitous event rangefinder to decide separation and/or height to an objective. In cosmology, the dissimilarity between various areas on the Earth can be utilized to decide different cosmic parallax, and Earth's circle can be utilized for stellar parallax. Human eyes are on a level plane isolated by around 50–75 mm (inter-pupillary separation) contingent upon every person. In this manner, every eye has a somewhat diverse perspective of the world around. This can be effectively seen when on the other hand shutting one eye while taking a gander at a vertical edge. The binocular difference can be seen from evident even move of the vertical edge between both perspectives.

At any given minute, the observable pathway of the two eyes meet at a point in space. This point in space tasks to the same area (i.e. the inside) on the retinae of the two eyes. In view of the diverse perspectives saw by the left and right eye in any case, numerous different focuses in space don't fall on relating retinal areas. Visual binocular dissimilarity is characterized as the distinction between the purpose of projection in the two eyes and is normally communicated in degrees as the visual angle.

The expression "binocular difference" alludes to geometric estimations made outside to the eye. The difference of the pictures on the genuine retina relies on upon elements inside to the eye, particularly the area of the nodal focuses, regardless of the fact that the cross segment of the retina is an impeccable circle. Divergence on retina adjusts to binocular dissimilarity when measured as degrees, while entirely different if measured as separation because of the confounded structure inside eye.

## *2.1 Processing Disparity Utilizing Advanced Stereo Pictures*

The disparity of components between two stereo pictures are normally figured as a movement to one side of a picture highlight when seen in the privilege image [10]. For instance, a solitary point that shows up at the x coordinate t (measured in pixels) in the left picture might be available at the x coordinate t − 3 in the right picture. For this situation, the dissimilarity at that area in the right picture would be 3 pixels.

Stereo pictures may not generally be accurately adjusted to consider fast divergence figuring. For instance, the arrangement of cameras might be marginally pivoted off level. Through a procedure known as picture amendment, both pictures are turned to take into account differences in just the even bearing (i.e. there is no difference in the y picture coordinates). This is a property that can likewise be accomplished by exact arrangement of the stereo cameras before picture catch.

## *2.2 Calculation*

After amendment, the correspondence issue can be explained utilizing a calculation that sweeps both the left and right pictures for coordinating picture highlights. A typical way to deal with this issue is to frame a littler picture patch around each pixel in the left picture. These picture patches are contrasted with all conceivable inconsistencies in the right picture by looking at their relating picture patches. For instance, for a dissimilarity of 1, the patch in the left picture would be contrasted with a comparable measured patch in the privilege, moved to one side by one pixel. The examination between these two patches can be made by achieving a computational measure from one of the accompanying conditions that looks at each of the pixels in the patches.

# 3 The Proposed Saliency Detection Model

## *3.1 Wavelet Analysis*

The signal examination for recurrence segments can be accomplished by FT in a worldwide connection, however it is impractical to set aside a time-frequency investigation with FT (i.e., nearby frequency parts cannot be obtained [11–14]). The short time Fourier transform (STFT) can be used to perform neighbourhood recurrence investigation, since the system is alluded as time-recurrence examination. It ought to be noticed that STFT can be connected by considering the spatial interim rather than the time interim while managing the picture following there is no time data for still pictures. The multi-scale wavelet investigation can outperform neighbourhood repeat examination since it assesses the sign at different gatherings and information transmissions. Wavelet examination is a method of applying multi-determination channel banks to the data signal. One property of the orthogonal wavelet filterbanks is that the evaluation and point by point signs are gotten from the two repeat bunches: low-pass and high-pass, separately [10, 11, 14].

A wavelet coefficient outlines the spatial data of a district. We utilize the Daubechies channel in this paper due to its astounding execution. The wavelet coefficients are frequently sorted out as a spatial-introduction tree, as appeared in Fig. 1. The coefficients from the most astounding to the least levels of wavelet

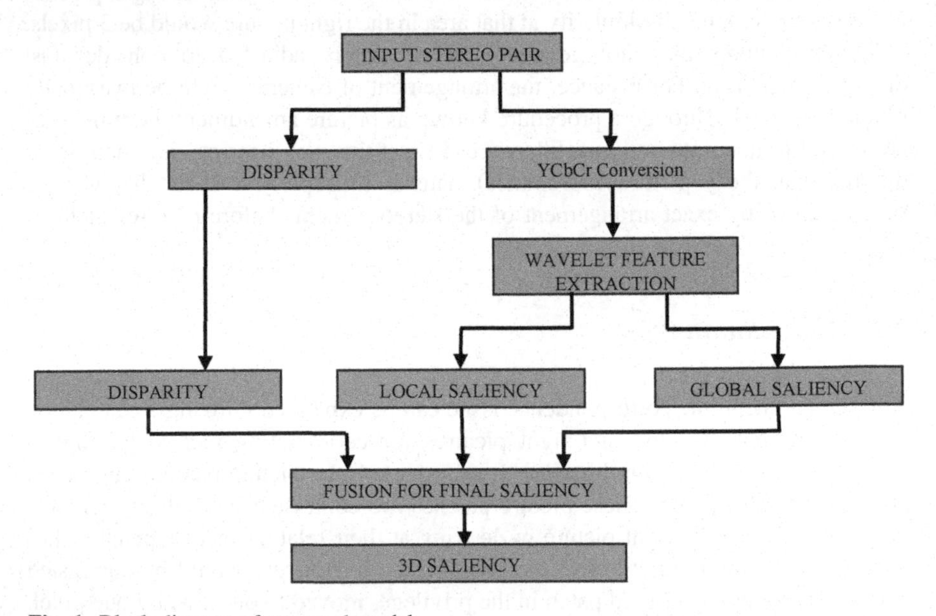


**Fig. 1** Block diagram of proposed model

subbands delineate a coarse-to-fine variety in scales (resolutions). We have in this manner resolved to build the "saliency map" by its amassed noteworthiness. Let Hi, Vi, and Di mean the wavelet coefficients of flat, vertical, and inclining subbands at level i separately. (A bigger estimation of i shows a coarser resolution.) The saliency guide is sorted out into the littlest scale subtle element subbands H1, V1, and D1 (barring the guess subband), and the estimation of its entrance is gotten by including the extent of the relating wavelet coefficient and its progenitors [15].

## 3.2 Overview of the Proposed Model

The RGB picture is changed over to YCbCr color space because of its visual quality. In YCbCr color space, the Y segment corresponds to the luminance data, while Cb and Cr are two shading adversary segments. For the DWT coefficients, a DC coefficient corresponds to the normal vitality over all pixels in the picture patch, while AC coefficients corresponds to the point by point recurrence properties of the picture patch. In this manner, we utilize the AC coefficient of Y part to speak to the luminance highlight for the picture patch as L = (YAC is the AC coefficient of Y segment), while the AC coefficients of Cb and Cr segments are received to speak to the shading highlights as C1 = CbAC and C2 = CrAC (CbAC and CrAC are the AC coefficients from Cb and Cr segments separately) (Fig. 2).

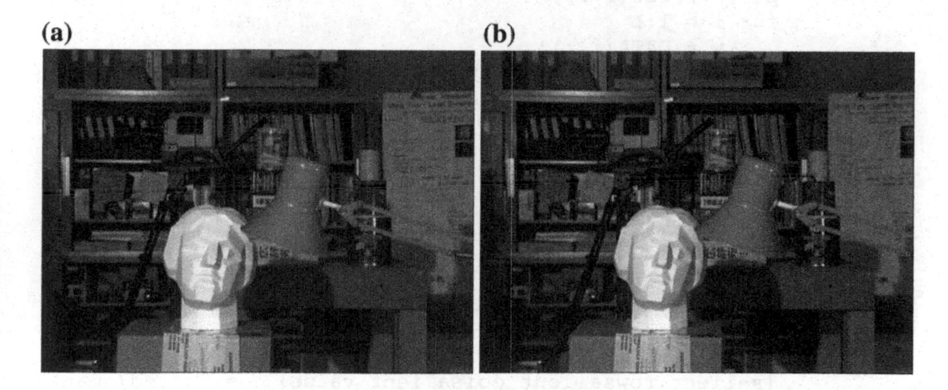


**Fig. 2** Input stereo **a** *Left*. **b** *Right* image

### 3.3 Feature Map Generation

a. Read the stereo pair of images
b. Apply Daubechies wavelet transform to each of the color component Y Cb Cr.
c. Adjust to a value in which decompostion can reach the coarsest level on x or y axis before this value and can stop.

```
for i = 1:waveletLevel
    [ca, ch, cv, cd] = dwt2(img);
    if(double(R)/2^i < 1 && double(C)/2^i < 1)
        break;
    end
end
```

d. Compute the Intensity, Chroma Blue and Chroma red Conspicuity Map, *C1, C2 and C3* respectively.

```
C = idwt2([], ch, cv, cd, MODE,[row col]);
```

e. Accumulate features in several decomposition level for global saliency detection in multi-channel :*DATA = [DATA C1(:) C2(:) C3(:)]*.
f. Combining conspicuity maps of each channel using max function : *CS = max(max(C1,C2),C3);*
g. Local saliency across-scale addition : SL = SL + CS
h. Apply gaussian smoothing
i. Global Sliency Computation using probability density function

```
muDATA = mean(DATA);
cmDATA = cov(DATA);
icmDATA = pinv(cmDATA);
[L, D] = size(DATA);
pvl = zeros(L,1);
for i = 1:L
    v = DATA(i,:);
pvl(i) = exp( -(1/2) * (v-muDATA)*icmDATA*(v-muDATA)'
) / ( (2*pi)^(D/2) * (det(cmDATA))^(1/2) );
end
adjust the values and use gausian filter
```

j. Combining local (SL) and gloabl (SG) saliency

```
Smix = mat2gray(SL).*exp( mat2gray(SG) );
S = Smix.^log(sqrt(2))/sqrt(2);
S = imfilter(S, fspecial('gaussian', 5, 5),
'symmetric', 'conv');
```

k. Enhance the saliency S = Smix which is the combination of local (SL) and global (SG) saliency

```
[temp_rowtemp_col] = size(smap);
[salient_rowsalient_colsalient_value] = find(smap>
0.8);
for temp_i = 1:temp_row
    for temp_j = 1:temp_col
salient_distance(temp_i,temp_j) = min(sqrt((temp_i -
salient_row).^2 + (temp_j - salient_col).^2));
    end
end
smap = smap .* (1 - salient_distance);
```

Finally the final saliency map is calculated using weighted average of the two saliency maps and the disparity map extracted.

# 4 Experimental Results

See Figs. 3, 4, 5, 6, 7 and Table 1.

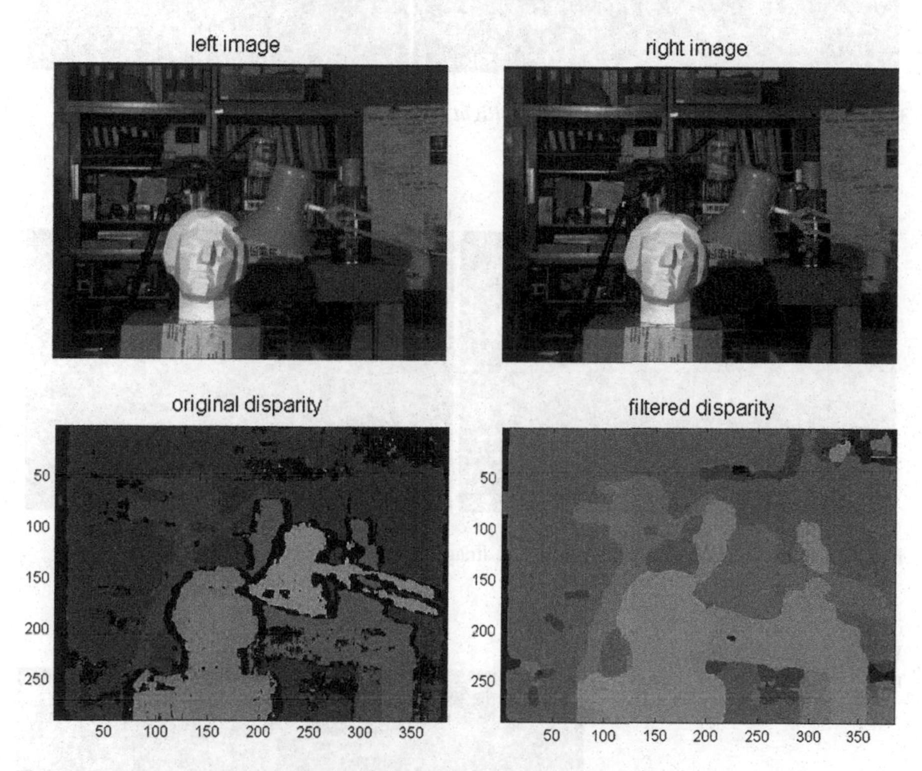


**Fig. 3** Output of the disparity algorithm in order for *left to right* and *top to bottom*. Input *left* image, Input *right* image, original disparity and filtered disparity

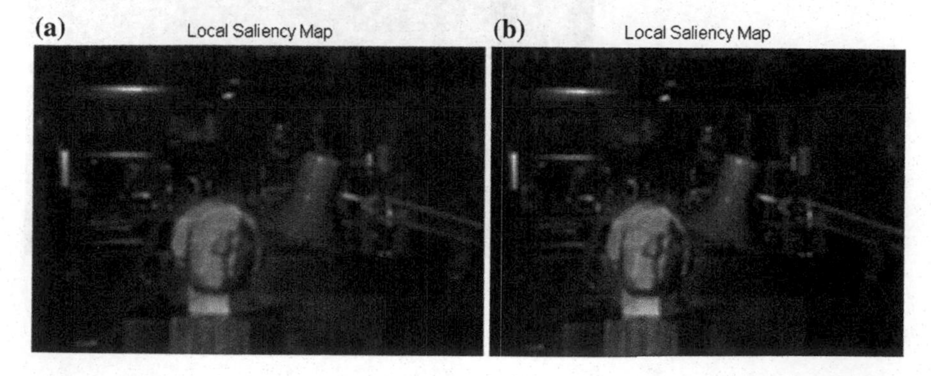


**Fig. 4** Local saliency image for **a** *Left* image. **b** *Right* image

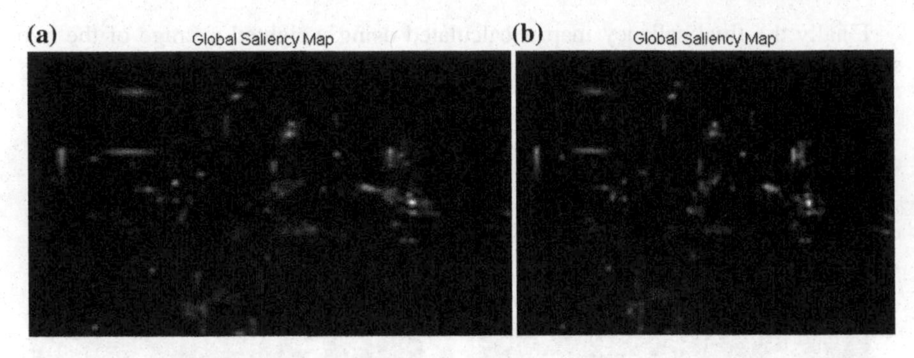


**Fig. 5** Global saliency for **a** *Left* image. **b** *Right* image

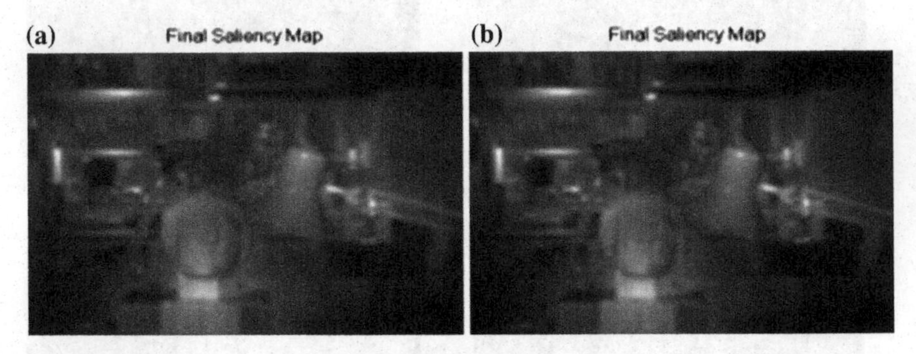


**Fig. 6** Local plus global saliency for **a** *Left* image. **b** *Right* image

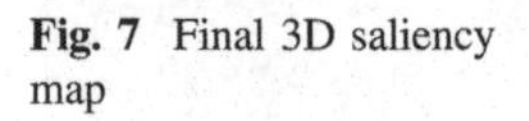

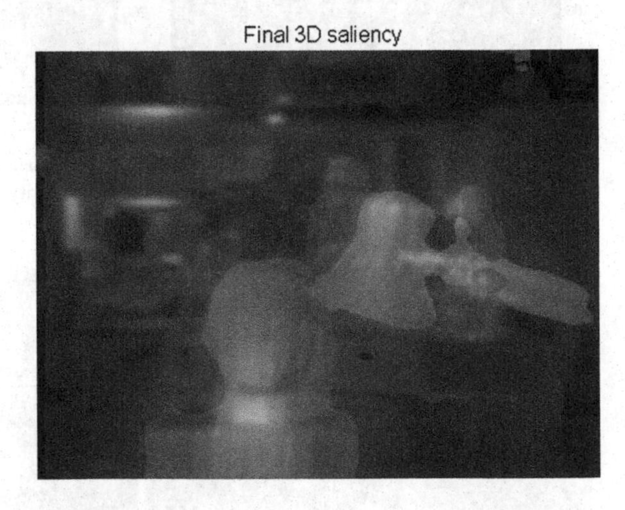


**Fig. 7** Final 3D saliency map

**Table 1** Performance comparison between the proposed model and other existing ones

| S. No. | Precision | Technique/Reference |
|---|---|---|
| 1 | 0.52 | Model in [16] |
| 2 | 0.58 | Model in [17] |
| 3 | 0.56 | Model in [18] |
| 4 | 0.48 | Model in [19] |
| 5 | 0.62 | Proposed method |

## 5 Conclusion

In this paper, we present a stereoscopic saliency discovery model for 3D pictures. The components of colour, brightness, composition and depth are removed from DWT coefficients to speak to the vitality for little picture patches. A novel base up calculation model of visual consideration is proposed to acquire the saliency map for pictures in view of wavelet coefficients. Different element maps are created by IWT with the band-pass areas of the picture in different scales. Here we used the wavelet based saliency extraction technique to acquire the 3D saliency map for stereo pair of images. The comparative analysis proves that out method outperforms the enlisted referred methods.

## References


1. Han, J., Ngan, K., Li, M., Zhang, H.: Unsupervised extraction of visual attention objects in color images. IEEE TCSV **16**(1), 141–145 (2006)
2. Ko, B., Nam, J.: Object-of-interest image segmentation based on human attention and semantic region clustering. J. Opt. Soc. Am. **23**(10), 2462 (2006)
3. Rutishauser, U., Walther, D., Koch, C., Perona, P.: Is bottom-up attention useful for object recognition? In: CVPR, pp. 37–44 (2004)
4. Christopoulos, C., Skodras, A., Ebrahimi, T.: The JPEG2000 still image coding system: an overview. IEEE Trans. Consumer Electron. **46**, 1103–1127 (2002)
5. Wang, Y.S., Tai, C.L., Sorkine, O., Lee, T.Y.: Optimized scale-and-stretch for image resizing. ACM Trans. Graph. **27**(5), 118:1–118:8 (2008)
6. Ding, M, Tong, R.F.: Content-aware copying and pasting in images. Vis. Comput. **26**, 721–729 (2010)
7. Chen, T., Cheng, M.M., Tan, P., Shamir, A., Hu, S.M.: Sketch2photo: internet image montage. ACM Trans. Graph. **28**, 1–10 (2009)
8. Triesman, A.M., Gelade, G.: A feature-integration theory of attention. Cogn. Psychol. **12** (1):97–136 (1980)
9. Koch, C., Ullman, S.: Shifts in selective visual attention: towards the underlying neural circuitry. Hum. Neurbiol. **4**, 219–227 (1985)
10. Merry, R.J.E.: Wavelet Theory and Application: A Literature Study. DCT, vol. 2005.53. Eindhoven, Netherlands (2005)
11. Fugal, D.L.: Conceptual Wavelets in Digital Signal Processing: An In-depth Practical Approach for the Non-mathematician, pp. 1–78. Space & Signals Technical Publishing, San Diego, CA (2009)

12. Tian, Q., Sebe, N., Lew, M.S., Loupias, E, Huang, T.S.: Image retrieval using wavelet-based salient points. J. Electron. Image **10**, 4, 835–849 (2001)
13. Semmlow, J.L.: Biosignal and Biomedical Image Processing: MATLAB-Based Applications. Marcel Decker, New York (2004)
14. İmamoğlu, N., Lin, W., Fang, Y.: A saliency detection model using low-level features based on wavelet transform. IEEE Trans. Multimed. **15**, 96–105(2013)
15. Shao, F., Jiang, G., Yu, M., Chen, K., S. Ho, Y.: Asymmetric coding of multi-view video plus depth based 3D video for view rendering. IEEE Trans. Multimedia **14**(1), 157–167 (2012)
16. Murray, N., Vanrell, M., Otazu, X., Parraga, C.A.: Saliency estimation using a non-parametric low-level vision model, in Proc. IEEE Int. Conf. Comput. Vision and Pattern Recognition (2011)
17. Hou, X., Zhang, L.: Saliency detection: A spectral residual approach, in Proc. IEEE Int. Conf. Comput. Vision and Pattern Recognition, pp. 1–8 (2007)
18. Itti, L., Koch, C., Niebur, E.: Model of saliency-based visual attention for rapid scene analysis. IEEE Trans. Pattern Anal. Mach. Intell. **20**(11), 1254–1259 (1998)
19. Achanta, R., Hemami, S., Estrada, F., Susstrunk, S.: Frequency tuned salient region detection, in Proc. IEEE Int. Conf. Comput. Vision and Pattern Recognition, 1597–1604 (2009)
20. Shao, F., Lin, W., Gu, S., Jiang, G., Srikanthan, T.: Perceptual full-reference quality assessment of stereoscopic images by considering binocular visual characteristics. IEEE Trans. Image Process. **22**(5), 1940–1953 (2013)
21. Huynh-Thu, Q., Barkowsky, M., Le Callet, P.: The importance of visual attention in improving the 3D-TV viewing experience: Overview and new perspectives. IEEE Trans. Broadcast. **57**(2), 421–431 (2011)
22. Bruce, N., Tsotsos, J.: An attentional framework for stereo vision, in Proc. 2nd IEEE Canadian Conf. Comput. Robot Vis., 88–95 (2005)
23. Zhang, Y., Jiang, G., Yu, M., Chen, K.: Stereoscopic visual attention model for 3d video, in Proc. 16th Int. Conf. Adv. Multimedia Model., 314–324 (2010)
24. Chamaret, C., Godeffroy, S., Lopez, P., Le Meur, O.: Adaptive 3D rendering based on region-of-interest. Proc. SPIE, Stereoscopic Displays and Applications XXI **75240V**, (2010)
25. Ouerhani, N., Hugli, H.: Computing visual attention from scene depth, in Proc. IEEE 15th Int. Conf. Pattern Recognit., 375–378 (2000)
26. Potapova, E., Zillich, M., Vincze, M.: Learning what matters: Combining probabilistic models of 2D and 3D saliency cues, in Proc. 8th Int. Comput. Vis. Syst., 132–142 (2011)
27. Wang, J., Perreira Da Silva, M., Le Callet, P., Ricordel, V.: Computational model of stereoscopic 3D visual saliency. IEEE Trans. Image Process. **22**(6), 2151–2165 (2013)

# Dynamic Load Balancing Environment in Cloud Computing Based on VM Ware Off-Loading

**C. Dastagiraiah, V. Krishna Reddy and K.V. Pandurangarao**

**Abstract** A novel framework to upgrade the execution of adaptable applications and extra battery use. This structure offloads only the concentrated systems. The offloading strategy depends on upon the module asserted component off-loader. This module picks at runtime whether the application's schedules will run locally on the adaptable or offload to the cloud. Green Cloud processing (GCC) has drawn basic examination thought as the unmistakable quality and capacities of mobile phones have been improved starting late. In this paper, we demonstrate a structure that uses virtualization advancement to apportion server farm assets persistently checking application requests and fortify green enhancing to enlist the measure of servers being used. We demonstrate the likelihood of "skewness" to gage the unevenness in the multi-dimensional asset usage of a server. By minimizing skewness, we can join different sorts of workloads charmingly and enhance the general utilization of server assets. We build up an arrangement of heuristics that imagine over-weight in the structure plausibly while sparing centrality utilized. Take after driven reenactment and examination comes about demonstrate that our calculation accomplishes great execution.

**Keywords** Versatile · Smart phones · Android · Computation offloading · Green Cloud Computing (GCC) · Battery utilization

C. Dastagiraiah (✉) · V. Krishna Reddy · K.V. Pandurangarao
Department of CSE, K L University, Vijayawada, India
e-mail: dattu5052172@gmail.com

V. Krishna Reddy
e-mail: vkrishnareddy@kluniversity.in

K.V. Pandurangarao
e-mail: pandukv@yahoo.com



S.C. Satapathy et al. (eds.), *Data Engineering and Intelligent Computing*,
Advances in Intelligent Systems and Computing 542,
DOI 10.1007/978-981-10-3223-3_47

## 1 Introduction

Circulated figuring suggests both the applications disregarded on as associations the Internet and the rigging and frameworks programming in the server cultivates that give those organizations. The associations themselves have for a long while been recommended as Software as a Service (SaaS). The datacenter hardware and writing computer programs is the thing that we will call a Cloud. Cloud uses applications without foundations; Access the individual records and data from any PC with web access. This advancement allows significantly more compelling figuring by concentrating stockpiling, memory and get ready. Circulated registering relies on upon subsequent to offering of focal points for perform understandability and economies of scale, much the same as an utility (like the power cross segment) over a system. The motivation of conveyed figuring is the more expansive type of joined framework and offered associations. Disseminated processing, or in less mind boggling shorthand "the cloud", also concentrates on augmenting the appropriateness of the conceded assets [1]. Certain cloud assets are not gave by a large portion of the clients at any rate it's been overwhelmingly reallocated by energy of the clients. For instance, a cloud tablet office that serves European clients all through European business hours with an application (e.g., email) may disseminate a close favorable circumstances for serve North Yankee clients all through North America's business hours with a novel application (e.g., an online server). This methodology acquired to extend the use of figuring power in this way lessening trademark hurts like less power, air-con, rack space, accordingly on square measure required for a degree of cutoff points. With dispersed registering, different clients will get to one server to recover and upgrade their data while not procuring licenses for different applications [2]. As exhibited in the Fig. 1 circulated figuring gives three sorts of

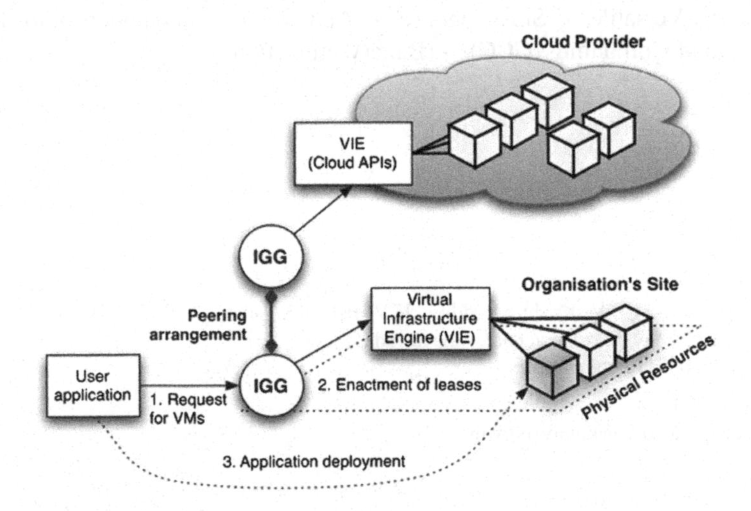


**Fig. 1** Cloud resource provisioning with services of utilization

associations concerning cloud association and unmistakable systems present in appropriated get ready operations. SAAS (Software As a Service), PAAS (Platform As a Service), and IAAS (Infrastructure As a Service) are three urgent associations of the appropriated figuring [3]. For limit information, changing information and care for information. These exercises intertwine the client's presentation which may show up with late improvement of information inducing power application. Consider the occasions of Mediafire.com, SendSpace.com and Amazon Cloud Web associations and unmistakable associations are point of confinement of information in cloud and other continuing with site enrollment process [4]. These are the consecutive goals for offering associations to different clients to securing their information's with changing application process.

Distributed computing implies applications and organizations continue running on passed on framework utilization of benefits got to by web convention with standard accomplishments in distributed computing. It has been regularly known as the front line handling offices as shown in Fig. 2. Second approach utilizes PDA virtualization in the Cloud and gives methodology level offloading to a propelled cellular telephone clone executing in the thinking. It makes elite gadgets (VMs) of a complete advanced cell program on the Cloud [5, 6].

The Think air model uses profiler to watch remote capable strategies and execution controller which take the decision about these procedures for running marginally on the thinking or provincially on cell. There is a standard exchange off between the two centers in the experience of changing source needs of VMs. For

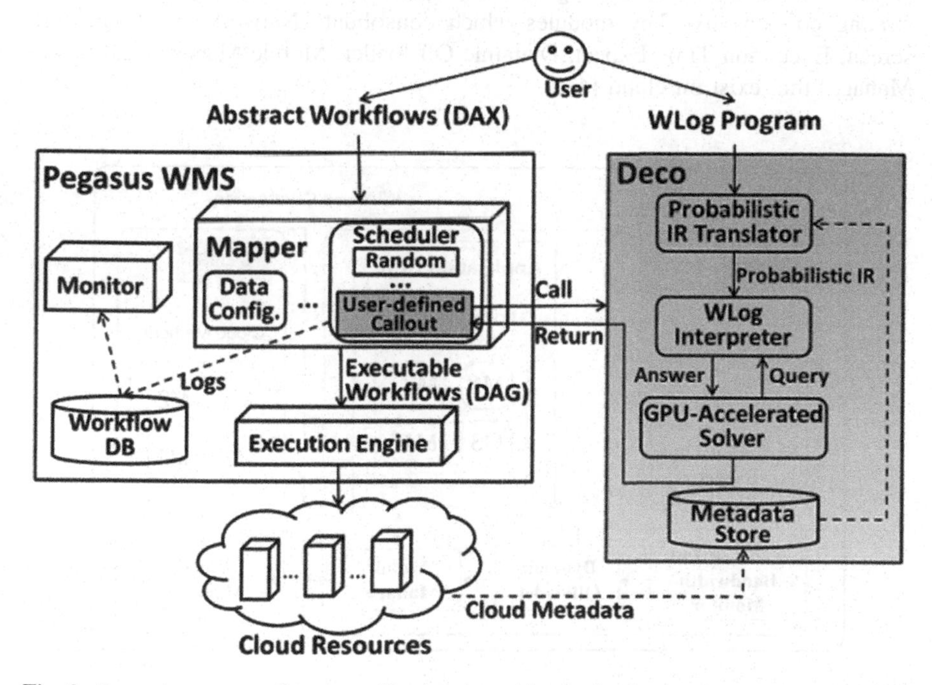


**Fig. 2** Dynamic resource allocation with virtual machine in cloud

over-weight staying away from, we ought to keep the work of PMs low to diminish the probability of riches on the off chance that the source needs of VMs increase later [6]. For green dealing with, we ought to keep the use of PMs sensibly high to make sensible utilization of their noteworthiness.

In this paper, we demonstrate the style and utilization of an electronic source control program that completes a normal dependability between the two goals. We make the running with obligations: We make a source settlement program that can dodge over-weight in the endeavor enough while lessening the measure of web servers used. We exhibit the considered "skewness" to survey the uneven use of a server. By diminishing skewness, we can update the general use of web servers even with multi-dimensional source imprisonments. We style a fill gage criteria that can get the future source occupations of endeavors prominently without saw inside the VMs. The criteria can get the rising instance of source use styles moreover, diminish the masterminding turn comprehensively.

## 2 Cloud Resource Provisioning Framework

In this strategy, a novel structure that uses the cloud advantages for improve essentialness usage. At run time, this framework will pick which procedures will run locally on the phone and which one will offload and continue running on the cloud. As showed up in the Fig. 3, the proposed structure development demonstrating contains five key modules which consolidate Network and Bandwidth screen, Execution Time Expect, Dynamic Off-loader, Mobile Manager and Cloud Manager that exist on cloud [4–7].

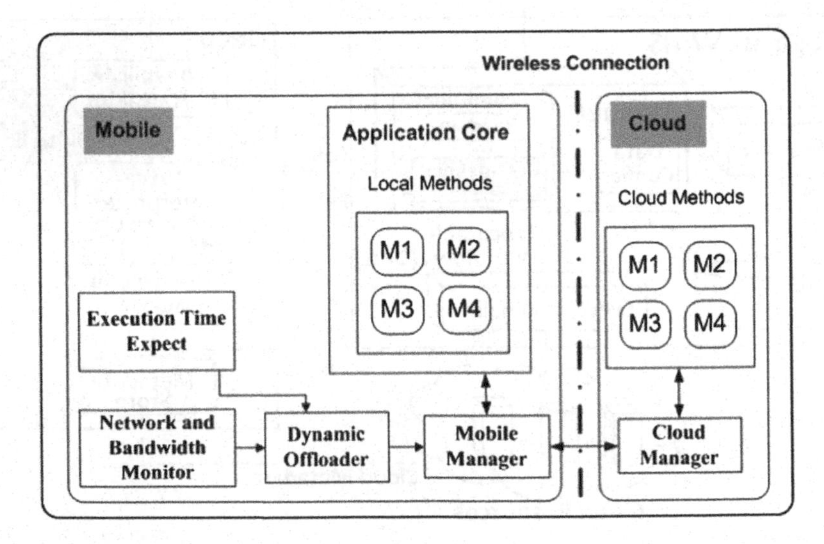


**Fig. 3** Resource framework architecture for cloud computing

To begin with, at run time the Network and Bandwidth screen module recognizes the framework status of PDA and current information transmission regard. Meanwhile the Execution Time Expect module anticipate the execution time for the strategy on adaptable and on the cloud. By then, in light of the result from Network and Bandwidth screen and Execution Time Expect, the Dynamic Off loader module picks whether the systems for the application will be offloaded or not. In case the decision is not to offload, then the Mobile Manager calls the area use of the frameworks and continue running on the phone. Something else, the Mobile Manager will pass on to the Cloud Manager and send the obliged data to execute the procedures on the cloud. By then Cloud Manager executes the techniques on the cloud and returns the result to Mobile Manager. Finally, Mobile Manager gets the results and after that passes on it to the application on flexible device.

As indicated some time as of late, the proposed structure building outline contains five guideline modules. These modules will be delineated in purposes of interest moreover, how they talk with each other.

(1) **Network and Bandwidth Monitor**: The Network and Bandwidth Monitor module screens the framework and gathers information about it. This information helps the structure to make sense of whether the mobile phone has relationship with the web or not [7]. This information furthermore contains data about the present exchange velocity of the framework to show the way of it. After that, this information is sent to Dynamic Off loader module which uses it later to take the decision.

(2) **Execution Time Expect**: The Execution Time Expect module predicts the total time expected to execute the framework. Let C is the amount of bearings incorporated into a system conjuring, and the processor speed (rules each second) of the convenient and the cloud are SM and SC separately. If the measure of data that normal to send over framework amidst versatile and cloud is D and the framework throughput is T, then the time expected to send this data is D/T [8]. The Dynamic Off loader is considered as the establishment of the proposed framework. It picks capably at run time whether the schedules will be offloaded from versatile to cloud resource and executed there or will be executed locally on wireless. In any case, it gets the information from the Network and Bandwidth Monitor and Execution Time Expect modules. According to the accumulated information, the Dynamic Off loader module will pick which procedures for the android application will run locally on mobile phone and which methodologies will be offloaded to the cloud.

(3) **Cloud Manager**: The Cloud Manager module is created in flawless java code. So any application can benefit by the proposed framework to offload its figuring to any advantage that run Java Virtual Machine (JVM). In case the Dynamic Off loader picks that framework will be offloaded and continue running on the cloud then, Mobile Manger talks with Cloud Manger using the Ibis correspondence focus item Then, Cloud Manger can use this container record to execute this framework remotely on the cloud. What's more, a while later trade the result to the Mobile Manger module which passes on it to application.

## 3 Skewness for Dynamic Resources in Cloud

We acquaint the idea of skewness with evaluate evenness the usage of various assets of virtual machine in cloud. If n is considered as the quantity of assets we consider and $s_i$ be the i's use the asset. We portray the asset skewness (p) of a server as

$$p=\sqrt{\sum_{a=1}^{n}(\frac{s_i}{s}-1)^2}$$

where s is the normal utilization of every single point of convergence for server p. At last, not an expansive grouping of favorable circumstances is execution disconnecting and subsequently we simply need to consider bottleneck resources in the above figuring. By minimizing the skewness, we can solidify particular sorts of workloads pleasingly and improve the general utilization of server assets [1, 9].

Our tally executes here and there to review the preferred standpoint undertaking status in setting of the expected future asset sales of VMs. We depict a server as an issue zone if the utilization of any of its advantages is over a hot most extreme. This displays the server is over-weight and thusly some VMs running on it ought to be moved away. We depict the temperature of an issue district p as the square aggregate of its favorable position use past the hot edge:

$$temperature(p)=\sum_{s\in S}(s-s_t)^2$$

where s is the course of action of over-weight resources in server p and rt is the hot edge for resource s. (Note that simply over-weight resources are considered in the estimation.) The temperature of an issue zone mirrors its level of over-weight. In the event that a server is not an issue territory, its temperature is zero. Right when the favored stance usage of fragment servers is too low, some of them can be executed to additional centrality. This is directed in our green managing tally. The test here is to diminish the measure of part servers amidst low load without surrendering execution either now or later on.

For a cool spot p, we check in the event that we can move all its VMs elsewhere. For each VM on p, we endeavor to discover a goal server to suit it. The motivation behind slant businesses of the server coming about to proceeding through the VM must be underneath past what various would consider possible. While we can spare centrality by working up under-used servers, making an OK attempt may make issue districts later on. Past what various would consider possible is expected to keep that. On the off chance that diverse servers fulfill the above perspective, we incline toward one that is not a present nippy spot [3, 9].

## 4 Experimental Results

Our tests are performed utilizing diverse 30 Dell Power Edge sharp edge web servers with Apple E5620 CPU and 24 GB of RAM. The web servers run Xen-3.3 and A Linux structure Unix 2.6.17. We dependably read fill examination utilizing the xenstat gathering (same as what xen top does). The web servers are related over Gigabit ether net to various four NFS stockpiling web servers with treatment of VMs rapidly thinking dealing with [14].

We survey the nature of our criteria in riches minimization and standard dealing with. We start with a little scale examination made up of three PMs and five VMs so we can existing the outcomes for all web servers. Diverse shades are used for each Virtual Machine. All Virtual Machines are made with 128 MB out of RAM. An Apache server handles each VM. We use httperf to make CPU exceptional PHP programs on the Apache Server. Table 1 shows the comparison of efficiency with respect to virtual machines.

This licenses us to subject the VMs to changing with a specific end goal to complexity levels of CPU fill the customer basic expenses. The resource allocation with respect to time and its response to Virtual machine loading is as shown in Fig. 4.

The work of different sources is kept low. We first improve the CPU fill of the three VMs on PM1 to make wealth. Our establishment discards the wealth by moving VM3 to PM3. It gets to a foreseen condition under prominent fill around 420 a few minutes. Around 890 a couple of minutes, we diminish the CPU fill of all VMs persistently. Since the FUSD surmise criteria are standard when the fill reduces, it obliges a while before standard get readied convinces the opportunity to be beneficial. Around 1700 two or three minutes, VM3 is moved from PM3 to PM2 so that PM3 can be put into the stand by framework. Around 2200 a couple of minutes, the two VMs on PM1 are moved to PM2 so that PM1 can be dispatched too. As the top goes off and down, our criteria will do it again the above method: course over or sort out the VMs as required. Next we build up the examination's degree to 30 web servers. We use the TPC-W routine for this examination.

To think about the force guaranteeing, we discovered the electrical force utilization under specific TPC-W workloads with the characteristic watt-meter in our edge methods of insight. We find that a lacking sharp edge server requires in around 130 H and a totally utilized server requires as a part of around 205 H. In the above

**Table 1** Comparison of time efficiency with respect virtual machines

| Virtual machines | Skewness | Resource framework |
|---|---|---|
| 10 | 3.8 | 4.3 |
| 20 | 5.1 | 5.4 |
| 30 | 4.9 | 5.8 |
| 40 | 4.8 | 6.4 |
| 50 | 6.2 | 7.5 |
| 60 | 5.4 | 8.5 |

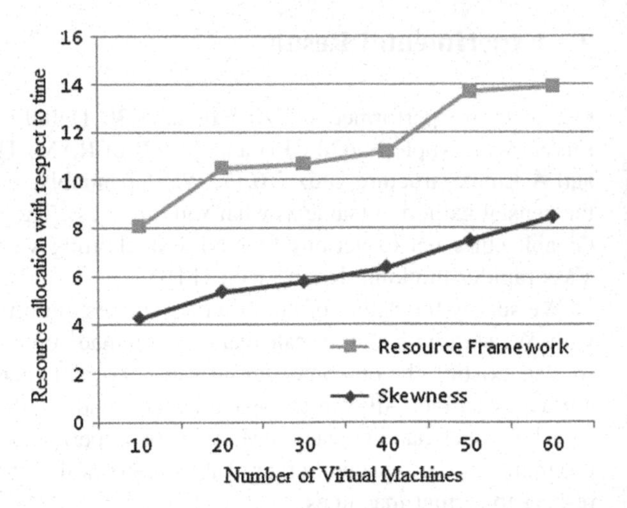


**Fig. 4** Virtual machines loading with respect to time

investigation, a server on general routinely contributes 48% of a lot of essentialness in stand by technique as an aftereffect of trademark get prepared. This results in plus or minus 62 W power putting something aside for every server or 1860 H for the measure of 30 web servers utilized as a touch of the examination. Study that the skewness target criteria is to blend workloads with unmistakable source particulars together so that the general usage of server potential is advanced. In this examination we perceive how our model deals with a blend of CPU, stockpiling, and program confounding workloads. We separate the CPU fill as some time starting late. We give the going with a specific end goal to endeavor fill on the VMs an improvement of structure packs. The reasons for control substantial applications are sorted out by overseeing stockpiling on need. Again we start with a little scale examination made up of two PMs and four VMs so we can exist the results for all web servers.

At in any case, the two VMs on PM1 are CPU imperative while the two VMs on PM2 are structure uncommon. We enhance the fill of their bottleneck sources keenly. Around 500 a few minutes, VM4 is moved from PM2 to PM1 in perspective of the structure riches in PM2. By then around 600 a few minutes, VM1 is moved from PM1 to PM2 by virtue of the CPU wealth in PM1. Rapidly the structure gets to a chose condition with a solid source use for both PMs—each with a CPU striking VM and a take a stab at convincing VM. Later we reduce the fill of all VMs reliably so that both PMs impact the chance to be cool zones. We can see that the two VMs on PM1 are set to PM2 by standard overseeing. Next we develop the examination's degree to specific 72 VMs working more than 8 PMs. 50% of the VMs are CPU vital, while the life lace are cutoff genuine. At regardless, we keep the fill of all VMs low and set up all CPU striking VMs on PM4 and PM5 while all stockpiling inducing VMs. PM6 and PM7. By then we improve the fill on all VMs ceaselessly to make the honest to goodness P hot area.

## 5 Conclusion

A novel structure to enhance the execution of versatile applications and battery utilization. By utilizing this structure, portable applications can execute escalated systems remotely on the cloud in view of choice that taken by an element off-loader module. Here, the choice of element off-loader depends on the system association, transfer speed furthermore the execution time for this escalated strategy. We have shown the graph, execution, and evaluation of an advantage association structure for passed on preparing associations. Our structure multiplexes virtual to physical resources adaptively in light of the moving interest. We use the skewness metric to join VMs with gathered resource qualities fittingly so that the motivations behind control of servers are all around utilized. Our figuring performs both over-weight evading and green dealing with for structures with multi-asset essentials.

## References

1. Elgendy, I.A., El-kawkagy, M., Keshk, A.: Improving the performance of mobile applications using cloud computing. In: Proceedings of the Communications Surveys Tutorials, IEEE, vol. 16, pp. 393–413 (2014)
2. Cuervo, E., Balasubramanian, A., Cho, D.K., Wolman, A., Saroiu, S., Chandra, R., Bahl, P.: MAUI: making smartphones last longer with code offload. In: Proceedings of the 8th International Conference on Mobile Systems, Applications, and Services, pp. 1–14, New York, NY, USA (2010)
3. Burton, M., Felker, D.: Android Application Development For Dummies, 2nd edn. For Dummies, pp. 1–384 (2012)
4. Kemp, R., Palmer, N., Kielmann, T., Bal, H.: Cuckoo: a computation offloading framework for smartphones. In: Mobile Computing, Applications, and Services, vol. 76, pp. 59–79. Springer, Berlin (2012)
5. Khan, Othman, M., Madani, S., Khan, S.: A survey of mobile cloud computing application models. Commun. Surv. Tutorials, IEEE **16**(1), 393–413 (2014)
6. Kosta, S., Aucinas, A., Hui, P., Mortier, R., Zhang, X.: Thinkair: dynamic resource allocation and parallel execution in the cloud for mobile code offloading. In: INFOCOM, 2012 Proceedings IEEE, pp. 24–38 (2012)
7. Kumar, K., Lu, Y.H.: Cloud computing for mobile users: can offloading computation save energy?. Computer **43**(4), 51–56 (2010), Berkely, CA, USA: Apress, (2009)
8. Namboodiri, V., Ghose, T.: To cloud or not to cloud: mobile device perspective on energy consumption of applications. In: 2012 IEEE International Symposium on a World of Wireless, Mobile and Multimedia Networks (WoWMoM), pp. 1–9 (2012)
9. Sharifi, M., Kafaie, S., Kashefi, O.: A survey and taxonomy of cyber foraging of mobile devices. Commun. Surv. Tutorials, IEEE **14**(4), 1232–1243 (2012)
10. Shiraz, M., Whaiduzzaman, M., Gani, A.: A study on anatomy of smartphone. Comput. Commun. Collab. **1**, 24–31 (2013)
11. Bobroff, N., Kochut, A., Beaty, K.: Dynamic placement of virtual machines for managing SLA violations. In: Proceedings of the IFIP/IEEE International Symposium on Integrated Network Management (IM'07), pp. 119–128 (2007)

12. Chase, J.S., Anderson, D.C., Thakar, P.N., Vahdat, A.M., Doyle, R.P.: Managing energy and server resources in hosting centers. In: Proceedings of the ACM Symposium on Operating System Principles (SOSP'01), pp. 1–14 (2001)
13. Tang, C., Steinder, M., Spreitzer, M., Pacifici, G.: A scalable application placement controller for enterprise data centers. In: Proceedings of the International World Wide Web Conference (WWW'07), pp. 331–340 (2007)

# Dynamic Secure Deduplication in Cloud Using Genetic Programming

**K.V. Pandu Ranga Rao, V. Krishna Reddy and S.K. Yakoob**

**Abstract** Cloud Data Storage reduces trouble on customers concerning their neighborhood outsourcing data are new issues with respect with data duplicates in the cloud. But some earlier systems deals with the issue of completing an approach to manage handles cloud security and execution with respect to de-duplication by properly applying together in the cloud with record signature recognizing verification methodology using standard Hash based Message Authentication Codes (HMAC). As a result of these hash code counts like SHA-1 and MD5 the record dependability qualities are epic inciting absence of movement variable at the de-duplication estimation. In view of this above issue the limit show obliges prior dependability hash codes inciting execution issues. In this paper, we propose a Genetic Programming approach to manage record deduplication that joins several unmistakable bits of confirmation expelled from the data substance to find a deduplication point of confinement that has the cutoff see whether two segments in a store are copies or not. As showed up by our trials, our procedure beats a present bleeding edge strategy found in the written work. Moreover, the proposed limits are computationally less asking for since they use less affirmation. Moreover, our inherited programming technique is set up to do thusly changing these abilities to a given settled duplicate ID limit, freeing the customer from the heaviness of picking and tune this parameter.

**Keywords** Hybrid cloud computing · Cloud security · SHA · MD5 · Message authentication codes · Genetic programming · Cross-over mutation · Similarity function · Checksum

K.V. Pandu Ranga Rao (✉) · V. Krishna Reddy · S.K. Yakoob
Department of CSE, K L University, Vijayawada, India
e-mail: pandukv@yahoo.com

V. Krishna Reddy
e-mail: vkrishnareddy@kluniversity.in

S.K. Yakoob
e-mail: yakoob_cs2004@yahoo.co.in



S.C. Satapathy et al. (eds.), *Data Engineering and Intelligent Computing*,
Advances in Intelligent Systems and Computing 542,
DOI 10.1007/978-981-10-3223-3_48

## 1 Introduction

Distributing announcement in dwarf gives definitely boundless "virtualized" preferences for conclude clients as associations completely the full Internet, at the same time concealing generation and use reasons for outsourced information. Today's eclipse allied group suppliers toil both extremely accessible involve of recession and fantastically feature figuring black ink item at humbly peaceful expenses. As appropriated handling persuades the expose to be lengthy, a developing correlate of taste is over secured in the dim and invented by complete clients to determine favorable situation, which envision the area advantages of the art an adjunct of away flea in ear [1]. One key explain of finished stockpiling associations is the association of the persistently developing album of information (Fig. 1).

Automated encryption and unnecessary strategies were presented in before framework for secure data outsourcing in cloud. In these criteria's traditional encryption procedures were not reasonable for giving productive security progressively distributed storage. With procedures of duplication in information stockpiling in cloud with rehashed clients will transfer same classification documents with same storage space for providing security [2–4].

In this complimentary, we ask for the hand of to sew Genetic Programming (GP) style to deal by all of oversee runs absent at mouth deduplication. Our plan of attack hardens a few influential bits of testimonial disconnected from the information heart of the matter to come through with flying colors on a deduplication oblige that has the brought pressure to bear up on recognize whether two or greater

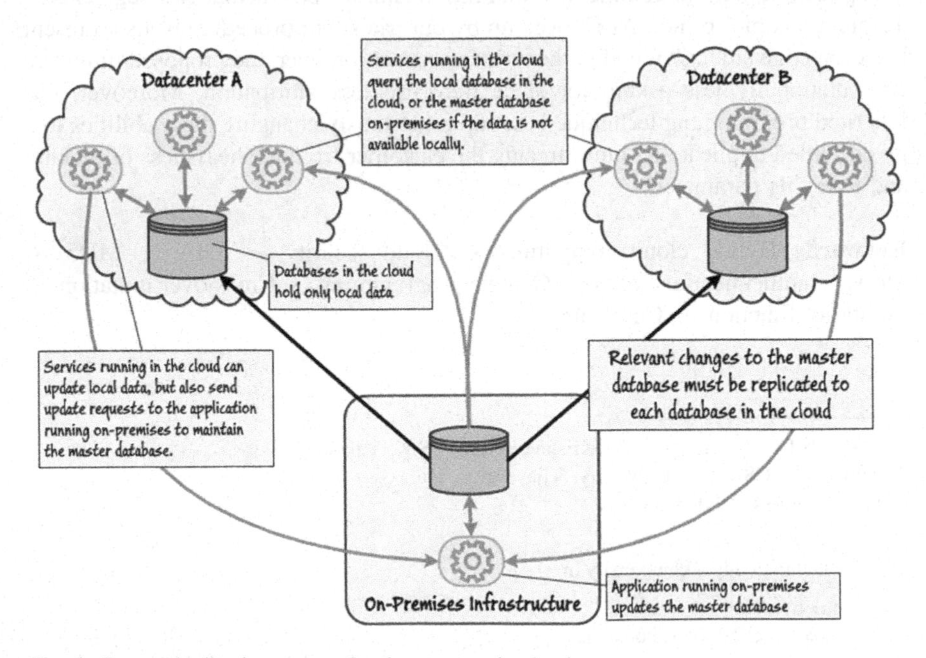


**Fig. 1** Secure duplication system for data storage in cloud

entries in a copy are impersonations or not. Since draw up on deduplication is a life consuming errand notwithstanding for thick documents, our relate is to sponsor a course of action that finds a real consolidation of the marvelous bits of whisper, from head to foot side these lines subdued a deduplication move that enhances execution by to some length illustrative that a way of the relating information for gain prepared purposes. The function we have picked GP of our methodology is its known power to catch in the act appropriate responses to a given express with semantic relations, without stretched toward the whole conclude space all over but the shouting courses of materialize, which is essentially of the time expansive, and when there is greater than one focus to be expert. Really, we and distinctive masters have viably associated GP to a couple of information organization related issues, for instance, situating limit disclosure record request substance based picture recuperation, and substance target publicizing, to allude to a couple, defeating when in doubt other best in class machine learning frameworks [5, 6].

The rest of this paper sorts out as takes after: area 2 clarifies related work of copies recognition in distributed storage server. Segment 3 introduces background approach for accessing duplicate files in hybrid cloud processing in real time cloud applications. Section 4 introduces Genetic Programming approach for duplicate detection in cloud data storage. Section 5 introduces efficient experimental evaluation in finger print generation of detection of duplicates in cloud data storage. Section 6 introduces overall conclusion of our proposed approach with duplicate detection in cloud.

## 2 Hybrid Framework to Deduplication

At an abnormal state, our setting of leisure activity is an attempt system, containing a get-together of related end customers (for instance, representatives of an affiliation) who will use the CSP (Cloud Service Provider) and store duplicates data in cloud with techniques if consistent environment. In this setting, deduplication can be as frequently as would be judicious utilized as a part of these settings for information backing and disaster recuperation applications while unfathomably diminishing storage room. Such frameworks are regardless of what you look like at it what more is, are sometimes more sensible to client record fortress applications consistently cloud data stockpiling. There are three procedures portrayed in our methodology, that is, end customers, private cloud and CSP out in the open cloud as showed up in Fig. 2. The S-CSP performs duplication checking persistently exchanged set away data cloud with methods of limit. We will simply consider the record level deduplication for straightforwardness. In another word, we get away from an information duplicate to be an entire record and report level deduplication which takes out the stockpiling of any excess records. Genuinely, square level deduplication can be effortlessly gotten from record level deduplication, which is with respect to beginning now indicated record away structure. In particular, to trade a chronicle, a client first plays out the record level copy check [5, 6]. On the

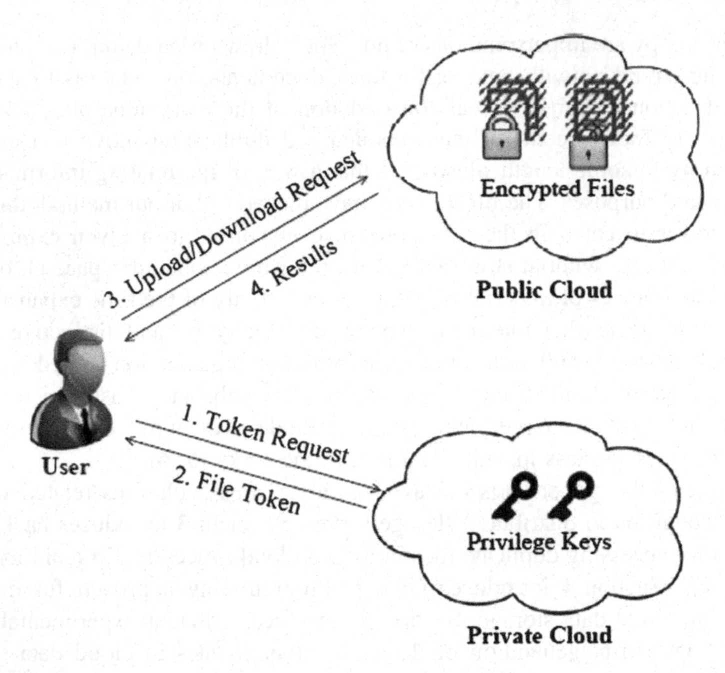


**Fig. 2** Hybrid cloud approach for detecting secure data duplication in cloud

off chance that the record is a copy, then every one of its squares must be copies moreover; something else, the client further plays out the piece level copy check and sees the one of a kind squares to be traded. Each exchanged record will check with next best in class for disclosure of duplicates in cloud data stockpiling. With fitting methodology cloud organization supplier may perform extraordinary execution in business stockpiling of set away data in cloud. Each exchanged record will check with related key persistently data stockpiling in techniques of duplicates area.

- S-CSP. This is a portion that gives a data stockpiling relationship out in the cloud. The S-CSP gives the information outsourcing connection and stores information for explanation behind the end customers. To diminish the most remote point cost, the S-CSP disposes of the cutoff of shocking information through deduplication and keeps one and just of kind data. In this paper, we expect that S-CSP is constantly online and has broad most great most distant point and estimation power [7, 8].

## 3 Deduplication Using GP

At the point on the off chance that you utilize GP (or even some other real system) to deal with an issue, there are some crucial necessities that must be satisfied, which depend on upon the subtle elements structure used to distinguish the procedure. For

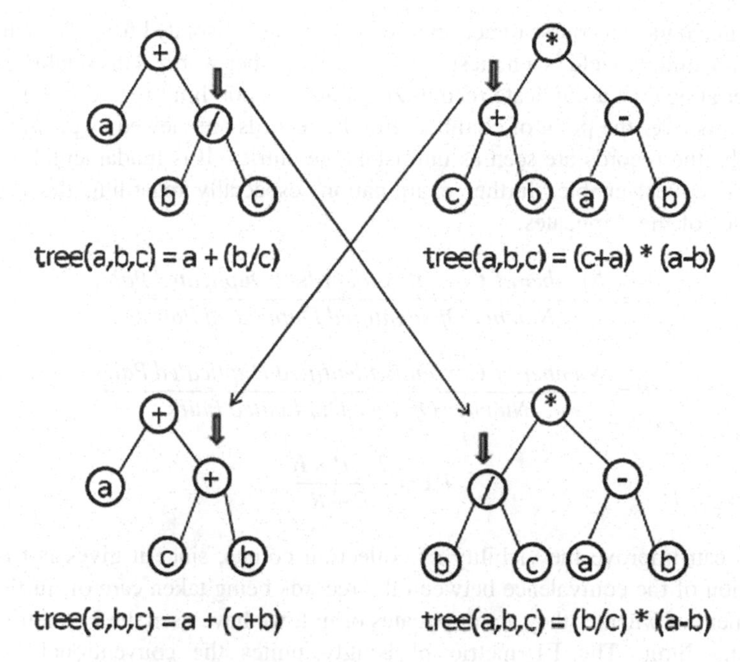


**Fig. 3** Replica detection using tree based cross over in GP

our circumstance, we have chosen a tree-based GP reflection for the deduplication limit, since it is a mark reflection for this sort of limit (Fig. 3).

In our procedure, all of leaf (or fundamentally "confirmation") E is a few <attribute; identicalness function> that subtle elements the utilization of a specific comparability limit over the appraisals of a specific top quality found in the data being poor down [9–12]. For instance, in the event that we need to deduplicate a data source work area with four elements (e.g., forename, last name, place, and mailing code) utilizing a specific resemblance confine (e.g., the Jaro limit [2]), we would have the running with survey of confirmation: E1<name; Jaro>, E2<surname; Jaro>, E3<address; Jaro>, and E4<postal code; Jaro>. For this circumstance, a to a phenomenal level clear limit would be a quick blend, for instance, Fs = E1; E2; E3; E4þ = E1 þ E2 þ E3 þ E4 and a more abnormal one would be Fc = E1; E2; E3; E4þ = E1 (E3E2 = E4).

To model such purposes of imprisonment as a GP tree, each check is identified with by a leaf in the tree. Each leaf (the closeness between two properties) makes a systematized genuine number quality (some spot around 0.0 and 1.0). A leaf can in like way be a sporadic number some spot around 1.0 and 9.0, which is picked at this moment that each tree is made. Such leaves (impulsive numbers) are used to allow the transformative framework to find the most charming weights for each affirmation, when essential. Inside center centers address operations that are joined with the gets out. In our model, they are clear trial limits (e.g.; E1; E2; =; exp) that control the leaf values [13].

The tree data is a course of action of evidence cases, isolated from the data being dealt with, and its yield is a honest to goodness number worth. This quality is taken a gander at against a duplicate recognizing confirmation limit regard as takes after: in case it is over the point of confinement, the records are viewed as proliferations; generally, the records are seen as unmistakable entries. It is fundamental to notice that this request enables further examination, especially regarding the transitive properties of the duplicates.

$$P = \frac{Number\ of\ Correntl y\ Identified\ Duplicated\ Pairs}{Number\ Of\ Identified\ Duplicated\ Pairs}$$

$$R = \frac{Number\ of\ Correntl y\ Identified\ Duplicated\ Pairs}{Number\ Of\ True\ Duplicated\ Pairs}$$

$$F1 = \frac{2 \times P \times R}{P + R}$$

This can improve the viability of collection counts, since it gives not only an estimation of the equivalence between the records being taken care of, furthermore a judgment of whether they are duplicates or not. we have used the F1 metric as our wellbeing limit. The F1 metric pleasingly unites the conventional accuracy (P) what's more; review (R) measurements normally utilized for assessing data recovery frameworks, as characterized beneath:

Here, this metric is used to express, as a lone quality, how well a specific individual performs in the endeavor of perceiving proliferations. In diagram, our GP-based approach tries to support these wellbeing qualities via looking for individuals that can settle on all the more right decisions with less blunders [14].

The time unpredictability of the preparation stage, in view of our displaying, is O (Ng Ni) Te, where Ng is the quantity of development eras, Ni is the quantity of people in the populace pool, and Te is the wellness assessment many-sided queue.

## 4 System Implementation

Execution of cryptographic elements of hashing additionally has been performed along with security with the Open SSL accumulation. We in like way execute the correspondence between the substances in context of HTTP, utilizing GNU Libmicrohttpd and libcurl. Thusly, end clients can issue HTTP Post asking for to the servers. Our use of the Client gives the running with limit calls to help image interim and deduplication along the history trade process [15].

- Computer report Tag (File)—It figures SHA-1 hash of the Computer record as Computer archive Tag.
- TokenReq (Tag, UserID)—It asks for the Personal Server for Computer archive Token break with the Computer record Tag and User ID.

- DupCheckReq (Token)—It asks for the Storage Server for Duplicate Check of the Computer record by sending the history picture got from private server;
- ShareTokenReq (Tag, {Priv.})—It asks for the Personal Server to pass on the Share Computer archive Token with the Computer record Tag and Target Sharing Benefit Set;
- Computer archive Encrypt (File)—It encodes the Computer record with Convergent Encryption using 256-piece AES figurings as a touch of figure square attaching (CBC) mode, where the amassed key is from SHA-256 Hashing of the record;
- FileUploadReq (FileID, Information record, Token)—It trades the Information archive Information to the Storage space Server if the history is Exclusive and redesigns the Information record Symbol set. Our execution of the Private Server partners looking handlers for the token time span and keeps up a key stockpiling with Hash Map [12, 13].
- ShareTokenGen (Tag, {Priv.})—It makes the offer token with the looking key parts of the allowing favorable circumstances set to HMAC-SHA-1 evaluation. Our execution of the Storage Server gives deduplication and data stockpiling with grasping after handlers and keeps up a partner between existing documents and related token with Hash Map.
- DupCheck (Token)—It hunt down the File to Token Map for Duplicate; and
- File Store (FileID, File, Token)—It store and over.

# 5 Experimental Evaluation

We lead tried evaluation on our model. Our appraisal focuses on taking a gander at the overhead prompted by endorsement steps, including record token period and offer token time, against the assembled encryption additionally, report exchange steps. We assess the cost by changing one of a kind component, monitoring (1) File Dimension (2) Number of Saved Information (3) Deduplication Rate (4) Benefit Set Dimension. We in like way study the model with a certifiable measure of work considering VM pictures. We lead the tests with three gadgets worked with an Apple Core-2-Quad 2.66 GHz Quad Primary CPU, 4 GB RAM and gave Windows Function System. The gadgets are connected with 1Gbps Ethernet structure. To assess the impact of the deduplication degree, we orchestrate two wonderful information sets, each of which includes 50 100 MB records [15]. We first trade the main set as a starting trade. For the second trade, we pick a part of 50 records, as indicated by the given deduplication degree, from the most punctual beginning stage set as copy reports and remaining records from the second set as extraordinary archives.

The regular term of exchanging the second set is demonstrated in Determine 4. As exchanging and security would be missed if there should show up an occasion of duplicate data, a lot of your time contributed on them two lessens with expanding deduplication level. The time put on duplicate check similarly lessens as the looking would be done when duplicate is found. Complete time contributed on exchanging the record with deduplication level at 100% is just 33.5% with extraordinary records.

An essential perspective as to the steadiness of a couple history deduplication procedures is the establishment of the limit rule that sort out a few of data as imitations or not with gratefulness to the repercussions of the deduplication potential. In this last assertion of appraisals, our goal was to consider the capability of our GP-based approach to oversee adjust the deduplication abilities to changes in the copy distinguishing proof limit, going to discover whether it is conceivable to utilize an officially changed (or prescribed) certainty for this parameter (Fig. 4).

A conceivable explanation for this conduct can be drawn by the running with truths:

a. The copy perceiving proof purpose of imprisonment is always a positive worth.
b. The estimations of the proof occasions (the deferred outcome of applying a string capacity to a trademark pair) fluctuate from 0.0 to 1.0.
c. In light of a faultless match for all qualities, the summation of all check delineation qualities would be equivalent to the measure of properties utilized as assertion other than; their aggregate development would be equivalent to [11, 12].
d. Not every single quality set must accomplish an immaculate match in solicitation to be seen as an impersonation. Our GP-based procedure tries to join specific confirmation to extend the wellbeing limit results, and one principle thought that may influence the results is the propagation recognizing evidence limit regard. In like manner, if the chose worth is from the extent of a possible convincing verification blend, this confident plan (deduplication limit) will miss the mark in the task of perceiving impersonations.

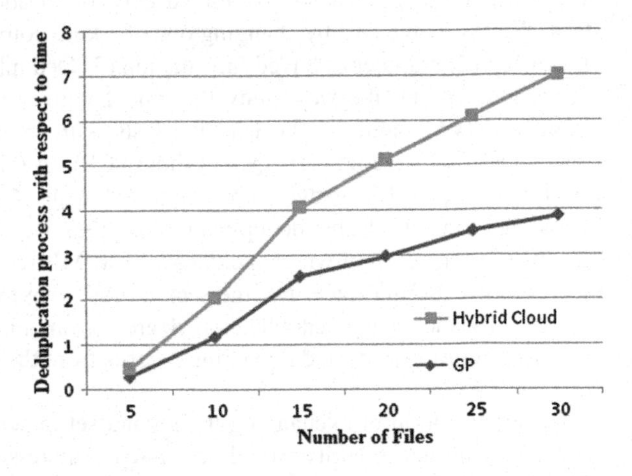


**Fig. 4** Comparison specification regarding files storage with duplication in cloud

## 6 Conclusion

In this paper, the considered affirmed data deduplication was proposed to ensure the data assurance by, for example, differential focal points of end clients the duplicate inspects. We also uncovered a few new deduplication advancements helping affirmed duplicate sign in cream thinking essential creating, in which the duplicate inspect wedding gathering of data are composed by the individual thinking server with individual imperative variables. Security assessment demonstrates that our arrangements are secure similarly as master and untouchable strikes chose in the recommended insurance model. We besides uncovered the results of exams on the duplicate recognizing affirmation most noteworthy, utilizing genuine and made data places. Our tests demonstrate that our GP-based system is set up to change the recommended deduplication capacities beyond what many would consider possible components used to speak to a few data as mimic or not. Additionally, the results suggest that the usage of a determined limit respect, as right around 1 as could be typical in light of the extraordinary circumstances, inspires the significant exertion besides rouses better contemplations. As future work, we would like to lead extra disclosure remembering the choosing reason to expand the level of use of our GP based way to deal with oversee record deduplication. For completing this, we arrange investigates diverse avenues as to data places from various territories.

## References


1. Li, J., Li, Y.K., Chen, X.F., Lee, P.P.C., Lou, W.: A hybrid cloud approach for secure authorized deduplication. IEEE Trans. Parallel Distrib. Syst. **99**, 1206–1216 (2014)
2. de Carvalho, G., Laender, A. H.F., Gonc¸alves, M.A., da Silva, A.S.: A genetic programming approach to record deduplication. IEEE Trans. Knowl. Data Eng. **24**, 766–770 (2012)
3. Bellare, M., Keelveedhi, S., Ristenpart, T.: Message-bolted encryption and secure deduplication. In: EUROCRYPT, pp. 296–312 (2013)
4. Bellare, M., Namprempre, C., Neven, G.: Security proofs for character based distinguishing proof and mark plans. J. Cryptol. **22**, 1–61 (2009)
5. http://www.peazip.org/copieshash-checksum.html
6. Halevi, S., Harnik, D., Pinkas, B., Shulman-Peleg, A.: Confirmations of proprietorship in remote stockpiling frameworks. In: ACM Conference on Computer and Communications Security, pp. 491–500. ACM, (2011)
7. Li, J., Chen, X., Li, M., Li, J., Lee, P., Lou. W.: Secure deduplication with proficient and solid merged key administration. IEEE Trans. Parallel Distrib. Syst. 456–468, (2013)
8. libcurl. http://curl.haxx.se/libcurl/
9. Ng, C., Lee, R.P.: A converse deduplication stockpiling framework advanced for peruses to most recent reinforcements. In: Proceedings of APSYS, pp. 74–88 (2013)
10. Ng, W.K., Wen, Y., Zhu, H.: Private information deduplication conventions in distributed storage. In: Proceedings of the 27th Annual ACM Symposium on Applied Computing, pp. 441–446. ACM, (2012)
11. Pietro, R.D., Sorniotti, A: Boosting productivity and security in evidence of proprietorship for deduplication. In: ACM Symposium on Information, Computer and Communications Security, pp. 81–82 (2012)

12. Quinla. S., Dorward, V.S.,: another way to deal with archiva stockpiling. In: Proceedings of USENIX FAST, pp. 554–568 (2002)
13. Rahumed, C.H.C.H., Tang, Y., Lee, P.P.C., Lui, J.C.S.: A protected cloud reinforcement framework with guaranteed cancellation and adaptation control. In: Third International Workshop on Security in Cloud Computing, pp. 1704–1712 (2011)
14. de Carvalho, M.G., Laender, A.H.F.,. Gonc¸alves, M.A., da Silva, A.S.: Copy identification using genetic programming. In: Proceedings of the 23rd Annual ACM Symposium Connected Computing (SAC), pp. 1801–1806 (2008)
15. Bilenko, M., Mooney, R., Cohen, W., Ravi Kumar, P., Fienberg, S.: Versatile name matching in information integration. IEEE Intell. Syst. **18**(5), 16–23 (2003)

# Synthesis of Sector Beams from Array Antennas Using Real Coded Genetic Algorithm

**Sudheer Kumar Terlapu and G.S.N. Raju**

**Abstract** In this paper, flat top far field radiation patterns known as sector beams are generated from linear arrays using Real Coded Genetic Algorithm. For all specified angular sectors, the pattern synthesis is carried out. The synthesis involves the determination of complex excitation functions for linear array for different beam widths. These patterns are numerically generated with the designed excitation levels. Controlling the ripples in the trade—in region maintaining the side lobe levels in the acceptable limits in the trade—off region is made possible in the present work.

**Keywords** Linear array • Genetic algorithm • Sector beam • Sidelobe level and beamwidth

## 1 Introduction

Design of antennas with high directive characteristics is often necessary in many of the radar and wireless systems which may not be possible with a single antenna. So the above requirement has necessitated the need for antenna arrays which can improve gain, directivity and radiation characteristics of the field pattern. The synthesis of antenna arrays to generate desired radiation patterns is highly a non-linear optimization problem. There are analytical methods such as Taylor method, Chebyshev method etc., have been proposed in the literature [1–4] Reduction of

S.K. Terlapu (✉)
Department of Electronics and Communication Engineering,
Shri Vishnu Engineering College for Women (Autonomous),
Vishnupur, Bhimavaram, Andhra Pradesh, India
e-mail: skterlapu@gmail.com

G.S.N. Raju
Department of Electronics and Communication Engineering,
AU College of Engineering (Autonomous), Andhra University,
Visakhapatnam, Andhra Pradesh, India



S.C. Satapathy et al. (eds.), *Data Engineering and Intelligent Computing*,
Advances in Intelligent Systems and Computing 542,
DOI 10.1007/978-981-10-3223-3_49

sidelobes with prescribed beamwidth is achieved with proper current distribution to resemble the Chebyshev pattern was solved by Dolph [5]. Synthesis of narrow beam with low sidelobe is proposed for linear arrays with continuous line source aperture by Taylor [6].

Synthesis of shaped beams using iterative sampling method has been introduced [7] using line source or discrete arrays. Synthesis techniques for shaped patterns have been proposed by several authors [8–11] in the literature. An alternative to the traditional methods, computational techniques like Genetic algorithms, a class of global optimization technique have been extensively used in the design of antenna arrays. An application of real-coded genetic algorithms for the design of reconfigurable array antennas is discussed in [12]. Use of evolutionary techniques provides methods that do not involve nonlinear equations and thus help in reducing mathematical complexity and computational time. Real Coded Genetic Algorithm (RCGA) is one such method that is used in the present work to synthesize the amplitude and phase excitation coefficients. Liu [13] proposed a method for the synthesis of linear array to obtain shaped power pattern.

In scan and non-scan radar and communication applications, it is required to generate sector beams over desired angular regions. This is to enhance more scan area by avoiding multiple scans of pencil beams. In the present work, an attempt is made to synthesize sector beams using evolutionary based real coded genetic algorithm. The fitness function is framed to obtain a flat-topped sector beam with specified beam widths, desired Sidelobe Levels and with minimum values of ripple in the trade-in region.

The paper is organized as follows. In Sect. 2, Linear Array and Fitness function formulation is discussed. Section 3 deals with Application of RCGA to the linear arrays. The synthesized radiation patterns with specified angular regions are presented in Sect. 4. Finally, conclusions are drawn in Sect. 5.

## 2 Formulation

### 2.1 Linear Array

An N element linear array radiating in the broadside which are positioned with equal spacing at half wavelength with radiating element as point source is considered. If the array is symmetrical around the axis with the central element or the midpoint of the array axis as reference, the elements are excited around the center of the linear array. The array factor corresponding to the far-field radiation is real. For even numbered linear array the array factor with the above consideration can be written as [14].

$$E(\theta) = \sum_{n=1}^{N} A_n e^{j\varphi_n} e^{j\pi n \sin\theta} \quad (1)$$

Here,

$A_n$ excitation weights of current for the nth element
$\varphi_n$ Phase weights of current for the nth element
$\theta$ angle between the line of observer and bore site
N Number of array elements

The above expression can be expressed in dB as

$$E(\theta) = 20\log(E_n(\theta)) \tag{2}$$

$E_n$ is the Normalized far field pattern and is given by

$$E_n(\theta) = \frac{E(\theta)}{E_{max}(\theta)} \tag{3}$$

## 2.2 *Fitness Function*

Keeping in view of achieving a flat top main beam, with reduced sidelobe level and admissible ripple in the trade in region, the corresponding fitness function is formulated (Fig. 1).

The proposed generalized fitness function corresponding to the above discussion is formulated as [15],

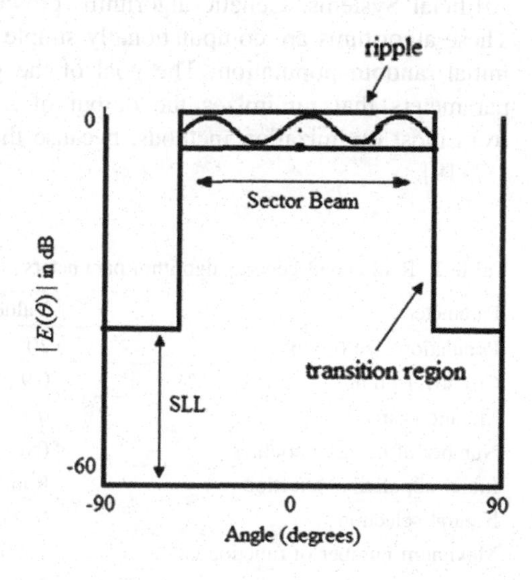


**Fig. 1** Desired radiation pattern

For $\theta_l \leq \theta \leq \theta_h$

$$f_1 = \sum_{\theta=\theta_l}^{\theta_h} |E(\theta)| \quad \text{if } E(\theta) \geq E_{sec} \tag{4}$$

For $\theta_h \leq \theta \leq \theta_l$

$$f_2 = \sum_{\theta=-\frac{\pi}{2}}^{\theta_l} (E(\theta) + SLL_{opt}) + \sum_{\theta=\theta_h}^{\frac{\pi}{2}} (E(\theta) + SLL_{opt}) \quad \text{if } E(\theta) \geq -SLL_{opt} \tag{5}$$

Finally, cost function is now written as

$$f_{sec} = f1 + f2 \tag{6}$$

where $SLL_{opt}$ and $E_{sec}$ represent the desired maximum sidelobe level and ripple in main beam respectively. The total sector pattern angular range is from $\theta_l$ to $\theta_h$. For $SLL_{opt} = 25$ dB and $E_{sec} = 1.5$ dB are used. The vector c consists of both amplitude and phase variables. $E_{sec}$ acts as a control parameter to achieve the desired flat beam.

## 3 Configuring RCGA for Linear Arrays

The main reference on genetic algorithm is Holland's [16] Adaption of Natural and Artificial Systems. Genetic algorithms provide robust search in complex spaces. These algorithms are computationally simple, and can be easily improved with an initial random population; The goal of the genetic algorithm is to find a set of parameters that minimizes the output of a function. Genetic algorithms differ from most optimization methods, because they have the following characteristics [17–19].

**Table 1** Real coded genetic algorithm parameters

| Parameter | Value |
|---|---|
| Population size (npop) | 20 |
| Crossover point | 0.9 |
| Mutation rate | 0.1 |
| Number of design variables | (No. of radiating elements)/2 |
| Initial population selection | Random |
| Natural selection | Npop/2 |
| Maximum number of function calls | 10000 |
| Maximum number of generations | 5000 |
| Termination criterion | Min cost or max number of generations |

- They work with a coding of the parameters, not the parameter themselves
- They search from many points instead of single point
- They don't use derivatives and use random transition rules, not deterministic rules

Real coded GA and its applications in the field of electromagnetic are discussed in the literature. The GA technique is applied for pattern synthesis in which the

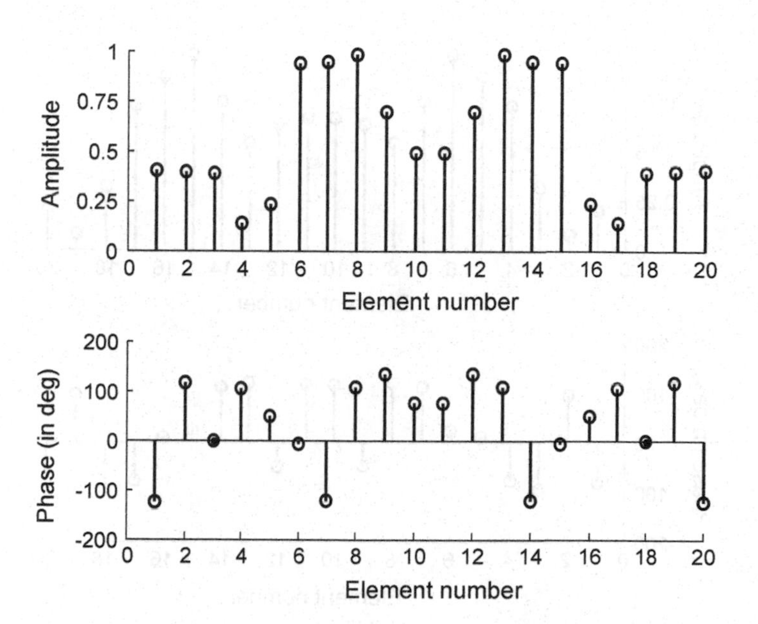


**Fig. 2** Amplitude and phase excitations for an array of 20 elements

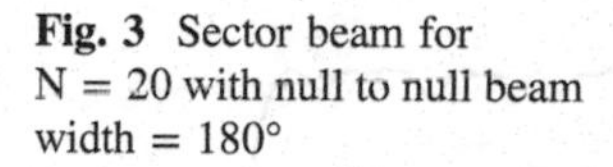

**Fig. 3** Sector beam for N = 20 with null to null beam width = 180°

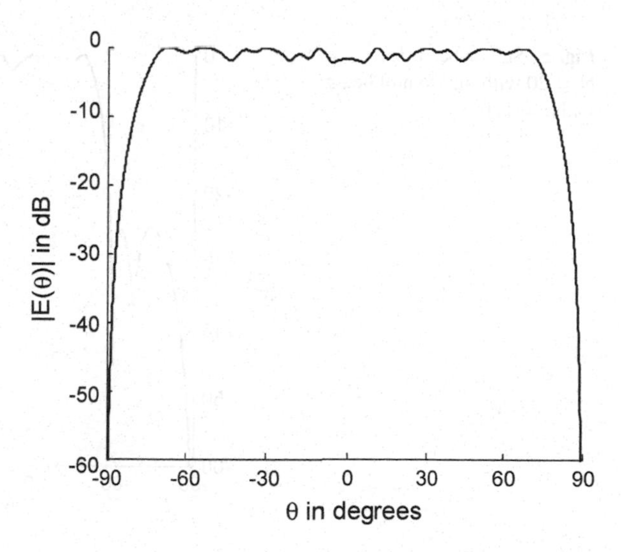

radiation patterns correspond to living beings and array weights correspond to chromosomes. GA is considered to manipulate a string of binary coding. Conventional GA encodes the parameters in Binary chromosomes and performs the binary operations. But the proposed approach does not consist of encoding or decoding and it uses decimal genetic operations treating the real or complex array element weights. Optimally large initial population is considered for small flat tops

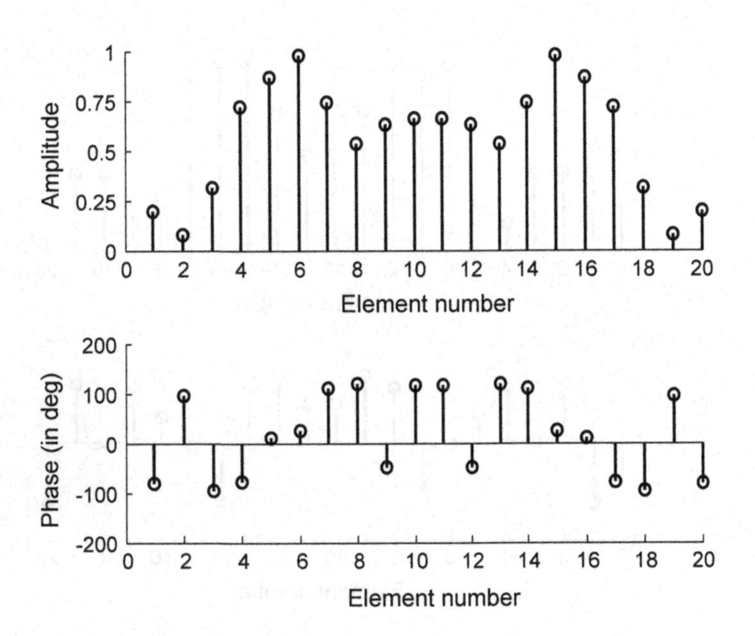


**Fig. 4** Amplitude and phase excitations for an array of 20 elements

**Fig. 5** Sector beam for N = 20 with null to null beam width = 120°

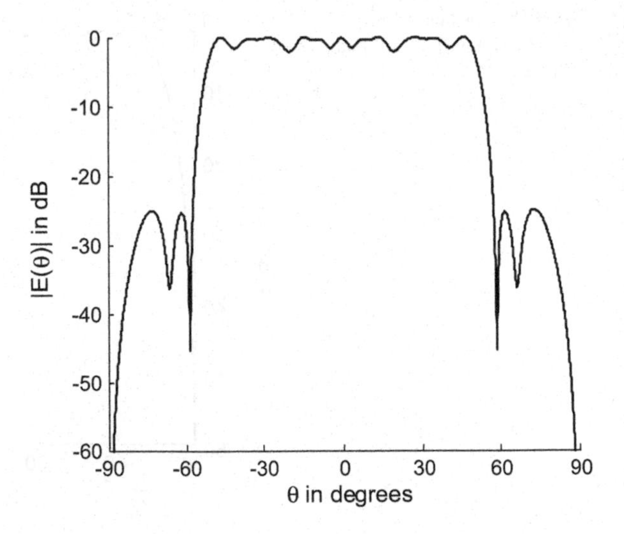

and vice versa. For the current problem, the selection of values of RCGA parameters is given in Table 1.

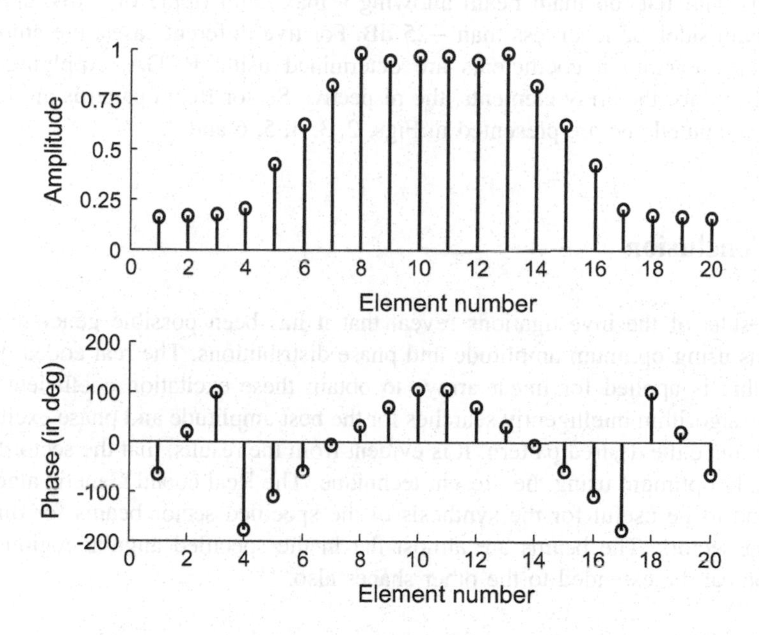


**Fig. 6** Amplitude and phase excitations for an array of 20 elements

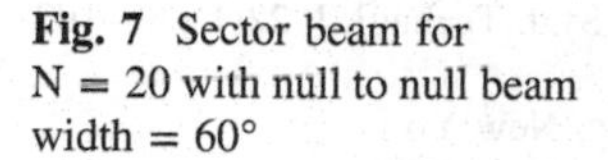

**Fig. 7** Sector beam for N = 20 with null to null beam width = 60°

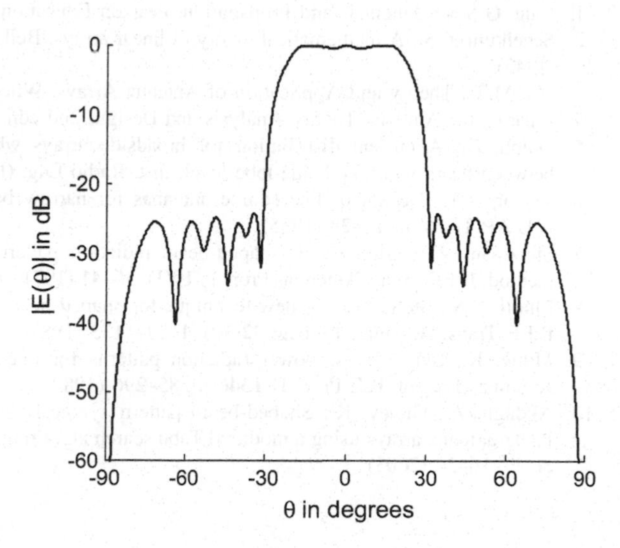

## 4 Results and Discussions

In the present work, Real Coded Genetic Algorithm (RCGA) is applied to determine the amplitude and phase excitation coefficients to obtain the optimized radiation patterns with flat top main beam allowing a maximum ripple of 1 dB, and with maximum sidelobe level less than −25 dB. For five different cases, the amplitude and phase excitation coefficients are determined using RCGA. Applying these coefficients for the array elements, the respective Sector Beam patterns are numerically computed and are presented in Figs. 2, 3, 4, 5, 6 and 7.

## 5 Conclusion

The results of the investigations reveal that it has been possible generate sector patterns using optimum amplitude and phase distributions. The real coded genetic algorithm is applied for linear arrays to obtain these excitation coefficients. The genetic algorithm intelligently searches for the best amplitude and phase excitations that produce the desired pattern. It is evident from the results, that the sector beams are really optimum using the present technique. The Real coded Genetic algorithm is found to be useful for the synthesis of the specified sector beams for different angular sectors. The beams are almost flat in the specified angular regions. The method can be extended to the other shapes also.

## References


1. Raju, G.S.N.: Antennas and Propagation. Pearson Education (2005)
2. Schelkunoff, S.: A mathematical theory of linear arrays. Bell Syst. Technol. J. **22**(1), 80–107 (1943)
3. Ma, M.T.: Theory and Application of Antenna Arrays. Wiley, New York
4. Balanis, C.: Antenna Theory Analysis and Design, 2nd edn. Wiley (1997)
5. Dolph, C.: A current distribution for broadside arrays which optimizes the relationship between beam width and side-lobe level. Inst. Radio Eng. (IRE) **34**(6), 335–348 (1946)
6. Taylor, T.T.: Design of line source antennas for narrow beam widths and low sidelobes. IRE AP Trans. **4**, 16–28 (1955)
7. Stutzman, W.: Synthesis of shaped-beam radiation patterns using the iterative sampling method. IEEE Trans. Antennas Propag. **19**(1), 36–41 (1971). doi:10.1109/TAP.1971.1139892
8. Elliott, R.S., Stern, G.J.: A new technique for shaped beam synthesis of equispaced arrays. IEEE Trans. Antennas Propag. **32**(10), 1129–1133 (1984)
9. Milne, K.: Synthesis of power radiation patterns for linear array antennas. Microwaves, Antennas Propag. IEE Proc. H **134**(3), 285–296 (1987)
10. Akdagli, A., Guney, K.: Shaped-beam pattern synthesis of equally and unequally spaced linear antenna arrays using a modified Tabu search algorithm. Microwave Opt. Technol. Lett. **36**(1), 16–20 (2003)

11. Azevedo, J.A.R.: Shaped beam pattern synthesis with non uniform sample phases. Prog. Electromagn. Res. B **5**, 77–90 (2008)
12. Mahanti, G.K., Chakrabarty, A., Das, S.: Phase-Only and amplitude-phase synthesis of dual-pattern linear antenna arrays using floating-point genetic algorithms. Prog. Electromagn. Res. **68**, 247–259 (2007)
13. Liu, Y., Nie, Z.-P., Liu, Q.H.: A new method for the synthesis of non-uniform linear arrays with shaped power patterns. Prog. Electromagn. Res. **107**, 349–363 (2010)
14. Balanis, C.A.: Antenna Theory: Analysis and Design. Wiley, New York (1997)
15. Blank, S.J., Hutt, M.F.: Antenna array synthesis using derivative, non-derivative and random search optimization. In: Sarnoff Symposium, pp. 1–4, Apr 2008
16. Holand, J.: Adaption of Natural and Artificial Systems. The University of Michigan Press, Ann Arbor (1975)
17. Goldberg, D.E.: Genetic Algorithms. Addison-Wesley, New York (1989), Ch. 1–4
18. Deb, K.: Multi-Objective Optimization using Evolutionary Algorithms. Wiley (2001)
19. Michalewicz, Z.: Genetic Algorithms + Data Structures = Evolution Programs. Springer, Berlin (1999)

# Performance Analysis of Encrypted Data Files by Improved RC4 (IRC4) and Original RC4

**Hemanta Dey and Uttam Kumar Roy**

**Abstract** In Cryptography, RC4 is one of the best known stream cipher from the last two decades in steam cipher family and it's already achieved the trust of many organizations after its code become public. Many Eminent researchers like Roos and et al. established that RC4 contains weakness and biasness in its internal stages. Most of the researcher feels that in KSA part of RC4 components contain most bias values related to the keys. If the KSA part generates strong output as for input of PRGA then resultant stream become more random. Already we proved and published that Improved RC4 (IRC4) gives better result compare to original RC4 in the generation of stream keys for encryption of any plaintext file. In this paper we generate two types of output stream files from the algorithm of IRC4 and original RC4 and implemented the both types of stream files to make encrypted files using same plaintext files and generates corresponding cipher text files. The two set of encrypted type files are testing through NIST Statistical package and from where we find out that IRC4 returns better result related to randomness of the cipher texts files, from where we can conclude that IRC4 algorithm generates more random stream and indirectly increase the security.

**Keywords** Improved RC4 · PRGA · KSA · SBox · NIST test suite package

H. Dey
Department of Computer Application, Techno India College of Technology,
Kolkata, India
e-mail: hemantadey13@gmail.com

U.K. Roy (✉)
Department of Information Technology, Jadavpur University, Kolkata, India
e-mail: royuttam@gmail.com



S.C. Satapathy et al. (eds.), *Data Engineering and Intelligent Computing*,
Advances in Intelligent Systems and Computing 542,
DOI 10.1007/978-981-10-3223-3_50

# 1 Introduction

In stream cipher family, RC4 is most popular, simple, efficient, and easy-to-implement stream ciphers [3]. Till now RC4 is successfully used in many popular products to maintain and customize the standards [1]. It is implemented to design for software and hardware products, since it become public in 1994. RC4 have two components, KSA and PRGA. In KSA part it initializes with identity s-boxes with a data-shuffling principle, which is basically a swap function. The most of the problem exists in the initial weak keys taken for the algorithm of RC4 first identified by Roos and et al. In Improved RC4 (IRC4) initial identity box values are initialized by the output, which are taken from popular Blum-Micali algorithms and another part of is random number generator i.e., Pseudo-Random Generation Algorithm (PRGA) in where initial values are taken from the last s-boxes of KSA and after that each two elements are shuffled for each byte generation.

IRC4 generates better result than original RC4 [1] but it not always true that generated streams XORed with plaintext makes the cipher text, which gives the better results. So here we encrypted both generated streams with same plain text file to produce respective cipher texts. Then we test all data files using NIST Statistical suite, *which* is basically a statistical package containing fifteen different tests and from which it is observed that IRC4 generated stream gives better result in the encrypted files rather than original RC4.

Original RC4 contains an array S for the S-Box of size 256, in which elements contain within it (0–255) are swapped, depending upon two parameters k and l where *k* is known as deterministic indicator and *l* is known as pseudo random indicator.

In this paper, we replaced the first part KSA using Blum-Micali algorithm to generate 256 random elements for the S-Boxes initial values. Hence the weakness and biasness of the initial stream sequence are overcome. So here starting value of parameter *l* in the part PRGA generated from return values of Blum-Micali algorithm functions, not like original RC4 algorithm, which gives more dynamic value to *l*.

NIST, USA recommended some minimum requirement for statistical tests using NIST *Statistical* package to find out the randomness of the stream sequences. This statistical package contains 15 different types of tests to analyze the randomness of binary sequences of order of $10^6$. We compared and analyzed IRC4 statistically with RC4, with proper guidelines Suggested by NIST in their package, which also again coded by us. It has been observed from last two decade that RC4 implementation in different security aspect is reliable, from last many years of its design, but now our new type IRC4 is proved more efficient and reliable than RC4.

**Table A**. The RC4 Algorithm

| KSA | PRGA |
|---|---|
| Taken Key *K*<br>for loop: *k* = 0, … *N* - 1<br>{<br>*SBox*[*k*] = *k*<br>}<br>*k* = 0<br>for loop: *k* = 0, …, *N* - 1<br>{<br>*l* = *l* + *SBox*[*k*] + *K*[*k*]<br>swap(*SBox*[*k*], *SBox*[*l*])<br>}<br><br>Generated S Box by *KSA* | Taken S-Box from KSA<br>*k* = 0; *l* = 0<br>while *TRUE*<br>{<br>*k* = *k* + *1*<br>*l* = *l* + *SBox*[*k*]<br>swap(*SBox*[*k*], *SBox*[*l*])<br>*z* = SBox[SBox[*k*] + *SBox*[*l*]]<br>}<br>Generated Stream *Z* |

## 2 Motivation

Stream cipher RC4 is very simple easy to understand few line codes but which have huge area of research after it's become public. As a result from 1994 it goes through tremendous different types of analysis on its code. In 1995 Roos and et al. [7] showed strong inter relation in the input secret key bytes with the obtained final key stream output. Their research group finds out classes of weak keys which are good in prediction for the output of KSA. And which will also shows the biasness on the final output stream sequences after its execution through PRGA.

In 2008, Maitra and Paul modified the RC4 stream cipher [4] to overcome the existing weakness in layer between KSA and PRGA, and named it as RC4+. They add an extra layer to resolve the non-uniformity in KSA. They prepared three-layer architecture in two parts KSA and PRGA, by adding extra phase as KSA+ and PRGA+.

In 2004, Paul with Preneel [6] add extra 2N swapping, where SBox size is N, to drop the initial bytes to solved statistical weakness related to the input key stream of RC4. They also proposed RC4A a new type of with less number of operations for generation of each output byte.

In the same year 2004, Tomasevic with Bojanic, also proposed a new technique of tree representation in RC4, to improve cryptanalytic attack on it.

In 2013, Sen Gupta et al., generate two key stream bytes at a time using a single clock cycle, implemented the whole logic to avoid the initial key weakness through hardware architecture. They also studied RC4 designing problem in throughput perspective.

In 2014, Dey Das [2] et al., replaced KSA and used BBS algorithms to generate the internal SBox of PRGA, and it observed that PRGA gives better result in the final output stream.

## 3 Improved RC4 (IRC4)

In 1995, Roos and et al. [7] show the weakness of KSA for some class of weak keys in RC4, after few months, when it become public. Roos argued that the chance of an element not to be swapped at all is near to 37% in KSA. Since in the RC4 algorithm the line of shuffling directly change the SBoxes in the time of exchange of two values and so the line $l = l + S[k] + K[k]$ is basically help to generate those indexes, therefore such exchange process may occur once, more than once, or not at all in between two array elements.

We proposed IRC4, where the initial key sets are taken from another set of random numbers generator of Blum-Micali algorithm using an input key of length 16 to generate the initial seed for the algorithms. So it contains the same flavor of RC4 algorithms but overcome the weakness and biasness of the KSA which are described by many researchers. We directly implemented the generated S-Boxes of size 256 into the Pseudo-random generation stage. So the new approach will be overcome all type of weakness related to non uniformity.

The random generator function using Blum-Micali algorithm is well known established function to the cryptographic society, so the generated 256 numbers for input element of PRGA will be reliable set of inputs for this stage. Since the deterministic variable $k$ and pseudo random variable j are evaluated through the index generated by line $S[k] + S[l]$ are taken for the input of PRGA, here remember that starting value of $l$ in the initial part of PRGA calculated from the return values of Blum-Micali algorithm functions, not like original RC4 algorithm which gives more dynamic value to $l$. The main aim is to make the algorithm more random as much possible which will be minimizing the chance of breaking through the intruder.

In this paper we generate two types of output stream files from the algorithm of IRC4 and original RC4 and implemented the both types of stream files to make encrypted files using same plaintext files and generates corresponding cipher text files of two sets. These two sets of encrypted files are going through NIST Statistical package and it is found that IRC4 generates better result related to randomness on the cipher texts files. In two sets, an exact same text file of size M is encrypted by at least 300 times using different generating streams of same size and obtained 300 cipher files with at least 1,342,500 bits in each file.

The both sets of cipher texts files are tested statistically to find out result of comparisons in respect to security variations in between original RC4 and IRC4.

## 4 NIST Statistical Test Suite

NIST, USA developed a Statistical package to find the measure of randomness of the generated stream cipher through random number and pseudorandom number generators, which are used for many areas like in cryptographic, modeling techniques and for simulation. NIST Statistical Test Suite is a popular established reliable package in the statistical society, containing of fifteen different tests which are used to measure of the various properties of randomness in the input long binary streams produced by random number and pseudorandom number generators [5].

For all fifteen statistical tests, we take an input bit stream sequence of at least $10^6$ size, and on which NIST implemented different mathematical concepts on it to calculate probability values (*p*-values) from both expected and observed results under the assumption of randomness [5]. The NIST package has been modified by us for windows application and Linux platform and this can execute any number of selected tests for any number of files individually or all at time. From which we can found the randomness measurements of RC4 and its improved variants.

## 5 Results and Discussions

In this paper we compare the outputs of the original RC4 and IRC4 over the encrypted cipher text files using the NIST Statistical package, as explained above, it has been observed that the improved algorithm gives better result compare to original RC4. Hence it enhances the security of encrypted cipher text files. The final analysis and comparison of both encrypted cipher files are displayed in three tables.

Table 1, displayed the comparisons of POP (Proportion of Passing) status with Uniformity Distribution of NIST Statistical Test suite for both types of algorithms

**Table 1** POP and uniformity distribution comparison between RC4 and IRC4

| Test number ↓ | POP status | | Uniformity distribution | |
|---|---|---|---|---|
| | Improved RC4 | RC4 | Improved RC4 | RC4 |
| T-01 | **0.993000** | 0.968000 | **5.671714$^{-01}$** | 4.122018$^{-01}$ |
| T-02 | 0.838000 | **0.929000** | 3.090909$^{-01}$ | **5.162834$^{-01}$** |
| T-03 | **0.972000** | **0.972000** | 6.300061$^{-01}$ | **8.520263$^{-01}$** |
| T-04 | **0.994000** | 0.982000 | 4.347634$^{-01}$ | **5.243211$^{-01}$** |
| T-05 | **0.987000** | 0.984000 | **3.154292$^{-01}$** | 2.432899$^{-01}$ |
| T-06 | **0.98000** | **0.98000** | **5.316178$^{-01}$** | 4.783081$^{-02}$ |
| T-07 | 0.968000 | **0.98000** | 4.145057$^{-01}$ | **8.905634$^{-01}$** |
| T-08 | 0.972000 | **0.982000** | **6.763452$^{-01}$** | 2.278544$^{-01}$ |
| T-09 | **0.98000** | 0.978000 | **3.762019$^{-01}$** | 3.208769$^{-02}$ |
| T-10 | 0.985000 | **0.99000** | **7.503285$^{-01}$** | 5.874090$^{-01}$ |
| T-11 | **0.992000** | 0.972000 | **3.719827$^{-01}$** | 1.613537$^{-01}$ |
| T-12 | 0.968000 | **0.982000** | **7.110938$^{-01}$** | 2.763018$^{-01}$ |
| T-13 | **0.992000** | **0.992000** | 9.241098$^{-03}$ | **8.209821$^{-01}$** |
| T-14 | **0.984000** | 0.981000 | 5.208720$^{-01}$ | **6.753185$^{-01}$** |
| T-15 | **0.986667** | 0.985567 | **5.263168$^{-01}$** | 4.835126$^{-02}$ |
| Total | **10** | *8* | **9** | *6* |

**Table 2** Histogram data values of the observed test of the architecture IRC4

| Test number | 0–0.01 | 0.01–0.1 | 0.1–0.2 | 0.2–0.3 | 0.3–0.4 | 0.4–0.5 | 0.5–0.6 | 0.6–0.7 | 0.7–0.8 | 0.8–0.9 | 0.9–1.0 |
|---|---|---|---|---|---|---|---|---|---|---|---|
| T-01 | 2 | 26 | 25 | 31 | 23 | 27 | 40 | 32 | 29 | 34 | 31 |
| T-02 | 1 | 27 | 29 | 29 | 40 | 29 | 33 | 31 | 24 | 28 | 29 |
| T-03 | 2 | 30 | 33 | 29 | 33 | 35 | 23 | 30 | 27 | 28 | 30 |
| T-04 | 0 | 34 | 22 | 38 | 31 | 31 | 30 | 30 | 29 | 26 | 29 |
| T-05 | 4 | 25 | 29 | 32 | 39 | 37 | 24 | 26 | 24 | 30 | 30 |
| T-06 | 5 | 21 | 48 | 26 | 19 | 31 | 19 | 36 | 41 | 24 | 30 |
| T-07 | 3 | 23 | 32 | 28 | 25 | 23 | 35 | 31 | 31 | 34 | 35 |
| T-08 | 4 | 34 | 35 | 37 | 30 | 32 | 23 | 28 | 25 | 32 | 20 |
| T-09 | 5 | 20 | 30 | 39 | 30 | 21 | 29 | 35 | 33 | 30 | 28 |
| T-10 | 0 | 33 | 31 | 31 | 33 | 27 | 22 | 30 | 34 | 25 | 34 |
| T-11 | 5 | 52 | 50 | 56 | 66 | 58 | 60 | 63 | 62 | 61 | 67 |
| T-12 | 4 | 22 | 22 | 36 | 35 | 29 | 27 | 27 | 34 | 28 | 36 |
| T-13 | 3 | 60 | 74 | 64 | 64 | 53 | 50 | 66 | 53 | 58 | 55 |
| T-14 | 30 | 213 | 270 | 240 | 243 | 216 | 244 | 248 | 246 | 217 | 233 |
| T-15 | 62 | 440 | 538 | 502 | 572 | 555 | 596 | 567 | 525 | 541 | 502 |

**Table 3** Uniformity distribution of probability values of (a) T-01 and (b) T-14

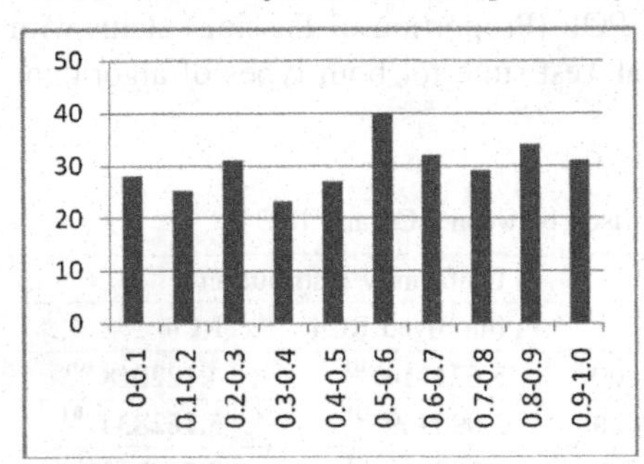


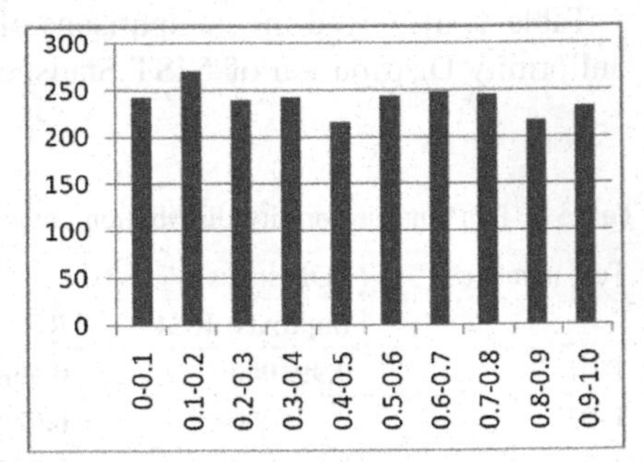


over encrypted 300 cipher files. The best values of all tests for both algorithm are shaded row wise and it has been found that the number of shaded cells for IRC4 is higher than original RC4, which indicate that IRC4 algorithm gives a better result compare to RC4 algorithm. Table 2 displays the histogram data values of the observed test of the architecture IRC4. Here 11 range set highlights the number of occurrence of different *p*-values from the total 300 files. Finally, it has been observed that a suitable mathematical model of random number generators like Blum-Micali can help to remove the biasness and weakness of the internal state array for RC4 and enhance the security of final output stream trough modified PRGA gives better randomization in the cipher texts (Table 3).

## 6 Conclusions

We found that Improved RC4 compare to the original RC4 is better random stream generated algorithm. It seems that security will be enhanced through the change of internal state array using Blum-Micali algorithm in KSA part of IRC4. So here as requirement of the user, they can create any random internal SBox by implementing secret input key to generate the initial seed of Blum-Micali and then corresponding output stream, which can taken as input for PRGA in IRC4. So from where we conclude that improved PRGA helps to generate more random output for IRC4.

## References

1. Akgün, M., Kavak, P., Demirci, H.: New results on the key scheduling algorithm of RC4. In: INDOCRYPT. Lecture Notes in Computer Science, vol. 5365, pp. 40–52. Springer (2008)
2. Dey, H., Roy, U.: An approach to find out the optimal randomness using parallel s-boxes in RC4, ICBIM-2016 NIT Durgapur, India. 978-1-5090-1228-2/16/$31.00 ©2016 IEEE
3. Foruzan, B.: Cryptography and Network Security. Tata McGraw-Hill, N. Delhi, Spl. Ind. Ed. (2007)
4. Maitra, S., Paul, G.: Analysis of RC4 and proposal of additional layers for better security margin. In: INDOCRYPT. Lecture Notes in Computer Science, vol. 5365, pp. 40–52. Springer (2008)
5. National Institute of Standard & Technology (NIST): A Statistical Test Suite for RNGs & PRNGs for Cryptographic Applications (2010)
6. Paul, S., Preneel, B.: A new weakness in the RC4 keystream generator and an approach to improve the security of the cipher. In: FSE 2004. LNCS, vol. 3017, pp. 245–259. Springer, Heidelberg (2004)
7. Roos, A.: A class of weak keys in the RC4 stream cipher. Post in sci. crypt (1995)

# A Novel Context and Load-Aware Family Genetic Algorithm Based Task Scheduling in Cloud Computing

**Kamaljit Kaur, Navdeep Kaur and Kuljit Kaur**

**Abstract** With the advent of web technologies and efficient networking capabilities, desktop applications are increasingly getting amalgamated with the touch of cloud computing. Most of the recent developments are dominated by consumer centric market, ensuring best quality of service and hence, greater customer base, leading to the rise of peaks in the profit charts. However, certain challenges in the field of cloud need to be dealt with, before peak performance is achieved and resource scheduling is one of these. This paper aims to present a context and load aware methodology for efficient task scheduling using modified genetic algorithm known as family genetic algorithm. Based on analysis of user characteristics, user requests are fulfilled by the right type of resource. Such a classification helps attain efficient scheduling and improved load balancing and will prove advantageous for the future of the cloud. Results show that the proposed technique is efficient under various circumstances.

**Keywords** Load balancing · Cloud · Task scheduling · Workload · Genetic algorithm

## 1 Introduction

Distributed computing was developed with the idea of enhancing the processing power of computers with visible benefits, making it evolve into the grid, cluster and cloud computing. Delivering computing resources as a utility became a very

K. Kaur (✉) · N. Kaur · K. Kaur
Department of Computer Engineering and Technology, Guru Nanak Dev University, Amritsar 143005, India
e-mail: kamal.aujla86@gmail.com

N. Kaur
e-mail: navdeep_binny@yahoo.com

K. Kaur
e-mail: kuljitchahal@yahoo.com



S.C. Satapathy et al. (eds.), *Data Engineering and Intelligent Computing*, Advances in Intelligent Systems and Computing 542,
DOI 10.1007/978-981-10-3223-3_51

welcomed idea all over the world, leading to the success of cloud [1]. Several authors have tried to define cloud, according to its characteristics, utility paradigm, structure or by comparing it with grid. In simple words, the cloud can be defined as a large pool of resources available to the users on pay per use model where scalability, reliability, security, heterogeneity, usability and other Qos parameters are taken care of. This makes cloud computing an enormously future oriented paradigm [2].

Talking about the structure and modeling of cloud, the layered architecture basically represents three main classes viz infrastructure, platform and software as a service. Virtualization augments this model by providing a logical interface to access cloud resources and makes it possible for cloud providers to modify this platform further into XaaS (everything as a service) [3]. Multi-tenancy, shared resource pooling, geo-distribution, self-organizing, dynamism in resource provisioning are some of salient features of cloud that ensure great response from its users. While all other architectures like grid, utility computing or other distributed computing environments, cloud computing with its fabulous architecture allows users with all kinds of needs to fit in. The SLA between customer and cloud provider ensures desirable Quality of Service (Qos). Various research issues have been explored in [4] out of which resource provisioning and allocation is a long known issue. The problem of resource management can be functionally divided into areas such as demand profiling, pricing, management of workload, application scaling and provisioning and scheduling both resource and workload. Cloud workload scheduling is one of the most prominent research areas. It consists of allocating suitable resources to the tasks so as to minimize or maximize a particular objective function that can be either makespan, execution time, resource utilization or throughput. This workload management is carried out by properly classifying the workload according to its characteristics as well as balancing this workload among various servers available [5]. As various research scholars have tried to devise scheduling algorithms to attain maximum efficiency in terms of response time or resource utilization. The problem taken up in this research paper involves workload management as well as resource scheduling for virtualized resources to match the needs of the workload by considering makespan as the prime parameter.

Section 2 discusses the work done by various authors in this domain with their advantages and loopholes. Section 3 describes a mathematical model for the proposed work which is elaborated in Sect. 4 along with the algorithm. Section 5 consists of the simulation results along with their interpretation.

## 2 Literature Survey

Last two decades have been completely evolutionary in the field of high speed computing. From distributed to the grid and then to cloud computing, purely demand based inventions have dominated IT.

The genetic algorithm proposed by Christina in [6] is a VM placement algorithm that aims to maximize physical resource utilization. The algorithm considers population divided into families with the dynamic mutation probability that prevents premature convergence [6]. Several other authors have tried to use GA and its hybrid versions in their works. For example, Kousik in [7] thrives to attain load balancing among VMs by using a fitness function that is based on latency, job arrival time, worst case completion time, etc. The research in this domain continues and another proposal by Tingting in [8] attempts to improve adaptive genetic algorithm by using a greedy approach to initialize population. Adaptive job spanning time and load balancing function has tremendously improved the performance of the algorithm. These meta-heuristic algorithms have the advantage of being flexible and can be hybridized in various ways as exemplified in [9, 10]. While the main topic of consideration is about genetic algorithm, various other meta heuristic techniques are available, including ant colony optimization, honey bee foraging behavior based techniques, particle swarm optimization and variable neighborhood search [11]. Efficient load balancing based on enhanced honey bee foraging behavior is presented in this paper. Task migration from overloaded to underloaded VM is carried out on the basis of task priority and while migration, underloaded VMs with highest capacity are chosen [12]. Hybrid approach for load balancing has been proposed by the author in [13] that uses improved max-min and ant colony algorithms. The max-min algorithm is modified to use execution time as the main parameter and then combined with an ant colony approach to obtain an optimum load balancing. A combination of fuzzy theory and GA is represented in [14] for job scheduling. Effort has been made to provide efficient scheduling for cost, energy and time constraint workloads in [15]. Focusing on providing load balancing, initial population is generated using max-min and min-min methods to provide efficient load balancing [16]. Resources can have different attributes like type, computability, storage needs and others. Keeping in mind these attributes, an efficient task scheduling algorithm is presented in [17] that aims to select a machine according to the resource requirements of the task. Throughput, execution time and resource utilization comparison with other algorithms represents the efficiency of the proposed method.

# 3 Proposed Methodology

## *3.1 Categorization*

This paper presents a fault tolerant-load balancing technique that uses a variation of genetic algorithm (FGA) along with considering various characteristics of workload submitted by the user. The foremost step in our research is to identify different characteristics of the workload that is being submitted. It is intended to divide the workload into different categories based upon these characteristics such as varying size, length, number of instructions in it etc. Likewise, Virtual resources (VM) are also classified depending upon their computing power, RAM, storage, etc. This makes scheduling and load management comparatively easy and better than the similar techniques found in the literature so far. The categories considered so far can be defined as computational intensive, network intensive and storage intensive as defined in [18].

### 3.2 Scheduling and Load Balancing

The categories defined are treated as families for Family Genetic Algorithm which is an improved version of GA. FGA module allows parallel processing of the different categories to assign tasks to the appropriate virtual resources (Virtual Machines). Fittness function of FGA is used to process requests to achieve fault tolerant load balancing. The fitness function is directly dependent on the category of tasks and uses bandwidth for network intensive, processing elements for computational intensive and size for storage intensive tasks.

Main objective of the proposed work:

1. Task scheduling to obtain minimum makespan and maximum efficiency.
2. To recognize the present and prospective upcoming requirements and expectations of the cloud customer.
3. To analyze the cloud workloads and group them using appropriate classification algorithm.
4. Provide load balancing.

## 4 Mathematical Model

A task scheduling problem can be formulated as a set of independent tasks represented by $\{t_1, t_2, t_3, t_4 \ldots\}$ being mapped on a set of resources (heterogeneous virtual resources, VMs in our case) represented as $\{v_1, v_2, v_3, v_4 \ldots\}$. The physical resources are represented as $\{p_1, p_2, p_3, p_4 \ldots\}$. As proposed, it is required to classify the tasks and the machines respectively based on certain characteristics. A task can be represented as a function of following parameters identified as id (task id), ST (starting time of task), length (number of instructions that comprise the task), size (filesize required to store the task), pe (number of processing elements required for the task) and bw (bandwidth requirement of the task).

This can be represented as:

$$t_i = f(\mathrm{id}, \mathrm{ST}, \mathrm{length}, \mathrm{size}, \mathrm{pe}, \mathrm{bw}) \quad (1)$$

Similarly, a virtual machine can also have certain characteristics that determine the idealness of a particular vm for a particular task. Using Eq. (1), a virtual machine set can be represented as:

$$\mathrm{vm_i} = f(\mathrm{id, st, r, pe, type, bw}) \tag{2}$$

As per the proposed methodology, the tasks (submitted in the form of cloudlet) need to be classified into three major categories—computational intensive, memory intensive and network intensive. Computational intensive (ci) tasks are those which require high computational power in terms of number of processing elements. Likewise, memory intensive (mi) have high storage requirements and network intensive (ni) with high bandwidth needs. Thus, the representation can be further modified as:

$$\text{Set of tasks t} = \left\{ \mathrm{t_i^{ci}, t_j^{mi}, t_k^{ni}}, \ldots\ldots \right\} \tag{3}$$

The scheduling process is improved if the characteristics of virtual machines match with those of the tasks. This necessitates the similar classification of vm to be done as:

$$\text{Set of vms vm} = \left\{ \mathrm{v_i^{ci}, v_j^{mi}, v_k^{ni}} \ldots \right\} \tag{4}$$

with vms having high mips and pe to be assigned to ci tasks, high storage to be assigned to mi tasks and high bandwidth for ni tasks.

FGA Algorithm: The real mapping of tasks and vms is carried out with the help of family genetic algorithm that uses a variation of genetic algorithm by dividing the initial population into families that run in parallel. Selection, crossover and mutation are applied to each of the families. In this paper, the families comprise three classes of tasks and vms as discussed above. The quality of schedule is assessed by calculating fitness function of each individual of each class. Using Eqs. (3) and (4):

Fitness function for computational intensive class:

$$\mathrm{t_i^{ci}/v_i^{ci}} = \mathrm{t(number\ of\ pes)/v(number\ of\ pes)} \tag{5}$$

Fitness function for storage intensive class:

$$\mathrm{t_i^{si}/v_i^{si}} = \mathrm{t(length)/v(size)} \tag{6}$$

Fitness function for network intensive class:

$$\mathrm{t_i^{ni}/v_i^{ni}} = \mathrm{t(bandwidth\ utilization)/v(bandwidth)} \tag{7}$$

Load balancing: Each assignment of task to vm is limited by a certain set of constraints depending upon the class to which it belongs. For example, in case of a set of computational intensive tasks, assignment of vm to that set can only be made if those vms can serve the request without getting overloaded. As a result, load balancing is implicitly taken care of.

$$\forall i \sum_{j=1}^{n} x_{ij} = 1 \tag{8}$$

Equations (9), (10) and (11) help ensure that the total processing element, size and bandwidth requirement of the tasks may not exceed the capacity of virtual resources (VMs). For example, Eq. (11) adds the constraint that summation of bandwidth as requested by users may not exceed the total processing bandwidth provided by Virtual Machines.

$$\forall i \sum_{j=1}^{n} x_{ij}.t_{i}^{pe} < v_{j}^{pe} \tag{9}$$

$$\forall i \sum_{j=1}^{n} x_{ij}.t_{i}^{\text{length}} < v_{j}^{\text{size}} \tag{10}$$

$$\forall i \sum_{j=1}^{n} x_{ij}.t_{i}^{bw} < v_{j}^{bw} \tag{11}$$

## 5 Pseudo-Code

The pseudo code for the proposed algorithm is as under:

1. Analyze user request and identify different parameters: no of Processing elements, RAM, file size, storage, network bandwidth.
2. Analyze the workload and group into three categories:

   Computational Intensive (ci), Storage Intensive (si) and Network Intensive (ni)

3. Initialize host machines and create VM instance for workload requests.
4. Select a VM to process a specific workload using FGA
5. Calculate makespan

   FGA Algorithm (Pseudocode):

1. Input lists of tasks and VMs
2. Randomly initialize the population
3. Define the number of families to be created
4. Repeat until all cloudlets are submitted

   a. for each cloudlet

      (i) if tasktype = ci

      for each vm

if (computational capacity of vm) > (computation needs of task)
add pair to pop_comp; comp_vm++

(ii) else if tasktype = si

for each vm
if (storage capacity of vm) > (storage requirement of task)
add pair to pop_mem; mem_vm++

(iii) else

for each vm
if (network capacity of vm) > (bandwidth requirement of task)
add pair to pop_net; net_vm++

b. for each class type

(i) if tasktype = ci

fitness function as in Eq. (5)

(ii) else if tasktype = si

fitness function as in Eq. (6)

(iii) else

fitness function in Eq. (7)

c. Perform single-point crossover and mutation with mutation probability 0.001
d. Select the task-vm pair with the best fitness function value.

end for

5. Select the fittest individual from the population to get the best solution.

# 6 Simulation and Results

This section comprises of experiments performed to guage the effectiveness of algorithm. Cloudsim [19] has been used as the simulation tool with Java 7, Window 10 running on i3 machine with 4 GB RAM. The experimentation done mainly focuses on two aspects, first to study the efficiency of proposed algorithm over certain existing algorithms and second to critically analyse the effect of various classifications on overall makespan.

Case1: Fig. 1 depicts comparison of proposed algorithm with default scheduling policy (RR) and ACO. The configuration of VM for this experiment is 6 processing elements, 300 mips, 6000 bw, 45 GB storage and 2 GB RAM size. A total of 100 cloudlets are run and depicted as set of 10 each, showing least execution time for FGA.

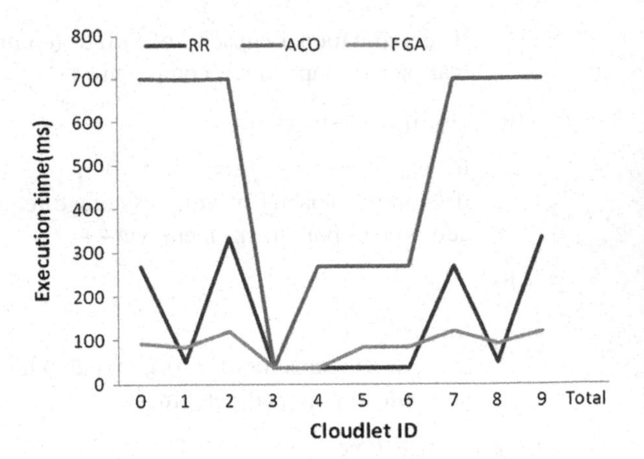


**Fig. 1** Effectiveness of proposed approach

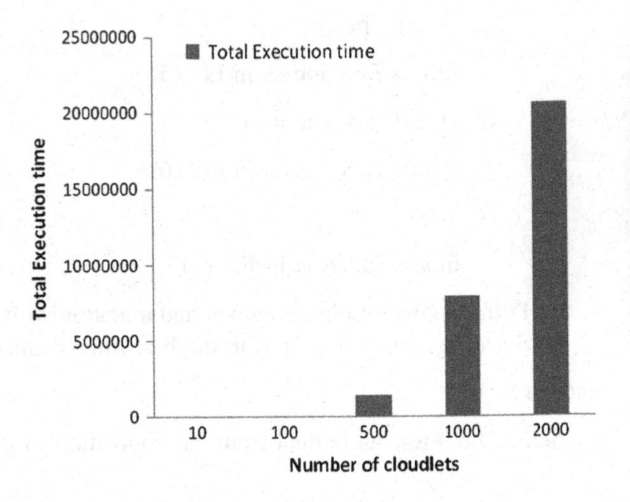


**Fig. 2** Execution time for varying workload (number of cloudlets)

Case 2: The algorithm performs efficiently for 500 cloudlets with the same configuration as mentioned in case 1. But as the number of cloudlets increase, the infrastructure needs to be scaled accordingly if efficient results are to be obtained. Efficient scheduling requires virtual resources with better configuration when load increases. This comparison results show peak rise in execution time for 2000 cloudlets (Fig. 2).

Case 3: Three different distributions are taken comprising of workloads that are dominant in storage type (80% storage intensive and 10% each network and 10% compute intensive), network type and computational type of tasks. For the same VM configuration, results are shown in Fig. 3. As expected, workload with maximum number of computational intensive tasks, are assigned VMs, with the maximum computational power decreasing the total execution time followed by storage intensive tasks.

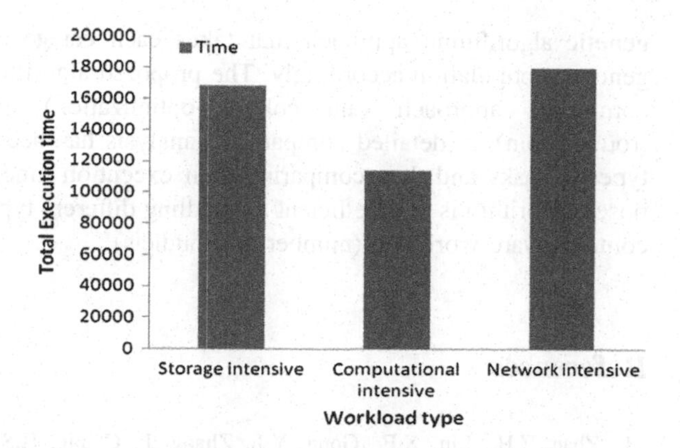


**Fig. 3** Total execution time for workloads dominating in computational, storage and network tasks

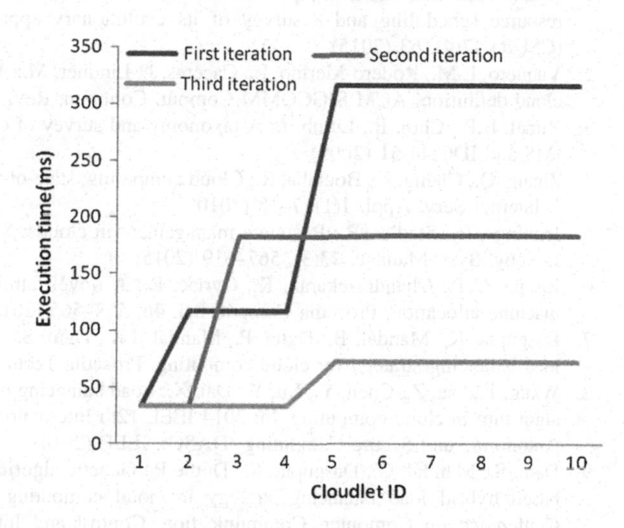


**Fig. 4** Execution time variation for different iterations of storage workload

Case 4: If different iterations of same type of workload (80% storage intensive, 10% compute and 10% network intensive, for example) are carried out, such iterative behaviour is depicted in Fig. 4. The variation of execution time can be explained due to the random assignment of virtual machines to the tasks.

## 7 Conclusion

Task scheduling is a critical aspect to developing most efficient systems. Load balancing, fault tolerance, context aware environment makes scheduling a complicated task. Therefore it is essential to carefully design all these algorithms to ensure performance doesn't lag at the cost of increased functionality. A similar attempt has been made in the proposed approach by categorization of workload before task scheduling can be done. Assignment of tasks to vms is done by hybrid genetic algorithmic approach that takes each category as individual input and generates population accordingly. The proposed algorithm is compared with a soft computing approach (ant colony optimization) and a default approach (round-robin). A detailed comparative analysis has been done by taking different types of tasks and then comparing their execution times. Results proves the proposed algorithm is more efficient at handling different types of workloads for varied context-aware workload (number of cloudlets).

## References


1. Zhan, Z.H., Liu, X.F., Gong, Y.J., Zhang, J., Chung, H.S.H., Li, Y.: Cloud computing resource scheduling and a survey of its evolutionary approaches. ACM Comput. Surv. (CSUR) **47**(4), 63 (2015)
2. Vaquero, L.M., Rodero-Merino, L., Caceres, J., Lindner, M.: A break in the clouds: towards a cloud definition. ACM SIGCOMM Comput. Commun. Rev. **39**(1), 50–55 (2008)
3. Rimal, B.P., Choi, E., Lumb, I.: A taxonomy and survey of cloud computing systems. INC, IMS and IDC:44-51 (2009)
4. Zhang, Q., Cheng, L., Boutaba, R.: Cloud computing: state-of-the-art and research challenges. J. Internet Serv. Appl. **1**(1), 7–18 (2010)
5. Jennings, B., Stadler, R.: Resource management in clouds: Survey and research challenges. J. Netw. Syst. Manage. **23**(3), 567–619 (2015)
6. Joseph, C.T., Chandrasekaran, K., Cyriac, R.: A novel family genetic approach for virtual machine allocation. Procedia Comput. Sci. **46**, 558–565 (2015)
7. Dasgupta, K., Mandal, B., Dutta, P., Mandal, J.K., Dam, S.: A genetic algorithm (ga) based load balancing strategy for cloud computing. Procedia Technol. **10**, 340–347 (2013)
8. Wang, T., Liu, Z., Chen, Y., Xu, Y., Dai, X.: Load balancing task scheduling based on genetic algorithm in cloud computing. In: 2014 IEEE 12th International Conference on Dependable, Autonomic and Secure Computing (DASC). IEEE (2014)
9. Dam, S., Mandal, G., Dasgupta, K., Dutta, P.: Genetic algorithm and gravitational emulation based hybrid load balancing strategy in cloud computing. In: 2015 Third International Conference on Computer, Communication, Control and Information Technology (C3IT). IEEE (2015)
10. Zhu, K., Song, H., Liu, L., Gao, J., Cheng, G.: Hybrid genetic algorithm for cloud computing applications. In: 2011 IEEE Asia-Pacific Services Computing Conference (APSCC). IEEE (2011)
11. Farrag, A.A.S., Mahmoud, S.A., El Sayed M.: Intelligent cloud algorithms for load balancing problems: a survey. In: 2015 IEEE Seventh International Conference on Intelligent Computing and Information Systems (ICICIS). IEEE (2015)
12. Babu, K.R., Samuel, P.: Enhanced bee colony algorithm for efficient load balancing and scheduling in cloud. In: Innovations in Bio-Inspired Computing and Applications. Springer International Publishing (2016)
13. Ghumman N.S. and Kaur R.: Dynamic combination of improved max-min and ant colony algorithm for load balancing in cloud system. In: 2015 6th International Conference on Computing, Communication and Networking Technologies (ICCCNT). IEEE (2015)
14. Shojafar, M., Javanmardi, S., Abolfazli, S., Cordeschi, N.: FUGE: a joint meta-heuristic approach to cloud job scheduling algorithm using fuzzy theory and a genetic method. Cluster Comput. **18**(2), 829–844 (2015)
15. Singh, S., Chana, I.: QRSF: QoS-aware resource scheduling framework in cloud computing. J. Supercomput. **71**(1), 241–292 (2015)

16. Zhan, Z.H., Zhang, G.Y., Gong, Y.J., Zhang, J.: Load balance aware genetic algorithm for task scheduling in cloud computing. In: Asia-Pacific Conference on Simulated Evolution and Learning. Springer International (2014)
17. Zhao, Y., Chen, L., Li, Y., Tian, W.: Efficient task scheduling for Many Task Computing with resource attribute selection. China Commun. **11**(12), 125–140 (2014)
18. Sandhu, R., Sood, S.K.: Scheduling of big data applications on distributed cloud based on QoS parameters. Cluster Comput. **18**(2), 817–828 (2015)
19. Calheiros, R.N., Ranjan, R., Beloglazov, A., De Rose, C.A., Buyya, R.: CloudSim: a toolkit for modeling and simulation of cloud computing environments and evaluation of resource provisioning algorithms. Softw. Pract. Experience **41**(1), 23–50 (2011)

# Indian Stock Market Analysis Using CHAID Regression Tree

**Udit Aggarwal, Sai Sabitha, Tanupriya Choudhury and Abhay Bansal**

**Abstract** Data mining is the technique which utilized to extract concealed "analytical" and "predictive" facts and figures from a big set of datasets and databases. It is being applied in various research areas by data scientists and analysts such as mathematics, marketing, genetics, cybernetics, etc. In this paper, A Chi Squared Automatic Interaction Detection (CHAID) regression tree model has been proposed to infer the volatility of the Stock Exchange Sensitive Index (SENSEX) data while explicitly accounting for dependencies between multiple derived attributes. Using real stock market data, dynamic time varying graphs are constructed to further analyze how the volatility depends on various factors such as Lok Sabha Elections, Domestic Riots, Union Budget of India, Indian Monsoon and Global factors. Factors have been analyzed to understand their role in the fluctuations seen over time in the market and how the SENSEX behave over these factors.

**Keywords** Data mining · CHAID regression · Prediction · Sensex index analysis · Comparative analysis · Decision tree

---

U. Aggarwal (✉) · S. Sabitha · T. Choudhury · A. Bansal
Amity University, Noida, Uttar Pradesh, India
e-mail: Udit.aggarwal015@gmail.com

S. Sabitha
e-mail: assabitha@amity.edu

T. Choudhury
e-mail: tchoudhury@amity.edu

A. Bansal
e-mail: abansal1@amity.edu



S.C. Satapathy et al. (eds.), *Data Engineering and Intelligent Computing*,
Advances in Intelligent Systems and Computing 542,
DOI 10.1007/978-981-10-3223-3_52

## 1 Introduction

Stock, equity or share market is an unrestricted marketplace where "mutual funds", "bonds", "stocks", etc. are traded. Although "Mutual fund market", "Stock market", etc. It offers enterprises right to use funds by offering or trading minor stake or possession to the investors. A SENSEX index is a "statistical" measure of variation in the country's economy or an equity market. All indexes have their own calculation procedures and commonly stated in the terms of alteration from the base worth.

In this paper certain factors have been taken and data are analyzed to find the trend followed by the index due to these selected factors. Stock Market Analysis is a major research area, and many techniques have been used in order to predict and analyze the market behavior. Some of important techniques include neural network, regression, lagged correlation, Fuzzy, ARIMA, hidden Markov model, data mining, etc. The research is carried on 26 years (1991 to 2016) data on Indian stock market index collected from BSE. What makes the stock market go up and down? There is no accurate answer to this question. Though fundamental analysis and technical analysis have been used to forecast market behavior, lots of research and study have been done in this regard, stock market experts have found it extremely difficult to predict the reasons which would cause a change in the stock market. The factors considered for the analysis of Indian stock market are Lok Sabha Election, Domestic Riots, Indian Union budget, Indian monsoon, Global Factors. The paper is further structured as follows: Theoretical Background, Methodology, Experimental setup, Analysis and Conclusion.

## 2 Theoretical Background

The most common factors that primarily affect stock market are industry performance, investor sentiments, economic factors, etc. These parameters are explained as follows:

- ***Industry Performance***:
  Mostly, the stock price of all of the companies in the same industry move aligned with each other because they behaved the same way to market conditions. But at times, the stock price of a company rises or falls due to bad or good results or decisions of other companies from the same industry, if they are competing for the same product or technology.
- ***Investor Sentiments***: Investor sentiment or confidence can cause up and down in the market, which result in fall or increase in share price. Investor sentiment can be analyzed in the market as Bull Market (shows high confidence among investors) and Bear Market (poor investor confidence):
- ***Economic Factors***: There are various economic factors which affect the market, some of the key factors are **interest rate** change by the RBI, Economic outlook

by different brokerage houses, Inflation, **Related markets** changes like more investment in share rather than government bonds, **Change in central government** may lead to change in investor sentiments and may also lead to changes in the economic outlook as these depends upon the factor that how stable the government is. Budget also plays a crucial role in economic policy.

- ***Seasonal/Environmental Factors***:
  There are certain other factors which directly or indirectly impact the market or stock price such as monsoon, cyclones, tsunami, earthquake, disasters, etc.

**Table 1** Data mining techniques used in researches

| S. no. | Literature references | Year | Area | Data mining techniques in stock market |
|---|---|---|---|---|
| 1 | Babu et al. [2] | 2014 | Indian stock prediction | Hybrid—ARIMA GARCH model |
| 2 | Saxena et al. [15] | 2013 | Prediction in stock market | Linguistic temporal approach |
| 3 | Kumar et al. [13] | 2013 | Performance analysis of indian stock market | Neural network with time series |
| 4 | Tirea et al. [16] | 2013 | Strongest performing stocks effects on the market | Neural networks |
| 5 | Abdullah et al. [1] | 2013 | Analysis of stock market using text mining | Natural language processing |
| 6 | Mr. Suresh et al. 2013 | 2013 | A study of fundamental and technical analysis | Statistical analysis |
| 7 | Bebarta et al. [3] | 2012 | Forecasting and classification of Indian market | Artificial neural network |
| 8 | Hendahewa et al. [6] | 2012 | Analysis of casuality in stock market | Regression |
| | Huang et al. | 2012 | Stock portfolio management | Cluster validity index |
| 9 | Troiano et al. [17] | 2010 | Predictive trend in next day market | Hidden Markov model. |
| 10 | Fathi Gasir | 2009 | An algorithm for creating fuzzy regression tree | Fuzzy and regression |
| 11 | Fonseka et al. [4] | 2008 | Stock market analysis | Lagged correlation |
| 12 | [12] | 2008 | A Phylogenetic Analysis for Stock Market Indices Using Time-Series String Alignments | Clustering |
| 13 | Mukherjee et al. [14] | 2007 | Comparative analysis of Indian stock market with international market | Statistical analysis |
| 14 | Zhongxing et al. [18] | 1993 | Stock market analysis | Neural network and fuzzy |

Table 1 given depicts the various Data mining techniques utilized in researches performed by various scholars where they predicted and analyzed stock market in different ways.

Predictive Analytics brings together advanced analytics capabilities spanning ad hoc statistical analysis, predictive modeling, data mining, text analytics, entity analytics, optimization, real-time scoring, machine learning and more. Predictive analytics have been shown to transform professions and industries through better use of data. Decision Tree is a "Predictive Model", it maps observation about the attribute to the conclusion about the target value. It is a predictive model which predicts the target values by applying simple decision rule which comes from the data features. The rule predicts an outcome in the target field. Viewing the decision rules (set of conditions derived from the data features) helps determine which conditions are likely to result in a specific outcome. The decision tree is of two main types in data mining, which are as follow by Han et al. [5]:

(1) Classification Tree: It is used when predicted or target outcome is related to the class to which dataset relates to.
(2) Regression Tree: It is used when we have to relate to any particular value or when the outcome is a real number.

There are different types of decision tree like ID3 (iterative dichotomiser3), C4.5, CART (Classification and regression tree),

CHAID ("Chi squared Automatic interaction detection"). We have used **CHAID** in this paper to predict the outcome of the SENSEX index in future. It stands for "Chi-Squared Automatic Interaction Detection". It is multi-way split decision tree technique developed by Kass [11]. It is an efficient technique to generate a regression tree and easy to understand. It uses dynamic branching method to determine the number of branches. It clubbed together or merge those attribute values which are similar in nature and kept the other which are heterogeneous in nature. The main motive behind this technique is to find the combination or interaction between two variables CHAID, a technique is original purpose is to detect interaction between variables (i.e., find "combination" variables). It partitions a population into separate and distinct groups, which are defined by a set of independent (predictor) variables, such that the CHAID objective is met: the variance of the dependent (target) variable is minimized within the groups, and maximized across the groups. A CHAID regression tree model has been utilized to infer the volatility of the SENSEX Index data which explicitly accounts for dependencies or interaction between different derived attributes. The CHAID algorithm consists of following stages: "Preparing" and "Merging" Stage, Splitting" Stage and Stopping" Stage.

- ***"Preparing" and "Merging" stage:*** *It* assigns a different category to each value of predictor and then creates contingency tables in order to identify most similar pairs of groups (the highest p-value) utilizing statistical significance test with respect to the dependent variable; F-test is used to identify significant predictors with regression problems. The algorithm merges together categories that have a significant level above a pre-specified merged threshold.

- ***"Splitting" stage***: The splitting stage chooses the predictor that has the smallest adjusted p-value to be used to best split the node. The terminal node of the split has the smallest pvalue that is above the split threshold.
- ***"Stopping" stage:*** Growth of the tree is stopped when it is not possible to generate more branches due to result of meeting any of the following conditions:-

a. All cases in a node have the same values for each predictor.
b. The node becomes pure, that is, all cases in a node have very similar values of the dependent variable.
c. One of the user specified CHAID regression parameter settings (tree depth, node size and child node size) has been reached.

# 3 Methodology

The SENSEX data is taken from BSE and the methodology as depicted in Fig. 1 is applied in this research. The various steps of adopted methodology are as follows:

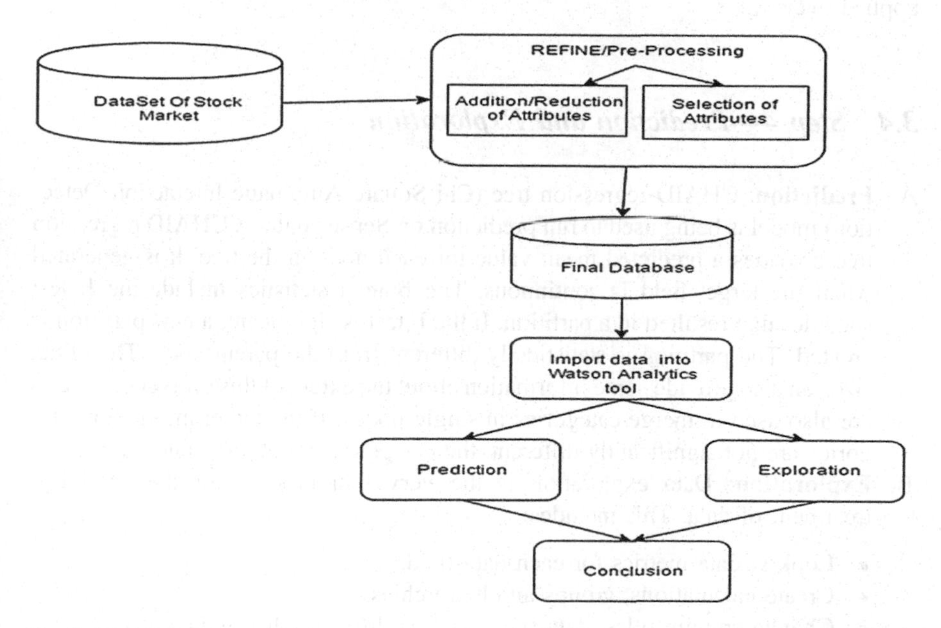


**Fig. 1** Methodology

### 3.1 Step 1—Dataset of Stock Market

The data set has been collected from the BSE stock exchange archived data publicly available on World Wide Web. The extracted dataset contains the following attributes: Date, Open, Low, Close and High.

### 3.2 Step 2—Pre-processing of Data

Data Set is preprocessed using "Refine" feature in the Watson Analytics. Using Refine, more attributes can be created or added to the dataset. So, new attributes have been added utilizing basic attributes. New derived attributes are as follows: prev_close, close-prev_close, max_gain, gain%, day_high, day_low, max_loss, and loss%. These attributes would be further utilized to run prediction and analyze various factors responsible for the trend in the stock market.

### 3.3 Step 3—Import Data into Watson Analytics Tool

Import dataset into the Watson Analytics tool, so that various techniques can be applied over it.

### 3.4 Step 4—Prediction and Exploration

A. **Prediction**: CHAID regression tree (Chi Square Automatic Interaction Detection) model is being used to run prediction on Sensex data. A CHAID regression tree provides a predicted mean value for each node in the tree. It is generated when the target field is continuous. The branch statistics include the F test statistics that resulted in a partition. If the F test is significant, a new partition is created. The partition is statistically different from the parent node. The effect size can also provide some information about the extent of this difference. F tests are also used to merge categories in single nodes. If the target means for categories are not significantly different, the categories are merged into a node.
B. **Exploration**: Data exploration is the very first process for the analytical treatment of data. This includes:
   - Look at data metrics for each data field.
   - Create calculations, groups and hierarchies.
   - Change column titles, data types and modify default aggregations.
   - Add in data points and explore things differently using the visualization menu, data filters, color pallets, etc.

### 3.5 *Step 5—: Result Analysis*

Results produced from the analysis and prediction are analyzed and concluded correspondingly.

## 4 Experimental Setup

A. *SENSEX Dataset*

The dataset, publilcy available for research was collected from the historical archive of BSE (Bombay Stock Exchange). Last 26 years (April, 1991 to Mar, 2016) of data has been taken and utilized for this research. In addition to existing attributes some new attributes have been added to give the edge to the prediction and analysis (Figs. 2 and 3).

B. Original Dataset and Preprocessed Dataset:

After pre-processing, the derived attributes are added to the original dataset. The attributes added are day high, day low, Close—prev_close, max gain, % max gain,

**Fig. 2** Original dataset snapshot

| Date | prev close | Open | High | Low | Close |
|---|---|---|---|---|---|
| 16-Apr-09 | 11355.01 | 11358.06 | 11367.23 | 10900.47 | 10947.4 |
| 17-Apr-09 | 10947.4 | 11067.71 | 11339.47 | 10946.25 | 11023.09 |
| 20-Apr-09 | 11023.09 | 11065.1 | 11209.66 | 10863.28 | 10979.5 |
| 21-Apr-09 | 10979.5 | 10765.39 | 11068.82 | 10764.08 | 10898.11 |
| 22-Apr-09 | 10898.11 | 10968.6 | 11036.24 | 10715.66 | 10817.54 |
| 23-Apr-09 | 10817.54 | 10841.57 | 11203.28 | 10758.97 | 11134.99 |
| 24-Apr-09 | 11134.99 | 11149.47 | 11362.88 | 11070.33 | 11329.05 |
| 27-Apr-09 | 11329.05 | 11237.42 | 11492.1 | 11176.55 | 11371.85 |
| 28-Apr-09 | 11371.85 | 11371.34 | 11375.97 | 10961.76 | 11001.75 |
| 29-Apr-09 | 11001.75 | 11091.56 | 11430.25 | 11091.56 | 11403.25 |
| 4-May-09 | 11403.25 | 11635.24 | 12161.9 | 11635.24 | 12134.75 |
| 5-May-09 | 12134.75 | 12159.74 | 12197.88 | 11985.88 | 12131.08 |
| 6-May-09 | 12131.08 | 12100.69 | 12272.1 | 11899.41 | 11952.75 |
| 7-May-09 | 11952.75 | 12064.51 | 12143.95 | 11981.13 | 12116.94 |
| 8-May-09 | 12116.94 | 12092.97 | 12180.07 | 11765.06 | 11876.43 |
| 11-May-09 | 11876.43 | 11997.37 | 12026.6 | 11621.3 | 11682.99 |
| 12-May-09 | 11682.99 | 11629.97 | 12194.63 | 11625.39 | 12158.03 |
| 13-May-09 | 12158.03 | 12201.93 | 12256.43 | 11934.44 | 12019.65 |
| 14-May-09 | 12019.65 | 11774.04 | 11935.86 | 11695.52 | 11872.91 |
| 15-May-09 | 11872.91 | 11948.7 | 12219.54 | 11948.7 | 12173.42 |
| 18-May-09 | 12173.42 | 13479.39 | 14284.21 | 13479.39 | 14284.21 |
| 19-May-09 | 14284.21 | 14757.82 | 14930.54 | 13824.13 | 14302.03 |

| Date | Prev Close | Open | Day High | Day Low | Close | close(minus) prev close | day high-open | daylow-open | max Gain | max loss | % max gain | % max loss |
|---|---|---|---|---|---|---|---|---|---|---|---|---|
| 11-Mar-16 | 24623.34 | 24620.39 | 24817.8 | 24552.26 | 24717.99 | 94.65 | 197.41 | -68.13 | 194.46 | -71.08 | 0.79 | -0.29 |
| 14-Mar-16 | 24717.99 | 24801.7 | 24960.51 | 24734.04 | 24804.28 | 86.29 | 158.81 | -67.66 | 242.52 | 16.05 | 0.98 | 0.06 |
| 15-Mar-16 | 24804.28 | 24832.04 | 24840.77 | 24517.28 | 24551.17 | -253.11 | 8.73 | -314.76 | 36.49 | -287 | 0.15 | -1.16 |
| 16-Mar-16 | 24551.17 | 24537.61 | 24706.85 | 24354.55 | 24682.48 | 131.31 | 169.24 | -183.06 | 155.68 | -196.62 | 0.63 | -0.8 |
| 17-Mar-16 | 24682.48 | 24852.18 | 24948.3 | 24576.52 | 24677.37 | -5.11 | 96.12 | -275.66 | 265.82 | -105.96 | 1.08 | -0.43 |
| 18-Mar-16 | 24677.37 | 24729.41 | 24986.94 | 24681.64 | 24952.74 | 275.37 | 257.53 | -47.77 | 309.57 | 4.27 | 1.25 | 0.02 |
| 21-Mar-16 | 24952.74 | 25007.56 | 25327.45 | 24988.27 | 25285.37 | 332.63 | 319.89 | -19.29 | 374.71 | 35.53 | 1.5 | 0.14 |
| 22-Mar-16 | 25285.37 | 25331.01 | 25381.33 | 25083.7 | 25330.49 | 45.12 | 50.32 | -247.31 | 95.96 | -201.67 | 0.38 | -0.8 |
| 23-Mar-16 | 25330.49 | 25322.1 | 25367.81 | 25156.82 | 25337.56 | 7.07 | 45.71 | -165.28 | 37.32 | -173.67 | 0.15 | -0.69 |
| 28-Mar-16 | 25337.56 | 25417.11 | 25432.94 | 24895.49 | 24966.4 | -371.16 | 15.83 | -521.62 | 95.38 | -442.07 | 0.38 | -1.74 |
| 29-Mar-16 | 24966.4 | 24957.24 | 25079.35 | 24835.56 | 24900.46 | -65.94 | 122.11 | -121.68 | 112.95 | -130.84 | 0.45 | -0.52 |
| 30-Mar-16 | 24900.46 | 25062.06 | 25358.84 | 25055.42 | 25338.58 | 438.12 | 296.78 | -6.64 | 458.38 | 154.96 | 1.84 | 0.62 |
| 31-Mar-16 | 25338.58 | 25364.75 | 25479.62 | 25223.22 | 25341.86 | 3.28 | 114.87 | -141.53 | 141.04 | -115.36 | 0.56 | -0.46 |
| 1-Apr-16 | 25341.86 | 25301.7 | 25354.94 | 25119.35 | 25269.64 | -72.22 | 53.24 | -182.35 | 13.08 | -222.51 | 0.05 | -0.88 |
| 4-Apr-16 | 25269.64 | 25333.98 | 25424.15 | 25223.49 | 25399.65 | 130.01 | 90.17 | -110.49 | 154.51 | -46.15 | 0.61 | -0.18 |
| 5-Apr-16 | 25399.65 | 25372.44 | 25372.44 | 24837.51 | 24883.59 | -516.06 | 0 | -534.93 | -27.21 | -562.14 | -0.11 | -2.21 |
| 6-Apr-16 | 24883.59 | 24978.86 | 25000.65 | 24834.16 | 24900.63 | 17.04 | 21.79 | -144.7 | 117.06 | -49.43 | 0.47 | -0.2 |
| 7-Apr-16 | 24900.63 | 24998.79 | 25013.13 | 24647.48 | 24685.42 | -215.21 | 14.34 | -351.31 | 112.5 | -253.15 | 0.45 | -1.02 |
| 8-Apr-16 | 24685.42 | 24665.8 | 24736.03 | 24608.51 | 24673.84 | -11.58 | 70.23 | -57.29 | 50.61 | -76.91 | 0.21 | -0.31 |

**Fig. 3** Pre-processed dataset of SENSEX data snapshot

**Table 2** Pre-processed dataset information

| Attributes of dataset | | | |
|---|---|---|---|
| Date | Date of the trading day | Close | The value at which SENSEX closed |
| Prev_close | The value at which SENSEX closed a day before | Close-prev_close | Total gain or loss in a trading day |
| Open | The value at which SENSEX index opened | Max gain | Total gain achieved in a trading day |
| Day high | Max value of SENSEX's index on a trading day | %max gain | Maximum percentage gain in a trading day |
| Day low | Min value of SENSEX's index on a trading day | Max loss | Total index point loss in a trading day |
| %max loss | Maximum percentage loss in a trading day | | |

max loss, % max loss. The attributes of preprocessed data set are explained in (Table 2).

C. *Tool Used—IBM Watson Analytics*

IBM Watson Analytics offers the benefits of advanced analytics without the complexity. It is a smart data discovery service available on the cloud which guides data exploration, automates predictive analytics and enables dashboard and infographic creation. IBM Watson Analytics provides guided data exploration, automated predictive analytics and automatic dashboard creation for exceptionally insightful discoveries, all on the cloud.

# 5 Analysis

## 5.1 *Case Study 1: Lok Sabha Elections Versus Indian Stock Market*

Lok Sabha elections have been identified as vital factor impacting stock market movement. Last few elections have been analyzed and compared in details and it has been found that elections results have direct impact on stock market. Figure 4 depicts hikes/trails in Index values on election results days, which denotes that stock market reacted on election results.

Below graph depicts point of time where political movements directly impact the stock market performance at-least in short term and then moved as per long term view and investors sentiments.

Table 3 shows summarized analysis of election results impacts on stock market trend.

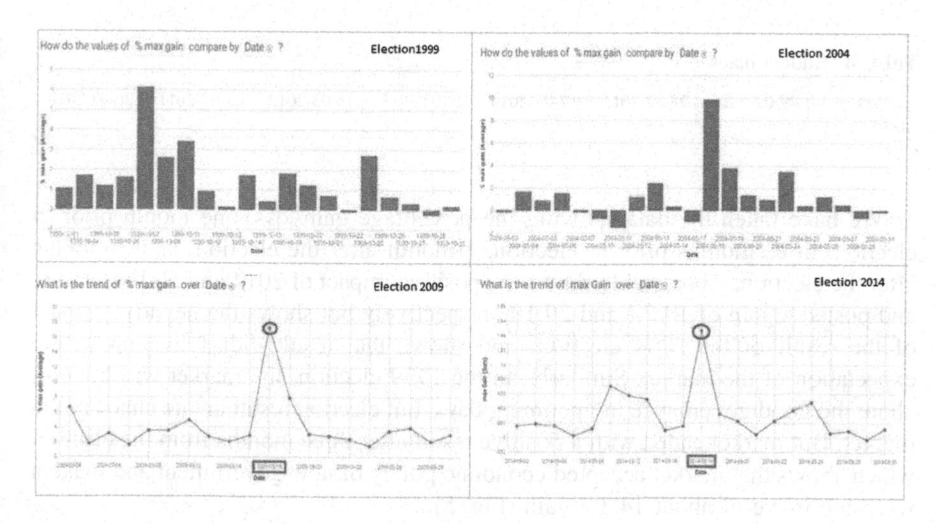


**Fig. 4** Election trends

**Table 3** Election year versus trends

| Year | Trend | Reason |
|---|---|---|
| May-14 | Positive | A huge gain on day of result, but could not retain it till day close |
| May-09 | Positive | Post biggest Sensex index gain of 2110 pts on results. All time highest gain in Sensex history |
| May-04 | Negative | Post a loss of 4.4% due to expectation of repetition of tenure of NDA government for next five years, but election result contradicts the market expectation |
| Oct-99 | Positive | Post a positive index gain due to majority of seats won by the leading party |

| SENSEX COMPARISON WITH ELECTION PERIOD IN INDIA | | | | | |
|---|---|---|---|---|---|
| | 1-month prior | 3-month prior | Election days | After 1-month | After 3-month |
| 2014 | 2.65 | 8.73 | 6.5 | 5.75 | 10.5 |
| 2009 | 26.2 | 21 | 6.5 | 26.8 | 29.1 |
| 2004 | -2.7 | 6.6 | -4.2 | -11 | -11 |
| 1999 | 2.4 | 16.5 | -0.2 | -5.2 | 14.3 |
| 1998 | -5.5 | -1.7 | 5 | 7.9 | 2.7 |

**Fig. 5** Comparative analysis of election period

**Table 4** Budget dates

| Budget dates | 29-02-2016 | 28-02-2015 | 17-02-2014 | 28-02-2013 | 16-03-2012 | 28-02-2011 | 26-02-2010 |
|---|---|---|---|---|---|---|---|

We have taken the data showing the percentage gain/loss, one month prior to election, three months prior to election, 1 month after the election, and 3 months after the election. Above table shows the positive impact of 2014 and 2009 election, and posted a gain of 10.5% and 29.1%, respectively but shows the negative impact of the result of the 2004 election, and shows that result didn't live up to the expectation of the market. Similarly, in the 1999 election, the market was hopeful about the good governance in upcoming days, but election result create chaos in the market. But market ended with a positive result after three months from the election which shows that market accepted economic policy of new government and make a welcome move of about 14.3% gain (Fig. 5).

## 5.2 *Case Study 2: Indian Union Budget Versus Indian Stock Market*

Union budget is one of the significant factors which bring volatility into the stock market. In this analysis, S&P BSE SENSEX is being utilized with 60 day's time span to check whether declaration of union budget has any impact on these or not. The period covered for analysis is from 2010 to 2016. This period includes total 7 budgets. Total sixty trading days are being considered to analyze the impact (Table 4).

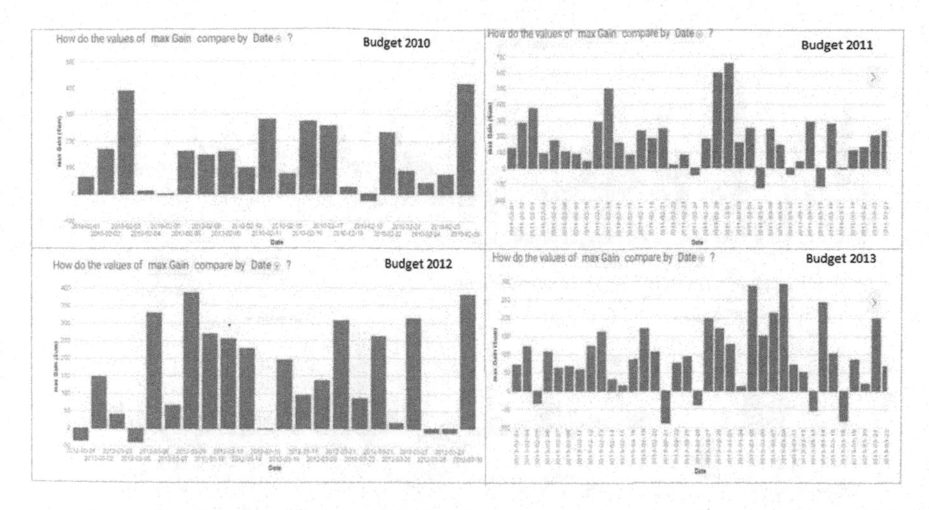


**Fig. 6** Budget days stock market performance

Based on the budget dates as depicted in the above table, February and March data are being analyzed budgets from 2010 to 2016. The market behaved differently on budget days and few days after budget day (Fig. 6).

**2010**: Market welcomed the budget because of the improved investment environment, number of steps had taken to simplify FDI (Foreign Direct Investment) regime etc. The market was positive on budget day and gained over the period of one month after budget.

**2011**: Market behaved positively because the budget was conservative one. The fiscal deficit was reined into the extent possible and money was spent towards growth. The market was positive on budget day and gained over the period of one month after budget.

**2012**: Market didn't like the budget because the standard rate of excise duty was raised from 10 to 12%, in addition to other reasons. The market was negative on budget day and depreciated over the period of one month after budget.

**2013**: Market behaved negatively on budget day because there were no tax rate cut, and other reasons. But market gained over the period of one month after budget due to positive moves such as Proposals such as the launch of Inflation Indexed Bonds or Inflation Indexed National Security Certificates to protect savings from inflation and other long term positive reasons (Fig. 7).

**2014**: Market welcomed the budget because of the composite cap of foreign investment which was raised to 49% with full Indian management and controlled through the FIPB route and other reasons. The market was positive on budget day and gained over the period of one month after budget.

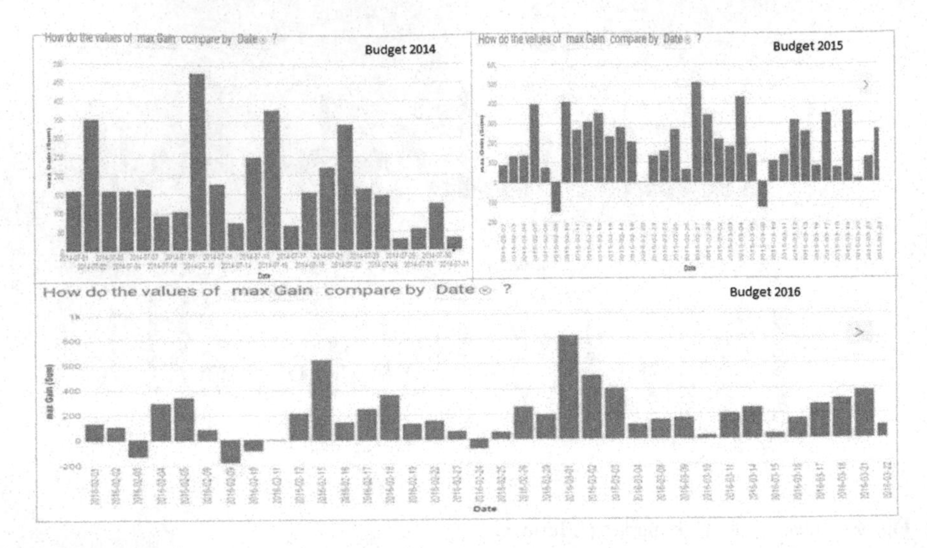


**Fig. 7** Budget days stock market performance

**Table 5** BUDGET versus trends

| Year | Trend | Reason | Year | Trend | Reason |
|---|---|---|---|---|---|
| 2010 | Positive | Step have been taken to simplify the FDI norms | 2014 | Positive | FDI limit is increased up to 49% |
| 2011 | Positive | Fiscal deficit is controlled by government | 2015 | Positive | Due to controlled fiscal deficit of 4.1% |
| 2012 | Negative | Excise duty was raised from 10 to 12% | 2016 | Negative | Increase in excise duty |
| 2013 | Negative | No tax rate cut as demanded by the market | | | |

**2015**: Market zoomed on the budget day but fall during the course of one month due to challenge of maintaining fiscal deficit of 4.1% of GDP met in 2014–15, despite lower nominal GDP growth due to lower inflation and consequent sub-dued tax buoyancy.

**2016:** Market moved down on the budget day, but gained during the course of one month due to Excise Duty on Branded Retail Garments at 6, 2.5% tax on diesel vehicles, proposed 100% deduction to undertakings for construction of affordable housing, etc. Below table shows supporting reasons of stock market trends, which was outcome of different budgets and became vital factors which influenced stock market directly as positive or negative trends (Table 5).

## *5.3 Case Study 3: Indian Monsoon Versus Indian Stock Market*

Monsoon plays a crucial role in Indian economy. As India constitutes 70% of the population dependent on the agriculture for their livelihood. So, rainfall plays a big role in the irrigation due to lack of water resources in India. Many sectors like fertilizers, automobile, housing, food, FMCG, and rural economy are closely dependent on the monsoon or good rainfall. Therefore, good or bad monsoon directly impacts the Indian economy. But in below comparative analysis, it is found that monsoon affects the Sensex for a while but does not have long lasting effect, even in the condition of drought. Sometime also in case of drought, Sensex fall affect have been counter support by the other factor like union election and government policies etc. (Table 6).

It has been found that deficient rainfall in 2002 posted a loss of 2.5% in the Sensex's index. But, 2009, a drought year, could not affect the Sensex growth and even made 80.5% rise. Same in 2014–15, Sensex have shown the rise of 24.88% in a drought year. Therefore, swings in the market were unpredictable, even in years when rainfall was normal. So, it affirms that snow and rainfall can't stop market and its participants to perform (Fig. 8).

It has been found that monsoon rainfall doesn't impact yearly returns and Sensex percentage moves, but it definitely affects some sectors such as fertilizers, automobile, FMCG, etc. It is found that during deficient rainfall, demand of fertilizer and pesticides increase, hence it opens new aspects for fertilizer and pesticide companies. Similarly, high demand patterns in cement industry have been observed in such scenarios. Below figure depicts FMCG sector's index trend due to monsoon rainfall. Here are some analysis on FMCG sector and auto sector.

**FMCG Versus Auto Sector:**

At the sector level, it seems to be some noticeable relationship. Such as, the returns of the BSE FMCG index demonstrate a positive relation with monsoon rainfall (Fig. 9). With a better monsoon, the returns on the BSE FMCG index improved,

**Table 6** Monson versus Sensex

| Year | Monsoon | Sensex change (%) | Year | Monsoon | Sensex change (%) |
|---|---|---|---|---|---|
| 2001–02 | Below normal | −4.48 | 2009–10 | Drought | 80.5 |
| 2002–03 | Drought | −2.26 | 2010–11 | Normal | 10 |
| 2003–04 | Near normal | 75.6 | 2011–12 | Normal | −10.49 |
| 2004–05 | Drought | 36.41 | 2012–13 | Below normal | 8.22 |
| 2005–06 | Near normal | 67.98 | 2013–14 | Normal | 18.84 |
| 2006–07 | Near normal | 25.71 | 2014–15 | Drought | 24.88 |
| 2007–08 | Above normal | 7.56 | 2015–16 | Drought | −9.08 |
| 2008–09 | Near normal | −37.9 | | | |

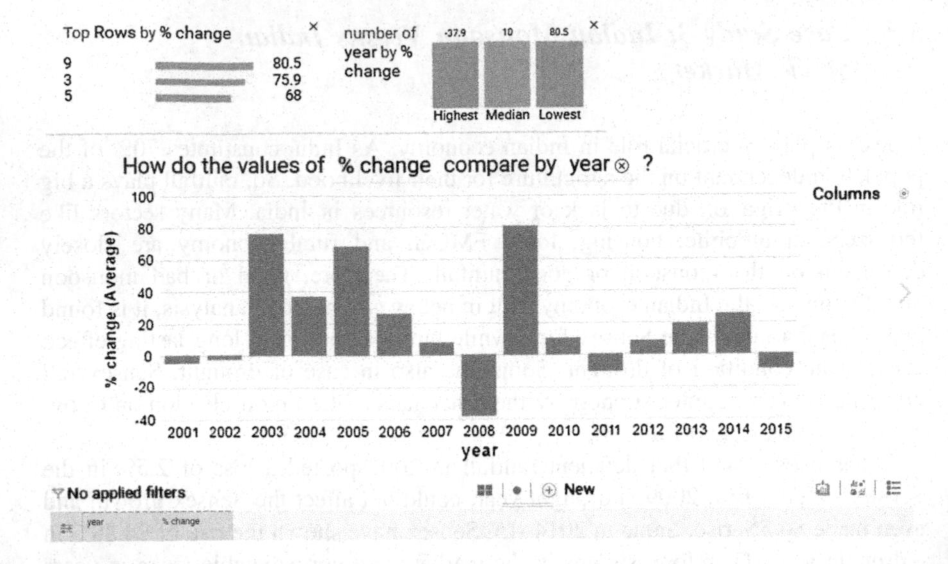


Fig. 8 Monson versus Sensex

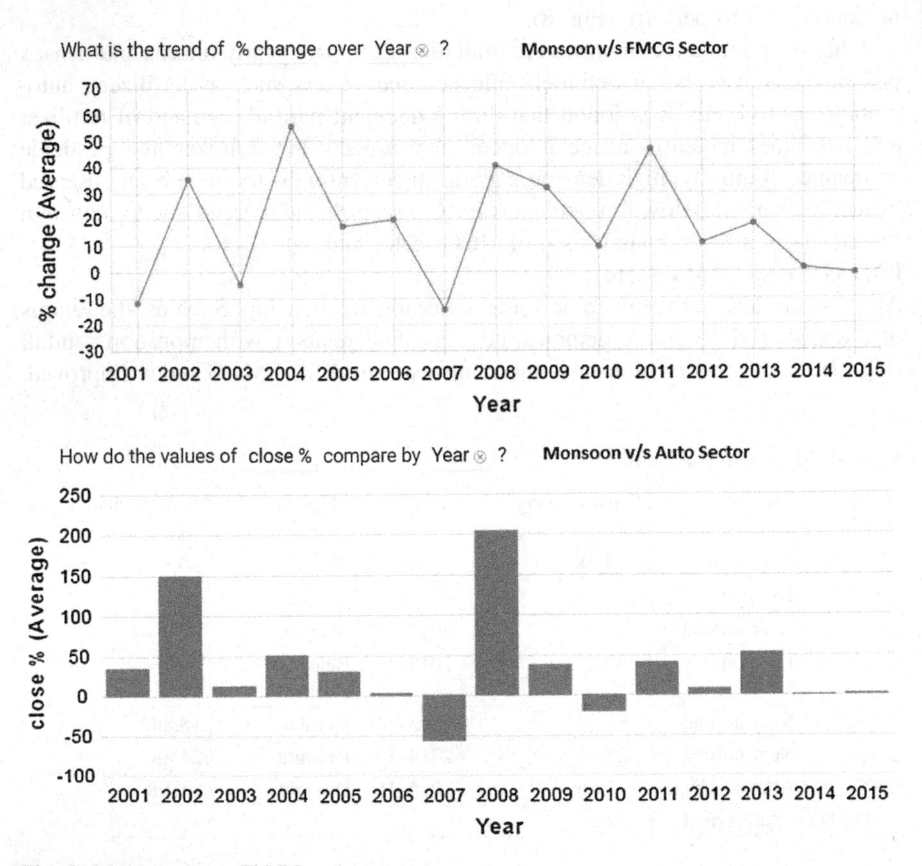


Fig. 9 Monson versus FMCG and auto sector

and vice versa. Yet, the high returns in an effective monsoon year on the BSE FMCG index are predominately balanced by the low returns the next year; the opposite happens during a bad monsoon year. It is expected that with the deficient rainfall, automobile sector feels much pressure due to less demand in rural area. It has been observed that the decrease in rainfall leads to less production in agriculture which results in less purchasing power of the rural area people. Hence post negative impact on BSE AUTO.

It shows that drought year (2009, 2014, 2015) posted less growth in the auto sector as compare to the growth in other years with normal monsoon rainfall.

## *5.4 Case Study 4: Domestic Riots Versus Indian Stock Market*

Riots are the biggest dent on any country's image. It creates a blow to both economic outlook or investor sentiments. We have analyzed different riots which happen in India, to infer whether riots have any direct impact on stock market or not Some of the findings are as follows:

**Mumbai Riots (10th Dec 1992 to 26th Jan 1993):**
There was a steep downfall of 150 points on 11th Dec which continued for 2–3 days but end with small gain. This shows that the market fluctuate within the small limits and did not show any great loss or gain.

**Bombay Blast (12th March 1993)**:
During Bombay blast, Bombay stock exchange (BSE) was among one of the blast sites. Stock trading remained closed on the day of the blast, but when the market opened on 15th March, Monday market moved up about 4.5%, which shows that investors were confident about economic system and stand with solidarity against terror attacks.

**Gujarat Riots (27th Feb 2002 to 2nd Mar 2002)**:
On the day of riots, there was a downward movement in the index, but that was not continued for long. It didn't end up in any big or huge loss to the market even after a few days. Hence market was mostly insensitive for riots (Fig. 10).

**Parliament Attack (13th Dec 2001)**:
During the parliament attack, the loss has been noticed till the next seven days of the attack, but ended with little fluctuations in gain or loss. But no major loss and gain have been seen due to the attack.

**Mumbai Attack (26th Nov 2008 to 29th Nov 2008)**:
The market was flat on 28th Nov but showed a steep fall in the next two sessions. It shows that attacks on Mumbai put some minor impact on Sensex but it was not long lasting effect.

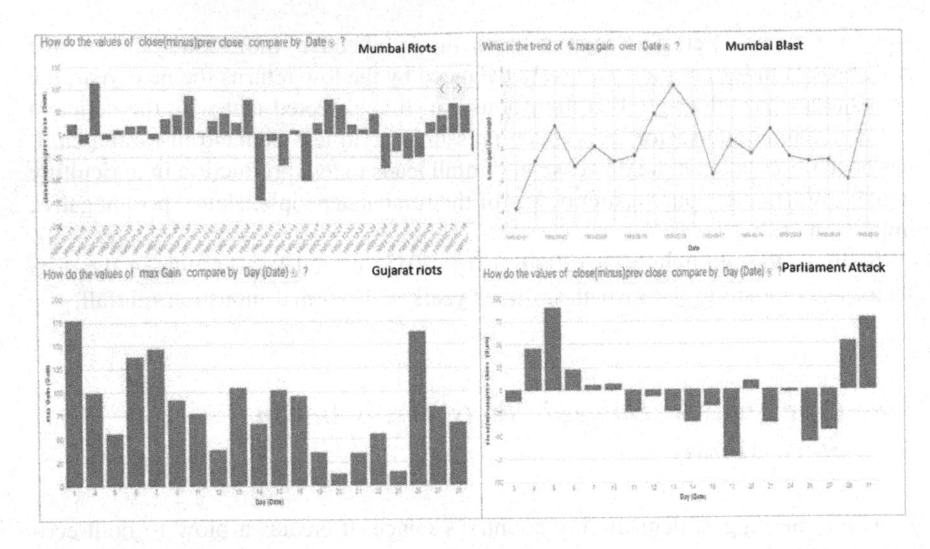


**Fig. 10** Riots versus Sensex

## 5.5 *Case Study 5: Global Factors Versus Indian Stock Market*

Global factors have a major impact on the stock market. We analyzed some dates on which rise and fall of the Sensex's index was more than 700 points. After analysis 39 date (days) were found where there is a huge difference in closing from the previous close and close and all are due to global factors. Below data shows a list of dates with huge fluctuation in day's gain or loss.

On May 18, 2006 Sensex registered a fall of 826 points (6.76%) to close at 11,391, its biggest ever, following heavy selling by FIIs, retail investors and a weakness in global markets. Oct 2007 was the most volatile month for the Sensex's index. On 9th Oct, market gained more than 700 points to create all time high of 18 K. On 18th Oct, Sensex fell more than 700 points because of uncertainty and conflict between SEBI and RBI on issuing of participatory notes issued for FII. But, index overcame and made a bull run of 800 points after intervention or clarification from SEBI chief.

Similarly, on 29th Oct, Sensex made a jump of 700 points to make a new high of 20 K but could not retain the spot till day's close.

Jan 2008 was the black month for the investors as on 21st and 22nd Jan market had a blood bath of more than 2200 points just in two trading days due to fear of recession in the U.S. but after reassurance from finance minister Sensex bounced back about 1900 pts in next two days. On 11th Feb 2008 Sensex's index soared 800 points due to poor performance by reliance industry share. As Reliance is a major contributor in the index. But index plunged more than 800 points on 14th Feb, due to heavy buying by the foreign investors and lowered the crude price over long time.

March 2008 month sheds more than 2700 points from 3rd March to 17th March due to fear of un-stability in U.S economy, the market tried to recover on 25th March but slides continued till 31st March due to heavy selling amid the global fear. July 2008 had some big upside moves, but in Oct 2008 Sensex had a great fall in index whole month due to recession in U.S, failing of the banking system across the world, poor industrial production data, lowering of GDP forecast by RBI.

During Jan 2009 Satyam fraud case gave a huge blow to the stock market and in May 2009 Sensex post all time high jump of more than 2100 pts, welcoming new government after union election. During July 2009, index showed a slide of 869 points due to higher fiscal deficit shown by government during the budget. The poor economy indication from the U.S government rattled the investor and which in turns lowered the index in Sep 2011. Sensex posted a fall of 769 points in August 2013 amid the fear of downgrade of the economic outlook and downgrade of currency value, and stern step by U.S Fed, but on 10th Sep 2013 posted a gain of 700 points and attained a level of 20 K on positive global cues. In 2015 also Sensex behaved as per global news one of the biggest fall was on 24 Aug 2015 when the market fell 1624 points as China equities fell over 8%. The similar trend has been followed up in 2016.

## 6 Prediction

Watson prediction includes three modes such as one field, two fields and combination. One field considers one input (attribute) to predict the target; two fields take multiple possible two inputs (attributes) to predict the target. A combination mode uses decision tree as a hierarchical classification of records to predict the value of a target field. The hierarchy is based on the values of one or more input fields. To create a tree, the records are segmented into the groups, which are called nodes. Each node contains records that are statistically similar to each other with respect to the target field. IBM Watson Analytics creates two types of decision trees: CHAID classification tree and CHAID regression tree. The predictions are based on combinations of values in the input fields. A CHAID regression tree provides a predicted mean value for each node in the tree. It is generated when the target field is continuous. The branch statistics include the F test statistics that resulted in a partition. If the F test is significant, a new partition is created. The partition is statistically different from the parent node. The effect size can also provide some information about the extent of this difference. F tests are also used to merge categories in single nodes. If the target means for categories are not significantly different, the categories are merged into a node.

Figure 11 shows top five decision rules predicting "close" to be high and five rule which predict the close to be the "low". Statistical details are as follows:

| Top five decision rules predicting close to be high | | | | | | |
|---|---|---|---|---|---|---|
| Day Low | Date (yrs) | daylow-open | %max loss | max loss | day high-open | Close |
| >17,852.73 | >41.37 | <=-139.94 | >-0.84 | <=-134.55 | NA | 25,602.94 |
| >17,852.73 | >41.37 | <=-139.94 | >-0.84 | >-134.55 | NA | 23,671.09 |
| >17,852.73 | >41.37 | <=-139.94 | <=-0.84 | NA | NA | 23,473.85 |
| >17,852.73 | >41.37 | >-139.94 | NA | NA | >110.95 | 23,442.01 |
| >17,852.73 | >41.37 | >-139.94 | >-0.35 | NA | <=110.95 | 22,171.36 |

| Top five decision rules predicting close to be low | | | | | | |
|---|---|---|---|---|---|---|
| Day Low | Date | daylow-open | %max loss | max loss | day high-open | Close |
| <=3,280.33 | <=26.66 | >-27.00 | NA | NA | <=8.82 | 2,016.53 |
| <=3,280.33 | <=26.66 | >-27.00 OR >-5.28 | NA | NA | "=8.82 to 26.74" | 2,024.62 |
| <=3,280.33 | <=26.66 | >=-27.00 AND <=-5.28 | NA | NA | "=8.82 to 26.74" | 2,344.50 |
| <=3,280.33 | <=26.66 | >-27.00 | NA | NA | >26.74 | 2,541.36 |
| <=3,280.33 | <=26.66 | <=-27.0 | NA | NA | NA | 2,647.71 |

**Fig. 11** Top 5 decision rules predicting close to be high

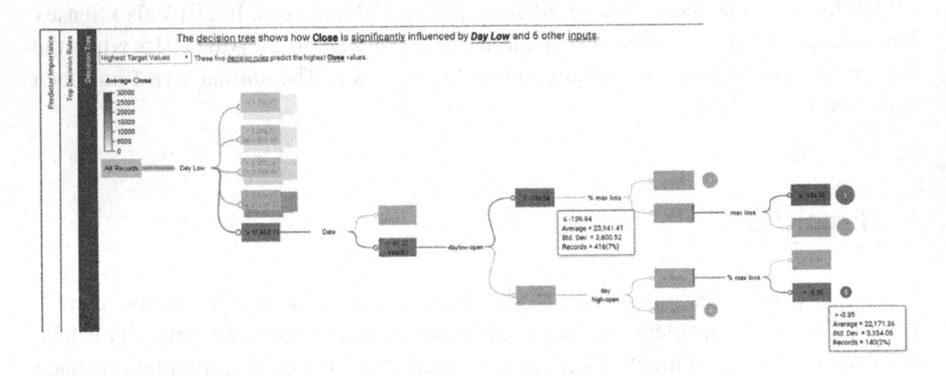


**Fig. 12** Decision tree predicting close to be high

Figure 12 depicts the decision tree how "close" is significantly influenced by "day low" and 6 other inputs. The five circles in the figure denotes corresponding rule for highest predicted value for Average Close.

The supporting rules produced as a result CHAID regression tree prediction methodology is as follows:

*IF (Day Low > 17,852.73) AND (Date > 41.37 year(s)) AND (daylow-open ≤ −139.94) AND (% max loss > −0.84) AND (max loss ≤ −134.55) THEN predicted Close = 25,602.94.*

*IF (Day Low > 17,852.73) AND (Date > 41.37 year(s)) AND (daylow-open = −139.94) AND (% max loss > −0.84) AND (max loss > −134.55) THEN predicted Close = 23,671.09.*

*IF (Day Low > 17,852.73) AND (Date > 41.37 year(s)) AND (daylow-open > −139.94) AND (day high-open = 110.95) AND (% max loss = −0.35) THEN predicted Close = 20,619.71.*
*IF (Day Low > 17,852.73) AND (Date > 41.37 year(s)) AND (daylow-open > −139.94) AND (day high-open = 110.95) AND (% max loss > −0.35) THEN predicted Close = 22,171.36.*
*IF (Day Low = 9,709.40 to 17,852.73) AND (Date = 36.48 year(s) to 41.37 year (s)) AND (% max gain = 1.66) AND (day high-open > 110.95) AND (% max loss > −0.84) THEN predicted Close = 15,748.78.*
*IF (Day Low = 9,709.40 to 17,852.73) AND (Date = 36.48 year(s) to 41.37 year (s)) AND (% max gain = 1.66) AND (day high-open > 110.95) AND (% max loss = −0.84) THEN predicted Close = 14,690.65.*

## 7 Conclusion

The research work uses CHAID regression and classification technique to understand the Indian stock market data of 26 years. From the analysis, the main influencing factors are India's Elections and Global Factors. The analysis also shows that budget is also another prominent aspect that affects the Indian Stock Market. External factors like monsoon and riots are vital parameters that influence stock market Sensex index. The work can be extended by considering the sector level (Auto, FMCG, IT, Metal) and enterprise level (TCS, HUL, BAJAJ). By Integrating factors like GDP, Indian currency as compared to USD, Gold Price etc., can be used in the prediction model to get more appropriate result. The other classification techniques like Naive Bayes, Support Vector Machine can also be used to compare different models and can check their performance and predicting capability.

## References


1. Abdullah, S.S., Rahaman, M.S., Rahman, M.S.: Analysis of stock market using text mining and natural language processing. In: 2013 International Conference on Informatics, Electronics and Vision (ICIEV), pp. 1–6. IEEE (2013)
2. Babu, C.N., Reddy, B.E.: Selected Indian stock predictions using a hybrid ARIMA-GARCH model. In: 2014 International Conference on Advances in Electronics, Computers and Communications (ICAECC), pp. 1–6. IEEE (2014)
3. Bebarta, D.K., Biswal, B., Rout, A.K., Dash, P.K.: Forecasting and classification of Indian stocks using different polynomial functional link artificial neural networks. In: 2012 Annual IEEE India Conference (INDICON), pp. 178–182. IEEE (2012)
4. Fonseka, C., Liyanage, L.: A Data mining algorithm to analyse stock market data using lagged correlation. In: ICIAFS 2008. 4th International Conference on Information and Automation for Sustainability, 2008, pp. 163–166. IEEE (2008)

5. Han, J., Kamber, M., Pei, J.: Data Mining: Concepts and Techniques. Elsevier (2011)
6. Hendahewa, C., Pavlovic, V.: Analysis of causality in stock market data. In: 2012 11th International Conference on Machine Learning and Applications (ICMLA), vol. 1, pp. 288–293. IEEE (2012)
7. Huang, K.Y., Wan, S.: Application of enhanced cluster validity index function to automatic stock portfolio selection system. Inf. Technol. Manage. **12**(3), 213–228 (2011)
8. IBM WatsonAnalytics (2016). https://watson.analytics.ibmcloud.com/catalyst/help/index.jsp?topic=/com.ibm.spss.analyticcatalysthelp/analytic_catalyst/chaid_regression_tree.html. Accessed 30 Mar 2016
9. Indian Budget. http://indiabudget.nic.in. Accessed 30 Mar 2016
10. Investopedia. http://www.investopedia.com/terms/s/stockmarket.asp. Accessed 30 Mar 2016
11. Kass, G.V.: An exploratory technique for investigating large quantities of categorical data. Appl. Stat. 119–127 (1980)
12. Kim, H.J., Ryu, C.K., Ryu, D.S., Cho, H.G.: A phylogenetic analysis for stock market indices using time-series string alignments. In: ICCIT'08. Third International Conference on Convergence and Hybrid Information Technology, 2008, vol. 1, pp. 487–492. IEEE (2008)
13. Kumar, D.A., Murugan, S.: Performance analysis of Indian stock market index using neural network time series model. In: 2013 International Conference on Pattern Recognition, Informatics and Mobile Engineering (PRIME), pp. 72–78. IEEE (2013)
14. Mukherjee, D.: Comparative analysis of Indian stock market with international markets. Great Lakes Herald **1**(1), 39–71 (2007)
15. Saxena, P., Pant, B., Goudar, R.H., Srivastav, S., Garg, V., Pareek, S.: Future predictions in Indian stock market through linguistic-temporal approach. In: 2013 7th International Conference on Intelligent Systems and Control (ISCO) (2013)
16. Tirea, M., Negru, V.: Stock market analysis-strongest performing stocks influence on an evolutionary market. In: Proceedings of the ITI 2013 35th International Conference on Information Technology Interfaces (ITI), pp. 263–270. IEEE (2013)
17. Troiano, L., Kriplani, P.: Predicting trend in the next-day market by hierarchical hidden Markov model. In: 2010 International Conference on Computer Information Systems and Industrial Management Applications (CISIM), pp. 199–204. IEEE (2010)
18. Zhongxing, Y., Liting, G.: A hybrid cognition system: application to stock market analysis. In: IJCNN'93-Nagoya. Proceedings of 1993 International Joint Conference on Neural Networks, 1993, vol. 3, pp. 3000–3003. IEEE (1993)

# A Novel Approach to Improve the Performance of Divisive Clustering-BST

P. Praveen and B. Rama

**Abstract** The traditional way of searching data has many disadvantages. In this context we propose Divisive hierarchical clustering method for quantitative measures of similarity among objects that could keep not only the structure of categorical attributes but also relative distance of numeric values. For numeric clustering, the quantity of clusters can be approved through geometry shapes or density distributions, in the proposed Divclues-T Calculate the Arithmetic mean it is called as a root node, the objects smaller than root node fall into left sub tree otherwise right sub tree this process is repeated until we find singleton object.

**Keywords** Agglomerative · Computational complexity · Clustering · Distance measure · Divclues-T

## 1 Introduction

Classification and cluster are important techniques that partition the objects that have many attributes into meaningful disjoint subgroups [1] so that objects in each group are more similar to each other in the values of their attributes than they are to objects in other group [2].

There is a serious distinction between cluster analysis and classification. In supervised classification, the categories are a unit outlined, the user already is aware of what categories there are a unit, and a few training data that's already tagged by their category membership is out there to training or build a model. In cluster analysis, one doesn't recognize what categories or clusters exist and also the downside to be resolved is to cluster the given data into purposeful cluster. Rather like application of supervised classification, cluster analysis has applications in

P. Praveen (✉) · B. Rama
Deparment of Computer Science, Kakatiya University, Warangal, Telangana, India
e-mail: prawin1731@gmail.com

B. Rama
e-mail: rama.abbidi@gmail.com



S.C. Satapathy et al. (eds.), *Data Engineering and Intelligent Computing*,
Advances in Intelligent Systems and Computing 542,
DOI 10.1007/978-981-10-3223-3_53

many various areas corresponding to in promoting, medicine, business. Sensible applications of cluster analysis have additionally been found in character recognition, internet analysis and classification of documents, classification of astronomical information. The first objective of cluster is to partition a collection of objects into homogenized teams. a good cluster wants an appropriate live of similarity or unsimilarity. Thus a partition structure would be known within the sort of natural teams [3].

Clustering has been exuberantly applied in various fields as well as care systems, client relationships management, producing, biotechnology and geographical data systems. Several algorithms that type clusters in numeric domain are planned; however few algorithms are appropriate for mixed knowledge like collective [4], the most aim of this paper a way to unify distance illustration schemes for numeric knowledge. Numeric cluster adopts distance metrics whereas emblematic uses a tally theme to calculate conditional probability estimates for defining the relationship between groups [5].

In this paper I would like to address the question of how to minimize computational complexity on synthetic data into natural groups efficiently and effectively, the proposed approach will be applied to identify the "optimal" classification scheme among that objects using Arithmetic mean. The extension of clustering to more general setting requires significant changes in algorithm techniques in several fundamental respected. Considering a data set, to measure the distance between objects, distance matrix for Hierarchical clustering method. The mean value for each iteration time complexity is O(n)/n is 1. We find the mean values n times, so the time complexity recursively O(n).The proposed algorithm takes O(n log n) time. It is smaller than agglomerative clustering algorithms.

## 2 Clustering Algorithms

Cluster analysis was first projected in numeric domains, where distance is clearly defined. Later it extended to categorical data. However, much of data in real world contains a mixture of categorical and unbroken facts; as a result, the demand of cluster analysis on the diverse data is growing.

Cluster analysis has been an area of research for several decades and there are too many different methods for all to be covered even briefly. Many new methods are still being developed. In this section we discuss some popular and mostly used clustering algorithm and present their complexity.

***K-means***: K-means is that the best and established agglomeration strategy that is clear to actualize. The traditional will exclusively be utilized if the data in regards to every one of the items is found inside the fundamental memory. The strategy is named K-implies subsequent to everything about K groups is depict by the mean of the objects inside it [6].

***Nearest Neighbor Algorithm***: Associate in nursing formula kind of like the one link technique is termed the closest neighbor formula. With this serial formula, things are iteratively united into the present clusters that are closet. during this formula a threshold, it's accustomed confirm if things are further to existing clusters or if a replacement cluster is formed [7].

***Divisive Clustering***: With dissentious agglomeration, all things are at the start placed in one cluster and clusters are repeatedly split in two till all things are in their own cluster. The concept is to separate up clusters wherever some parts don't seem to be sufficiently near to alternative parts [7, 8].

***BIRCH Algorithm***: BIRCH is meant for agglomeration an oversized quantity of numerical information by integration of stratified agglomeration (at the initial small agglomeration stage) and alternative agglomeration ways equivalent to reiterative partitioning (at the later macro agglomeration stage). It overcomes the 2 difficulties of clustered agglomeration methods: (1) measurability and (2) the lack to undo what was wiped out the previous step [9, 10].

***ROCK (Robust agglomeration mistreatment links***) may be a stratified agglomeration formula that explores the thought of links (the variety of common neighbors between 2 objects) for information with categorical attributes [11].

***CURE formula***: One objective for the CURE agglomeration algorithm is to handle outliers well. It's each a stratified element and a partitioning element [12].

***Chameleon***: Chameleon may be a stratified agglomeration formula that uses dynamic modeling to work out the similarity between pairs of clusters [13]. It absolutely was derived supported the determined weakness of the Two stratified agglomeration algorithms: ROCK and CURE [14]. Distance based HC methods are widely used in unsupervised data analysis but few authors fake account of uncertainty in the distance data [5].

Distance between P1 to P2 = 5, d(P1, P2) = 5.

$(p_1, q_1)$ $l_2$ $(p_2, q_2)$

$$l_2 = \left((p_2 - p_1)^2 + (q_2 - q_1)^2\right)^{1/2}$$

$$D\,(A, B) = \min \sum\nolimits_{i=1}^{n} \|x_i - x_j\|^q$$

where A and B are pair of elements considered as cluster, d (a,b) denotes the distance between the two elements. Distance function nature is defined by an integer q (q = 2) for a data set of numeric values [15].

# 3 Agglomerative Versus Divclues-T

## *3.1 Agglomerative Methods*

The basic plan of the clustered methodology is to begin with n clusters for n data focuses, that is, each group comprising of one data point [16]. Utilizing a live of distance, at each progression of the methodology, the technique combines two closest groups, hence decreasing the number of groups and building in turn larger clusters. The method continues till the desired range of clusters has been obtained or all the information points are in one cluster. The clustered methodology ends up in hierarchical clusters during which at every step we have a tendency to build larger and bigger clusters that embrace more and more dissimilar objects [3].

**Algorithm of Agglomerative method**

1. The clustered methodology is largely a bottom-up approach that involves the subsequent steps. Associate degree implementation but could embrace some variation of those steps [15].
2. Assign every purpose to a cluster of its own. Therefore we have tendency to begin with n clusters for n objects.
3. Produce a distance matrix by generating distance between the clusters either victimization, as associate instance. These distances are in ascending order [13].
4. Find the smallest distance among the groups.
5. Merge clusters with nearby clustered objects.
6. The above steps are repeated to produce a single group.

### 3.1.1 Computational Complexity (SLINK)

The basic rule for hierarchical agglomeration [5] isn't terribly economical. At every step, we have a tendency to work out the distances between every pair of clusters, to begin with to search an object is O ($n^2$) time, but upcoming steps take time of $(n - 1)^2$, $(n - 2)^2$ ……. The squares of n are up to n is O ($n^3$). By computing the space between all pairs of objects consumes O ($n^2$) times. Their distances are stored into a priority queue, thus we are able to continually realize the smaller distance on one step. This operation takes O ($n^2$).

We calculate distances between the emerging cluster and remaining clusters. This work take O (n log n), Steps 5 and 6 above steps are executed n times, Steps 1 and 2 are executed only for one time. It consumes O ($n^2$log n) time.

## *3.2 Divclues-T*

DIVCLUS-T is divisive hierarchical agglomeration algorithms supported a monothetic bipartitional approach permitting the dendrogram of the hierarchy to be browse as a call tree. It's designed for either numerical categorical information.

Divisive hierarchical clustering reverse method of clustered hierarchical agglomeration [17].

The divisive methodology is that the opposite of the clustered methodology therein the strategy starts with the total information set together cluster then divide the cluster into two sub-clusters repeatedly until each cluster has one object. There square measure two forms of discordant ways that [18].

***Mono_Thetic***: It divides a cluster victimization just one element at an instant. Associate degree attribute that has the foremost dissimilarity may be well chosen.

***Poly_Thetic***: It divides cluster victimization for all attributes. Two clusters may have distant well designed supported distance among items [17]. A typical poly-thetic divisive methodology works just like the following [15]:

1. Choose a way of activity the distance between two objects. Additionally decide a threshold distance.
2. A distance matrix is computed among all pairs of objects in the cluster.
3. Find the pair which has the biggest distance between the objects. They're foremost dissimilar items.
4. If the distance between the two objects is smaller than the pre-specified threshold and there's no different cluster that has to be divided then stop, otherwise continue [3].
5. Use the pair of objects as seeds of a K-means methodology to make two new clusters.
6. If there's just one object in every cluster then stop otherwise continue with step a pair of.

In this the higher than methodology, we want to resolve the subsequent two issues:

- Which cluster to split next?
- How to split a cluster?

#### 3.2.1 Divclues-T Algorithm

1. Initially all objects are in single cluster.
2. Find arithmetic mean of distance matrix.
3. The mean value is stored on tree it is called as root.
4. Object distance value is less then mean value create a new cluster and place the objects in new cluster.
5. If object distance value is greater than mean value creates a new cluster and place the objects in new cluster.
6. Continue step 4 and 5 until single elemented cluster.

In the event that the sought key (object separation) is not found after an invalid sub tree is achieved then the item is not present in all clusters.

### 3.2.2 Computation Complexity for Search

To start most pessimistic scenario this calculation must hunt from the base of the tree to the leaf O (n log n), The pursuit operation takes relative time to the tree's height, on a normal paired inquiry trees with n hubs have O(log n). The mean value for each iteration time complexity is 0(n)/n i.e. 1. We find the mean values n times, so the time complexity recursively O (n). The proposed algorithm takes O (n log n) time. It is smaller than agglomerative clustering algorithms. The binary search tree that has lack of load balancing. This method guarantees that object or element can be found less or equal than O (log n).

## 4 Experimental Evaluation

**Exercise 1**: the Agglomerative algorithm using Euclidean distance to cluster and Divclues-T algorithm using mean value for the following 6 objects (Table 1).

We have 6 objects i.e. [A, B, C, D, E, F] and we put every item into one group. Presently every item choices them as a group. In this way, at first we have 6 groups are a unit objective is to cluster those 6 cluster determined at the highest point of the cycles, we are going to have exclusively single group comprises of the complete six unique articles. In each progression of the cycle, we find the most elevated join groups. Amid this case, the most elevated group is between clusteres. D and F with most limited separation of 0.5. In this manner we tend to overhaul the space grid, separation between ungrouped clusteres won't transform from the first separation lattice, now the issue is the way to compute separation between recently assembled groups (D,F) and different groups these show on Table 2:

That is precisely where the linkage principle happens. Utilizing single linkage, we determine least distance between unique objects of the two groups. Utilizing the information separation lattice, distance among group (D, F) and cluster.

$$d\,(D,F) \rightarrow A \ \min\,(d_{DA},\, d_{FA}) = \min\,(3.61,\, 3.20) = 3.20 \tag{1}$$

**Table 1** Matrix D1 of pair wise distance

| Dist | A | B | C | D | E | F |
|---|---|---|---|---|---|---|
| A | 0.0 | 0.71 | 5.66 | 3.61 | 4.24 | 3.21 |
| B | | 0.0 | 4.95 | 2.92 | 3.54 | 2.50 |
| C | | | 0.0 | 2.24 | 1.41 | 2.50 |
| D | | | | 0.0 | 1.00 | 0.50 |
| E | | | | | 0.0 | 1.12 |
| F | | | | | | 0.0 |

Similarly, distance between cluster (D, F) and cluster B,

$$d(D, F) \rightarrow B \min(d_{DA}, d_{FB}) = \min(2.92, 2.50) = 2.50.$$

Similarly, distance between cluster (D, F) and cluster C—2.24, Similarly, distance between cluster (D, F) and cluster ∈—1.00 Then, the updated distance matrix becomes in Table 3.

From Eq. (1) Table 3.

In above distance matrix, e discovered that the shut separation between group An and B is presently 0.71. So we amass cluster An and group B into a solitary cluster name (A, B) using the input distance matrix (6 × 6) C and cluster (A, B) we found out the minimum distance is 4.95 it is similarly (1).

Cluster (A, B) and cluster (D, F) minimum distance is 2.50. Cluster (A, B) and cluster ∈ minimum distance is 3.54 then updated distance matrix (Table 4).

In above distance matrix the cluster is area by cluster (D, F) we merge ((D, F), E) continue this process for cluster C, so cluster C is nearby cluster ((D, F), E), C finally all clusters are combined together (Fig. 1).

**Divclues-T using mean**

Mean value of $\overline{X} = \overset{N}{\underset{i=1}{\in}} \frac{d(x_i, y_i)}{N}$ Now $\overline{X}$ is root—2.67 (Fig. 2; Table 5).

**Table 2** Matrix D2 join (D, F) of pair wise distance

| Dist | A | B | C | D, F | E |
|---|---|---|---|---|---|
| A | 0.0 | 0.71 | 5.66 | ? | 4.24 |
| B | | 0.0 | 4.95 | ? | 3.54 |
| C | | | 0.0 | ? | 1.41 |
| D, F | ? | ? | ? | 0.0 | ? |
| E | | | | | 0.0 |

**Table 3** Updated matrix D2

| Dist | A | B | C | D, F | E |
|---|---|---|---|---|---|
| A | 0.0 | 0.71 | 5.66 | 3.61 | 4.24 |
| B | | 0.0 | 4.95 | 2.50 | 3.54 |
| C | | | 0.0 | 2.24 | 1.41 |
| D, F | | | | 0.0 | 1.0 |
| E | | | | | 0.0 |

**Table 4** Final distance matrix D3 join (A, F) of pair wise distance

| Dist | A, B | C | (D, F) | E |
|---|---|---|---|---|
| A, B | 0 | 4.95 | 2.50 | 3.54 |
| C | | 0 | 2.24 | 1.41 |
| D, F | | | 0 | 1.00 |
| E | | | | 0 |

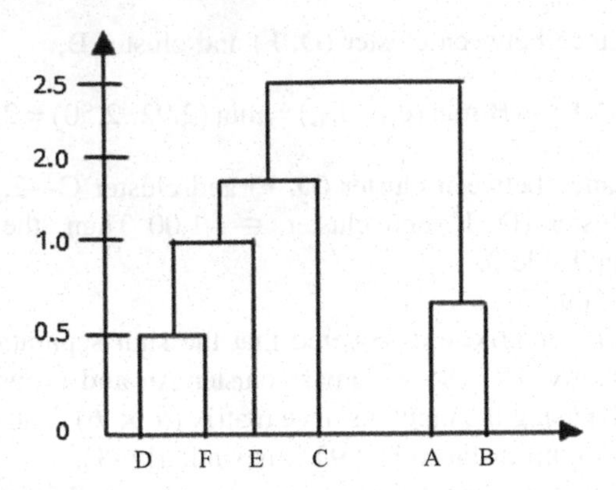


Fig. 1 Dendrogram for single link hierarchical clustering

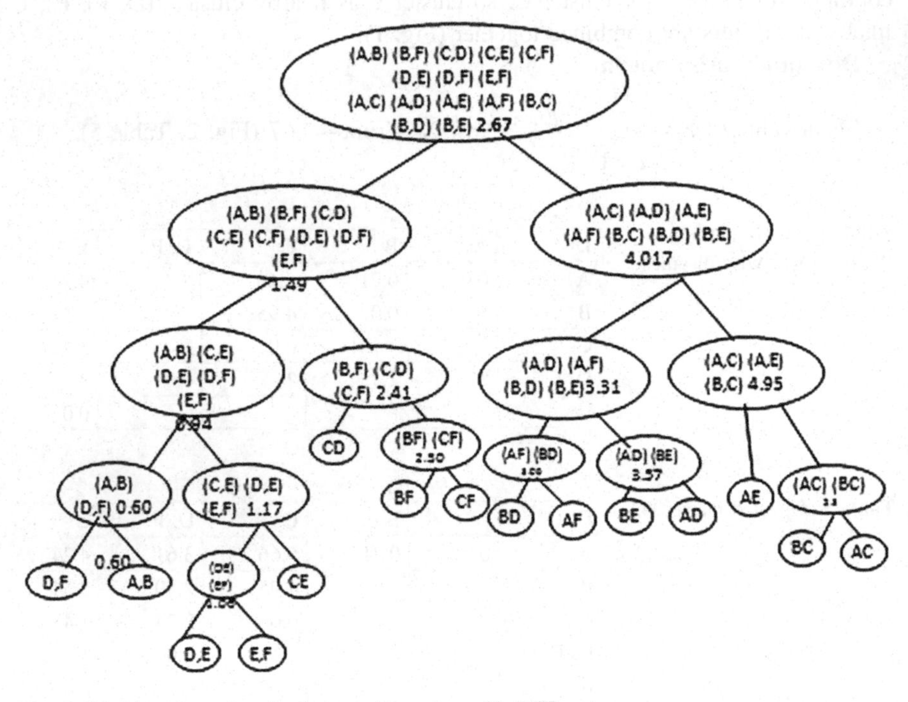


Fig. 2 Divisive clustering implementation using with BST

Searching an object in clusters by using mean value. Here the mean value is a key. We begin by examining the root node. In the event that the key is equivalents that of the root the pursuit it effective and we give back the hub i.e. object. On the off chance that the key is not as much as that of the root we seek the left sub tree. Correspondingly in the event that the key is more prominent than that of the root.

**Table 5** Adjacency matrix for Divclues

| Dist | A | B | C | D | E | F |
|---|---|---|---|---|---|---|
| A | 0.0 | 0.71 | 5.66 | 3.61 | 4.24 | 3.21 |
| B | | 0.0 | 4.95 | 2.92 | 3.54 | 2.50 |
| C | | | 0.0 | 2.24 | 1.41 | 2.50 |
| D | | | | 0.0 | 1.00 | 0.50 |
| E | | | | | 0.0 | 1.12 |
| F | | | | | | 0.0 |

We look the right sub tree this procedure is rehashed until the key is found or remaining sub tree is invalid.

## 5 Conclusion

We have examined single link hierarchical clustering and Divisive clustering algorithm for pure numeric synthetic data set. So far we have examined distance measure on single link for numeric data sets and mean value of hierarchical clustering methods. The results are achieved highly encouraging and more optimal performance. Divisive algorithm by using the mean value of objects can also examined. This paper, an agglomerative and Divisive clustering method designed for numerical data. Compared or classical method like K-means and hierarchical clustering. To conclude this section we have that the running time of our Divisive algorithms are faster than agglomerative (SLHC) algorithm for above dataset. The future scope of this work is to store clusters in a n dimensional array which reduces the time complexity.

## References


1. MacQuuen, J.B.: Some methods for classification and analysis of multivariate observation. In: Proceedings of the 5th Berkley Symposium on Mathematical Statistics and Probability, pp. 281–297 (1967)
2. Lance, G.N., Williams, W.T.: A general theory of classificatory sorting strategies. Comput. J.
3. Langfelder, P., Zhang, B., Horvath, S.: Defining clusters from a hierarchical cluster tree: the dynamic tree cut package for r. Bioinform. Appl. Note **24**(5), 719–720 (2008)
4. Ahmad, A., Dey, L.: A k-mean clustering Algorithm for mixed numeric and categorical data. Data Knowl. Eng. **63**(2), 503–527 (2007)
5. Apresjan, J.D.: An algorithm for constructing clusters from a distance matrix. Mashinnyi perevod: Prikladnaja lingvistika **9**, 3–18 (1966)
6. Charalampidis, D.: A modified K-means algorithm for circular invariant clustering. IEEE Trans. Pattern Anal. Mach. Intell. **27**(12), 1856–1865 (2005)
7. Han, J., Kamber, M.: Data Mining Concepts and Techniques, 2nd edn
8. Galluccio, L., Michel, O., Comon, P., Kligee, M., Hero, A.O.: Inform. Sci. **2**(51), 96–113 (2013)

9. Yamaguchi, T., Ichimura, T., Mackin, K.J.: Analysis using adaptive tree structured clustering method for medical data of patients with coronary heart disease
10. Castro, R.M., Member, S., Coates, M.J., Nowak, R.D.: Likelihood based hierarchical clustering. IEEE Trans. Signal Process. **52**, 2308–2321 (2004)
11. Jain, A.K., Murty, M.N., Flynn, P.J.: Data clustering: a review. ACM Comput. Surv. **31**(3), 264–323 (1999)
12. Xu, R., Wunsch, D.: Survey of clustering algorithms. IEEE Trans. Neural Netw. **6**(3), 645–672 (2005)
13. Karypis, G., Han, E.H., Kumar, V.: CHAMELEON: a hierarchical clustering algorithm using modeling. IEEE Comput. **32**(8), 68–75 (1999)
14. Kuchaki Rafsanjani, M., Asghari Varzaneh, Z., Emami Chukanlo, N.: A survey of hierarchical clustering algorithms. J. Math. Comput. Sci. 229–240 (2012)
15. Frawley, W., Piatetsky, G., Matheus, S.C.: Knowledge discovery in databases: an overview. AI Mag. 213–228 (1992)
16. Karypis, G., Han, E.H., Kumar, V.: CHAMELEON: a heiararchical clustering algorithm using dynamic modeling. IEEE Comput. **32**(8), 68–75 (1999)
17. Yuruk, N., Mete, M., Xu, X., Schweiger, T.A.J.: A divisive hierarchical structural clustering algorithm for networks. In: Proceedings of the 7th IEEE International Conference on Data Mining Workshops, pp. 441–448 (2007)
18. Li, M., Ng, M.K., Cheung, Y.M., Huang, Z.: Agglomertive fuzzy K-means clustering algorithm with selection of number of clusters. IEEE Trans. Knowl. Eng. **20**(11), 1519–1534 (2008)

# Prediction and Analysis of Pollution Levels in Delhi Using Multilayer Perceptron

Aly Akhtar, Sarfaraz Masood, Chaitanya Gupta and Adil Masood

**Abstract** Air Pollution is a major problem faced by humans worldwide and is placed in the top ten health risks. Particulate Matter (PM10) is one of the major parameters to measure the air quality of an area. These are the particulate matter of the size 10 μm or less suspended in the air. PM10 occur naturally from volcanoes, forest fire, dust storms etc., as well as from human activities like coal combustion, burning of fossil fuels etc. The PM10 value is predicted by multilayer perceptron algorithm, which is an artificial neural network, Naive Bayes algorithm and Support Vector Machine algorithm. Total of 9 meteorological factors are considered in constructing the prediction model like Temperature, Wind Speed, Wind Direction, Humidity etc. We have then constructed an analysis model to find the correlation between the different meteorological factors and the PM10 value. Results are then compared for different algorithms, which show MLP as the best.

**Keywords** Artificial neural network · Naive Bayes · Support vector machine · Multilayer perceptron (MLP) · Correlation · PM 10 · Air quality index · Delhi

A. Akhtar · S. Masood (✉) · C. Gupta
Department of Computer Engineering, Jamia Millia Islamia, New Delhi, India
e-mail: smasood@jmi.ac.in

A. Akhtar
e-mail: alyakhtar94@gmail.com

C. Gupta
e-mail: cgupta319@gmail.com

A. Masood
Department of Civil Engineering, Al-Falah University, Faridabad, India
e-mail: adil.engg.cvl@gmail.com





S.C. Satapathy et al. (eds.), *Data Engineering and Intelligent Computing*,
Advances in Intelligent Systems and Computing 542,
DOI 10.1007/978-981-10-3223-3_54

## 1 Introduction

The capital of India—Delhi, according to a survey by World Health Organization (WHO) in 1600 cities around the world, has the worst air quality among the major cities [1]. About 1.5 million people are estimated to die every year because of air pollution and it is the fifth largest killer in India. India tops the list of death rate from chronic diseases and asthma in world. In Delhi, poor air quality is damaging the lungs of 2.2 million or 50% of all children. The major sources of pollution include Primary Pollutants e.g. CO, $SO_x$, PM etc., Secondary Pollutants e.g. $H_2SO_4$, $O_3$ etc., Vehicular Emissions and Industrial Pollution.

Among all the pollutants, Particulate Matter is a significant pollutant due to its effect on people's health when found in higher concentration than normal. The Study of the relationship of various meteorological parameters and the corresponding air pollution levels may not be very encouraging, as it does not affirm any interrelation between the parameters. However, the air pollutant levels usually respond to all metrological variables representing an air mass. This study attempts to examine the relation between the various meteorological features and urban air pollutant (Particulate Matter), and understand the relation between the local meteorology, atmospheric circulation and the concentrations of tropospheric air pollutants.

## 2 Literature Review

Air Quality Forecasting is a hot topic nowadays and we have more and more people working on it every coming day so as to come up with results that would in turn facilitate the government to have a greater control over the levels of Air Pollution. Some of the recent works in this field done globally have been discussed in this section.

### 2.1 *Related Work*

In [2], there is a study on the relationship between air pollution and individual meteorological parameters for predicting the Air Quality Index. However, they do not explain the interrelation between the parameters as the level of the air pollutants usually respond to all of meteorological variables representing an air mass. Studying individual meteorological factors is not sufficient as there are numerous meteorological variables that can have an effect on air quality.

The work done in [3] considers the cumulative effects of meteorological factors on PM 2.5. However this work includes only a few of the factors and misses out factors like wind speed and wind direction which are pretty important meteorological factors and should be considered while investigating the trends with PM 10 values.

The work in [4] extensively studies the cumulative effects of a range of meteorological factors on the value of PM 1.0 and ultrafine particles (UFP). In this work PM 2.5 or PM 10 were not taken into account. UFP are particulate matter of nano scale size (less than 100 nm in diameter). As these particles are far smaller than PM 2.5 and PM 10 hence regulations do not exist for this size class of ambient air pollution particles. Due to this, UFP are not considered to be true indicatives of the Air Quality index. For a better understanding of an area's pollution, it is necessary to study either PM 2.5 or PM 10 along with UFP and PM 1.0.

In [5], the effects of meteorological conditions have been studied on the content of Ozone in the atmosphere. Global greenhouse effects are a proof that Carbon dioxide is one of the most harmful gases in the world.

The above discussed factors indicate that there exist some loopholes in the existing studies which account for incomplete or partial factors, so it is important to consider as much input factors as possible so as to increase the universal acceptability of the designed model and to achieve accurate results. The goal of this work is to design a neural network model that would accurately predict the level of PM 10 for Delhi and consequently analyze the pattern of PM 10 with variation of the input factors by studying the correlations between each individual factor and a combination of input factor and PM 10.

# 3 Design of Experiment

The data set collected for this work, consists of meteorological data for 396 days for the Delhi city. The idea is to use various machine learning techniques such as Multilayer Perceptron, Support Vector Machines and Naïve Bayes Classifier for the purpose of model construction and analysis.

## *3.1 Dataset*

The Meteorological data was collected from Control Pollution Central Board (CPCB) [6]. The CPCB is the division of government of India under the ministry of Environment and Forests which tracks and researches on air quality in India. CPCB does the monitoring of meteorological parameters along with the monitoring of air quality at various centers in Delhi state as shown in the Fig. 1. For this work the R.K. Puram center was selected, as it is a densely populated area with many residential complexes, schools and various government buildings in its vicinity.

The data consisted of values for 396 days starting from 1st May 2015–1st June 2016 which had a total of 9 parameters: Average Temperature, Minimum Temperature, Maximum Temperature, Humidity, Pressure, Max Wind Speed, Visibility

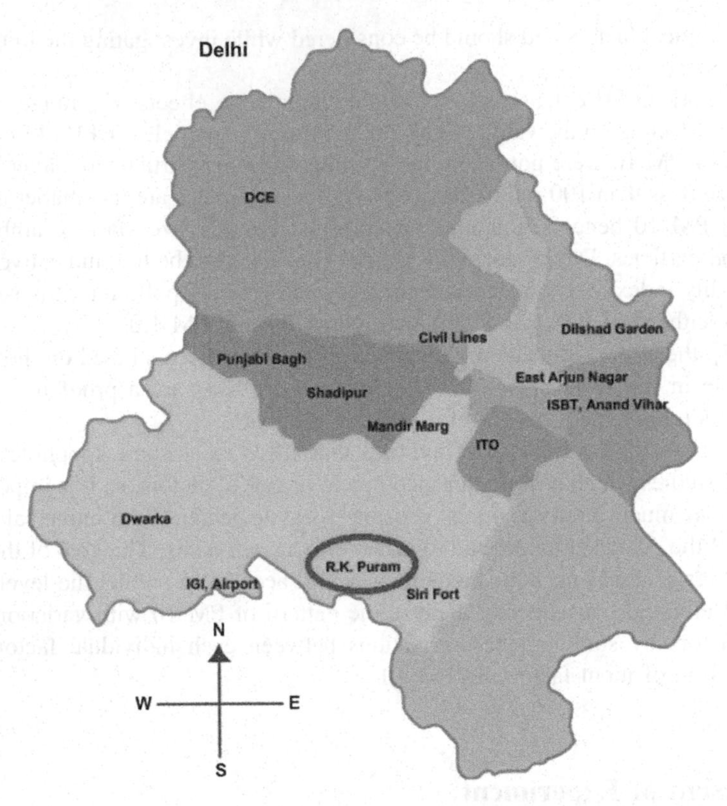


**Fig. 1** Map of Delhi [1] with marked location of data collection center

and Wind Speed, wind direction and PM 10. The data was fragmented into two parts: training dataset and the test dataset. Training data set has 316 sample observations and the test data has 80 sample observations. Each point represents the meteorological condition of a specific day in Delhi City at the selected center. A brief description of some of the parameters used in this work is as follows:

**Average Temperature**: Temperature affects the air quality due to the temperate inversion: the warm air above cooler air acts as a lid, thereby trapping the cooler air at the surface and suppressing vertical mixing. As pollutants from the vehicles, the industries, and the fireplaces are emitted into the air, this inversion phenomenon entraps the pollutants along the ground.

**Average Wind Speed**: Wind speed is a major player in diluting pollutants. Generally, scattering of the pollutants takes place due to strong winds, whereas light winds generally cause stagnant conditions, making pollutants to build up over an area.

**Average Relative Humidity**: Relative humidity (RH) is the ratio of the partial water vapor pressure to the equilibrium water vapor pressure at the same temperature.

**Atmospheric Pressure**: Atmospheric Pressure has a positive correlation with value of PM 10, this means that if the atmospheric pressure is high in an area then it is bound to be more polluted than an area having less atmospheric pressure.

**Wind Direction**: Wind Direction plays a major role in dispersion of the pollutants from one place to another. The major pollutants that enter Delhi are from Rajasthan i.e. South-West Direction.

**Average Visibility**: In meteorology, visibility is a measure of the distance at which a light or an object can be clearly distinguished. Average visibility is inversely proportional to pollution, this means that if visibility is high then it is very likely that value of PM 10 is low and vice versa.

## *3.2 Prediction Models*

We used 3 Machine Learning techniques i.e. the Multilayer Perceptron (MLP), the Support Vector Machine (SVM) and the Naive Bayes method, for the purpose of building a predictor. The PM10 level were classified into "High" (>100 μg/m$^3$) and "low" (≤ 100 μg/m$^3$) categories on the basis of Air Quality Level standard in Delhi [7], which is 100 μg/m$^3$, and is considered to be a mild level pollution.

**Multilayer Perceptron**: It is a feed forward artificial neural network model that takes in a set of input data and maps it onto a set of outputs. MLP utilizes back-propagation for training a network, which is a supervised learning technique. In this work, the MLP consists of four layers, with 2 hidden layers (Fig. 2).

**Support Vector Machines**: SVM's are models of supervised learning technique, which can be used for data analysis using the classification or regression

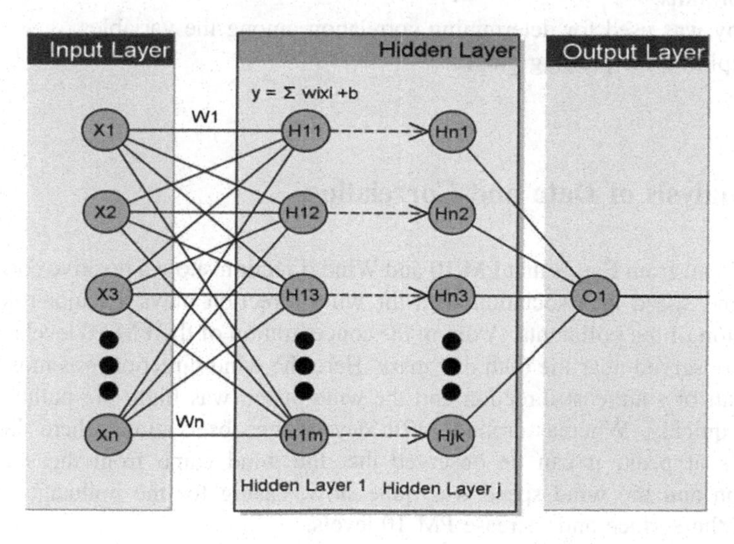


**Fig. 2** Multilayer perceptron architecture

analysis. The model of SVM created is such that, all the data points are considered points in a space and these points are divided into clear categories as wide as possible.

**Naïve Bayes**: In machine learning, naive Bayes classifiers are a group of simple probabilistic based classifiers, which use the concept of Bayes Theorem.

Naïve Bayes is a set of algorithms, rather than a single algorithm whose aim is to consider each feature of the data set as an individual and independent entity, rather that every feature being connected or sharing it with each other.

**Correlation**. It is the dependence of any two or more variable on each other in a statistical relationship [8]. In this work the Pearson Coefficient has been used to find the correlation between each input factor and PM 10 separately. The Pearson correlation coefficient is the linear correlation whose value +1 refers to a total positive correlation, 0 refers to no correlation, and −1 refers to a total negative correlation.

### *3.3 Tools Used*

For this work, the Python programming language was selected along with the following libraries:

- Pandas for performing data processing tasks such as data cleaning and normalization.
- Theano served as the base library for implementation of multilayer perceptron.
- Sklearn was used for implementing Naïve Bayes and SVM algorithm.
- Keras library that runs on top of theano, which helps us execute different algorithms.
- Scipy was used for determining correlation among the variables.
- Matplotlib for plotting graphs.

## 4 Analysis of Data and Correlation

It is evident from Fig. 3 that PM 10 and Wind direction show a positive correlation. The wind speed in association with the wind direction plays a major role in the dispersion of the pollutants. A dip in the concentration of the PM 10 levels in Delhi can be observed near the 60th day mark. Here the wind direction was mostly from the south or southeast direction and the wind speed was high, the pollutants dispersed quickly. Whereas around 217th day of the observation, where the PM10 value is at peak, it can be observed that the wind came from the south-west direction and the wind speed was quite slow, easing for the pollutants to settle around the surface and increase PM 10 levels.

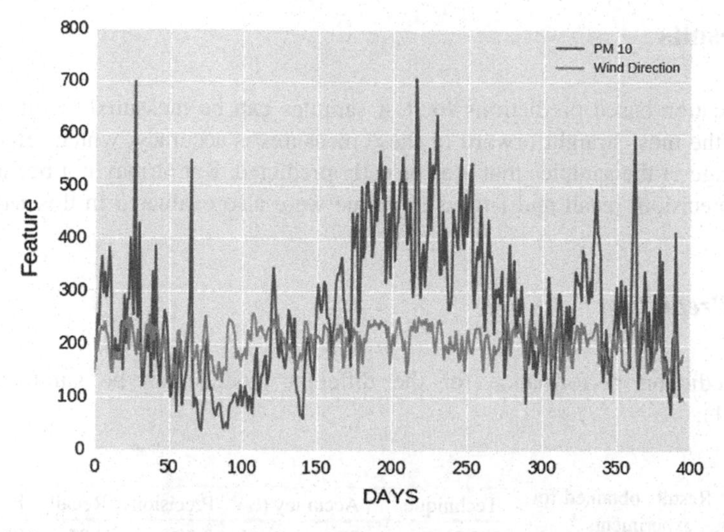


**Fig. 3** Maximum positive correlation observed between PM10 and wind direction

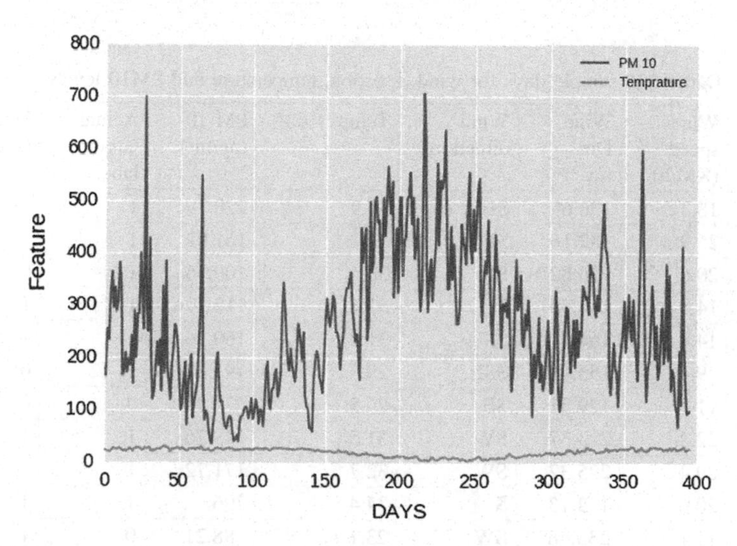


**Fig. 4** Maximum negative correlation observed between PM10 and temperature

In Fig. 4, the clear effect of negative correlation between PM 10 and Temperature can be observed. As the temperature increases, the value of PM 10 will decrease and vice versa. Hence we can say that during summers the value of PM10 is less than what we observe during the winters, where the temperature is low.

## 5 Results

Classification-based predictions for test samples can be measured in many ways. One of the most straightforward of these measures is accuracy, which refers to the percentage of the samples that are correctly predicted. But, it may not be sufficient, hence precision, recall and f-measure value were also evaluated in this work.

### 5.1 Prediction

The prediction performance for the different models can be summarized as (Table 1):

**Table 1** Results obtained for the various experiments

| Technique | Accuracy (%) | Precision | Recall | F-measure |
|---|---|---|---|---|
| MLP | 98.1 | 0.98 | 0.95 | 0.97 |
| SVM | 92.5 | 0.92 | 0.90 | 0.91 |
| Naive Bayes | 91.25 | 0.90 | 0.87 | 0.89 |

**Table 2** Data of 20 sample days for wind direction, temperature and PM10 levels

| Day No | Wind speed (KM/h) | Wind Dir. (in °) | Wind direction | Temp. (°C) | PM 10 (μg/m$^3$) | Actual target label | Predicted labels |
|---|---|---|---|---|---|---|---|
| 50 | 18.3 | 126.69 | SE | 35.9 | 220.79 | 1 | 1 |
| 51 | 25.9 | 132.16 | SE | 33 | 161.92 | 1 | 1 |
| 52 | 20.6 | 126.82 | SE | 30.4 | 100.46 | 1 | 0 |
| 53 | 14.8 | 134.56 | SE | 32.1 | 112.59 | 1 | 1 |
| 54 | 14.8 | 164.5 | SSE | 31.6 | 160.36 | 1 | 1 |
| 55 | 9.4 | 153.48 | SSE | 29.8 | 191.47 | 1 | 0 |
| 56 | 11.1 | 129.31 | SE | 26.5 | 72.24 | 0 | 0 |
| 57 | 22.2 | 223.37 | SW | 31.5 | 148.76 | 1 | 1 |
| 58 | 29.4 | 225.32 | SW | 32.7 | 171.72 | 1 | 1 |
| 59 | 20.6 | 182.13 | S | 34.4 | 206 | 1 | 1 |
| 64 | 11.1 | 233.48 | SW | 23.1 | 88.21 | 0 | 1 |
| 103 | 14.8 | 163 | SSE | 25.5 | 42.5 | 0 | 0 |
| 110 | 11.5 | 155.7 | SSW | 27.4 | 94 | 0 | 0 |
| 142 | 13.3 | 219.67 | SE | 19.7 | 45 | 0 | 0 |
| 212 | 7.6 | 198.47 | SSW | 19.3 | 357.69 | 1 | 1 |
| 213 | 11.1 | 228.53 | SW | 17.9 | 423.06 | 1 | 1 |
| 214 | 18.3 | 246.49 | WSW | 17.6 | 432.28 | 1 | 1 |
| 215 | 14.8 | 251.09 | WSW | 17.7 | 506.44 | 1 | 1 |
| 216 | 7.6 | 222.65 | SW | 17.4 | 509.14 | 1 | 1 |
| 217 | 9.4 | 213.22 | SSW | 17.2 | 555.18 | 1 | 1 |

**Multilayer Perceptron** gave the best accuracy of 98.1%. The results are achieved after the removal of all the outliers. Therefore, we can say that MLP is the best technique among the three for the prediction of the PM 10 values.

These results, which were obtained on MLP, were achieved after a number of explorative experiments in which the various network parameters were varied to get their optimal values.

Table 2 does not show the entire dataset. Instead it contains a small sample describing only the highly positively and a few highly negatively correlated factors with PM 10 levels, their actual class labels and their corresponding predicted class labels. The high quality of the prediction of PM 10 levels which the model is successfully able to perform can be observed. As per the regulations stated in [7] the PM 10 levels above 100 $\mu g/m^3$ is considered as high (1) and levels below 100 $\mu g/m^3$ are considered low (0). It can be seen from the table that, out of the 20 selected instances the model is able to correctly predict 17 instances while only 03 are incorrectly predicted in this small sample space.

## 6 Conclusion

In this work, the task of constructing a pollution prediction model for Delhi was successfully accomplished. Of the various machine learning techniques used, we observed that Multilayer Perceptron gave the best results of all, with an overall accuracy of 98%.

Also, for the analysis part we found that out of all the meteorological factors—Wind Direction has the maximum positive correlation with PM 10 out of all the input factors considered individually. Temperature has the maximum negative correlation with PM 10 out of all the input factors considered individually. This means that if value of Temperature decreases then the value of PM 10 will increase and vice versa. Using this newly constructed model, highly accurate results can be predicted based on the current trends of the meteorological data, which can be, used abetment of particulate pollution in that area and help to develop pollution control strategies.

However this is a work attempts to identify a correlation between various metrological data and PM10 values, but doesn't consider vehicular traffic data. A further work can be seen as relevant where the vehicular traffic data may also be considered and correlation must be established between them and the particulate pollution. Also as the dataset used in this work is of about just over a year, hence the use of a dataset of a larger time span will help to establish the correlations and predictions better. A larger time span would mean repetition of weathers conditions which would help in asserting the reasons for pollution level changes at a certain time in the year.

## References

1. World Health Organization. http://www.who.int/phe/health_topics/outdoorair/databases/cities/en/
2. Li, Y., Wang, W., Wang, J., Zhang, X., Lin, W., Yang, Y.: Impact of air pollution control measures and weather conditions on asthma during the 2008 Summer Olympic Games in Beijing. Int. J. Biometeorol. **55**(4), 547–554 (2011)
3. Wei, D.: Predicting Air Pollution Level in a Specific City. Stanford Publication (2014)
4. Pandey, G., Zhang, B., Jian, L.: Predicting submicron air pollution indicators: a machine learning approach. Environ. Sci.: Process. Impacts **15**(5), 996–1005 (2013)
5. Krupa, S., Nosal, M., Ferdinand, J.A., Stevenson, R.E., Skelly, J.M.: A multi-variate statistical model integrating passive sampler and meteorology data to predict the frequency distributions of hourly ambient ozone (O 3) concentrations. Environ. Pollut. **124**(1), 173–178 (2003)
6. Central Pollution Control Board. http://www.cpcb.gov.in
7. Central Pollution Control Board. http://cpcb.nic.in/National_Ambient_Air_Quality_Standards.php
8. Slini, T., Karatzas, K., Moussiopoulos, N.: Correlation of air pollution and meteorological data using neural networks. Int. J. Environ. Pollut. **20**(1–6), 218–229 (2003)

# An Architectural View Towards Autonomic Cloud Computing

**Ravi Tomar, Abhirup Khanna, Ananya Bansal and Vivudh Fore**

**Abstract** Cloud computing is causing significant transformations in the world of information technology. It continues to be the hot favorite of people from both within and outside the IT industry. One of the key factors for which the Cloud is know is its accessibility to never ending resources. This is a perception which the Cloud has been able to maintain since long but due to extensive user involvements and humungous amounts of data this perception seems to fade away. In present day Cloud seems to face challenges when it comes to over utilization of resources, fault tolerance and dynamic monitoring and management of services. In order to address these problems, human intervention continues to increases thus causing a negative impact on the QoS and generic nature of cloud. In order to overcome these challenges we propose an Autonomic Cloud Computing Environment which provides dynamic allocation and monitoring of resources along with orchestration of cloud services based upon VM migration. The system that we propose is SLA complaint and automates the user experience depending upon the clauses mentioned in the SLA.

**Keywords** Cloud computing · Autonomic cloud computing · VM migration · Cloudsim

R. Tomar (✉) · A. Khanna · A. Bansal
University of Petroleum and Energy Studies, Dehradun, India
e-mail: rtomar@ddn.upes.ac.in

A. Khanna
e-mail: abhirupkhanna@yahoo.com

A. Bansal
e-mail: bansal.ananya14@stu.upes.ac.in

V. Fore
Gurukul Kangri Vishwavidyalaya, Haridwar, India
e-mail: vivudh.fore@gmail.com



S.C. Satapathy et al. (eds.), *Data Engineering and Intelligent Computing*,
Advances in Intelligent Systems and Computing 542,
DOI 10.1007/978-981-10-3223-3_55

# 1 Introduction

Cloud Computing is one of the greatest establishments of internet computing. It is tantamount to internet. It is a computing paradigm wherein resources are rendered on a utility basis just like electricity. Cloud provides on call virtualized IT resources that can been subscribed by the user on a pay as you use basis. More apps will be built and maintained on cloud. According to recent Forbes studies [1] worldwide spending on public cloud services will grow at a 19.4% compound annual growth rate (CAGR) from nearly $70B in 2015. It is estimated that there will be more utilization of cloud services in enterprises as all the applications will be available on the cloud. The struggle of ownership of the infrastructure will be over soon, the clash of applications is about to begin. Since Cloud computing is a utility and does not require hardware but only a remote network therefore the demand of infrastructure will go down.

We all are aware that Cloud Computing reduces human effort but the next big progressive step in its technological advancements is Autonomic Cloud Computing (ACC). Autonomic Cloud Computing is the fabrication, execution, management of a cloud system that requires bare minimum human inputs and interrupts along with providing the best available outputs. This Autonomic Computing system is in the manner of an autonomous nervous system in human body as both of them are built to be self-dependent, self-healing and self-configuration. ACC is an evolved cloud computing architecture wherein the solutions for certain situations and conditions are saved in the system. The cloud ecosystem gets automatically managed and optimized as and when autonomous methods are applied onto it. Since all the work will be performed by the machine therefore if there are any faulty programs or any viruses be it software or a hardware bug it will be found quickly and fixed. Even though the same job is done by a team of highly skilled humans then also they stand no chance in competing with an autonomic cloud engine. One of the key features which ACC includes self-configuration. In case of self- configuration the system could configure itself according to the environment and its needs. In this way the user is well satisfied as all the work is done in the most efficient manner. An ACC system should be able to find the most optimum and feasible solution to any problem that arises, that too in a short span of time at magnificent speed. To do so the system should be well aware of all its resources, capabilities and its shortcomings. Autonomic cloud computing is one of the most beneficial and sophisticated ideas of today's times as its helps in reducing both energy consumption & time which are the two most paramount commodities today's world. we have no time to do tasks manually so if the jobs are done in a computerized manner then a lot of time and effort can be saved.

The rest of the paper is categorized as follows: Sect. 2 talks about the related work in the field of autonomic cloud computing. Section 3 elucidates the algorithm that depicts the workflow of the entire system, whereas its complete layered architecture is discussed in Sect. 4. Finally, Sect. 5 exhibits the implementation and simulation of our proposed system.

## 2 Related Work

Since the past couple of years, researchers have formulated various algorithms and techniques for creating autonomous clouds. Continuous work is still going on so as to device more optimized and efficient ways in which cloud environments could be automated. In this section, we have presented some of the research works pertaining to this area.

- Cloud security is one the biggest challenges when it comes to cloud deployment. There are many independent prerequisites that are needed to be fulfilled before securing our cloud applications. There can be a number of ways in securing a cloud environment like vulnerability scan, configuration management, source code analysis etc. Autonomic Resilient Cloud Management (ARCM) is one such methodology based on moving target defense [2]. It uses two algorithms namely, Cloud Service Behavior Obfuscation and Autonomic Management for securing the cloud. The aim of the ARCM is to accomplish fundamental security for a cloud environment. This is achieved through constant change in the virtual environment of the cloud. So that, by the time the hacker figures out the system's vulnerabilities and its security is comprised the environment is formerly altered. The unique attribute which distinguishes this methodology from the others is the diversity technique which unsystematically changes the versions and resources of the cloud environment in order to make it difficult for hackers to attack the system.
- The work of a Resource Management System is to manage and coordinate all the available resources pertaining to a cloud environment. RMS is a manifestation of a central manager who manages the resources at levels both physical and virtual. In order to do so there is steady communication between the manager and the nodes to make decisions relating to allocation of resources. Role Based Self Appointment (RBSA) management system [3] is a Resource Management System that is based on the principles of autonomic computing in order to provide enhanced scalability and prevention from server bottlenecks. The concept involves the creation of an autonomic node that can manage all the resources without any external intervention from a central manager. The node would be able to take decisions, keeping in mind the current status in the network.

## 3 Algorithm

See Fig. 1.

```
RU: Resource Utilization                          MA1: Mobile Agent at VM level
MA2: Mobile Agent at server level                 AS: Application Scheduler
VS: VM Scheduler                                  AC: Autonomic Controller

1.      AC monitors and controls the federated cloud
2.      If(RU factor of VM >= threshold value)
3.      Notify MA1
4.      while(MA1 searches for appropriate VM)
5.      If (search = = success)
6.      AS notifies source and destination VM
7.      Cloudlet created at the destination VM with same configuration of the source VM
8.      AS notifies AC
9.      Print ("migration successful")
10.     End if
11.     Else (search! =success)
12.     MA1 will inform the server and goto step 11
13.     End while
14.     If (RU factor of server >= threshold value)
15.     while(MA2 searches for appropriate server for VM migration)
16.     If (search = = success)
17.     VS notifies AC
18.     If(AC refers to the SLA)
19.     VS notifies source and destination server
20.     VM migration takes place in running state
21.     New VM created at the destination server with the same configuration of the source VM
22.     Cloudlets created of same configuration at the destination VM
23.     Source VM deleted
24.     Print("migration successful")
25.     End if
26.     Else
27.     goto step 36
28.     End if
29.     Else (search! =success)
30.     Server reached in compromising state
31.     while(AC searches for an appropriate server residing on a different Cloud)
32.     If (search = = success)
33.     goto step 17
34.     End if
35.     Else(search! =success)
36.     AC stops or removes the abnormally high resource intensive cloudlet
37.     End while
```

**Fig. 1** Algorithm

# 4 Proposed Work

In this section, we discuss our proposed Autonomic Cloud Computing Environment and all the various actors that constitute it. We present a layered architecture that depicts our autonomic environment and showcases all the different entities which it comprises. As the core of our proposed system is based upon VM migration and dynamic monitoring and management of resources thus we specify the nature of our proposed environment by elaborating the conditions that would lead to VM or Application migration.

## *4.1 Conditions for Migration*

In this subsection, we elucidate of our autonomic cloud environment and discuss its functionalities and working by explaining the conditions that cause VM or Application migration. Following are the conditions wherein our system would undertake VM or Application migration.

- *Load Balancing*: It is one of the core areas where VM migration finds its application. The main aim of the system would be to facilitate load balancing between different servers of various data centers. With the help of load balancing migration, the load on a particular server can be reduced and distributed among rest of the servers [4]. It involves effective provisioning and scheduling of VMs thereby leading to the optimal distribution of load among the servers in order to enhance the scalability of the cloud environment.

- *Energy Optimization*: The main source of power consumption in a data center is from its servers. Servers continue to consume 70% of their maximum power needed even in cases of their minimum utilization [5]. In order to avoid such situations the system ensures energy efficient migration that would help achieve optimal power consumption of servers by migrating VMs from one server to another. If the total number of VMs running on a server are much less than its maximum capacity, then in such case these VMs are migrated to another server, making the source server free and allowing it to be powered off, thereby reducing the energy consumption of a data center.
- *Autonomous Application Scaling*: It is a condition wherein there is a sudden rise in terms of resource requirements of an application(s). This sudden rise in resource requirements may lead to increased Resource Utilization factor (RU factor) of a VM thus causing it to reach its threshold value. In such conditions the system would either look for an appropriate VM for application migration or may also consider VM migration to some another server. This server may be present within the same or different cloud computing environment. All migrations undertaken by the system are done post referring to the SLA.
- *Data Privacy*: In order to achieve higher levels of data privacy and security the end user may desire to switch its computing environment from a Public to a Private Cloud. The system accomplishes this change in computing environment in an autonomic fashion by migrating the respective VM(s) from Public to a Private Cloud.
- *Fault Tolerance*: Whenever there is some fault in the system be it at the server or datacenter level it causes huge losses. In such situations, the system commences VM migration as it seems to be very prolific as all the VMs residing on the faulted server or datacenter would be migrated to another server without causing any interruption in the services being provided to the end users. Fault tolerant migration allows VMs to migrate from one server to another depending upon the failure that has occurred using proactive fault tolerance mechanism. It helps to increase the availability of resources and also prevents performance degradation for the cloud environment.
- *IT maintenance*: With the advent of VM migration, IT maintenance has become very convenient as all the VMs of a server could be migrated to another server. In cases the system migrates all the VMs residing on a server thus, allowing the source server to shut down for maintenance.
- *Adaptive Pricing Models*: Cloud computing since its inception has been known for its lucrative and flexible pricing models. As our proposed system focuses on the role of the SLA thus it allows its users to alter its pricing models according to their convenience. Alteration of pricing models could lead to either enhanced or reduced consumption of cloud services and resources. This change in services and resources would pilot change in computing environments of a VM. Such

changes are very well incorporated and managed by our system in an autonomic fashion.

## 4.2 Layered Architecture

In this subsection we would be discussing the layered architecture of our proposed autonomic cloud computing environment. We would also be talking about the various actors along with their functionalities that constitute the system. Following is the diagram that depicts the layered architecture for our proposed system (Fig. 2).

- *Application Scheduler*: The work of the application scheduler is to migrate an application from one VM to the other in consultation with MA1 and Autonomic Controller. It is the application scheduler which receives orders from the autonomic controller with regard to application migration. Once the source and destination VMs have been identified by the application scheduler it subsequently collects application information from the source hypervisor and transmits the same to the destination hypervisor.
- *VM Scheduler*: The work of the VM scheduler is to migrate a VM from one server to the other in consultation with MA2 and Autonomic Controller. It is the VM scheduler which receives orders from the autonomic controller with regard to VM migration. Once the source and destination servers have been identified by the VM scheduler it subsequently collects VM information from the source hypervisor and transmits the same to the destination hypervisor. This information comprises of CPU state, Memory content and Storage content. The system makes use of a Network Attached Storage (NAS) so as to avoid any

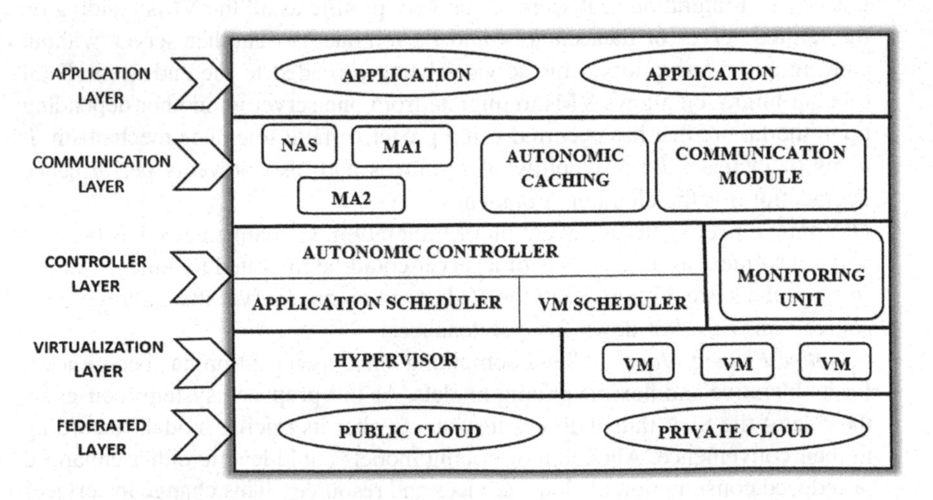


**Fig. 2** Autonomic cloud computing architecture

network latency in transfer of storage contents. The system also incorporates the concept of autonomic caching which is archived through use of surrogate servers and prevent middle mile bottlenecks as far as possible.

- *Autonomic Controller*: It is acts as the heart of the system. It is the Autonomic Controller which is responsible for orchestrating all the other actors and various cloud services in an efficient manner. It plays a very important role in the system as all decisions from application migration to server consolidation are taken by the Autonomic Controller. The Autonomic Controller works on the principle of Monitor-Analyze- Act. It maintains continuous communication with the mobile agents and acts in an autonomic fashion depending upon the inputs it gets. Prior to taking any decision the Autonomic Controller refers to the SLA and goes through all the liberties and services that are being offered to the end user. It is the Autonomic Controller which is accountable for all the proceedings in a cloud environment for which it is in continuous contact with the monitoring unit which continues to update it regarding all the various parameters of the cloud environment.
- *Monitoring Unit*: The work of the monitoring unit is to monitor all the various parameters of a cloud computing environment. These parameters range from memory utilization, power consumption, processing capabilities, bandwidth consumption, network topologies, server and VM uptime, network latencies, etc. The monitoring unit over regular intervals of time continues to update the autonomic controller regarding the status of these parameters.
- *Mobile Agents*: They are software components which travel autonomously across different machines with very less overhead. These are very small computational entities that majorly focus on moving randomly around hosts so as to collect information which can be used to carry out other processes. They use Remote Procedure Calls to facilitate one to one transition from one host to another. In case of deploying multiple mobile agents at different levels, one needs to establish a shared system in order to carry out a single process in a parallel. The main characteristics of mobile agents is their mobility, low network latency, robustness, light weight, parallel processing, easy and flexible management, dynamic adaptation and fault-tolerant capabilities [6]; amongst all mobility is of prime importance.

# 5 Implementation and Simulation

The above mentioned algorithm is implemented on the CloudSim framework. In CloudSim [7] there are various predefined classes that provide a simulation environment for cloud computing. It is a java bases simulation tool and can be implemented either with Eclipse or NetBeans IDE. For our proposed algorithm we have used the eclipse IDE. To run CloudSim in eclipse, we first need to download the eclipse IDE and install it. After successful installation of eclipse download the latest CloudSim package, extract it and import it in eclipse. Talking of our proposed work we have created our own classes in CloudSim and have portrayed our algorithm in form of java code. The following are the screenshots that depict the working of our algorithm on CloudSim framework.

The following figure shows creation of three Hosts (Servers) on different Datacenters. Each datacenter represents a Cloud in itself thus creating a federation of clouds. For our experiment we have created two datacenters, Datacenter #0 and Datacenter #1 representing Public and Private Cloud environments respectively. The aim of this experiment is to depict how our system reacts to spontaneous resource requirements and undergoes load balancing. Each Host has certain set of VMs being hosted on them, which subsequently have Cloudlets (applications) running. After certain amount of time the RU factor of Host #0, Datacenter #1 surpasses its threshold value. Within no time MA2 gets to know about this situation and starts searching for a Host where VM(s) can be migrated. Host #1, Datacenter #1 is selected by MA2 as the destination server for VM #1. Host #1 is notified by MA2 and a new VM is created of having same configurations of that of VM #1, Host #0 thus resulting in Successful migration (Fig. 3).

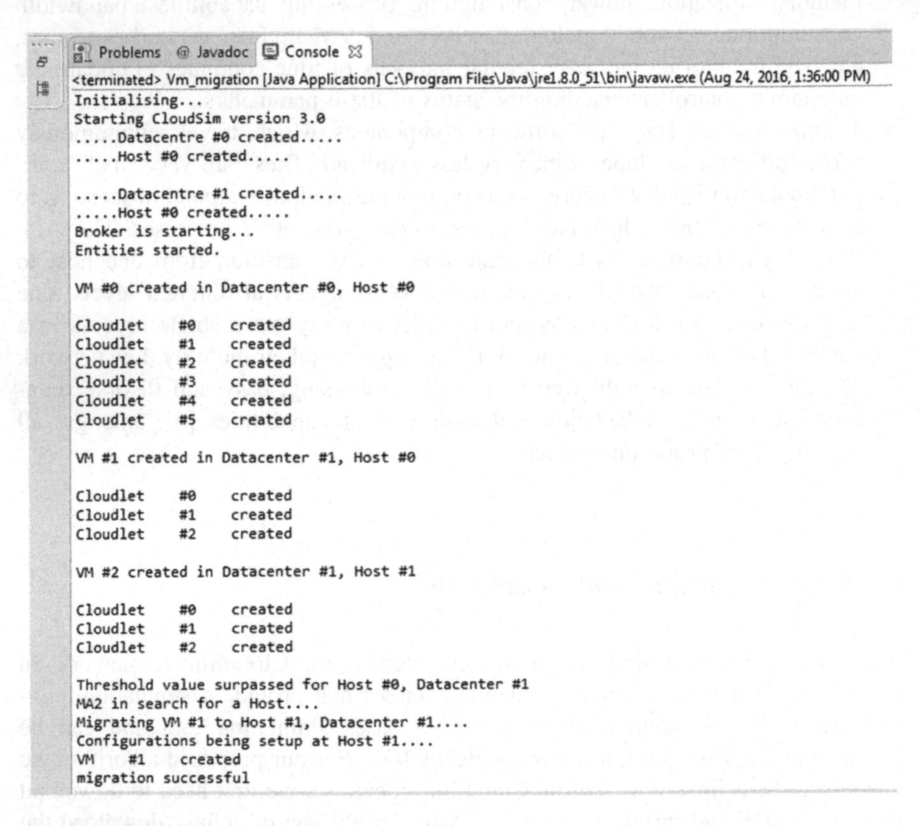


**Fig. 3** CloudSim simulation

In our next experiment, we demonstrate how our autonomous system responds to failure in a datacenter and how inter-cloud migration takes place. The pictorial depiction of this experiment is expressed in Fig. 4. For our experiment we have created two datacenters, Datacenter #0 and Datacenter #1 representing Public and Private Cloud environments respectively. Each Host has certain set of VMs being hosted on them, which subsequently have Cloudlets (applications) running. At a particular instance of time a fault occurs causing failure of Datacenter #1. The Autonomic Controller in response to the situation starts to search for an alternate cloud environment wherein the VM(s) of Datacenter #1 can be migrated. It also refers to the SLA of every VM as it comprises of application constraints such as data privacy, cost model, resource utilization levels and performance parameters. Soon after getting a consent from the SLA, the Autonomic Controller starts to undergo the process of VM migration from source to destination server. In this way VM #1 and VM#2 are created on Datacenter #0 thus resulting in Successful migration.

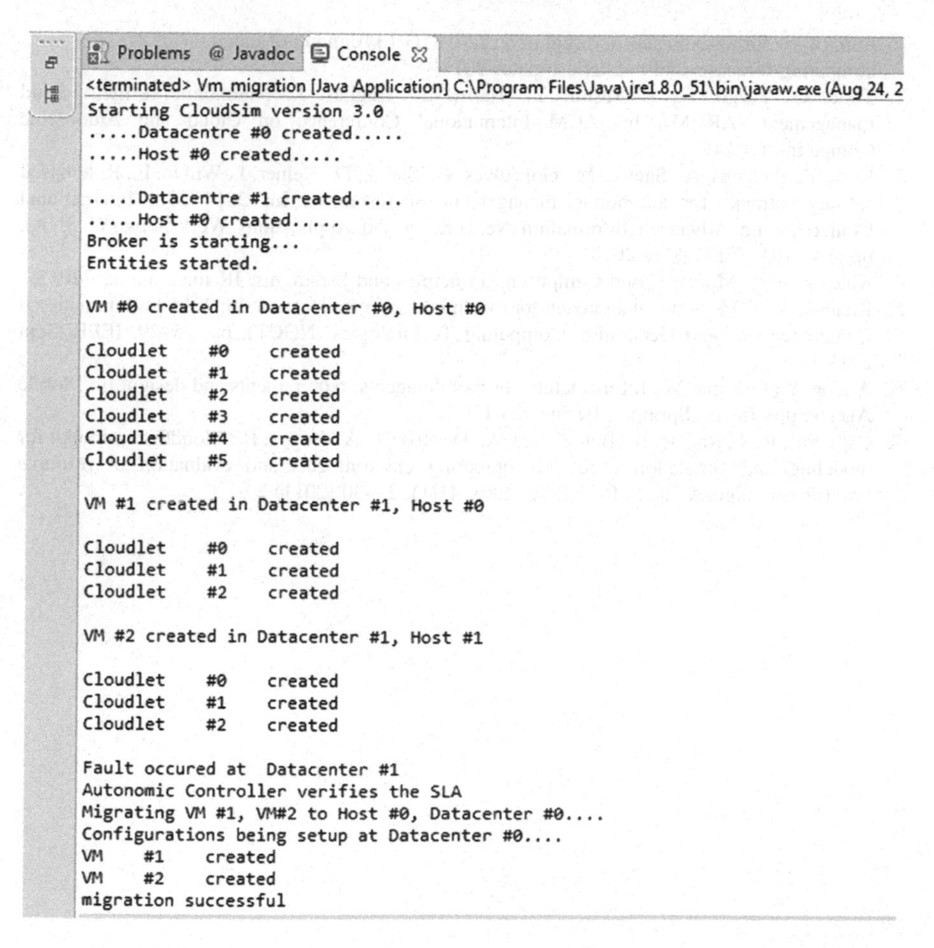


**Fig. 4** CloudSim simulation

## 6 Conclusion

Autonomic Cloud Computing is one of the many domains of Cloud wherein significant research can be seen. It is a concept which has answers to many current day problems which the Cloud faces. We propose an Autonomic Cloud Computing which facilitates dynamic monitoring and management of resources through VM migration thus automating the entire working of the Cloud. We also discuss the layered architecture for the proposed system and explain the working and functionalities of every actor involved in it. The aim of our proposed work is to maintain QoS and automate the entire user experience.

## References

1. http://www.forbes.com/sites/louiscolumbus/2016/03/13/roundup-of-cloud computing-forecasts-and-market-estimates-2016/
2. Tunc, C., Fargo, F., Al-Nashif, Y., Hariri, S., Hughes, J.: Autonomic resilient cloud management (ARCM). In: ACM International Conference on Cloud and Autonomic Computing (2014)
3. Endo, P., Palhares, A., Santos, M., Gonçalves, G., Sadok, D., Kelner, J., Wuhib, F.: Role-based self-appointment for autonomic management of resources. In: 2014 28th International Conference on Advanced Information Networking and Applications Workshops (WAINA), pp. 696–701. IEEE, May 2014
4. Khanna, A.S.: Mobile Cloud Computing: Principles and Paradigms. IK International (2015)
5. Khanna, A.: RAS: A novel approach for dynamic resource allocation. In: 2015 1st International Conference on Next Generation Computing Technologies (NGCT), pp. 25–29. IEEE, Sept 2015
6. Aridor, Y., Oshima, M.: Infrastructure for mobile agents: requirements and design. In: Mobile Agents, pp. 38–49. Springer, Berlin, Jan 1998
7. Calheiros, R.N., Ranjan, R., Beloglazov, A., De Rose, C.A., Buyya, R.: CloudSim: a toolkit for modeling and simulation of cloud computing environments and evaluation of resource provisioning algorithms. Softw.: Pract. Exp. **41**(1), 23–50 (2011)

# Implementation and Analysis of TL-SMD Cryptographic Hash Algorithm

**Manish Grewal, Nihar Ranjan Roy and Twinkle Tiwari**

**Abstract** With an advent of technological innovative tools and technology, Data security has become a major challenge in today's world. The solution to this challenge comes out in the way of Cryptographic hash function which is used in various security applications. It is a one-way hash function which is designed to provide data security. One-way hash functions are those hash functions which cannot be reverted back i.e. we cannot find the input or the actual message bits using the hexadecimal output value. TL-SMD is a cryptographic hash function having two layers of encryption. This paper is an extension to the TL-SMD work, here the algorithm is implemented using MATLAB and results are analysed and discusses on various steps for the further improvement in data security.

**Keywords** TL-SMD • Merkle-Demgard construction • Avalanche effect • Fast wind pipe construction • Block chaining technique

## 1 Introduction

A hash function is a function that converts an arbitrary input bits known as plain text into compressed fixed output bits known as cipher text. The input to the hash function is of arbitrary length that varies according to the application requirement i.e. depending on the usage. For example if we are using this algorithm for the password purpose, then we can fix the input length to 8 or 9 or may be bit more than

M. Grewal (✉) · N.R. Roy
Department of Computer Science and Engineering, School of Engineering, GD Goenka University, Sohna, Gurgaon, Haryana, India
e-mail: manishgrewal86@gmail.com

N.R. Roy
e-mail: nihar.ranjanroy@gdgoenka.ac.in

T. Tiwari
Information Technology Department, KIET, Ghaziabad, Uttar Pradesh, India
e-mail: twinkle.tiwari@kiet.edu



S.C. Satapathy et al. (eds.), *Data Engineering and Intelligent Computing*, Advances in Intelligent Systems and Computing 542,
DOI 10.1007/978-981-10-3223-3_56

that. So the input may vary as per the usage of this algorithm. But the cipher text generated by this algorithm is fixed to 64 bits and that to a hexadecimal code.

Cryptographic Hash Functions plays an important role in the security application of today's digital world. There are many hash function which has already been in use, for example MD5 [1, 2], SHA [3, 4] etc. Among all the other algorithms, there is an algorithm TL-SMD [5, 6], which provides two-layer encryption. TL-SMD [5, 6] stands for Two Layered-Secured Message Digest. In comparison to other hash functions like MD5 and SHA series algorithm, TL-SMD also provides encryption of message bits, but here there are two layers defined for the encryption of message bits in order to make it more secure. The main idea behind adding additional layer of encryption is to achieve avalanche effect [7, 8].

Horst Feistel uses the term avalanche effect for the first time. The avalanche effect says that a small change in input bits (for example, a change in single bit) should bring a significant change in the output bits i.e. small change in plain text brings considerably change in the cipher text. The algorithm, that achieves the avalanche effect at max, is considered as the best algorithm in cryptographic world [4], As it will make it very difficult for hackers to guess the password.

TL-SMD hash function is a one-way hash function. This makes it widely used for the application such as digital signature, digital timestamp [9], message authentication code (or MAC) [10], public key encryption, tamper detection of files etc. The only way to recreate the original plain text from the output Hexa-decimal bits (or cipher text) is to try brute force attack method [11]. This method is very time consuming, and become ineffective if limited number of attempts is there for providing input password.

## 2 Basic Properties of TL-SMD

TL-SMD uses the modified version of the combination of all the construction models. For example, instead of using the electronic codebook (ECB) or cipher block-chaining (CBC) mode [12, 13], TL-SMD uses the combination of both. The word size used is 64 bits in all the stages of processing. The padding technique used here is a standard one i.e. binary bit “zero” is used for the padding. This is separated from the plain text by binary bit “one” and the 128 bits in 1024 bits encoded message is kept for identifying the length of message in plain text.

It uses the block size of 1024 having 256 (4 × 64) bits internal block size. The input stream also known as the plain text can be of arbitrary size but the output generated i.e. the cipher text is a 64 bit hexa-decimal code. So TL-SMD [5, 6] converts an arbitrary size input to a fixed size output which can not be reversed back to obtain the plain text. The details of TL-SMD algorithm are shown in Table 1.

**Table 1** Shows the details of TL-SMD algorithm

| TL-SMD details | |
|---|---|
| Block size | 1024 |
| Internal state size | 256 (4 × 64) |
| Output size bit | 256 |
| Maximum message size | $2^{128} - 1$ |
| Rounds | 24 |
| Security (bits) | 128 |
| Operations used | AND, OR, NAND, EX-NOR, EX-NOR, left rotate, right rotate |

# 3 Architecture of TL-SMD

## *3.1 Layer One Architecture*

The basic architecture of layer one of TL-SMD is shown in Fig. 1. The variables A, B, C, D and $K_i$ are independent variables initialized using the 64 bit binary conversion of the fraction part of the cube root and the square root of the prime numbers. It consists of four main functions, which operates on the words made up of plain text along with the padding, and generates output (out1, out2, out3, out4) each of 64 bits. Feedback mechanism used here is adopted from fast wind pine construction model, but with some modifications. Notations used in this paper are shown in Table 2.

The data is taken from the user and then converted to ASCII code. This converted message is then concatenated and extended using the standard padding technique to make it 1024 bits. Then this 1024 bits are divided into 16 words of 64 bits each. In the next stage, using the Independent variables and these 16 words of 64 bits each new words are created using this function:

$$W_{(i+1)} = \text{Lshft}^{24}[(W_{(i-16)} X_{nor} W_{(i-11)}) X_{or} W_{(i-1)}]$$

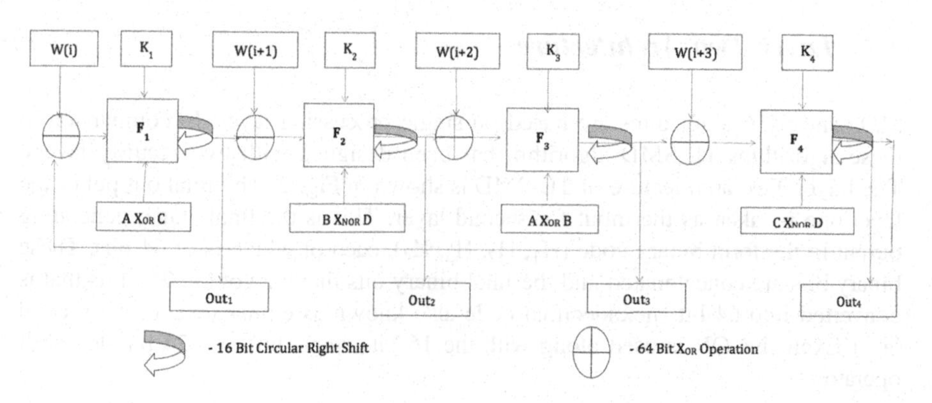


**Fig. 1** Shows the basic architecture of layer one of TL-SMD

**Table 2** Notations

| | |
|---|---|
| Lshftn | n bits circular left rotation |
| Rshftn | n bits circular right rotation |
| Xnor | Exclusive NOR operation |
| Xor | Exclusive OR operation |
| Nnd | Bitwise NAND operation |
| && | Bitwise AND operation |
| \|\| | Bitwise OR operation |
| Wi | 64 bit word |

where i is initialized to 16 and it will go till 23. This will generate 8 new words of 64 bits each. The architecture of Layer one is shown in Fig. 1.

The words processed by the functions are as follows:

Function $F_1$ will process the following words: $W_0$, $W_4$, $W_8$, $W_{12}$, $W_{16}$, $W_{20}$.
Function $F_2$ will process the following words: $W_1$, $W_5$, $W_9$, $W_{13}$, $W_{17}$, $W_{21}$.
Function $F_3$ will process the following words: $W_2$, $W_6$, $W_{10}$, $W_{14}$, $W_{18}$., $W_{22}$.
Function $F_4$ will process the following words: $W_3$, $W_7$, $W_{11}$, $W_{15}$, $W_{19}$., $W_{23}$.

The output function designed for layer one are:

$$\text{output}_1 = [\{W_i\, X_{or}(\text{Rshft}^{16}(\text{output}_4))\}\, N_{nd}\{(K_1 X_{or}(A\, X_{or} C))\}]$$
$$\text{output}_2 = \left[\{W_i\, X_{or}\text{Rshft}^{16}(\text{output}_1)\}\&\&\{K_2 || (B\, X_{nor} D)\}\right]$$
$$\text{output}_3 = [\{(W_i\, X_{or}(\text{Rshft}^{16}(\text{output}_2))\, X_{nor} K(j)\}\, X_{nor}(A\, X_{or}\, B)\}]$$
$$\text{output}_4 = [\{(W_i\, X_{or}(\text{Rshft}^{16}(\text{output}_3))||(C\, X_{nor} D)\}\&\&\{(K_4||(W_i X_{or}(\text{Rshft}^{16}(\text{output}_3)))\, X_{or}\{K_4||(C\, X_{nor} D)\}]$$

All this functions uses basic binary logic gates to process the plain text entered by the user in order to generate an output. The output of function $F_1$ i.e. Out1 is first passed through the 16 bit circular right shift and then used as one of the input to function $F_2$ and so on. The final outputs of layer one act as an input for the layer two, which further process this and gives the final cipher text of 64 bits.

## 3.2 Layer Two Architecture

MD5 and SHA algorithms are based on single processing layer. In comparison to these algorithms TL-SMD algorithm has been designed with two security layers. The Layer Two architecture of TL-SMD is shown in Fig. 2. The final out put of the layer one is taken as the input for second layer. This is the final stage, generating output in the form binary code ($H_1$, $H_2$, $H_3$, $H_4$), each of which is of 64 bits. These binary bits are concatenated and the final binary bits that we got are 256 bits that is converted into 64 bits hexadecimal code also known as cipher text. In the second layer Exclusive-OR is used along with the 16 bits right shift and 24 bits left shift operators.

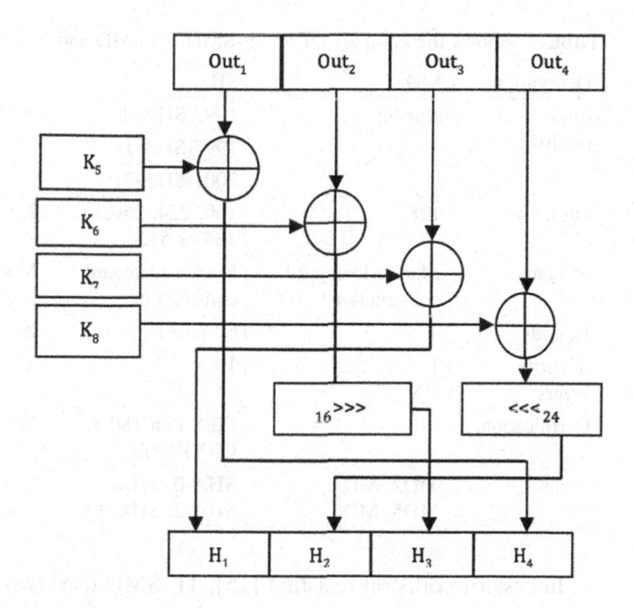


**Fig. 2** Shows the layer two architecture of TL-SMD

The variables $K_5$–$K_8$ used here are independent variables initialized with the 64 bit binary conversion of the fraction part of the cube root and the square root of the next consecutive prime numbers. At the last step the final output is shuffled before concatenation. The main idea of using this layer two is to have more and more variation in the output in case of a small change in the input.

Final cipher text is represented with the following shuffling:

$H_1 = \text{output}_3\ X_{or}\ K_7$
$H_2 = \text{Lshft}^{24}\ (\text{output}_4\ X_{or}\ K_8)$
$H_3 = \text{Rshft}^{16}\ (\text{output}_2\ X_{or}\ K_6)$
$H_4 = \text{output}_1\ X_{or}\ K_5$

The cipher text is the concatenation of $H_1$, $H_2$, $H_3$ and $H_4$. These are of 16 bits each in hexadecimal format and finally led to the cipher text of 64 bits. So what ever may be the size of the plain text, the cipher text will always be of 64 bits.

# 4 Comparison of TL-SMD with Others

Other cryptographic hash algorithms are designed using the Merkle-Damgard construction method [14] only. If two prefixes with similar value of hash is constructed, a common suffix can be added to both to make the collision accepted as actual plain text. Where as in case of TL-SMD the basic construction includes the modified combination of Merkle-Damgard construction [14] and the fast wind pine construction methodology.

**Table 3** Shows the comparison of TL-SMD with MD and SHA [6]

| Decryption | MD | SHA | TL-SMD |
|---|---|---|---|
| First publish | Apr-92 | 1993(SHA-0) | 2015 |
| | | 1995(SHA-1) | |
| | | 2001(SHA-2) | |
| Digit size | 128 | 160, 224, 256, 384 or 512 | 256 |
| Structure | Markle-Demgard construction | Markle-Demgard construction | Modified combination of MDC and fast wind pipe construction |
| Rounds | 4 | 60 or 80 | 24 |
| Security layers | 1 | 1 | 2 |
| Certification | | FIPS Pub 180-4, CRYPTREC | N/A |
| Series | MD2, MD4, MD5, MD6 | SHA-0, SHA-1, SHA-2, SHA-3 | TL-SMD |

In case of collision resistant [15], TL-SMD uses two levels of encryption where as all other algorithm uses only single layer of encryption of plain text. Therefore chances of collision are less in TL-SMD. Table 3 shows the comparison among the TL-SMD, SHA algorithms [16, 17, 6] and the MD5 algorithms. Both MD i.e. Message Digest and SHA i.e. Secure Hash Algorithm have many versions [18, 15, 3], but for the comparison purpose, we have used the widely used MD5 and SHA series along with TL-SMD.

## 5 Analysis of TL-SMD

The Two Layered Secure Message Digest had been implemented in Matlab R2015a. On the basis of the designed parameters of the algorithm, we have analyzed the variation in the hexadecimal code or the cipher text generated by the algorithm when there is a change in one bit, two bits and so on up to five bits change in the input or the plain text. Figures 3, 4, 5, 6 and 7 show these variations.

Along with this, hamming distances in each test case have been calculated. Hamming Distance is the number of bits that should be changed in order to change the first code to second code or vice versa. In TL-SMD, we have used the following coding to calculate the hamming distance for each test case:

Bit 0: If there is no variation between the cipher text generated from original plain text and the cipher text generated from the different plain text.
Bit 1: If there is a variation between the cipher text generated from original plain text and the cipher text generated from the different plain text.

The hamming distance for each test case has been shown in Fig. 8. While calculating hamming distance for each test case, we have considered the plain text

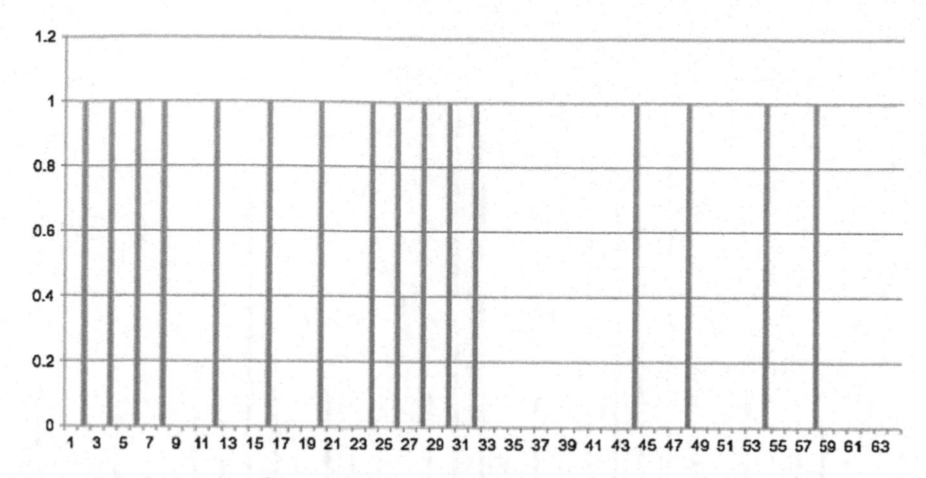

Fig. 3 Shows the variation in each bit in case 1 hamming distance: 16

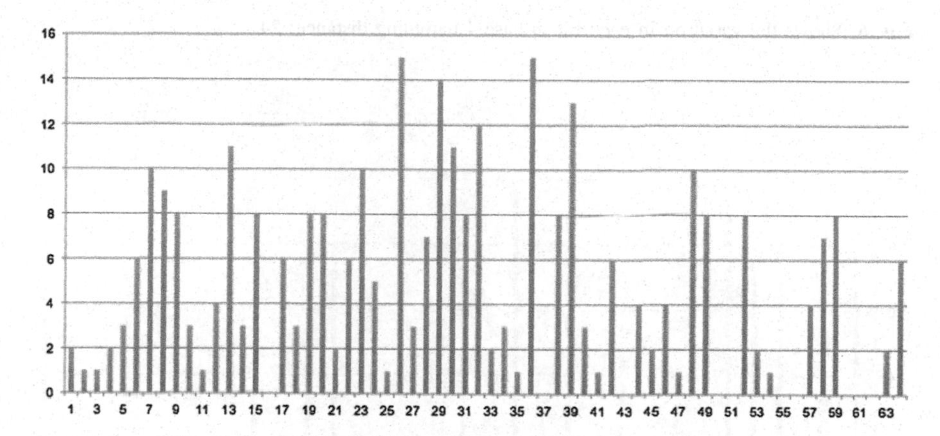

Fig. 4 Shows the variation in each bit in case 2 hamming distance: 54

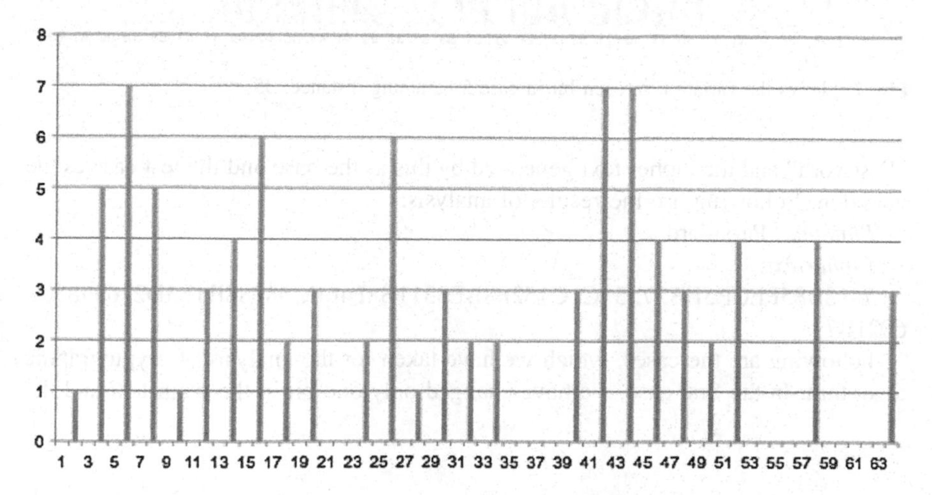

Fig. 5 Shows the variation in each bit in case 3 hamming distance: 25

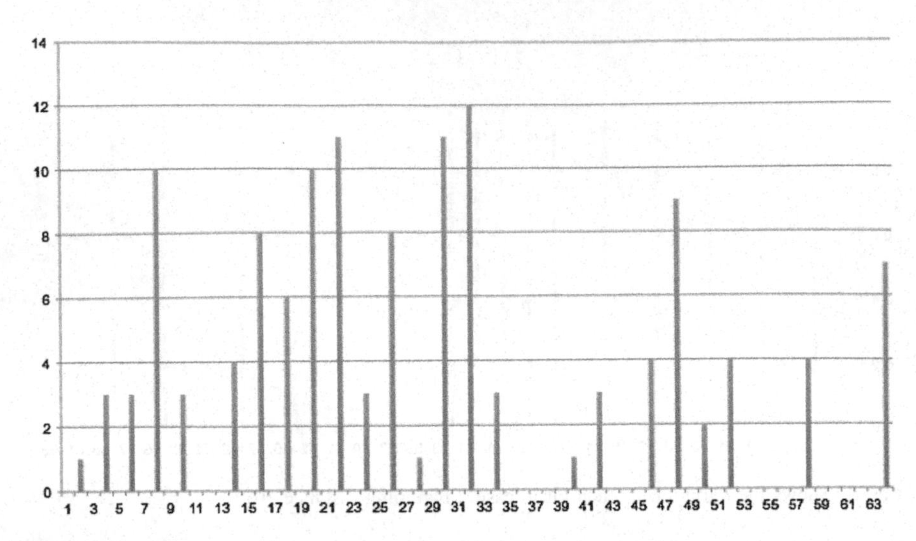


**Fig. 6** Shows the variation in each bit in case 4 hamming distance: 24

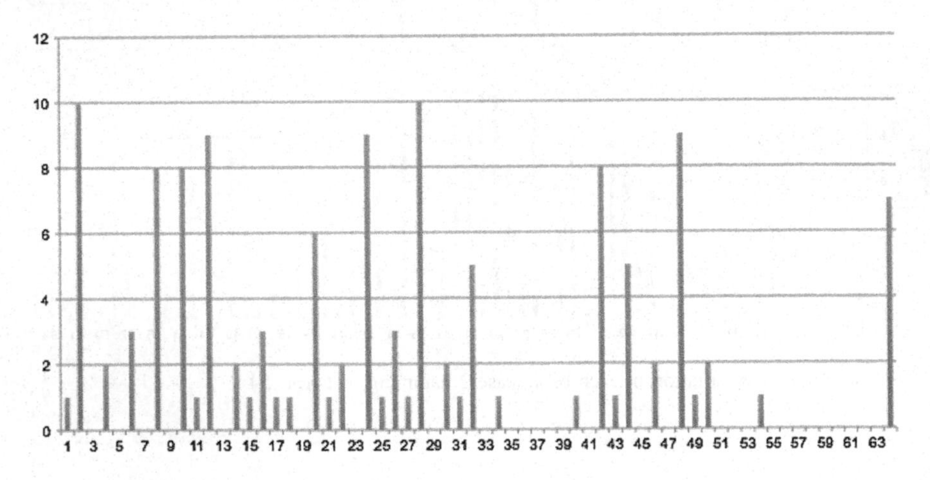


**Fig. 7** Shows the variation in each bit in case 5 hamming distance: 35

"Password" and the cipher text generated by this as the base and the test case as the variation. Following are the results of analysis:

*Plaintext*: Password

*Ciphertext*:

7418B85FECE5EB27251C1C152F8BE3311B8F46FC58388B1F3D27639850-6921B7

Following are the cases, which we have taken for the analysis of cryptographic algorithm. In the first case, we have changed only one bit in the plain text and the

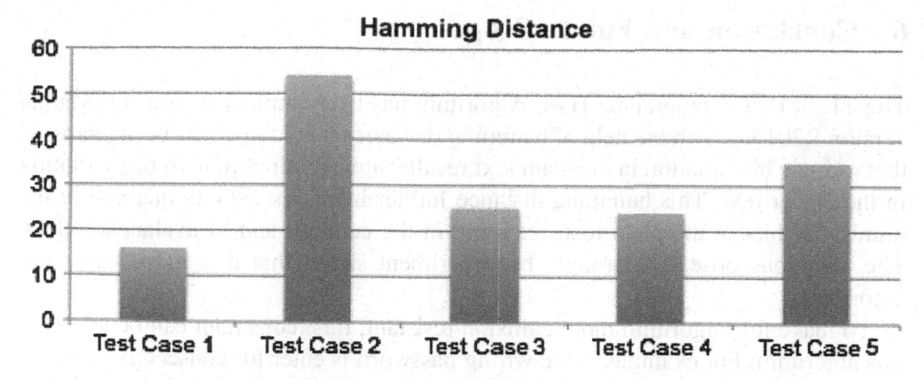


**Fig. 8** Shows the hamming distance in each test case

variation in each bit are shown in the Fig. 3. In case 2, two bits are changed in the plain text and the variation in each bit is shown in the Fig. 4 and so on.

**Test Case 1**: one bit variation in plaintext

*Plaintext*: passwork

*Ciphertext*:

7519B95EECE4EB26251D1C142E8AE2301B8F46FC58398B1E3D27629851-6921B7

**Test Case 2**: Two bits variation in plaintext

*Plaintext*: passwoab

*Ciphertext*:

530A8EF669F138A7889436BA10B40EBD38904E296234A725BD2F429817-E9219D

**Test Case 3**: Three bits variation in plaintext

*Plaintext*: passwxyz

*Ciphertext*:

751DBF5AEBE2EF21231E1C17298AE633198F46FF5F3F8F1C3F23639854-6921B4

**Test Case 4**: Four bits variation in plaintext

*Plaintext*: passpass

*Ciphertext*:

7315B555EFE5E72F2B121112278CEE3D188F46FD5B3887163F236398546-921BE

**Test Case 5**: Five bits variation in plaintext

*Plaintext*: pasklmno

*Ciphertext*:

6E1AB557E4FEE93436160A1E3C91E1261A8F46FD502389162F276298506-921BE

## 6 Conclusion and Future Scope

The TL-SMD Cryptographic Hash Algorithm has been implemented using Matlab version R2015a. With the help of hamming distance technique, it has been analyzed that a single bit variation in the plain text results into a minimum of 16 bits variation in the cipher text. This hamming distance further increases with an increase in the number of bits in the plain text, resulting in the achievement of avalanche effect. The variations observed through the experiment shows that it is indeed a secure algorithm.

To make this algorithm more collision resistant, time constraint can be added to this algorithm. For example, if the wrong password is enter for consecutive 3 times then the system will automatically block the input screen for 30 min and further more after 30 min if again the user entered the wrong password for the next 3 consecutive times then the input screen will be blocked for the next 60 min and so on. By this we can reduce the number of attempts made by the hacker in a particular time limit.

## References

1. Rivest, R.L.: The MD4 message digest algorithm, request for comments (RFC) 1320. Internet Activities Board, Internet Privacy Task Force, April 1992
2. Rivest, R.L.: The MD5 message digest algorithm, request for comments (RFC) 1321. Internet Activities Board, Internet Privacy Task Force, April 1992
3. NIST Interagency Report 7764, Status Report on the Second Round of the SHA-3 Cryptographic Hash Algorithm Competition, pp. 1–38 (2011). http://csrc.nist.gov/groups/ST/hash/sha3/Round2/documents/Round2_Report_NISTIR_7764.pdf
4. Cryptographic Algorithm Hash Competition. http://csrc.nist.gov/groups/ST/hash/sha-3/
5. Bajaj, S.B., Grewal, M.: TL-SMD: two layered secure message digest algorithm. In: Advance Computing Conference (IACC), 2015 IEEE International, 12–13 June 2015
6. Grewal, M., Chaudhari, S., Bajaj, S.B.: A survey on different cryptographic algorithms. In: REDSET 2015 2nd International Conference, October 30–31 (2015)
7. Manda, A.K., Tiwari, A.: Analysis of Avalanche effect in plaintext of DES using binary codes, vol. 1, issue 3, September–October 2012
8. Henriquez, F., Saqib, N., Prez, D., Kaya Koc, C.: Cryptographic Algorithms on Reconfigurable Hardware. Springer, (2006)
9. Haber, S., Stornetta, W.S.: How to timestampting a digital document. J. Cryptol. **3**(2), 99–111 (1991)
10. Krawczyk, H., Bellare, M., Canetti, R.: HMAC: keyed-hashing for message authentication. Internet RFC 2104, February 1997
11. Owens, J., Matthews, J.: A study of passwords and methods used in brute-force SSH attacks
12. El-Fishawy, N., Zaid, O.M.A.: Quality of encryption measurement of bitmap images with RC6, MRC6, and Rijndael block cipher algorithms. Int. J. Netw. Secur. **5**(3), 241–251 (2007)
13. Rijmen, V., Oswald, E.: Update on SHA-1. In: Menezes, A. (ed.) Topics in Cryptology—CT-RSA 2005, The Cryptographers' Track at the RSA Conference 2005, San Francisco, CA, USA, volume 3376 of LNCS, pp. 58–71 (2005)

14. Coron, J.S., Dodis, Y., Malinaud, C., Puniya, P.: Merkle Damgard revisited: how to construct a hash function. In: Advances in Cryptology—CRYPTO 2005, vol. 3621 of the series Lecture Notes in Computer Science, pp. 430–448
15. Wang, X., Yin, Y.L., Yu, H.: Collision Search Attacks on SHA-1 (2005). http://www.c4i.org/erehwon/shanote.pdf
16. Sklavos, N., Koufopavlou, O.: Implementation of the SHA-2 hash family standard using FPGAs. J. Supercomput. **227248** (2005)
17. McEvoy, R.P., Crowe, F.M., Murphy, C.C., Marnane, W.P.: Optimisation of the SHA-2 family of hash functions on FPGAs. In: IEEE Computer Society Annual Symposium on Emerging VLSI Technologies and Architectures (ISVLSI'06), pp. 317–322 (2006)
18. Rao, M., Newe, T., Grout, I.: Secure hash algorithm-3(SHA-3) implementation on Xilinx FPGAs, suitable for IoT applications. In: 8th International Conference on Sensing Technology, September 2–4, Liverpool, UK (2014)

# Contrast Enhancement of an Image by DWT-SVD and DCT-SVD

**Sugandha Juneja and Rohit Anand**

**Abstract** In this paper a novel contrast stretching technique is proposed that is based on two methods: (a) Discrete Wavelet Transform (DWT) followed by SVD and (b) Discrete Cosine Transform (DCT) followed by SVD where SVD refers to Singular Value Decomposition. In DWT-SVD technique, DWT is applied on an image resulting in the conversion of that entire image into four distinct frequency subbands (i.e. LL, LH, HL and HH subbands) followed by the application of SVD on LL sub-band of DWT processed image (because LL subband contains illumination coefficients). In this way, the values of illumination coefficients are normalized and LL subband is reformed using updated coefficients. Afterwards, image is reconstructed by using inverse DWT. In 2nd method, image is processed with DCT followed by SVD and after the reconstruction of the image in frequency domain, finally image is reconstructed by taking inverse DCT. This paper provides modified technique of DWT-SVD technique as DWT-SVD technique alone cannot produce appreciable results for some images having low contrast. Based upon the quality of contrast within an image, DWT-SVD or DCT-SVD can be used.

**Keywords** Contrast enhancement • Discrete wavelet transform • Singular value decomposition • Discrete cosine transform

## 1 Introduction

Before discussing the contrast enhancement, there are some important terms like contrast stretching (or enhancement), discrete wavelet transform, singular value decomposition and discrete wavelet transform.

---

S. Juneja
N.C. College of Engineering, Israna, Panipat, India
e-mail: sugandha.juneja05@gmail.com

R. Anand (✉)
G.B. Pant Engineering College, New Delhi, India
e-mail: roh_anand@rediffmail.com



S.C. Satapathy et al. (eds.), *Data Engineering and Intelligent Computing*,
Advances in Intelligent Systems and Computing 542,
DOI 10.1007/978-981-10-3223-3_57

## 1.1 Contrast Enhancement

Image contrast may be defined as the difference in the intensity between the highest gray level and lowest gray level in an image. Some of the images are of low contrast which may be due to poor illumination or due to poor dynamic range of the sensor or due to the improper setting of a lens aperture. More generally, an image with low dynamic range can have a very dull look or washed-out look. So, contrast enhancement (or contrast stretching) refers to expanding the range of intensity levels in an image so as to cover almost the full range of the display device [1].

## 1.2 Discrete Wavelet Transform

Similar to the expansion of Fourier series, there exists the expansion of wavelet series that converts a continuous function into a string of coefficients. If the expanded function is discontinuous or discrete, the coefficients are referred to as the Discrete Wavelet Transform (DWT).

Wavelet transform is based on very small waves (called wavelets) which can be of varying frequencies and very limited duration (unlike Fourier transform which is based on the different sinusoids). Using Wavelet transform, it is very much easy to compress the images, to transmit them and to analyze them. Wavelets can be shown to be basis of the different novel approaches to the signal processing and analysis called as 'Multiresolution Theory' [1].

In 2-D discrete wavelet transform (DWT) technique, image is first decomposed along rows and then along columns. In this way, DWT fragments an image into four distinct frequency bands LL, LH, HL, HH subbands as shown in Fig. 1. The LL sub-band contains illumination coefficients (i.e. low frequency content) and high frequency content is contained by other three sub-bands LH, HL and HH. So for improving contrast, basically illumination of LL band will be changed [2].

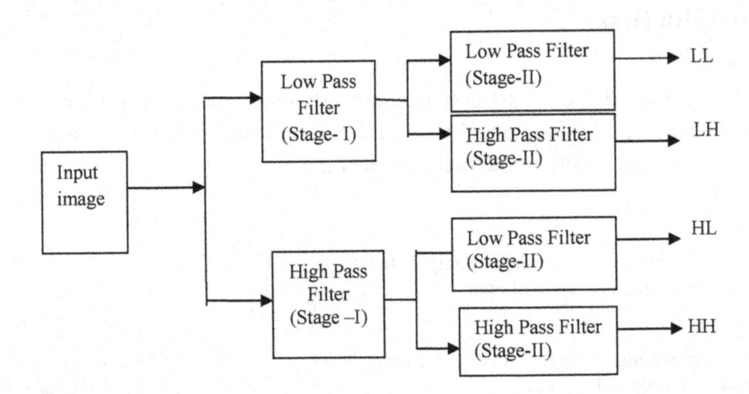


**Fig. 1** Block diagram showing DWT filter

### 1.3 Singular Value Decomposition

For singular Value Decomposition (SVD), an image matrix is decomposed by Singular Value Decomposition (SVD) technique [3]. In general SVD for any image is represented as

$$A = U_A \sum_A V_A^T \tag{1}$$

where A is input image matrix, $U_A$, $V_A$ are orthogonal square matrices and $\Sigma_A$ contains sorted singular values on its principal diagonal and it represents the illumination values of the pixels [3].

$U_A$ and $V_A$ are basically the eigen vectors of $AA^T$ and $A^TA$ respectively. If matrix A is real, the singular values will be real and hence U and V will also be real [1].

### 1.4 Discrete Cosine Transform

Discrete cosine transform (DCT) converts a signal from spatial domain to frequency domain (i.e. spectral information). It provides quite better approximation of the signal with less no. of coefficients. Low frequency DCT coefficients contain important information because of which applying illumination enhancement with low frequency coefficients can produce enhanced image. The discrete cosine transform has outstanding and huge energy constriction for the highly correlated data [1].

The purpose of contrast enhancement is the improvement of image so that it looks better than the original for a specific application. One of the most important reasons for poor contrast image is poor illumination. If intensity levels are highly concentrated within a particular small range, then information might be lost in those areas that are heavily and equally concentrated. Global Histogram Equalization (GHE) is the most widely used method for the improvement of contrast. But disadvantage of this technique is that it does not preserve the Probability Density Function (PDF) of curve [2]. Preserving PDF is a key aspect in image processing. There are some techniques such as Brightness Preserving Dynamic Histogram Equalization (BPDHE) [4], Dynamic Histogram Specification (DHS) [1] etc. which are used for contrast enhancement as well as for preserving PDF of the curve [5]. Although these techniques help in preserving shape of the PDF but these techniques use interpolation methods as a result of which loss of high frequency components (e.g. at edges) may be there [6, 7]. DWT-SVD technique cannot produce better results for some low contrast image [8]. This paper provides modification of this technique based on some checking parameter € (to be discussed later). The wavelet techniques are used presently in a number of image processing applications.

## 2 Proposed Contrast Enhancement Scheme

Let us introduce a new input image A of dimensions (M × N). The complete proposed scheme is shown in Fig. 2.

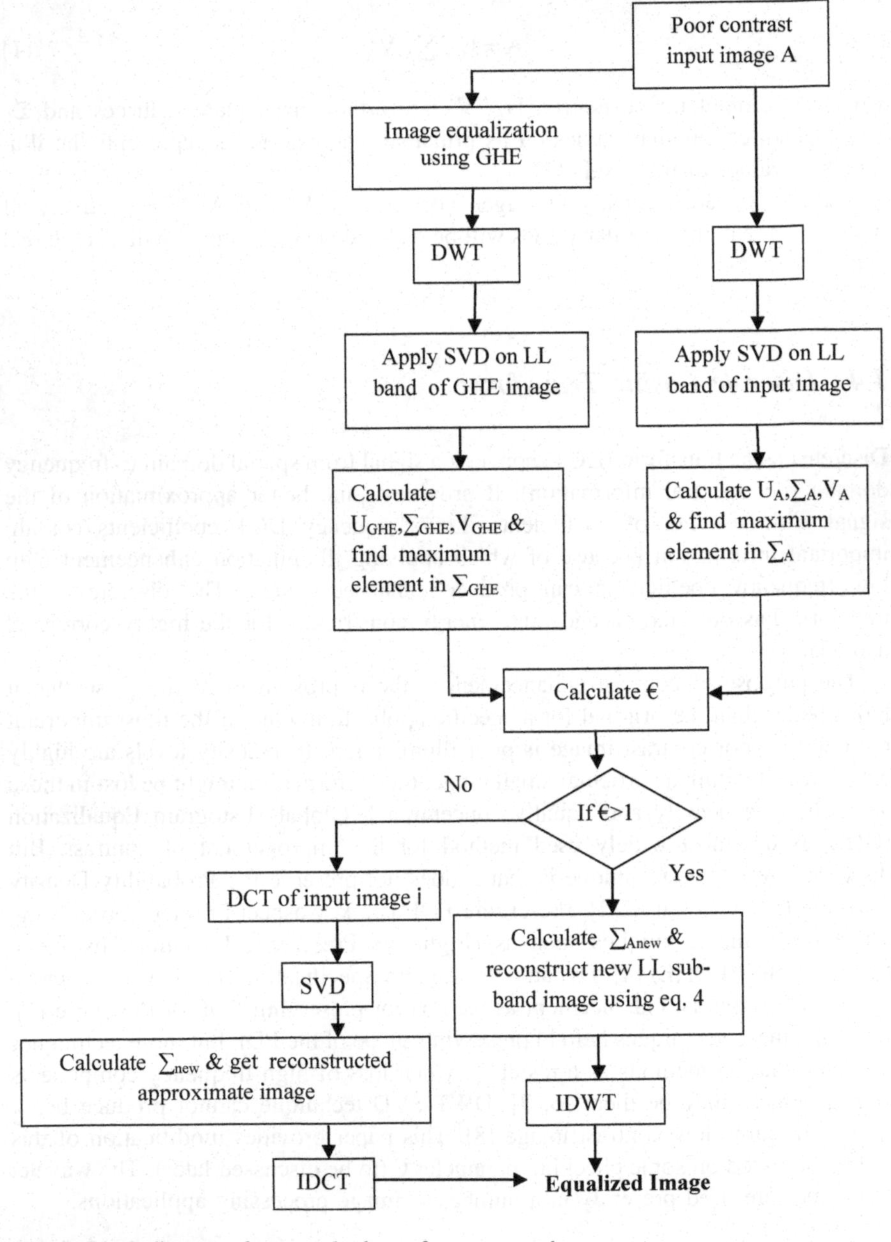


**Fig. 2** Block diagram of proposed scheme for contrast enhancement

Apply DWT on this input image and GHE version of this input image. To produce GHE version of image, a global histogram equalization (GHE) technique is applied on the entire input image.

The purpose of DWT is to separate low and high frequency components into the four subbands (as discussed earlier). Now, apply SVD on LL sub-band of both images (DWT version of input image and DWT version of GHE based input image) and find the three parameters (namely U, $\sum$ and $V^T$) in both cases.

Thereafter, find out the maximum element (i.e. the maximum illumination coefficient) contained in $\sum$ for each case. Introduce a checking parameter € defined as the ratio of maximum illumination coefficient of GHE-DWT-SVD version image and the maximum illumination coefficient of DWT-SVD version of image A.

$$€ = \max(\Sigma_{GHE}) / \max(\Sigma_A) \quad (2)$$

Now, if checking parameter € is more than unity, do perform singular value equalization by multiplying illumination coefficients with € parameter to calculate $\sum_{A(new)}$.

$$\Sigma_{A(new)} = €\Sigma_A \quad (3)$$

At this step, we get transformed illumination coefficients. For getting new LL band, apply the following transformation:

$$\bar{A} = U_A \Sigma_{A(new)} V_A \quad (4)$$

$\bar{A}$ is updated LL band.

Now take IDWT and get equalized image.

If the value of € parameter is less than 1, then above method does not produce satisfactory results. In this case, apply DCT on input image and then apply SVD and hence using (3), calculate the transformed illumination cocfficients and thereafter use (4) to find A' (where A' is the reconstructed approximate image of A in frequency domain) and after that, take IDCT to get the equalized image in spatial domain.

## 3 Experimental Results

The left column represents the original images and right column represents the enhanced images in Fig. 3 using the proposed enhancement scheme discussed in the previous section. In all the images (1–7) shown, the left column (i.e. left image) shows the original image and the right column (i.e. right image) shows the enhanced image.

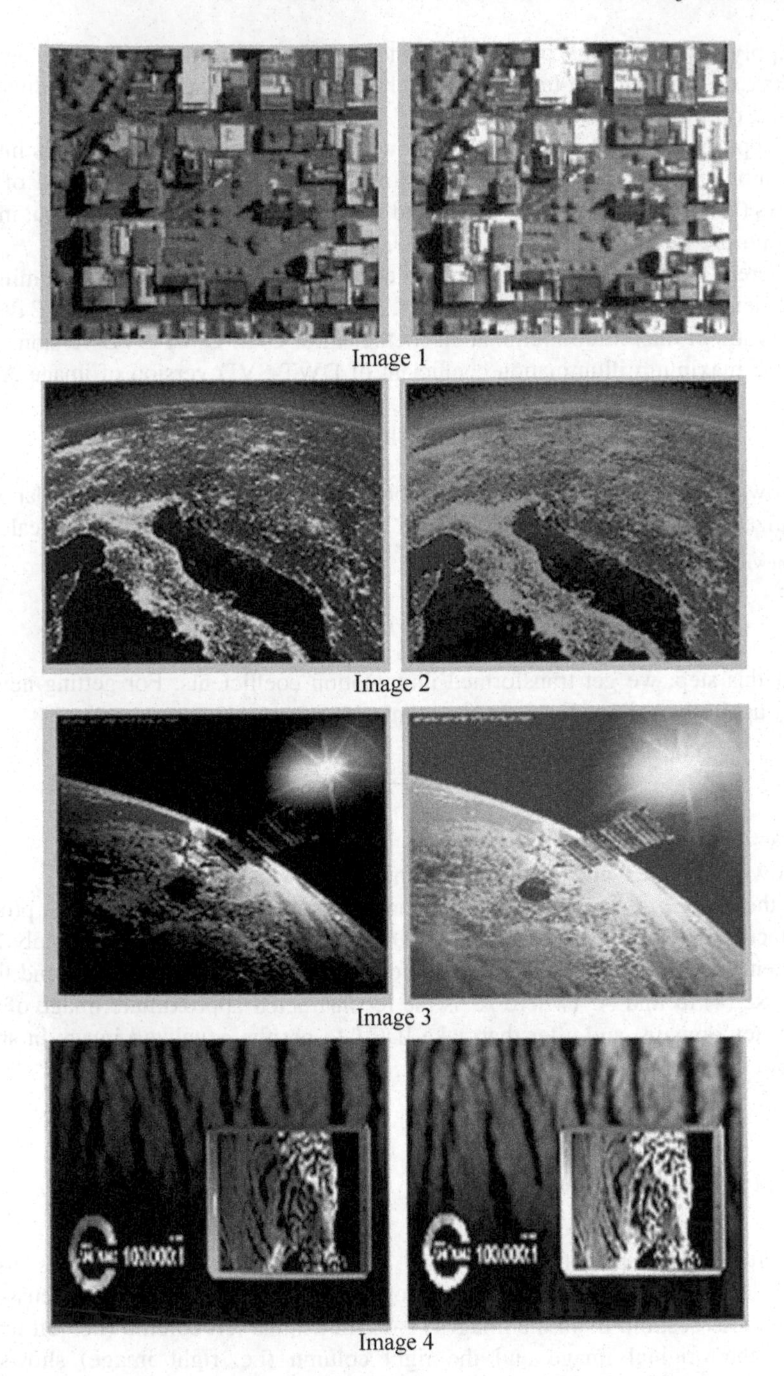


**Fig. 3** *Left Column* shows original images and *right column* shows enhanced images

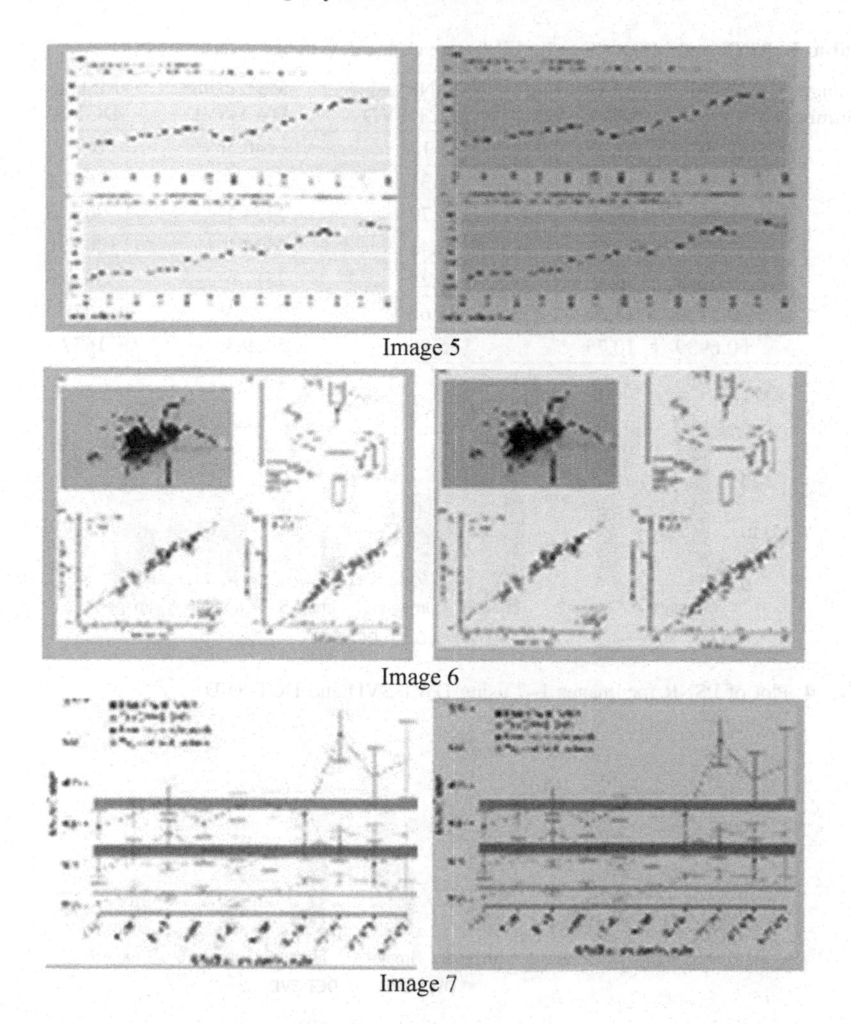

Image 5

Image 6

Image 7

**Fig. 3** (continued)

It may be seen that Images 1–4 (left column) represent the images having most of the intensity levels close to the lower intensity levels. Hence, for these images, € parameter will be more than unity (as (2) indicates) and hence, first technique discussed (i.e. DWT-SVD) will be applied in a satisfactory way.

Unlike first four images, images 5–7 (left column) have most of the intensity levels close to the higher intensity levels. So, for these images, € parameter will be less than unity (as (2) indicates) and hence, second technique discussed (i.e. DCT-SVD) will give the satisfactory results.

So, images 1–4 (right column) represent the enhanced images by DWT-SVD approach and images 5–7 (right column) represent the enhanced images by DCT-SVD approach.

**Table 1** PSNR and MSE values for all the seven images (image 1–7)

| Image number | Ratio | PSNR using DWT-SVD | PSNR using DCT-SVD | MSE using DWT-SVD | MSE using DCT-SVD |
|---|---|---|---|---|---|
| 1 | 1.09 | 27.23 | 18.17 | 146.56 | 1697.2 |
| 2 | 1.66 | 17.49 | 15.31 | 3171.5 | 7607.1 |
| 3 | 1.97 | 17.28 | 15.79 | 4754.5 | 9633.5 |
| 4 | 2.37 | 17.48 | 16.41 | 6534.1 | 12051.0 |
| 5 | 0.6438 | 9.3 | 13.25 | 7517.6 | 3071.6 |
| 6 | 0.7750 | 13.54 | 23.68 | 2873.45 | 278.48 |
| 7 | 0.6939 | 10.75 | 15.98 | 5459.3 | 1637.8 |

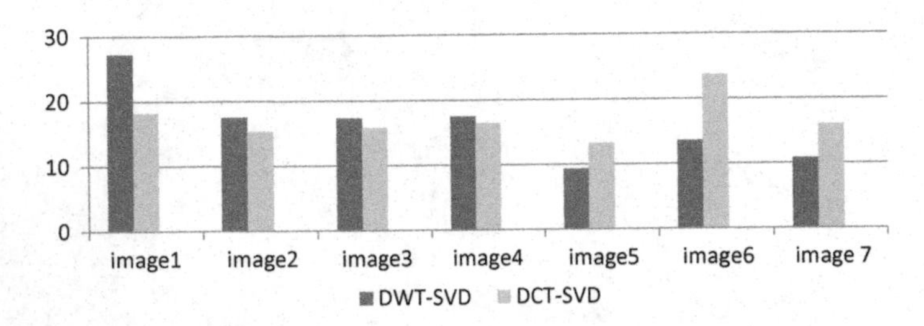


**Fig. 4** Plot of PSNR for Images 1–7 using DWT-SVD and DCT-SVD

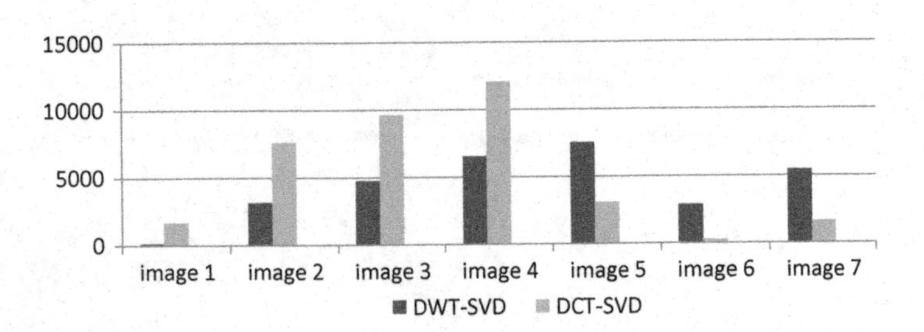


**Fig. 5** Plot of MSE for Images 1–7 using DWT-SVD and DCT-SVD

As shown in Table 1, Peak Signal to Noise Ratio (PSNR) is higher for the images 1–4 when they are enhanced using DWT-SVD (right column) and higher for the images 5–7 when they are enhanced using DCT-SVD (right column). The table also shows that Mean Square Error (MSE) is lower for the images 1–4 when they are enhanced using DWT-SVD (right column) and lower for the images 5–7 when they are enhanced using DCT-SVD (right column).

All the calculated values of PSNR and MSE have been plotted in the graph shown in Figs. 4 and 5 respectively.

## 4 Conclusion

In this paper, a new contrast enhancement technique has been proposed that calculates the ratio (€) of maximum illumination coefficient in the global histogram equalized image to the maximum illumination coefficient in input image. If ratio is more than 1, image is enhanced using DWT- SVD and if ratio is less than 1, it is enhanced using DCT-SVD. Final image is reconstructed by IDWT (for € > 1) or IDCT (for € < 1). Visual results are shown and the comparison between PSNR and MSE values for both the approaches (DWT-SVD and DCT-SVD) has been made with the help of a table and a graph.

## References

1. Gonzalez, R.C., Woods, R.E.: Digital Image Processing. Prentice Hall (2009)
2. Demirel, H., Ozcinar, C., Anbarjafari, G.: satellite image contrast enhancement using discrete wavelet transform and singular value decomposition. IEEE Geosci. Remote Sens. Lett. **7**(2), 333–337 (2010)
3. Demirel, H., Anbarjafari, G., Jahromi, M.N.S.: Image equalization based on singular value decomposition. In: 23rd International Symposium on Computer and Information Sciences, pp. 1–5, Turkey (2008)
4. Ibrahim, H., Kong, N.S.P.: Brightness preserving dynamic histogram equalization for image contrast enhancement. IEEE Trans. Consum. Electron. **53**(4), 1752–1758 (2007)
5. Kim, T.K., Paik, J.K., Kang, B.S.: Contrast enhancement system using spatially adaptive histogram equalization with temporal filtering. IEEE Trans. Consum. Electron. **44**(1), 82–87 (1998)
6. Rakesh, D., Sreenivasulu, M., Chakrapani, T., Sudhakar, K.: Robust contrast and resolution enhancement of images using multi-wavelets and SVD. Int. J. Eng. Trends Technol. **4**(7), 3016–3021 (2013)
7. Harikrishna, O., Maheshwari, A.: Satellite image resolution enhancement using DWT technique. Int. J. Soft Comput. Eng. **2**(5) (2012)
8. Atta, R.: Low contrast satellite images enhancement using discrete cosine transform pyramid and singular value decomposition. IET Image Proc. **7**(5), 472–483 (2013)

# A Fuzzy Logic Based Approach for Data Classification

**Shweta Taneja, Bhawna Suri, Sachin Gupta, Himanshu Narwal, Anchit Jain and Akshay Kathuria**

**Abstract** In this paper, we have developed a new algorithm to handle the classification of data by using fuzzy rules on real world data set. Our proposed algorithm helps banks to decide whether to grant loan to customers by classifying them into three clusters—accepted, rejected and those who have probability to get loan. To handle third cluster, fuzzy logic based approach is appropriate. We have implemented our proposed algorithm on standard bank of England data set. Our algorithm makes prediction for getting loan on basis of various attributes like job status, applicant is the chief loan applicant or not, source of income, weight factor etc. Fuzzy rules generated from the numerical data give output in linguistic terms. We have compared our algorithm with the state of the art algorithms—K-Means, Fuzzy C-means etc. Our algorithm has proved to be more efficient than others in terms of performance.

**Keywords** Fuzzy C-means (FCM) algorithm · Fuzzy logic · Classification technique

S. Taneja (✉) · B. Suri · S. Gupta · H. Narwal · A. Jain · A. Kathuria
Department of Computer Science, BPIT, GGSIPU, New Delhi, India
e-mail: shweta_taneja08@yahoo.co.in

B. Suri
e-mail: suri_bhawna@yahoo.com

S. Gupta
e-mail: guptasachin579@gmail.com

H. Narwal
e-mail: himanshunarwal@yahoo.co.in

A. Jain
e-mail: anchitjain1994@gmail.com

A. Kathuria
e-mail: akshay.kathuria@gmail.com

 
S.C. Satapathy et al. (eds.), *Data Engineering and Intelligent Computing*,
Advances in Intelligent Systems and Computing 542,
DOI 10.1007/978-981-10-3223-3_58

## 1 Introduction

Data Mining [1] is the process of extracting useful information or knowledge from huge amount of data. This information is used to perform strategic operations and thus helps in effective decision making. Decision making generally needs different and new methodologies in order to make sure that decision made is accurate and valid. There are many problems that arise in data analysis. Sometimes, the real world data is uncertain. Thus, it is necessary to develop new techniques to handle the ever increasing and uncertain data.

To handle uncertainty in data, we have used the concept of fuzzy logic with classification technique of data mining. As suggested by Zadeh [2], fuzzy set theory deals with ambiguity, uncertainty and seeks to overcome most of the problems generally found in classical set theory. We have taken a data set of loans applied by various applicants to a bank. According to classical set theory, an applicant will either be granted loan or rejected. But, in the real world, there is a probability that he or she might get the loan. Thus, to handle such ambiguous situation, we have defined the membership functions for granting loan to applicants. Our algorithm can be used to perform fuzzy classification where uncertainty or fuzziness can be resolved using membership functions. We have implemented this algorithm using Canopy (python based tool) and various python libraries like Peach 0.3.1 [3], NumPy [4], Matplotlib 1.5.1 [5], SciPy [6].

This paper is organized as follows. In Sect. 2, we give the description of the approaches that are currently in practice. The problems with these approaches and how the proposed method overcomes them are highlighted. The proposed algorithm is stated and explained in Sect. 3. Section 4 gives the implementation results by applying the proposed algorithm on bank loans data set. In the next section, comparison of the proposed algorithm with other existing algorithms is done. The last section concludes the paper.

## 2 Related Work

The classification technique [7] is one of the data mining techniques which classify data into various classes. We have used the concept of fuzzy logic to classify the uncertain or ambiguous data.

There are some approaches in literature to handle uncertainty in data. In [8], the author uses linguistic terms for database queries and shows the advantages of using linguistic terms as well as the difference between classical and fuzzy approaches. But in case of many-valued logic, this is not sufficient. A new approach to fuzzy classification has been discussed by the authors in [9]. They state that by using fuzzy discretization the results can be represented in linguistic terms which is superior to other classification techniques. However, it cannot be applied for all kinds of data. Another technique is proposed in [10], where the authors have suggested the use of Data Envelopment Analysis for graduate admissions using

GMAT scores and GPA. The advantage of this technique is that it does not require expert participation for obtaining membership functions. In [11], authors define a new function based on fuzzy logic but the method is not tested on large datasets. Another work is done in [12], authors suggest a new approach for decision making on university admissions using fuzzy logic.

In this paper, we have tried to develop a better data classification algorithm to handle real world problems in terms of efficiency as well as universal applicability. Our algorithm uses Fuzzy logic [2], as data for granting loans is not crisp and we need to analyze it carefully to get more accurate and efficient results, which is one of the main benefits. The proposed algorithm helps both banks and applicants as it predicts the probability of getting the loan in linguistic terms, which cannot be easily done by most other methods mentioned above.

# 3 Proposed Algorithm

There are certain assumptions to be made for the application of the proposed algorithm. These are as follows:

- Every applicant has a valid job status.
- He/she must be the chief loan applicant.
- He/she must have a regular and valid source of income.
- Applicant must not have taken other loans.

The following Fig. 1 shows the proposed algorithm.

The detailed version of algorithm is given in Figs. 2 and 3 of Sect. 3.1.

STEPS:

1. Divide the dataset into training, validation & test set.
2. Generate Fuzzy Rules using Association rules and classification techniques of data mining.
3. Identify Acceptance, Fuzzy and Rejection regions on the basis of rules.
4. For Fuzzy region, apply fuzzy c-means clustering and obtain outliers which are used to calculate the weight factors.
5. Apply Quantifiers on fuzzy region by using following formula of $LP_i$ (Loan Points).

$$LP_i = [a*(J) + b*(I) + c*(O) + d*(W) + e*(A)]. \tag{1a}$$

$$LP_{adjusted} = LP_i / LP_{i\,Max}. \tag{1b}$$

Where,

$LP_i$ =Loan Points

J = Job Status of applicant

I = Income of applicant

O = whether the applicant has taken any loan or not

W = Weight Factor assigned by the bank

A = Age of applicant

a, b, c, d, e = Constants that can vary as they depend on the bank giving weightage to all these factors.

**Fig. 1** Proposed algorithm

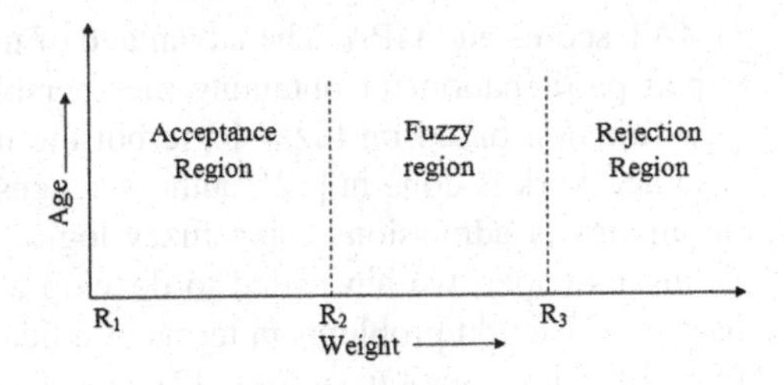


**Fig. 2** Regions obtained from rules. On X-axis: weight, on Y-axis: age

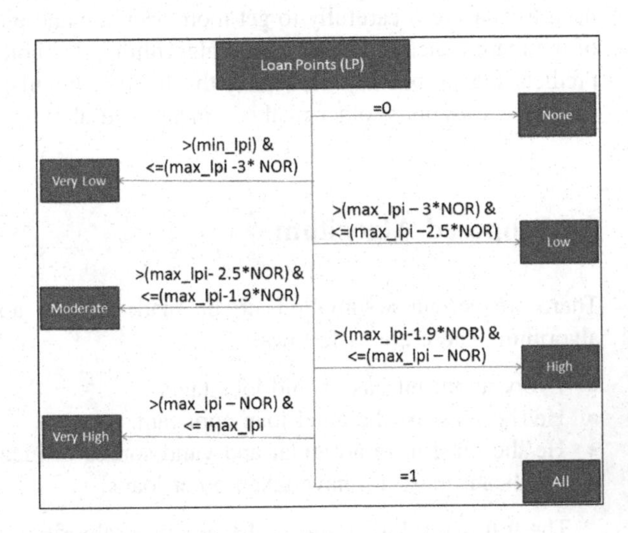


**Fig. 3** Probability of getting loan

## *3.1 Detailed Algorithm*

Input: Dataset D

Output: Probability of getting loan in terms of quantifiers

Method:

1. Suppose we have a dataset D, which is divided into 3 parts that are: training dataset ($d_t$), Validation dataset ($d_v$) and test dataset ($d_{test}$) such that they follow following properties:

$$d_t \cap d_v = \varnothing \tag{2}$$

$$d_v \cap d_{test} = \varnothing \tag{3}$$

$$d_t \cap d_{test} = \varnothing \tag{4}$$

$$d_t \cup d_v \cup d_{test} = D \tag{5}$$

2. Fuzzy Rules are generated using association rules or classification technique using WEKA tool.

**Table 1** Weight factors assigned

| Weight | Weight factor (W) |
|---|---|
| Cluster 1 | $W_1$ |
| Cluster 2 | $W_2$ |
| ⋮ | ⋮ |
| Cluster n | $W_n$ |

n = number of cluster
$W_n$ = Weight factor for nth cluster

**Table 2** Quantifiers assigned to range assigned

| $LP_i$ | Quantifier |
|---|---|
| [0] | None |
| [0.074, 0.261) | Very low |
| [0.261, 0.308) | Low |
| [0.308, 0.364) | Moderate |
| [0.364, 0.448) | High |
| [0.448, 0.542] | Very high |
| [1] | All |

- Generated fuzzy rules are in form of if-else.
- Eg. If weight factor < 1.24, then loan request is declined.

3. On the basis of the rules generated above, Acceptance region, Fuzzy region and Rejection region are Fig. 2.
4. For Fuzzy Region:
   4.1. Fuzzy C-Means (FCM) Algorithm is applied, clusters are assigned values according to the factors as shown in Table 1.
   Similarly different attributes such as Job Status, Income, Age and Other Loan are assigned values (also the ranges) according to the clusters made.
   4.2 Apply quantifiers on fuzzy region by using the formulae of $LP_i$ (Loan Point) given in Eqs. 1a and 1b.
   4.3 For General Case,

$$NOR = \left(LP_{i\,(max)} - LP_{i\,(min)}\right)/5 \tag{6}$$

   where, NOR = Number of Regions, $LP_{i\,(max)}$ = maximum value of $LP_i$, $LP_{i\,(min)}$ = minimum value of $LP_i$

5. On the basis of the values calculated above, following quantifiers are assigned to $AP_i$ values as shown in Table 2.
   The probability of an applicant being granted a loan is shown in Fig. 3.

# 4 Experiments Conducted and Results Obtained

In this section, we illustrate the working of our algorithm for applicants seeking for loan from a bank. Loan granting decisions are taken by the bank which considers several factors including applicant's job status, secure and source of income and other taken loans.

## 4.1 Dataset

We have used a standard Bank of England dataset (collected by NMG surveys from the period 2004–2011) [13]. The bank committee uses a weighing function mentioned in Eqs. (1a) and (1b) to decide whether to grant loan to the applicant or not. Candidates with a weight of more than or equal to 1.25 are checked by the bank committee one-by-one and in case of supporting evidences (job status, income status, other loans etc.) applicant is granted loan, otherwise the application is rejected. The loan data contains attributes like ID, weight, Job status, Income etc. A subset of dataset having some tuples is shown in Table 3.

## 4.2 Weka Tool

We have used Weka tool [14]. The Weka3.6 is the latest version of Weka used nowadays. It contains algorithms for implementation of different data mining techniques like classification, prediction, clustering, association rules etc.

**Table 3** Dataset used

| ID | Weight | Job status | Income | Loan |
|---|---|---|---|---|
| 65 | 0.549 | Unemployed | Don't know | Yes |
| 231 | 3.233 | In paid employ | 75000–99999 | No |
| 1543 | 0.886 | In paid employ | 25000–29999 | Yes |
| 644 | 1.332 | In paid employ | 40000–49999 | No |
| 108 | 0.877 | Retired | Refused | No |
| 1081 | 1.091 | Long-term sick | Don't know | Yes |
| 522 | 0.713 | Retired | Refused | No |
| 1454 | 0.646 | Retired | 7500–9499 | Yes |
| 364 | 0.708 | Unemployed | Refused | No |
| 110 | 1.29 | In paid employ | 50000–74999 | Yes |

### 4.3 Results Obtained

The loan data set fed as input to weka tool. By using Apriori algorithm [15], following association rules are generated.

if weight factor < 1.24, then reject     else if weight factor > 2.105, then accept

According to rules, obtained data is divided into 3 regions which are shown in Fig. 4.

After division into regions, the fuzzy region is considered to remove the uncertainty. The FCM algorithm [16] is applied to it and clusters are made to assign different weight factor values. Each data point is assigned membership corresponding to each cluster center on the basis of gap between the cluster center and the data point. The more the data is near to any specific cluster center, more is its membership towards that particular cluster center. Clearly, results obtained on adding memberships of each data point should be equal to one. After each successful iteration, membership and cluster centers are updated according to the formula given below:

$$\mu_{ij} = 1/\sum_{k=1}^{c} \left(d_{ij}/d_{ik}\right)^{\left(\frac{2}{m}-1\right)} \tag{7}$$

$$\upsilon_j = \frac{\sum_{i=1}^{n} \left(\mu_{ij}\right)^m x_i}{\sum_{i=1}^{n} \left(\mu_{ij}\right)^m}, \forall_j = 1, 2, \ldots, c \tag{8}$$

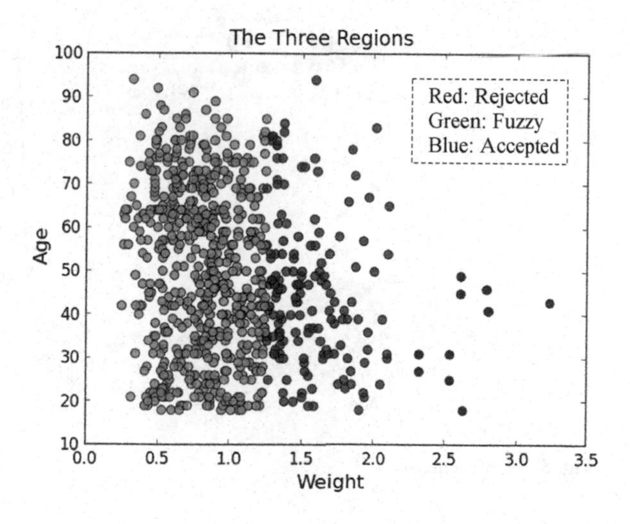


**Fig. 4** The regions obtained shown in 2-D

where,

'n' is the number of data points.
'$\upsilon_j$' is *j*th cluster center.
'M' is the fuzziness index m € [1, ∞].
'c' is the number of cluster center.
'$\mu_{ij}$' is the membership of *i*th data to *j*th cluster center.
'$d_{ij}$' is the Euclidean distance between *i*th data and *j*th cluster center.

The main objective of fuzzy c-means algorithm is to minimize:

$$J(U,V) = \sum_{i=1}^{n}\sum_{j=1}^{c}(\mu_{ij})^m \left\|x_i - v_j\right\|^2 \tag{9}$$

where, '$\left\|x_i - v_j\right\|$' is the Euclidean distance between *i*th and *j*th cluster center

After the FCM algorithm, we got 5 centers shown as black dots in Fig. 5, thus using them further in our algorithm. After calculating the centers of clusters and distance of each data point from it the following graph shown in Fig. 6 is obtained.

From Fig. 6 the range of age for the age factor (A) is calculated as shown in Table 4. Similarly other factors are also assigned values based on their ranges (Tables 5, 6, 7 and 8).

After obtaining various factors from FCM algorithm, we calculate the $LP_i$ by using formula (1) and NOR is calculated using formula (6). According to the range of $LP_i$ obtained, results are obtained in a linguistic manner, as shown in Table 9.

For general case, the range of $AP_i$ can be calculated as displayed in the Fig. 7.

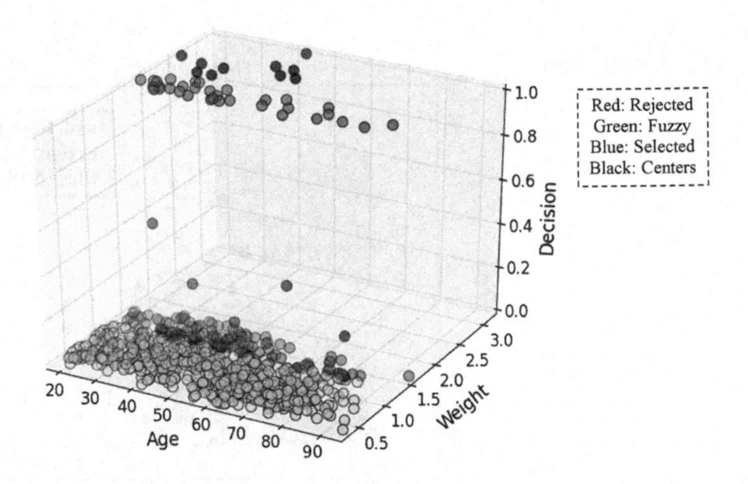


**Fig. 5** FCM centers

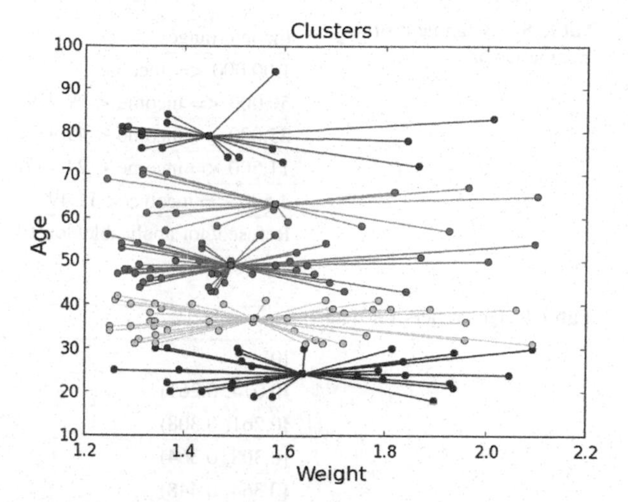


**Fig. 6** Clusters with different age values

**Table 4** Assignment of age factor

| Age range | Age factor |
|---|---|
| 31 <= age < 43 | 5 |
| 43 <= age < 57 | 4 |
| 18 <= age < 31 | 3 |
| 57 <= age < 72 | 2 |
| 72 <= age < 93 | 1 |
| 94 < age < 18 | 0 |

**Table 5** Assignment of weight factor

| Weight range | Weight factor |
|---|---|
| 2.21 < weight | 5 |
| 1.96 <= weight < 2.21 | 4 |
| 1.72 <= weight < 1.96 | 3 |
| 1.48 <= weight < 1.72 | 2 |
| 1.24 <= weight < 1.48 | 1 |
| Weight < 1.24 | 0 |

**Table 6** Assignment of other loan factor

| Other loan | Other loan factor |
|---|---|
| No | 2 |
| Don't know/refused/NA | 1 |
| Yes | 0 |

**Table 7** Assignment of job factor

| Job status | Job factor |
|---|---|
| In paid | 5 |
| Self employed | 4 |
| Other | 3 |
| Sick/disable | 2 |
| Student/retired | 1 |
| Unemployed | 0 |

**Table 8** Assignment of income factor

| Income range | Income factor |
|---|---|
| 1,00,000 <= Income | 5 |
| 50,000 <= income < 99,999 | 4 |
| 25,000 <= income < 49,999 | 3 |
| 11,500 <= income < 24,999 | 2 |
| 4,500 <= income < 11,499 | 1 |
| Refused/not applicable/don't know | 0 |

**Table 9** Quantifiers for data

| $LP_i$ | Quantifier |
|---|---|
| [0] | None |
| [0.074, 0.261) | Very low |
| [0.261, 0.308) | Low |
| [0.308, 0.364) | Moderate |
| [0.364, 0.448) | High |
| [0.448, 0.542] | Very high |
| [1] | All |

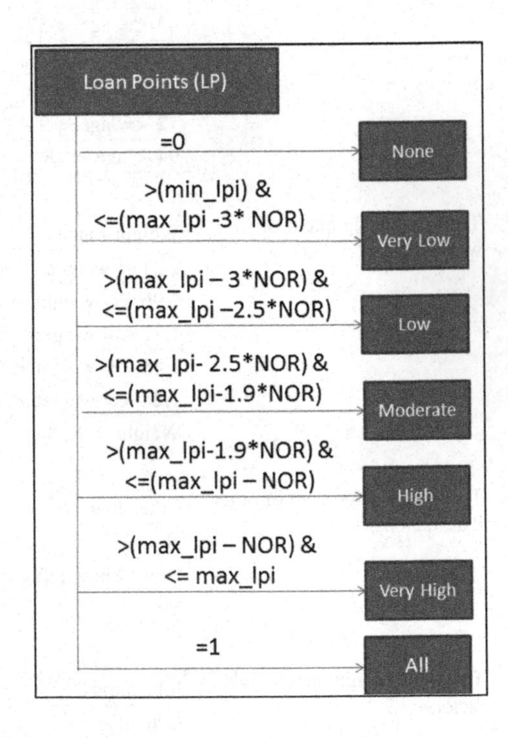


**Fig. 7** Quantifiers for general case

Table 10 Comparison based on accuracy

| Data Set | K-means | FCM | Proposed |
|---|---|---|---|
| Bank of England (%) | 76.32 | 78.72 | 83.49 |

## 5 Comparison of Proposed Algorithm with Other Classification Algorithms

We have compared our proposed algorithm on Bank of England data set with K-Means [17] and Fuzzy C-Means algorithm. Table 10 shows the accuracy obtained. Our proposed algorithm obtains the best accuracy amongst the other two.

## 6 Conclusion and Future Directions

Data mining is an intelligent technique to extract knowledge from a large set of data. We have proposed a new classification algorithm using the concept of fuzzy logic. The algorithm deals with the granting of loans to different applicants by banks. The possible options could be the applicant would get the loan or get rejected or there is a probability that he/she might be granted or rejected the loan. We have considered the third case where the probability of getting loan is to be handled. In order to do so we have defined the degree of membership to the respective clusters using quantifiers. This is done by using fuzzy rules and various factors.

The efficiency of our proposed algorithm is proved to be better by comparing it with standard algorithms—K-Means, Fuzzy C-Means. In future, our proposed algorithm can be extended and applied on other real world applications like signal processing etc. Moreover, the proposed algorithm can be applied on large datasets and its performance can be compared in terms of execution time as well as accuracy with other algorithms of similar nature.

## References

1. Han, J., Kamber, M., Pei, J.: Data Mining Concepts and Techniques. Morgan Kaufmann Publishers, San Francisco (2012)
2. Zadeh, L.A.: Fuzzy sets. In: Information and Control, issue 3 (1965)
3. Peach library. https://pypi.python.org/pypi/Peach/0.3.1
4. NumPy library. http://www.numpy.org
5. Matplotlib library. http://matplotlib.org
6. SciPy library. https://www.scipy.org
7. Kromer, P., Platos, J., Snasel, V., Abraham, A.: Fuzzy classification by evolutionary algorithms. In: Systems, Man, and Cybernetics (SMC), 2011 IEEE International Conference (2011)

8. Hudec, M., Vujošević, M.: Integration of data selection and classification by fuzzy logic. In: Expert Systems with Applications, issue 10 (2012)
9. Mehta, R.G., Rana, D.P., Zaveri, M.A.: A novel fuzzy based classification for data mining using fuzzy discretization. In: Computer Science and Information Engineering, 2009 WRI World Congress (2009)
10. Pendharkar, P.: Fuzzy classification using the data envelopment analysis. In: Knowledge-Based Systems (2012)
11. Gupta, Y., Saini. A., Saxena, A.K.,: A new fuzzy logic based ranking function for efficient information retrieval system. In: Expert Systems with Applications, issue 3 (2015)
12. Taneja, S., Suri, B., Narwal, H., Jain, A., Kathuria, A., Gupta, S.: A new approach for data classification using fuzzy logic. In: 6th International Conference—Cloud System and Big Data Engineering (Confluence), IEEE (2016)
13. Bank of England NMG household survey from 2004 to 2011. http://www.bankofengland.co.uk/research/Documents/onebank/ nmgface.xlsx
14. Witten, I.H., Frank, E.: Data mining: Practical Machine Learning Tools and Techniques—Tutorial Exercises for the Weka Explorer. Morgan Kaufmann Publishers, San Francisco (2011)
15. Chai, S., Yang, J., Cheng,Y.: The research of improved apriori algorithm for mining association rules. In: Service Systems and Service Management, 2007 International Conference (2007)
16. Bezdek, J.C., Ehrlich, R., Full, W.: FCM: The fuzzy C-means clustering algorithm. In: Computers and Geosciences, issue 2 (1984)
17. K-means clustering algorithm. https://sites.google.com/site/dataclusteringalgorithms/k-means-clustering-algorithm

# Facebook Like: Past, Present and Future

**Kumar Gaurav, Akash Sinha, Jyoti Prakash Singh and Prabhat Kumar**

**Abstract** As a social networking website, Facebook has a huge advantage over other sites: the emotional investment of its users. However, such investments are meaningful only if others respond to them. Facebook provides a way to its users for responding to posts by writing comments or by pressing a Like button to express their reactions. Since its activation on February 9, 2009, the Facebook Like button has evolved as an essential part of users' daily Facebook routines and a popular tool for them to express their social presence. However, the inadequacy of the Like button in expressing the original sentiments of a user towards a post has raised serious discussions among the users. It is an apparent deduction that Facebook Like disappoints at addressing the wide spectrum of emotions that an online human communication entails. It does not let the post creator ascertain that the sentiment behind his post has been perceived in its true essence. Even after the collaboration with emotions, the Like button still has a wide range of issues that needs to be addressed. The paper considers these pros and cons associated with the current Facebook Like button. The paper also provides novel technique to improve the efficiency of the Like feature by associating it with an intelligent engine for generating recommendations to the users. This, in turn, shall improve the user-posted content on Facebook.

**Keywords** Social media · Social network · Facebook · Like · Privacy

K. Gaurav (✉) · A. Sinha · J.P. Singh · P. Kumar
Department of Computer Science and Engineering,
National Institute of Technology Patna, Patna, India
e-mail: kumargaurav.nitp@gmail.com

A. Sinha
e-mail: akash.cse15@nitp.ac.in

J.P. Singh
e-mail: jps@nitp.ac.in

P. Kumar
e-mail: prabhat@nitp.ac.in



S.C. Satapathy et al. (eds.), *Data Engineering and Intelligent Computing*,
Advances in Intelligent Systems and Computing 542,
DOI 10.1007/978-981-10-3223-3_59

# 1 Introduction

Social networking has brought the world closer in dimensions which besides being fascinating in comprehension are quite pragmatic in implementation. Just few years after Mark Zuckerberg found "The Facebook" in his dorm room as a way for the students of Harvard University to create and maintain social ties, the company joined the ranks of the Web's great superpowers. With 1.44 billion monthly active users in early 2015 [1], Facebook has turned up as the leading player in the social networking domain in most of the countries. Surviving in an arena where consistent innovation is persistently required, Facebook has reflected commendable improvements over time. The changes in the News Feed algorithm, transforming Facebook into a "perfect personalized newspaper", have added yet another feather to its cap. However, in the attempt to maneuver new ways for ceaseless updating of features like Adding a Legacy Contact, On This Day Facebook, history browsing, etc., a key feature has been left unattended since long, the iconic Facebook Like button.

The Facebook Help feature mentions that clicking the Like button available underneath a Facebook post "is an easy way to let someone know that you enjoy it, without leaving a comment. Just like a comment though, the fact that you liked the post is visible below it" [2]. However, as a broader portion of the population becomes Facebook members, the website will be used in increasingly varied ways [3], which might also lead to the disparities in people's perception and use of the like button. Figure 1 shows the iconic Like button of Facebook.

Recently, the application of Like button has raised a few questions originating from the dilemma it often leads its users into. Numerous instances have been witnessed by the users of Facebook irrespective of their geographical existence where they were reluctant at clicking on the Facebook Like button because it hardly expressed their original sentiment.

# 2 Yesterday of Like

When was the last time you were going through your news feed on Facebook and happened to come across a post pertaining to a grave social issue with a hint of melancholy like a post which mentions about lives lost in a terrorist attack or about how a friend lost someone from his/her family and felt uncomfortable at clicking the Facebook Like button. You wish to console your friend by doing something that would convey your regret for the trauma he is going through. But again all you are left with is a button that fails at reciprocating the plausible gesture. The more

**Fig. 1** Facebook Like button

optimistic mediators among our lot may come up with a very convenient solution saying that we can always choose to comment on things we can't like. But that's like doing another thing when we don't have the one thing we would like to do in the first place.

One of the crucial advantages of an online social network is that one gets a medium of communication with a comparatively greater scalability than the conventional offline medium of communication [4, 5]. The reach of any message is wider and the propagation of the message scales up to meet the definition of virality. Such a medium that has the potential to address a large audience offers a platform conducive for conducting and propagating a mass movement. An instance for the aforesaid situation in the context of Facebook could be a post that has the objective of spreading awareness about a missing child. The parents lose contact with their child in an unfortunate incident and therefore decide to take the matter to the online social network. They create a post on Facebook containing physical description about the child, the information concerning the missing status of the child and a request for an earnest intimation to the parents in case if anyone who has been notified about the incident happens to spot the missing child. Their Facebook friends wish to convey their support to the parents in their hour of need. Are they supposed to do that by clicking on the Facebook Like button? The logical answer would arise that they won't use the Like button. But what if some of the online friends do? Be it ignorance or haste, the culprit could be any factor but down the line the fact remains that the faux pas could be committed.

There are times when the context of a post plays a more prominent role than the content itself. There are numerous instances where approximately same contents provide varied meanings some of which could even be in contradiction to each other. Suppose there is a scenario where a Facebook user shares a post: "I came fourth in the 100 m race finishing only a second later than the one who came third." And there is another scenario where a user shares a post: "I came fourth in the National Science Olympiad." Now, even a cursory glance will suffice in leading anyone to the conclusion that the content of both the posts are in line provided that only the opening three words are taken into consideration. But the same verdict cannot be passed on for the context.

There is a considerable contrast in the sentiments invoked or pursued by the aforementioned posts. The first scenario reveals about a result submerged in the colors of regret. He is sorry that he could not achieve at least the third position. In the best case scenario what he will be receiving should comprise of empathetic responses. At best, he needs encouragement from that section of the online society with which he is sharing his near-success-turned-failure episode. On the contrary, he receives only 'likes' of those interested in responding to his post.

In the second scenario, the user has posted about securing the same rank as in the first scenario but the platform for the performance has altered. Securing a fourth rank in the National Science Olympiad is an achievement in a general sense. Of course, sense of achievement or failure is a question of perception and varies from person to person. But that triggers a different chain of studies. Generally speaking, the person securing such a rank, as mentioned in the aforesaid scenario, has earned a sense of general appreciation from the online social community with whom he has shared the post. Here the Like button reciprocates the kind of reaction expected by the creator/sharer of the post and thus helps in building the relationship on the online social network.

In the former case the social network lost in providing the kind of reciprocation inducted from the post and which would have met the expectations of the creator/sharer of the post. The person receiving likes on the post that was meant to represent his conveyance of regret shall feel offended or disappointed for the fact that the readers of the post failed to comprehend his true sentiments. The same might not to be true for the readers (Facebook users) who had no other (read quick) option than to Like the post. They might have comprehended the true intent of the post but were swayed in the action of clicking the Like button subject to the absence of any other alternative. The goal of any social network is building relationships but in scenarios like the latter one Facebook fails to meet its most cardinal goal.

To counteract the anomalies discerned in resorting to Facebook Like button, new alternatives have to emerge. And these alternatives need to be devised with diligence of a higher order because online social platforms as such are found to be potent tools that mirror and magnify the good, the bad and the ugly [6].

## 3 The Counterpart of Like: *Dislike*

Provision for a Dislike button has been proposed and explored on various occasions since a long period of time. Instances of users raising concerns for implementation of such an alternative have been witnessed often. The story behind the quest for such a button on Facebook is of considerable length and spun out of viral campaigns, false notions, skeptical marketing strategies and spam. The button has never been considered for implementation even in any of the test forms of Facebook. However, the idea of such a button has been widely discussed and debated over within the headquarters of Facebook in light of the creation of countless profiles, communities, groups and pages demanding for provision of such a social plug-in. The Facebook development team doesn't seem to reflect much regard for any argument in favor of such a modification. They are convinced that disapproval is not a beneficial gesture for its users since it runs the risk of generating negativity on the social platform. What would often feel playful to the user in 'disliking' a post could inflict pain to the recipient of the dislike by it being overly critical or judgmental.

A Dislike button is also counterproductive to the financial gains of Facebook. Facebook is a platform where we log in after returning brain-dead from work, where we scroll through the wall posts or newsfeed aimlessly just before falling asleep and after waking up. We wander there in our leisure simply to take a break from the monotony of our lives. If we were ever to humanize Facebook then pleasant, amusing or congenial would likely be the first few appropriate characteristics that we would consider attributing to its personality. It is quite evident from all of the above that a Dislike button is in clear contradiction to the popular image of the social network. It will eventually institute negativity in an otherwise positive platform and will eventually repel its own users leading to a clear downfall in the traffic.

The users can still convey a wide spectrum of responses through comments representing general appreciation as well as criticism. However, consideration and practice of less affirmative or rather negative shades of comments more prominently in mainstream communication would facilitate and encourage negativity in an amount that might prove detrimental to the inclination of the users to such social media platforms.

## 4 Today of Like

Recently, Facebook has introduced a new feature called 'Reactions', as shown in Fig. 2. Whereas liking tells your friends that you enjoyed their post, reacting allows you to specify your response [2].

The inclusion of five more buttons has reinforced the ability to emote better but it has also brought certain complications latent in its operation. In the pre-'Reactions' era, the users of the online social network had only to take a decision about whether they were going to Like something or not. Now they are supposed to not just take a decision about whether they wish to like something or not but also to choose the degree of their liking or disliking it. Loving some post in the context of communication on Facebook is a superlative manifestation of liking it. The user needs to discern between two posts in order to justify his decision of whether liking it or loving it. Similarly, in case of disliking a post a user has to decide whether he wishes to express simply his sadness over the post or escalate to a level higher and express indignation. Thus, in all cases the indulgence considering the time and effort in regard of the clicks infer at its manifold increment.

There is another issue with the operation of the feature 'Reactions'. Theory predicts that conforming behavior occurs when status is signaled through publicly observed actions and individuals' concern about social status is sufficiently high [7]. In a hypothetical scenario, each user is supposed to reciprocate his individual reaction on a certain post. The reaction is supposed to be a representation of his

**Fig. 2** Facebook Like with emotions

original sentiment independent of any constraints. On the contrary, a user's choice of reaction in the aura of the 'Reactions' may be less of his individual reaction and coincide more with the average opinion of his peer group. He may eschew from making a reaction that is in contradiction with the overall reaction of the group of individuals with whom he is in a close proximity or related in a manner which may subject him to any form of intimidation. 'Reactions' tends to be less of a statement of an individual's perception and more of the average of the reactions of individual clusters of the society. Hence, the feature of 'Reactions' has an inherent drawback from psychological perspective of a user and his social background.

## 5 Tomorrow of Like

Several studies have shed light on the application of analyzing online behavior for user profiling [8]. Mining of online social behavior data for efficient prediction of real-life conditions like depression [9] and stress [10] has witnessed identification of common behavior patterns and astonishing revelation.

Considering the dominance of common patterns in online social interaction, an intelligent engine could be proposed as the future of Facebook content-'liking' or 'reacting'. The function of this intelligent engine would be to pose intelligent alerts or recommendations on detection of semantic anomalies in content-'liking' or 'reacting'. The engine would monitor the trending topics and analyze the contents of the post being shared. It may then choose to suggest intelligent alerts/recommendations selectively when users under ignorance or negligence happen to 'like' or 'react' on a sensitive issue in a manner detrimental to global peace and harmony. The engine shall analyze the post at the time it is being posted. An additional flag corresponding to the post can be utilized that shall indicate whether to provide alert to the users reacting to the post. This intelligent alert system shall provoke the users to reconsider their expression of emotions on the social media portal. However, a general dilemma can pose as an ineludible limitation in determining whether the recommendation provided is in contradiction to the general opinion or inflicts any offence to a particular sect or belief.

Another enhancement may involve assigning weights to different Reactions. Different Reactions can be weighted differently by News Feed to do a better job of showing everyone the stories they most want to see. This enhancement is driven by the fact that a user may wish to glance at the stories or posts with which he associates positive reactions. In light of the fact that more emphasis is being laid on paid social media marketing [11], which is similar to interruption-based marketing, the aforementioned enhancement is the need of the hour. The enhancement aims to reinstate the elements of a permission-based marketing which is the key feature of an organic social media.

## 6 Privacy of Like

Various studies have confirmed that peer pressure plays a crucial role in altering the decisions of a particular individual or a group [12]. Peer pressure, in the context of social network, refers to the kind of influence a peer group, an individual or certain observers exert on others that manipulates them to alter their opinions, reactions or social behavior in order to comply or conform to those of the dominant individual or peer group. The groups that are affected over the social network include membership groups in which the individuals are actual members such as in case of trade unions, political parties etc. as well as cliques where there is absence of any clear definition of the membership of the individuals. However, it is not mandatory for an individual to seek membership or pose as a member of a group in order to be

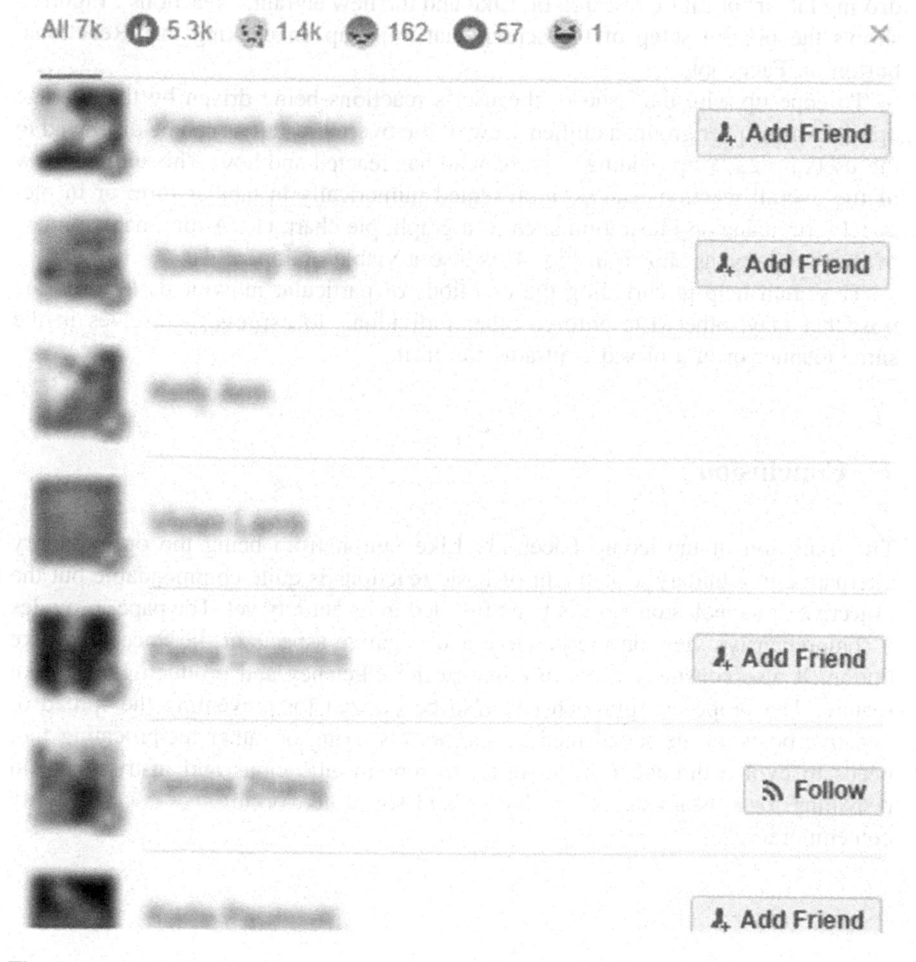


**Fig. 3** Facebook 'Reactions' screen

All 6.1k 5.5k 420 141 44 3 1

**Fig. 4** Minimalistic version of Facebook 'Reactions' screen

influenced under peer pressure. There are also dissociative groups with which one chooses to avoid any linkage or association and, thus, maintains a behavior which is in contradiction to that of the group(s).

It is only natural to assume that in such cases it is better to conceal the individuality of those 'liking' or 'reacting' on the social media platform. However, there is a very fine line of demarcation between privacy and anonymity. The concealment has to be in such a measure that does not violate or intrigue the key characteristic of social network: sharing. Therefore, privacy of its users and their contributions is a major concern for social network that is supposed to pose as a driving factor for future research on Like and the new entrant, 'Reactions'. Figure 3 shows the present setup of the screen that pops up on clicking the 'Reactions' button on Facebook.

To cope up with the issue of the user's reactions being driven by the average opinion of the peer group, a unified view of the overall reactions can be displayed to the users instead of providing a list of who has reacted and how. This unified view of the overall reactions can be represented numerically in tabular form or in pictorial form using an illustration such as a graph, pie chart, etc. A minimalistic view of the reactions, as shown in Fig. 4, is also a viable option.

This shall help in curtailing the emotions of particular individuals towards the post that may, otherwise, enforce other individuals to express themselves in the same manner or in a biased contradiction to it.

## 7 Conclusion

The transition of the iconic Facebook Like button from being the only solitary alternative to a unitary assortment of basic reactions is quite commendable but the objective of its inclusion has not been fulfilled in its entirety yet. The paper provides a comprehensive view on the positive and negative aspects of the Facebook Like button. It also considers ways to enhance the efficiency and productivity of Like feature. The proposed approach can also be utilized for preventing the spread of negative posts on the social media. Facebook's liking or rather reciprocating tool needs to evolve through time in order to remain efficacious and instrumental in providing for a means of online sharing and social networking that is consistently contemporary.

## References

1. Protalinski, E.: Facebook passes 1.44B monthly active users and 1.25B mobile users; 65% are now daily users (2015). http://venturebeat.com/2015/04/22/facebook-passes-1-44b-monthly-active-users-1-25b-mobile-users-and-936-million-daily-users/
2. Facebook. https://www.facebook.com/help/933093216805622
3. Brandtzæg, P.B.: Social networking sites: their users and social implications—a longitudinal study. J. Comput. Mediat. Commun. **17**(4), 467–488 (2012)
4. Ross, C., Orr, E.S., Sisic, M., Arseneault, J.M., Simmering, M.G., Orr, R.R.: Personality and motivations associated with facebook use. Comput. Hum. Behav. **25**(2), 578–586 (2009)
5. Bargh, J.A., McKenna, K.Y.: the internet and social life. Annu. Rev. Psychol. **55**, 573–590 (2004)
6. Hamburger, E.: The area of Facebook is an anomaly (2014). http://www.theverge.com/2014/3/13/5488558/danah-boyd-interview-the-era-of-facebook-is-an-anomaly
7. Bernheim, B.D.: A theory of conformity. J. Polit. Econ. **102**(5), 841–877 (1994)
8. Devineni, P., Koutra, D., Faloutsos, M., Faloutsos, C.: If walls could talk: patterns and anomalies in Facebook wallposts. In: Proceedings of the 2015 IEEE/ACM International Conference on Advances in Social Networks Analysis and Mining 2015 (ASONAM '15), pp. 367–374, New York (2015)
9. Choudhury, M.D., Counts, S., Horvitz, E.J., Hoff, A.: Characterizing and predicting postpartum depression from shared facebook data. In: Proceedings of the 17th ACM Conference on Computer Supported Cooperative Work and Social Computing (CSCW '14), pp. 626–638, New York (2014)
10. Wang, R., Chen, F., Chen, Z., Li, T., Harari, G., Tignor, S., Zhou, X., Ben-Zeev, D., Campbell, A.T.: StudentLife: assessing mental health, academic performance and behavioral trends of college students using smartphones. In: Proceedings of the 2014 ACM International Joint Conference on Pervasive and Ubiquitous Computing (UbiComp '14), pp. 3–14, New York (2014)
11. Neher, K.: The future of social media: paid vs. organic (2014). https://www.clickz.com/the-future-of-social-media-paid-vs-organic/29078/
12. Durkin, K.: Peer Pressure. The Blackwell Encyclopedia of Social Psychology (1996). http://www.blackwellreference.com/public/tocnode?id=g9780631202899_chunk_g978063120289918_ss1-4

# Criminal Policing Using Rossmo's Equation by Applying Local Crime Sentiment

**Fuzail Ahmad, Simran Syal and Mandeep Singh Tinna**

**Abstract** The paper explains discusses criminal policing by applying Rossmo's equation and local crime sentiment approach is proposed where nine types of crimes are discussed which is put into four categories. Each category weight is calculated from the criminal database based on his/her crimes. Rapid miner tool is used to generate graph using results generated through Rossmo's equation and local crime sentiment approach. The resultant graph is then analyzed to predict the most probable criminal. Gurgaon proclaimed offender case is used for case study. The experimental proves that the approach presented in the paper will give accurate results.

## 1 Introduction

In this era of information technology boom, the science can be very effectively put to use in criminal policing. Investigative police work is mostly about the recovery, analysis of information from previous crime history [1]. These days a lot of predictive tools are used in various fields and police has already started using this branch of study in their investigative cases [2]. IT sector proves to be an important field of criminal investigations because they facilitate creation, storage, retrieval, transfer, and application of investigation-related information [3]. With the advancement of technology, the criminal records are also centrally managed and the record can be opened from anywhere by any investigating department. This data is analyzed through proper equation for fast and better policing. Criminal policing can be done

F. Ahmad
Oracle India Private Limited, Hyderabad, India
e-mail: fuzail.a.ahmad@oracle.com

S. Syal (✉)
CGC Group of Colleges, Mohali, Punjab, India
e-mail: simusyal@gmail.com

M.S. Tinna
Chandigarh University, Mohali, Punjab, India
e-mail: mandeep.tinna@gmail.com



S.C. Satapathy et al. (eds.), *Data Engineering and Intelligent Computing*,
Advances in Intelligent Systems and Computing 542,
DOI 10.1007/978-981-10-3223-3_60

by learning the pattern in which a crime can be committed by criminal [4]. Criminal policing is inhibiting crime activity by actively resolving ongoing crime committed in an area and try to prevent any further crime to take place by applying real-time analysis of situation of area. Rossmo's equation [5] was developed and patented by criminologist Kim Rossmo and it widely used in various researches and software tools to do geographic profiling based on intensity of crimes and various other crime related studies. The mathematical formula of Rossmo's equation is:

$$p_{i,j} = k \sum_{n-1}^{total\,crime} \left[ \frac{\phi_{ij}}{(|X_i - x_n| + |Y_j - y_n|)^f} + \frac{(1-\phi_{ij})(B^{g-f})}{(2B - |X_i - x_n| - |Y_j - y_n|)^g} \right] \tag{1}$$

where,

$P_{i,j}$ denotes Probability (location or criminal depends on use-case).
$X_i$, $Y_j$ denotes the mean location of crime by criminal.
$x_n$, $y_n$ denotes the actual crime location.
$f$, $g$ denotes Constants determined through experimental results.

Buffer Zone (B): It is area for which probability is to be generated. Particularly, it is the area kept for monitoring. More discrete is the Buffer zone, more the accuracy of probability generated by equation.

$$\phi_{ij} = \begin{cases} 1 & if\,(|X_i - x_n| + |Y_i - y_n|) > B \\ 0 & otherwise \end{cases} \tag{2}$$

The first term deals with *decreasing probability with increasing distance*. The second term deals with the concept of a *buffer zone*. The analysis of formula is that the probability of crimes first increases as one moves through the buffer zone away from the *hot zone*, but decreases afterwards.

The proposed local crime sentiment approach using Rossmo's equation explained in this paper uses large datasets, however due to space constraint small data sets are used. There has been study which proves that based on previous crime trends of a region, we can predict future crime trend and hence we have incorporated this theory in our approach and verified through both abstract data and experimental data.

## 2 Proposed Methodology

Local databases will be setup in the defined regions for which the criminal policing has to be done. Normally, it may contain 9–10 police stations depending on the regions covered by the police stations.

Criminal data is taken from crime repository and Rossmo's equation is then used to calculate the probability.

The data is segregated with criminal's name and father's name. Each criminal can be assigned an auto-incremental identification number also.

Mean sector (area) is being calculated using below equation.

$$Mean\ sector(x) = \sum (Criminal\ past\ crime\ sectors)/Total\ number\ of\ Crimes \quad (3)$$

Now, the Rossmo's equation is applied on the data to calculate probability of crime for each criminal.

## 2.1 Probability Calculation

$$D = \sum (|Actual\ sector\ where\ crime\ is\ committed - Mean\ sector\ of\ particular\ criminal|) \quad (4)$$

where sector is calculated in *x,y* co-ordinates and D is mean distance of finding the criminal.

$$B = |Analysis\ Sector\ which\ is\ farthest\ among\ all - Mean\ sector\ of\ particular\ criminal| \quad (5)$$

where B is the Buffer Zone area in *x,y* co-ordinates.

$$\text{Probabilty of each Criminal} = \begin{cases} \frac{1}{D^f} & \text{if } D > B \\ \frac{B^g}{2*(B-D)^{(f-g)}} & \text{if } D \le B \end{cases} \quad (6)$$

where, f = 0.8 and g = 0.5 are constants and taken from experimental results calculated in [6].

## 2.2 Sentiment Approach

Future crime trend can be predicted using data of criminals over a period of time (say, 4 years record). In this paper, nine types of crimes are studied which are defined into 4 categories based on its severity (however the categorization can be changed which is not a part of study in the paper):

Category 1: Murder, Rape
Category 2: Assault, Kidnapping
Category 3: Burglary, Theft, Extortion
Category 4: Fraud, Bribery

Weight calculation equation,

$$Individual\ Category\ weight = \frac{Number\ of\ crimes\ committed\ in\ one\ category}{Total\ number\ of\ crimes\ committed\ in\ an\ year} \quad (7)$$

Mean value of individual category weight is used to predict the crime trend of current year.

Example:

$$Category1\ weight = \frac{Category1}{Category1 + Category2 + Category3 + Category4}$$

$$Category1\ weight\ (2016) = \frac{Category1\ weight\ (2013) + Category1\ weight\ (2014) + Category1\ weight\ (2015)}{3}$$

## 2.3 Assumptions

1. The distance taken in the research is taken to be absolute in miles with respect to city Sheffield to ease the calculation since probability calculated is relative to other location and criminal.
2. f and g constants' value used in Rossmo's equation was taken from the experimental results of "A Mathematical Modeling Approach for Geographical Profiling and Crime Prediction" research paper's Yorkshire ripper result.
3. Only 9 types of crimes are analyzed for weight based approach proposed to ease the calculation and database data.
4. Crime clusters were made based on types of crime and clustered into four types for calculation ease. Furthermore, it can be divided into more parts if needed. As the base of research was to propose a new methodology to improve policing, hence this assumption can be made to ease the approach.

## 2.4 Existing Research

Predpol (full-fledged software made based on research and prototype by MIT scholars). The system produces red boxes, 500 feet on each side, which suggests or predicts where property crime, thefts are more likely to happen. It collects data over a period of time and using Rossmo's equation presents the red boxes which are high probable areas of such crimes (Figs. 1 and 2).

In [7] paper, data mining approach is followed to determine hotspots by analyzing the previous records of crimes.

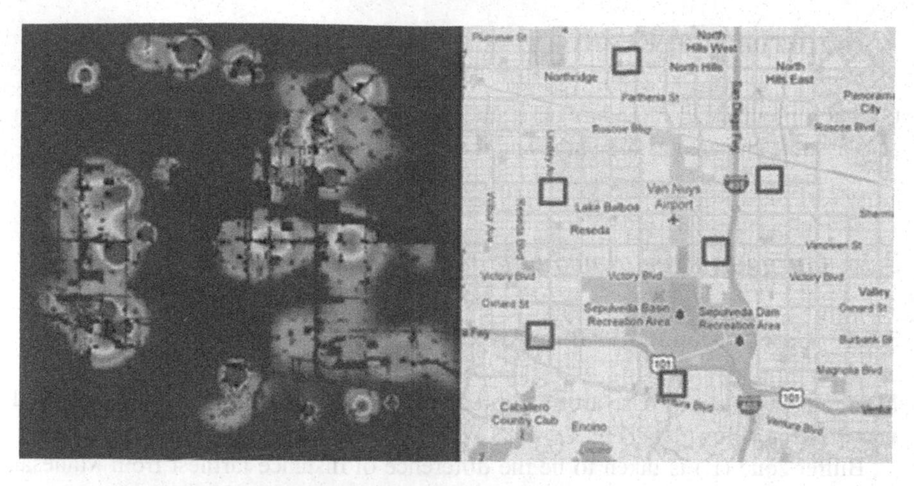


**Fig. 1** Existing Predictive analysis based probable crime prone areas [6]

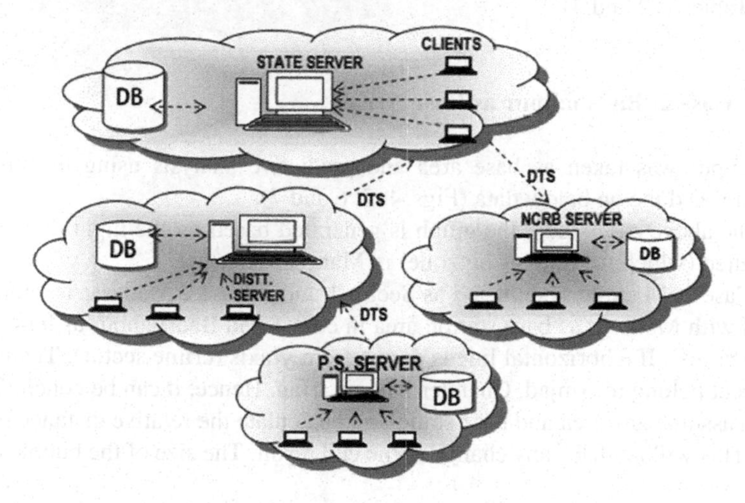


**Fig. 2** Architecture of crime criminal information system [8]

The police can get real time analysis of any region of the possible criminals (usually history sheeters having previous crime records). To back our research's feasibility, it is known that currently NCRB has over 100000000 criminal's data in its database (numbers published on NCRB website) [8].

# 3 Experimental Result

Gurgaon proclaimed offenders case study proves the convict of crime and we can also assume any area as base area and it won't affect the analytics result.

## 3.1 *Gurgaon Proclaimed Offender Case Study*

### 3.1.1 Case 1: Manesar as Base Area

Manesar was taken as base area and predictive analysis using the proposed method was done on below data (Fig. 3).

Buffer zone (B) is taken to be the difference of distance farthest from Manesar and mean value of distance for a particular criminal. The average constitutes buffer zone (Tables 1, 2 and 3).

### 3.1.2 Case 2: Badshahpur as Base Area

Badshahpur was taken as base area and predictive analysis using the proposed method was done on below data (Figs. 4, 5, 6 and 7).

In the above case study, the graph is generated based on the input that crime is committed (which we refer as hotzone) in Manesar Area.

In Case 1, Manesar is depicted as Sector 1 and in case 2 Manesar is depicted as Area 8 with Manesar as base station area in case 1 and Badshahpur as base station area in case 2. If a horizontal line is drawn from y-axis (crime sector). The bubbles which cut belong to Amjad, Chander Pal and Irfan. Hence, it can be concluded that we can assume any area and base station and calculate the relative distance of other areas. This will not have any change in the end result. The size of the bubble depicts

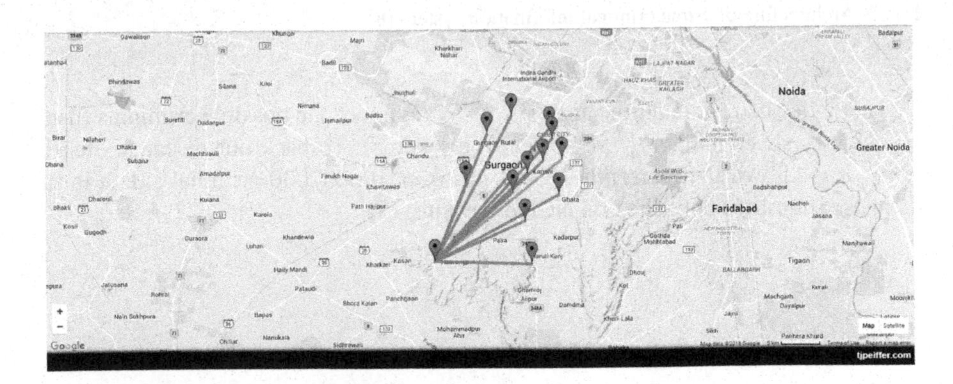


**Fig. 3** Map distance from base area Manesar (*Source* http://www.tjpeiffer.com)

**Table 1** Input data from internet sources (*Source* http://www.gurgaon.haryanapolice.gov.in)

| Criminal | Crime type | Crime area |
|---|---|---|
| Amjad | Theft, assault | PS manesar |
| Anil | Kidnapping | PS DLF PH-1 |
| Anil | Smuggling | PS DLF PH-2 |
| Anup | Assault | PS Badshahpur |
| Anvar | Theft, burglary | PS Sadar |
| Bisambar | Extortion | PS Sushant Lok |
| Chander Pal | Kidnapping, extortion, fraud | PS Manesar |
| Deena | Smuggling | PS Sushant Lok |
| Dharambir | Fraud, theft | PS Palam Vihar |
| Firey Gujjar | Murder, assault | PS Sushant Lok |
| Irafan | Burgalry, theft | PS Manesar |
| Mahesh | Fraud, murder | PS Sec-40 |
| Manoj | Burglary, theft | PS Sec-40 |
| Manoj Kumar | Murder | PS Sadar |
| Mota | Theft | PS Rajendra Park |
| Nution | Fraud | PS SEC-56 |
| Pardeep | Theft | PS SEC-56 |
| Parkash | Murder | PS Sadar |
| Pintu Mandal | Theft | PS DLF PH-2 |

**Table 2** Areas with distance in miles from Manesar (Manesar is taken with base value 1 miles)

| City | Distance (Miles from Manesar) |
|---|---|
| Manesar | 1 |
| Bhondsi | 7 |
| Sushant Lok | 11 |
| Udyog Vihar | 13 |
| Palam Vihar | 12 |
| Badshahpur | 8 |
| Sadar | 8 |
| Pataudi | 6 |
| Rajendra Park | 10 |
| Sec-40 | 10 |
| Sec-56 | 11 |
| DLF PH-2 | 13 |
| DLF PH-1 | 12 |

**Table 3** Areas with distance in miles from Badshahpur (Badshahpur is taken with base value 1 miles)

| City | Distance (Miles from Badshahpur) |
|---|---|
| Manesar | 8 |
| Bhondsi | 3 |
| Sushant Lok | 5 |
| Udyog Vihar | 7 |
| Palam Vihar | 7 |
| Badshahpur | 1 |
| Sadar | 2 |
| Pataudi | 17 |
| Rajendra Park | 7 |
| Sec-40 | 4 |
| Sec-56 | 3 |
| DLF PH-2 | 7 |
| DLF PH-1 | 5 |

| Name of Criminal | Mean Crime Sector | Probability Value by Rossmo's equation | Mean Cluster Weight |
|---|---|---|---|
| Amjad | 1 | 0.812491 | 0.666666 |
| Anil | 12 | 0.420235 | 0.208333 |
| Anil | 13 | 0.411973 | 0 |
| Anup | 8 | 0.47227 | 0.208333 |
| Anvar | 8 | 0.47227 | 0.458333 |
| Bisambar | 11 | 0.429891 | 0.458333 |
| Chander Pal | 1 | 0.812491 | 0.833333 |
| Deena | 11 | 0.429891 | 0 |
| Dharambir | 12 | 0.420235 | 0.625 |
| Firey Gujjar | 11 | 0.429891 | 0.375 |
| Irfan | 1 | 0.812491 | 0.458333 |
| Mahesh | 10 | 0.441347 | 0.333334 |
| Manoj | 10 | 0.441347 | 0.458333 |
| Manoj Kumar | 8 | 0.47227 | 0.166667 |
| Mota | 10 | 0.441347 | 0.458333 |
| Nution | 11 | 0.429891 | 0.166667 |
| Pardeep | 11 | 0.429891 | 0.458333 |
| Parkash | 8 | 0.47227 | 0.166667 |
| Pintu Mandal | 13 | 0.411973 | 0.458333 |

**Fig. 4** Data result after applying input from raw data collected from http://www.gurgaon.haryanapolice.gov.in/proclaimed-offenders.html

the real time crime trend (which we refer as crime weight in the graph) calculated using total crimes committed by all the criminals in the Buffer zone. Hence, the probability order of most probable criminals can be sorted out and prioritized. In reality all these criminals have committed the crime in those areas and the trend also matches with the data.

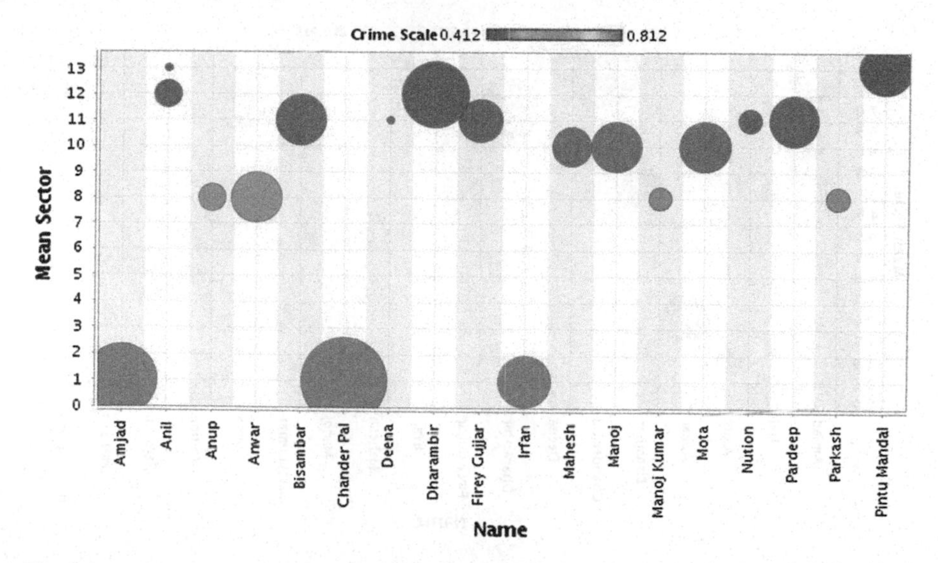


**Fig. 5** Trend plot for data result using rapid miner tool

| Name of Criminal | Mean Crime Sector | Probability Value by Rossmo's equation | Mean Cluster Weight |
|---|---|---|---|
| Amjad | 8 | 0.915101 | 0.666666 |
| Anil | 5 | 0.620884 | 0.208333 |
| Anil | 7 | 0.915101 | 0 |
| Anup | 1 | 0.47227 | 0.208333 |
| Anvar | 2 | 0.493967 | 0.458333 |
| Bisambar | 5 | 0.915101 | 0.458333 |
| Chander Pal | 8 | 0.915101 | 0.833333 |
| Deena | 5 | 0.620884 | 0 |
| Dharambir | 7 | 0.915101 | 0.625 |
| Firey Gujjar | 5 | 0.620884 | 0.375 |
| Irfan | 8 | 0.915101 | 0.458333 |
| Mahesh | 4 | 0.562099 | 0.333334 |
| Manoj | 4 | 0.562099 | 0.458333 |
| Manoj Kumar | 2 | 0.493967 | 0.166667 |
| Mota | 7 | 0.915101 | 0.458333 |
| Nution | 3 | 0.522545 | 0.166667 |
| Pardeep | 3 | 0.915101 | 0.458333 |
| Parkash | 2 | 0.493967 | 0.166667 |
| Pintu Mandal | 7 | 0.915101 | 0.458333 |

**Fig. 6** Data result after applying input from raw data collected from http://www.gurgaon.haryanapolice.gov.in/proclaimed-offenders.html

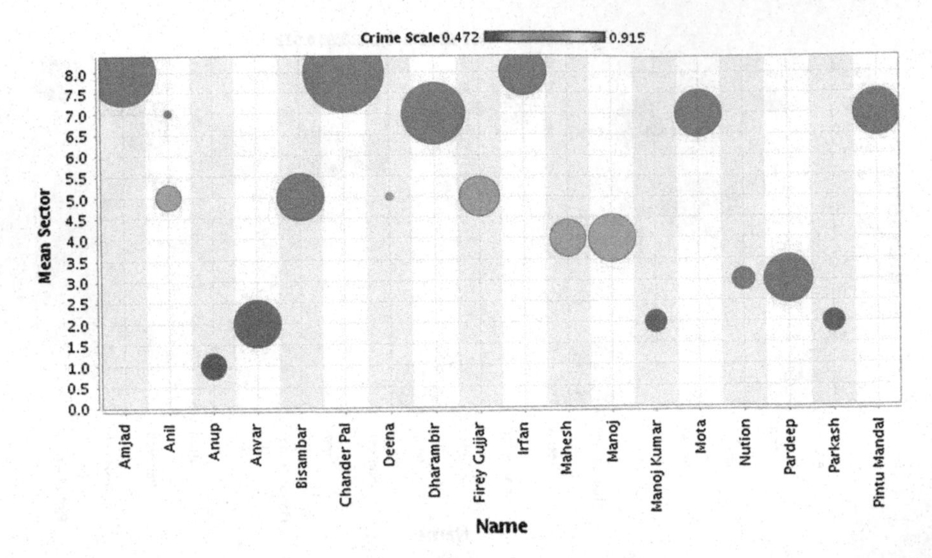


**Fig. 7** Trend plot for data result using rapid miner tool

## 4 Conclusion

This paper, tries to solve the big challenge of fighting crime wherein distributed server model will be used in regions and each cluster set of servers constitute a Buffer area.

However, the results will not be always accurate in predicting the actual criminal but it certainly helps in narrowing down the most predictive criminals.

The proposed approach proves that crimes can be reduced by effectively learning the behavioral pattern of previous crimes or criminals. Moreover, further studies can be done on the proposed approach and it can be made more accurate by working on the type of crime to be categorized together which falls in the study of a professional criminologist. The idea of this paper was to prove the weight based approach used with Rossmo's equation which came out to be accurate in the case study of Gurgaon Proclaimed offenders case.

## References

1. Osterburg, J., Ward, R., Miller, L.: Criminal Investigation: A Method for Reconstructing the Past. LexisNexis (2007)
2. Luen, T., Al-Hawamdeh, S.: Knowledge management in the public sector: principles and practices in police work. J. Inf. Sci. **27**, 311–318 (2001)
3. Dean, G., Gottschalk, P.: Knowledge management in policing and law enforcement: foundations, structures, applications. Oxford University Press, Oxford (2007) For more information about this book please refer to the publisher's website (see link) or contact the author

4. Malathi, A., Baboo, S.S.: Algorithmic crime prediction model based on the analysis of crime clusters. In: GJCST, pp. 47–51 (2011)
5. Rossmo, D.K.: Place, space and police investigations: hunting serial violent criminals (1995). https://www.ncjrs.gov/App/Publications/abstract.aspx?ID=160740
6. Zheng, X., Cao, Y., Ma, Z.: A mathematical modeling approach for geographical profiling and crime prediction. In: 2011 IEEE 2nd International Conference on Software Engineering and Service Science, pp. 500–503 (2011)
7. Yu, C.H., Ward, M.W., Morabito, M., Ding, W.: Crime forecasting using data mining techniques. In: Proceedings of the 2011 IEEE 11th International Conference on Data Mining Workshops. ICDMW'11, Washington, DC, USA, pp. 779–786. IEEE Computer Society (2011)
8. Gupta, M., Chandra, B., Gupta, M.P.: Crime data mining for Indian police information system. Commun. ACM **46**, 28–34 (2011)

# Moving Object Tracking and Detection Based on Kalman Filter and Saliency Mapping

**Priyanka Prasad and Ashutosh Gupta**

**Abstract** There are many applications like video surveillance, object detection and tracking which require the processing of video to extract the desired result out of it. In this paper, we use saliency mapping to extract the interested regions of a video after its successful detection and tracking using Kalman filter. The saliency mapping uses the concept of temporal saliency mapping and spatial saliency mapping to distinguish between the various regions of a video. The high motion region is detected with the help of temporal mapping, while region consisting of regular movement is identified by spatial mapping. The effective saliency map is created using the combination of both spatial and temporal saliency mapping of the salient object. The experimental results obtained using public dataset, shows that our method performed well in detection, tracking and saliency mapping of the object.

**Keywords** Spatial mapping · Temporal mapping · Kalman filter · Motion vectors · Spatial temporal saliency mapping

## 1 Introduction

Human beings have an inherited ability to visualise and understand the interested or salient regions of a scene. The scene can either be a still picture or a video produced using a camera. Since in the digital world, the result from any system is expected to

P. Prasad (✉) · A. Gupta
Department of Electronics and Communication Engineering,
Ambedkar Institute of Advanced Communication Technologies and Research,
Geeta Colony, New Delhi, India
e-mail: manavpriyanka1992@gmail.com

A. Gupta
e-mail: ashutosh14139@iiitd.ac.in

P. Prasad · A. Gupta
Department of Electronics and Communication Engineering, IIIT-Delhi Okhla,
New Delhi, India



S.C. Satapathy et al. (eds.), *Data Engineering and Intelligent Computing*,
Advances in Intelligent Systems and Computing 542,
DOI 10.1007/978-981-10-3223-3_61

be flawless and close to the perfection, therefore various researches are ongoing to develop enhanced computational model that could extract useful features of salient region in a video. Video processing is a method that uses various video filters to produce video output when the input itself is video signal. When video processing is accompanied by saliency mapping, the results become better. Saliency mapping is a process in which a region is mapped, exploiting the salient features. This method typically based on heat equation that shows the intensity of the core region of the frame extracted from the video.

A better review of saliency based models for segmentation is explained by Li et al. [1]. However such feature information obtained using saliency model has restricted its scope in case of lighting variations and background complexity. The use of spatial and temporal coherency for salient object detection is given by Mahapatra et al. [2]. This concept is well used for saliency detection in the segmented video. Borji et al. in [3] has explained visual attention modelling to detect visual salient regions using high level information. In [4], Guo et al. incorporate a spatiotemporal saliency detection model with multiresolution. Though this approach has some limitation for real time video but it effectively improve the coding efficiency for image and video compression. A computational model for object detection integrating the low level visual features with image saliency to obtain the salient region is given in [5]. The selection of saliency features and generation of complex features for object detection is precisely explained by Gao et al. in [6]. In [7], the fusion of two constructed spatial and temporal coherency maps is used to create a spatio-temporal attention model.

Chen et al. in [8] propose a method for video adaptation based on content reposition. It has successfully provided small size video emphasizing on only the important aspects of the scene keeping background context unaltered. The framework to detect the interested region in both images and video is proposed by Kim et al. in [9] where the problem of minimizing the false positives is reduced to a considerable level. It uses the precise differences between center regions and surround regions to extract the temporal information. Mahadevan et al. [10] incorporate some computation to obtain motion saliency map. It makes the spatiotemporal analysis of dynamic scenes of video easier by using centre surround principle. Chen et al. in [11] propose dynamic visual saliency modelling. It is based on spatiotemporal analysis and uses time domain Harris corner detector. In [12], another method of saliency called as context aware saliency method is proposed for effective detection. It only detects the dominant object of a scene, after emphasizing on both high and low level features. Li et al. in [13] detect the saliency by reconstructing the dense and sparse features of a video.

The relatedness of objectness and saliency is shown by Chang et al. in [14]. It effectively minimize the energy by simultaneously improving the quality of saliency and objectness estimations. The histogram based contrast and contrast based on spatial information of enhanced region are two approaches in [15] used to generate high quality saliency maps for object detection. Jiang et al. in [16] uses a regression based approach to compute the saliency score for constructing saliency maps. An efficient way of detecting the salient region can be done through soft image abstraction as described by Cheng et al. in [17]. In [18], a search mechanism based on saliency is incorporated. This method is for covert and overt shifts of visual mechanism. Kim et al. in [19] finds an approach to retargeting of video sequences using Fourier analysis. It uses adaptive scaling and partitioning concept to get the reliable performance. Nie et al. in [20] suggests the use of Adaptive Rood Pattern Search method for computation and estimation of motion vectors. A saliency model is proposed by Itti et al. in [21] with feed forward feature extraction mechanism. It gives a strong performance in rapid scene analysis like detection of salient traffic signs of different shapes. We explored the importance of Kalman filter and saliency mapping for detection and tracking of a salient object in a video. Our method extracts low and high level features of video and gives more accurate saliency maps reducing the probability of false positives. The paper is organised in different sections. Section 2 presents our proposed method for salient object detection, tracking and saliency mapping. Section 3 shows the obtained experimental results using public dataset SegTrack. Section 4 presents the conclusion of this paper.

## 2 Proposed Work

The detection and tracking of moving object is the foremost step in the video processing. In our process, we used Kalman filter for this purpose. Kalman filter performs this operation in two steps where first is the detection of moving salient objects and second is tracking of desired object from frame to frame.

Figure 1 depicts the sequential stages of our method. The coherency maps are obtained using various number of frames of a video.

Kalman filter uses various mathematical equations to estimate the current state by minimising the mean squared error. It employs an estimator consisting of predictor and corrector. In the prediction stage, time update is considered where the anticipation of the next future value is done and a priori estimate is obtained after analysing the current estimated value. While in correction stage, the measurement update is taken and the correction of the current estimated state is calculated which minimises the estimated error covariance and produce the result close to the original input. In our system, after the averaging of previous and current frames, the

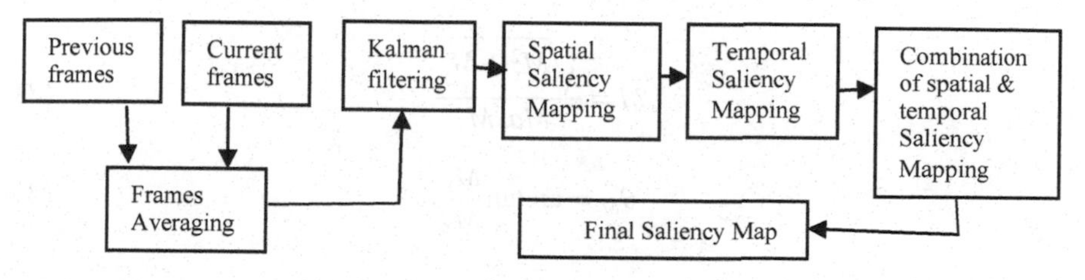


**Fig. 1** Block diagram of various steps of our proposed method

averaged frames are passed through Kalman filter. The Kalman filter keeps on estimating the future value from the frames resulting in detection and tracking of salient object. This cycle continues for all the frames taken.

The next step of our method is spatial mapping of salient object. It is done for identification of regions belonging to a regular object in a video. Spatial coherency requires the colour information and entropy of the salient of the object. Though colour information might increase the computation if analysed deeply but we only extract the intensities values from it. The variation in the intensities of colour in the frames create patches of different entropies deciding the spatial features such as coherency for spatial mapping. Higher the variation in intensities, more is the information contained in the patch and better will be the coherency. The coherency of spatial mapping can lead to qualitative result and its equation is obtained from [2], the spatial coherency C for a patch x is given as

$$C(x) = -\sum_{\theta} p(\theta)\log p(\theta) \tag{1}$$

where p(Θ) is the probability of gradient angles Θ. The spatial saliency mapping is done in three stages. First is extraction of multiple low level visual features like intensities, orientation and colour. Second is activation of these multiple low level features maps to construct the activation maps. Finally is the normalisation of the normalised combination of activation maps produces the spatial saliency map.

After the spatial mapping of the desired region, our next step is to obtain the temporal coherency map to identify the region with highly coherent motion unlike the irregular motion. Centre surround maps [21] and entropy of the object are the basic parameters for temporal mapping. To distinguish the pattern of motion of salient object from the random motion, motion vectors or a technique called optical flow can be used. The motion vectors are used to estimate the displacement between the consecutive blocks. But optical flow is not preferred because of its inability to process blocks at a time which is possible through motion vectors. The variation in the displacement of block is effectively evaluated for obtaining the two main components of temporal coherency- motion coherency and direction coherency. The motion coherency is evaluated with the help of motion entropy $M_e$ and motion centre surround map $M_{CSM}$. Similarly, the direction coherency depends on direction entropy $\Theta_e$ and direction centre surround map $\Theta_{CSM}$. The motion magnitude M and direction $\Theta_M$ responsible for motion and direction coherency respectively are derived from [2] as given below

$$M = \frac{\sqrt{M_x^2 + M_y^2}}{MaxM} \tag{2}$$

$$\theta_M = \arctan\frac{M_y}{M_x} \tag{3}$$

MaxM denotes the maximum motion magnitude. $M_y$ and $M_x$ are the components of motion vector of an image. The motion magnitude is stored for each block. The set of maps from motion coherency are obtained for blocks from [2] as $M_e$ and $M_{CSM}$

$$M_e(x) = -\sum_n p_n \log p_n \tag{4}$$

$$M_{CSM} = \frac{1}{Y}\sum_{i=1}^{|N_x|} |M_x - M_i| \tag{5}$$

where $p_n$ is the probability of motion magnitude n and Y is the number of neighbouring patches. Here $N_x$ denotes the neighbours. A similar set of maps from direction coherency is obtained from [2] as $\Theta_e$ and $\Theta_{CSM}$

$$\theta_e(x) = -\sum_{\theta_M} p(\theta_M) \log p(\theta_M) \tag{6}$$

$$\theta_{CSM} = \frac{1}{Y}\sum_{i=1}^{|N_x|} (\theta_x - \theta_i) \tag{7}$$

These quantities are evaluated to map the salient object using temporal coherency. Higher value of $\Theta_e$ indicates incoherent motion which is undesired. This problem can be overcome by combining the spatial and temporal saliency maps. The final step of our method is obtaining the combined map of spatial and temporal mapping. This combination is done to determine the relative contribution of both the mapping in effective detection and tracking of the salient object. It performs the dual function of observing whether the salient region is from a regular object and secondly if there is any high coherent motion of the object different from random motion. We have combined the two techniques for obtaining the final saliency map independent of variation in any control factor. Thus final spatial temporal mapping produces the desired output highlighting the well define salient object being tracked.

## 3 Experimental Results

The performance of our method is evaluated using a public dataset, SegTrack [22]. The SegTrack datasets consists of set of images with length varying from 20–50 frames and image dimensionalities of different sizes. Various parameters such as precision and recall are calculated which are given as

$$\text{Recall} = |G \cap A| / |G|; \quad \text{Precision} = |G \cap A| / |A|$$

Where G is the ground truth salient pixels as denoted by the experts and A is denoting the salient pixels obtained from automated method. $G \cap A$ is the intersection result of these two quantities.

Figure 2 shows the result of various stages of our proposed method on SegTrack parachute dataset. Figure 2a shows the original frame of dataset video. Figure 2b is showing the salient object being detected in the processed frame. The tracking of moving object is shown in Fig. 2c. The detection and tracking are done with the help of kalman filter. Figure 2d shows the masking of object in the frame. The temporal mapping of salient object is shown in Fig. 2e. Figure 2f is showing the final combined spatial temporal mapping. Figure 3 shows spatial temporal mapping of salient object in different frames of same video dataset.

The Table 1 shows various measures obtained for object tracking and detection and saliency mapping from SegTrack dataset. The parameters calculated are Recall and Precision. The recall and Precision values of our method is quite higher than other two methods. The parameter values of PQFT becomes lower due to its limitation of creating false positives while CSM performs better than PQFT but our method gives appreciating result for tracking, detection and saliency mapping of object.

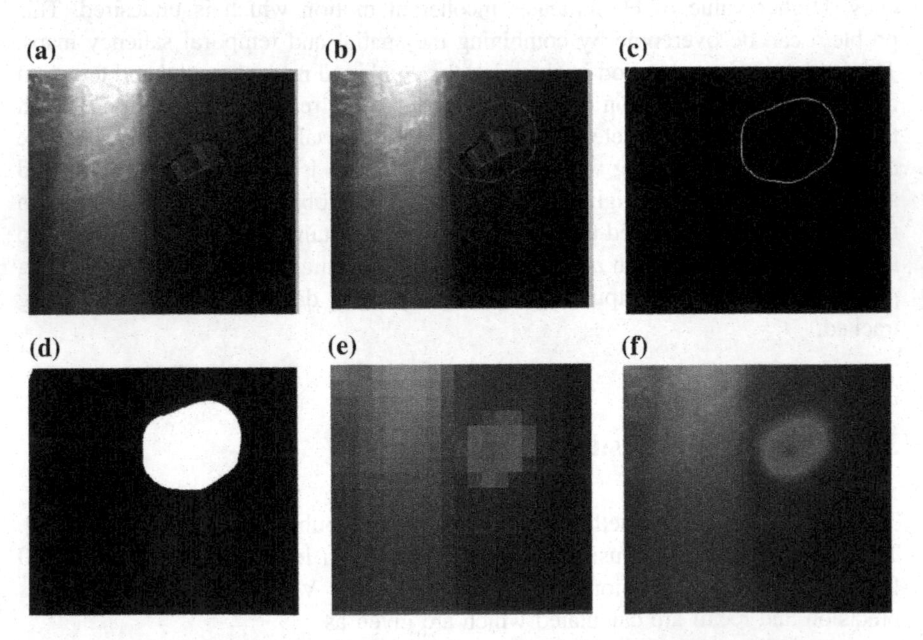


**Fig. 2** Results of various stages obtained from SegTrack (parachute) dataset. **a** Original frame of a video; **b** and **c** Salient object detection and tracking using Kalman filter respectively; **d** masking of salient object; **e** temporal mapping of salient object; **f** Combined spatial temporal mapping of salient object

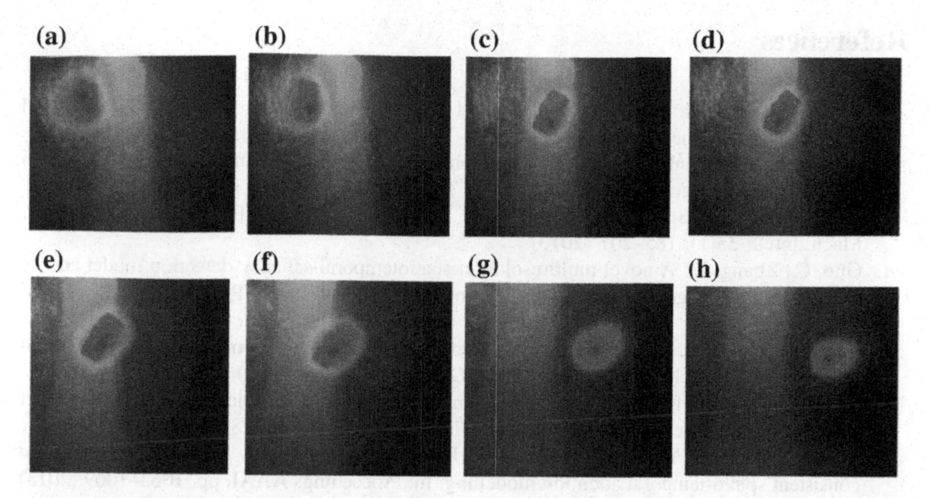


**Fig. 3** Final combination of spatial temporal saliency mapping results on SegTrack parachute dataset; **a**, **b**, **c**, **d**, **e**, **f**, **g** and **h** are showing the moving object saliency mapping using different frames from same video

**Table 1** Quantitative measures of performance for various methods using SegTrack dataset

| Methods used | SegTrack | |
|---|---|---|
| | Recall | Precision |
| Proposed method | 0.81 | 0.95 |
| PQFT [4] | 0.54 | 0.76 |
| CSM [2] | 0.79 | 0.94 |

# 4 Conclusion

We have proposed an efficient method for detection and tracking of salient object in a video using Kalman filter and saliency mapping concept. The saliency mapping is done spatially and temporally. The spatial mapping identifies whether the region belongs to a regular object or not. And the temporal coherency mapping helps in distinguishing the region with small regular motion to the region with regular motion. The combination of spatial and temporal mapping gives final saliency map to map the salient object. Our method is very much successful in detection of moving object, detection of tumors in human body, detection of foreign object in prohibited territory, tracking the path of a flight, video segmentation etc. The limitation of proposed method is its reduced efficiency in case where multiple smaller moving objects are present. The experimental results are obtained using public database namely SegTrack. Our whole process is implemented in MATLAB on 2.20 GHz processor.

## References

1. Li, H., Ngan, K.N.: Saliency model-based face segmentation and tracking in head and shoulder video sequences, J. Vis. Commun. Image Represent. **19**(5), 320–333 (2008)
2. Mahapatra, D., Gilani, S.O., Saini, M.K.: Coherency based spatio-temporal saliency detection for video object segmentation. IEEE J. Sel. Top. Signal Process. **8**(3) (2014)
3. Borji, A., Itti, L.: State-of-the-art in visual attention modeling. IEEE Trans. Pattern Anal. Mach. Intell. **35**(1), 185–207 (2013)
4. Guo, C., Zhang, L.: A novel multiresolution spatiotemporal saliency detection model and its applications in image and video compression. IEEE Trans. Image Process. **19**(1), 185–198 (2010)
5. Olivia, A., Torralba, A., Castelhano, M., Henderson, J.: Top-down control of visual attention in object detection. In: Proceedings IEEE ICIP, pp. 253–256 (2003)
6. Gao, D., Vasconcelos, N.: Integrated learning of saliency, complex features, and object detectors from cluttered scenes. In: Proceedings IEEE CVPR, pp. 282–287 (2005)
7. Zhong, S.-H., Liu, Y., Ren, F., Zhang, J., Ren, T.: Video saliency detection via. dynamic consistent spatio-temporal attention modelling. In: Proceedings AAAI, pp. 1063–1069 (2013)
8. Chen, W.-H., Wang, C.-W., Wu, J.-L.: Video adaptation based for small display based on content recomposition. IEEE Trans. Circuits Syst. Video Technol. **17**(1), 43–58 (2007)
9. Kim, W., Jung, C., Kim, C.: Spatiotemporal saliency detection and its applications in static and dynamic scenes. IEEE Trans. Circuits Syst. Video Technol. **21**(4), 446–456 (2011)
10. Mahadevan, V., Vasconcelos, N.: Spatiotemporal analysis in dynamic scenes. IEEE Trans. Pattern Anal. Mach. Intell. **32**(1), 71–77 (2010)
11. Chen, D.-Y., Tyan, H.-R., Hsiao, D.-Y., Shih, S.-W., Liao, H.-Y.M.: Dynamic visual saliency modeling based on spatio temporal analysis. In: Proceedings IEEE ICME, pp. 1085–1088 (2005)
12. Goferman, S., Zelnik-Manor, L., Tal, A.: Context aware saliency detection. IEEE Trans. Pattern Anal. Mach. Intell. **34**(10), 1915–1926 (2012)
13. Li, X., Lu, H., Zhang, L., Ruan, X., Yang, M.-H.: Saliency detection via dense and sparse reconstruction. In: Proceedings IEEE ICCV, pp. 2976–2983 (2013)
14. Chang, K.-Y., Liu, T.-L., Chen, H.-T., Lai, S.-H.: Fusing generic objectness and visual saliency for salient object detection. In: Proceedings IEEE ICCV, pp. 914–921 (2011)
15. Cheng, M.-M., Zhang, G.-X., Mitra, N.J., Huang, X., Hu, S.-M.: Global contrast based salient region detection. In: Proceedings IEEE CVPR, pp. 409–416 (2011)
16. Jiang, H., Wang, J., Yuan, Z., Wu, Y., Zheng, N., Li, S.: Salient object detection: a discriminative regional feature integration approach. In: Proceedings IEEE CVPR, pp. 2083–2090 (2013)
17. Cheng, M.-M., Warrell, J., Lin, W.-Y., Zheng, S., Vineet, V., Crook, N.: Efficient salient region detection with soft image abstraction. In: Proceedings IEEE ICCV, pp. 1–8 (2013)
18. Itti, L., Koch, C.: A saliency-based search mechanism for overt and covert shifts of visual attention. Vis. Res. **40**, 1489–1506 (2000)
19. Kim, J.-S., Kim, J.-H., Kim, C.-S.: Adaptive image and video retargeting technique using Fourier analysis. In: Proceedings IEEE CVPR, pp. 1730–1737 (2009)
20. Nie, Y., Ma, K.-H.: Adaptive rood pattern search for fast block matching motion estimation. IEEE Trans. Image Process. **11**(12), 1442–1448 (2002)
21. Itti, L., Koch, C., Niebur, E.: A model of saliency-based visual attention for rapid scene analysis. IEEE Trans. Pattern Anal. Mach. Intell. **20**(11), 1254–1259 (1998)
22. Tsai, D., Flagg, M., Rehg, J.M.: Motion coherent tracking with multi-label mrf optimization. Int. J. Comp. Vis. **100**(2), 190–202 (2012)

# XUBA: An Authenticated Encryption Scheme

**R. Neethu, M. Sindhu and Chungath Srinivasan**

**Abstract** In this paper, we propose a stream cipher based authenticated encryption scheme, XUBA capable of achieving standard security. It is a bit based stream cipher with a key and initialization vector of 128 bits. Authenticity is provided by generating a tag of 128 bits irrespective of the input message. The new cipher is resistant to algebraic, differential and time-memory-data trade-off attacks. 128-bits of the tag makes the cipher resistant to forgery attack and guessing of MAC.

**Keywords** Authenticated encryption · Stream cipher · MAC

## 1 Introduction

Authenticated encryption is a mechanism for simultaneously providing confidentiality, authenticity and integrity on the data with a single key [2]. The need for authenticated encryption schemes emerged from the security issues faced by the communication channels in the networks. There exist so many authenticated encryption schemes with different design approaches. But most of them failed due to the lack of security. In this paper, a new stream cipher based authenticated encryption scheme is designed which have the ability to resist cryptographic attacks. Symmetric ciphers provide authenticated encryption using either stream ciphers or block ciphers [2]. High throughput is obtained when we use stream cipher mode of authenticated encryption. In order to design a new cipher, the structure and security analysis of existing ciphers should be analyzed. Some of the existing stream cipher based

R. Neethu · M. Sindhu (✉) · C. Srinivasan
TIFAC-CORE in Cyber Security Amrita School of Engineering,
Amrita Vishwa Vidyapeetham Amrita University, Coimbatore, India
e-mail: m_sindhu@cb.amrita.edu

R. Neethu
e-mail: neethurajankk@gmail.com

C. Srinivasan
e-mail: c_srinivasan@cb.amrita.edu



S.C. Satapathy et al. (eds.), *Data Engineering and Intelligent Computing*,
Advances in Intelligent Systems and Computing 542,
DOI 10.1007/978-981-10-3223-3_62

authenticated encryption schemes like Grain-128a, Snow3G, Trivia-ck, Acorn-128, HS1-SIV and Sablier are analyzed. In 2013, NIST announced a competition for submitting authenticated encryption schemes suitable for widespread adoption [1]. Among the ten stream cipher based proposals, only Trivia-ck and HS1- SIV are currently competing in the CEASAR competition. Other schemes got rejected due to lack of security [1].

This paper comprises of basically four sections including introduction and related works. We describe the design of the proposed cipher in Sect. 2. Security analysis of the cipher is given in Sect. 3 and we conclude this paper with Sect. 4.

## 1.1 Related Works

In this section, some existing stream cipher based authenticated encryption schemes are described. Snow3G [2] is the updated version of Snow 2.0 stream cipher with authentication feature. Snow3G is mainly used in 3GPP technology for wireless communication with a key and *IV* of 128 bits. Encrypt-then-MAC mode is used for this 32 bit word based cipher. Generation of session keys makes the cipher resistant to fault attacks, but later the cipher was failed due to differential resynchronization and multiset collision attacks.

Grain-128a is the successor of Grain-128 stream cipher with more security enhancement and optional authentication [6]. Keystream and tag bits are generated from 128 bit key and 96 bit *IV*. The throughput of Grain-128a can be increased by using some additional hardware which makes it more expensive. The design is resistant to time-memory-data trade-off, side channel, and fault attacks. But the cipher is declined due to a probabilistic algebraic attack which retrieves LFSR bits in the internal state. Moreover, authentication will slow down the encryption process and later this issue is patched in Sablier [12], which is a word based scheme resistant to differential and cube attacks. Sablier is a hardware efficient stream cipher whose encryption speed in hardware is 16 times faster than Trivium. But a full key recovery on Sablier leads to its rejection from CEASAR competition. Acorn-128 v2 [11] and Trivia-ck v2 [4] are round two candidates in CEASAR competition. Even though Acorn is a lightweight cipher which is resistant to linear and differential attacks, its weakness in the state update function leads to internal state collision and results in rejection from CEASAR competition. But in Trivia-ck, the large number of initialization rounds and internal state makes the cipher resistant to cube attack and guess-then-find attack. But complex EHC-hash computation prevents it from lightweight applications. HS1-SIV v2 [7] is another round two candidate in CEASAR competition which uses Chacha20 stream cipher and HS1-Hash to achieve confidentiality and integrity. The use of HS1-Hash in tag computation made it resistant to collision attack. The resistance to related-key attack is achieved using pseudo-random functions in the core of Chacha20 and security improvements made in Salsa to launch Chacha20 cipher [7]. Like Trivia-ck, HS1 also uses complex computations.

Our objective is to construct a simple stream cipher based authenticated encryption which is resistant to cryptographic attacks. In this paper, we propose a new stream cipher based authenticated encryption scheme XUBA, which achieves a high level of security.

## 2 Design

XUBA is a new stream cipher based authenticated encryption scheme with more security enhancements. Authenticated encryption is done using Encrypt-and-MAC approach (E&M). XUBA generates keystream from a 128 bit secret key and 128 bit initialization vector ($IV$). Two non-linear feedback shift registers (NFSR) and a tag buffer are the main components of XUBA. It has an internal state of 280 bits which comprises of two registers of size 136 bits and 144 bits.

The internal state $S = (S_0, S_1, ..., S_{278}, S_{279})$ is represented in two registers as: $R_1 = (S_0, S_1, ..., S_{134}, S_{135})$ and $R_2 = (S_{136}, S_{137}, ..., S_{278}, S_{279})$. The structure and working of XUBA is depicted in the Fig. 1.

### 2.1 *Key-IV Loading*

Initially, both registers will be filled with zeros. Later the 280 bits of internal state of XUBA is loaded with 128 bit key, $K = (K_0, ..., K_{127})$ and initialization vector, $IV = (IV_0, ..., IV_{127})$ in the following way:

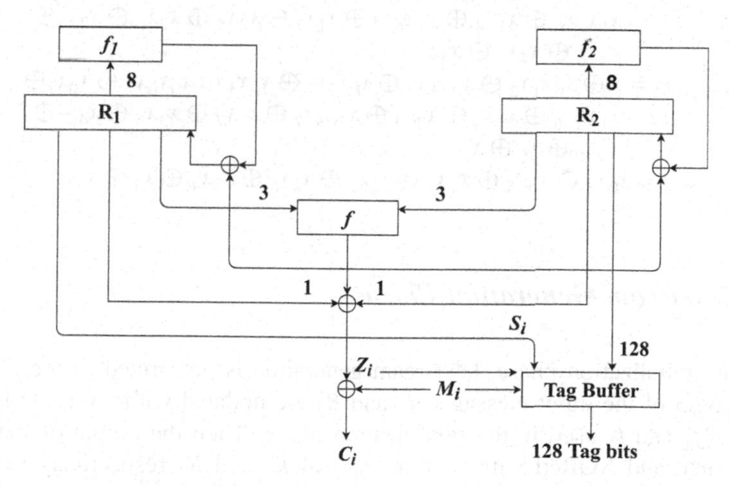


**Fig. 1** Block diagram representation of XUBA

$$
\begin{array}{ll}
S_i \leftarrow K_i & 0 \le i \le 55 \\
S_i \leftarrow IV_{i-56} & 56 \le i \le 132 \\
S_i \leftarrow 1 & 133 \le i \le 135 \\
S_i \leftarrow K_{(i-80)} & 136 \le i \le 207 \\
S_i \leftarrow IV_{i-131} & 208 \le i \le 258 \\
S_i \leftarrow 1 & 259 \le i \le 268 \\
S_i \leftarrow K_{(i-269)} \oplus IV_{(i-269)} & 269 \le i \le 279
\end{array}
$$

## 2.2 *Initialization Phase*

After loading key and *IV* to the internal state, the cipher is clocked 280 times. The purpose of initialization is to mix the key and *IV* completely into the state so that an attacker will not be able to get the secret key directly from the state. During initialization, complete state bits are only updated without taking any output.

Two eight variable non-linear Boolean functions $f_1$ and $f_2$ are used for updating $R_1$ and $R_2$ registers. $f_1$ takes $S_0, S_1, S_9, S_{11}, S_{40}, S_{78}, S_{94}$ and $S_{114}$ as inputs from $R_1$. Similarly, $f_2$ takes $S_{136}, S_{137}, S_{144}, S_{177}, S_{179}, S_{206}, S_{228}$ and $S_{260}$ as inputs from $R_2$. Then output of a six variable non-linear Boolean function $f$ with three inputs from both registers is computed for giving feedback to both registers. The inputs for $f$ are: $S_{15}, S_{60}, S_{90}, S_{188}, S_{208}$ and $S_{276}$. A single bit left shift is applied to all bits except the last one in both registers. Then output of $f$ is XORed with output of $f_1$ and fed back to last bit of $R_1$. Similarly, output of $f$ is XORed with outputs of $f_2$ and fed back to last bit of $R_2$. These functions are chosen with good cryptographic properties. The functions used for initializing the proposed scheme are:

$$
\begin{aligned}
f_1(x_0, \ldots, x_7) = {} & 1 \oplus x_0x_1x_2 \oplus x_0x_1x_7 \oplus x_0x_2x_6 \oplus x_0x_3 \oplus x_0x_6x_7 \oplus x_0x_6 \oplus \\
& x_1x_2x_6 \oplus x_1x_4 \oplus x_1x_6x_7 \oplus x_1x_7 \oplus x_2x_5 \oplus x_2x_6 \oplus x_2x_7 \oplus \\
& x_3x_6 \oplus x_4x_6 \oplus x_5x_7 \\
f_2(x_0, \ldots, x_7) = {} & 1 \oplus x_0x_1x_2 \oplus x_0x_1x_6 \oplus x_0x_2x_7 \oplus x_0x_3 \oplus x_0x_6x_7 \oplus x_0x_7 \oplus \\
& x_1x_2x_6 \oplus x_1x_4 \oplus x_2x_5 \oplus x_2x_6x_7 \oplus x_2x_7 \oplus x_3x_6 \oplus x_4x_7 \oplus \\
& x_5x_6 \oplus x_6 \oplus x_7 \\
f(x_0, \ldots, x_5) = {} & x_0x_1 \oplus x_0x_5 \oplus x_1x_4 \oplus x_2x_3 \oplus x_2x_5 \oplus x_3x_4 \oplus x_4 \oplus x_5.
\end{aligned}
$$

## 2.3 *Keystream Generation Phase*

After the initialization phase, keystream generation is performed $l$ times, where $l$ is the length of the input message. $R_1$ and $R_2$ are updated with the same Boolean functions $f_1$ and $f_2$ used in the initialization phase. Then the output of function $f$ is computed and XORed with $S_{55}$ and $S_{171}$ of $R_1$ and $R_2$ respectively to get the keystream bit, $Z_i$. Like the initialization phase, the output of $f$ function is used for computing the feedback bit. The keystream bit is XORed with plaintext bit $M_i$ to form the corresponding ciphertext $C_i$ and also given as an input to the tag computation.

### 2.4 Tag Computation

XUBA uses a tag buffer of size 128 bits for generating message authentication tag corresponding to the input message. Initially, the buffer is filled with zeros. After initialization phase, the first 128 bits of $R_2$ is loaded to tag buffer.

Tag computation is performed simultaneously with the keystream generation. After generating each keystream bit, the message bit $M_i$, $0 \leq i \leq l-1$ is taken as input along with corresponding keystream bit for tag computation. Here each bit of message and keystream is used to update the whole 128 bits of tag. A four variable Boolean function, $f_3$ is used to update tag buffer in each round of keystream generation. Let $x_0, x_1, x_2$ and $x_3$ are the four variables corresponding to $Z_i, M_i$, tag bit $T_j$ and state bit $S_j$ where $0 \leq i \leq l-1$ and $0 \leq j \leq 127$. The function $f_3(x_0, x_1, x_2, x_3) = x_0x_1 \oplus x_0x_2 \oplus x_1x_3 \oplus x_2x_3 \oplus x_2 \oplus x_3$ is used for updating each bit of the tag.

## 3 Security Analysis

The resistance of XUBA against algebraic, differential and time-memory-data trade off attack along with MAC forgery and guessing of MAC are analyzed in this section. The randomness of the proposed scheme is checked using cryptographic randomness tests.

### 3.1 Differential Attack

Differential cryptanalysis is an attack developed by Biham and Shamir [3]. In this attack, an attacker tries to find the influence of input difference in the corresponding output difference. Some differences in key, initial state or plaintext are used to predict the corresponding differences in keystream or internal state in order to retrieve the secret key.

Here some bits in the internal states are flipped to find the difference propagation through the internal state. Changes were applied to internal state bits $S_2, S_8, S_{79}, S_{95}, S_{138}, S_{191}, S_{205}, S_{259}, S_{275}$ and the number of differences that propagated into the 280-bit internal state for selected number of rounds are analyzed [8] and given in Table 1. The difference is propagated into the whole 280-bit internal state within 243 rounds which is less than 280 initialization rounds. Since the cipher is initialized for 280 rounds, the attack is infeasible in XUBA.

**Table 1** Differential propagation

| Rounds | Number of differences propagated | | | | | | | | | |
|---|---|---|---|---|---|---|---|---|---|---|
| | $S_2$ | $S_8$ | $S_{41}$ | $S_{79}$ | $S_{95}$ | $S_{138}$ | $S_{191}$ | $S_{205}$ | $S_{259}$ | $S_{275}$ |
| 100 | 96 | 84 | 92 | 124 | 122 | 160 | 168 | 140 | 77 | 70 |
| 150 | 194 | 182 | 191 | 223 | 217 | 252 | 264 | 240 | 176 | 169 |
| 200 | 274 | 268 | 263 | 275 | 271 | 280 | 280 | 279 | 258 | 240 |
| 230 | 280 | 280 | 280 | 280 | 280 | 280 | 280 | 280 | 280 | 272 |
| 243 | 280 | 280 | 280 | 280 | 280 | 280 | 280 | 280 | 280 | 280 |

### 3.2 Algebraic Attack

The algebraic attack is an attack in which attacker tries to retrieve the secret key by solving a system of non-linear equations within the polynomial time [10]. It is one of the recent cryptanalytic technique that deals with breaking a system of polynomial equations. Here the 280 bit internal state of XUBA is considered as 280 unknown variables. During each initialization round, three new equations of degree three and two new unknowns are generated. After 32 rounds, $m = 96$ equations in $n = 344$ variables of degree 3 are generated. The XL algorithm consists of multiplying these $m$ equations by all possible monomials of degree up to $D - d$. Let $R$ be the number of equations and $T$ be the number of monomials of degree $D$ of this newly generated system of equations. Here $R$ is computed as $m \times \frac{n^{D-2}}{(D-2)!}$ and $T = \frac{n^D}{D!}$ where $D \geq (n/\sqrt{m})$ [10]. If most of the equations are linearly independent, we expect to succeed with a complexity $\mathcal{O}(T^3)$ when the number of equations exceeds the number of variables. In the proposed cipher, we need to take $D \geq 77$ to make the value of $R$ greater than or equal to $T$ ($>2^{273}$). Since the complexity of the attack is $\mathcal{O}(T^3)$, this large value of $T$ makes the attack infeasible.

### 3.3 Time-Memory-Data Trade-off Attack

In time-memory-data trade-off attack, the attacker takes advantage over the small internal state of the cipher [6]. If the internal state size of a cipher is $n$, then the complexity of attack is $\mathcal{O}(2^{n/2})$. In XUBA, internal state size is 280 bits which make a complexity of $\mathcal{O}(2^{140})$, which is larger than the complexity of the exhaustive key search.

### 3.4 Forgery Attack on MAC

Forgery attack is a common attack performed on integrity component. In this attack, the attacker will be able to prejudge the value of MAC for any message without knowing the secret key [9]. If the nonce bits or length of plaintext is changed, then the probability of forgery is $2^{-t}$ where $t$ is the length of tag bits. $2^{-128}$ attempts make the forgery attack infeasible in XUBA.

### 3.5 Guessing of MAC

Guessing of MAC is done by choosing an arbitrary fraudulent message and to append a randomly chosen MAC value to it. Then attacker tries to guess the correct MAC value. The probability that this MAC value is correct is equal to $1/2^t$. When a cipher produces a tag of size $t = 32, 33, ..., 64$, then the attack is not possible due to the low probability value of guessing [9]. The 128 bit tag makes the attack infeasible in the proposed scheme.

### 3.6 SAC Test

Strict Avalanche Criterion (SAC) is an important property satisfied by ciphers. When one input bit is changed, every output bit should change with a probability of $p = \frac{1}{2}$. Using this test, the effect of a single bit flip in the input is studied in the output of tag bits and keystream bits [5].

In SAC test, a SAC matrix of size $m \times n$ ($m$ is input size and $n$ is output size) is created as follows: the SAC matrix entries are initialized to zeros. A random input is taken from a set of $2^{10}$ samples and the corresponding output is calculated. The $a$th bit ($1 \leq a \leq m$) of the this input is flipped and the output is calculated. Then XOR operation is performed on both the outputs and for each non-zero bit $b$ ($1 \leq b \leq n$) of the difference, $(a, b)$th entry of the SAC matrix is incremented by 1. This process is repeated for each input bit $a$. Then $\chi^2$ Goodness of Fit Test is used to check whether the SAC matrix values follows binomial distribution or not. Then a probability value is obtained for the corresponding $\chi^2$ value. If this probability value is less than 0.01, then the cipher is not considered as random. The parameters for SAC test are given below:

- Size of SAC Matrix (Tag): $128 \times 128$
- Size of SAC Matrix (Keystream): $280 \times 100$
- Degrees of Freedom: 4
- Significance Level: 0.01
- Number of Inputs Tested: $2^{10}$

**Table 2** Results of SAC test

| Range | Keystream | | MAC | |
|---|---|---|---|---|
| | Expected count | Observed count | Expected count | Observed count |
| 0–498 | 5584 | 5766 | 3267 | 3354 |
| 499–507 | 5316 | 5207 | 3111 | 3123 |
| 508–516 | 6201 | 6126 | 3628 | 3596 |
| 517–525 | 5316 | 5350 | 3111 | 3060 |
| 526–1024 | 5583 | 5551 | 3267 | 3251 |

The ranges, expected count and observed count of SAC Test for $2^{10}$ trials on tag bits and keystream computation is given in the Table 2. The *p*-value and $\chi^2$ value for SAC test on tag bits are 0.46 and 3.55, respectively. In the case of keystream, we got the $\chi^2$ value as 9.47 and it corresponds to the *p*-value, 0.04. According to SAC test, we obtained *p*-values greater than the significance level indicating randomness.

### 3.7 Collision Test

Collision resistance is an important property of hash functions. A function is said to be collision resistant if it is infeasible to find any two messages having the same tag. The number of collisions in the whole 128 bits of tag is computed in this test and thus checks its resistance to collision [5].

According to collision test, a random input message is selected and its tag is computed. Then $2^{12}$ input messages are formed by changing the first 12 bits and corresponding tags are calculated. Tag for all input messages is stored in a hash table of size $2^{12} \times 128$. If tag bits of more than one message is same, then it is considered as a collision. As we are computing tag for $2^{12}$ inputs, the number of collision in a hash table ranges from 0 to 4096. This process is repeated for $2^{10}$ input samples and an average collision of 67 is obtained. As this count is a very small percentage of 4096, the cipher is considered to be collision resistant.

## 4 Conclusion

We have proposed a new authenticated encryption scheme with more security enhancement. The cipher is resistant to algebraic, differential and time-memory-data trade-off attacks and MAC is resistant to forgery and guessing attack. The randomness of the cipher has been tested using collision and SAC test. Comparison of XUBA with existing ciphers is given in the Table 3.

**Table 3** Comparison of XUBA with existing ciphers

| Ciphers | Key & IV (bits) | MAC (bits) | Algebraic attack | Differential attack | Time-memory-data trade-off | MAC Forgery |
|---|---|---|---|---|---|---|
| Grain-128a | 128 & 96 | 32 | Yes | Yes | No | Yes |
| Snow3G | 128 & 128 | 32 | No | No | No | Yes |
| Trivia | 128 & 128 | 128/160 | No | No | No | No |
| Acorn-128a | 128 & 128 | 64–128 | No | Yes | No | No |
| HS1-SIV | 128 & 128 | 64–256 | No | No | No | No |
| Sablier | 80 & 80 | 32 | No | No | No | Yes |
| XUBA | 128 &128 | 128 | No | No | No | No |

# References


1. Abed, F., Forler, C., Lucks, S.: General overview of the first-round caesar candidates for authenticated encryption. Technical report, Cryptology ePrint report 2014/792 (2014)
2. ALMashrafi, M.J.:Analysis of stream cipher based authenticated encryption schemes. Ph.D. thesis, Queensland University of Technology (2012)
3. Biham, E., Shamir, A.: Differential cryptanalysis of des-like cryptosystems. J. Cryptol. **4**(1), 3–72 (1991)
4. Chakraborti, A., Nandi, M.: Trivia-ck v2 (2015)
5. Doganaksoy, A., Ege, B., Koçak, O., Sulak, F.: Cryptographic randomness testing of block ciphers and hash functions. IACR Cryptology ePrint Archive 2010, 564 (2010)
6. Johansson, T., Agren, M., Hell, M., Meier, W.: Grain-128a: a new version of grain-128 with optional authentication. Int. J. Wirel. Mob. Comput. **5**(1), 48–59 (2011)
7. Krovetz, T.: Hs1-siv (v2). CAESAR Second Round Submission (2015). https://competitions.cr.yp.to/round2/hs1sivv2.pdf
8. Megha, P., Sindhu, M., Srinivasan, C., Sethumadhavan, M.: Hash-one: a lightweight cryptographic hash function. IET Inf. Secur. (2016)
9. Preneel, B., Mercierlaan, K.: Cryptanalysis of Message Authentication Codes. Department Electrical Engineering, Katholieke Universiteit Leuven, Belgium (2004)
10. Vörös, M.: Algebraic attack on stream ciphers. Master's thesis, Comenius University, Faculty of Mathematics, Physics and Informatics, Department of Computer Science (2007)
11. Wu, H.: Acorn: a lightweight authenticated cipher (v1). CAESAR Second Round Submission (2015). https://competitions.cr.yp.to/round2/acornv1.pdf
12. Xu, C., Yao, Y., Zhang, B., Shi, Z., Li, Z.: Sablier v1. CAESAR 1st Round (2014). https://competitions.cr.yp.to/round1/sablierv1.pdf

# Secure Speech Enhancement Using LPC Based FEM in Wiener Filter

**Kavita Bhatt, C.S. Vinitha and Rashmi Gupta**

**Abstract** Speech enhancement is a process which cultivates the quality of speech signal in noisy environment. It refers to removing or reducing the background noise in order to obtain an improved quality of original speech signal. Degradation of speech signal is most common problem in speech communication, so enhancement of the speech plays a vital role in improving the quality of speech signal. A number of methods are used for speech enhancement. Here we are using LPC based FEM in Wiener filter for speech enhancement. This method is then compared with several speech enhancement algorithms to obtain a better result or better speech quality. On comparing this method with other speech enhancement methods we obtain better speech performance. Here we are using NOIZEUS speech database in order to compare different speech enhancement methods. The experimental result shows that our proposed method provides better result and there is no information loss in original speech signal.

**Keywords** Speech enhancement · Wiener filtering · Linear prediction coefficients (LPCs) · Formant enhancement method (FEM)

## 1 Introduction

Speech is a common form of human to human communication. Basically speech is the physical production of sound by using our lips, tongue, palate and respiratory system to communicate ideas. In many situations it can be noted that some

K. Bhatt (✉) · C.S. Vinitha · R. Gupta
Department of Electronics and Communication Engineering, AIACTR
(Affiliated to GGSIPU), Geeta Colony, New Delhi, India
e-mail: kavitabhatt57@gmail.com

C.S. Vinitha
e-mail: vinithavinod@rediffmail.com

R. Gupta
e-mail: rashmig71@yahoo.com



S.C. Satapathy et al. (eds.), *Data Engineering and Intelligent Computing*,
Advances in Intelligent Systems and Computing 542,
DOI 10.1007/978-981-10-3223-3_63

additional noise corrupts the original speech signal and its presence affects the intelligibility of speech. Ex. Conversation of a pilot and a controller in air traffic control tower, where speech is usually degraded by the addition of engine noise. In such situations the basic necessity is the improvement in intelligibility and quality of speech signal.

Enhancement plays a significant role in increasing the quality as well as improving the performance of something and when it comes to speech it uses some processing tools for improving the quality of the noise corrupted speech. The basic purpose of enhancing the speech is to bring improvement in the performance of speech communication in noisy environment without affecting the original speech signal. The speech enhancement algorithm reduces or eliminates the background noise without distorting the speech signal. Enhancement of the speech which is degraded or corrupted by noise, is one of the important parts of speech enhancement, and thus used in multiple applications like hearing aids, mobile phones, speech recognition, and teleconferencing systems [1]. The speech enhancement techniques are divided on the basis of number of channel of data available. According to this criterion we have, single and multi channel speech enhancement techniques [2]. In single channel speech enhancement techniques only a single channel is present and these techniques are less expensive than multichannel techniques. The multichannel speech enhancement techniques are those in which multiple inputs are present.

In this paper we have used single channel speech enhancement techniques. Over the past four decades a number of speech enhancement techniques are used. The methods used for speech enhancement are MMSE estimation methods [3], spectral subtraction [4], Wiener filter [5], and Kalman filter [6].

The MMSE spectrum estimation method includes two parts: 1. A priori SNR, 2. MMSE spectral amplitude estimation. In order to obtain MMSE estimator, coefficients of a priori probability distribution and noise Fourier expansion in speech should be known. In this the speech and possibly the noise are neither stationary nor ergodic process. The spectral subtraction method is considered as a nonparametric method. This method requires an estimation of the noisy spectrum and it is considered that the estimation of noisy spectrum is done while the speaker is silent. In subspace method the noise subspace is removed and the original speech signal is estimated from the noisy environment. Basically subspace methods of speech enhancement are preferred for white noise environment. Spectral information needs to describe the noise spectrum which is obtained from the signal measured during non speech activity. Kalman filters are widely used in speech enhancement. The Kalman filter is a model based method which models speech signal as an autoregressive (AR) process and then recovers the speech signal [7]. The Kalman filtering method of speech enhancement has no assumption of stationary speech signals. The Kalman filtering offers the following advantages: it is designed to work with finite data sets, it makes use of models and it can be made to work with non-stationary signals. The methods used for the purpose of speech enhancement provide low SNR so a number of new methods are used in order to increase SNR value. Tarek Mellahi and Rachid Hamdi proposed a new method LPC based FEM in Kalman filter for enhancing the quality of speech signal [8]. In this method the new coefficient obtained by using LPC-FEM is applied to Kalman filter which provides better speech output. In order to obtain better perceptual quality and higher SNR values, here we are using LPC based FEM in Wiener filter.

In this paper we are comparing and evaluating four different speech enhancement methods. Here we are using the NOIZEUS database. The noisy dataset includes 30 IEEE sentences which were pronounced by three male and female speakers, and was interrupted by eight different noises at different SNRs [9]. This paper contains different sections: the second section describes the proposed method used for speech enhancement. Different comparison graphs and spectrogram are described in the third section. The fourth section contains conclusion.

## 2 Proposed Work

Here we are using LPC based FEM in Wiener filter for improving speech quality. Figure 1 shows the basic block diagram of speech enhancement system. In various stages degradation might occur to the speech signal. Most common degradation at the source of the speech includes background noise, reverberation and speech of competing speakers. Background noise is due to the presence of noise sources in the ambiance, whereas reverberation occurs due to the reflection from various surfaces. The speech enhancement system reduces the unwanted noise present in the background and channel, and passes the desired clean speech signal.

In this proposed method we are using the linear predictive coding based formant enhancement method with Wiener filter. The LPC methods provides extremely accurate estimates of speech parameters, basic idea of linear prediction is that the present samples of speech can be closely approximated as a linear combination of previous samples [10, 11].

$$\hat{a}(n) = \sum_{k=1}^{p} \alpha_k a(n-k) \tag{1}$$

where $\propto_k$ is the coefficient of LPC and $\hat{s}$ is the prediction of noisy speech s. The samples are calculated easily for clear speech but in practice the clear sound is not

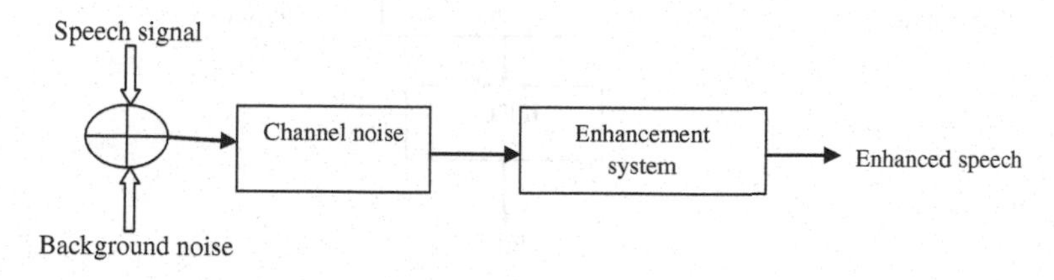


**Fig. 1** Block diagram of speech enhancement system

present. So a number of methods are used but the SNR obtained by those methods are low. In the linear predictive coding analysis the speech frames are extracted by Hamming or Hanning window.

$$w_H(n) = \begin{cases} 0.5 - 0.5\cos\frac{2\prod n}{M}, 0 \le n \le M - 1 \\ 0, \; otherwise \end{cases} \tag{2}$$

The prediction error y(n) is given as

$$u(n) = a(n) - \hat{a}(n) \tag{3}$$

So

$$a(n) = y(n) - \sum_{k=1}^{p} \alpha_k a(n-1) \tag{4}$$

Now the main motive is to find the coefficient of noisy speech b(n). The Levension-Durbin method is used to calculate the coefficients. The vocal tract response has a number of resonance frequencies because of vibration of vocal cords. These resonance frequencies are called as formant frequency or formants [12]. Figure 2 shows the basic block diagram of the proposed method. A(t) is the combination of clean speech signal and white noise.

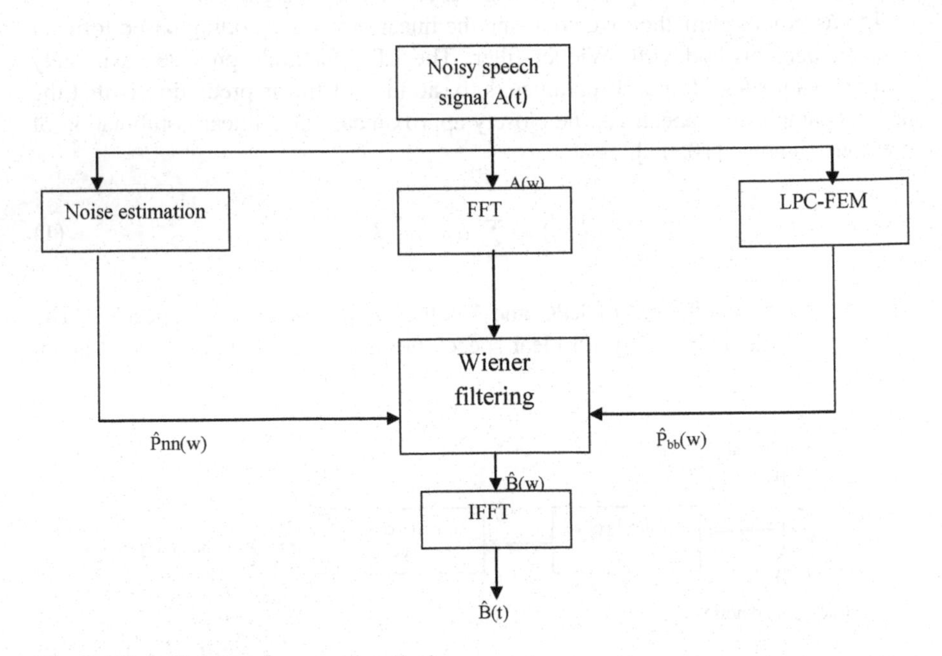


**Fig. 2** Block diagram of proposed method

$$a(t) = b(t) + n(t) \tag{5}$$

Figure 2 shows that the noisy speech is applied to LPC-FEM. The LPC based FEM depends on the modification of power spectrum of the linear prediction coding model and then re-evaluation of new LPCs. The algorithm for LPC-FEM is given as [13, 14]:

i. Evaluation of power spectrum from linear predictive coding, which uses FFT.
ii. Modifying the power spectrum by decreasing the low energy parts.
    a. Find formants and valleys from the smooth power spectrum.
    b. Multiply low energy regions by a small coefficient.
    c. Formants are left unmodified.
iii. Construct autocorrelation from the new power spectrum, which uses IFFT.
iv. Re-evaluation of new LPC from autocorrelation functions by using Yule-Walker.

The noisy speech is estimated and the new coefficients of speech obtained by LPC based Fem is applied to the Wiener filter. Basically Wiener filter is optimum filter. The Wiener filter method of speech enhancement is the most popular method. For Eq. 5, the Wiener filter can be defined as

$$B(w) = H(w)A(w) \tag{6}$$

Here w represents the frequency index. B(w) represents the discrete Fourier transform of clean speech, A(w) represents the transform of noisy speech and H(w) represents the transfer function of Wiener filter. The Wiener filter can be obtained by

$$H(w) = \frac{P_{bb}(w)}{P_{bb}(w) + P_{nn}(w)} \tag{7}$$

Here $P_{bb}(w)$ is power spectrum of speech b(t) and $P_{nn}(w)$ is power spectrum of noise n(t). By using Eqs. (6) and (7), the enhanced speech can be estimated in the frequency domain as

$$B(w) = \frac{P_{bb}(w)}{P_{bb}(w) + P_{nn}(w)} .A(w) \tag{8}$$

Now by using inverse FFT we obtained the enhanced speech for B(w). Basically Wiener filter performance depends on accuracy of $P_{bb}(w)$. The LPC analysis estimates the power spectrums, $P_{bb}(w)$ and $P_{nn}(w)$, for noisy corrupted speech i.e. a(n). The Wiener filter is operated in order to obtain more refined output. The proposed method provides the following advantages: this method provides good noise reduction effect, the muteness present in the output of wiener filter algorithm is reduced by using LPC-FEM, and compared with other methods this method gives more refined output speech enhancement.

# 3 Speech Enhancement Experiment and Result

In this experiment we are using NOIZEUS database. Here we are comparing the four different method of speech enhancement. The comparison is done by using tabular, graphical method, spectrogram and evaluation of the performance of speech is done by using several methods like perceptual evaluation of speech quality (PESQ), SNR, normal correlation, and PSNR. Here we have used four different methods for comparison purpose. Higher the value of PESQ, SNR, NC, PSNR, better the performance which means better the quality of speech signal.

The following Table 1 shows the comparison of different speech enhancement method based on PESQ, SNR, NC and PSNR. For the comparison purpose we used the speech signal corrupted with Gaussian noise (5 db).

**Table 1** Performance comparison of different speech enhancement techniques

| Methods | Objective evaluation | | | |
|---|---|---|---|---|
| | PESQ | SNR(db) | PSNR(db) | NC |
| Noise | 1.80 | 5 | 55 | 0.65 |
| MMSE | 1.98 | 5.4 | 76.5 | 0.8 |
| Kalman | 2.15 | 11.16 | 77.47 | 0.84 |
| Kalman + LPC based FEM | 3.2 | 11.53 | 82.36 | 0.98 |
| Wiener + LPC based FEM | 3.54 | 12.25 | 99.18 | 0.97 |

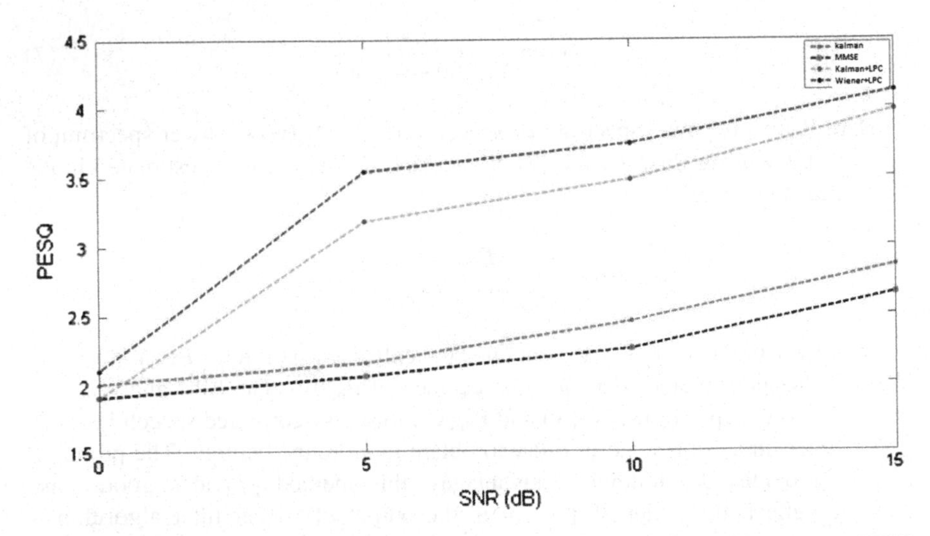


**Fig. 3** Comparative performance of various speech enhancement methods with respect to PESQ

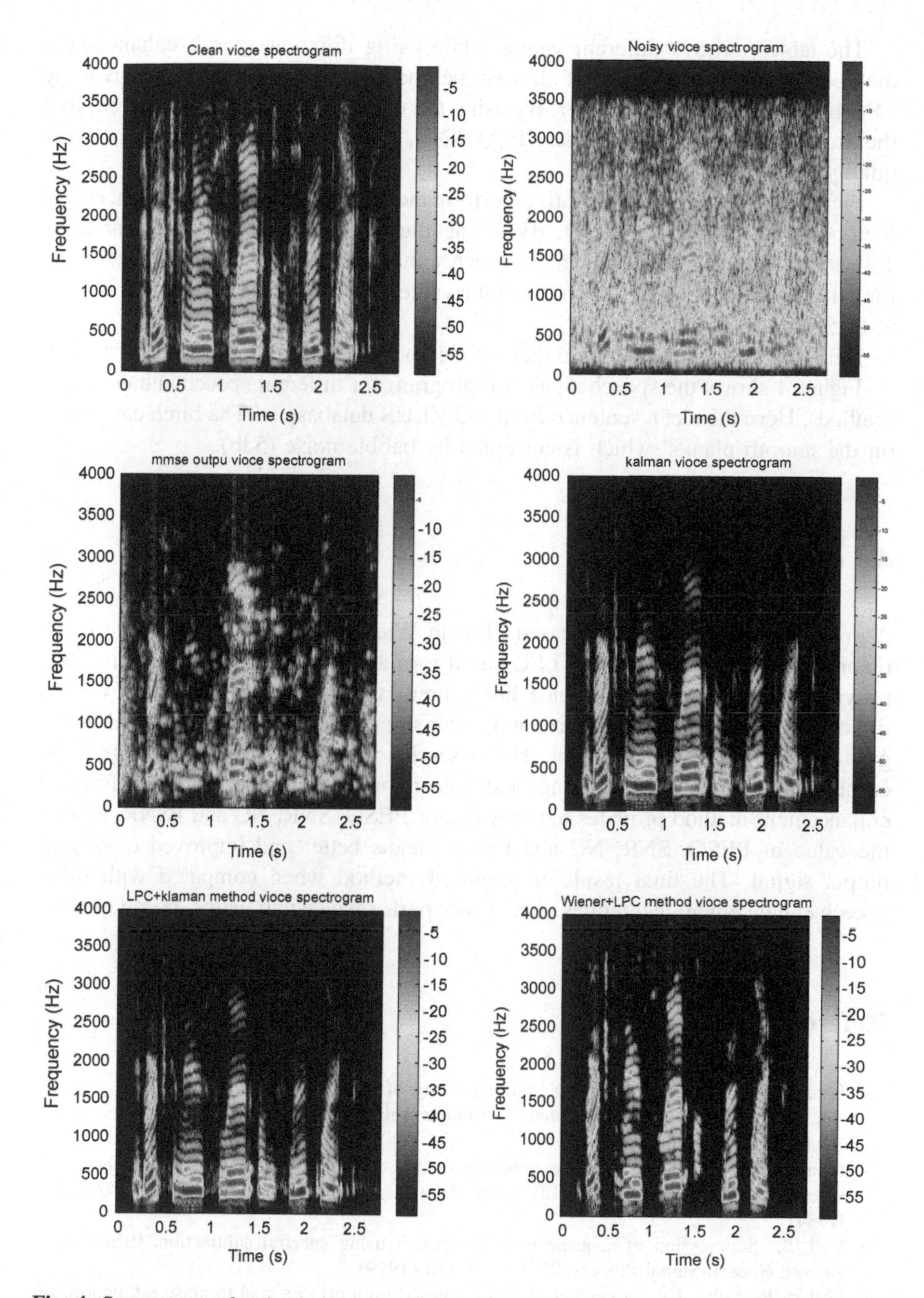


**Fig. 4** Spectrogram of **a** clean speech from NOIZEUS data, **b** Babble noise corrupted sentence at 5db SNR, **c** output of MMSE method **d** output of Kalman method, **e** output of Kalman +LPC based FEM method, **f** output of Wiener + LPC based FEM method

The table portrays different values while using different speech enhancement methods. We can see that out of all used methods the best performance is given by LPC based FEM in Wiener filter. By using the comparison table it is apparent that the proposed method gives higher PESQ, SNR, PSNR and NC which means better quality result is obtained.

Figure 3 shows the comparative performance of various speech enhancement methods with respect to PESQ. By taking the Gaussian noise at different SNR values the comparison of different speech enhancement techniques is performed then the graph is plotted on the basis of the values of PESQ. The experiment is done at 0, 5, 10 and 15 db.

From the line graph it is vivid that the method used gives higher value of PESQ.

Figure 4 shows the speech signal spectrogram for different speech enhancement methods. Here the clean sentence from NOIZEUS database is 'The birch canoe slid on the smooth planks' which is corrupted by babble noise (5db).

## 4 Conclusion

A new method based on LPC based FEM in Wiener filter is presented. The result obtained in this paper by using LPC based formant enhancement method in filters provides better performance since linear predictive coding technique in Wiener filter improves the speech quality and it considerably reduces the loss and ambiguities in original speech signal. The observed values and result shows that the combination of Wiener filtering and linear predictive coding based formant enhancement method provides higher value of PESQ, SNR, NC and PSNR. Higher the value of PESQ, SNR, NC and PSNR means better and improved quality of output signal. The final result of proposed method when compared with other speech enhancement techniques gives better performance and better speech quality.

## References

1. Rabiner, L.R., Schaffer, R.W.: Digital processing of speech signal
2. Xia, Y., Wang, J.: Low dimensional recurrent neural network based Kalman filter for speech enhancement. Neural Netw. (2015)
3. Ephraim, Y., Malah, D.: Speech enhancement using a minimum mean square error short time spectral amplitude estimator. IEEE Trans. Acoust. Speech Signal Process. **32**, 1109–1121 (1984)
4. Boll, S.: Suppression of acoustic noise in speech using spectral subtraction. IEEE Trans. Acoust. Speech Signal Process. **27**(2), 113–120 (1979)
5. Scalart, P., Filho, J.V.: Speech enhancement based on a priori signal to noise estimation. In: IEEE International Conference on Acoustics, Speech, and Signal Processing, pp. 629–632 (1996)
6. Paliwal, K.K., Basu, A.: A speech enhancement method based on Kalman filtering. In: IEEE International Conference on Acoustics, Speech, and Signal Processing, pp. 177–180 (1987)

7. Gannot, S., Burshtein, D., Weinstein, E.: Iterative and sequential Kalman filter based speech enhancement algorithms. IEEE Trans. Speech Audio Process. **6**(4), 373–385 (1998)
8. Mellahi, T., Hamdi, R.: LPC based formant enhancement method in Kalman filter for speech enhancement. AEU Int. J. Electron. Commun. (2015)
9. Hu, Y., Loizou, P.: Subjective evaluation and comparison of speech enhancement algorithm. Speech Commun. **49**, 588–601 (2007)
10. Tierney, J.: A study of LPC analysis of speech in additive noise. IEEE Trans. Acoust. Speech Signal Process. **28**(4), 389–479 (1980)
11. Kalman, R.E.: A new approach to linear filtering and prediction problems. J. Basic Eng. Trans. ASME **82**, 35–45 (1960)
12. Welling, L., Ney, H.: A model for efficient formant estimation. In: IEEE International Conference on Acoustics, Speech, and Signal Processing, pp. 797–800 (1996)
13. Kabal, P., Wang, F., O'Shaughnessy, D., Ramachandran, R.: Adaptive post filtering for enhancement of noisy speech in the frequency domain. IEEE Int. Symp. Circuits Syst. 312–315 (1991)
14. Chen, B., Loizou, P.C.: Formant frequency estimation in noise. In: IEEE International Conference on Acoustics, Speech, and Signal Processing, pp. 581–4 (2004)

# Author Index





S.C. Satapathy et al. (eds.), *Data Engineering and Intelligent Computing*,
Advances in Intelligent Systems and Computing 542 ,
DOI 10.1007/978-981-10-3223-3

Printed in the United States
By Bookmasters